r   Indicates volumes in Continuous Revision.

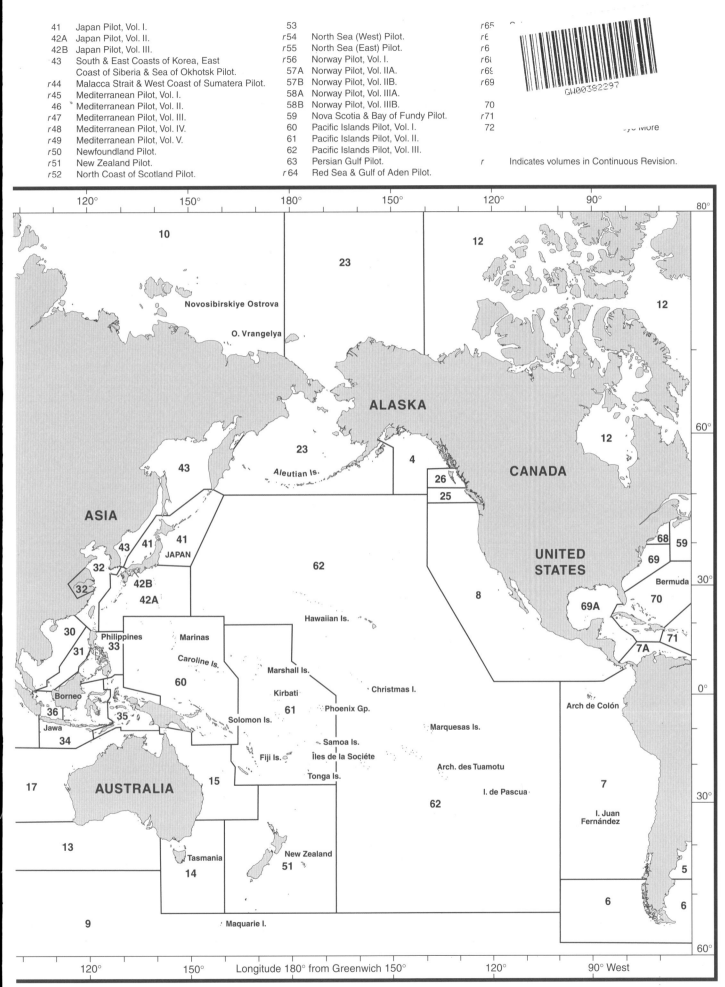

Admiralty Sailing Directions

RECORD OF CORRECTIONS

The table below is to record Section IV Notice to Mariners corrections affecting this volume.
Sub paragraph numbers in the margin of the body of the book are to assist the user with corrections to this volume from these amendments.

Weekly Notices to Mariners (Section IV)

2004	2005	2006	2007

NP 136

OCEAN PASSAGES FOR THE WORLD

FIFTH EDITION
2004

PUBLISHED BY THE UNITED KINGDOM HYDROGRAPHIC OFFICE

Previous editions:

First published 1895
1st Edition 1923
2nd Edition 1950
3rd Edition 1973
4th Edition 1987

Oh God be good to me,
Thy sea is so wide and my ship is so small

Breton fisherman's prayer

PREFACE

The Fifth Edition of Ocean Passages for the World has been prepared by Captain P.J. Pritchard, Master Mariner, and contains the latest information received in the United Kingdom Hydrographic Office to the date given below.

This edition supersedes the Fourth Edition (1987), which is cancelled.

Meteorological Information has been supplied by the Meteorological Office, Exeter.

In addition to UK Hydrographic Office Publications and Ministry of Defence papers, certain volumes of *United States Sailing Directions (Passage Planning Guides)* have been consulted:

Dr D W Williams
United Kingdom National Hydrographer

The United Kingdom Hydrographic Office
Admiralty Way
Taunton
Somerset TA1 2DN
England
January 2004

CONTENTS

CHAPTER 5

CHAPTER 6

CHAPTER 7

CHAPTER 8

CHAPTER 9

Sailing Passages for Indian Ocean, Red Sea and Eastern Archipelago

CHAPTER 10

Sailing Passages for Pacific Ocean

TABLES, APPENDIX AND INDEX

EXPLANATORY NOTES

Ocean Passages of the World will be kept up-to-date by supplements, each new supplement cancelling the previous one. In addition important amendments which cannot await the next supplement are published in Section IV of the weekly editions of *Admiralty Notices to Mariners*. A list of such amendments and notices in force is published in the last weekly edition for each month. Those still in force at the end of the year are reprinted in the *Annual Summary of Admiralty Notices to Mariners*.

This book should not be used without reference to the latest supplement and Section IV of the weekly editions of Admiralty Notices to Mariners.

References to hydrographic and other publications

The Mariner's Handbook gives general information affecting navigation and is complementary to this volume.

Admiralty List of Lights should be consulted for details of lights, lanbys and fog signals, as these are not fully described in this volume.

Admiralty List of Radio Signals should be consulted for information relating to coast and port radio stations, radio details of pilotage services, radiobeacons and direction finding stations, meteorological services, radio navigational aids, Global Maritime Distress and Safety System (GMDSS) and Differential Global Positioning System (DGPS) stations, as these are only briefly referred to in this volume.

Annual Summary of Admiralty Notices to Mariners contains in addition to the temporary and preliminary notices, and amendments and notices affecting Sailing Directions, a number of notices giving information of a permanent nature covering radio messages and navigational warnings, distress and rescue at sea and exercise areas.

The International Code of Signals should be consulted for details of distress and life-saving signals, international ice-breaker signals as well as international flag signals.

Remarks on subject matter

Geographical positions are taken from the largest scale Admiralty chart. Those of the arrival and departure positions given in the Routes Index are precise. Positions given in the Gazetteer are approximate and are given to enable the reader to identify the location on his passage planning chart.

Names have been taken from the most authoritative source. When an obsolete name still appears on the chart, it is given in brackets following the proper name at the principal description of the feature in the text and where the name is first mentioned.

Units and terminology used in this volume

Bearings and directions are referred to the true compass and when given in degrees are reckoned clockwise from 000° (North) to 359°
Bearings used for positioning are given from the reference object.
Bearings of objects, alignments and light sectors are given as seen from the vessel.
Courses always refer to the course to be made good over the ground.

Winds are described by the direction from which they blow.

Tidal streams and currents are described by the direction towards which they flow.

Distances are expressed in sea miles of 60 to a degree of latitude and sub-divided into cables of one tenth of a sea mile.

Depths are given below chart datum, except where otherwise stated.

Heights of objects refer to the height of the structure above the ground and are invariably expressed as "... m in height".

Elevations, as distinct from heights, are given above Mean High Water Springs or Mean Higher High Water whichever is quoted in *Admiralty Tide Tables*, and expressed as, "an elevation of ... m". However the elevation of natural features such as hills may alternatively be expressed as "... m high" since in this case there can be no confusion between elevation and height.

Metric units are used for all measurements of depths, heights and short distances, but where feet/fathoms charts are referred to, these latter units are given in brackets after the metric values for depths and heights shown on the chart.

Time is expressed in the four-figure notation beginning at midnight and is given in local time unless otherwise stated. Details of local time kept will be found in the relevant *Admiralty List of Radio Signals*.

Conspicuous objects are natural and artificial marks which are outstanding, easily identifiable and clearly visible to the mariner over a large area of sea in varying conditions of light. If the scale is large enough they will normally be shown on the chart in bold capitals and may be marked "conspic".

Prominent objects are those which are easily identifiable, but do not justify being classified as conspicuous.

ABBREVIATIONS

The following abbreviations are used in the text.

Directions

N	north (northerly, northward, northern, northernmost)	S	south
		SSW	south-south-west
NNE	north-north-east	SW	south-west
NE	north-east	WSW	west-south-west
ENE	east-north-east	W	west
E	east	WNW	west-north-west
ESE	east-south-east	NW	north-west
SE	south-east	NNW	north-north-west
SSE	south-south-east		

Navigation

DGPS	Differential Global Positioning System	Satnav	Satellite navigation
GPS	Global Positioning System	TSS	Traffic Separation Scheme
Lanby	Large automatic navigation buoy	VTS	Vessel Traffic Services
ODAS	Ocean Data Acquisition System	VTMS	Vessel Traffic Management System

Offshore Operations

ALC	Articulated loading column	FSO	Floating storage and offloading vessel
ALP	Articulated loading platform	SALM	Single anchor leg mooring system
CALM	Catenary anchor leg mooring	SALS	Single anchored leg storage system
ELSBM	Exposed location single buoy mooring	SBM	Single buoy mooring
FPSO	Floating Production Storage and Offloading vessel	SPM	Single point mooring

Organizations

IALA	International Association of Lighthouse Authorities	IMO	International Maritime Organization
		NATO	North Atlantic Treaty Organization
IHO	International Hydrographic Organization	RN	Royal Navy

Radio

DF	direction finding	RT	radio telephony
HF	high frequency	UHF	ultra high frequency
LF	low frequency	VHF	very high frequency
MF	medium frequency	WT	radio (wireless) telegraphy
Navtex	Navigational Telex System		

Rescue and distress

AMVER	Automated Mutual Assistance Vessel Rescue System	MRCC	Maritime Rescue Co-ordination Centre
		MRSC	Maritime Rescue Sub-Centre
EPIRB	Emergency Position Indicating Radio Beacon	SAR	Search and Rescue
GMDSS	Global Maritime Distress and Safety System		

Tides

HAT	Highest Astronomical Tide	MHWS	Mean High Water Springs
HW	High Water	MLHW	Mean Lower High Water
LAT	Lowest Astronomical Tide	MLLW	Mean Lower Low Water
LW	Low Water	MLW	Mean Low Water
MHHW	Mean Higher High Water	MLWN	Mean Low Water Neaps
MHLW	Mean Higher Low Water	MLWS	Mean Low Water Springs
MHW	Mean High Water	MSL	Mean Sea Level
MHWN	Mean High Water Neaps		

Times

ETA	estimated time of arrival	UT	Universal Time
ETD	estimated time of departure	UTC	Co-ordinated Universal Time

Units and Miscellaneous

°C	degrees Celsius		km	kilometre(s)
dwt	deadweight tonnage		kn	knot(s)
feu	forty foot equivalent unit		kw	kilowatt(s)
fm	fathom(s)		m	metre(s)
ft	foot(feet)		mb	millibar(s)
g/cm^3	gram per cubic centimetre		MHz	megahertz
GRP	glass reinforced plastic		mm	millimetre(s)
grt	gross register tonnage		MW	megawatt(s)
gt	gross tonnage		No	number
hp	horse power		nrt	nett register tonnage
hPa	hectopascal		teu	twenty foot equivalent unit
kHz	kilohertz			

Vessels and cargo

HMS	Her (His) Majesty's Ship		POL	Petrol, Oil & Lubricants
LASH	Lighter Aboard Ship		RMS	Royal Mail Ship
LNG	Liquefied Natural Gas		Ro-Ro	Roll-on, Roll-off
LOA	Length overall		SS	Steamship
LPG	Liquefied Petroleum Gas		ULCC	Ultra Large Crude Carrier
MV	Motor Vessel		VLCC	Very Large Crude Carrier
MY	Motor Yacht			

OCEAN PASSAGES FOR THE WORLD

CHAPTER 1

ROUTE PLANNING

DETAILS OF THE BOOK

Coverage
1.1

1   Ocean Passages for the World is written for use in planning deep-sea voyages. It contains notes on the weather and other factors affecting passages, directions for a number of selected commonly-used routes and distances and dangers affecting those routes.

2   Chapters 2 - 7 describe climatic conditions and give routes recommended for full-powered vessels within the areas described.

3   Chapters 8 - 10 give the usual routes which were used by sailing vessels, however these routes may have to be adjusted to reflect current regulations and changed conditions. These chapters also give details of routes recommended for low-powered or hampered vessels.

4   The book should be used in conjunction with the appropriate charts and publications described below.

Routes
1.2

1   Power vessels are regarded as of moderate draught and belonging to two categories:

Full-powered and able to maintain a sea-going speed in excess of 15 knots.

Low-powered or hampered by damage or towing and unable to meet the requirements for full-powered vessels.

2   Moderate draught refers to vessels of 12 m. It is appreciated that there is a wide variety of modern vessels currently operating with draughts considerably in excess of 12 m. However, by setting a deeper draught as standard a substantial number of useful and well frequented routes would be excluded. The special requirements of deep-draught vessels are therefore not covered.

3   In most instances fully-powered vessels would be expected to take the shortest navigable route between ports. In certain cases, however, such vessels would find considerable saving in wear and tear, time and fuel by following the alternative recommended routes given.

4   Masters of low-powered, or hampered vessels, would do well to consult the appropriate passages for sailing vessels, described in Chapters 8 to 10, where they will find more detail regarding weather, tidal stream and currents than is given in the chapters for full-powered vessels.

Directions
1.3

For each route, the directions provide a guide for planning, however, conditions will rarely be precisely as predicted and the advice should be reviewed in the light of existing circumstances.

Diagrams are provided for each route in the text. These, however, should only be taken as an indication of the general direction of the route and should not be relied on to show all details. The scale of these diagrams only permits large land masses to be included, but wherever possible points of reference and dangers, mentioned in the directions, are shown. Ocean and coastal charts, of a scale sufficient for the safe navigation of the vessel, must always be used for passage planning and navigation. See *The Mariner's Handbook* (1.12.11) for advice on the use of charts and other navigational aids.

Departure and arrival positions
1.4

Distances, where given for routes, are between positions used in *Admiralty Distance Tables* (1.12.8). For ports and harbours the position quoted is usually the pilot ground or anchorage and may be a considerable distance from the port itself.

Distances
1.5

1   The Figure of the Earth used for computing distances is the International Spheroid.

2   Distances are given in International Nautical Miles of 1852 m to the nearest:

10 miles for passages of over 1000 miles:

5 miles for passages less than 1000 miles.

Gazetteer

1.6

1   The geographical positions of places and features named in the text are given in the Gazetteer. They are also given throughout the text, when mentioned, in order to assist the mariner in plotting the various routes when using electronic charting systems.

2   The positions, except Departure and Arrival positions (1.4), are approximate and only intended to assist in locating the place or feature on the chart.

Indexes

1.7

1   General subjects are listed alphabetically in the General Index.

2   Routes are listed in the Route Index, by their place of departure, if this point is precise. its latitude and longitude is given.

ADMIRALTY CHARTS AND PUBLICATIONS

Routeing charts

1.8

The series of Routeing charts are an important aid to passage planning. Each chart is published in 12 monthly versions and shows weather and ice conditions, ocean currents, load line zones and some recommended tracks and distances.

Limits of the charts are shown on Diagram 1.8.

Ocean charts

1.9

1   For the selection of suitable charts, which can be obtained from Admiralty Chart Agents, the *Catalogue of Admiralty Charts and Publications* (1.12.13) should be consulted. The catalogue, as well as the *Annual Summary*

of Admiralty Notices to Mariners (1.12.10), gives full details of all Admiralty Chart Agents throughout the world.

2   It is emphasised that ocean charts are still mainly based on sparse and inadequate sounding data obtained from a wide variety of sources of varying reliability and accuracy. Sounding coverage is best along well-frequented routes, but even in these waters undetected dangers may still exist, especially for modern deep-draught vessels.

3   For further remarks on such dangers and for the reliance that can be placed on charts, see *The Mariner's Handbook* (1.12.11).

Gnomic ocean charts

1.10

1   Gnomic ocean charts are convenient for laying off great circle courses. Details of those published to cover the Atlantic, Pacific and Indian Oceans can be found in the *Catalogue of Admiralty Charts and Publications* (1.12.13).

2   *Chart 5097 North Pacific* and *Chart 5098 South Pacific* are drawn with a greater range of latitude than the gnomic charts for the other oceans. By suitably renumbering the meridians of the Pacific Ocean charts, they can be readily adapted for similar latitudes in other oceans.

3   *Chart 5029 Great Circle Diagram* can also be used for laying off great circle courses particularly those between positions outside the limits of the Gnomic Ocean Charts.

4   For the computing of positions for the laying off a great circle course, see *Admiralty Distance Tables* (1.12.8).

Miscellaneous charts

1.11

The following Admiralty charts, which are referred to in this book, will prove useful when planning passages:

    Chart 5006 The World—Time Zone Chart, reproduced in this book at Tables D and E.

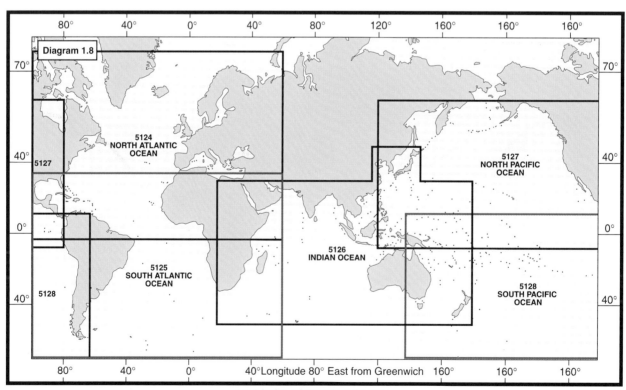

1.8 Limits of Routing Charts

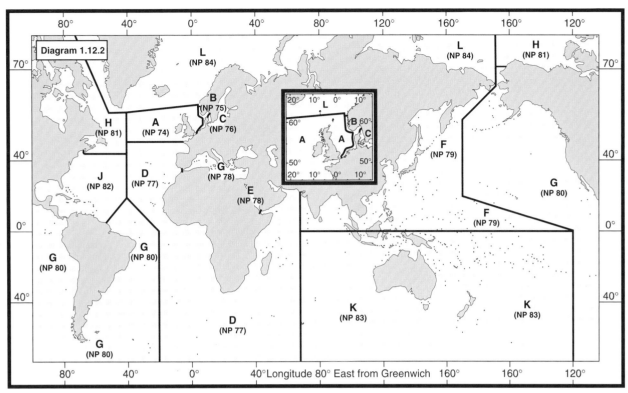

1.12.2 Limits of Volumes of Admiralty List of Lights and Fog Signals

Chart 5307 The World—Main Ocean Routes: Power Vessels.

Chart 5308 The World—Sailing ship routes, sections of which are reproduced in this book at the start of Chapters 8, 9, and 10.

Chart 5309 The World—Tracks followed by Sailing and Auxiliary Powered Vessels.

Chart 5310 The World—General Surface Current Distribution.

Publications
1.12

For the coastal sections of the routes given herein, for port arrival information and for other information affecting passages the following publications should be consulted.

All these publications, both digital and paper products, are kept updated by weekly Admiralty Notices to Mariners (1.12.9).

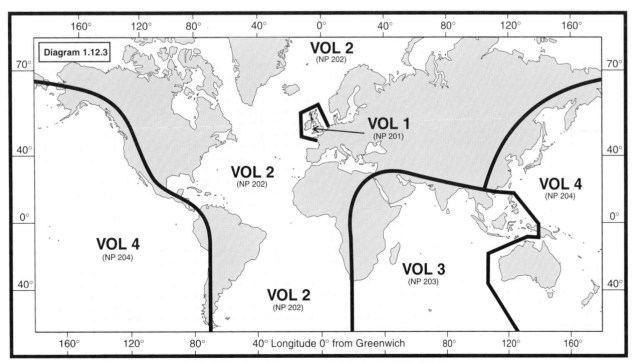

1.12.3 Limits of Volumes of Admiralty Tide Tables

1.12.1

Admiralty Sailing Directions give world-wide details of local conditions, directions, regulations and port information. Limits of the various volumes are shown in Diagram 1.12.1 on the front end-papers of this book.

1.12.2

Admiralty List of Lights give details of navigational lights and fog signals world-wide.

The series is available in both paper and digital form and meets SOLAS Carriage Requirements (1.13.4).

Limits of the paper volumes are shown in Diagram 1.12.2.

1.12.3

Admiralty Tide Tables give predictions of tidal heights throughout the world.

Limits of the volumes are shown in Diagram 1.12.3.

Digital tide prediction systems are also available, see 1.13.2 and 1.13.3

1.12.4

Admiralty Tidal Stream Atlases give details of tidal streams in North-west Europe.

Limits of the volumes are shown in Diagram 1.12.4.

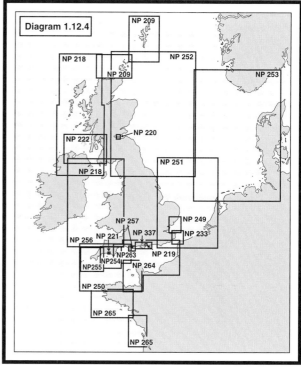

1.12.5

Admiralty List of Radio Signals comprehensive and authorative information on maritime radio communication systems.

They are published in 6 volumes:

Volume 1 (2 parts) Coast Radio Stations:

Volume 2 (1 part) Radio Aids to Navigation, Satellite Navigation Systems, Legal Time, Radio Time Signals, and Electronic Position Fixing Systems and:

Volume 3 (2 parts) Maritime Safety Information Services:

Volume 4 (1 part) Meteorological Observation Stations:

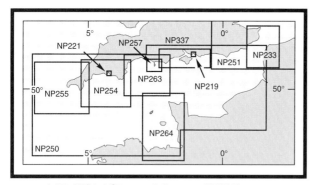

1.12.4 Tidal Stream Atlases of NW Europe

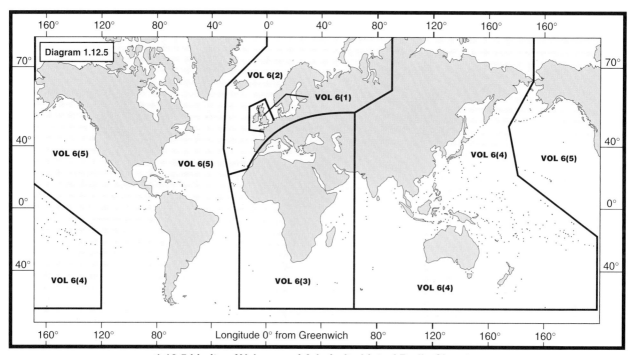

1.12.5 Limits of Volumes of Admiralty List of Radio Signals

Volume 5 (1 part) Global Maritime Distress and Safety System (GMDSS):

Volume 6 (5 parts) Pilot Services, Vessel Traffic Services and Port Operations. See Diagram 1.12.5 for the limits of coverage of these five publications.

1.12.6

1   **Admiralty Maritime Communications** are published in three volumes and are designed to provide easy reference to essential information on all aspects of maritime communications for both leisure and commercial small craft. See Diagram 1.12.6 for the limits of coverage of these three publications.

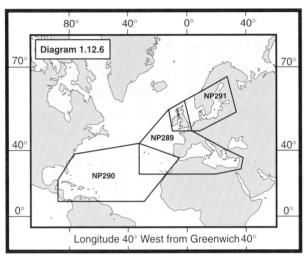

1.12.6 Limits of Volumes of Admiralty Maritime Communications

1.12.7

1   **Automatic Identification Systems (AIS)**. A description of this system, for the United Kingdom, is published in the *Admiralty List of Radio Signals, Volume 1*. See also *The Mariner's Handbook* (1.12.11) for a summary of this system.

1.12.8

Admiralty Distance Tables can conveniently be used for comparing distances by different routes through archipelagos or round land masses and for connecting distances by routes in *Ocean Passages for the World* to other ports.

1.12.9

Admiralty Notices to Mariners are produced weekly and should be consulted for recent, imminent and temporary changes of navigational information. They can be obtained from Admiralty Chart Agents (1.9).

1.12.10

Annual Summary of Admiralty Notices to Mariners contains reprints of all Temporary and Preliminary Notices affecting Admiralty charts and amendments to Admiralty Sailing Directions, in force on 1st January of the year of publication.

It also amplifies information given on a number of subjects mentioned in this book and contains Annual Notices covering other matters of interest to the mariner, including:

Suppliers of Admiralty Charts and Publications.
Availability of Notices to Mariners.
Distress and Rescue at Sea.
National Claims to Maritime Jurisdiction.
Information concerning Submarines.

1.12.11

The Mariner's Handbook gives general information on the supply, upkeep and use of Admiralty charts and publications and also on weather, ice, navigational hazards and other matters.

1.12.12

Chart 5011 Symbols and Abbreviations used on Admiralty Charts gives the meanings of all symbols and abbreviations used on Admiralty charts.

1.12.13

Catalogue of Admiralty Charts and Publications is a fully comprehensive reference in graphical and textual form detailing the world-wide inventory of all Admiralty Charts and Publications available for purchase.

Digital products

1.13

A growing range of digital products to complement paper charts and publications are available to the mariner and can be obtained from Admiralty Distributors.

For details of the coverage and availability of the following digital products the *Catalogue of Admiralty Charts and Publications* (1.12.13) should be consulted.

1.13.1

Admiralty Raster Chart Service (ARCS) is an official electronic charting service developed from the Admiralty paper chart. ARCS charts are exact reproductions of the corresponding paper charts in digital form for use with electronic navigational systems.

1.13.2

Admiralty TotalTide is a comprehensive tidal prediction system which provides instant tidal height and stream predictions for the world's commercial shipping routes.

Annual updates are recommended and are available from Admiralty Distributors each January.

1.13.3

Simplified Harmonic Method of Tidal Predictions is Windows-based and available on CD-ROM. The user inputs Harmonic Constants from Admiralty Tide Tables to obtain a graph of heights against time.

1.13.4

Admiralty Digital List of Lights is designed to reduce time and effort spent on manual corrections and provide exactly the same data as the paper versions, but with a graphic interface. Limits of the areas covered are shown in Diagram 1.13.4.

Weekly updates by weekly updates are available on CD-ROM, via the UKHO website or by e-mail.

NATURAL CONDITIONS

Climatic conditions

1.14

1   World climatic conditions are shown on Charts *5301 World Climatic Chart - January* and *5302 World Climatic Chart - July*, which are reproduced in this book as Diagrams 1.14a and 1.14b at the end of this chapter.

2   The diagrams give the general distribution of atmospheric pressure, wind, sea surface temperature, fog, currents and ice. Their accompanying notes should be read carefully.

3   The general principles of maritime meteorology, of ocean current circulation, of formation and distribution of ice and the formation of fog are given in *The Mariner's Handbook* (1.12.11). A more detailed treatment is given in *Meteorology for Mariners* produced by the Meteorological Office and published through Her Majesty's Stationary Office.

Admiralty Digital List of Lights - Area Limits

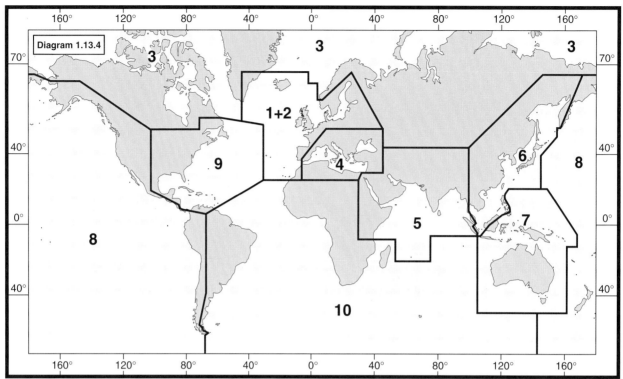

© Crown Copyright 2004

AREAS 1+2.	Northern Europe and the Baltic
AREA 3.	Northern waters
AREA 4.	Mediterranean and Black Seas
AREA 5.	Indian Ocean (northern part) and Red Sea
AREA 6.	Singapore to Japan
AREA 7.	Australia, Borneo and Philippines
AREA 8.	Pacific Ocean
AREA 9.	North America (east coast) and Caribbean
AREA 10.	South Atlantic and Indian Ocean (southern part)

4   The UK Meteorological Office can be contacted for Marine Weather Services on:

Website: www.metoffice.com.
e-mail: sales@metoffice.com.
Tel: +44 1344 855680
Fax: +44 1344 855681

5   The climatic features of particular oceanic areas are given at the beginning of the appropriate chapters of this book.

Seasonal winds
1.15

1   Over certain parts of the oceans the general distribution of pressure and winds, such as the Trade Winds, is greatly modified by the seasonal heating and cooling of adjacent large land masses. The result is a seasonal reversal of the prevailing wind over the adjacent oceans and known in low latitudes as "monsoons".

2   The seasons of the principal monsoons and their typical strengths over the area concerned are shown in Table B.

Tropical Storms
1.16

Signs of Tropical Storms, their typical behaviour and precautions which can be taken to avoid them, are described in *The Mariner's Handbook* (1.12.11). When affecting particular areas they are described in the appropriate volume of *Admiralty Sailing Directions* (1.12.1). The areas usually affected and in which period of the year storms can be expected are shown in Table C.

Depressions
1.17

In the middle latitudes, depressions which usually sweep in an E direction across the oceans of the N and S hemispheres, are the dominant influence on the weather. Of immense power, they cover great areas of the oceans with uninterrupted winds of long duration and build up extensive areas of high seas and heavy swell. They are an important factor in deciding the route of a passage.

Sea and swell
1.18

Areas where rough seas or heavy swell can be expected are shown on Diagrams 1.18a to 1.18d, at the end of this chapter. They indicate the percentage frequency of waves equal to, or greater than, 3·5 m and 6·0 m for the different seasons.

1.18.1

1   **Sea** is the name given to waves in, what is known as, the generating area where the waves are formed by the wind.

2   The following terms are used to describe the height of the sea waves:

Description	Height in metres
Calm - glassy	0·0
Calm - rippled	0·0 – 0·1
Smooth wavelets	0·1 – 0·5
Slight	0·5 – 1·25
Moderate	1·25 – 2.5

Rough	2·5 – 4·0
Very rough	4·0 – 6·0
High	6·0 – 9·0
Very high	9·0 – 14·0
Phenomenal	Over 14·0

1.18.2

1   **Swell** is the wave motion caused by a meteorological disturbance, which persists after the disturbance has died down or moved away. Swell often travels for considerable distances out of its generating area.

2   The following terms are used to describe swell waves:

Length of wave		Height of wave	
Short	0 – 100 m	Low	0 – 2 m
Average	100 – 200 m	Moderate	2 – 4 m
Long	Over 200 m	Heavy	Over 4 m

Abnormal waves
1.19

1   Wherever sea or swell waves encounter a seabed rising steeply from deep water, a strong opposing tidal stream or current, or are reinforced by waves from another wave system, they may be distorted to form large abnormal waves. Where waves are normally large, such abnormal waves may be massive, steep-fronted and capable of causing severe structural damage to the largest of ships.

2   Places where abnormal waves have been reported, usually near the 200 m depth contour at the steep edge of a continental shelf, include:

       Parts of the Norwegian Sea;
       Off the entrance to the Chesapeake Bay;
       Off the NW coast of Spain;
       Off the SE coast of South Africa.

3   It is off the SE coast of South Africa that most reports have been received and most research has been carried out on these waves.

4   For further information, see *The Mariner's Handbook* (1.12.11).

Currents
1.20

1   The principal currents of each ocean are described at the beginning of the appropriate chapter of this book and in greater detail in the appropriate volume of *Admiralty Sailing Directions* (1.12.1).

2   *Chart 5310* (1.11) depicts the predominant direction of the currents over all the world's oceans. More detailed information can be found in the series of Routeing Charts (1.8) which depict, on a three monthly basis, the predominant direction, rate and constancy of currents in the various oceans.

3   It should be appreciated that the number of current observations available varies enormously, being greatest in the areas of the principal shipping routes. In other parts of the oceans, particularly in polar regions, observations are sparse.

1.20.1

    Predominant direction is the mean direction of the 90° sector containing the greatest number of vector representations of all the current observations in the area.

1.20.2

1   **Constancy**, as indicated by the thickness of the arrow on the diagrams, is a measure of the persistence of a current. Low constancy, for example, implies marked variability in rate and, particularly, in direction.

2   Change in currents is continuous and most marked in tropical waters. Variability of direction occurs in the most constant currents and in many other areas the predominance of the indicated direction is often minimal.

    On average, over the greater part of the oceans the proportion of current observations of less than ½ kn is between 50 and 60 percent.

Ice
1.21

1   Ice limits and drift and ice reporting services in particular areas are given at the beginning of the appropriate chapters of this book. More detailed information regarding ice reports can be found in *Admiralty List of Radio Signal, Volume 3.*

2   Over recent years, with increasing availability of information from satellites, it has become apparent that the variability of ice extent from year to year is probably much greater than previously thought. There may also be a tendency for some trends to continue for a number of years.

Coral waters
1.22

1   Coral reefs are often steep-to, and depths of more than 200 m may be found within 1 cable of the edge of a reef. Soundings are therefore of little value as a warning of their proximity. The soundings furthermore shoal so rapidly that it is sometimes difficult to follow the echo sounder trace and the echo itself is often weak owing to the steep bottom profile.

2   Navigation among coral reefs is therefore almost entirely dependent upon the eye and in ocean areas where these reefs abound the greatest care is required. Wherever possible, passage through the worst parts of such areas should be made in daylight, while every precaution should be taken to keep an accurate check on the ship's position.

3   For further information on navigation in coral waters, see *The Mariner's Handbook* (1.12.11).

Local magnetic anomalies
1.23

1   In parts of the oceans, particularly where volcanic action has taken place, local magnetic anomalies have been reported.

2   For further information on local magnetic anomalies, see *The Mariner's Handbook* (1.12.11), and for details of reports, see the appropriate volume of *Admiralty Sailing Directions* (1.12.1).

PASSAGE PLANNING

Track selection
1.24

    The selection of the best track for a passage demands skilled evaluation of a number of factors, the principal of which are the sea conditions, winds and currents which it is expected to encounter and the way in which the vessel will react to them. Such factors as the likelihood of damage to ship or cargo, as well as fuel economy and time on passage will also need to be considered. Some cargoes, such as those carried on deck and livestock are likely to be more susceptible to the weather than others and may therefore affect the choice of route or speed.

Weather routeing
1.25

1 Having planned the passage taking into account the normal conditions, consideration should be given to the actual conditions which it is expected will be encountered.

2 With the aid of the latest weather forecasts, weather maps and ice charts, a system of weather routeing enables the original route to be modified to make best use of the actual weather pattern and the alterations expected to take place within it. This will produce the greatest economy in fuel expenditure and reduce the risk of heavy weather damage to the ship and her cargo.

3 Weather Routeing Services are provided by certain foreign governments and private firms; details of which can be obtained from *Admiralty List of Radio Signal, Volume 3.*

These services apply latest weather reports and long range forecasts to determine the best route for a particular vessel. On passage, modifications to the route are passed to the vessel to enable early action to avoid developing areas of adverse conditions.

1.25.1

1 The World Meteorological Organization (WMO) has established a global service for the transmission of high seas weather warnings and routine weather bulletins.

2 Meteorological service areas (METAREAS) are identical to the 16 NAVAREAS (1.32)

3 For details of the service, see *Admiralty List of Radio Signals, Volume 3* (1.12.5) and *Annual Summary of Admiralty Notices to Mariners* (1.12.10).

Load Line Rules
1.26

1 The zones, areas and seasonal periods defined in Statutory Instruments 1968 No 1053 — *The Merchant Shipping Load Line Rules, 1968*, as amended, are shown on Diagram 1.26 on the back end-papers of this book.

2 These rules apply to all ships, except ships of war, ships engaged solely in fishing and pleasure yachts.

Offshore installations
1.27

1 In some parts of the world oil and gas fields are found many miles offshore. The rigs, platforms and associated moorings and pipelines used in operating these fields form hazards to navigation. Many fields lie within Prohibited or Restricted Areas and most of the installations are protected by Safety Zones.

2 For descriptions of installations and Safety Zones, see *The Mariner's Handbook* (1.12.11), and *Annual Summary of Admiralty Notices to Mariners* (1.12.10).

3 Offshore windfarms, for generating electricity, are increasingly being developed in various parts of the world and could pose a hazard to shippng.

Traffic Separation Schemes
1.28

1 Where main shipping lanes converge when entering straits, channels or round headlands, or in other areas where traffic is congested, Traffic Separation Schemes (TSS) have been established.

2 All such schemes are listed in *Annual Summary of Admiralty Notices to Mariners* (1.12.10) which indicates which schemes have been adopted by the IMO; they are also shown on appropriate charts and referred to in *Admiralty Sailing Directions* (1.12.1).

3 Rule 10 of *International Regulations for Preventing Collisions at Sea 1972* applies to all IMO-adopted schemes. Regulations for unadopted schemes are given in *Admiralty Sailing Directions*.

4 In this book reference is made to Traffic Separation Schemes in waterways through which the various routes described pass, but not to those in port approaches or inner coastal waters.

Archipelagic Sea Lanes
1.29

Within this book Archipelagic Sea Lanes have been established in Indonesian waters. For full details, see Appendix A.

Admiralty charts show all adopted archipelagic sea lanes, including the axis lines and the lateral limits of the sea lanes.

Areas to be Avoided
1.30

1 Areas which are to be avoided for various reasons have been established in some parts of the world.

2 All such areas, which are IMO-approved, are shown on Admiralty Charts and described in *Admiralty Sailing Directions* (1.12.1).

Maritime Safety Information (MSI)
1.31

1 MSI is defined as "navigational and meteorological forecasts and other urgent safety-related messages", of vital importance to all ships at sea.

2 It is a sub-system of the Global Maritime Distress and Safety System (GMDSS).

3 Full details of MSI and GMDSS can be found in *Admiralty List of Radio Signals, Volumes 3* and *5* respectively (1.12.5).

World-Wide Navigational Warning Service (WWNWS)
1.32

1 Navigational warning messages to give early information of important incidents or changes which may constitute a danger to navigation are promulgated by the WWNWS. The service consists of three warning systems:

1.32.1

NAVAREA warnings which are issued for each of the 16 geographical areas covering the sea areas of the world. They normally provide sufficient information to enable vessels to pass in safety through main shipping lanes clear of the coast.

1.32.2

Coastal warnings which are used for information of importance only in a particular region. They are broadcast by the country concerned.

1.32.3

Local warnings usually refer to inshore waters and do not carry information needed for ocean passages. They are often issued by coastguards, port or pilotage authorities.

For details of the service, see *Admiralty List of Radio Signals, Volumes 3 and 5* (1.12.5) and *Annual Summary of Admiralty Notices to Mariners* (1.12.10).

Ship Reporting Systems
1.33

1   The purpose of such systems are either to:

Enable the positions of ships in oceanic areas to be maintained ashore in order to facilitate search and rescue operations when they are required. Ships of all nations are encouraged to participate in these systems. For details see *Admiralty List of Radio Signals, Volume 1* (1.12.5).

Regulate traffic in ports and their approaches. For details of these systems see *Admiralty List of Radio Signals, Volume 6* (1.12.5).

2   AMVER (Automated Mutual-assistance Vessel Rescue System) is such a system and is operated on a world wide basis by the US Coast Guard. Certain other countries operate similar systems for particular areas off their coasts.

For AMVER and some of the systems participation is voluntary while for others it is mandatory.

3   For further details of these systems, see *Annual Summary of Admiralty Notices to Mariners* (1.12.10).

Pollution of the Sea
1.34

1   International regulations concerning pollution of the sea by oil or other substances are contained in *International Convention for the Prevention of Pollution from Ships, 1973* (MARPOL 1973) as amended by the Protocol of 1978 and known as *MARPOL 73/78*. Further information can be found in *The Mariner's Handbook* (1.12.11).

2   Details of "Pollution Reports by Radio" are given in *Admiralty List of Radio Signals, Volume 1* (1.12.5).

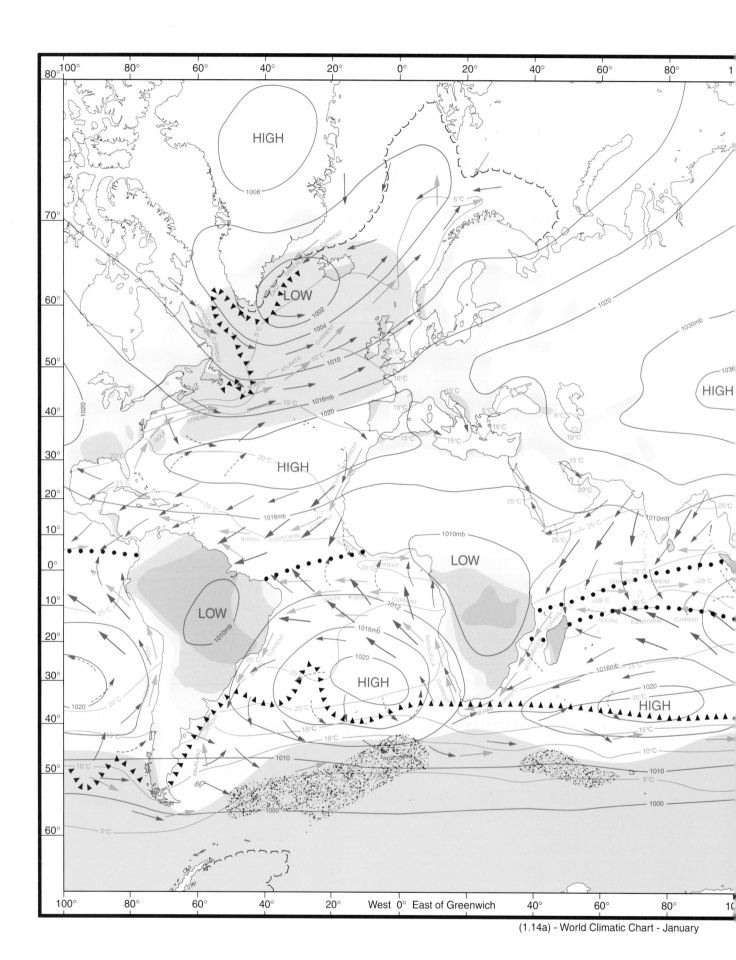

(1.14a) - World Climatic Chart - January

JANUARY

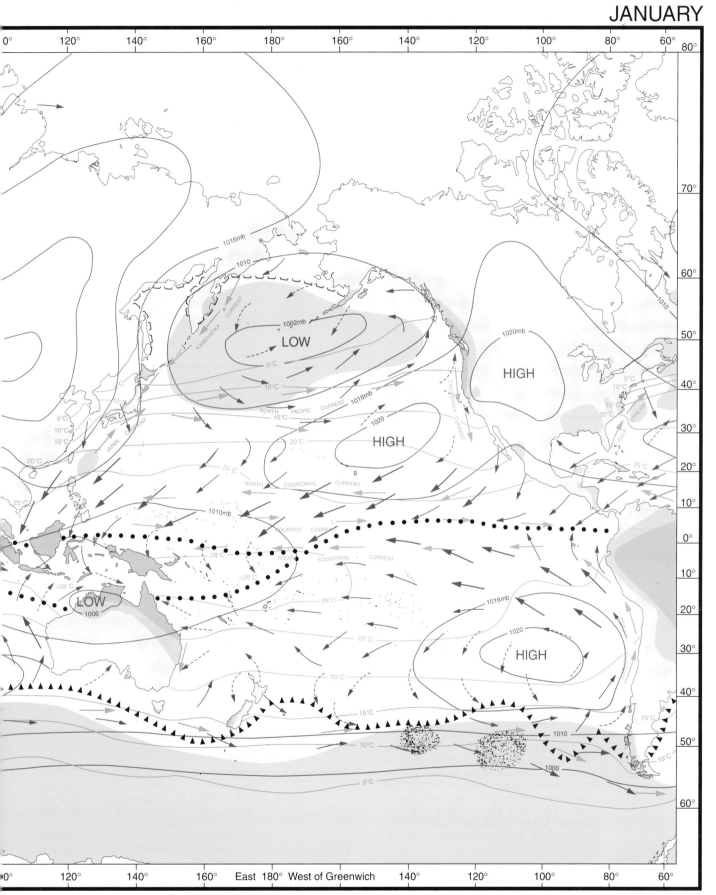

(Diagram 1.14a)

Explanatory Notes for diagram 1.14a (January)

GENERAL

The information incorporated in this chart deals with the average, not actual, conditions and this is therefore a Climactic, and not a Weather Chart.

Unless otherwise stated, the chart represents the characteristic state of affairs for January (approximately the height of the northern winter and southern summer).

For detailed information about any particular part of the world, reference should be made to the appropriate Weather Handbooks, Climactic and Current Atlases, Admiralty Pilots and Ice Charts.

PRESSURE

——— 1016mb ———

Pressure, like wind, is variable in quantity. On any given day in January the actual pressures recorded and the arrangement of lines of equal pressure(= isobars) may be quite different from those shown on this chart. This is especially likely to be the case with the "travelling Lows" (or depressions) which usually move from West to East across the temperature zone. The "Highs" (or anticyclones) are much less mobile and their day-to-day variations of pressure much smaller. This is the main reason why the Trades, for instance, are steadier in force and direction than the Westerlies of the North Atlantic.

WINDS

The arrows give the general picture of the prevailing winds over the oceans. Where these winds are unusually steady, as in the Trades, the arrows have been strengthened: in such cases it can be assumed that the winds blow in the direction indicated on more than 2/3rds of all occasions. Where the arrows are broken or omitted altogether, as in the heart of the High and Low pressure areas, there the winds are variable.

GALE FREQUENCY

5 - 10 days/month

> 10 days/month

Gales and low pressures usually go together: this is the main reason why they are mostly confined to the extra-tropical depression zones of the world. Winds of force 7 or more occur on more than 10 days a month in the darker-tinted areas. Such winds occur on 5-10 days a month in the lighter tinted areas.

SWELL

No attempt has been made to show the distribution of swell, but it can be inferred broadly from information on the chart. Low or moderate swell is liable to occur to the leeward of all the main wind belts; there is, for instance, a persistent short and low - at times moderate - swell along the E. African coast during the N.E. monsoon. Heavy swell is generally only produced in the extra-tropical gale zones; however, in favourable circumstances it generates up to 3,000 miles beyond the confines of these zones. These circumstances are most likely to arise in the rear of the eastward-moving depressions where, in the Northern Hemisphere, the winds are N. Westerly and in the Southern Hemisphere, S. Westerly. From this it follows (i) that heavy swell is more likely to out-run the gale zones on their tropical, than on their poleward flanks, and (ii) that it is more likely to be experienced in the eastern than in the western sides of the oceans.

TROPICAL STORMS

In low latitudes gales are almost entirely confined to those areas frequented by tropical storms. In a normal year the number of such storms seldom exceeds two a month in any one area. In the South Indian and South Pacific Oceans the cyclone season usually lasts from December to April; in the South Atlantic and north of the equator there are generally no tropical storms in these months.

INTERTROPICAL CONVERGENCE ZONE (EQUATORIAL TROUGH) (DOLDRUMS)

• • • • • •

The chart shows the mean monthly position of the Intertropical Convergency Zone. The severity of the weather near the Zone depends upon the degree of wind convergence occurring there and varies both in space and time from clear skies, when there is no convergence, to squalls, with heavy rain and thunderstorms, when the convergence is marked.

CURRENTS

Only the location and direction of the main ocean currents are indicated. Except where the winds are generally light and/or variable, as in the vicinity of the Equatorial Counter Currents, there is close agreement between the direction of the current and that of the prevailing wind. A direct result of this is that, according to whether the wind blows from warmer to colder latitudes, or vice versa, the surface water temperatures will be above, or below, the average for the latitude. Some idea of the relative warmth or coldness of the currents is given by the extent to which the sea isotherms in the vicinity of the current arrows bend polewards or equatorwards respectively; thus the Benguella Current is obviously cold, the Agulhas Current warm.

SEA TEMPERATURE

——— 25°C ———

Unlike pressure and winds, the day-to-day variation of sea temperature is very small; monthly changes of more than 1°- 2°C are unusual, and year-to-year departures from the mean monthly value are of the same low order.

In the open ocean, and along leeward shores, sea and air temperatures agree fairly closely (within 1° or so); thus February sea temperatures off Scilly is approximately 9°-10°C. and the average air temperature for the same place 8°C. Along windward shores, especially in high latitudes, differences up to 10°- 20°C. are not uncommon. Sometimes the air will be much colder than the sea (e.g. in high latitudes in winter); at others (e.g. in summer along the west coast of S. Africa) the sea will have the lower temperature.

FOG

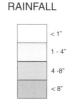 at least 5 days/month

Sea fog is likely to form wherever warm air passes over cold water; it is most likely to be persistent where the sea isotherms lying athwart the track of a wind blowing from warmer to colder waters are packed closely together. Areas in which sea fog may be encountered on at least 5 days a month are indicated by a light stipple.

ICE

Ice is about the most variable of all the elements depicted on this chart. The year-to-year fluctuations in the limit of pack-ice (or icebergs for that matter) are often very considerable. For instance, in some seasons Jan Mayen remains almost completely ice-free; in others it cannot be approached by a ship not specially strengthened, until July. Icebergs (formed by the "calving" of ice from continental ice-caps and valley glaciers) are liable to be encountered beyond the pack-ice limits at all seasons of the year, but mostly in early summer. They melt fast once they have drifted into comparatively warm water e.g. the Gulf Stream.

NORTHERN HEMISPHERE

ᴗ ᴗ ᴗ ᴗ ᴗ Mean limit of 4/8 pack-ice at time of greatest extent (March, but February in Gulf of St. Lawrence)

▼▼▼▼▼▼▼▼ Approximate limit of Icebergs - October to November (mean least extent)

SOUTHERN HEMISPHERE

◠ ◠ ◠ ◠ ◠ Mean Limit of 4/8 pack-ice at time of least extent (February, March)

▲▲▲▲▲▲▲▲ Extreme limit of iceberg sightings (at all times of the year)

RAINFALL

< 1"

1 - 4"

4 - 8"

< 8"

This is another uncertain quantity, and the amount falling in a particular January may bear little relation to the averages shown on the chart; especially is this the case with areas lying near the lower limit of the 1" - 4" zone. Over the oceans the only habitually rainy areas are the Doldrums and the storm (depression) belts of middle and high latitudes. The Trades and N.E. Monsoon blowing, it will be noticed, off arid lands, are practically rainless.

Explanatory Notes for diagram 1.14b (July)

GENERAL

The information incorporated in this chart deals with average, not actual, conditions and this is therefore a Climactic, and not a Weather Chart.

Unless otherwise stated, the chart represents the characteristic state of affairs for July (approximately the height of the northern summer and the southern winter).

For detailed information about any particular part of the world, reference should be made to the appropriate Weather Handbooks, Climactic and Current Atlases, Admiralty Pilots and Ice Charts.

PRESSURE

——— 1016mb ———

Pressure, like wind, is variable in quantity. On any given day in July the actual pressures recorded and the arrangement of lines of equal pressure(= isobars) may be quite different from those shown on this chart. This is especially likely to be the case with the "travelling Lows" (or depressions) which usually move from West to East across the temperature zone. The "Highs" (or anticyclones) are much less mobile and their day-to-day variations of pressure much smaller. This is the main reason why the Trades, for instance, are steadier in force and direction than the Westerlies of the North Atlantic.

WINDS

The arrows give the general picture of the prevailing winds over the oceans. Where these winds are unusually steady, as in the Trades, the arrows have been strengthened: in such cases it can be assumed that the winds blow in the direction indicated on more than 2/3rds of all occasions. Where the arrows are broken or omitted altogether, as in the heart of the High and Low pressure areas, there the winds are variable.

GALE FREQUENCY

5 - 10 days/month

> 10 days/month

Gales and low pressures usually go together: this is the main reason why they are mostly confined to the extra-tropical depression zones of the world. Winds of force 7 or more occur on more than 10 days a month in the darker-tinted areas. Such winds occur on 5-10 days a month in the lighter tinted areas.

SWELL

No attempt has been made to show the distribution of swell, but it can be inferred broadly from information on the chart. Low or moderate swell is liable to occur to the leeward of all the main wind belts; there is, for instance, a persistent moderate - at times heavy - swell along the N.W. coast of India during the S.W. monsoon. Heavy swell is generally only produced in the extra-tropical gale zones; however, in favourable circumstances it penetrates up to 3,000 miles beyond the confines of these zones. These circumstances are most likely to arise in the rear of the eastward-moving depressions where, in the Northern Hemisphere, the winds are North Westerly and in the Southern Hemisphere, South Westerly. From this it follows (i) that heavy swell is more likely to out-run the gale zones on their tropical, than on their poleward flanks; and (ii) that it is more likely to be experienced in the eastern than in the western side of the oceans.

TROPICAL STORMS

In low latitudes gales are almost entirely confined to those areas frequented by tropical storms. In a normal year the number of such storms seldom exceeds two a month in any one area. In the North Indian and North Pacific Oceans the cyclone season usually lasts from May to December; in the North Atlantic form May to November. South of the Equator there are no tropical storms in these (winter) months.

INTERTROPICAL CONVERGENCE ZONE (EQUATORIAL TROUGH) (DOLDRUMS)

• • • • • •

The chart shows the mean monthly position of the Intertropical Convergency Zone. The severity of the weather near the Zone depends upon the degree of wind convergence occurring there and varies both in space and time from clear skies, when there is no convergence, to squalls, with heavy rain and thunderstorms, when the convergence is marked.

CURRENTS

———→

Only the location and direction of the main ocean currents are indicated. Except where the winds are generally light and/or variable, as in the vicinity of the Equatorial Counter Currents, there is close agreement between the direction of the current and that of the prevailing wind. A direct result of this is that, according to whether the wind blows from warmer to colder latitudes, or vice versa, the surface water temperatures will be above, or below, the average for the latitude. Some idea of the relative warmth or coldness of the currents is given by the extent to which the sea isotherms in the vicinity of the current arrows bend polewards or equatorwards respectively; thus the Benguella Current is obviously cold, the Agulhas Current warm.

SEA TEMPERATURE

——— 25°C ———

Unlike pressure and winds, the day-to-day variation of sea temperature is very small; monthly changes of more than 1°- 2°C are unusual, and year-to-year departures from the mean monthly value are of the same low order.

In the open ocean, and along leeward shores, sea and air temperatures agree fairly closely (within 1° or so); thus August sea temperature off Scilly is approximately 15°-16°C. and the average air temperature for the same place 17°C. Along windward shores, especially in high latitudes, differences up to 10°- 20°C. are not uncommon. Sometimes the air will be much colder than the sea (e.g. in high latitudes in the winter hemisphere); at others (e.g. in summer along Trade wind coasts) the sea will have the lower temperature.

FOG

 at least 5 days/month

>10 days/month

Sea fog is likely to form wherever warm air passes over cold water; it is most likely to be persistent where the sea isotherms lying athwart the track of a wind blowing from warmer to colder waters are packed closely together. Areas in which sea fog may be encountered on at least 5-10 days a month are indicated by a light stipple; 10-20 days a month areas are shown by a darker stipple.

ICE

Ice is about the most variable of all the elements depicted on this chart. The year-to-year fluctuations in the limit of pack-ice (or icebergs for that matter) are often very considerable. For instance, in some seasons Jan Mayen remains almost completely ice-free; in others it cannot be approached by a ship not specially strengthened, until July. Icebergs (formed by the "calving" of ice from continental ice-caps and valley glaciers) are liable to be encountered beyond the pack-ice limits at all seasons of the year, but mostly in early summer. They melt fast once they have drifted into comparatively warm water e.g. the Gulf Stream.

NORTHERN HEMISPHERE

⌣ ⌣ ⌣ ⌣ ⌣ ⌣　Mean limit of 4/8 pack-ice at time of least extent (September)

▼▼▼▼▼▼▼▼　Approximate limit of Icebergs - March to June (mean greatest extent)

This does not include exceptional sightings in the Eastern part of the North Atlantic

SOUTHERN HEMISPHERE

⌐ ⌐ ⌐ ⌐ ⌐ ⌐　Mean Limit of 4/8 pack-ice at time of greatest extent (September and October)

▲▲▲▲▲▲▲▲　Extreme limit of iceberg sightings (at all times of the year)

RAINFALL

< 1"

1 - 4"

4 -8"

< 8"

This is another uncertain quantity, and the amount falling in a particular July may bear little relation to the averages shown here; especially is this the case with areas lying near the lower limit of the 1" - 4" zone. Over the oceans the only habitually rainy areas are the Doldrums and the storm (depression) belts of middle and high latitudes. The Trades and N.E. and S.E. Trades, blowing, it will be noticed, off arid lands, are practically rainless.

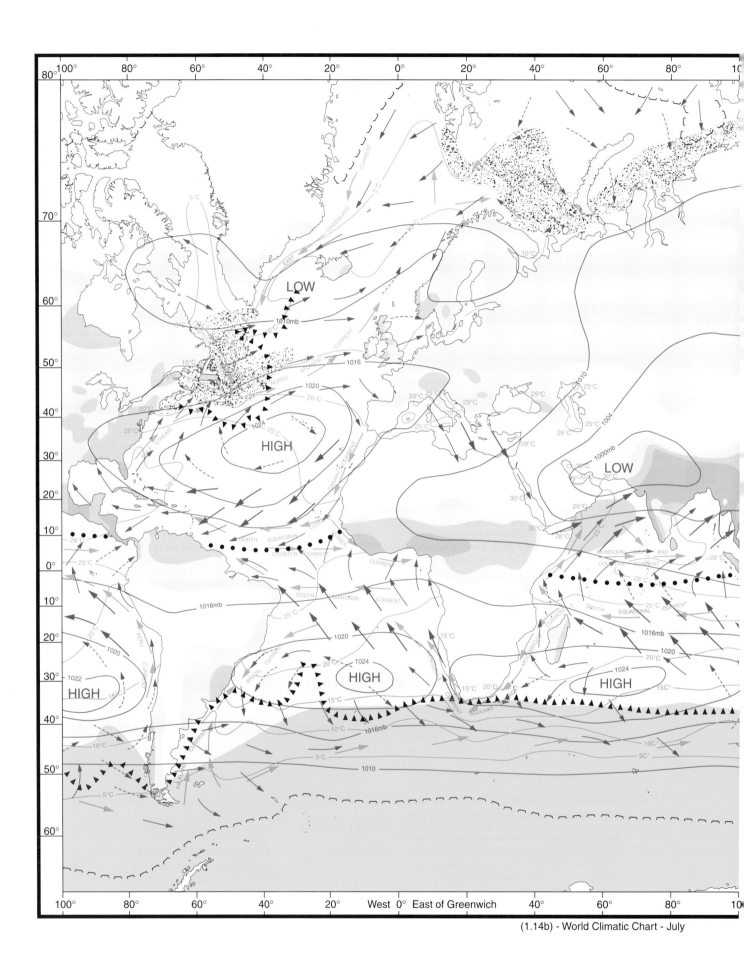

(1.14b) - World Climatic Chart - July

JULY

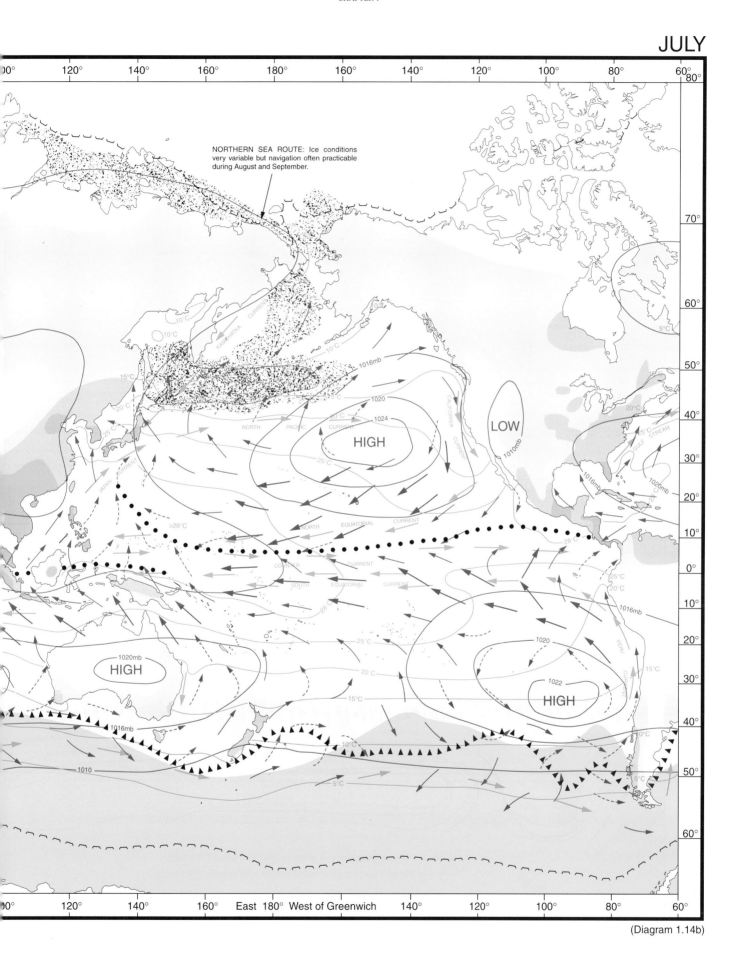

NORTHERN SEA ROUTE: Ice conditions very variable but navigation often practicable during August and September.

(Diagram 1.14b)

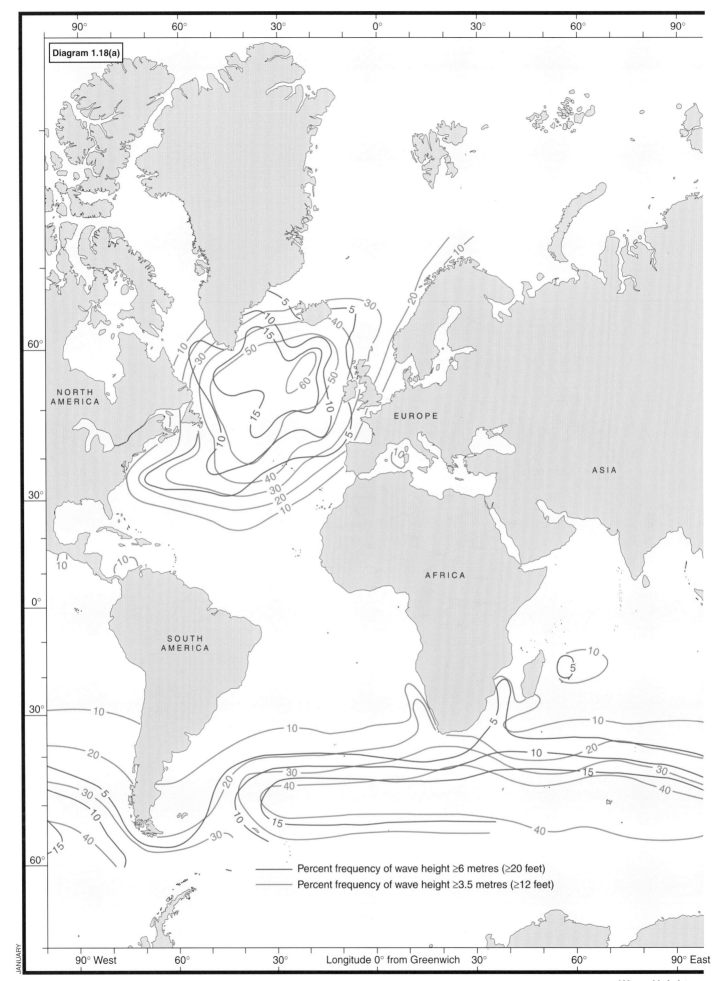

Diagram 1.18(a)

Percent frequency of wave height ≥6 metres (≥20 feet)
Percent frequency of wave height ≥3.5 metres (≥12 feet)

NORTH
AMERICA

EUROPE

ASIA

AFRICA

SOUTH
AMERICA

JANUARY

90° West   60°   30°   Longitude 0° from Greenwich   30°   60°   90° East

Wave Heights –

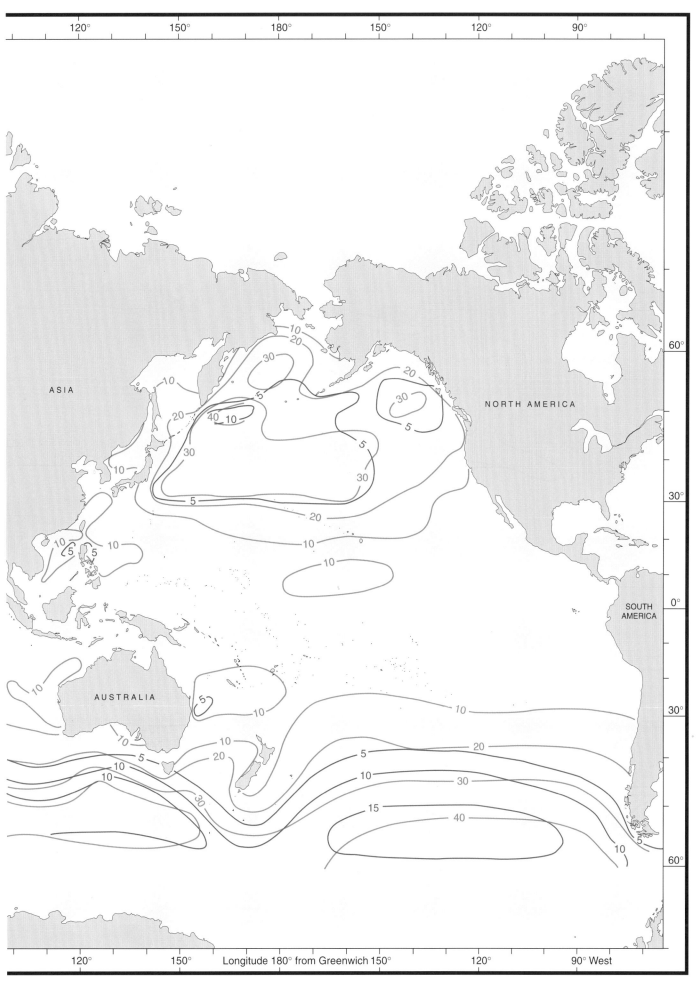

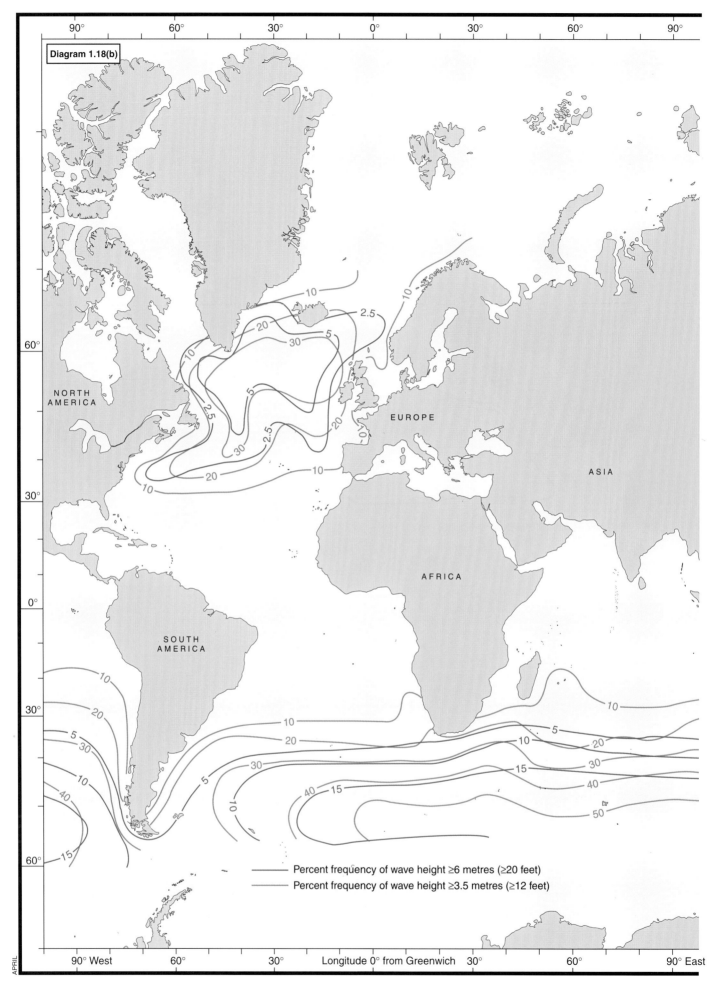

Diagram 1.18(b)

Percent frequency of wave height ≥6 metres (≥20 feet)
Percent frequency of wave height ≥3.5 metres (≥12 feet)

90° West   60°   30°   Longitude 0° from Greenwich   30°   60°   90° East

Wave Heights -

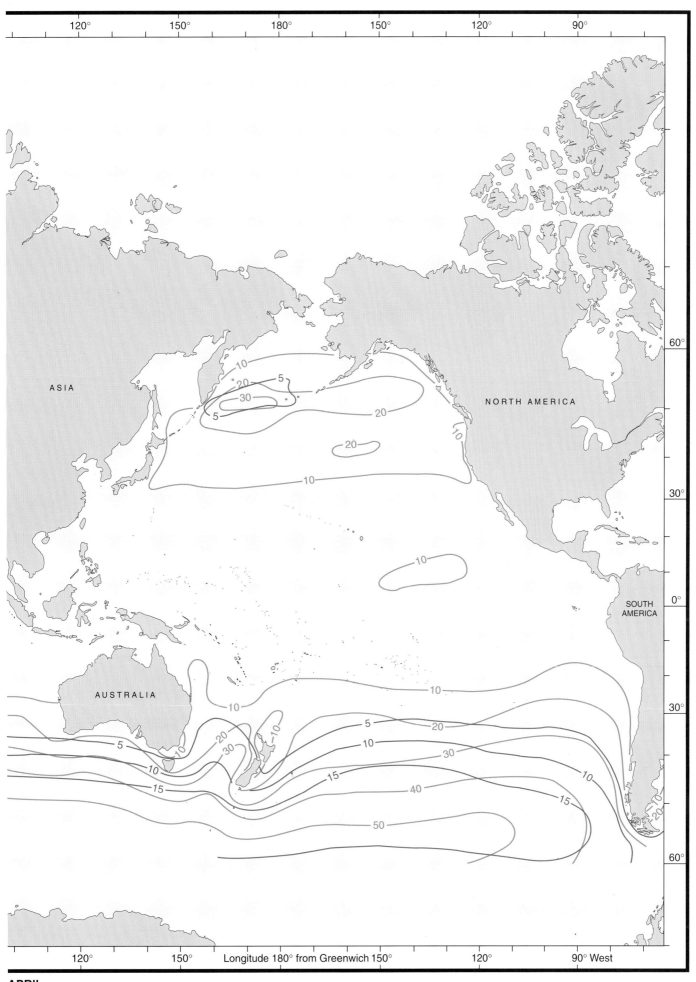

APRIL

19

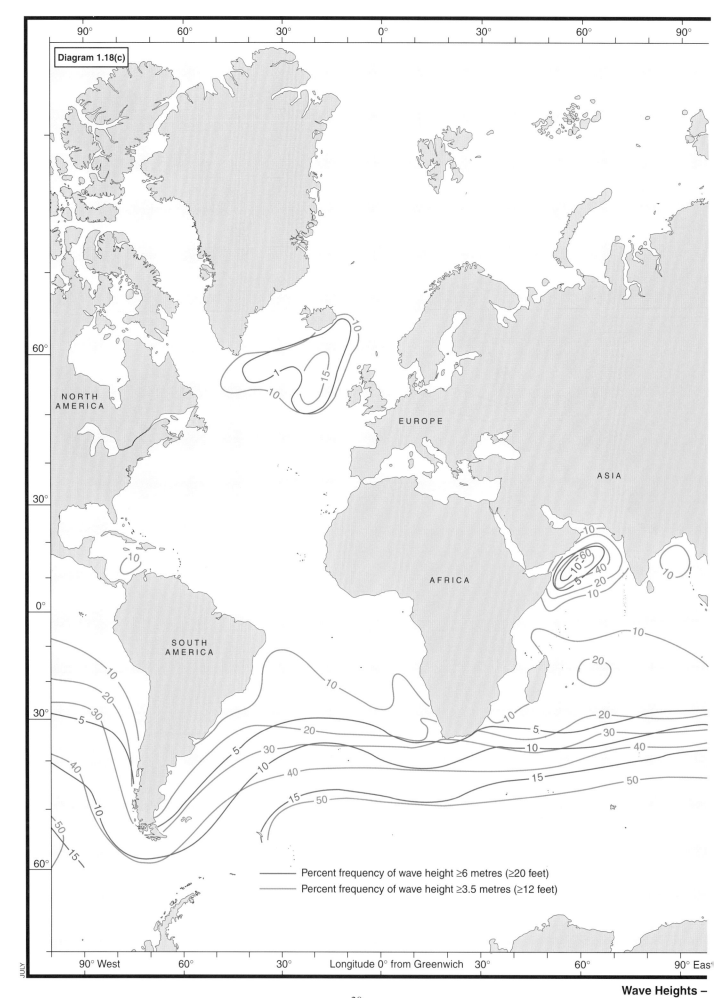

Diagram 1.18(c)

90°   60°   30°   0°   30°   60°   90°

60°

NORTH
AMERICA

EUROPE

ASIA

30°

AFRICA

0°

SOUTH
AMERICA

30°

60°

——— Percent frequency of wave height ≥6 metres (≥20 feet)
——— Percent frequency of wave height ≥3.5 metres (≥12 feet)

90° West   60°   30°   Longitude 0° from Greenwich   30°   60°   90° East

JULY

20

Wave Heights –

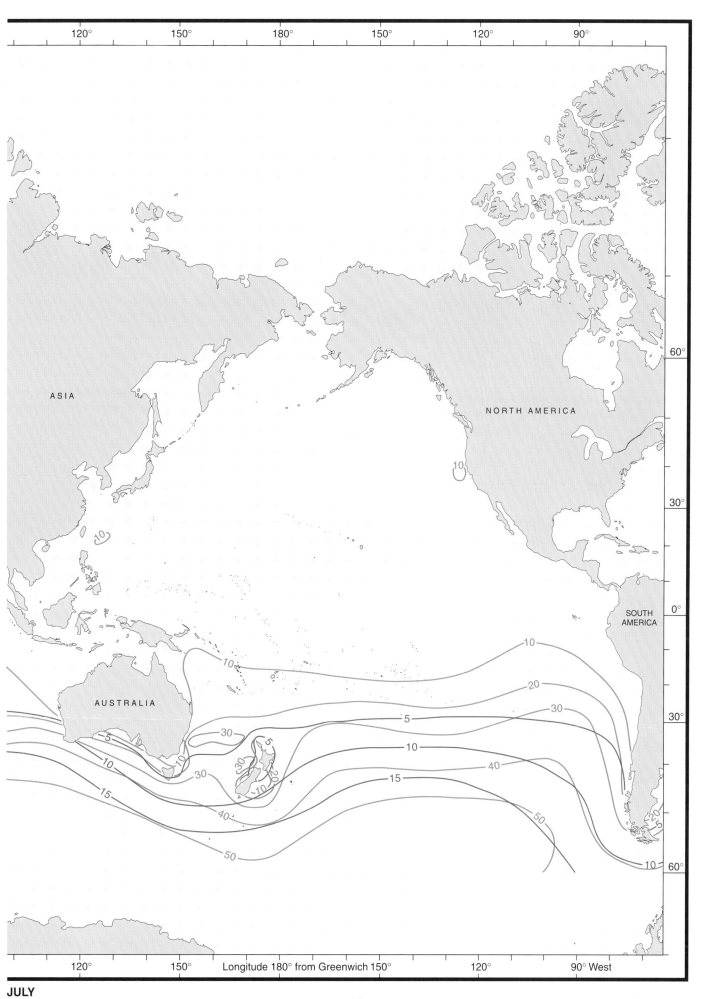

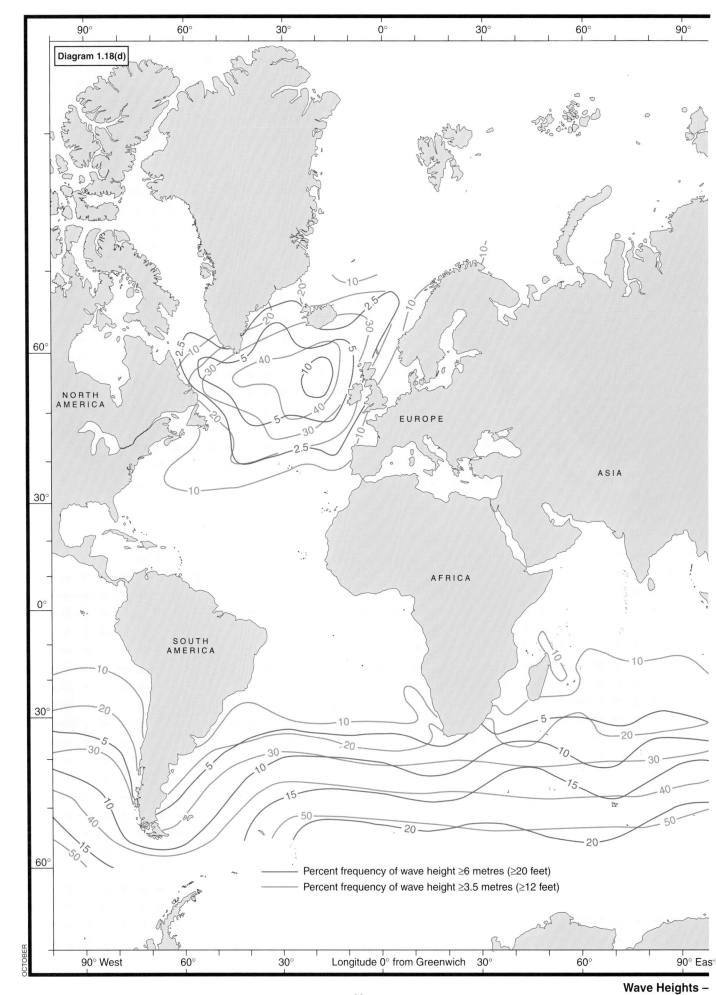

Diagram 1.18(d)

Percent frequency of wave height ≥6 metres (≥20 feet)
Percent frequency of wave height ≥3.5 metres (≥12 feet)

90° West 60° 30° Longitude 0° from Greenwich 30° 60° 90° East

OCTOBER

22

Wave Heights –

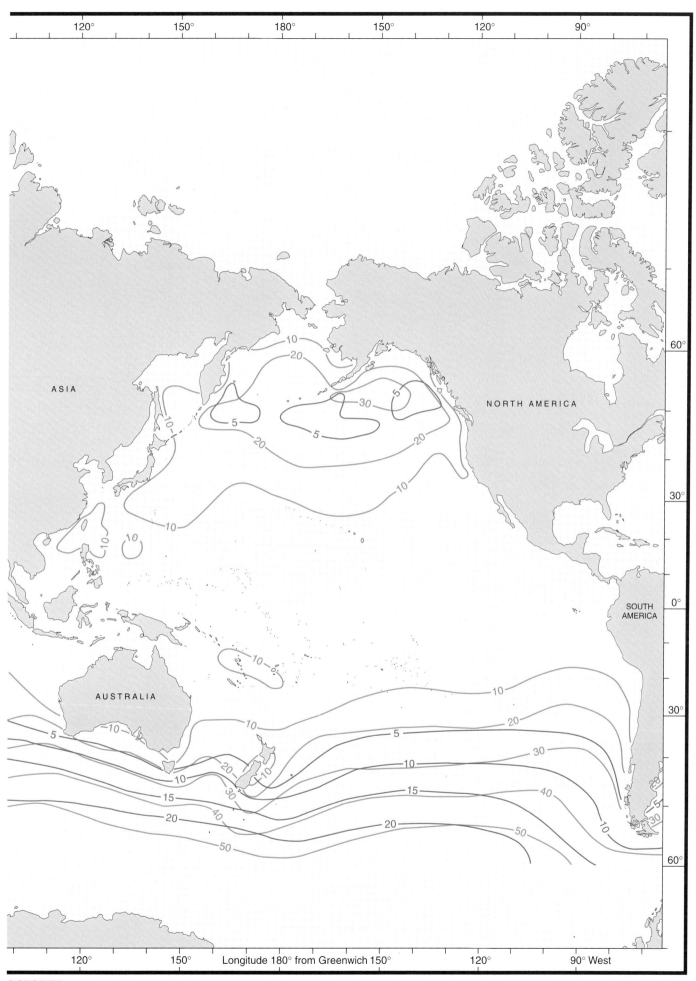

ASIA

NORTH AMERICA

AUSTRALIA

SOUTH AMERICA

60°

30°

0°

30°

60°

120°   150°   Longitude 180° from Greenwich 150°   120°   90° West

OCTOBER

METEOROLOGICAL CONVERSION TABLE AND SCALES

Fahrenheit to Celsius
°Fahrenheit

	0	1	2	3	4	5	6	7	8	9
°F					Degrees Celsius					
-100	-73·3	-73·9	-74·4	-75·0	-75·6	-76·1	-76·7	-77·2	-77·8	-78·3
-90	-67·8	-68·3	-68·9	-69·4	-70·0	-70·6	-71·1	-71·7	-72·2	-72·8
-80	-62·2	-62·8	-63·3	-63·9	-64·4	-65·0	-65·6	-66·1	-66·7	-67·2
-70	-56·7	-57·2	-57·8	-58·3	-58·9	-59·4	-60·0	-60·6	-61·1	-61·7
-60	-51·1	-51·7	-52·2	-52·8	-53·3	-53·9	-54·4	-55·0	-55·6	-56·1
-50	-45·6	-46·1	-46·7	-47·2	-47·8	-48·3	-48·9	-49·4	-50·0	-50·6
-40	-40·0	-40·6	-41·1	-41·7	-42·2	-42·8	-43·3	-43·9	-44·4	-45·0
-30	-34·4	-35·0	-35·6	-36·1	-36·7	-37·2	-37·8	-38·3	-38·9	-39·4
-20	-28·9	-29·4	-30·0	-30·6	-31·1	-31·7	-32·2	-32·8	-33·3	-33·9
-10	-23·3	-23·9	-24·4	-25·0	-25·6	-26·1	-26·7	-27·2	-27·8	-28·3
-0	-17·8	-18·3	-18·9	-19·4	-20·6	-20·6	-21·1	-21·7	-22·2	-22·8
+0	-17·8	-17·2	-16·7	-16·1	-15·6	-15·0	-14·4	-13·9	-13·3	-12·8
10	-12·2	-11·7	-11·1	-10·6	-10·0	-9·4	-8·9	-8·3	-7·8	-7·2
20	-6·7	-6·1	-5·6	-5·0	-4·4	-3·9	-3·3	-2·8	-2·2	-1·7
30	-1·1	-0·6	0	+0·6	+1·1	+1·7	+2·2	+2·8	+3·3	+3·9
40	+4·4	+5·0	+5·6	6·1	6·7	7·2	7·8	8·3	8·9	9·4
50	10·0	10·6	11·1	11·7	12·2	12·8	13·3	13·9	14·4	15·0
60	15·6	16·1	16·7	17·2	17·8	18·3	18·9	19·4	20·0	20·6
70	21·1	21·7	22·2	22·8	23·3	23·9	24·4	25·0	25·6	26·1
80	26·7	27·2	27·8	28·3	28·9	29·4	30·0	30·6	31·1	31·7
90	32·2	32·8	33·3	33·9	34·4	35·0	35·6	36·1	36·7	37·2
100	37·8	38·3	38·9	39·4	40·0	40·6	41·1	41·7	42·2	42·8
110	43·3	43·9	44·4	45·0	45·6	46·1	46·7	47·2	47·8	48·3
120	48·9	49·4	50·0	50·6	51·1	51·7	52·2	52·8	53·3	53·9

Celsius to Fahrenheit
°Celsius

	0	1	2	3	4	5	6	7	8	9
°C					Degrees Fahrenheit					
-70	-94·0	-95·8	-97·6	-99·4	-101·2	-103·0	-104·8	-106·6	-108·4	-110·2
-60	-76·0	-77·8	-79·6	-81·4	-83·2	-85·0	-86·8	-88·6	-90·4	-92·2
-50	-58·0	-59·8	-61·6	-63·4	-65·2	-67·0	-68·8	-70·6	-72·4	-74·2
-40	-40·0	-41·8	-43·6	-45·4	-47·2	-49·0	-50·8	-52·6	-54·4	-56·2
-30	-22·0	-23·8	-25·6	-27·4	-29·2	-31·0	-32·8	-34·6	-36·4	-38·2
-20	-4·0	-5·8	-7·6	-9·4	-11·2	-13·0	-14·8	-16·6	18·4	-20·2
-10	+14·0	+12·2	+10·4	+8·6	+6·8	+5·0	+3·2	+1·4	-0·4	-2·2
-0	32·0	30·2	28·4	26·6	24·8	23·0	21·2	19·4	+17·6	+15·8
+0	32·0	33·8	35·6	37·4	39·2	41·0	42·8	44·6	46·4	48·2
10	50·0	51·8	53·6	55·4	57·2	59·0	60·8	62·6	64·4	66·2
20	68·0	69·8	71·6	73·4	75·2	77·0	78·8	80·6	82·4	84·2
30	86·0	87·8	89·6	91·4	93·2	95·0	96·8	98·6	100·4	102·2
40	104·0	105·8	107·6	109·4	111·2	113·0	114·8	116·6	118·4	120·2
50	122·0	123·8	125·6	127·4	129·2	131·0	132·8	134·6	136·4	138·2

HECTOPASCALS TO INCHES

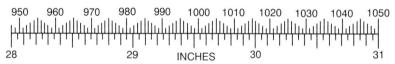

MILLIMETRES TO INCHES

(1) (for small values)

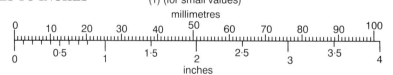

(2) (for large values)

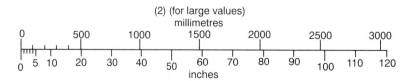

CHAPTER 2

NORTH ATLANTIC OCEAN

GENERAL INFORMATION

COVERAGE

Chapter coverage
2.1

This chapter contains details of the following passages:

To and from Cape Farvel and Davis Strait (2.46).

To and from Strait of Belle Isle (2.56).

To and from St John's Harbour, Newfoundland (2.63).

Between Cabot Strait or N American ports and Europe (2.66).

Between North American ports, West Indies or Caribbean Sea and Africa (2.73).

Between West Indies or Bermuda and Europe (2.78).

Between English Channel and Strait of Gibraltar or intermediate ports (2.87).

Between English Channel or Strait of Gibraltar and west coast of Africa, Cape Town or Indian Ocean (2.89).

Between North America and places between Gulf of Mexico and Cabo Calcanhar (2.95).

Between North-east coast of South America and eastern part of North Atlantic Ocean (2.103).

Between eastern part of North Atlantic Ocean and Recife (2.106).

To and from Arquipélago dos Açores (2.113).

To and from Arquipélago de Cabo Verde (2.116).

Admiralty Publications
2.2

Relevant navigational publications should be consulted, in addition to this book, when planning and conducting passages. Details of these, and the areas covered, can be found, as follows:

Admiralty Sailing Directions; see 1.12.1

Admiralty List of Lights and Fog Signals; see 1.12.2

Admiralty Tide Tables; see 1.12.3

Admiralty Tidal Stream Atlas; see 1.12.4

Admiralty List of Radio Signals; see 1.12.5

Admiralty Distance Tables; see 1.12.8

2.2.1

See also *The Mariner's Handbook* for notes on electronic products.

NATURAL CONDITIONS

Charts:

North Atlantic Routeing 5124 (1–12)

World Surface Current Distribution 5310

WINDS AND WEATHER
Intertropical Convergence Zone
2.3

1   In the North Atlantic Ocean, the belt of calms and light variable winds, known as the Intertropical Convergence Zone, Equatorial Trough or Doldrums which lies between the Trade Winds of the two hemispheres remains N of the equator throughout the year.

2   The actual position is subject to much day-to-day variation, as is also the width of the zone, which averages about 200 to 300 miles but may at times be reduced to almost nothing by an increase in strength of one or both Trade Winds.

3   There is evidence to show that showers, squalls or thunderstorms are more common within 200 to 300 miles from the African coast than in the W part of the area.

4   Visibility in the Intertropical Convergence Zone is invariably good except in rain.

South-west Monsoon
2.4

1   In summer the intense heating of the land mass of N Africa lowers the atmospheric pressure over that area and distorts the Equatorial Trough towards N. The south-east Trade Wind (3.4) is drawn across the equator and is forced to veer by the earth's rotation, so that it arrives off the W coast of Africa between the equator and about 15°N, to the

E of about 20°W, as a SW wind which is known as the south-west monsoon.

1   This monsoon, which is accompanied by cloudy weather and considerable rainfall, lasts from about June until the middle of October; the rainfall is heavy on the coast between The Gambia and Liberia. Visibility is good at this season except in rain.

2   During the rest of the year winds in this area are mainly N between Liberia and Mauretania, but are mostly from between S and W in the Gulf of Guinea; in both cases they are generally light. Between November and February a dry, dust laden wind known as the "Harmattan" occurs at times. Weather at this season is generally fine, but visibility is often only moderate due to haze and it may become poor while the Harmattan is blowing.

3   Towards the beginning and end of the rainy season, that is April to May and October to November, violent thunderstorms accompanied by severe squalls, generally from the E, occur at times near the coast. These are known locally as "Tornadoes", but they should not be confused with storms of that name which occur in the interior of the United States and of Australia, to which they bear no relation.

North-east Trade Wind
2.5

1   The North-east Trade Wind forms the SE and equatorial sides of the clockwise circulation round the oceanic anticyclone situated in about 30°N. This Trade Wind belt extends from the African coast as far W as the Caribbean Sea and the Gulf of Mexico, blowing from the NNE on the E side of the ocean and from a little N of E in the W part of the zone. The S limit of the North-east Trade Wind is marked by the Intertropical Convergence Zone.

2   The winds blow permanently with an average strength of force 4, though on rare occasions they may increase to force 7 or decrease to force 2. In the Gulf of Mexico (4.3) they are more variable both in direction and strength; between October and April they are sometimes interrupted in that area by strong or gale force N winds, known as "Northers".

3   In the NE part of the Trade Wind zone the weather is generally fair or fine with small amounts of detached cumulus and little or no rain. Cloud cover and showers increase towards the Intertropical Convergence Zone and towards the W part; in the latter area rain is comparatively frequent, particularly in summer.

4   Haze occurs frequently in the E part of the Trade Wind zone; it is caused by the dust or sand carried seaward by the prevailing offshore wind. Sea fog forms at times in the NE part of the zone over the cold water of the Canary Current (2.14). In the W part of the zone, visibility is good except in rain.

The Variables (Horse Latitudes)
2.6

1   A belt of generally light or variable winds over the oceanic area of high pressure extends across the ocean in about 30°N, oscillating from about 28°N in winter to 32°N in summer. The predominant winds in this area, E of about 20°W in winter and 30°W in summer, are from between N and NE and form an extension of the North-east Trade Winds, particularly in summer.

2   Weather in the E part of the zone is fine with little cloud; in the W part there is more cloud and rain is fairly common. Visibility in the E part is is often reduced by haze and sometimes by fog, see 2.4.

Hurricanes
2.7

1   Hurricanes occur in the W part of the North Atlantic Ocean. They affect in particular the Caribbean Sea, the Gulf of Mexico, Florida, the Bahamas and Bermuda, with the adjacent sea areas. They occur from June to November and sometimes in May and December, with their greatest frequency from August to October.

2   More detailed information on storm frequencies will be found in the relevant volumes of *Admiralty Sailing Directions*. Notes of precursory signs and avoiding action are published in *The Mariner's Handbook*.

Westerlies
2.8

1   The N part of the Atlantic Ocean experiences predominantly unsettled weather on the polar side of the oceanic anticyclone. As a result of the almost continuous passage of depressions across this zone in an E or NE direction, the wind varies greatly in both direction and strength and there is a high frequency of strong winds. Gales are common especially in winter. The stormiest belt extends roughly from the vicinity of Newfoundland to the channel between Iceland and Føroyar. The central and E sections of this belt are especially stormy and winds of force 7 and over may be expected on 16 to 20 days per month in January and February. In July, which is the quietest month, the stormiest area remains SW of Iceland but the frequency of winds of force 7 and above is only about 7 days a month in it. Close to the coasts of Greenland, Iceland and Norway, katabatic winds are common.

Fog and visibility
2.9

1   The frequencies of fog and poor visibility are indicated on the Routeing Charts and the subject is treated at length in the relevant *Admiralty Sailing Directions*.

2   In the region of the Westerlies (2.8), overcast skies, with periods of rain or snow, alternate with brief fine spells. Cloud amounts are generally large. The part of the North Atlantic Ocean most affected by fog lies E and S of Newfoundland.

3   In the vicinity of the coast between Long Island and Nova Scotia, and the Newfoundland Banks, fog is very prevalent in late spring and early summer, being due to the movement of warm, moist air from S or SW over the cold Labrador Current (2.14 and 2.16); over a large part of this area fog is experienced on more than 10 days a month. It is also liable to occur at times in other parts of this zone; usually in spring and early summer and in association with winds from between S and SW. Visibility is good with NW winds except in showers.

The North Polar regions
2.10

1   The greater part of the region lying on the polar side of the Westerlies is denied to navigation on account of ice. The prevailing wind is from some E point, though, as in the case of the Westerlies (2.8), great variations in direction and strength are caused by the passage of depressions across the area. Gales are common but less so than in the Westerlies.

2   Weather is generally very cloudy, and precipitation, usually in the form of snow, may occur at any time.

3   Fog, often of the Arctic sea-smoke type, is prevalent in summer.

4   Further information is published in the relevant *Admiralty Sailing Directions*.

SWELL

Height and direction
2.11

1   Between the equator and 30°N, frequencies of swell greater than 4 m in height rarely exceed two to four percent. One of the most persistent swells is from the NE, between Islas Canarias and the NE coast of South America. In the extreme SE, off Freetown, S and E swells prevail.

2   Between 30°N and 40°N, frequencies of swell greater than 4 m in height are:

April	10%
May to August	5 to 10%
September to November	10%
December to March	20%

The predominant direction is from between W and NW.

3   Between 40°N and 60°N, frequencies of swell greater than 4 m in height are:

April	20%
May to July	10%
August and September	20%
October to March	30%

In December and January a maximum of 40% is reached in an area centred on 55°N, 22°W. Throughout the year swell comes mainly from between SW and NW, with swell from W predominating.

Length of swell
2.12

1 Swell in the Atlantic Ocean is generally short (less than 100 m) or average (100 to 200 m) in length. However long swells may be found from time to time, though they are less frequent than in the Pacific Ocean.

CURRENTS

Chart 5310 World Surface Current Distribution
North and South Atlantic Oceans
2.13

1 The Atlantic Ocean is dominated in equatorial regions by the semi-permanent N and S sub-tropical anticyclones centred in latitudes 25° to 30° N and S respectively.

2 Under the influence of the North-east and South-east Trade Winds the respective clockwise circulation of the N gyre and counter-clockwise circulation of the S gyre ensure a steady transport of water to the W between 25°N and 25°S, the N and S boundaries varying by a few degrees with the season. Just N of the equator a counter-current sets to the E between the North and South Equatorial Currents over a narrow latitude band for part of each year.

3 Polewards and across the belt of variable winds near the centre of the gyres weaker E-going flows are generated under the influence of W winds blowing around the Icelandic and Sub-Antarctic low pressure belts. Again, in both hemispheres, there is a weak return flow to the W adjacent to the land masses in polar latitudes.

North Atlantic Ocean
2.14

1 The **North Equatorial Current** setting W to the S of the main sub-tropical gyre provides a continuous supply of warm water into the Caribbean Sea and Gulf of Mexico. This is supplemented through the year by a S equatorial flow that has crossed the equator, and been diverted to the WNW by the prominent NE coast of Brazil. However from May to November a good deal of this water turns E into the Equatorial Counter-current (2.15), which then splits, some continuing E, the rest turning back NW to join the North Equatorial Current. In the N winter an anti-clockwise circulation develops at about 7°N, 47°W, probably because of interaction between the North Equatorial Current and a weakened Equatorial Counter-current. The WNW-going sets along the Brazilian coast are of large constancy with average rates of between 2 and 3 kn.

2 The combination of warm surface waters and a steady inflow into the Gulf of Mexico allows a noticeable "head of water" to develop. The resulting outflow through the Straits of Florida is further enhanced by recurving equatorial waters being topographically accelerated through Old Bahama Channel onto its E flank. This combined flow is initially called the **Florida Current** and later the **Gulf Stream**. Average rates in the highly constant main stream reach 3 to 3½ kn during the summer months between 25°N and 30°N at 78° to 80°W. Rates decrease to the N to near 1 kn in the vicinity of Cape Hatteras where the current tends to turn away from the shore and set in a broad E to ENE direction towards W Europe. In this general area, to the NE of Cape Hatteras and S of Newfoundland, the current is joined on its W flank by cooler water that has originated in the N part of Baffin Bay and steadily moved S gathering melt water from the Canadian mainland and island coasts. South of Newfoundland more fresh and cool water from the Gulf of Saint Lawrence adds to the formation of a marked temperature discontinuity both at the surface and at depth as the cold and warm currents converge. The cold **Labrador Current** is turned through S to E on the N side of the of the warmer Gulf Stream. The combined flow then continues ENE at a reduced rate and greater variability as the **North Atlantic Current**. Approaching the coasts of W Europe the S part of the flow is turned gradually through SE to S and the N part continues to the W of the British Isles and on to the N coast of Norway and the Arctic Basin The relatively weak but general S-going drift E of 30°W and between 25°N and 45°N is known as the **Azores Current** in the W, the **Portugal Current** off the Iberian coast and the **Canary Current** in the SE. Average rates are ½ to ¾ kn and constancies low, increasing to moderate or more in the Canary Current.

3 The N part of the North Atlantic Current continues toward Iceland but to the SW of the island part is turned NW as the relatively warm **Irminger Current**. The W part forms a weak counter-clockwise eddy between S Greenland and Iceland and the E part continues clockwise around Iceland rejoining the NE-setting North Atlantic Current. There is a diffuse counter-clockwise circulation between the N coast of Norway, Spitsbergen and NE Greenland. This is complex and probably consists of a number of minor eddies formed along the boundary between the predominant warm N Atlantic water in the SE and the cold outflow from the Arctic Basin in the N and W.

4 To the NW of Norway the warm N-setting current diverges, part passing to the W of Spitsbergen and part continuing around Nordkapp, into the S part of the Barents Sea and along the W coast of Novaya Zemlya. As the warm more saline water meets the colder fresher Arctic water it sinks but continues into the Arctic Basin as a sub-surface warm current.

5 The **East Greenland Current** emerges from the Arctic as an extension of the Transpolar Drift and continues SSW, the cold temperature being maintained by melt water from the ice edge, glaciers and fjords. Rates are mostly near ½ kn but increase to near 1 kn at times during the summer months, particularly S of Denmark Strait. Currents are mainly variable in the central regions of the main gyres and of near moderate constancy along the E coast of Greenland. Off Kap Farvel the combined East Greenland and recurving Irminger Currents turn N along the SW coast of Greenland. The circulation in Davis Strait and Baffin Bay is broadly counter-clockwise with a tendency for a N and S gyre. The strongest of these generally weak currents are on the W side. Further S the flow becomes more organised continuing SSE as the Labrador Current thus completing the gyre.

Equatorial Counter-current
2.15

1 South of about 10°N the currents are more complex and show a marked seasonal variation. Lying between the North and South Equatorial Current, which both set W, the E-setting Equatorial Counter-current fluctuates throughout the year. It is a minimum during March and April when there is little evidence of any E-going sets in the W, and it is only evident from 20°W in the E. During the late spring a small area of E-going sets appears off the NE coast of Brazil between the equator and about 4°N. This gradually moves NW and expands E over the following few months emerging with an extension to the W of the **Guinea Current**. By the late summer and early autumn the Counter-current is at its maximum extent from near 50°W to its confluence with the Guinea Current in 20°W. During

November and December the Equatorial Currents tend to merge near 30°W. A slow contraction of both the E and W sections of the Counter-current takes place over the following months to the spring minimum. Constancies in the Counter-current are mostly moderate or high and average rates are between 1 and 1½ kn. The width of the Counter-current varies continuously but is usually between 4° and 6° of latitude and is centred about 7°N.

Newfoundland Banks
2.16

1 After passing the Strait of Belle Isle and the E coast of Newfoundland, the Labrador Current (2.14) covers the whole of the Grand Banks except, during summer, the extreme S part. A large branch of the current follows the E edge of the bank; this is the part which carries the ice farthest S to reach the transatlantic shipping routes. Another branch rounds Cape Race and sets SW. Although some of the water that has passed on to the Grand Banks continues in a more S direction, especially during August to October, the bulk of it sets SW and continues, as a SW-going set, to fill the region between Newfoundland, Nova Scotia and the Gulf Stream.

2 The Labrador Current subsequently continues S along the coast of the United States as a cold current as far as about 36°N from November to January, 37°N from February to April, 38°N from May to July and 40°N from August to October. Between the S limit of the Labrador Current and the Tail of the Bank, the warm and cold waters converge on a line which is known as the "Northern Edge" (or sometimes the North Wall) of the Gulf Stream.

3 The E end of the Northern Edge presents the greatest hydrographic contrasts to be found in the world, the water changing from the olive or bottle green of the Arctic side to the indigo blue of the Gulf Stream: a temperature change of 12° to 0° Celsius has been recorded within a ship's length.

4 The currents off the coasts of Labrador and Newfoundland are complex; for details, reference should be made to the relevant volumes of *Admiralty Sailing Directions*.

North Sea
2.17

1 A branch of the North Atlantic Current (2.14) diverges from the main flow NE of the Shetland Isles and flows S, fanning out E towards the S part of the Skagerrak, along the E coast of Britain as far as the Thames Estuary. It is there joined by a branch of the North Atlantic Current which passes through the English Channel and the Strait of Dover, the combined currents then flowing along the Netherlands and Jutland coasts. This current then flows around Skagerrak in a counter-clockwise direction and finally sets N along the W coast of Norway.

2 The outflow from the North Sea forms the **Norwegian Coastal Current** and is probably the most constant part of the circulation. In about 62°N this current rejoins the main branch of the North Atlantic Current flowing towards Nordkapp.

3 In most parts of the North Sea, except in Skagerrak, these currents are small and mostly insignificant to navigation compared with the predominant tidal influence and the effect of wind drift currents.

Western approaches to English Channel
2.18

1 After SW or W gales, a set towards the mouth of the channel may be expected, at a rate depending on the locality, strength and duration of the gale. In winter, sets of up to 1½ kn are sometimes recorded, mainly in directions between ENE and SE, but the tidal streams are responsible for most of the water movement within the 200 m contour.

Bay of Biscay
2.19

1 Off the mouth of the Bay of Biscay the current is trending SE to S to form the beginning of the Portugal Current (2.14). A branch enters the bay and recurves W along the N coast of Spain, but over most of the bay the currents are highly variable with a tendency for directions between E and S to predominate. The rates for the most part do not exceed 1 kn and very rarely reach 2 kn.

2 Following W or NW gales, E-going sets occur off the N coast of Spain, sometimes attaining a rate of 3 kn off Bilbao and 4 to 5 kn at the head of the bay particularly where current and tidal stream are in the same direction.

ICE

General remarks
2.20

1 The following brief account of ice in the North Atlantic is by no means comprehensive. Before undertaking voyages through areas in which ice is likely to be met, *The Mariner's Handbook* and the relevant *Admiralty Sailing Directions* should be studied, as well as the monthly Routeing Charts, which show the ice limits. These limits are shown approximately on *Diagrams 1.14a* and *1.14b* but they may not always agree with the Routeing Charts which endeavour to show the extreme limits on a monthly basis as far as this is possible with the limited and variable data available.

2 Details of Ice Warning and facsimile ice chart broadcasts are also available and are described in the relevant volumes of *Admiralty List of Radio Signals*.

3 A factor always to be borne in mind where ice conditions are concerned is their great variability. For this reason, and on account of the sparsity of observations in many areas, the charted positions of ice limits must be regarded as approximate. The dates which follow refer to average conditions.

Ice limits and drift
2.21

1 The Routeing Charts show the influence of the ocean currents (2.14 and 2.16) in setting the pack ice over much of the area of the Grand Banks of Newfoundland from the latter part of January until May, while the E part of the ocean remains ice-free to high latitudes.

2 Almost all the icebergs which menace the North Atlantic routes originate in the glaciers of the W coast of Greenland where they are calved at a rate of several thousand a year. Most are carried N by the West Greenland Current, round the head of Baffin Bay, and then S by the Canadian and Labrador Currents (2.14), and when they finally reach the shipping routes they may be several years old. The bergs calved on the E coast of Greenland also drift S, and may be met off Kap Farvel. Some drift across the East Greenland Current (2.14) and may be met throughout the year on the E flank of that current, extending SW from the W extremity of Iceland. Others drift round Kap Farvel, but they do not survive the relatively warm waters of the Davis Strait and are not a source of danger on the regular transatlantic routes.

3 Icebergs may be found beyond the limits of the pack ice at all seasons, but mostly in early summer, in winter many are frozen into the pack ice.

Ice in specific localities
2.22

1 **Kap Farvel.** The greatest distance at which bergs are met S of Kap Farvel is generally about 120 miles. This usually occurs in May when they may be encountered as far S as 66°N and as far E as 32°W. Their least extent is in December. Bergs are not usually met S of 48°N between September and December, but may well be encountered in any month N of 52°N.
2.23

1 **Saint Lawrence River.** Below Montreal the river is closed by ice between early December and mid-April. Commercial navigation ceases in most parts of the Gulf of Saint Lawrence by mid-December; in the S part navigation is not considered safe between early December and mid-April.
2.24

1 **Strait of Belle Isle.** The Strait is generally not navigable from late December until June.
2.25

1 **Cabot Strait.** The Strait is usually navigable from mid-April until February. Pack ice arrives from N of Cape Race about the end of January in an ordinary season, extending round the coasts of the Avalon Peninsular in February, until early May.
2.26

1 **The Grand Banks of Newfoundland.** The Grand Banks are entirely free of pack ice between July and December inclusive. Pack ice reaches the banks in January and extends farthest S in March and April, on the E edge of the banks. In very rare seasons, dangerous pack ice may extend to the Tail of the Bank and even S of it but, on average, the floes begin to break up on reaching 45°N.

2 In the region of the Grand Banks, the worst season for icebergs is between March and July, with April, May and June as the months of greatest frequency. Bergs are not often found S of 40°N or E of 40°W, though occasionally they may be considerably outside these limits. They are particularly prevalent around the E flanks of the banks, on which many of them ground. More detail is given in the relevant volume of *Admiralty Sailing Directions*.
2.27

1 **Denmark Strait.** The strait is normally free of ice on its E side throughout the year, but on rare occasions, as in the spring of 1968, the ice spreads across from Greenland to close the strait. Icebergs may be met throughout the year on both sides of the Denmark Strait.
2.28

1 **White Sea.** The White Sea is normally closed to navigation from about mid-December to mid-May.
2.29

1 **Kol'skiy Zaliv.** The N part remains open throughout the year but, from December to April ice forms along the shore and at times breaks away, to be carried out to sea. It may be a hindrance for three or four days at a time in exceptionally cold winters.

2.30

1 **Norwegian coast.** None of the main ports on the W coasts is ever closed by ice, and the closure of Oslo is rare.
2.31

1 **North Sea.** Serious ice conditions in the entrances to German, Netherlands and Danish ports, lasting from 1 to 4 weeks, occur about two or three times in ten years at some time between mid-January and early March.

Ice Information Services
2.32

1 Ice information, comprising up-to-date reports and forecasts from the Gulf of Saint Lawrence, the Grand Banks of Newfoundland, Greenland, Iceland and the NW approaches to Europe are transmitted from the coast radio stations listed in the relevant volume of *Admiralty List of Radio Signals*.
2.33

1 **International Ice Patrol.** This service is operated by the US Coast Guard with the primary object of collecting data and warning shipping of the amount and extent of icebergs and sea ice in the vicinity of the Grand Banks. The service operates principally between the parallels of 39° and 50°N and the meridians of 42° and 60°W during the ice season from February or March until about the end of June.

2 In spite of the efforts of the International Ice Patrol bergs are known to drift unnoticed into the usual routes in the vicinity of the Grand Banks. For details of the International Ice Patrol see the relevant volumes of *Admiralty Sailing Directions* and *Admiralty List of Radio Signals*.
2.34

1 **Ice Advisory Service.** This service, maintained by the Canadian Coast Guard during the winter navigational season, is based on aerial reconnaissance. Reports of existing and forecast ice conditions are broadcast from certain Canadian radio stations. For details of the service see the relevant volumes of *Admiralty Sailing Directions* and *Admiralty List of Radio Signals*. The volume *Ice Navigation in Canadian Waters*, issued by the Canadian Coast Guard, should also be consulted.
2.35

1 **Caution.** Tests conducted by the International Ice Patrol have shown that radar cannot provide positive assurance for iceberg detection. Sea-water is a better reflector than ice. this means that unless a berg or growler is observed outside the area of 'sea return' or 'clutter' it will not be detected by radar. The average range of detection of a dangerous growler, if detected at all, is only 4 miles.

2 Radar is a valuable aid, but its use cannot replace the traditional caution exercised during a passage passing near the Grand Banks during the ice season.

NOTES AND CAUTIONS
Meteorological and Oceanographic Data Buoys
2.36

1 Automated buoys are deployed in the world oceans under the auspices of the World Meteorological Organisation. They make routine measurements of wind speed and direction, air temperature and humidity, atmospheric pressure, currents and sea surface temperature, which data is transmitted via satellite.

2 Those automated buoys established in permanent or semi-permanent positions are shown on Admiralty charts.

3 Further details are given in *Annual Summary of Admiralty Notices to Mariner*.

Western approaches to the English Channel
2.37

1 When navigating in these waters it is essential to assess the surface drift current caused by lately prevailing winds as well as the present winds. See 2.18 and relevant *Admiralty Sailing Directions*.

2 Traffic Separation Schemes have been established in the English Channel and its approaches and off Fastnet Rock.

3 *Chart 5500—Mariner's Routeing Guide—English Channel and Southern North Sea*, displays much information useful for passage planning, routeing and radio and pilotage services in the area it covers.

Île d'Ouessant
2.38

1 Mariners approaching Île d'Ouessant Traffic Separation Scheme must guard against the danger of being set E of their reckoning. Unless certain of the position, this scheme should be given a wide berth and a depth exceeding 110 m maintained.

Bay of Biscay and west coasts of Spain and Portugal
2.39

1 There may be a strong E-going set off the N coast of Spain after a W or NW gale (2.19).

2 An onshore wind brings cloud that develops into fog or thick mist when it reaches the elevated land at both the N and S points of the Bay of Biscay.

3 Traffic Separation Schemes have been established off Berlenga, Cabo de Roca and Cabo de São Vicente.

Strait of Gibraltar
2.40

1 In thick weather when approaching the Strait from the Atlantic Ocean, soundings should be taken until the position is certain. Caution is necessary since the currents, tidal streams and eddies between Cabo de São Vicente and Isla da Tarifa are very variable. In the Strait itself with a strong W wind, the flow of water (the resultant of tidal stream and current) is liable to reach a rate of 6 kn.

2 A Traffic Separation Scheme has been established in the Strait.

Strait of Belle Isle
2.41

1 Approaching from the E in low visibility, soundings on the banks E of Newfoundland and Labrador will be found of great assistance if not certain of the position.

2 A Traffic Separation Scheme has been established in the Strait.

Newfoundland coasts
2.42

1 Fog is exceedingly prevalent off the S coast of Newfoundland, especially in summer. The set of the current and the indraughts into the deep bays, particularly on their E sides, should be guarded against.

2 Approaching from the E in thick weather, radiobeacons on the E coast of Newfoundland or other radio aids, and soundings over the Grand Banks and Ballard Bank should indicate the position with enough accuracy to round Cape Race in safety.

3 Although the current between the Grand Banks and Newfoundland ordinarily sets SW at a rate which may slightly exceed 1 kn, it is not unusual, particularly for a short period before a gale, for the current to be so disturbed as to set across its ordinary direction or even to be reversed on the surface. Close inshore it is affected by the tidal streams.

4 The currents between Cape Race and Saint Pierre are irregular, with a greatest rate of 1 kn, and are influenced by the wind, and, near the shore, by the tidal streams. See relevant volumes of *Admiralty Sailing Directions*.

5 Approaching from the W, Cape Pine and Cape Race should not be closed in depths of less than 55 m unless certain of the position.

Grand Banks of Newfoundland
2.43

1 The principal shipping routes from N European ports to ports on the E coast of the United States, and to the Gulf of Saint Lawrence through Cabot Strait, lead over or near the Grand Banks.

2 They are among the busiest routes in the world. At the same time they are amongst the most dangerous.

3 Icebergs, growlers and pack ice are common in this region notorious for the frequency and density of its fogs. Many depressions pass close to the area so that gales are frequent and severe. In addition, many fishing vessels are found throughout the year on the Grand Banks, as well as vessels and platforms used to exploit oil, gas and mineral deposits.

4 In view of these hazards the *International Convention for the Safety of Life at Sea, (1974)* advises that all ships proceeding on voyages in the vicinity of the Grand Banks should avoid as far as practical, the fishing banks of Newfoundland N of 43°N and to pass outside regions known or believed to be endangered by ice.

5 The International Ice Patrol (2.33) also advises against venturing into pack ice N of 45°30′N before the middle of April.

Bermuda
2.44

1 Extreme caution is necessary in the vicinity of Bermuda due to the extensive and fringing reefs, skirted on their N and W sides by an "Area to be avoided" by certain ships. See the relevant volume of *Admiralty Sailing Directions*.

Penedos de São Pedro e São Paolo and Ilha de Fernando de Noronha
2.45

1 When approaching these islets caution is necessary as the South Equatorial Current (3.9) sets to the WNW past them at a rate of 1 to 2 kn.

PASSAGES TO AND FROM KAP FARVEL AND DAVIS STRAIT

GENERAL INFORMATION

Coverage
2.46
This section contains details of the following passages:
> To and from Kap Farvel and Nordkapp (2.49).
> To and from Kap Farvel and W Coast of Norway and North Sea (2.51).
> To and from Kap Farvel and the British Isles, Biscay and ports on the W coasts of Spain and Portugal (2.53).
> To and from the Davis and Hudson Straits to ports in Europe (2.55).

Davis Strait
2.47
1   For directions and ice conditions in Davis Strait, Hudson Strait and Hudson Bay, see *NP12 Arctic Pilot Volume III*. For Ice Information Services see 2.32 to 2.34.

Kap Farvel
2.48
1   In view of the weather and ice conditions off the coast of Greenland the routes which follow are taken from 58°30′N, 44°00′W (**K**), about 75 miles S of Kap Farvel.

PASSAGES

Kap Farvel ⇐ ⇒ Nordkapp

Diagram 2.50
Notes
2.49
1   **Hazards.** The hazards of navigation in high latitudes are described in *The Mariner's Handbook*. For remarks on ice see 2.20-2.31.

2   **Magnetic field.** The directive force of the earth's magnetic field is weak and the values of the magnetic variation change rapidly along this and other routes from Kap Farvel. Furthermore, local magnetic anomalies have been reported in the vicinities of Jan Mayen and Iceland, particularly in depths of less than 135 m, and off the Norwegian coast.

Routes
2.50
1   The shortest route is through the Denmark Strait, however ice or weather conditions may make a passage S of Iceland preferable.
2.50.1
1   **Via Denmark Strait.** The route from 58°30′N, 44°00′W (**K**), about 75 miles S of Kap Farvel, is by great circle, passing through Denmark Strait and 33 miles S of Jan Mayen, to 71°15′N, 25°40′E, 5 miles N of Nordkapp.
> Distance   1810 miles.
2.50.2
1   **South of Iceland.** If Denmark Strait is not navigable, passage must be made S of Iceland, by great circle from 58°30′N, 44°00′W (**K**), about 75 miles S of Kap Farvel to 63°20′N, 16°00′W (**C**), 15 miles S of Surtsey (63°20′N, 20°28′W) and 12 miles S of the S coast of Iceland, thence by great circle to 71°15′N, 25°40′E, 5 miles N of Nordkapp.
> Distance   1910 miles.

Kap Farvel ⇐ ⇒ West coast of Norway and North Sea

Diagram 2.52
Notes
2.51
1   **Oil and gasfields,** with platforms, drilling rigs and associated pipelines and buoyage, are situated in the North Sea extending as far N as the 200 m depth contour.

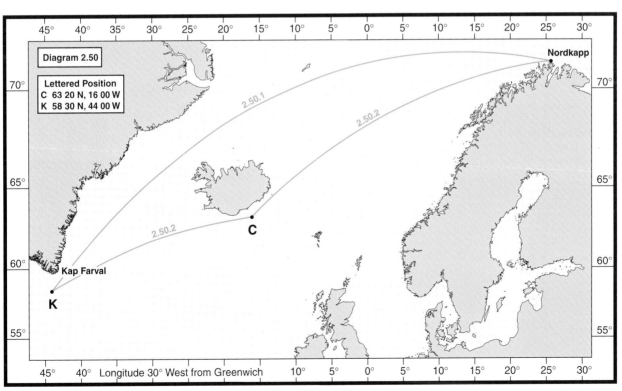

2.50 Kap Farval ← → Nordkapp

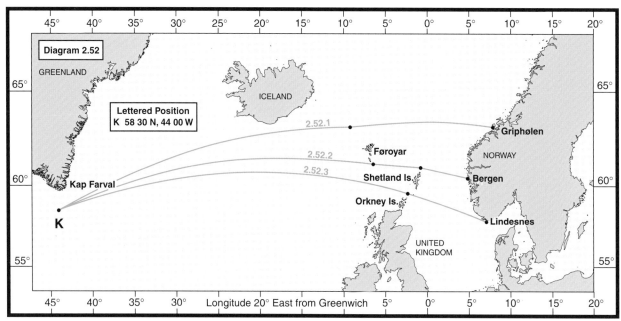

2.52 Kap Farval ← → West Coast of Norway and North Sea

Routes

2.52

1   From 58°30′N, 44°00′W **(K)**, about 75 miles S of Kap Farvel, the routes to Trondheim, Bergen or Lindesnes are as direct as navigation permits.

2.52.1

1   To **Trondheim** (63°27′N, 10°23′E). The route passes between Iceland and Føroyar to the pilot ground in Griphølen (63°15′N, 7°37′E), thence through Indreleia.

Distance   1590 miles.

2.52.2

1   To **Bergen** (60°24′N, 5°18′E). The route passes between Føroyar and Shetland Isles.

Distance   1490 miles.

2.52.3

1   To position 57°55′N, 7°03′E, 4 miles S of **Lindesnes**. The route passes between Fair Isle and Orkney Islands.

Distance   1580 miles.

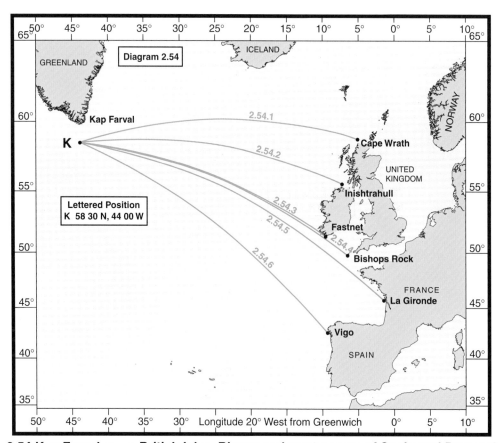

2.54 Kap Farvel ← → British Isles, Biscay and west coasts of Spain and Portugal

Kap Farvel ⇐ ⇒ British Isles, Biscay and ports on the west coasts of Spain and Portugal

Diagram 2.54

Notes

2.53

1   For information on W approaches to English Channel, Île d'Ouessant, Bay of Biscay and W coasts of Spain and Portugal, see 2.37 to 2.39.

Routes

2.54

1   From 58°30′N, 44°00′W (**K**), about 75 miles S of Kap Farvel the routes are as follows.

2.54.1

1   To 58°43′N, 5°00′W, 5 miles N of **Cape Wrath**, the route is by great circle.

Distance   1210 miles.

2.54.2

1   To 55°31′N, 7°15′W, 5 miles N of **Inishtrahull**, the route is by great circle.

Distance   1200 miles.

2.54.3

1   To 51°18′N, 9°36′W, 5 miles S of **Fastnet Rock**, the route is by great circle.

Distance   1250 miles.

2.54.4

1   To 49°47′N, 6°27′W, 5 miles S of **Bishop Rock**, the route is by great circle.

Distance   1400 miles.

2.54.5

1   To **La Gironde** (45°40′N, 1°28′W), 70 miles from Bordeaux, the route is by great circle. Vessels for Biscay ports should pass at least 10 miles SW of Chaussée de Sein (48°00′N, 5°00′W).

Distance   1710 miles.

2.54.6

1   To **Vigo** (42°13′N, 8°50′W) the route is by great circle.

Distance   1640 miles.

Davis and Hudson Straits ⇐ ⇒ European ports

Diagram 2.55

Routes

2.55

1   **Davis Strait** (60°00′N, 56°00′W). From the N and E parts of the Davis Strait, routes are as navigation permits to a position 58°30′N, 44°00′W (**K**), S of Kap Farvel, thence as at 2.50, 2.52 and 2.54 to destinations in Europe.

2.55.1

1   **Hudson Strait.** From Hudson Strait (61°00′N, 64°50′W) routes, to those destinations in Europe N of Leixões, are by great circle to a position 58°30′N, 44°00′W (**K**), S of Kap Farvel, thence as at 2.50, 2.52 and 2.54 to destinations in Europe.

Distance   650 miles.

(to Kap Farvel)

2.55.2

1   To **Leixões** (41°10′N, 8°42′W) the route is by great circle.

Distance   2340 miles.

2.55.3

1   To **Lisboa** (38°36′N, 9°24′W) the route is by great circle.

Distance   2440 miles.

2.55.4

1   To **Strait of Gibraltar** the route is by great circle to Cabo de São Vicente (37°01′N, 9°00′W), thence as navigation permits to a position (36°00′N, 5°21′W), 6 miles S of Europa Point.

Distance   2720 miles.

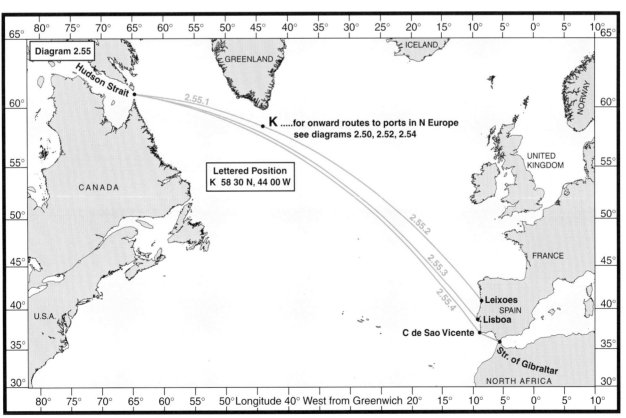

2.55 Davis and Hudson Strait ← → European Ports

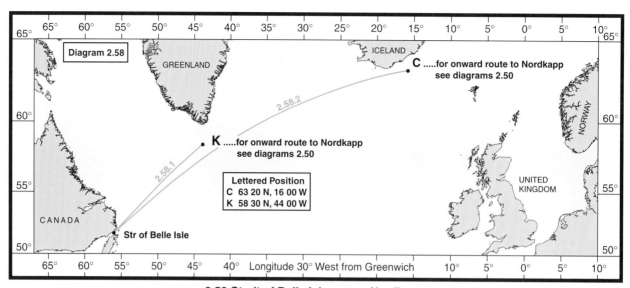

2.58 Strait of Belle Isle ← → Nordkapp

PASSAGES TO AND FROM STRAIT OF BELLE ISLE

GENERAL INFORMATION

Coverage
2.56

This section contains details of the following routes:
    To and from Strait of Belle Isle and Nordkapp (2.58).
    To and from Strait of Belle Isle and W Coast of Norway and North Sea (2.59).
    To and from Strait of Belle Isle and the British Isles and Bay of Biscay (2.61).

Belle Isle
2.57

1   **Ice.** For details of ice conditions see *2.22 to 2.31* and the relevant volumes of *Admiralty Sailing Directions.* For Ice Information Services see 2.32 to 2.34.

2   **Cautions.** See 2.41.

PASSAGES

Strait of Belle Isle ⇐ ⇒ Nordkapp

Diagrams 2.58 and 2.50

Routes
2.58

1   The shortest route is through the Denmark Strait, however ice or weather conditions may make a passage S of Iceland preferable, see also notes at 2.49.

2.58.1

1   **Via Denmark Strait.** The route from Strait of Belle Isle (51°44′N, 56°00′W) is to a position 58°30′N, 44°00′W **(K)**, about 75 miles S of Kap Farvel, thence by great circle, passing through Denmark Strait and 33 miles S of Jan Mayen, to 71°15′N, 25°40′E, 5 miles N of Nordkapp. See Diagram 2.50, Route 2.50.1.
    Distance   2390 miles.

2.58.2

1   **South of Iceland.** If Denmark Strait is not navigable, passage must be made S of Iceland, by great circle from

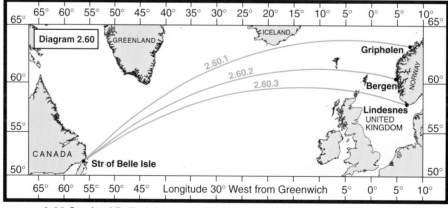

2.60 Strait of Belle Isle ← → West coast of Norway and North Sea

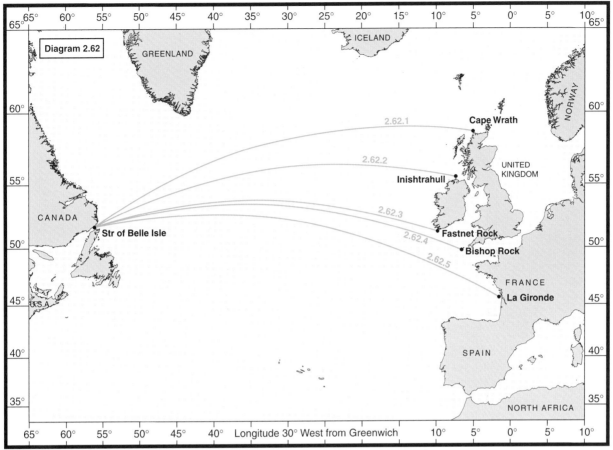

2.62 Strait of Belle Isle ← → British Isles and Biscay

Strait of Belle Isle (51°44′N, 56°00′W) to 63°20′N, 16°00′W (**C**), which passes 15 miles S of Surtsey (63°20′N, 20°28′W) and 12 miles S of the S coast of Iceland, thence by great circle to 71°15′N, 25°40′E, 5 miles N of Nordkapp. See Diagram 2.50, Route 2.50.2.

Distance   2490 miles.

Strait of Belle Isle ⇐ ⇒ West coast of Norway and North Sea

Diagram 2.60
Notes
2.59

1   **Oil and gasfields,** see 2.51.

Routes
2.60

1   From Strait of Belle Isle (51°44′N, 56°00′W) the routes to Trondheim, Bergen or Lindesnes are as follows.
2.60.1

1   To **Trondheim** (63°27′N, 10°23′E). The route is by great circle to the pilot ground in Griphølen (63°15′N, 7°37′E), thence through Indreleia.

Distance   2160 miles.
2.60.2

1   To **Bergen** (60°24′N, 5°18′E). The route is by great circle passing between Føroyar and Shetland Isles.

Distance   2050 miles.
2.60.3

1   To position 57°55′N, 7°03′E, 4 miles S of **Lindesnes**. The route is by great circle passing between Fair Isle and Orkney Islands.

Distance   2140 miles.

Strait of Belle Isle ⇐ ⇒ British Isles and Biscay ports

Diagram 2.62
Notes
2.61

1   For information on W approaches to English Channel, Île d'Ouessant, Bay of Biscay and W coasts of Spain and Portugal, see 2.37 to 2.39.

Routes
2.62

1   The routes from Strait of Belle Isle (51°44′N, 56°00′W) are as follows.
2.62.1

1   To 58°43′N, 5°00′W, 5 miles N of **Cape Wrath**, the route is by great circle.

Distance   1760 miles.
2.62.2

1   To 55°31′N, 7°15′W, 5 miles N of **Inishtrahull**, the route is by great circle.

Distance   1720 miles.
2.62.3

1   To 51°18′N, 9°36′W, 5 miles S of **Fastnet Rock**, the route is by great circle.

Distance   1710 miles.
2.62.4

1   To 49°47′N, 6°27′W, 5 miles S of **Bishop Rock**, the route is by great circle.

Distance   1860 miles.
2.62.5

1   To **La Gironde** (45°40′N, 1°28′W), 70 miles from Bordeaux the route is by great circle.

Distance   2150 miles.

PASSAGES TO AND FROM ST JOHN'S HARBOUR, NEWFOUNDLAND

GENERAL INFORMATION

Coverage
2.63
This section contains details of the passages to and from St John's Harbour, Newfoundland and ports in Europe from Cape Wrath in the N to Strait of Gibraltar in the S.

References
2.64
1   **Ice.** For details of ice conditions see *2.22 to 2.31* and the relevant volumes of *Admiralty Sailing Directions*. For Ice Information Services see 2.32 to 2.34.

2   **Cautions:**
      Western approaches to the English Channel, see 2.37.
      Île d'Ouessant, see 2.38.
      Bay of Biscay and west coasts of Spain and Portugal, see 2.39.
      Strait of Gibraltar, see 2.40.
      Coast of Newfoundland, see 2.42.

PASSAGES

St John's Harbour ⇐ ⇒ Positions between Cape Wrath and Strait of Gibraltar

Diagram 2.65
Routes
2.65
1   The routes from St John's Harbour, Newfoundland (47°34′N, 52°38′W) are as follows.
2.65.1
1   To 58°43′N, 5°00′W, 5 miles N of **Cape Wrath**, the route is by great circle.
    Distance   1800 miles.
2.65.2
1   To 55°31′N, 7°15′W, 5 miles N of **Inishtrahull**, the route is by great circle.
    Distance   1730 miles.
2.65.3
1   To 51°18′N, 9°36′W, 5 miles S of **Fastnet Rock**, the route is by great circle.
    Distance   1680 miles.

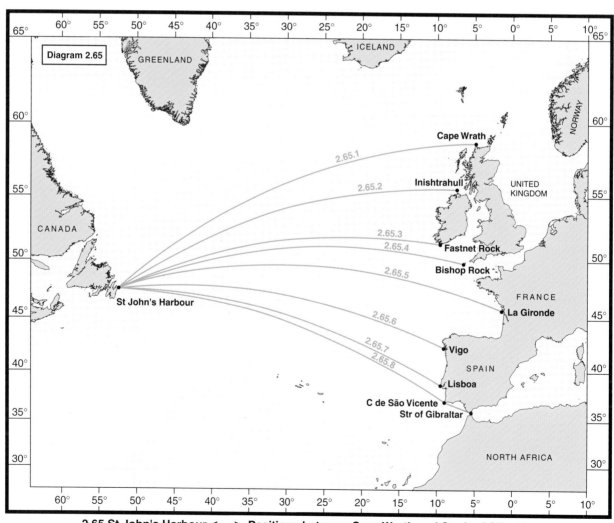

2.65 St John's Harbour ← → Positions between Cape Wrath and Strait of Gibraltar

2.65.4

1   To 49°47′N, 6°27′W, 5 miles S of **Bishop Rock**, the route is by great circle.

Distance   1810 miles.

2.65.5

1   To **La Gironde** (45°40′N, 1°28′W), 70 miles from Bordeaux, the route is by great circle.

Distance   2080 miles.

2.65.6

1   To **Vigo** (42°13′N, 8°50′W) the route is by great circle.

Distance   1870 miles.

2.65.7

1   To **Lisboa** (38°36′N, 9°24′W) the route is by great circle.

Distance   1950 miles.

2.65.8

1   To **Strait of Gibraltar** the route is by great circle to Cabo de São Vicente (37°01′N, 9°00′W), thence as navigation permits to a position (36°00′N, 5°21′W), 6 miles S of Europa Point.

Distance   2220 miles.

PASSAGES BETWEEN CABOT STRAIT OR NORTH AMERICAN PORTS AND EUROPE

GENERAL INFORMATION

Coverage

2.66

This section contains details of the following assages:

Recommended passages on the E seaboards of Canada and the United States avoiding dangers on the Grand Banks (2.68).

To and from position (42°30′N, 50°00′W) (**BS**) and Europe (2.69).

To and from Cape Race and Europe (2.70).

West-bound alternative routes (2.71).

References

2.67

1   **Traffic Separation Schemes** see 1.28.

2   **Weather Routeing** see 1.25.

3   **Ice.** For details of ice conditions see *2.22 to 2.31* and the relevant volumes of *Admiralty Sailing Directions*. For Ice Information Services see 2.32 to 2.34.

4   **Cautions**:

Western approaches to the English Channel, see 2.37.

Île d'Ouessant, see 2.38.

Bay of Biscay and west coasts of Spain and Portugal, see 2.39.

Strait of Gibraltar, see 2.40.

Coast of Newfoundland, see 2.42.

Grand Banks of Newfoundland, see 2.43.

PASSAGES

Passages avoiding the Grand Banks of Newfoundland

Diagram 2.68

Recommended routes

2.68

1   Recommended routes between Cabot Strait, Halifax or the N part of the E coast of the United States and Europe skirt the S side of the Grand Banks, passing through, or S of, position 42°30′N, 50°00′W (**BS**), to avoid the hazards of crossing the Grand Banks (2.43).

2   There are alternative routes to and from Biscay ports and places farther N, passing 20 miles S of Cape Race. They are usable between May and November.

3   In recommending routes to and from ports SW of Cape Race, account must be taken of the seasonal movement of ice in the Grand Banks area (2.26). It can never be assumed that a particular route will be clear of ice; constant study of ice reports, and the utmost vigilance at sea, are essential.

2.68.1

1   From **Cabot Strait** (47°32′N, 59°30′W), routes from the Traffic Separation Scheme are either:

  a) S of the Grand Banks to position 42°30′N, 50°00′W (**BS**), or

  b) Along the S coast of Newfoundland to Cape Race (46°39′N, 53°04′W).

2.68.2

1   From **Halifax** (44°31′N, 63°30′W), routes are either:

  a) Through 43°00′N, 60°00′W (**C**) (50 miles S of Sable Island), thence to position 42°30′N, 50°00′W (**BS**), or

  b) Direct to Cape Race (46°39′N, 53°04′W).

2.68.3

1   From **Boston** (42°20′N, 70°46′W), routes are direct to either:

  a) Position 42°30′N, 50°00′W (**BS**), or:

  b) Cape Race (46°39′N, 53°04′W).

2.68.4

1   From places between **Boston** (42°20′N, 70°46′W) and **Chesapeake Bay** (36°56′N, 75°58′W), routes are either direct to position 42°30′N, 50°00′W (**BS**), or to position 43°00′N, 60°00′W (**C**) and thence to Cape Race (46°39′N, 53°04′W).

Routes from:

  (a) **New York** (40°28′N, 73°50′W) and its vicinity pass S of Nantucket TSS (40°30′N, 69°15′W).

  (b) **Delaware Bay** (38°48′N, 75°02′W) to Cape Race (46°39′N, 53°04′W) pass S of Nantucket TSS (40°30′N, 69°15′W).

  (c) **Chesapeake Bay** (36°56′N, 75°58′W) to Strait of Gibraltar, the route is by great circle to Cabo de São Vicente (37°01′N, 9°00′W), passing S of position 42°30′N, 50°00′W (**BS**) thence as navigation permits to a position (36°00′N, 5°21′W), 6 miles S of Europa Point.

Position BS ⇐ ⇒ Europe

Diagram 2.69

Routes

2.69

1   The routes from position (42°30′N, 50°00′W) (**BS**) (2.68) are as follows (distances are given in Table 2.72).

2.69.1

1   To 71°15′N, 25°40′E, 5 miles N of **Nordkapp**, the route is by great circle to 57°50′N, 18°00′W (**A**), thence as navigation permits.

2.69.2

1   To **Trondheim** (63°27′N, 10°23′E) the route is by great circle to 61°14′N, 6°40′W (**B**) (10 miles S of Føroyar),

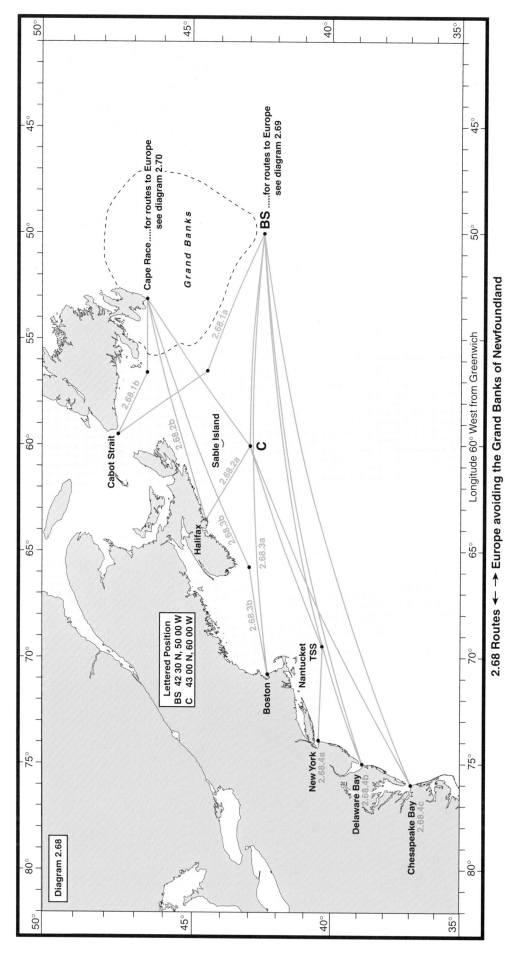

2.68 Routes ← → Europe avoiding the Grand Banks of Newfoundland

Longitude 60° West from Greenwich

Diagram 2.68

Lettered Position
BS 42 30 N, 50 00 W
C 43 00 N, 60 00 W

Cape Race.....for routes to Europe
see diagram 2.70

Grand Banks

BS .....for routes to Europe
see diagram 2.69

Cabot Strait

Sable Island

Halifax

Boston

Nantucket
TSS

New York

Delaware Bay

Chesapeake Bay

2.68.1a
2.68.1b
2.68.2b
2.68.2a
2.68.3b
2.68.3a
2.68.3b
2.68.4a
2.68.4b
2.68.4c

C

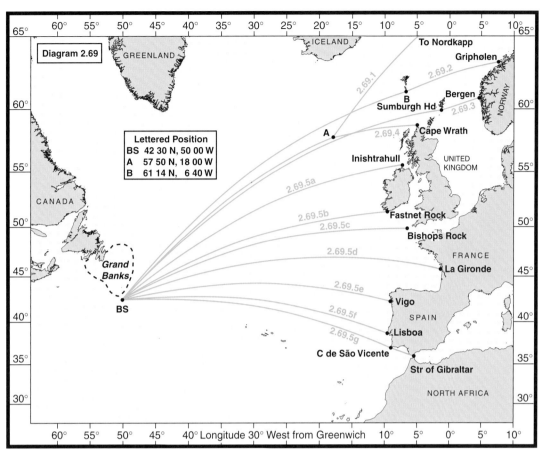

2.69 Position BS (42° 30'N. 50° 00'W) ←→ Europe

thence by rhumb line to the pilot ground at Griphølen (63°15′N, 7°37′E), thence through Indreleia.

2.69.3

1　To **Bergen** (60°24′N, 5°18′E) the route is by great circle to a landfall off Sumburgh Head (59°51′N, 1°16′W), thence as navigation permits.

2.69.4

1　To 58°43′N, 5°00′W, 5 miles N of **Cape Wrath**, the route is by great circle to 57°50′N, 18°00′W (**A**), thence direct.

2.69.5

1　To To places **S of Cape Wrath** the routes are by great circle to destination.

Cape Race ⇐ ⇒ Europe

Diagram 2.70
Routes
2.70

1　The routes from Cape Race (46°39′N, 53°04′W) are as follows (distances are given in Table 2.72).

2.70.1

1　To 71°15′N, 25°40′E, 5 miles N of **Nordkapp**, the route is by great circle to 57°50′N, 18°00′W (**A**), thence as navigation permits.

2.70.2

1　To **Trondheim** (63°27′N, 10°23′E) the route is by great circle to the pilot ground at Griphølen (63°15′N, 7°37′E), thence through Indreleia.

2.70.3

1　To **Bergen** (60°24′N, 5°18′E) the route is by great circle to a landfall off Sumburgh Head (59° 51′N, 1°16′W), thence as navigation permits.

2.70.4

1　To 58°43′N, 5°00′W, 5 miles N of **Cape Wrath**, the route is by great circle to 57°50′N, 18°00′W (**A**), thence direct.

2.70.5

1　To places S of **Cape Wrath** the routes are by great circle to destination.

West-bound alternative routes

Diagram 2.71
Routes
2.71

1　West-bound trans-atlantic vessels can gain some advantage in weather and currents by proceeding along the following route.

2　From **Strait of Gibraltar** (36°00′N, 5°21′W) bound for **New York** (40°28′N, 73°50′W) pass 20 miles S of São Miguel (37°45′N, 25°30′W), Arquipélago dos Açores, thence by rhumb line to Nantucket TSS (40°30′N, 69°15′W), thence as navigation permits.

Distance　　　　　　　3260 miles.

2.71.1

Alternative W-bound routes, recommended for low-powered vessels, can be found at 8.29 and 8.32.

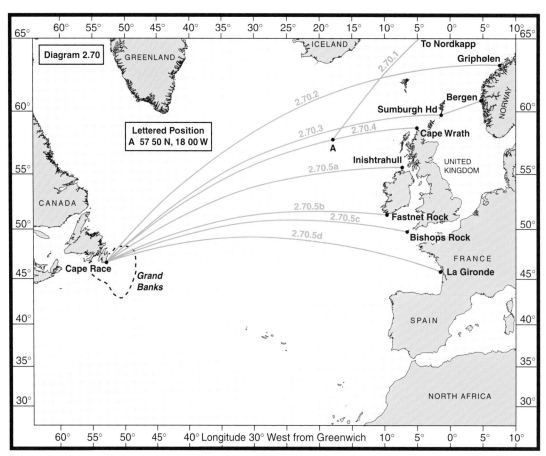

2.70 Cape Race ← → Europe

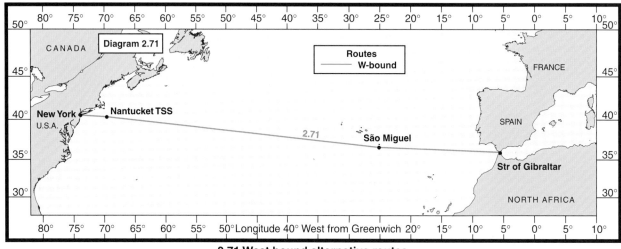

2.71 West bound alternative routes

Distances in miles
2.72

	Cabot Strait		Halifax		Boston		New York		Delaware Bay		Chesapeake Bay	
via	BS	C Race	BS	C Race	BS	C Race	BS	C Race	BS	C Race	BS	C Race
Nordkapp	3380	3070	3480	3240	3780	3590	3940	3800	4020	3880	4100	3970
Trondheim	2940	2600	3040	2770	3340	3120	3500	3330	3580	3410	3660	3500
Bergen	2780	2480	2880	2650	3190	3000	3350	3210	3430	3290	3510	3380
Cape Wrath	2450	2140	2550	2320	2860	2660	3010	2870	3100	2950	3180	3040
Inishtrahull (a)	2340	2060	2440	2240	2750	2590	2910	2800	2990	2870	3070	2970
Fastnet Rk (b)	2240	2000	2340	2180	2640	2520	2800	2730	2880	2810	2960	2900

	Cabot Strait		Halifax		Boston		New York		Delaware Bay		Chesapeake Bay	
Bishop Rk (c)	2360	2130	2460	2310	2760	2660	2920	2870	3000	2940	3080	3040
La Gironde†(d)	2590	2410	2690	2570	3000	2920	3160	3130	3240	3210	3320	3300
Vigo (e)	2330		2430		2740		2900		2980		3060	
Lisboa (f)	2370		2470		2780		2940		3020		3100	
Strait of Gibraltar** (g)	2620		2720		3020		3180		3260		3340*	

* Great circle direct to Cabo de São Vicente   † 70 miles from Bordeaux
** 6 miles S of Europa Point

PASSAGES BETWEEN NORTH AMERICAN PORTS, WEST INDIES OR CARIBBEAN SEA AND AFRICA

GENERAL INFORMATION

Coverage
2.73
This section contains details of the following passages:
  To and from ports in North America or West Indies and the W coast of Africa (2.74).
  To and from Caribbean Sea and Cape Town or Indian Ocean (2.76).
For details of the following passages:
  To and from North American ports to Gulf of Guinea, SW coast of Africa or Cape Town see 3.38, 3.54 and 3.55.
  To and from Colón or Tobago and SW coast of Africa see 3.35 and 3.36.

North America or West Indies ⇐ ⇒ West coast of Africa

Diagram 2.74
Routes
2.74
1   Subject to the ordinary requirements of navigation, all routes, except those passing through Arquipélago dos Açores, are by great circle. Where great circle routes pass through, or close to, Arquipélago dos Açores see 2.115.
2.74.1
Routes to and from **Casablanca** (33°38′N, 7°35′W), for distances see Table 2.75.
2.74.2
Routes to and from **Dakar** (14°41′N, 17°24′W), for distances see Table 2.75.
2.74.3
Routes to and from **Freetown** (8°30′N, 13°19′W). Some of these great circle routes, as indicated at 2.75, are to Position 10°40′N, 17°40′W (**A**), thence direct, for distances see Table 2.75.

Distances in miles
2.75

	Casa-blanca	Dakar	Freetown
New York (a)	3140 *	3320	3750 †
Delaware Bay (b)	3220	3360	3780 †
Chesapeake Bay (c)	3290	3390	3800 †
NE Providence Ch (d)	3590	3400	3700 †
St Lucia/ St Vincent Ch (e)	3150	2540	2830
Tobago (f)	3190	2530	2800

* Via Nantucket TSS
† Via 10°40′N, 17°40′W (**A**)

Caribbean Sea ⇐ ⇒ Cape Town or Indian Ocean

Routes
2.76
1   Routes through the Caribbean Sea are as at 4.13 and 4.14 from Gulf of Mexico to its E end, thence direct as at 2.102 (in reverse) to 4°40′S, 34°35′W (**E**) off Cabo Calcanhar, thence by great circle to destination, see *Diagram 3.50*.

Distances in miles
2.77

	Colón	St Lucia/ St Vincent Channel
Cape Town	6450	5370
Cape Agulhas (15 miles S of)	6540	5470
Cape of Good Hope (145 miles S of)	6530	5450

PASSAGES BETWEEN WEST INDIES OR BERMUDA AND EUROPE

GENERAL INFORMATION

Coverage
2.78
This section contains details of the following passages:
  Approaches to Gulf of Mexico (2.80).
  From Straits of Florida to Vigo and European coast farther North (2.81).
  From Straits of Florida to Lisboa and Strait of Gibraltar (2.82).
  From Bishop Rock to North-east Providence Channel (2.83).

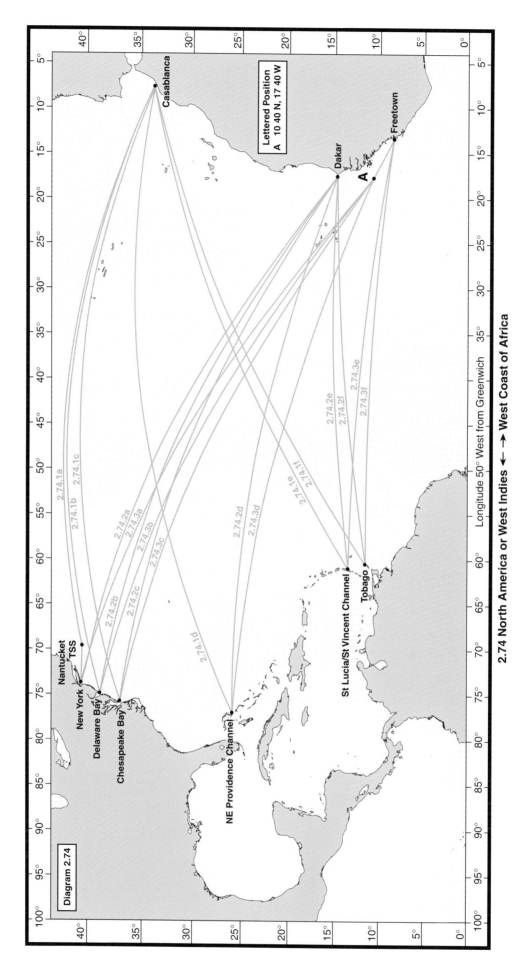

Lettered Position
A 10 40 N, 17 40 W

Diagram 2.74

2.74 North America or West Indies ← → West Coast of Africa

Longitude 50° West from Greenwich

Casablanca

Dakar

Freetown

A

St Lucia/St Vincent Channel

Tobago

NE Providence Channel

Chesapeake Bay

Delaware Bay

New York

Nantucket TSS

2.74.1a
2.74.1b
2.74.1c
2.74.2a
2.74.3a
2.74.3b
2.74.3c
2.74.2b
2.74.2c
2.74.1d
2.74.2d
2.74.3d
2.74.1e
2.74.1f
2.74.2e
2.74.2f
2.74.3e
2.74.3f

To and from West Indies passages or Bermuda and Europe, including alternative routes for W-bound vessels (2.84).

References
2.79
1 **Cautions**:
Western approaches to the English Channel, see 2.37.
Île d'Ouessant, see 2.38.
Bay of Biscay and west coasts of Spain and Portugal, see 2.39.
Strait of Gibraltar, see 2.40.
Grand Banks of Newfoundland, see 2.43.
Bermuda, see 2.44.
Caribbean Sea entrance channels, see 4.7.

PASSAGES

Diagram 2.80
Approaches to Gulf of Mexico
2.80
1 East-bound passages from Gulf of Mexico are recommended through Straits of Florida, thence N to take full advantage of the Florida Current, the Gulf Stream and North Atlantic Current (2.14), as well as the predominantly W winds in the N part of the North Atlantic Ocean.

2 For directions for passage through Straits of Florida to the departure point (27°00′N, 79°49′W), off Jupiter Inlet, see *West Indies Pilot, Volume I*.

3 West-bound passages are recommended either through Providence Channels or through Old Bahama Channel and Nicholas Channel.

4 Passages in Gulf of Mexico and Caribbean Sea, and the approaches thereto, are continued in Chapter 4.

Straits of Florida ⇒ Vigo and European coast farther North

Diagram 2.80
Routes
2.81
1 From the departure point, off Jupiter Inlet, the route leads to position 42°30′N, 50°00′W (**BS**) thence to destinations in N Europe (2.69).
2.81.1
1 From 27°00′N, 79°49′W (**A**), at the N end of the Straits of Florida, the routes are:
Through 30°00′N, 79°40′W (**G**), keeping in the main body of the Gulf Stream, thence:
Through 35°30′N, 72°40′W (**B**), thence:
Great circle to position (42°30′N, 50°00′W) (**BS**), thence:
As at 2.69 to destination.

2.81.2
1 Distances from Straits of Florida:

Nordkapp	4660 miles.
Trondheim	4220 miles.
Bergen	4060 miles.
Cape Wrath	3730 miles.
Inishtrahull	3620 miles.
Fastnet Rock	3520 miles.
Bishop Rock	3640 miles.
La Gironde*	3870 miles.
Vigo	3610 miles.

* 70 miles from Bordeaux

Straits of Florida ⇒ Lisboa and Strait of Gibraltar

Diagram 2.80
Routes
2.82
1 From 27°00′N, 79°49′W (**A**), at the N end of the Straits of Florida, the routes are:
2.82.1
Lisboa:
Through 30°00′N, 79°40′W (**G**), keeping in the main body of the Gulf Stream, thence:
By great circle to Lisboa (38°36′N, 9°24′W).
Distance   3630 miles.
2.82.2
Strait of Gibraltar:
Through 30°00′N, 79°40′W (**G**), keeping in the main body of the Gulf Stream, thence:
By great circle to Strait of Gibraltar (36°00′N, 5°21′W), adjusting course to avoid the islands of Arquipélago dos Açores, see 2.115.
Distance   3840 miles.

Bishops Rock ⇒ North-East Providence Channel

Diagram 2.80
Routes
2.83
1 From a position (49°47′N, 6°27′W) 5 miles S of Bishop Rock the route is:
By great circle to position (42°30′N, 50°00′W) (**BS**), thence:
By great circle to destination (25°50′N, 77°00′W).
Distance   3500 miles.
2.83.1
Alternative recommended routes for low-powered vessels can be found at 8.17.

West Indies passages or Bermuda ⇐ ⇒ Europe

Diagram 2.84
Routes
2.84
1 Subject to the ordinary requirements of navigation, including, on certain routes, avoidance of Bermuda (2.44) and passage through the islands of Arquipélago dos Açores, see 2.115, the routes are by great circles.

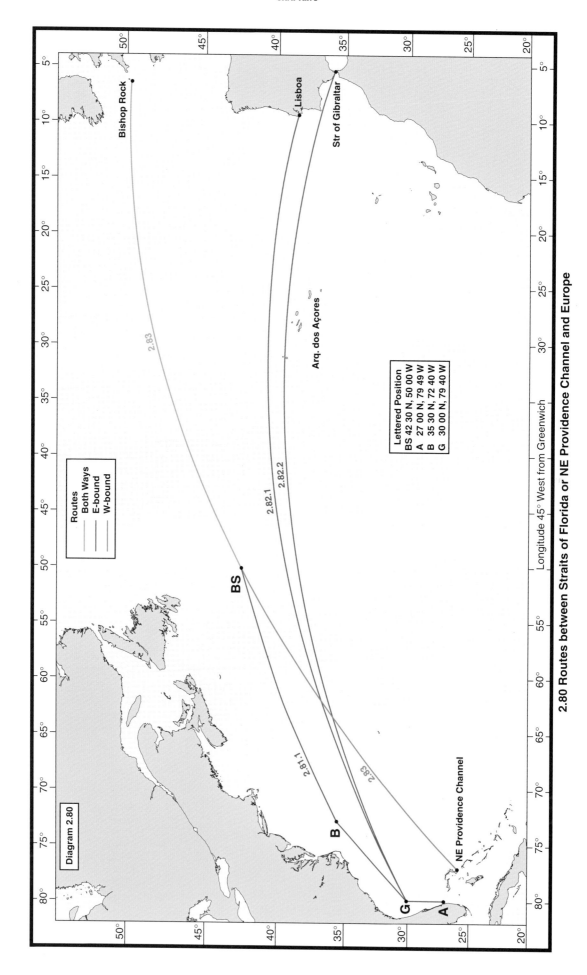

2.80 Routes between Straits of Florida or NE Providence Channel and Europe

Diagram 2.80

Routes
Both Ways
E-bound
W-bound

Lettered Position
BS 42 30 N, 50 00 W
A 27 00 N, 79 49 W
B 35 30 N, 72 40 W
G 30 00 N, 79 40 W

Longitude 45° West from Greenwich

Bishop Rock

Lisboa

Str of Gibraltar

Arq. dos Açores

BS

B

G

A

NE Providence Channel

2.83

2.82.2

2.82.1

2.81.1

2.83

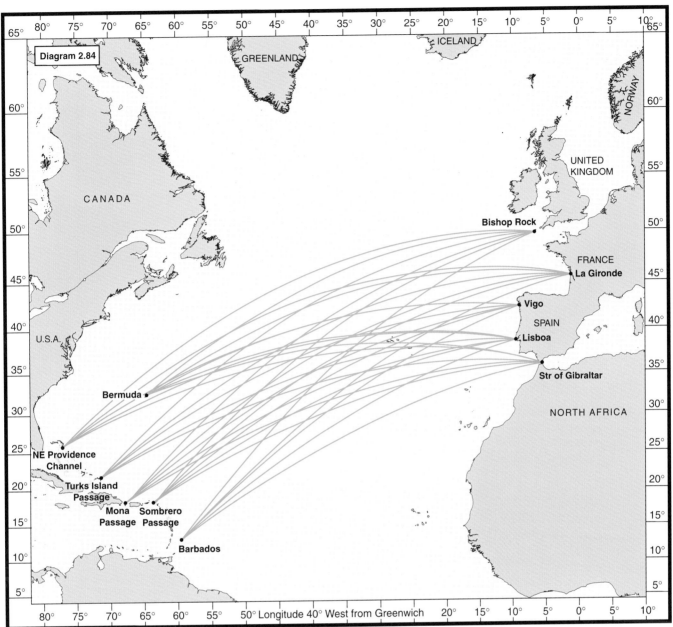

2.84 West Indies passages or Bermuda ← → Europe

Diagram 2.85

Alternative routes for west-bound vessels

2.85

1   For W-bound vessels from the Bay of Biscay and places farther N the following routes may be preferable in order to avoid the effects of the E-going North Atlantic Current and the predominance of W winds and head seas in the N part of the North Atlantic Ocean.

2.85.1

1   From the S part of the Bay of Biscay:

Rhumb line to 44°15′N, 8°30′W (**F**), thence:
Rhumb line to 36°40′N, 24°45′W (**D**), thence:
Great circle to 30°00′N, 45°30′W (**E**), thence:
By great circle to destination.

2.85.2

1   From the N part of the Bay of Biscay and places farther N:

Great circle to join Route 2.85.1 in 36°40′N, 24°45′W (**D**), thence:
As for Route 2.85.1 to destination.

Distances in miles

2.86

	Great circles (2.84)					Alternative W-bound routes (2.85)	
	Bishop Rock	La Gironde	Vigo	Lisboa	Strait of Gibraltar	Bishop Rock	La Gironde
NE Providence Channel	See 2.83	3730	3450	3450	3670		3980
Turks Island Passage	3450	3650	3330	3310	3510	3710	3770

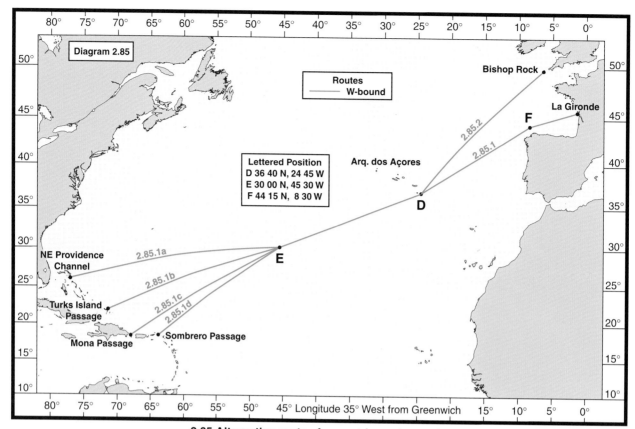

2.85 Alternative routes for west-bound vessels

	Great circles (2.84)					Alternative W-bound routes (2.85)	
Mona Passage	3470	3640	3310	3260	3450	3640	3700
Sombrero Passage	3310	3470	3120	3070	3250	3450	3510
Barbados	3410	3530	3170	3080	3230		
Bermuda	2760	2980	2690	2690	2930		

PASSAGES BETWEEN ENGLISH CHANNEL AND STRAIT OF GIBRALTAR OR INTERMEDIATE PORTS

GENERAL INFORMATION

Notes and cautions
2.87

1   **Traffic Separation Schemes** exist in the entrance to the English Channel, Île d'Ouessant and off the Spanish and Portuguese coasts. For regulations governing them see *Admiralty Sailing Directions*.

2   **Fishing.** For details of fishing grounds see relevant volume of *Admiralty Sailing Directions*.

3   **Île d'Ouessant.** When rounding Île d'Ouessant in uncertain weather great care should be taken, see 2.18, 2.37, 2.38 and *Channel Pilot*.

4   The incidence of fog in the vicinity is high, it is important to remember, when in fog, that it is not always possible from the land to determine the existence of fog banks in the offing and that fog signals may not therefore be in operation.

5   **Bay of Biscay.** Between Île d'Ouessant (48°28′N, 5°05′W) and Cabo Finisterre (42°53′N, 9°16′W) a general E-going set may be experienced. Onshore winds bring clouds, which may develop into low visibility near the coast.

6   The coast between Cabo Ortegal (43°46′N, 7°52′W) and Cabo Finisterre (42°53′N, 9°16′W) is a dangerous landfall except in good weather, owing to the E-going set of the current, the tidal streams and the risk of poor visibility with low cloud which may obscure the lights. Approaching across the Bay of Biscay, a landfall should be made at Cabo Villano (43°10′N, 9°13′W) which is high and easily recognised. In poor visibility, soundings will give a good indication of the distance off the shore. See also *Bay of Biscay Pilot*.

7   **West coasts of Spain and Portugal.** Although in general a good offing is advisable off the coasts of Spain and Portugal, when coasting the normal route for N-bound traffic follows the coast to Cabo Finisterre (42°53′N, 9°16′W) more closely than that for S-bound traffic.

8   The channel between Ilha Berlenga (39°25′N, 9°30′W) and Cabo Carvoeiro (39°21′N, 9°24′W) is clear and deep and may be taken in clear weather. When uncertain of the position near Os Farilhões (39°29′N, 9°33′W) and Ilha Berlenga (39°25′N, 9°30′W), it is vital to gain sea room

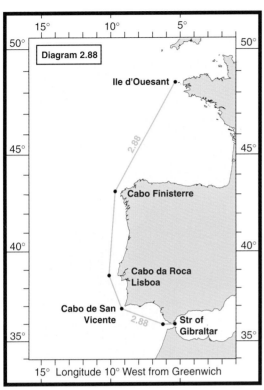

2.88 English Channel and Strait of Gibraltar intermediate ports

since sounding gives little indication of the vicinity of these islands. For general remarks on depths off these coasts see *West Coasts of Spain and Portugal Pilot.*

9   In the vicinity of Cabo de São Vicente (37°01'N, 9°00'W), the currents set strongly along the coast and have a tendency towards the cape. S-going currents predominate; N-going currents are especially likely to occur during SW gales.

10   **Strait of Gibraltar.** For the approach to the Strait of Gibraltar (36°00'N, 5°21'W) see 2.40.

PASSAGES

Diagram 2.88
Routes
2.88

1   Routes are direct and coastal, subject to the ordinary requirements of navigation. For details see *Bay of Biscay Pilot* and *West Coasts of Spain and Portugal Pilot.*

Distances in miles
2.88.1

Île d'Ouessant

430	**Vigo**		
640	225	**Lisboa**	
930	520	295	**Strait of Gibraltar****

**6 miles S of Europa Point

PASSAGES BETWEEN ENGLISH CHANNEL OR STRAIT OF GIBRALTAR AND WEST COAST OF AFRICA, CAPE TOWN OR THE INDIAN OCEAN

GENERAL INFORMATION

Coverage
2.89

This section contains details of the following passages:
   To and from English Channel and Cape Town or Indian Ocean (2.91).
   To and from English Channel and West Africa or Gulf of Guinea (2.92).
   To and from Strait of Gibraltar to West Africa, Gulf of Guinea, Cape Town or Indian Ocean (2.93).

References
2.90

1   **Cautions:**
   Bay of Biscay. For passage across the Bay of Biscay, see 2.87.
   West coasts of Spain and Portugal. For passage off the coasts of Spain and Portugal, see 2.87
   West Africa. The charting of the coast between Cabo Bojador (26°07'N, 14°30'W) and Feuve Senegal (16°00'N, 16°30'W) is reported to be inaccurate, and at night, as there are few lights, it should be given a wide berth. For general remarks on the reliability of charts of the W coast of Africa, see *Africa Pilot, Volumes I and II.*
   Abnormal refraction occurs at times off the African coast.
   Indian Ocean. For remarks on routes to and from the Indian Ocean see 3.16.

PASSAGES

English Channel ⇐ ⇒ Cape Town or Indian Ocean

Diagrams 2.91 and 3.29
Routes
2.91

1   From a position 48°28'N, 5°23'W, 10 miles W of Île d'Ouessant (2.38) the routes are:
2.91.1
   Cape Town.
      Rhumb line to 43°00'N, 10°00'W **(B)**, 30 miles off Cabo Finisterre, thence:
      Great circle to 20°50'N, 18°10'W **(C)**, off Cap Blanc, passing between Tenerife and Gran Canaria, thence:
      Rhumb line to 10°40'N, 17°40'W **(A)** to clear the breakers off Arquipélago dos Bijagós, thence:
      Great circle to Cape Town (33°53'S, 18°26'E).
2.91.2
   Cape Agulhas.
   If rounding Cape Agulhas then:
      as 2.91.1 to 10°40'N, 17°40'W **(A)** to clear the breakers off Arquipélago dos Bijagós, thence:
      Great circle to round Cape Point, thence:
      Direct to a position 35°05'S, 20°00'E, 15 miles S of Cape Agulhas.
2.91.3
   Indian Ocean.
   If entering the Indian Ocean (3.16) then:
      as 2.91.1 to 10°40'N, 17°40'W **(A)** to clear the breakers off Arquipélago dos Bijagós, thence:

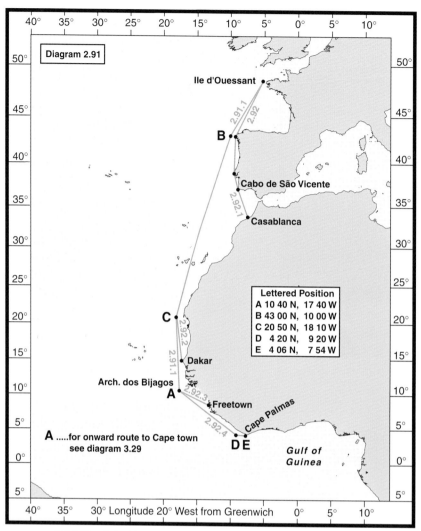

**2.91 English Channel ← → West Africa, Gulf of Guinea,
Cape Town or Indian Ocean**

Great circle to position 36°45′S, 19°00′E, 145 miles S
of Cape of Good Hope.

2.91.4

Routes beyond Arquipélago dos Bijagós are shown on
Diagram 3.29. For distances see Table 2.94.

English Channel ⇐ ⇒ West Africa and Gulf of Guinea

Diagram 2.91
Routes
2.92

1   From a position 48°28′N, 5°23′W, 10 miles W of Île
d'Ouessant (2.38) the routes are:

2.92.1

Casablanca.

As at 2.88 along the coasts of Spain and Portugal to
Cabo de São Vicente (37°01′N, 9°00′W), thence:
As navigation permits to Casablanca (33°38′N,
7°35′W).

2.92.2

Dakar.

As at 2.91.1 to 20°50′N, 18°10′W (**C**), off Cap
Blanc, passing between Tenerife and Gran Canaria,
thence:
As navigation permits to Dakar (14°41′N, 17°24′W).

2.92.3

Freetown.

As at 2.91.1 to 10°40′N, 17°40′W (**A**) to clear the
breakers off Arquipélago dos Bijagós, thence:
Direct to the outer end of the channel leading to
Freetown (8°30′N, 13°19′W).

2.92.4

Gulf of Guinea.

As at 2.91.1 to 10°40′N, 17°40′W (**A**) to clear the
breakers off Arquipélago dos Bijagós, thence:
Through 4°20′N, 9°20′W (**D**), thence:
Through 4°06′N, 7°54′W (**E**) (20 miles SSW of Cape
Palmas), thence:
As navigation permits to destination.

2.92.5

For distances see Table 2.94.

Strait of Gibraltar ⇐ ⇒ West Africa, Gulf of Guinea, Cape Town or Indian Ocean

Diagram 2.91 and 3.29
Routes
2.93

1   From Strait of Gibraltar (36°00′N, 5°21′W) the routes
are:

2.93.1

West Africa and Gulf of Guinea.
Direct as navigation permits.

2.93.2

Cape Town or Indian Ocean.
Coastwise to 10°40′N, 17°40′W **(A)** to clear the

breakers off Arquipélago dos Bijagós, thence:
As at 2.91 to destination.

2.93.3

Routes beyond Arquipélago dos Bijagós are shown on Diagram 3.29. For distances see Table 2.94.

Distances between English Channel, Strait of Gibraltar, West Africa and Cape of Good Hope
2.94

Île d'Ouessant	Strait of Gibraltar**					
950	190†	Casablanca				
1310	700	530	Las Palmas			
2160	1500‡	1330‡	835	Dakar		
2670	2040‡	1870‡	1370	530†	Freetown	
2380	1750‡	1570‡	1070	240	295	**10°40′N, 17°40′W (A. dos Bijagós)**
3090	2450‡	2280‡	1780	945	530	**4°06′N, 7°54′W (Cape Palmas)**
5750	5110‡	4940‡	4440	3610	3200	**Cape Town**
5860	5220‡	5050‡	4550	3720	3310	**Cape Agulhas (15 miles S)**
5890	5260‡	5080‡	4580	3750	3340	**Cape of Good Hope (145 miles S)**

** 6 miles S of Europa Point
‡ E of Islas Canarias, and joining route from Île d'Ouessant in 20°50′N, 18°10′W.
† Coastwise

PASSAGES BETWEEN NORTH AMERICA AND PLACES BETWEEN GULF OF MEXICO AND CABO CALCANHAR

GENERAL INFORMATION

Coverage
2.95

This section contains details of the following passages:
From Straits of Florida to North American ports (2.97).
To and from North American ports and West Indies (2.98).
To and from North American ports and Río Pará (2.99).
To and from North American ports and Cabo Calcanhar (2.100).
To and from Caribbean Sea, Río Pará and Cabo Calcanhar (2.102).

Notes and cautions
2.96

1 **Traffic separation scheme.** In Cabot Strait, a Traffic Separation Scheme has been established.

2 **Gulf of Mexico.** For recommended approaches to the Gulf of Mexico see 2.80.

3 **West Indies channels and E coast of USA.** For directions for passages through the West Indies Channels or along the E coast of the United States, see relevant volume of *Admiralty Sailing Directions*.

A strong N-going current will be encountered for 200 miles in the N approaches to North-East Providence Channel.

4 **Bahamas.** The Gulf Stream (2.14) is the main factor affecting voyages in the part of the ocean between Gulf of Saint Lawrence and the Bahamas.

5 **Bermuda.** For routes passing near Bermuda Islands, see caution at 2.44.

PASSAGES

Straits of Florida ⇒ North American ports

Diagram 2.97

Routes
2.97

1 From 27°00′N, 79°49′W **(A)**, at the N end of the Straits of Florida, the routes are:

2.97.1

To 30°25′N, 79°40′W **(B)**, thence:
To 31°11′N, 79°15′W **(C)**, thence:
Continuing in the main body of the Gulf Stream to 34°00′N, 75°49′W **(D)**, thence:
As navigation permits to destination, passing E of Diamond Shoal Light-buoy (35°09′N, 75°18′W).

2.97.2

For distances see Table 2.101.

North American ports ⇐ ⇒ West Indies

Diagram 2.98.1

Routes
2.98

1 The routes shown are between Cabot Strait, Halifax, New York, Delaware Bay, Chesapeake Bay and:

2.98.1

West Indies passages. The routes are by rhumb lines. The routes between Mona Passage or Cabot Strait and Halifax pass near Bermuda Islands, see caution at 2.44.

Diagram 2.98.2

2.98.2

Barbados. The routes are by rhumb lines passing E of Barbuda and the Lesser Antilles.

2.98.3

For distances see Table 2.101.

North American ports ⇐ ⇒ Río Pará

Diagram 2.99
Routes
2.99

1   Routes from N of Cape Hatteras (35°14′N, 75°31′W) are by great circle to 5°00′N, 47°30′W (**F**), then as navigation permits to the pilot ground off Salinopolis (0°30′S, 47°23′W).

2   The route between New York and 5°00′N, 47°30′W (**F**) passes near Bermuda Islands, see caution at 2.44.

2.99.1

1   North of Bermuda, the Gulf Stream will be felt; squally weather is frequent within its limits, and fog is prevalent along its N border.

2   On the S part of these routes, full allowance is required for the effects of the W-going North Equatorial Current and the E-going Equatorial Counter-current (2.14 and 2.15). Great care must be taken not to make a landfall W of Ponta da Atalaia (4 miles ENE of Salinopolis) because, in

that region, fresh to strong ESE winds and rough seas may be expected, with occasional poor visibility.

2.99.2

For distances see Table 2.101.

North American ports ⇐ ⇒ Cabo Calcanha

Diagram 2.100
Routes
2.100

1   Routes are by great circle to 4°40′S, 34°35′W (**E**), 60 miles off Cabo Calcanhar. The routes to and from New York and Delaware Bay pass near Bermuda Islands, see caution at 2.44.

2   For cautions regarding Atol das Rocas (3°52′S, 33°49′W) and dangers N and NW of Cabo Calcanhar (5°10′S, 35°29′W), see *South America Pilot Volume I*.

2.100.1

For distances see Table 2.101.

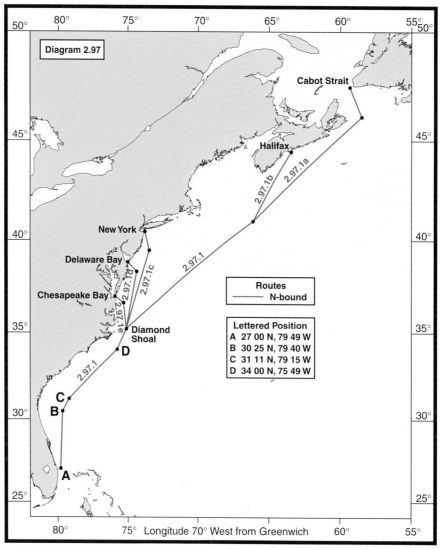

2.97 Straits of Florida → North American Ports

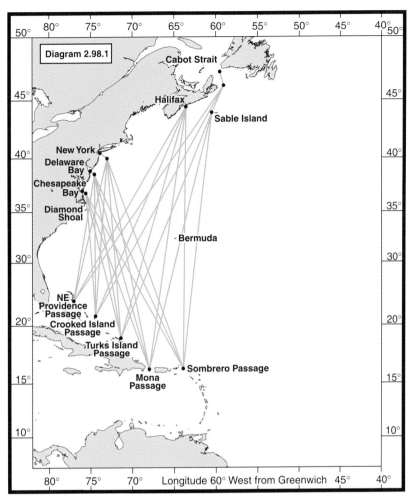

2.98.1 North American Ports ← → West Indies

Distances in miles
2.101

From	Cabot Strait	Halifax	New York	Delaware Bay	Chesapeake Bay
Strait of Florida	1630*	1360‡	900	800	690
NE Providence Channel	1570*	1300	890	790	680
Crooked Island Passage§	1580*	1330	970	880	770
Turks Island Passage	1650*	1420	1130	1040	940
Mona Passage	1800*	1590	1360	1290	1190
Sombrero Passage	1760*	1560	1420	1360	1280
Barbados	2070**	1900	1800	1750	1680
Río Pará†	2950**	2840	2800	2850	2790
Cabo Calcanhar (60 miles ENE)	3390**	3330	3460	3440	3400

§   10 miles NE of San Salvador
†   Río Pará to Belem: 125 miles.
‡   E of Georges Shoal

*   W of Sable Island
**   E of Sable Island

Caribbean Sea ⇐ ⇒ Río Pará ⇐ ⇒ Cabo Calcanhar

Diagram 2.102
Routes
2.102

1   The routes between Caribbean Sea, Río Pará and Cabo Calcanhar are as follows:

2.102.1
Saint Lucia/Saint Vincent Channel to Cabo Calcanhar.

From the channel between Saint Lucia (14°02′N, 61°01′W) and Saint Vincent (13°15′N, 61°15′W) the route is:

Direct to 5°00′N, 45°00′W (**D**), or even farther N, to avoid the strength of the Guiana and South Equatorial Currents (3.9), thence:

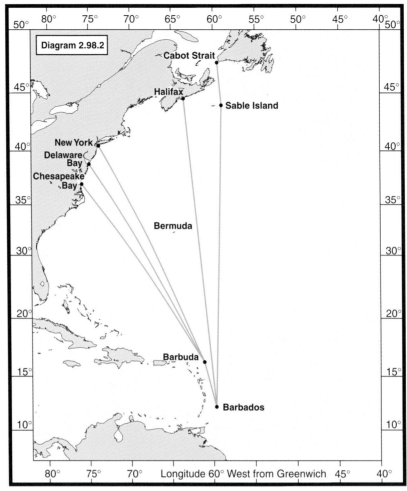

2.98.2 North American Ports ← → Barbados

Direct to 4°40′S, 34°35′W **(E)**, off Cabo Calcanhar, making due allowance for the W-going set.

Distance   1930 miles.

2.102.2

Tobago to Cabo Calcanhar.

From Tobago (11°15′N, 60°35′W) the route is:

Direct to 11°00′N, 56°20′W **(B)** where it joins Route 2.102.1.

Distance   1870 miles.

2.102.3

1 **Río Pará to Cabo Calcanhar**.

From Río Pará (0°30′S, 47°23′W) it is possible to take advantage of the tidal streams and an E-going counter-current by keeping close inshore if conditions permit, see *South America Pilot Volume I*.

2 An oilfield, with platforms and associated pipelines and buoyage, is situated close within the 100 m depth contour 40 miles NNW of Fortaleza (3°43′S, 38°31′W).

3 For cautions regarding Atol das Rocas (3°52′S, 33°49′W) and dangers N and NW off Cabo Calcanhar, see *South America Pilot Volume I*.

  Distance   820 miles.

  † Río Pará to Belem: 125 miles.

2.102.4

1 **Caribbean Sea to Río Pará**.

The route follows Route 2.102.1 to 6°20′N, 47°30′W **(C)**, thence:

As navigation permits to the pilot ground off

Salinopolis (0°30′S, 47°23′W).

Alternatively, a curving track, keeping about 100 miles offshore will shorten the distance by about 100 miles, but the adverse current will be stronger.

2 **Distances**.

Río Pará† from:

St Lucia/St Vincent Channel	1320 miles.
Tobago	1260 miles.

† Río Pará to Belem: 125 miles.

2.102.5

1 **Cabo Calcanhar to Río Pará and Carribean Sea**.

For caution regarding dangers N and NW of Cabo Calcanhar and off Fortaleza, see 2.102.3 and *South America Pilot Volume I*.

Approaching Río Pará (0°30′S, 47°23′W) from the E (route 2.102.3 in reverse), attention must be paid to Recife Manoel Luis (0°50′S, 44°15′W). The coast is low and devoid of prominent features, so that reliance must therefore be placed on non-visual methods for fixing the position.

For the Caribbean Sea routes are direct.

2 **Distances**:

From Cabo Calcanhar

St Lucia/St Vincent Channel	1910 miles.
Tobago	1830 miles.
Galleons Passage	1830 miles.
Río Pará†	820 miles.

† Río Pará to Belem: 125 miles.

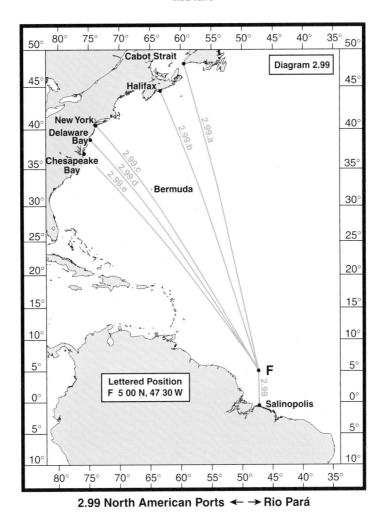

2.99 North American Ports ← → Rio Pará

PASSAGES BETWEEN NORTH-EAST COAST OF SOUTH AMERICA AND EASTERN PART OF NORTH ATLANTIC OCEAN

GENERAL INFORMATION

Coverage
2.103
This section contains details of passages between Río Pará (0°30′S, 47°23′W) and ports on the E side of the Atlantic Ocean.

General notes and references
2.104

1   **North-east coast of South America**. When approaching the coast between Trinidad (10°38′N, 61°34′W) and Cabo Calcanhar (5°10′S, 35°29′W), the effects of the Equatorial Current and Counter-current will be felt. See *Admiralty Sailing Directions.*

2   **Cautions**:

Western approaches to the English Channel, see 2.37.
Île d'Ouessant, see 2.38.
Bay of Biscay and west coasts of Spain and Portugal, see 2.39.
Strait of Gibraltar, see 2.40.

PASSAGES

Río Pará ⇐ ⇒ East part of North Atlantic

Diagram 2.105

Routes
2.105

1   From Río Pará (0°30′S, 47°23′W), in general, great circle tracks are recommended.

2.105.1

Distances from Río Pará†:

Bishop Rock	3670 miles	(a).
Île d'Ouessant	3650 miles	(b).
La Gironde*	3690 miles	(c).
Vigo	3300 miles	(d).
Lisboa	3140 miles	(e).
Strait of Gibraltar**	3210 miles	(f).
Casablanca	3040 miles	(g).
Dakar	2000 miles	(h).
Freetown	2100 miles	(i).
Ponta Delgada	2580 miles	(j).
Porto Grande	1680 miles	(k).
Las Palmas	2530 miles	(l).

*   70 miles from Bordeaux

** 6 miles S of Europa Point

†   Río Pará to Belem: 125 miles

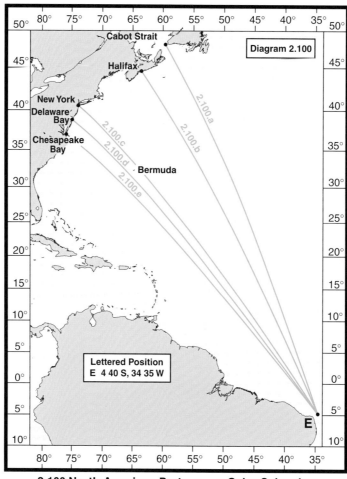

2.100 North American Ports ← → Cabo Calcanhar

PASSAGES BETWEEN EASTERN PART OF NORTH ATLANTIC OCEAN AND RECIFE

GENERAL INFORMATION

Coverage
2.106

This section contains details of the following passages:
To and from English Channel and Recife (2.108).
To and from Lisboa and Recife (2.109).
To and from Strait of Gibraltar and Recife (2.110).
To and from Las Palmas and Recife (2.111).
To and from Porto Grande and Recife (2.112).

General notes and references
2.107

1   **Cautions**

Western approaches to the English Channel, see 2.37.
Île d'Ouessant, see 2.38.
Bay of Biscay and west coasts of Spain and Portugal, see 2.39.
Strait of Gibraltar, see 2.40.

2   **Offshore hazards**

Penedos de São Pedro e São Paolo (2.45) and Atol das Rocas, which are both low-lying and dangerous, as well as Arquipélago de Fernando de Noronha (2.45), lie close to the routes to Recife, and in the main stream of the W-going South Equatorial Current.

The coast S of Cabo de São Roque should be approached with caution. Currents, which often set onshore should be guarded against.

Landfall off Recife should normally be made in 8°00′S, 34°40′W, 10 miles E of Ponta de Olinda, or N of this position from October to January. See *South America Pilot, Volume I*.

For destinations S of Recife see 3.18—3.24.

PASSAGES

English Channel ⇐ ⇒ Recife

Diagram 2.108
Routes
2.108

1   From the English Channel the routes are by great circle passing:
W of Madeira (32°40′N, 17°00′W) to 17°00′N, 25°30′W **(A)** (6 miles W of Santo Antão in Arquipélago de Cabo Verde), thence:
Direct to landfall off Recife (8°00′S, 34°40′W), passing between Atol das Rocas (3°52′S, 33°49′W) and Ilha de Fernando de Noronha (3°50′S, 32°27′W) (2.45).

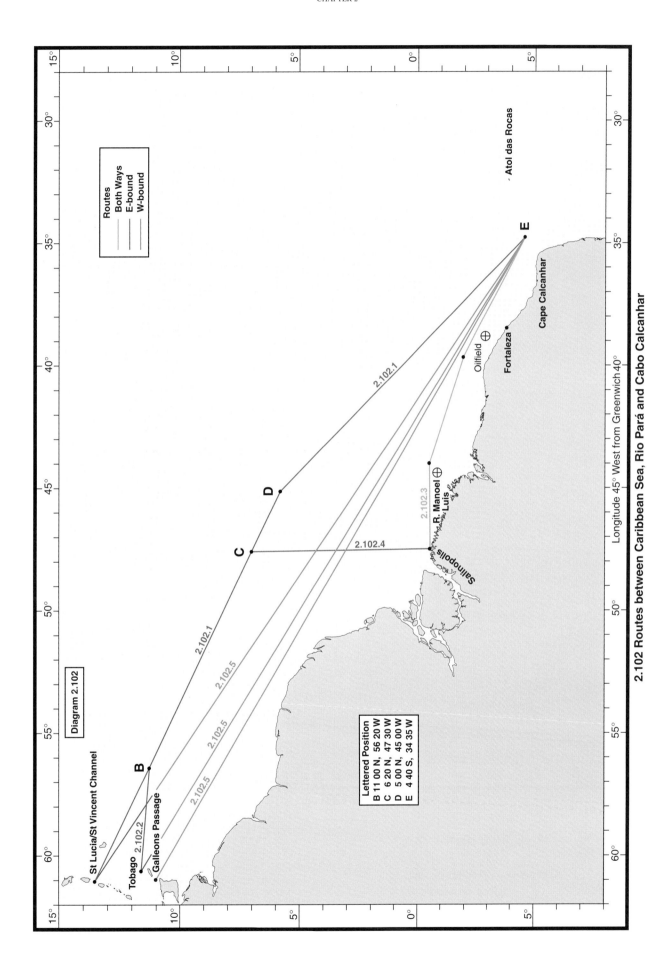

Diagram 2.102

Routes
Both Ways
E-bound
W-bound

St Lucia/St Vincent Channel

Tobago 2.102.2

Galleons Passage

B

2.102.1

2.102.5

2.102.5

2.102.5

C

2.102.4

D

2.102.1

2.102.3

R. Manoel Luis ⊕

Salinopolis

Oilfield ⊕

Fortaleza

Cape Calcanhar

E

· Atol das Rocas

Lettered Position
B 11 00 N, 56 20 W
C 6 20 N, 47 30 W
D 5 00 N, 45 00 W
E 4 40 S, 34 35 W

Longitude 45° West from Greenwich 40°

2.102 Routes between Caribbean Sea, Rio Pará and Cabo Calcanhar

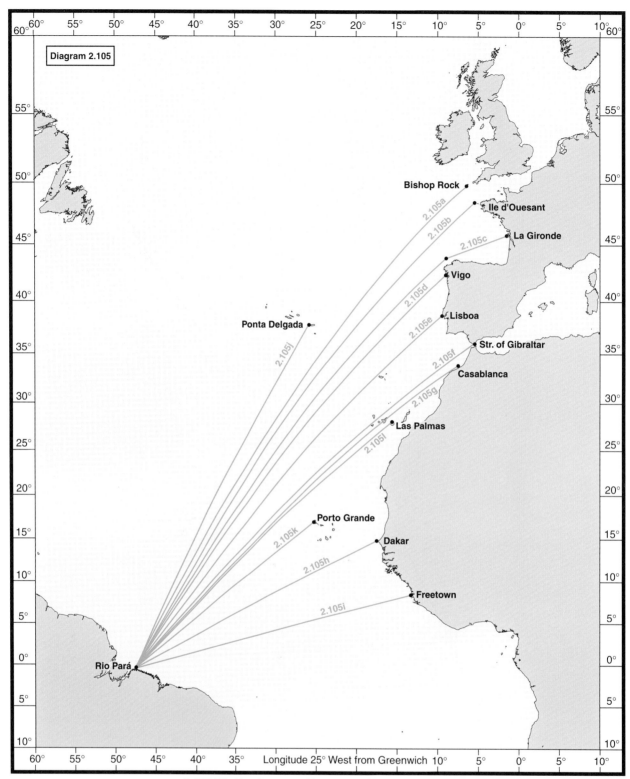

Diagram 2.105

2.105 Rio Pará ← → East part of North Atlantic

2.108.10

Distance to landfall off Recife, from:
Île d'Ouessant   3720 miles.

Lisboa ⇐ ⇒ Recife

Diagram 2.108

Routes

2.109

1   From Lisboa (38°36′N, 9°24′W) the route is direct
passing:

Through the channel between Fuerteventura (28°20′N,
14°00′W) and Gran Canaria (28°00′N, 15°30′W),
thence:

By great circle to landfall off Recife (8°00′S,
34°40′W), passing E of Arquipélago de Cabo
Verde (17°00′N, 24°00′W), and W of Penedos de
São Pedro e São Paolo (0°55′N, 29°21′W) and Ilha
de Fernando de Noronha (3°50′S, 32°37′W) (2.45).

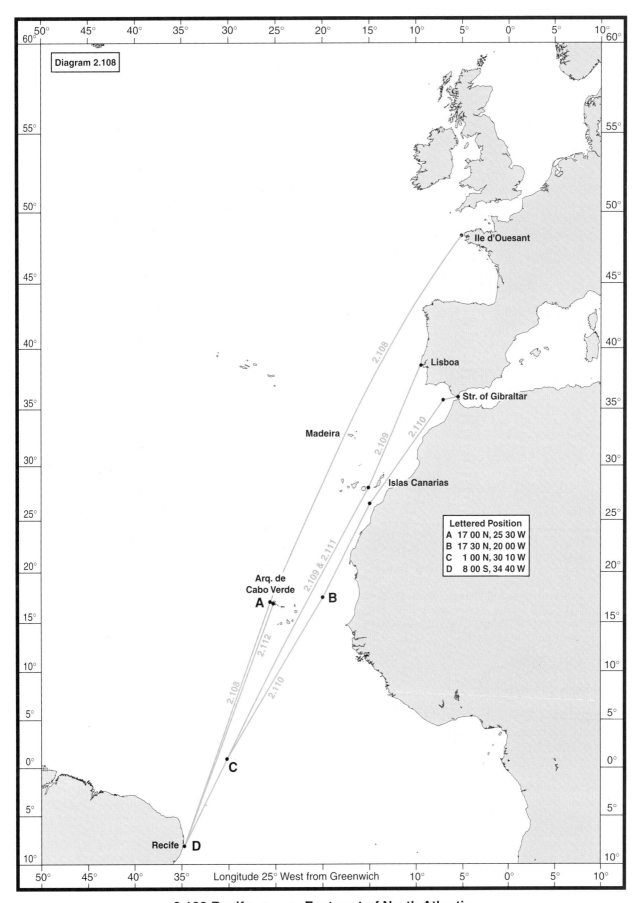

Diagram 2.108

Ile d'Ouesant

Lisboa

Str. of Gibraltar

Madeira

Islas Canarias

Lettered Position
A 17 00 N, 25 30 W
B 17 30 N, 20 00 W
C   1 00 N, 30 10 W
D   8 00 S, 34 40 W

Arq. de
Cabo Verde

A

B

C

Recife D

Longitude 25° West from Greenwich

2.108 Recife ⟵ ⟶ East part of North Atlantic

2.109.1

Distance to landfall off Recife, from:
Lisboa   3130 miles.

Strait of Gibraltar ⇐ ⇒ Recife

Diagram 2.108
Routes
2.110

1   The route from Strait of Gibraltar (36°00′N, 5°21′W) is:
Along the African coast as far as Dakhla (23°42′N, 15°56′W), thence:
Direct to 17°30′N, 20°00′W (**B**), thence:
Direct to 1°00′N, 30°10′W (**C**), 45 miles W of Penedos de São Pedro e São Paolo, thence:
Direct to landfall off Recife (8°00′S, 34°40′W), passing W of Ilha de Fernando de Noronha (3°50′S, 32°27′W) (2.45).

2.110.1

Distance to landfall off Recife, from:
Strait of Gibraltar**   3130 miles.
** 6 miles S of Europa Point.

Las Palmas ⇐ ⇒ Recife

Diagram 2.108
Routes
2.111

1   The route from Las Palmas (28°07′N, 15°24′W) passes:
Around the E side of Gran Canaria (28°00′N,

15°30′W), thence:
Joining the route from Lisbon (2.109) S of that island.

2.111.1

Distance to landfall off Recife, from:
Las Palmas   2440 miles.

Porto Grande ⇐ ⇒ Recife

Diagram 2.108
Routes
2.112

1   The route from Porto Grande (16°54′N, 25°02′W) passes:
Through Canal de São Vicente (16°56′N, 25°06′W), thence:
Direct to landfall off Recife (8°00′S, 34°40′W), passing between Atol das Rocas (3°52′S, 33°49′W) and Ilha de Fernando de Noronha (3°50′S, 32°27′W) (2.45).

2.112.1

Distance to landfall off Recife, from:
Porto Grande   1600 miles.

PASSAGES TO AND FROM ARQUIPÉLAGO DOS AÇORES

GENERAL INFORMATION

Coverage
2.113

This section contains details of passages between the Arquipélago dos Açores and the English Channel, Bay of Biscay, W coasts of Spain and Portugal, Strait of Gibraltar, Arquipélago de Cabo Verde, Islas Canarias, W coast of Africa, NE coast of South America, West Indies, Caribbean Sea, or E coast of North America.

General notes and references
2.114

Cautions:
Western approaches to the English Channel, see 2.37.
Île d'Ouessant, see 2.38.
Bay of Biscay and west coasts of Spain and Portugal, see 2.39.
Strait of Gibraltar, see 2.40.
Strait of Belle Isle, see 2.41
Coast of Newfoundland, see 2.42.
Grand Banks of Newfoundland, see 2.43
Bermuda, see 2.44.
Caribbean Sea entrance channels, see 4.7
Traffic Separation Scheme. A Traffic Separation Scheme has been established in Cabot Strait.
Currents. The islands lie in the main flow of the Azores Current (2.14), and S of the predominantly W winds of the North Atlantic Ocean.

PASSAGES

Arquipélago dos Açores ⇐ ⇒North Atlantic ports

Diagram 2.115
Routes
2.115

1   Subject to the ordinary requirements of navigation, Arquipélago dos Açores (38°00′N, 27°00′W) can be approached by great circle routes from all directions.
2.115.1

Distances from Ponta Delgada (37°44′N, 25°39′W):

Bishop Rock	1110 miles (a).
Île d'Ouessant	1100 miles (b).
La Gironde*	1190 miles (c).
Vigo	825 miles (d).
Lisboa	770 miles (e).
Strait of Gibraltar**	985 miles (f).
Casablanca	920 miles (g).
Las Palmas	775 miles (h).
Dakar	1460 miles (i).
Porto Grande	1250 miles (j).
Río Pará†	2580 miles (k).
Barbados	2350 miles (l).
Sombrero Passage	2310 miles (m).
Mona Passage	2500 miles (n).
Turks Island Passage	2540 miles (o).
N-E Providence Ch.	2690 miles (p).
Bermuda	1930 miles (q).
Straits of Florida¤	2790 miles (r.i).
Delaware Bay	2310 miles (s).
New York	2230 miles (t).
Halifax‡	1770 miles (u).

58

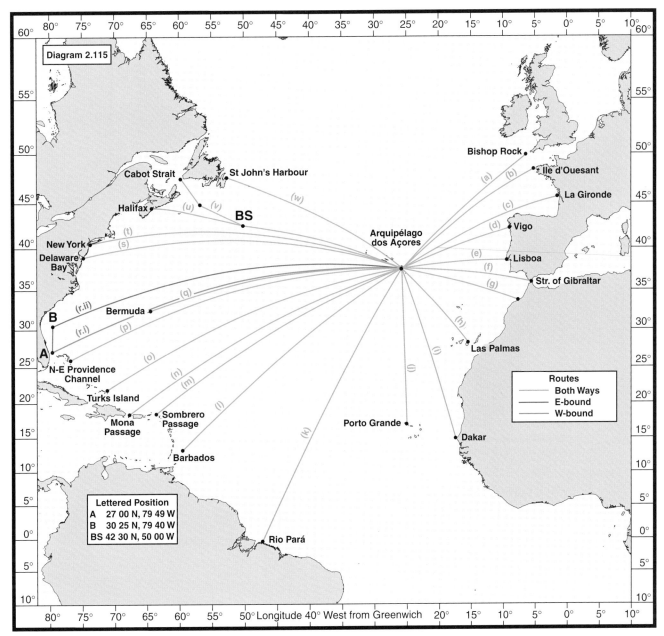

2.115 Routes to and from Arquipélago dos Açores

Cabot Strait#	1670 miles (v).	†	Río Pará to Belem: 125 miles
St.John's, Nfndland	1330 miles (w).	¤	Direct. If E-bound (see 2.82), 2880 miles (r.ii).
* 70 miles from Bordeaux		‡	Via Position BS as at 2.68.2
** 6 miles S of Europa Point		#	Via Position BS as at 2.68.1

PASSAGES TO AND FROM ARQUIPÉLAGO DE CABO VERDE

GENERAL INFORMATION

Coverage
2.116

This section contains details of the following passages:

To and from Arquipélago de Cabo Verde and English Channel or western side of North Atlantic Ocean (2.118).

To and from Arquipélago de Cabo Verde and Strait of Gibraltar or west coast of Africa (2.119).

General notes and references
2.117
Cautions:

Western approaches to the English Channel, see 2.37.

Île d'Ouessant, see 2.38.

Bay of Biscay and west coasts of Spain and Portugal, see 2.39.

Strait of Gibraltar, see 2.40.

Strait of Belle Isle, see 2.41.

Coast of Newfoundland, see 2.42.

Grand Banks of Newfoundland, see 2.43.

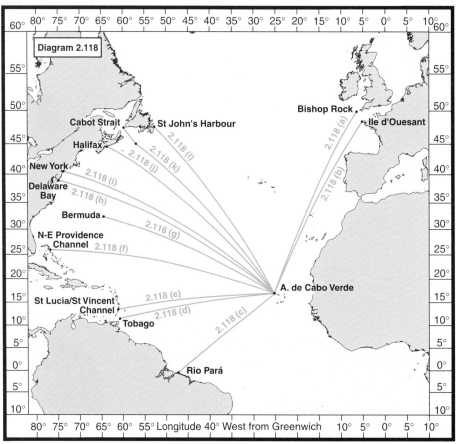

2.118 Porto Grande ← → English Channel and Western side of North Atlantic Ocean

Bermuda, see 2.44.

Caribbean Sea entrance channels, see 4.7.

Traffic Separation Scheme. A Traffic Separation Scheme has been established in Cabot Strait.

Currents. The E islands of Arquipélago de Cabo Verde more especially feel the force of the Canary Current (2.14) setting to the SW; several wrecks have been caused by disregarding it. The currents between the islands of the group are frequently strong, irregular and influenced by the wind. For further details of currents see *Africa Pilot, Volume 1*.

Local magnetic anomalies have been reported in the vicinity of Arquipélago de Cabo Verde, especially off the W side of Sal, off the E side of Boavista and near Fogo and Brava.

PASSAGES

Porto Grande ⇐ ⇒ English Channel and western side of North Atlantic Ocean

Diagram 2.118
Routes
2.118

1 Subject to the ordinary requirements of navigation, Arquipélago de Cabo Verde (17°00'N, 24°00'W) can be approached by great circle from the English Channel and from the W.

2.118.1

Distances from Porto Grande (16°54'N, 25°02'W):

Bishop Rock	2170 miles	(a).
Île d'Ouessant	2120 miles	(b).
Río Pará†	1680 miles	(c).
Tobago	2090 miles	(d).

2.119 Porto Grande ← → Strait of Gibraltar and West Africa

St Lucia/St Vincent   2090 miles   (e).
N-E Providence Ch.   2940 miles   (f).
Bermuda   2340 miles   (g).
Delaware Bay   2950 miles   (h).
New York   2890 miles   (i).
Halifax   2560 miles   (j).
Cabot Strait   2510 miles   (k).
St.John's, Nfndland   2290 miles   (l).
†   Río Pará to Belem: 125 miles

Porto Grande ⇐ ⇒ Strait of Gibraltar and West Africa

Diagram 2.119
Routes
2.119

1   For Dakar (14°41′N, 17°24′W) and Banjul (13°32′N, 16°54′W), the routes pass N of Boavista (16°05′N, 22°55′W).

2   For Freetown (8°30′S, 13°19′W)) and ports further S, the routes pass S of Santiago (15°06′N, 23°40′W).

3   For ports in the Gulf of Guinea (see 3.26), routes pass:
Through 4°20′N, 9°20′W **(D)**, thence:
Through 4°06′N, 7°54′W **(E)**, 20 miles SSW of Cape Palmas.

2.119.1
Distances from Porto Grande (16°54′N, 25°02′W):

Strait of Gibraltar**   1560 miles   (a).
Casablanca   1390 miles   (b).
Las Palmas   870 miles   (c).
Dakar   470 miles   (d).
Banjul   515 miles   (e).
Freetown   885 miles   (f).
Monrovia   1070 miles   (g).
Cape Palmas*   1300 miles   (h).

*   20 miles SSW of Cape Palmas
**   6 miles S of Europa Point

NOTES

CHAPTER 3

SOUTH ATLANTIC OCEAN

GENERAL INFORMATION

COVERAGE

Chapter coverage

3.1

This chapter contains details of the following passages:
Passages off the E coast of South America (3.17).
Passages off the W coast of Africa (3.26).
Transatlantic passages (3.31).

Admiralty Publications

3.2

Relevant navigational publications should be consulted, in addition to this book, when planning and conducting passages. Details of these, and the areas covered, can be found, as follows:
Admiralty Sailing Directions; see 1.12.1
Admiralty List of Lights and Fog Signals; see 1.12.2
Admiralty Tide Tables; see 1.12.3
Admiralty List of Radio Signals; see 1.12.5
Admiralty Distance Tables; see 1.12.8

3.2.1

See also *The Mariner's Handbook* for notes on electronic products.

NATURAL CONDITIONS

Charts:

South Atlantic Routeing 5125 (1-12)
World Surface Current Distribution 5310

WINDS AND WEATHER

General information

3.3

1   The wind system of the South Atlantic Ocean resembles that of the North Atlantic, except that the circulation round the oceanic anticyclone is anti-clockwise, and there is no wind corresponding to the south-west monsoon off West Africa. There is no Intertropical Convergence Zone, and there are no tropical storms.

South-east Trade Wind

3.4

1   The South-east Trade Wind forms the equatorial side of the circulation around the oceanic anticyclone, which is centred in about 20°S to 28°S. It is the counterpart of the North-east Trade Wind (2.5) and blows with equal persistence and constancy of direction, from about SSE on the E side of the ocean to almost E on the W side. It extends as far N as the equator in winter (July) and to within two or three degrees of it in summer (January).

2   The average strength of the South-east Trade Wind is is similar to that of the North-east Trade Wind except N of 10°S, E of about 10°W, where it averages only force 2 to 3.

3   Weather is similar to that of the zone of the North-east Trade Wind, except that fog is frequent over the cold waters of the Benguela Current (3.9) close to the coast of South-west Africa between 20° and 30°S.

The Variables (Horse Latitudes)

3.5

1   The Variables, a belt of light and generally variable winds in the neighbourhood of the oceanic areas of high pressure, extend across the ocean in about 29°S, oscillating from about 26°S in winter to about 31°S in summer. Conditions are similar to those in the corresponding zone (2.6) of the North Atlantic Ocean.

2   East of the prime meridian winds are predominantly from between S and ESE, being in fact an extension of the South-east Trade Wind. In the W part of the zone NE winds are commonest, particularly in summer.

The Westerlies (Roaring Forties)

3.6

1   South of about 35°S, W winds predominate. As in the North Atlantic Ocean, the almost continuous passage of depressions from W to E causes the wind to vary greatly both in direction and strength, and winds from any direction can be experienced; the centres of the depressions generally move from the vicinity of Cabo de Hornos in the direction of South Georgia and then approximately along 50°S.

2   Gales are very prevalent; S of about 40°S, even at midsummer, winds reach force 7 on from 7 to 9 days per month, and S of about 43°S and S and E of about 40°W the frequency rises to about 15 days per month. In winter this latter frequency is generally S of a line joining the Falkland Islands and Cape of Good Hope, while most of the area between this line and 30° S has from 5 to 10 days per month with winds of this force.

3   Weather is of a similarly variable nature to that experienced in the corresponding zone of the N hemisphere.

4   Fog is not uncommon in summer, and is generally associated with winds from warmer latitudes.

SWELL

Zones

3.7

1   Fewer observations of ocean swells are available from the South Atlantic Ocean than from the North Atlantic Ocean. For purposes of swell the South Atlantic Ocean can conveniently be divided into three zones. For length of swell see 2.12.

2   **0° to 20°S.** Slight to moderate swell, rarely heavy; from SE in the E part of the zone, and from between SE and E in the W part.

3   **20° to 40°S.** The swell is mainly moderate, but sometimes heavy. In the E part of the zone it is from S; the direction is variable in the W part, with a high proportion from between NE and N. Reports of confused swell are frequent.

4   **40° to 60°S.** The swell is mainly moderate, but in the extreme S it is often heavy.

5   Throughout the year, the worst conditions are likely to occur between 40°S and 50°S. The depressions, which are of much the same size as those that produce the North Atlantic winter storms, move in continual succession from W to E,

usually along tracks S of 50°S. The strongest winds blow from NW with a heavy overcast sky on a falling barometer; they are followed by SW winds as the barometer rises and the sky clears.

6 Heavy swell is present for between 30% and 70% of the time between 50°S and 60°S. In summer, the frequency of high seas and swell decreases towards the circumpolar trough which generally lies in about 64°S, where the mean wind speeds are less than farther N, although relatively small strong gales occur from time to time.

7 Most of the very high seas and swell appear to be raised by the Westerlies. Abnormal waves, which are almost certainly due to a number of component wave trains becoming momentarily in step, are a very real possibility which appears to be increased in the vicinity of shoal water and when the wave train is moving against the current. See 1.19 for further details.

CURRENTS

North and South Atlantic Oceans
3.8

1 The current circulation systems of the whole Atlantic region are described at 2.13.

South Atlantic Ocean
3.9

1 Just S of the **Equatorial Counter-current**, the **South Equatorial Current** sets W to the N of the sub-tropical gyre which circulates water in a counter-clockwise direction between West Africa in the E and South America in the W.

2 Off SW Africa the **Benguela Current** sets NW with moderate constancy fanning out to the W from its left flank as it progresses. Two main factors combine to form the Benguela Current. Offshore the main circulation carries relatively cool water ENE from the sub-Antarctic and then turns N as it approaches the coastal shelf of SW Africa. Inshore the South-east Trade Wind blowing persistently from the land, removes the surface water seaward and allows cooler sub-surface waters to rise as replacement. This enhances the temperature gradient with the Benguela Current appearing as a cool current. On the left flank of the S part of the Benguela Current a narrow mixing zone, approximately 3° to 5° of latitude wide, extends from about 30°S, 0° through 34°S, 10°E to near 38°S, 20°E. Along this axis there is continuous interaction between the E-going sets of the **South Atlantic Current**, the ENE-going flow of the **Southern Ocean Current** and the general NW-going sets of the combined Agulhas (6.28) and Benguela Currents. Areas of local convergence and divergence occur with marked variation of both sea temperature and sea state.

3 The Benguela Current gradually turns W towards South America as the South Equatorial Current. This current is stronger on its equatorial flank where its N boundary extends across the equator. To the S towards the axis of the gyre, constancies and rates both decrease, becoming variable and mostly less than ½ knot.

4 Approaching the Brazil coast near Recife the current diverges, the N part continuing WNW towards the Caribbean Sea (see 2.14) and the S part turning S as the warm **Brazil Current**. The latitude at which this division occurs varies at the coast from about 8°S in summer (December) to near 12°S in winter (June). In central longitudes near 20°W the axis of the gyre also fluctuates by about 5° being farthest N near 20°S in the winter months.

5 The Brazil Current extends S to the latitude of Río de La Plata where interaction with the N-flowing cold **Falkland**

Current occurs. Fanning out from its left flank into the open ocean leads to a belt of E-going sets between 30°S and 40°S. This flow is sometimes referred to as the South Atlantic Current to distinguish it from the much cooler Southern Ocean Current which extends ENE from Drake Passage and continues to the E, passing S of South Africa into the South Indian and Pacific Oceans as a circumpolar current.

6 The Falkland Current derives from the Southern Ocean Current as it enters the South Atlantic Ocean. Most of it passes W of the Falkland Islands but some continues S of the islands before turning N, joining the main stream in about 45°S. There are indications of marked eddies associated with patches of warm water penetrating well S in longitudes 50°W to 55°W from the S part of the Brazil Current.

ICE

General remarks
3.10

1 For general remarks and references, see 2.20, which are also applicable to the South Atlantic Ocean.

2 For facsimile ice charts and ice forecasts from Antarctica, see the relevant volume of *Admiralty List of Radio Signals*.

Pack ice
3.11

1 The approximate mean limits of pack ice are indicated on Routeing Charts, Climatic Charts and *US Marine Climatic Atlas of the World, 1995 Edition (available on CD-ROM)*.

2 The main shipping routes of the S hemisphere are not affected by pack ice, but in the South Atlantic Ocean its presence prevents the use of a great circle track between Cape of Good Hope and Cabo de Hornos except during March, April and May.

3 The long-term average position of the pack ice (4/8 concentration) in August to September, at its greatest extension, runs from about 60°S, 60°W to a position just E of South Georgia in about 54°S, 30°W. Thence it extends E while gradually increasing in latitude to about 55°S on the meridian of Greenwich and about 58°S in 50°E. This is an average position of the edge of the pack ice which, in severe years, can be appreciably farther N.

4 The average position of the edge of the pack ice in the months of least average extension (February to March) is well S of the foregoing positions. In those parts where the Antarctic continent extends continuously to lower latitudes (ie from longitudes 10°W through 0° to 160°E) the average ice edge at this season does not extend much beyond 100 miles from the coast and in some places retreats to the coast. Off the Weddel and Ross Seas the ice is more extensive, reaching its farthest N on the parallel of about 62°S between 20°W and 60°W. It is stressed that these are average positions.

Icebergs
3.12

1 Antarctic icebergs, unlike those of the North Atlantic Ocean, are not usually calved from glaciers, but consist of portions that have broken away from the great ice shelves which surround parts of the Antarctic continent. They are consequently flat-topped and they may be of immense size.

2 The extreme limit of icebergs, irrespective of season, is illustrated on *Diagrams 1.14a* and *1.14b*, and for each month on *US Pilot Charts — Monthly Weather Hazards*. In the S hemisphere, icebergs are more liable to be encountered in lower latitudes in the South Atlantic Ocean than in the

other oceans of this hemisphere. Near the coasts of Argentina and S Brazil, icebergs may be found as far N as 31°S. Abnormally, one has even been reported in about 26°S, 26°W. In the rest of the South Atlantic Ocean, icebergs are largely confined to latitudes S of 35°S.

3   The relatively simple nature of Antarctic geography, with an almost symmetrical flow of currents round a nearly circular continent, means that there is less cause here than in Arctic waters for a great concentration of icebergs in a few comparatively narrow "lanes". Some concentration does occur due to the deflection and concentration of the E-going circumpolar stream by the projection to the N of Graham Land. Some of the icebergs in the resulting NE flow between South America and Graham Land are carried into the Falkland Current which takes them N as far as, or even beyond, the estuary of Río de La Plata. Another branch of the NE-going flow through Drake Passage continues NE and passes E of the Falkland Islands, carrying icebergs to similar latitudes in the more central parts of the South Atlantic Ocean.

4   Due notice should be taken of the caution at 2.35 regarding the use of radar for detecting icebergs.

NOTES AND CAUTIONS

Penedos de São Pedro e São Paolo and Ilha de Fernando de Noronha.
3.13

1   Caution is necessary in the vicinity of these islands as the South Equatorial Current (3.9) sets WNW past them at a rate of 1 to 2 kn.

Rounding Cabo de Hornos
3.14

1   Off Cabo de Hornos, W winds predominate with at least 30% of force 7 or more throughout the year. The current off the cape is usually E-going. Swell is mainly from the W throughout the year, heavy on 25% of occasions. Icebergs may be encountered.

Estrecho de Magallanes
3.15

1   Estrecho de Magallanes provides an alternative route between the E and W coasts of South America, it is regularly navigated by ocean-going vessels. Although caution is required in making the passage, the difficulties and dangers in navigating the strait, in either direction, are the same as those experienced in other narrow channels and harbours, but they are accentuated by the prevalence of bad weather and by the generally foul and rocky character of the anchorages.

2   Manoeuvrability of a vessel is important as an unhandy or low-powered vessel is at a disadvantage in those parts of the strait where rapid action may be required to counter strong cross-tidal streams, or where there is a risk of meeting another vessel simultaneously with one of the violent and unpredictable squalls which are common.

3   W-bound vessels avoid the adverse currents and gales and the heavy head seas so commonly experienced in rounding Cabo de Hornos and off the archipelago NW of the cape. The risk of encountering icebergs is avoided and there is usually a saving in distance.

4   E-bound vessels in bad weather may find it difficult to make a good landfall which is essential before a vessel enters the strait. Tidal streams in the E part of the strait tend to be less favourable to an E-bound ship than to a W-bound one.

5   Some masters favour the passage of the strait both W-bound and E-bound, particularly for vessels W-bound in ballast.

6   For further information and details of pilotage requirements, see *South America Pilot, Volume II*.

Rounding Cape of Good Hope
3.16

1   The SW-going Agulhas Current flows off the SE coast of South Africa (6.35). It is strongest in the vicinity of the 200 m depth contour and, at times, reaches a rate of 5 kn.

2   To the S of the cape W winds predominate from April to September and winds of force 7 or more can be expected for about 30% of the time. From October to March winds are more variable in direction and the frequency of strong winds is less.

3   Loaded tankers navigating off the coast of South Africa are governed by regulations regarding the offing to be kept. See *Africa Pilot, Volumes II and III*.

3.16.1
Entering the Indian Ocean. There is a choice of either making Cape Agulhas and thence keeping inshore of the Agulhas Current (see Route 6.35.1) or passing S of the current through 36°45′S, 19°00′E (145 miles S of Cape of Good Hope).

3.16.2
Leaving the Indian Ocean. The Agulhas Current should be sought, but avoiding the dangerous seas and abnormal waves (1.19) found at times in its vicinity.

PASSAGES OFF THE EAST COAST OF SOUTH AMERICA

GENERAL INFORMATION

Coverage
3.17

This section contains details of the following passages:

    To and from Cabo Calcanhar or Recife and Río de la Plata and intermediate ports (3.19).

    To and from Río de la Plata and Cabo de Hornos and intermediate ports (3.20).

    To and from Falkland Islands and Estracho de Magallanes (3.22).

    To and from Falkland Islands and east coast of South America (3.23).

    To and from Falkland Islands and Cabo de Hornos (3.24).

Notes and cautions
3.18

1   Between ports on the E coast of South America, routes are coastwise and are not described in this book; the relevant *Admiralty Sailing Directions* should be consulted.

3.18.1

1   **Off-lying shoals**. Generally speaking, all passages are as direct as prudent navigation permits, but off-lying shoals make wide divergence from the coast necessary in some places, notably near Arquipélago dos Abrolhos (18°00′S, 38°40′W), Cabo de São Tomé (22°02′S, 41°03′W) and Banco do Albardão (33°10′S, 52°25′W). All ships, particularly deep-draught vessels, should note the dangers which extend offshore between the parallels of 15°S and 22°S, particularly the ridge extending 600 miles E from the

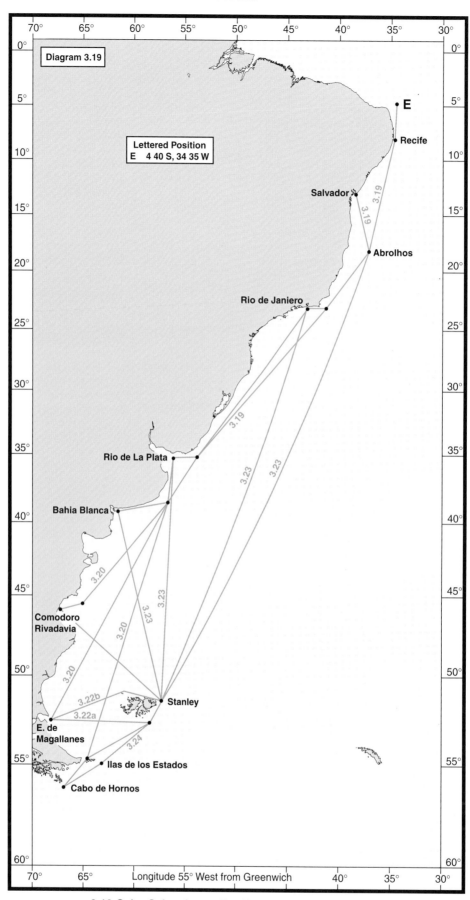

Diagram 3.19

Lettered Position
E   4 40 S, 34 35 W

E

Recife

Salvador

Abrolhos

Rio de Janiero

Rio de La Plata

Bahia Blanca

Comodoro
Rivadavia

Stanley

E. de
Magallanes

Ilas de los Estados

Cabo de Hornos

3.19 Cabo Calcanhar or Recife ←→ Rio de La Plata
3.20 Rio de La Plata ←→ Cabo de Hornos
3.22 Falkland Islands ←→ Estrecho de Magallanes
3.23 Falkland Islands ←→ East Coast of South America
3.24 Falkland Islands ←→ Cabo de Hornos

2 coast which terminates in Ilhas Martin Vaz (20°31′S, 28°51′W); this region is largely unsurveyed.

2 Known dangers are described in *Admiralty Sailing Directions,* but it should be borne in mind that modern deep-draught ships could well ground on shoals over which their predecessors passed safely.

3.18.2

An oilfield, with platforms, pipelines and associated operations, is situated about 50 miles SE of Cabo de São Tomé (22°02′S, 41°03′W).

3.18.3

1 **Currents** in general move water towards Río de la Plata (35°10′S, 56°15′W); S from Cabo Calcanhar (5°10′S, 35°29′W) and N from Cabo de Hornos (56°04′S, 67°15′W), though the latter set is well away from the coast. Off the coast S of Peninsula Valdes (42°30′S, 64°00′W) currents are very variable, and within 20 miles of the shore tidal influences only are felt. Onshore currents are prevalent at any time of the year between Cabo de São Roque (5°29′S, 35°16′W) and Cabo Frio (23°01′S, 42°00′W).

2 Seasonal changes in the coastal currents should be noted. See 3.9 and the relevant volumes of *Admiralty Sailing Directions.*

3 Advantage may be taken of these currents by making N-bound passages between Río de la Plata and Cabo Frio closer inshore during May, June and July than at other times, but with due regard to the possibility of onshore sets.

PASSAGES

Cabo Calcanhar or Recife ⇐ ⇒ Río de la Plata and intermediate ports

Diagram 3.19
Routes
3.19

Routes are as direct as navigation permits, taking due precautions against onshore currents.

For distances, see Table 3.25.

Río de la Plata ⇐ ⇒ Cabo de Hornos and intermediate ports

Diagram 3.19
Notes and cautions
3.20

1 **Currents**. The N-going Falkland Current (3.9) will affect voyages between Río de la Plata (35°10′S, 56°15′W) and the Falkland Islands (51°40′S, 57°49′W) or Estrecho de Le Maire (54°50′S, 65°00′W) or, to a lesser extent, Estrecho de Magallanes (52°27′S, 68°26′W).

2 **Tidal streams**. Approaching Estrecho de Magallanes, special attention is required as the range of the tide is great, and the tidal streams at the entrance run with great strength, causing, at times, an indraught towards Banco Sarmiento (52°30′S, 68°04′W) and the dangers extending from Cabo Vírgenes (52°20′S, 68°21′W). The tidal streams in the strait are a controlling factor in the choice of the time of arrival.

3 **Icebergs**. For dangers from icebergs see 3.12.

Routes
3.21

For choice of route around Cabo de Hornos or through Estrecho de Magallanes, see 3.14 and 3.15. For directions for approaching the Falkland Islands see *South America Pilot, Volume II.*

Routes are as direct as navigation permits.

For distances see Table 3.25.
3.21.1

For low-powered, or hampered, vessels the following routes are advised:

South-bound, see 8.87.1.
North-bound, see 8.92.1.

Falkland Islands ⇐ ⇒ Estrecho de Magallanes

Diagram 3.19
Routes
3.22

The route is as direct as navigation permits, passing either S or N of the Falkland Islands (51°40′S, 57°49′W).

Distances between Stanley and Estrecho de Magallanes (52°27′S, 68°26′W) (3 miles S of Dungeness):
(a) S of Falkland Islands   420 miles.
(b) N of Falkland Islands   445 miles.

Falkland Islands ⇐ ⇒ East coast of South America

Diagram 3.19
Routes
3.23

Routes are as direct as navigation permits.
For distances see Table 3.25.

Falkland Islands ⇐ ⇒ Cabo de Hornos

Diagram 3.19
Routes
3.24

The route is as direct as navigation permits, either:
Through Estrecho de Le Maire (54°50′S, 65°00′W), or
E of Isla de Los Estados (54°50′S, 64°10′W).
Distance by either route   435 miles.

Distances on the E coast of South America
3.25

Cabo Calcanhar*						
210	Recife**					
600	395	Salvador				
1270	1070	740†	Rio de Janiero			
2260	2060	1730†	1020	Rio de la Plata		
2620	2420	2090†	1390	410	Bahía Blanca	
3050	2850	2510†	1810	840	510	Comodoro Rivadavia

3340	3130	2810	2120	1170	865	445	**Est. de Magallanes** (3′ S of Dungeness)
3470	3270	2940	2260	1340	1070	675	**Cabo de Hornos** (5 miles S of)
3060	2850	2520	1850	1000	785	530	**Falkland Islands** (Stanley)

*   60 miles ENE of (4°40′S, 34°35′W).
** For distances from places S of Recife to landfall off Recife given in 2.107, add 5 miles.
†   E of Parcel dos Abrolhos.

PASSAGES OFF WEST COAST OF AFRICA

GENERAL INFORMATION

Coverage
3.26

This section covers the passages to and from Arquipélago dos Bijagós and Cape Town with intermediate ports or the Indian Ocean.

Notes and cautions
3.27

Caution. The coast between Rio Cunene (17°15′S, 11°45′E) and Walvis Bay (22°54′S, 14°30′E) is known as the Skeleton Coast, having been the scene of innumerable shipwrecks, not only from the imperfect nature of the surveys, but because currents setting onshore are frequently experienced and in addition fog is prevalent, especially in winter. See *Africa Pilot, Volume II.*

3.27.1

Abnormal refraction, liable to cause appreciable error in sights, occur at times near the coast, especially off SW Africa.

3.27.2

Oilfields, with numerous structures, pipelines and submerged obstructions, sometimes marked by buoys, exist within the 200 m depth contour between Benin River (5°45′N, 5°00′E) and River Congo (6°00′S, 12°30′E).

3.27.3

Tankers. Loaded tankers navigating off the coast of South Africa are governed by regulations regarding the offing to be kept. See *Africa Pilot, Volumes II and III.*

PASSAGES

West Africa
3.28

Voyages between ports on the W coast of Africa are mainly coastwise, and passages are as direct as prudent navigation permits. For details of coastal routes see *Africa Pilot, Volumes I and II.*

Arquipélago dos Bijagós ⇐ ⇒ Cape Town and intermediate ports or the Indian Ocean

Diagram 3.29
Routes
3.29

The route between 10°40′N, 17° 40′W **(A)**, 75 miles SW of Arquipélago dos Bijagós and Gulf of Guinea (2°00′N, 3°00′E) is given at 2.92.4.

Between ports in Gulf of Guinea and Cape Town or the Indian Ocean, routes are as direct as navigation permits. For distances see Table 3.30, below.

Distances between positions in West Africa, South Africa and Indian Ocean
3.30

Arq.dos Bijagós*											
295	**Freetown**										
705	530	**Cape Palmas†**									
1080	900	380	**Takoradi**								
1400	1220	695	320	**Lagos**							
1600	1420	895	525	270	**Bonny River**						
1760	1570	1050	685	426	160	**Douala**					
1760	1580	1060	715	500	270	220	**Libreville**				
2000	1820	1300	1000	840	630	610	410	**Pointe Noire**			
2220	2140	1620	1380	1270	1070	1040	840	465	**Lobito**		
3370	3200	2720	2590	2560	2380	2350	2150	1810	1400	**Cape Town**	
3480	3310	2830	2700	2670	2490	2470	2270	1930	1520	**Cape Agulhas** (15 miles S)	
3510	3340	2870	2750	2720	2540	2520	2320	1980	1570	**Cape of Good Hope** (145 miles S)	

*   10°40′N, 17°40′W.
†   4°06′N, 7°54′W, 20 miles SSW of Cape Palmas

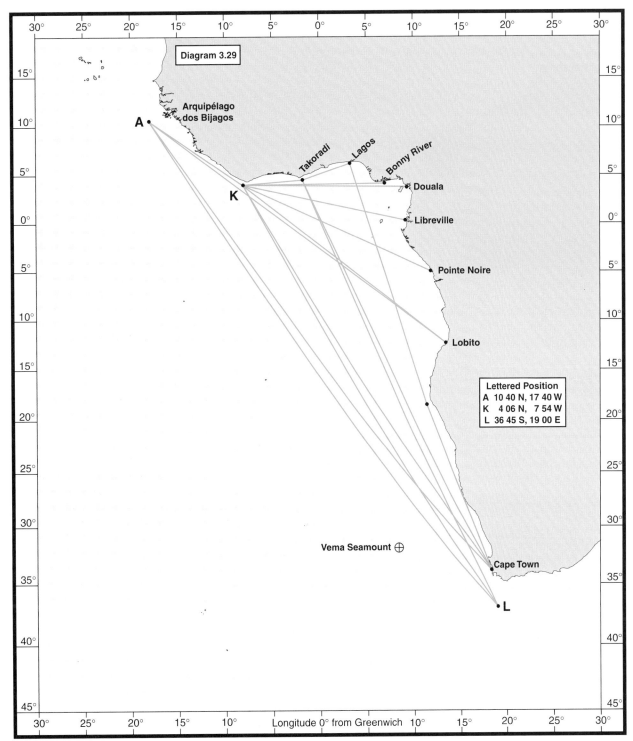

3.29 Arquipélago dos Bijagos ← → Cape Town or Indian Ocean

Within the diagram:

Diagram 3.29

Arquipélago dos Bijagos

A

Takoradi

Lagos

Bonny River

K

Douala

Libreville

Pointe Noire

Lobito

Lettered Position
A 10 40 N, 17 40 W
K 4 06 N, 7 54 W
L 36 45 S, 19 00 E

Vema Seamount ⊕

Cape Town

L

Longitude 0° from Greenwich

TRANSATLANTIC PASSAGES

GENERAL INFORMATION

Coverage
3.31

1 This section contains details of the following passages:
   Northern Passages (3.33)
   To and from W African ports and South American ports (3.33).

To and from SW coast of Africa or Gulf of Guinea and Colón (for Panama Canal) (3.35).
To and from Gulf of Guinea and North American ports (3.37).
Southern Passages (3.39)

2 From Rio de Janiero to Cape Town or Indian Ocean (3.40).

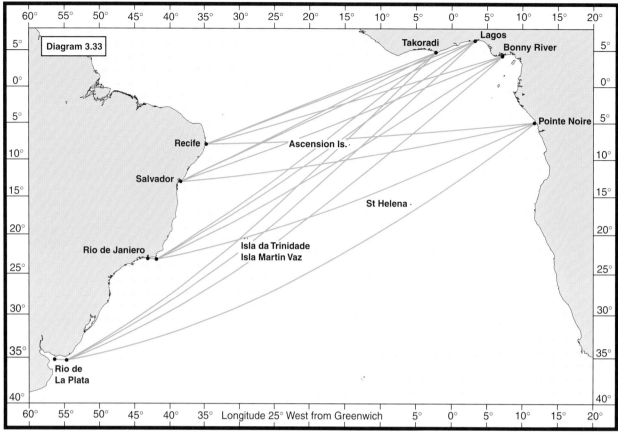

3.33 West African ports ← → South American ports

From Cape Town or Indian Ocean to Rio de Janiero (3.41).

From Río de La Plata to Cape Town or Indian Ocean (3.42).

From Cape Town or Indian Ocean to Río de La Plata (3.43).

To and from Estrecho de Magallanes and Gulf of Guinea (3.44).

To and from Estrecho de Magallanes and Pointe Noire (3.45).

From Estrecho de Magallanes to Cape Town or Indian Ocean (3.46).

From Cape Town or Indian Ocean to Estrecho de Magallanes, Falkland Islands or Cabo de Hornos (3.47).

From Falkland Islands to Cape Town or Indian Ocean (3.48).

From Cabo de Hornos to Cape Town or Indian Ocean (3.49).

From Cape Town to Galleons Passage and Colón (for Panama Canal) (3.50).

From Cape Town to N part of Caribbean Sea and Gulf of Mexico (3.51).

To and from Cape Town and Recife or Salvador (3.52).

To and from Cape Town and Cabot Strait or Halifax (3.53).

To and from Cape Town and New York (3.55).

Notes and cautions
3.32

1   **Caution**. Vema Seamount lies in 31°48′S, 8°20′E, about 500 miles WNW of Cape Town. It has a depth of 11 m over it and constitutes a danger to deep-draught vessels.

NORTHERN PASSAGES

West African ports ⇐ ⇒ South American ports

Diagram 3.33
Routes
3.33

Routes are by great circle in both directions between Río de la Plata (35°10′S, 56°15′W) and ports farther N and ports on the African coast N of about 25°S.

Distances
3.34

	Recife	Salvador	Rio de Janiero	Río de la Plata
Takoradi	2130	2460	2960	3900
Lagos	2450	2770	3260	4190
Bonny River	2610	2920	3370	4270
Pointe Noire	2790	3040	3370	4170

South-west coast of Africa and Gulf of Guinea ⇐ ⇒ Colón (for Panama Canal)

Diagram 3.35
General information
3.35

West-bound, benefit may be obtained by making use of the W-going South Equatorial Current (3.9) and North Equatorial Current (2.14 and 4.5).

East-bound, currents in general are unfavourable, but it may be possible to make use of the E-going Equatorial Counter-current and Guinea Current (2.15).

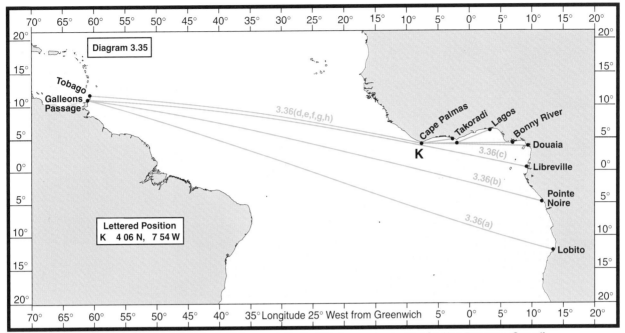

3.35 South West Coast of Africa and Gulf of Guinea ← → Colón (for Panama Canal)

Routes
3.36

From ports S of Douala (3°54′N, 9°32′E), routes are by great circle to Galleons Passage (10°57′N, 60°55′W), thence as at 4.15 to Colón (9°23′N, 79°55′W).

From places between Douala and Takoradi (4°53′N, 1°44′W) routes are:

Round position (4°06′N, 7°54′W) **(K)** 20 miles SSW of Cape Palmas, thence:

By great circle to 11°35′N, 60°35′W (15 miles N of Tobago), thence:

As at 4.15 to Colón.

Distances
3.36.1

Distances to Colón from:

(a)	Lobito	5830 miles
(b)	Pointe Noire	5620 miles
(c)	Libreville	5410 miles
(d)	Douala	5400 miles
(e)	Bonny River	5240 miles
(f)	Lagos	5040 miles
(g)	Takoradi	4730 miles
(h)	Cape Palmas*	4350 miles

* (20 miles SSW of)

Gulf of Guinea ⇐ ⇒ North American ports

Diagram 3.37

General information
3.37

In Cabot Strait a Traffic Separation Scheme has been established.

Routes
3.38

Routes are:

Direct to position (4°06′N, 7°54′W) **(K)** 20 miles SSW of Cape Palmas, thence:

Direct to 4°20′N, 9°20′W, thence:

Direct to 14°40′N, 24°55′W **(H)**, to the SW of Arquipélago de Cabo Verde, thence:

By great circle to destination.

3.38.1

From Lobito (12°19′S, 13°35′E), however, the route is direct to 14°40′N, 24°55′W **(H)**.

3.38.2
Distances:

	Cabot Strait	Halifax	New York
Lobito	5430	5470	5780
Pointe Noire	5120	5160	5480
Libreville	4880	4920	5240
Douala	4870	4910	5230
Bonny River	4720	4760	5080
Lagos	4520	4560	4880
Takoradi	4200	4240	4560
Cape Palmas*	3820	3860	4180

* (20 miles SSW of)

SOUTHERN PASSAGES

General information
3.39

South of about 25°S, E-bound routes are, in general, by great circle, but W-bound routes are by rhumb line to reduce headwinds and adverse currents.

For information for rounding Cape of Good Hope, see 3.16.

Rio de Janiero ⇒ Cape Town or the Indian Ocean

Diagram 3.40

Routes
3.40

Routes are by great circle.

3.40.1

Distances:

Cape Town	3280 miles
Cape Agulhas*	3350 miles
Cape of Good Hope†	3290 miles

* (15 miles S of) † (145 miles S of)

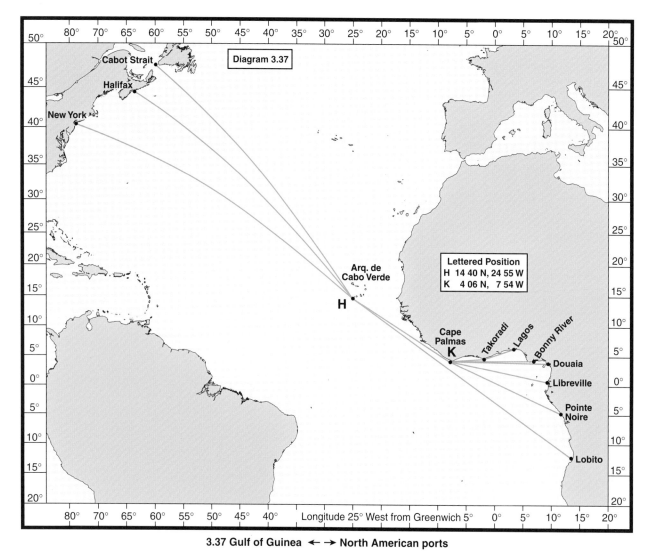

Lettered Position
H 14 40 N, 24 55 W
K 4 06 N, 7 54 W

Diagram 3.37

3.37 Gulf of Guinea ← → North American ports

Cape Town or the Indian Ocean ⇒ Rio de Janiero

Diagram 3.40
Routes
3.41

Routes are direct by rhumb line.

Caution. Vema Seamount (31°48′S, 8°20′E) (3.32) lies close N of the route from Cape Town (33°53′S, 18°26′E).
3.41.1

Distances:

Cape Town	3320 miles
Cape Agulhas*	3390 miles

* (15 miles S of)
3.41.2

An alternative route for low-powered vessels is given at 8.63.3.

Río de la Plata ⇒ Cape Town or the Indian Ocean

Diagram 3.40
Routes
3.42

Routes are by great circle.
3.42.1

Parts of the tracks lie within the extreme iceberg limits. The most S points on the routes are:

For Cape Town	41°00′S, 20°00′W
For Cape Agulhas	41°50′S, 18°15′W
For Cape of Good Hope†	42°30′S, 17°30′W

† (145 miles S of)
3.42.2

Distances:

Cape Town	3610 miles
Cape Agulhas*	3650 miles
Cape of Good Hope†	3570 miles

* (15 miles S of)
† (145 miles S of)
3.42.3

An alternative route for low-powered vessels is given at 8.85.

Cape Town or the Indian Ocean ⇒ Río de La Plata

Diagram 3.40
Routes
3.43

Routes are by rhumb line:
To 35°00′S, 40°00′W (**Z**), thence:
Rhumb line to Río de la Plata.

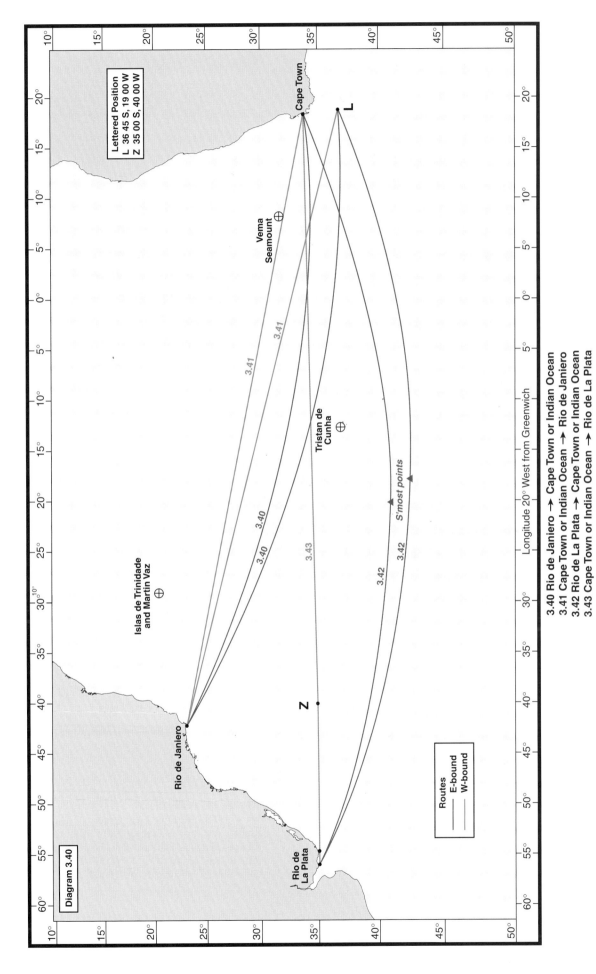

Diagram 3.40

Lettered Position
L 36 45 S, 19 00 W
Z 35 00 S, 40 00 W

Routes
E-bound
W-bound

Longitude 20° West from Greenwich

3.40 Rio de Janiero → Cape Town or Indian Ocean
3.41 Cape Town or Indian Ocean → Rio de Janiero
3.42 Rio de La Plata → Cape Town or Indian Ocean
3.43 Cape Town or Indian Ocean → Rio de La Plata

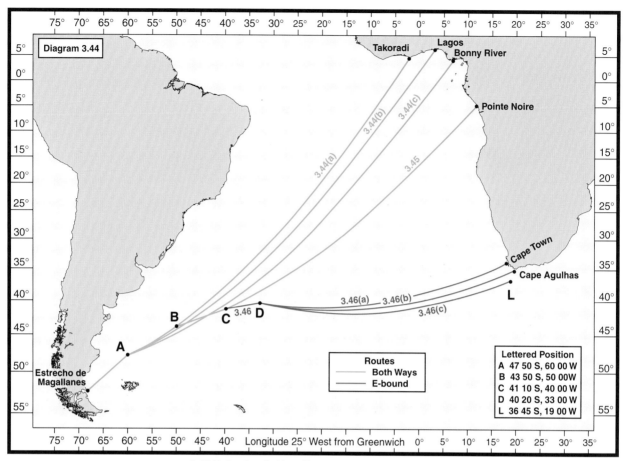

3.44 Estrecho de Magallanes ← → Gulf of Guinea
3.45 Estrecho de Magallanes ← → Pointe Noire
3.46 Estrecho de Magallanes → Cape Town or Indian Ocean

3.43.1

Distances:

Cape Town	3700 miles
Cape Agulhas*	3760 miles

* (15 miles S of)

3.43.2

An alternative, but longer, route for low-powered vessels is given at 8.63.1.

Estrecho de Magallanes ⇐ ⇒ Gulf of Guinea

Diagram 3.44

Routes

3.44

Routes are by rhumb line:
  To 47°50'S, 60°00'W (**A**), thence:
  Great circle to destination.

3.44.1

Distances from Estrecho de Magallanes (3 miles S of Dungeness):

(a)	Takoradi	4820 miles
(b)	Lagos	5080 miles
(c)	Bonny River	5110 miles

Estrecho de Magallanes ⇐ ⇒ Pointe Noire

Diagram 3.44

Routes

3.45

Route is by rhumb line:
  To 47°50'S, 60°00'W (**A**), thence:
  Rhumb line to 43°50'S, 50°00'W (**B**), thence:
  Great circle to Pointe Noire.

3.45.1

Distance from Estrecho de Magallanes (3 miles S of Dungeness)   4910 miles

Estrecho de Magallanes ⇒ Cape Town or the Indian Ocean

Diagram 3.44

Routes

3.46

Routes are by rhumb line:
  To 47°50'S, 60°00'W (**A**), thence:
  Rhumb line to 43°50'S, 50°00'W (**B**), thence:
  Rhumb line to 41°10'S, 40°00'W (**C**), thence:
  Rhumb line to 40°20'S, 33°00'W (**D**), thence:
  Great circle, passing S of Gough Island (40°20'S, 9°55'W), to destination.

3.46.1

Distances from Estrecho de Magallanes (3 miles S of Dungeness):

(a)	Cape Town	4170 miles

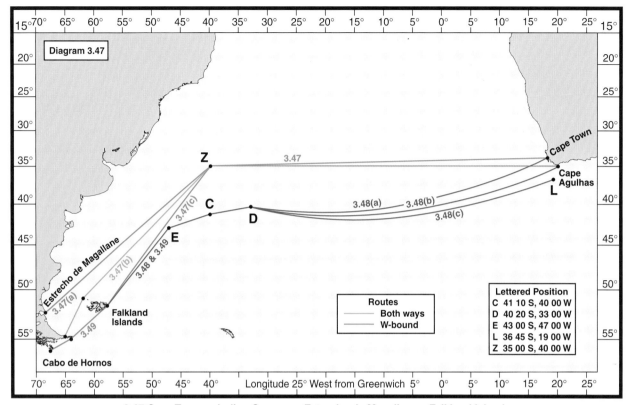

**3.47 Cape Town or Indian Ocean → Estrecho de Magallanes, Falkland Islands
or Cabo de Hornos**

(b) Cape Agulhas*   4210 miles
(c) Cape of Good Hope†   4130 miles
* (15 miles S of)
† (145 miles S of)
3.46.2
An alternative route for low-powered vessels is given at 8.94.

Cape Town or the Indian Ocean ⇒ Estrecho de Magallanes, Falkland Islands or Cabo de Hornos

Diagram 3.47
Routes
3.47
Routes are by rhumb line:
To 35°00′S, 40°00′W (**Z**), thence:
Rhumb line to destination.
3.47.1
Distances:

	Estrecho de Magallanes†	Cabo de Hornos‡	Falkland Islands
Cape Town	4510	4620	4170
Cape Agulhas*	4570	4670	4220

* (15 miles S of)
† (3 miles S of Dungeness)
‡ W of Falkland Islands and through Estrecho de Le Maire; if E of Falkland Islands and Islas de los Estados, subtract 25 miles.
3.47.2
An alternative, but longer, route for low-powered vessels is given at 8.63.2.

Falkland Islands ⇒ Cape Town or the Indian Ocean

Diagram 3.47
Routes
3.48
The great circle distance to Cape Town (33°53′S, 18°26′E) is 3370 miles but the track reaches 53°S and cannot therefore be recommended on account of ice and weather.
The normal route is:
Rhumb line to 43°00′S, 47°00′W (**E**), thence:
Rhumb line to 41°10′S, 40°00′W (**C**), thence:
Rhumb line to 40°20′S, 33°00′W (**D**), thence:
Great circle to destination, passing S of Gough Island.
3.48.1
Distances:
(a) Cape Town   3800 miles
(b) Cape Agulhas*   3840 miles
(c) Cape of Good Hope†   3760 miles
* (15 miles S of)
† (145 miles S of)
3.48.2
An alternative route for low-powered vessels is given at 8.94.

Cabo de Hornos ⇒ Cape Town or the Indian Ocean

Diagram 3.47
Routes
3.49
Routes are either through Estrecho de Le Maire (54°50′S, 65°00′W) or E of Islas de los Estados (54°50′S, 64°10′W), as at 3.24, to join the routes from the Falkland Islands (3.48) off Stanley (51°40′S, 57°49′W).

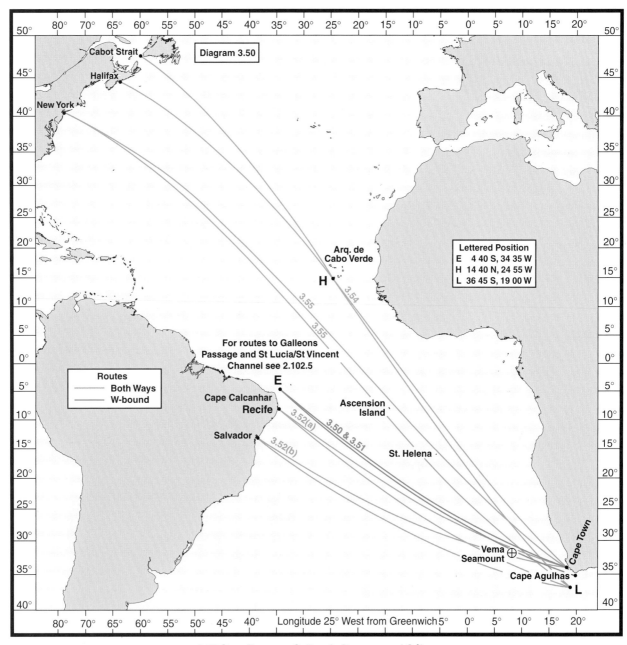

50° 80° 70° 65° 60° 55° 50° 45° 40° 35° 30° 25° 20° 15° 10° 5° 0° 5° 10° 15° 20° 50°

Diagram 3.50

Cabot Strait

Halifax

New York

Arq. de
Cabo Verde

H

Lettered Position
E 4 40 S, 34 35 W
H 14 40 N, 24 55 W
L 36 45 S, 19 00 W

For routes to Galleons
Passage and St Lucia/St Vincent
Channel see 2.102.5

E

Routes
— Both Ways
— W-bound

Cape Calcanhar
Recife

Salvador

Ascension
Island

St. Helena

3.52(a)
3.50 & 3.51
3.52(b)

Cape Town

Vema
Seamount

Cape Agulhas

L

80° 70° 65° 60° 55° 50° 45° 40° Longitude 25° West from Greenwich 5° 0° 5° 10° 15° 20°

3.50 Cape Town → Galleon's Passage and Cólon
3.51 Cape Town → North part of Caribbean Sea and Gulf of Mexico
3.52 Cape Town ← → Recife or Salvador
3.53 Cape Town ← → Cabot Strait or Halifax
3.55 Cape Town ← → New York

3.49.1

Distances:

Cape Town	4230 miles
Cape Agulhas*	4270 miles
Cape of Good Hope†	4190 miles

* (15 miles S of)

† (145 miles S of)

Cape Town ⇒ Galleons Passage and Colón
(for Panama Canal)

Diagram 3.50 and 4.12

Routes
3.50

From Cape Town (33°53′S, 18°26′E) or Cape Agulhas (35°05′S, 20°00′E) routes are:

Great circle to 4°40′S, 34°35′W **(E)**, off Cabo Calcanhar, thence:

Direct to Galleons Passage (10°57′N, 60°55′W) (2.102.5), thence:

As at 4.15 to Colón (9°23′N, 79°55′W).

3.50.1

Distances:

	Cape Town	Cape Agulhas*
Galleons Passage	5260	5360
Colón	6430	6530

* (15 miles S of)

Cape Town ⇒ Northern part of Caribbean Sea and Gulf of Mexico

Diagram 3.50 and 4.12
Routes
3.51

From Cape Town (33°53′S, 18°26′E) or Cape Agulhas (35°05′S, 20°00′E) routes are:

Great circle to 4°40′S, 34°35′W (**E**), off Cabo Calcanhar, thence:

Direct to St Lucia/St Vincent Channel (13°30′N, 61°00′W) (2.102.5), thence:

As at 4.13.2 to Yucatan Channel (21°30′N, 85°40′W), or to Mona Passage (18°20′N, 67°50′W) and through Old Bahama Channel (21°30′N, 75°45′W) to Gulf of Mexico.

3.51.1

Distances to St Lucia/St Vincent Channel:

From Cape Town   5340 miles
From Cape Aghulas*   5440 miles

* (15 miles S of)

Cape Town ⇐ ⇒ Recife or Salvador

Diagram 3.50
Routes
3.52

Routes between Recife (8°04′S, 34°51′W) or Salvador (13°00′S, 38°32′W) and Cape Town (33°53′S, 18°26′E), Cape Aghulas (35°05′S, 20°00′E) or Cape of Good Hope (36°45′S, 19°00′E) are by great circle in both directions.

3.52.1

Distances

	Recife (a)	Salvador (b)
Cape Town	3320	3340
Cape Agulhas*	3410	3430
Cape of Good Hope†	3390	3390

* (15 miles S of)
† (145 miles S of)

Cape Town ⇐ ⇒ Cabot Strait or Halifax

Diagram 3.50
General information
3.53

In Cabot Strait a Traffic Separation Scheme has been established.

Routes
3.54

The routes are:

3.54.1

From Cape Town (33°53′S, 18°26′E):

Great circle to 14°40′N, 24°55′W (**H**), thence
Great circle to destination.

3.54.2

From Cape Agulhas (35°05′S, 20°00′E):

As navigation permits to round Cape Point (34°21′S, 18°30′E), thence:
Great circle to 14°40′N, 24°55′W (**H**), thence
Great circle to destination.

3.54.3

Distances:

	Cabot Strait	Halifax
Cape Town	6450	6490
Cape Agulhas*	6560	6600
Cape of Good Hope†	6580	6630

* (15 miles S of)
† (145 miles S of)

Cape Town ⇐ ⇒ New York

Diagram 3.50
Routes
3.55

The routes between Cape Town (33°53′S, 18°26′E) or Cape Agulhas (35°05′S, 20°00′E) and New York (40°28′N, 73°50′W) are:

Great circle to destination.

3.55.1

Distances:

From Cape Town   6790
From Cape Agulhas*   6900
From Cape of Good Hope†   6910

* (15 miles S of)
† (145 miles S of)

NOTES

CHAPTER 4

CARIBBEAN SEA AND GULF OF MEXICO

GENERAL INFORMATION

COVERAGE

Chapter coverage

4.1

This chapter contains details of the following passages:
To and from English Channel and Caribbean Sea or Gulf of Mexico (4.8).
From Bermuda to Habana (4.9).
From Habana to Bermuda (4.10).
From Bermuda to Kingston (4.11).
To and from Colón and Gulf of Mexico (4.12).
To and from Yucatan Channel and E part of Caribbean Sea (4.13).
From E part of Caribbean Sea to South American ports (4.14).
To and from Colón and Trinidad, Galleons Passage or Tobago (4.15).

Admiralty Publications

4.2

Relevant navigational publications should be consulted, in addition to this book, when planning and conducting passages. Details of these, and the areas covered, can be found, as follows:
Admiralty Sailing Directions; see 1.12.1
Admiralty List of Lights and Fog Signals; see 1.12.2
Admiralty Tide Tables; see 1.12.3
Admiralty List of Radio Signals; see 1.12.5
Admiralty Distance Tables; see 1.12.8

4.2.1

See also *The Mariner's Handbook* for notes on electronic products.

NATURAL CONDITIONS

Charts:

North Atlantic Routeing 5124 (1-12)
World Surface Current Distribution 5310

WINDS AND WEATHER

General description

4.3

1   Over the Caribbean Sea, NE to E winds prevail throughout the year, while over the Gulf of Mexico the wind is generally lighter and more variable in direction, though frequently from between NE and SE. In coastal waters, strong N winds may reach gale force at times over the Gulf. For the whole area, wind speeds are mainly light or moderate, except for occasional hurricanes, see 2.7 for details, which may affect the area from June to November. Most hurricanes track N of Cuba and they rarely occur S of 15°N.

2   The weather over the area is generally partly cloudy with scattered showers. Sunny spells are frequent and, from May to December, periods of heavy rain and thunderstorms are frequent. Squalls may occur at any time, but fog seldom occurs at sea.

3   Visibility is generally good throughout the year though it may at times be drastically reduced by heavy rain.

4   See 2.5 for details of "Northers".

SWELL

General description

4.4

1   Swells are generally lower in the Gulf of Mexico than in the Caribbean Sea.

2   In the Caribbean Sea the prevailing direction is from NE to E; in the Gulf of Mexico, from March to September it is from E to SE and from October to February it is from NE.

3   Highest swells occur in the area around 13°N, 77°W in the Caribbean Sea, especially in June and July, when the frquency of swell greater than 4 m is 20%. These swells are invariably short or average in length.

CURRENTS

General description

4.5

1   The **North Equatorial Current** (2.14) flows WNW throught the Caribbean Sea with little change of direction until it approaches the Yucatan Channel where it turns to the N. It leaves an anti-clockwise eddy in the S part of the sea, S of about 12°N. There is also an E-going counter-current close to the S coast of E and central Cuba.

2   In the Gulf of Mexico, part of the N-going flow from Yucatan Channel fans out in directions between SW and NW. Currents setting in these directions occupy most of the Gulf W of a line from Cabo Catoche to close W of the Mississippi delta. From the NW flow along this line, water fans out NE and then shortly recurves to join the SE flow extending from the Mississippi delta to the W approaches to the Strait of Florida. This SE-going stream joins the NE-going stream which emerges from Yucatan Channel and the combined flow continues E, and through the Straits of Florida as the Florida Current (2.14). The emerging stream meeting the NW flowing water of **North Sub-tropical Current**, turns N off the E coast of Florida and forms the beginning of the Gulf Stream (2.14).

3   Along the W coast of Florida there is a N-going current which, with the SE flow coming from the Mississippi delta, forms an anti-clockwise eddy in the E part of the Gulf of Mexico.

4   There is little seasonal variation in the pattern of the currents.

5 The average current rates in most of the Caribbean Sea are about 1 kn, increasing on the W side of Yucatan Channel to about 4 kn. The strongest currents are observed in the Straits of Florida in about 25°00′N, 80°00′W and for about 300 miles N from that position. Here the average rate is nearly 3 kn in summer and 2½ kn in winter.

6 Over most of the Gulf of Mexico the average rates are about 1 kn, but N-going sets of about 1½ kn are reported in summer near the Mexican coast N of Tampico and SE-going sets of a similar magnitude for much of the year between the Mississippi delta and Cuba.

NOTES AND CAUTIONS

Navigation

4.6

1 In the Caribbean Sea and Gulf of Mexico, and in the channels leading thereto, great care is necessary near the cays and banks, as some of the charts are based on old and imperfect surveys.

2 Furthermore depths over the shoals may be less than those charted owing to the growth of the coral of which many of them are composed or to the imprecise nature of the least depths reported over them. Shoal water should be approached with caution at all times and given a wide berth when conditions for fixing are poor; many of the banks are steep-to.

4.6.1

Caution. Strong currents can be expected in the entrances and channels leading to the Caribbean Sea and Gulf of Mexico, particularly in Straits of Florida.

Having chosen the route, the mariner should consult the relevant volumes of *Admiralty Sailing Directions* for details of currents and tidal streams affecting it.

Entrance channels

4.7

1 The Caribbean Sea may be approached through Crooked Island Passage (23°15′N, 74°25′W), Caicos Passage (22°15′N, 72°20′W) or Turks Island Passage (21°48′N, 71°16′W), all of which lead to Windward Passage (20°00′N, 74°00′W).

Crooked Island Passage is the most frequently used.

Caicos Passage is not well lighted and Turks Island Passage, not lighted in its S approach, is not recommended for N-bound vessels at night.

2 Other entrances in common use are:

Mona Passage (18°20′N, 67°50′W), which is much frequented and presents no difficulty. Although subject to heavy squalls it is safer than Turks Island Passage.

Sombrero Passage (18°25′N, 63°45′W), which is not lighted in its S approach.

The Channel between Saint Lucia (14°02′N, 61°01′W) and Saint Vincent (13°15′N, 61°15′W).

The passages N and S of Tobago (11°15′N, 60°35′W).

3 In many cases routes through the various passages differ little in distances, and selection will depend principally on the ship's particular requirements.

4 For distances from these passages, see 4.16.

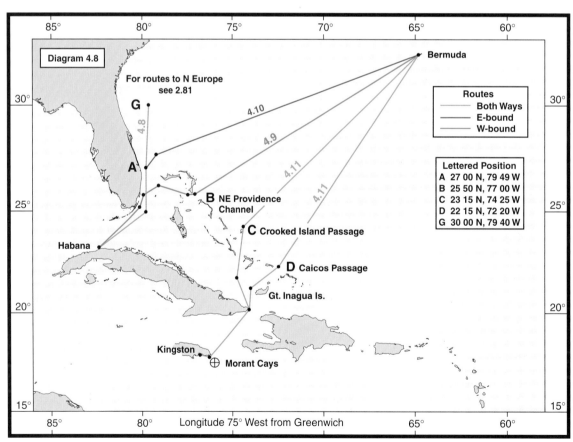

4.8 English Channel ← → Caribbean Sea and Gulf of Mexico
4.9 Bermuda → Habana
4.10 Habana → Bermuda
4.11 Bermuda ← → Kingston

PASSAGES

English Channel ⇐ ⇒ Caribbean Sea and Gulf of Mexico

Diagram 4.8
Recommended Passages
4.8

Recommended Passages from the North Atlantic to ports in the Caribbean Sea and Gulf of Mexico are as follows:

Belize (17°20′N, 88°01′W). Providence Channels (2.83) or Turks Island Passage (2.84) and Windward Passage (20°00′N, 74°00′W) are suitable.

Kingston (17°54′N, 76°44′W) **or Colón** (9°23′N, 79°55′W). Turks Island Passage (2.84) and Windward Passage (20°00′N, 74°00′W) are suitable.

Colón (9°23′N, 79°55′W) **or Curaçao** (12°06′N, 68°56′W). Mona Passage (2.84) or Sombrero Passage (2.84) are suitable.

Gulf of Mexico. Approach may be made either N of Cuba (22°30′N, 80°00′W), through Providence Channels (2.83) or Old Bahama Channel (21°30′N, 75°45′W) and Nicholas Channel (23°15′N, 80°00′W); or S of Cuba through Caribbean Sea and Yucatan Channel (21°30′N, 85°40′W). For routes within Gulf of Mexico see the relevant volumes of *Admiralty Sailing Directions*. Straits of Florida, through which the current runs strongly to the N, is best for departure from Gulf of Mexico (2.81).

4.8.1

For distances through these passages, see 4.16 and 2.86.

Bermuda ⇒ Habana

Diagram 4.8
Routes
4.9

From Bermuda (32°23′N, 64°38′W) the route is through North-East Providence Channel (25°50′N, 77°00′W), North-West Providence Channel (26°10′N, 78°35′W) and Straits of Florida (27°00′N, 79°49′W).

    Distance   1130 miles.

For low-powered vessels an alternative route is described at 8.17.1

Habana ⇒ Bermuda

Diagram 4.8
Route
4.10

From Habana (23°10′N, 82°22′W) the route is:
Through Straits of Florida (27°00′N, 79°49′W), thence:
Direct from NW end of Little Bahama Bank (27°00′N, 78°30′W).

Caution. For dangers in approaches to Bermuda (32°23′N, 64°38′W), see 2.44.

    Distance   1150 miles.

Bermuda ⇐ ⇒ Kingston

Diagram 4.8
Route
4.11

From Bermuda (32°23′N, 64°38′W) the route is:
Through either Crooked Island Passage (23°15′N, 74°25′W) or Caicos Passage (22°15′N, 72°20′W) (see 4.7), thence:

W of Great Inagua Island (21°00′N, 73°20′W), and:
Through Windward Passage (20°00′N, 74°00′W).

Cautions:

Morant Cays (17°25′N, 76°00′W) have been the scene of many wrecks. Currents in their vicinity vary greatly, both in direction and rate. If passing them at night, ships should keep well N of them.

For dangers in the approaches to Bermuda, see 2.44.

4.11.1

Distances:

Crooked Island Passage	1180 miles.
Caicos Passage	1110 miles.

Colón (for Panama Canal) ⇐ ⇒ Gulf of Mexico

Diagram 4.12
Routes
4.12

The routes are between the entrance to the Panama Canal, between the breakwaters off Colón (9°23′N, 79°55′W), and the Yucatan Channel (21°30′N, 85°40′W).

4.12.1

N-bound from the canal entrance at Colón the route runs:
NNE to round Bajo Nuevo (15°54′N, 78°40′W) at a prudent distance, thence:
Direct to Yucatan Channel.

4.12.2

S-bound from Yucatan Channel the route is:
To a position outside the 200 m depth contour off the South-West Point of Grand Cayman Island (19°20′N, 81°15′W), thence:
To a point off the SW extremity of Pedro Bank (17°10′N, 78°45′W), thence:
Direct to Colón, passing between Bajo Nuevo (15°54′N, 78°40′W) and Alice Shoal (16°05′N, 79°20′W).

Or, if unsure of the position, the route continues:
SE until Bajo Nuevo has been cleared, thence
SSW to Colón.

4.12.3

Cautions:

1   Bajo Nuevo and Alice Shoal are charted mainly from a leadline survey of 1835.

2   Due allowance must be paid to the prevailing W-going current between Colón and Jamaica.

3   A vessel entering the W part of the Gulf of Mexico from the Caribbean Sea should be kept in the deep water in Yucatan Channel, all precautions being taken against the N-going current and the position constantly checked in order to identify the point of crossing the edge of the Banco de Campeche (22°30′N, 90°00′W).

4   The edge of the bank is generally marked by rippling and only a short distance within it the water becomes discoloured. New reports of shoal patches are often received, so soundings should be continuous when crossing it. Dangers on the bank are steep-to and sometimes indicated by discoloured water.

5   For routes W of Thunder Knoll (16°30′N, 81°20′W) and from Yucatan Channel to Ports in the Gulf of Mexico, see the relevant volumes of *Admiralty Sailing Directions*.

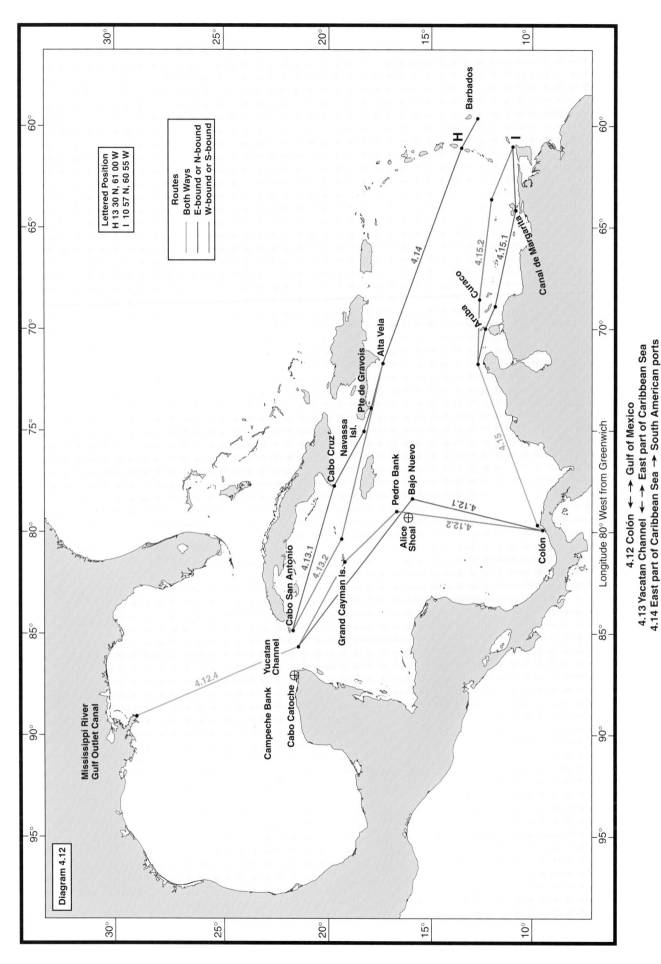

Diagram 4.12

Lettered Position
H 13 30 N, 61 00 W
I 10 57 N, 60 55 W

Routes
Both Ways
E-bound or N-bound
W-bound or S-bound

Mississippi River
Gulf Outlet Canal

4.12.4

Campeche Bank

Cabo Catoche

Yucatan Channel

Cabo San Antonio

4.13.1

4.13.2

Grand Cayman Is.

Cabo Cruz

Navassa Isl.

Pte de Gravois

Alta Vela

4.14

Pedro Bank

Alice Shoal

Bajo Nuevo

4.12.1

4.12.2

Colón

4.15

Aruba

Curaco

4.15.2

4.15.1

Canal de Margarita

Barbados

H

I

Longitude 80° West from Greenwich

4.12 Colón ←→ Gulf of Mexico
4.13 Yacatan Channel ←→ East part of Caribbean Sea
4.14 East part of Caribbean Sea → South American ports
4.15 Colón ←→ Trinidad, Galleons Passage and Tobago

4.12.4

Distance from Colón, passing W of Bajo Nuevo*:

Cabo Catoche (28 miles ENE of)	965 miles.
Cabo San Antonio (10 miles W of)	900 miles.
New Orleans**	1380 miles.
Balboa (Panama Canal S Entrance)	45 miles.

*   If passing E of Bajo Nuevo add 15 miles

** Mississippi River South Pass

Yucatan Channel ⇐ ⇒ eastern part of Caribbean Sea

Diagram 4.12

Routes

4.13

Cabo Catoche (21°36′N, 87°04′W) should be given a wide berth on account of the shoals N of it.

4.13.1

E-bound from 7 miles S of Cabo San Antonio (21°52′N, 84°57′W) the route is:

Along the S coast of Cuba to pass 5 miles S of Cabo Cruz (19°51′N, 77°44′W), thence:

5 miles S of Navassa Island (18°24′N, 75°01′W), thence:

5 miles S of Pointe de Gravois (18°01′N, 73°54′W), thence:

To a position 5 miles S of Alta Vela (17°28′N, 71°39′W).

This track, mostly under the lee of the land, makes use of the E-going counter-current (4.5). Off the coast of Cuba, between Cabo San Antonio and Cabo Cruz, special caution is required because onshore sets sometimes run strongly.

4.13.2

W-bound from Alta Vela (17°28′N, 71°39′W) the route passes:

N of Jamaica, thence:

As direct as navigation permits to Cabo San Antonio (21°52′N, 84°57′W).

Eastern part of Caribbean Sea ⇒ South American ports

Diagram 4.12

Routes

4.14

For destinations E of the Caribbean Sea, the best route from Alta Vela (17°28′N, 71°39′W) passes between Saint Lucia (14°02′N, 61°01′W) and Saint Vincent (13°15′N, 61°15′W) and S of Barbados (13°06′N, 59°39′W).

Colón (for Panama Canal) ⇐ ⇒ Trinidad, Galleons Passage and Tobago

Diagram 4.12

Routes

4.15

For directions for these routes, see *South America Pilot, Volume IV.*

4.15.1

E-bound from Colón (9°23′N, 79°55′W) passages are best made keeping close inshore, passing:

S of Aruba (12°26′N, 69°56′W) and Curaçao (12°06′N, 68°56′W), and:

Through Canal de Margarita (10°51′N, 64°05′W).

4.15.2

W-bound, advantage can be taken of the W-going North Eqtorial Current, by passing:

N of the outlying islands and dangers.

4.15.3

Distances from Colón by either route:

Trinidad (Port of Spain)	1140 miles.
Galleons Passage	1170 miles.
Tobago*	1180 miles.

*   15 miles N of North Point.

Distances from entrances to Caribbean Sea

Diagram 4.16

4.16

The distances to ports within the Caribbean Sea and Gulf of Mexico from the main entrances to the Caribbean Sea are as follows:

4.16.1

A Straits of Florida (N end) (27°00′N, 79°49′W):

Dry Tortugas (24°30′N, 83°05′W)	285 miles.
New Orleans (Mississippi River Gulf Outlet Canal) (29°25′N, 88°58′W)	720 miles.
Habana (23°10′N, 82°22′W)	285 miles.

4.16.2

B North-East Providence Channel (25°50′N, 77°00′W):

Dry Tortugas (24°30′N, 83°05′W)	380 miles.
New Orleans (Mississippi River Gulf Outlet Canal) (29°25′N, 88°58′W)	805 miles.
Habana (23°10′N, 82°22′W)	370 miles.

4.16.3

C Crooked Island Passage (23°15′N, 74°25′W) (from landfall on San Salvador (24°00′N, 74°30′W))

via Old Bahama Channel (21°30′N, 75°45′W):

Habana (23°10′N, 82°22′W)	585 miles.
New Orleans (Mississippi River Gulf Outlet Canal) (29°25′N, 88°58′W)	1060 miles.

via Windward Passage (20°00′N, 74°00′W):

Kingston (17°54′N, 76°44′W)	480 miles.
Belize (17°20′N, 88°01′W)	1080 miles.
Colón (9°23′N, 79°55′W)	980 miles.

4.16.4

D Caicos Passage (22°15′N, 72°20′W):

via Old Bahama Channel (21°30′N, 75°45′W):

New Orleans (Mississippi River Gulf Outlet Canal) (29°25′N, 88°58′W)	1070 miles.
Habana (23°10′N, 82°22′W)	590 miles.
Kingston (17°54′N, 76°44′W) (passing W of Great Inagua Island) (21°00′N, 73°20′W)	390 miles.

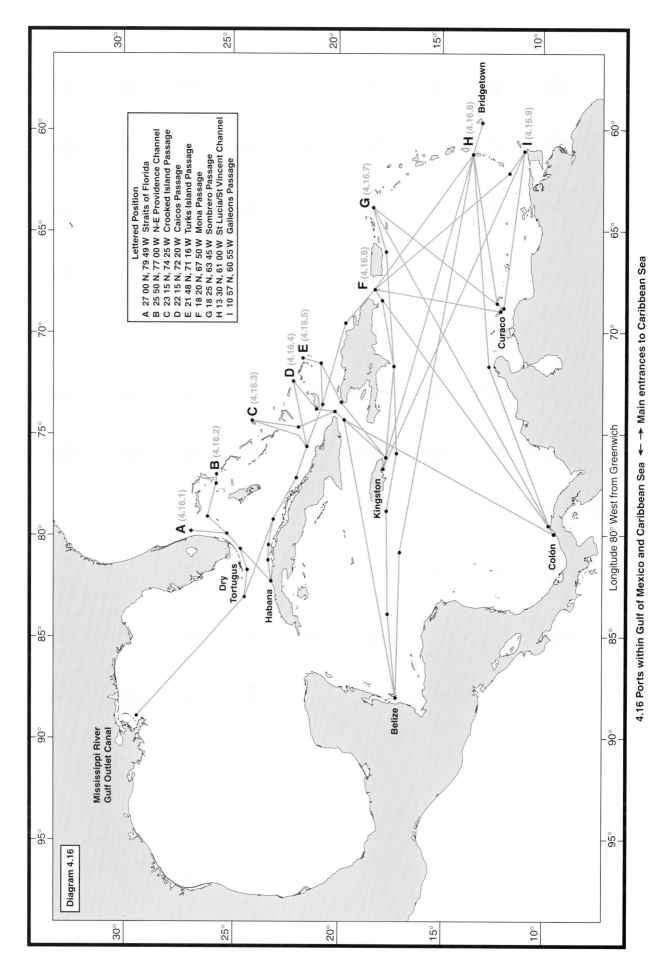

Diagram 4.16

Lettered Position

A 27 00 N, 79 49 W Straits of Florida
B 25 50 N, 77 00 W N-E Providence Channel
C 23 15 N, 74 25 W Crooked Island Passage
D 22 15 N, 72 20 W Caicos Passage
E 21 48 N, 71 16 W Turks Island Passage
F 18 20 N, 67 50 W Mona Passage
G 18 25 N, 63 45 W Sombrero Passage
H 13 30 N, 61 00 W St Lucia/St Vincent Channel
I 10 57 N, 60 55 W Galleons Passage

4.16 Ports within Gulf of Mexico and Caribbean Sea ⟵ ⟶ Main entrances to Caribbean Sea

E Turks Island Passage (21°48′N, 71°16′W) (from landfall on San Salvador (24°00′N, 74°30′W))

  via Old Bahama Channel (21°30′N, 75°45′W):

    New Orleans (Mississippi River Gulf Outlet Canal) (29°25′N, 88°58′W) . . . . 1160 miles.

    Habana (23°10′N, 82°22′W) . . . . 685 miles.

  via Windward Passage (20°00′N, 74°00′W):

    Belize (17°20′N, 88°01′W) . . . . 1030 miles.

    Colón (9°23′N, 79°55′W) . . . . 925 miles.

    Kingston (17°54′N, 76°44′W) . . . . 420 miles.

Yucatan Channel (21°30′N, 85°40′W) gives the shortest route to Tampico (22°17′N, 97°44′W) and ports in Gulf of Mexico farther S.

Old Bahama Passage gives the shortest route to ports N and E of Tampico.

4.16.5

F Mona Passage (18°20′N, 67°50′W):

  Kingston (17°54′N, 76°44′W) . . . . 520 miles.

  Curaçao (12°06′N, 68°56′W) . . . . 390 miles.

  Colón (9°23′N, 79°55′W) . . . . 885 miles.

  Belize (17°20′N, 88°01′W) . . . . 1180 miles.

4.16.6

G Sombrero Passage (18°25′N, 63°45′W):

  Kingston (17°54′N, 76°44′W) . . . . 755 miles.

  Curaçao (12°06′N, 68°56′W) . . . . 500 miles.

  Colón (9°23′N, 79°55′W) . . . . 1090 miles.

  Belize (17°20′N, 88°01′W) . . . . 1410 miles.

4.16.7

H Saint Lucia/Saint Vincent Channel (13°30′N, 61°00′W):

  Barbados (Bridgetown) (13°06′N, 59°39′W) . . . . 85 miles.

  Kingston (17°54′N, 76°44′W) . . . . 950 miles.

  Curaçao (12°06′N, 68°56′W) . . . . 480 miles.

  Colón (9°23′N, 79°55′W) . . . . 1150 miles.

  Belize (17°20′N, 88°01′W) . . . . 1590 miles.

  via Mona Passage (18°20′N, 67°50′W) and Old Bahama Channel (21°30′N, 75°45′W):

    New Orleans (Mississippi River Gulf Outlet Canal) (29°25′N, 88°58′W) . . . . 1860 miles.

    Habana (23°10′N, 82°22′W) . . . . 1370 miles.

4.16.8

I Galleons Passage (10°57′N, 60°55′W):

  Kingston (17°54′N, 76°44′W) . . . . 1010 miles.

  Curaçao (12°06′N, 68°56′W) . . . . 480 miles.

  via Mona Passage (18°20′N, 67°50′W) and Old Bahama Channel (21°30′N, 75°45′W):

    New Orleans (Mississippi River Gulf Outlet Canal) (29°25′N, 88°58′W) . . . . 1960 miles.

NOTES

CHAPTER 5

MEDITERRANEAN SEA

GENERAL INFORMATION

COVERAGE

Chapter coverage
5.1

This chapter contains details of the following Passages:

From Strait of Gibraltar to W part of the Mediterranean Sea except S shore (5.11).

From W part of the Mediterranean Sea except S shore to Strait of Gibraltar (5.12).

From Strait of Gibraltar to E part of the Mediterranean Sea and Algerian coast (5.13).

From E part of the Mediterranean Sea and Algerian coast to Strait of Gibraltar (5.14).

To and from Bûr Sa'îd (Port Said) (5.15).

Admiralty Publications
5.2

Relevant navigational publications should be consulted, in addition to this book, when planning and conducting passages. Details of these, and the areas covered, can be found, as follows:

Admiralty Sailing Directions; see 1.12.1
Admiralty List of Lights and Fog Signals; see 1.12.2
Admiralty Tide Tables; see 1.12.3
Admiralty List of Radio Signals; see 1.12.5
Admiralty Distance Tables; see 1.12.8

5.2.1

See also *The Mariner's Handbook* for notes on electronic products.

NATURAL CONDITIONS

Charts:

North Atlantic Routeing 5124 (1-12)
World Surface Current Distribution 5310

WINDS AND WEATHER

General description
5.3

1   The weather of the Mediterranean Sea is markedly seasonal, being characterised by hot dry summers with mainly light or moderate winds and mild rainy winters with a rather high frequency of strong winds and gales. The situation of this sea, surrounded by land, much of which is either mountainous or desert, together with the indented nature of parts of the coast, leads to the occurrence of a large number of local winds, many with special names and characteristics. Some of these can produce extremely violent conditions. Information concerning these local winds can be found in the relevant volume of *Admiralty Sailing Directions*.

2   Over the greater part of the open waters of the Mediterranean Sea, winds from between N and W are the most frequent, though the passage of depressions across the area causes great variations in both the direction and force of the wind. From about November to March these depressions are frequent and often vigorous, while from about May to September they are less common and much less intense.

3   For convenience in describing the winds and weather of the Mediterranean Sea, the area has been divided up into into the W part lying W of the Sicilian Channel and the E part E of the Sicilian Channel.

4   It is emphasised that the statements which follow apply to the open sea away from the influence of the land, in the vicinity of which marked differences are to be found.

Western part of the Mediterranean Sea
5.4

For convenience the following descriptions are divided into seasonal periods.

5.4.1

1   **November to March**. In the more confined part of the area W of about 1°W, winds blow mainly parallel with the coast, westerlies being somewhat more common than easterlies from January to March and very considerably more frequent in November and December.

2   Over the remainder of the area as far E as the longitude of Sardegna the most frequent wind directions are from between N and W, with a bias towards the latter direction in the S part of the area.

3   In the N part of the Tyrrhenian Sea there is no clearly predominant wind direction, though winds from some N point are more common than from a S point. In the S part of this sea and in the Sicilian Channel the prevailing direction is NW.

4   In January, the stormiest month in most of the W part of the Mediterranean Sea, winds reach force 7 or above on six to nine days per month in the NW and on three to seven days per month elsewhere. Most of the winter gales are from between N and W, though NE gales are not uncommon and gales from other directions may occasionally occur. Weather at this season is subject to rapid changes due to the passage of depressions with their associated frontal belts of cloud and rain. The rain is usually heavy but of relatively shorter duration than in the British Isles.

5   Visibility over the open sea is generally good except when reduced by rain, but it may at times be only moderate with winds from a S quarter.

5.4.2

1   **May to September**. In that part of the area S of about 40°N and W of the longitude of Sardegna, winds are most frequent from between E and NE or from between W and SW, the former being slightly more common. Elsewhere from June to August the most frequent wind directions are from between N and W, but in May and September there is no clearly predominant wind direction.

2   Winds are likely to reach force 7 or above on one to three days per month in the NW part of the area; elsewhere, winds of that strength are rare at this season, but in September the mariner should be alert to the

possibility of rapid development of depressions and squalls which can give unexpected onsets of strong winds.

3   Weather in July and August is generally fine with little or no rainfall, especially in the S and E. Cloud amounts are larger and rain is somewhat more common in May and September, especially in the latter month and NE of a line joining the Gulf of Lions, Sardegna and Sicilia.

4   Visibility is generally good, though occasional patches of sea fog may be experienced in early summer and, with winds from a S quarter, haze is sometimes prevalent.

5.4.3

1   **April and October**. In the transitional months of April and October conditions can be taken as intermediate between winter and summer, though it should be realised that considerable variations are likely from year to year.

Eastern part of the Mediterranean Sea
5.5

For convenience the following descriptions are divided into seasonal periods.

5.5.1

1   **November to March**. South of about 35°N, winds are most often from between SW and N, while to the N of that parallel between Sicilia and Greece there is no clearly prevailing wind direction.

2   In the greater part of the Adriatic Sea and the N part of the Aegean Sea winds from between N and E are the most frequent, though these are often interrupted by winds from a S quarter blowing in advance of an approaching depression.

3   In the S part of the Aegean Sea, S winds occur more frequently than in the N. Winds, however, blow mainly as in the N part, from between N and E.

4   The confined nature of the Adriatic Sea gives rise to many local effects, details of which can be found in *Mediterranean Sea Pilot, Volume III*.

5   At the height of the season, winds are likely to reach force 7 or above on six to nine days per month in the Aegean Sea and the E part of the Ionian Sea and on three to six days per month elsewhere in the area.

6   Weather at this season, as in the W part of the Mediterranean Sea, is subject to rapid changes caused by moving depressions and the statements made in paragraph 5.4.1 for that area apply equally to the E part of the Mediterranean Sea.

7   Visibility is generally good except, when reduced by rain, but with winds from a S quarter, which are experienced in advance of a depression, it is often only moderate.

5.5.2

1   **May to September**. Over the whole of the E part of the Mediterranean Sea, other than the Aegean Sea, the prevailing winds are NW throughout the period, and particularly persistent in July and August and E of 20°E, where winds from directions, other than between N and W, are uncommon.

2   Over the Aegean Sea the prevailing wind is N; here also, the degree of persistence is particularly high in July and August, during which months the great majority of winds are from between NE and NW.

3   From May to August winds are likely to reach force 7 only on rare occasions, except over the Aegean Sea in July and August, where winds of this strength may be expected on one or two days per month. In September the frequency of these winds is one to three days per month over most of the E part of the Mediterranean Sea, but the mariner should

be particularly alert to the possibility of rapid development of depressions and squalls which can give unexpected onsets of strong winds and violent conditions.

4   Over the greater part of the open waters of the area, weather at this season is fine with small amounts of cloud and little or no rain, especially in the S and E of the area in July and August. Over the N parts of the Aegean and Adriatic Seas some rain is likely throughout the period.

5   Visibility is generally good, though occasional patches of sea fog may be experienced in early summer, most often in the N part of the area; with winds from a S quarter haze is sometimes prevalent.

5.5.3

1   **April and October**. In the transitional months of April and October conditions can be taken as intermediate between winter and summer, though it should be realised that considerable variations are likely from year to year.

SWELL

General remarks
5.6

1   Heavy swells are more frequent in the W part than in the E part of the Mediterranean Sea.

2   In the W part of the Mediterranean Sea, between Corse and Islas Baleares, the percentage frequency of swell greater than 4 m is:

June to September	1 to 2 %.
October	2 to 5 %.
November to March	10%.
April and May	2 to 5 %.

These swells are invariably short or average in strength.

CURRENTS

Mediterranean Sea, Adriatic Sea and Aegean Sea
5.7

1   In the Mediterranean Basin the rate of evaporation is about three time as great as the inflow from the rivers which discharge into it. In consequence, there is a continuous inflow of water through the Strait of Gibraltar from the Atlantic Ocean.

2   Evaporation causes the Mediterranean water to increase its salinity; this dense water sinks and its excess emerges through the Strait of Gibraltar as a W-going sub-surface current, a smaller quantity similarly reaches the Black Sea.

3   The main body of water entering the Strait of Gibraltar flows E along the N coast of Africa; this is the most constant part of the main circulation, but it gradually loses its strength as it penetrates E. On reaching Malta Channel part of it turns N to circulate counter-clockwise in the Western Mediterranean; the remainder continues through the Malta Channel and along the African coast turning N at the E end of the Mediterranean Sea and returning W along its N shores until it reaches the Ionian Sea where it turns S to rejoin the main E-going flow. Branches of this current enter the Aegean and Adriatic Seas, giving rise to counter-clockwise circulations in those areas.

ICE
5.8

No ice occurs in the Mediterranean Sea.

PASSAGES

GENERAL INFORMATION

Detail
5.9

Passages from the Strait of Gibraltar and Bûr Sa'îd (Port Said) to principal ports in the Mediterranean Sea are given below.

Main routes through the Adriatic and Aegean Seas are given in the appropriate volumes of *Admiralty Sailing Directions*.

Dhiórix Korínthou
5.10

1   The saving of distance by using Dhiórix Korínthou (Corinth Canal) instead of entering the Aegean Sea through Stenó Elafonísou is most marked in the case of passages between ports in the Adriatic Sea or on the W coast of Greece and ports in the Aegean or Black Seas. To a lesser degree there is a saving of distance between Stretto di Messina and ports in the Aegean or Black Seas.

2   Thus the saving from:

Brindisi to Piraiévs is about 130 miles and to İstanbul about 100 miles:

Messina to Piraiévs is about 80 miles and to İstanbul about 50 miles.

3   For a description and for regulations regarding the canal, including pilotage, maximum permissible draughts and times of opening see *Mediterranean Pilot, Volume III* and for approaches to its SE entrance see *Mediterranean Pilot, Volume IV*.

ROUTES

Strait of Gibraltar ⇒ West part of Mediterranean Sea except south shore

Diagram 5.11
Routes
5.11

Full advantage should be taken of the E-going current by keeping well away from the Spanish coast and passing about 20 miles off Cabo de Gata (36°43′N, 2°11′W).

For ports between Barcelona and Genoa
5.11.1

Routes for ships passing N of Islas Baleares (39°45′N, 3°00′E) continue about 20 miles off the Spanish coast until departure can be taken for destination from abreast of Cabo de San Antonio (38°48′N, 0°12′E).

For Napoli
5.11.2

Routes pass close S of Capo Spartivento (38°53′N, 8°51′E) thence as navigation permits.

For Palermo or Messina
5.11.3

Routes are direct to the N coast of Sicilia, passing:
N of the dangers extending N from Îles de la Galite (37°30′N, 8°55′E) and giving Keith Reef (37°49′N, 10°55′E) a wide berth.

West part of Mediterranean Sea except south shore ⇒ Strait of Gibraltar

Diagram 5.11
5.12

All routes make Cabo de Gata (36°43′N, 2°11′W), thence follow the coast as closely as navigation permits to Strait of Gibraltar (36°00′N, 5°21′W), to avoid the full strength of the E-going current.

From Marseille or Barcelona
5.12.1

Routes make the Spanish coast at Cabo de San Antonio (38°48′N, 0°12′E) and thence follow it to Cabo de Gata (36°43′N, 2°11′W).

From North Italian ports
5.12.2

Routes passing:
N of Islas Baleares (39°45′N, 3°00′E) make the Spanish coast at Cabo de Palos (37°58′N, 0°41′W) and thence follow the coast to Cabo de Gata (36°43′N, 2°11′W).
S of Islas Baleares make Cabo de Gata direct.

From Napoli
5.12.3

Routes pass close S of Capo Spartivento (38°53′N, 8°51′E) thence as navigation permits to Cabo de Gata (36°43′N, 2°11′W).

From Messina or Palermo
5.12.4

Routes are direct to Cabo de Gata (36°43′N, 2°11′W), giving Keith Reef (37°49′N, 10°55′E) and the banks extending N from Îles de la Galite (37°30′N, 8°55′E) a wide berth.

From other ports
5.12.5

Routes are as direct as navigation permits to Cabo de Gata (36°43′N, 2°11′W).

Strait of Gibraltar ⇒ East part of Mediterranean Sea and Algerian coast

Diagram 5.11
Routes
5.13

From Strait of Gibraltar (36°00′N, 5°21′W) full advantage should be taken of the E-going current by keeping well away from the Spanish coast.

For Isola di Pantelleria
5.13.1

The route passes:
Ten miles N of Alborán (35°56′N, 3°02′W), thence:
Along the African coast, keeping 10 to 20 miles off the salient points, and:
Through Canal de la Galite (37°18′N, 9°00′E), passing 5 miles N of Cap Serrat (37°14′N, 9°13′E), Les Fratelli (37°18′N, 9°24′E) and Ras Ben Sekka (37°21′N, 9°45′E), thence:
Through the Traffic Separation Schemes off Îles Cani (37°21′N, 10°07′E) and Cap Bon (37°05′N, 11°03′E), thence:

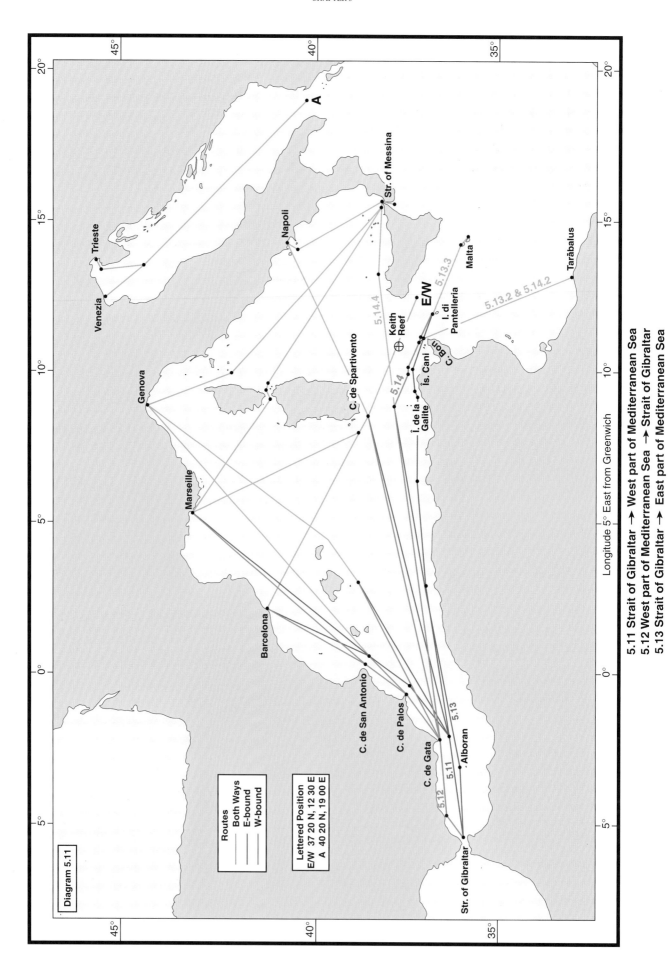

Diagram 5.11

Routes
—— Both Ways
—— E-bound
—— W-bound

Lettered Position
E/W 37 20 N, 12 30 E
A 40 20 N, 19 00 E

5.11 Strait of Gibraltar → West part of Mediterranean Sea
5.12 West part of Mediterranean Sea → Strait of Gibraltar
5.13 Strait of Gibraltar → East part of Mediterranean Sea

Longitude 5° East from Greenwich

Five miles N of Isola di Pantelleria (36°47'N, 12°00'E).

For North African ports depart from the above route as appropriate.

For Tarābulus
5.13.2

The route continues direct to Tarābulus (32°56'N, 13°12'E) from the end of the Traffic Separation Scheme off Cap Bon, described at 5.13.1.

For Malta
5.13.3

The route from 5 miles N of Isola di Pantelleria, described at 5.13.1, continues passing 5 miles N of Għawdex (36°03'N, 14°15'E), thence to Malta (35°55'N, 14°33'E).

Diagram 5.14
For other destinations
5.13.4

Routes continue from 5 miles N of Isola di Pantelleria, described at 5.13.1, as direct as navigation permits to destination.

East part of Mediterranean Sea and Algerian coast ⇒ Strait of Gibraltar

Diagram 5.11
Routes
5.14

All routes make Cabo de Gata (36°43'N, 2°11'W), thence follow the coast as closely as navigation permits to Strait of Gibraltar (36°00'N, 5°21'W), to avoid the full strength of the E-going current.

From East Mediterranean ports
5.14.1

From E Mediterranean ports, except Tarābulus, routes are as navigation permits:
To a position 5 miles N of Isola di Pantelleria (36°47'N, 12°00'E), thence:
Through the Traffic Separation Schemes off Cap Bon (37°05'N, 11°03'E) and Îles Cani (37°21'N, 10°07'E) to a position 25 miles N of Îles de la Galite (37°30'N, 8°55'E), thence:
To Cabo de Gata (36°43'N, 2°11'W), avoiding the full strength of the E-going current, thence:
Along the Spanish coast to the Strait of Gibraltar (36°00'N, 5°21'W), keeping as close to the coast as navigation permits.

From Tarābulus
5.14.2

From Tarābulus (32°56'N, 13°12'E) the route is direct to join the route described at 5.14.2 at the end of the Traffic Separation Scheme, off Cap Bon (37°05'N, 11°03'E).

From Algerian coast
5.14.3

Routes join that described at 5.14.2 off Cabo de Gata (36°43'N, 2°11'W).

From Adriatic Sea or Dhiórix Korínthou
5.14.4

Slightly shorter routes are:
Through Stretto di Messina (38°12'N, 15°36'E), where a Traffic Separation Scheme is in force, thence:
Along the N coast of Sicilia, thence:
Joining route described at 5.14.2 N of Îles de la Galite (37°30'N, 8°55'E).

Routes to and from Bûr Sa'îd (Port Said)

Diagram 5.14
Routes
5.15

For descriptions of routes through the Adriatic and Aegean Seas see the appropriate volumes of *Admiralty Sailing Directions*.
5.15.1

Between places in the W part of Mediterranean Sea S and W of Barcelona (41°20'N, 2°10'E) and Bûr Sa'îd (31°21'N, 32°33'E):
E-bound routes should join the routes described at 5.13, as appropriate:
W-bound routes are as described at 5.14 until N of Îles de la Galite (37°30'N, 8°55'E), thence either:
Following that route for W ports, or:
As navigation permits to other destinations.
5.15.2

Between Bûr Sa'îd (31°21'N, 32°33'E) and other places in Mediterranean Sea, including Barcelona, routes are direct.

Distances through Mediterranean Sea
5.16

The list gives the E-bound distances. Where E-bound and W-bound routes differ appreciably in distance a note indicates the difference.

	Strait of Gibraltar	Bûr Sa'îd
Algiers (36°46'N, 3°05'E)	415	1500
Barcelona (41°20'N, 2°10'E)	525 [a]	1590 [b]
Marseille (43°18'N, 5°20'E)	705 [a]	1510 [c]
Genova (44°23'N, 8°53'E)	865 [ad]	1420
Napoli (40°50'N, 14°17'E)	980	1110
Stretto di Messina (38°12'N, 15°36'E)	1030	935
Venezia (45°24'N, 12°29'E)	1720 [ef]	1300
Trieste (45°39'N, 13°44'E)	1720 [ef]	1300
Stenó Elafonísou (36°25'N, 22°57'E)	1380 [e]	565
Piraiévs (37°56'N, 23°37'E) (via Dhiórix Korínthou)	1430 [e]	—
Piraiévs (via Aegean Sea)	1490 [e]	590
Thessaloníki (40°37'N, 22°56'E)	1700 [e]	730 [g]
Çanakkale Boğazı (SW entrance) (40°09'N, 26°24'E)	1650 [e]	650 [g]
İzmir (38°25'N, 27°00'E)	1640 [e]	610 [g]
Beirut (33°55'N, 35°31'E)	2020 [e]	220

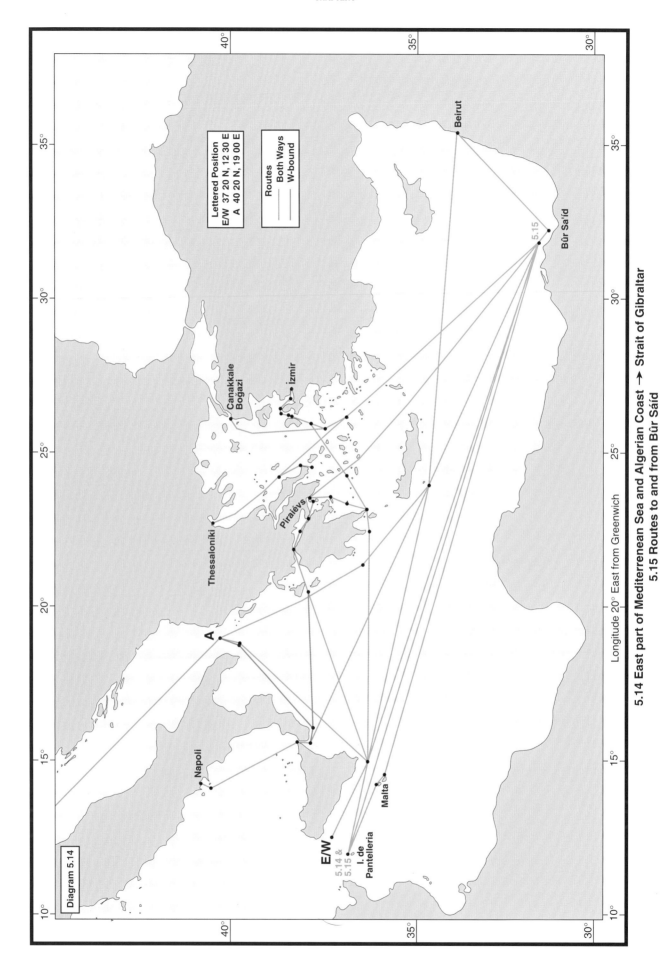

5.14 East part of Mediterrenean Sea and Algerian Coast → Strait of Gibraltar

5.15 Routes to and from Bûr Sa'îd

	Strait of Gibraltar	Bûr Sa'îd
Bûr Sa'îd (31°21′N, 32°33′E)	1910 [e]	—
Tarābulus (32°56′N, 13°12′E)	1080 [e]	990
Malta (35°55′N, 14°33′E)	980 [e]	935

Notes

[a]   W-bound subtract 10 miles

[b]   N or S of Sicilia

[c]   N of Sicilia. If S of Sicilia, add 55 miles

[d]   N or S of Islas Baleares

[e]   W-bound add 15 miles

[f]   S of Sicilia

[g]   Via Stenó Elafonísou

Suez Canal
5.16.1

Distance from the Fairway light-buoy off Bûr Sa'îd to the seaward end of Newport Rock Channel in the entrance to Bahr el Qulzum (Suez Bay) is 99 miles.

NOTES

CHAPTER 6

INDIAN OCEAN

GENERAL INFORMATION

COVERAGE

Chapter coverage
6.1

This chapter contains details of the following passages:
To and from Red Sea and Persian Gulf (6.33).
To and from East coast of Africa, Arabian Sea and Bay of Bengal (6.34).
To and from Mauritius (Port Louis) (6.63).
To and from Mahé Island (Seychelles Group) (6.76).
Western approaches to and passages off the Australian coast (6.83).
Passages on the Eastern side of the Indian Ocean (6.93).
Trans-Ocean Passages (6.97).

Admiralty Publications
6.2

Relevant navigational publications should be consulted, in addition to this book, when planning and conducting passages. Details of these, and the areas covered, can be found, as follows:
Admiralty Sailing Directions; see 1.12.1
Admiralty List of Lights and Fog Signals; see 1.12.2
Admiralty Tide Tables; see 1.12.3
Admiralty List of Radio Signals; see 1.12.5
Admiralty Distance Tables; see 1.12.8

6.2.1

See also *The Mariner's Handbook* for notes on electronic products.

NATURAL CONDITIONS

Charts:
Indian Ocean Routeing 5126 (1-12)
World Surface Current Distribution 5310

WINDS AND WEATHER

General information
6.3

The following description of the winds and weather of the region of the Indian Ocean amplifies the general statement given in *The Mariner's Handbook*. More precise information about oceanic winds and weather, or detailed information about specific localities, should be sought in the appropriate volumes of *Admiralty Sailing Directions*.

North Indian Ocean

Monsoons
6.4

The winds and weather of the whole North Indian Ocean are dominated by the alteration of the Monsoons, which are seasonal winds generated by the changes in pressure resulting from the heating and cooling of the land mass of Asia.

South-west Monsoon
6.5

1   From June to September the heating of the Asiatic land mass results in the establishment of a large area of low pressure centred approximately over the NW part of India. The South-east Trade Wind of the South Indian Ocean is drawn across the equator, deflected to the right by the effects of the earth's rotation, and joins the cyclonic circulation round the area of low pressure mentioned above. The resulting SW wind, felt in the North Indian Ocean, the Arabian Sea and the Bay of Bengal, from June to September, is known as the south-west monsoon. The general distribution of pressure and wind at this time is shown on *Diagram 1.14b World Climatic Chart (July)*, from which it will be noted that in the E part of the

Arabian Sea the prevailing wind direction is more nearly W than SW.

2   The strength of the wind varies considerably between different parts of the ocean. It is strongest in the W part of the Arabian Sea where, over a considerable area, the wind averages force 6 at the height of the season and reaches force 7 or above on more than 10 days per month; the worst area is some 250 miles E of Suquṭrá (12°30′N, 54°00′E), where in July about half the observations report winds of force 7 or above.

3   In the extreme N part, and in the E part of the Arabian Sea in July and August, the monsoon wind averages about force 4, although it often freshens to force 5 or 6, and attains force 7 or above on 5 to 10 days per month N of about 10°N.

4   In the Bay of Bengal the average strength of the monsoon wind is force 4 to 5; over the greater part of the bay the wind reaches force 7 or above on 5 to 10 days per month in July and August.

5   Between the equator and about 5°N and E of 60°E, winds are generally lighter and only average about force 3; they are also considerably more variable in direction, though generally from between S and W.

6   In Malacca Strait the wind is mostly light and is subject to considerable variation in direction and strength due to land and sea breezes and other local influences. In the N part of the strait the winds are most often SW, while in the S the most frequent direction is SE. Although the Monsoon is generally light, there are often periods of stronger winds accompanied by squalls which sometimes reach gale force. Of particular note are the squalls, widely known as 'Sumatras', which blow from some W point and occur most frequently at night; they are described in *Malacca Strait and West Coast of Sumatera Pilot*.

7   The weather over most of the North Indian Ocean during the south-west monsoon season is cloudy and unsettled, with considerable rainfall, especially off the W coasts of India and Mayanmār (Burma), where it is very heavy. In the W part of the Arabian Sea, however, cloud amount and rainfall decrease towards the N and W and both are generally small in the vicinity of the African and

Arabian coasts. Rainfall is also small at this time in the immediate vicinity of the E coast of Sri Lanka and India as far N as about 15°N.

8   Visibility is good in most parts of the area except when reduced by rain. In the N and W parts of the Arabian Sea it is often only moderate and sometimes poor within about 200 miles of the coast particularly during the south-west monsoon period when, although the sky may be clear, the surface visibility may be reduced because of dust haze. In this latter zone in July and August visibility is likely to be less than 5 miles on about 50% of occasions.

9   The waters off the S Arabian coast in the vicinity of the Juzur al Ḥalāniyāt (Kuria Muria Islands) are frequently affected by sea fog during the south-west monsoon.

North-east Monsoon
6.6

1   From November to March, a NE wind is experienced in the North Indian Ocean, the Arabian Sea and the Bay of Bengal. This wind is known as the north-east monsoon. The general distribution of pressure and wind at this season is shown on *Diagram 1.14a World Climatic Chart (January)* from which it will be observed that over the E part of the Arabian Sea, and towards the equator, the prevailing wind direction is more nearly N than NE.

2   There are two areas in which the Monsoon is subject to considerable interruption, or in which the wind is rather variable in direction. The first is in the Arabian Sea N of about 20°N, where the variations in the strength and direction of the wind are caused by the passage of depressions across Iran or along the Makrán coast. The second is between the equator and about 5°N and E of about 90°E where winds, though mostly N, are generally light and somewhat variable in direction.

3   Over the greater part of the North Indian Ocean the strength of the north-east monsoon averages force 3 to 4 at the height of the season, though towards the equator it averages force 2 to 3, except W of about 55°E. It is also only light in the Malacca Strait. Winds are likely to reach force 7 only on rare occasions.

4   The weather in the Arabian Sea and Bay of Bengal is generally fine with small amounts of cloud and little or no rain. Cloudiness and rainfall increase towards the S and E, especially in December and January when considerable rain occurs in the S part of Bay of Bengal, S of a line joining the N extremities of Sri Lanka and Sumatera.

5   Visibility over the open ocean away from the effects of land is generally good or very good at this season and fog is unknown. In the N and E parts of the Arabian Sea, however, visibility is often reduced by dust haze, especially in the latter part of the season, while in the N part of the Bay of Bengal it may be reduced by smoke haze and land mists carried seawards by the prevailing N winds.

Inter-monsoon seasons
6.7

1   The months of April, May and October are characterised by the N and S shift across the area of the Intertropical Convergence Zone (6.11) and by the progressive replacement of the north-east monsoon by the south-west monsoon in April and May, and vice versa in October. The south-west monsoon becomes established in the S earlier than in the N and the reverse is true for the north-east monsoon. The width of the Equatorial Trough, however, varies greatly from day to day and its movements are irregular, consequently the whole area can be regarded primarily as one of light winds (apart from squalls and tropical storms) with a rather high frequency of calms, and

with the oncoming monsoon becoming gradually established.

2   Except in squalls, which are common, or in association with tropical storms, winds over the open ocean are likely to reach force 7 or above only on rare occasions. In the W part of the Arabian Sea between 5°N and 10°N and W of 55°E, however, winds of this strength may be expected on about 2 days in May. 'Sumatras' (6.5) occur occasionally in the Malacca Strait.

3   The weather varies considerably, fair or fine conditions alternating with cloudy, squally weather, with frequent heavy showers and thunderstorms. These conditions spread N during April and May and retreat S during October. In the N part of the Arabian Sea, however, fine weather predominates during these inter-monsoon months.

4   Visibility over the open ocean is good except when reduced by heavy rain. Near the shores of the N and E parts of the Arabian Sea, however, it is sometimes reduced by dust haze in April and May.

Gulf of Aden
6.8

1   The winds in the Gulf of Aden form part of the monsoon circulation of Asia. The north-east monsoon prevails but winds are constrained by topography as they enter the Gulf of Aden and blow generally from an ENE or E direction, veering sharply to SE on nearing the Straits of Bab el Mandeb. In May, wind direction is variable, while from June to September SW winds prevail.

2   In the Gulf of Aden wind speeds average force 2 to 4 from December to March, but winds are funnelled in the approaches to the Straits of Bab el Mandeb (12°40′N, 43°30′E) to reach force 5 to 6 and, occasionally, force 7 to 8 in the Straits; otherwise gales are rare. From June to September in the main part of the Gulf, the strength of the south-west monsoon averages about force 4 and winds reach force 7 or above on 1 to 2 days per month. The average strength of the wind and the frequency of gales, however, increase rapidly towards the E end of the Gulf and E of Raas Caseyr winds are likely to reach force 7 or above on 10 to 15 days in July. Tropical cyclones are very rare in the Gulf, only 3 or 4 being experienced in the last 50 years.

3   The weather over the whole of the Gulf of Aden is generally fine with small amounts of cloud; when rainfall does occur it is in the form of showers and may be heavy. Total rainfall is slight.

4   Over the open sea, fog and mist are rare except in the extreme E part of the Gulf of Aden during the south-west monsoon season. Sand and dust haze is, however, widespread from June to August, visibility at this time of year being less than 5 miles on about 1 day in 4 or 5 on the African side of the Gulf of Aden, and 1 day in 2 on the Arabian side of the Gulf. In September the frequency of haze decreases greatly, while from December to February it is not usual. Sandstorms which mostly occur in the Gulf of Aden from May to August or September may occasionally reduce visibility to 50 m or less.

Gulf of Oman
6.9

1   The following remarks apply to open water away from the local effects of land. Near land, both land and sea breezes and other local effects are likely to cause considerable modification. Detailed information about specific localities will be found in *Persian Gulf Pilot*.

2   From December to February, winds are mainly from some N point with NW as the most frequent direction.

Occasionally, south-easterlies may occur ahead of depressions advancing across the Persian Gulf. From March to May winds are very variable, with north-westerlies decreasing and south-westerlies increasing in frequency until, by May, the latter winds predominate. From June to August the prevailing wind is SE, being an offshoot of the south-west monsoon of the Arabian Sea. From September to November the frequency of SE winds decreases and that of northerlies increases, but wind direction is, in general, very variable.

3   Winds reach force 7 on about 1 to 2 days per month from December to March, but rarely attain force 8. Squalls are common. On rare occasions the Gulf may be affected by a tropical storm originating in the Arabian Sea.

4   Rain and large amounts of cloud are practically confined to the period November to April, and are associated with E-moving depressions, in the intervals between which fine weather, with small amounts of cloud, prevails. In summer, the influence of the monsoon causes an increase in cloudiness in July and August.

5   Visibility is for the most part good or very good from November to February; after this dust haze causes a progressive deterioration until in June and July visibility is less than 5 miles on 10 to 12 days per month over the open sea and more often near the coast. Dust haze decreases considerably after July. Duststorms and sandstorms occur in all seasons, but are most frequent during June and July and least so during winter; during their occurrence they often reduce visibility to less than 500 metres.

South Indian Ocean

General remarks
6.10

The winds and weather of the South Indian Ocean are governed by the advance of the North Indian Ocean monsoon into the S hemisphere from November to February and its retreat from June to September; the result is the establishment, in this zone, of alternating seasonal winds. South of this zone the normal wind and pressure distribution, as outlined in *The Mariner's Handbook*, prevails.

Intertropical Convergence Zone
6.11

1   This region is known variously as the Intertropical Convergence Zone (ITCZ), the Intertropical Front (ITF), the Doldrum Belt, the Equatorial Trough, the Equatorial Front or the Shearline.

2   It is, in the Indian Ocean, S of the equator from about November to April and reaches its most S position in February. The winds and weather are similar to those encountered in the Intertropical Convergence Zone in other oceans and consist of fair weather, calms and light variable winds alternating with squalls, heavy showers and thunderstorms. Both the width of the belt and its position vary considerably from day to day; the width averages 200 miles but may, at times, be much more, while at others it may be reduced to almost nothing by a strong burst of the South-east Trade Wind. Visibility in this zone is good except in heavy rain.

North-west Monsoon
6.12

1   During the period from November to March, when the Equatorial Trough is situated in the S hemisphere, the north-east monsoon of the North Indian Ocean is drawn across the equator, deflected to the left by the effect of the earth's rotation, and is felt in the N part of the South Indian Ocean as a NW wind known as the north-west monsoon. See *Diagram 1.14a World Climatic Chart (January)*.

2   Winds are in general light, and vary considerably in direction, but in the W part of the zone the prevailing direction is more nearly N than NW, and becomes NE close to the African coast and N of about 10°S. In the Moçambique Channel a N wind prevails as far as 15°S to 17°S; it is known here as the Northern Monsoon.

3   In the E part of the ocean, just S of Jawa (7°00′S, 110°00′E) and in the Timor and Arafura Seas, the prevailing wind direction is between W and NW.

4   Except in squalls, which are common, or in association with tropical storms (6.16) winds over the greater part of the zone are likely to reach force 7 or above only on rare occasions.

5   The weather is generally rather cloudy and unsettled and rain, mostly in the form of heavy showers, is frequent. Visibility is good except in rain.

South-east Trade Wind
6.13

1   This wind blows on the equatorial side of the anti-clockwise circulation round the oceanic high-pressure area situated about 30°S. in this ocean, however, the oceanic anticyclone seldom consists of a single cell; more frequently it contains a more or less regular succession of E-moving anticyclones, from the N sides of which blow the Trade Wind, permanently and with little variation in direction throughout the year.

2   In summer, the South-east Trade Wind extends from about 30°S to the Equatorial Trough, the general direction of the wind being from between E and SE over most of the area, but becoming S off the W coast of Australia and mainly SW off its NW coast, though in the latter area the direction is much more variable than in the Trade Wind proper. In the S part of the Moçambique Channel an extension of the Trade Wind gives prevailing S to SE winds. The *Diagram 1.14a World Climatic Chart (January)* shows the area covered by the Trade Winds at this season.

3   In winter, the South-east Trade Wind extends from about 27°S to the equator, though N of about 5°S and E of 70°E the winds are weak and though, generally from some S point, they vary considerably in direction. Elsewhere over the greater part of the open ocean winds are almost exclusively from between SSE and ESE, but in the E part of the area and in the Timor Sea the predominant connection is somewhat more E. In the Timor and Arafura Seas the South-east Trade Wind is sometimes referred to as the south-east monsoon in contradistinction to the north-west monsoon (6.12), which prevails there in summer. In the Moçambique Channel an extension of the South-east Trade Winds gives prevailing S to SE winds over the whole length of the channel from about April to September. These winds are known as the Southern Monsoon in contradistinction to the Northern Monsoon (6.12) which prevails in the N part of the channel in summer. The *Diagram 1.14b World Climatic Chart (July)* shows the area covered by the South-east Trade Winds at this season.

4   The average strength of the South-east Trade Wind is force 3 to 4 in summer and 4 to 5 in winter; it reaches a mean of force 5 between about 10°S and 20°S and 65°E and 100°E when at its strongest during the winter. In summer, winds are likely to reach force 7 or above on 1 to 3 days per month over the greater part of the zone, rising to 3 to 6 days per month over the central part of the area.

In winter, winds of this strength are likely to be encountered on 1 to 3 days per month in the E and W parts of the zone, while over a considerable area between about 65°E and 90°E their frequency rises to 6 to 9 days per month as shown on the *Diagram 1.14b World Climatic Chart (July)*. In the Timor and Arafura Seas winds are unlikely to reach force 7 on more than 1 or 2 days per month.

5   The weather over the open ocean is mostly fair or fine with skies about half covered, but belts of cloudy showery weather occur at intervals. To the NW and N of the Australian continent, between the NW part of Australia and Jawa, in the Timor Sea and, to a lesser extent, in the Arafura Sea, cloud amounts and rainfall are small from April to September, while the South-east Trade Wind prevails in these regions. Extensive dust haze prevails here, especially in the Timor Sea and towards the end of the season. Elsewhere in the South-east Trade Wind zone visibility is good except in rain.

Variables
6.14

1   To the S of the S limit of the South-east Trade Wind, there is a zone of light variable winds in the area of the oceanic high-pressure region. In winter the centre of the high-pressure region is located about 30°S, while in summer it moves to about 35°S over the greater part of the ocean, dipping somewhat farther S near the SW part of Australia.

2   The weather also varies considerably in this zone, alternating between fair or fine conditions near the centres of the E-moving anticyclones and cloudy, showery weather in the intervening troughs of low pressure. Visibility is generally good except in rain.

The Westerlies (Roaring Forties)
6.15

1   To the S of the high pressure region described in 6.13 and 6.14, W winds predominate. As in the Westerlies of other oceans, the almost continuous passage of depressions from W to E causes the wind to vary greatly both in direction and strength; the centres of most of these depressions pass S of 50°S. Gales are very prevalent in the zone of the Westerlies especially in winter, during which season winds reach force 7 or above on 12 to 16 days per month S of the 36°S. During summer, winds of this force are likely to be encountered on 6 to 12 days per month S of about 40°S. The Routeing Charts 5126(1) to 5126(12) show the regions in which gales are most common.

2   As in the Westerlies of other oceans, the weather is very variable, periods of overcast skies and rain or snow associated with the fronts of E-moving depressions alternating with fairer conditions. Fine weather is, however, seldom prolonged, and cloud amounts are generally large throughout the year.

3   Visibility varies considerably; with winds from a S point it is generally good, while N winds are often associated with moderate or poor visibility. South of the 40th parallel visibility of less than 2 miles may be expected on perhaps 5 days per month, while fog is not uncommon during the summer it is usually associated with winds from a N point.

Tropical storms
6.16

1   Tropical storms occur in the Arabian Sea, the Bay of Bengal and in parts of the South Indian Ocean. They are described, and advice on avoiding them is given, in *The Mariner's Handbook*. Information regarding storm frequencies and tracks will be found the appropriate volume of *Admiralty Sailing Directions*, or on the Routeing Charts 5126(1) to 5126(12).

2   Tropical storms are known as 'Cyclones' in the area covered by this chapter, except off the N, NW and W coasts of Australia where they are known as 'Hurricanes'.

3   In the Arabian Sea, cyclones occur in May, June, July, October and November, the periods of gretest frequency being June and November. Although they have been recorded, they are extremely rare in July, September and December. They are unknown from January to March and in August.

4   In the Bay of Bengal most cyclones occur from May to November, with May, June, October and November as the months of greatest frequency. They occur very occasionally in March, April and December and are almost unknown in January and entirely so in February.

5   In the South Indian Ocean, cyclones occur from December to April, the period of greatest frequency being between January and March; they also occur occasionally in November and May.

6   In the Timor and Arafura Seas and off the W coast of Australia the hurricane season and the period of greatest frequency are the same as for the cyclones of the South Indian Ocean, except that hurricanes are not known in May.

7   For the effects of tropical storms upon currents see, *The Mariner's Handbook*.

SWELL

Arabian Sea and the Bay of Bengal
6.17

1   The swell is governed by the direction and strength of the monsoon winds.

2   In the Arabian Sea a swell from the SW becomes established during May and persists until September. A swell from the NE becomes established during November and persists until March. There is no predominant direction in April and October and the swell is normally low or moderate in the changeover months and mainly moderate once the monsoon is established, though from June to September a heavy swell may be encountered.

3   In the Bay of Bengal a swell from the SW becomes established during March and persists until October. A swell from the NE becomes established during November and persists until February. Swell is normally low or moderate except for the period from May to August when it is moderate or heavy.

4   In Malacca Strait there is no predominant direction of swell. Throughout the year swell is normally low and only on rare occasions does it become moderate.

5   Swell in the Arabian Sea and the Bay of Bengal is normally short or average in length. However, on about 10% of occasions, swells of over 200 m in length may be encountered; such swells are almost invariably low in height.

Gulf of Aden
6.18

A swell from the SW occurs from June to September and a swell from between E and NE from November to March. These swells are low or moderate. There is no predominant direction in April, May or October, when the swell is mainly low. The length of swell is generally short, though a small number of average swells do occur.

Gulf of Oman
6.19

Swell is from the NW from December to February and from the SE from June to August. At other times there is no predominant direction. The swell is normally low or moderate and only rarely heavy. Most swells in the Gulf of Oman are short and have periods of between 3 and 6 seconds.

South Indian Ocean
6.20

1   Swell is a regular feature. The swell generated by the depressions S of 50°S often travels to all parts of the North and South Indian Ocean; more than one swell is frequently present and confused swell is often reported. As shown on the following table (6.21), it is normally moderate to heavy.

2   In length it covers the complete range from short to long; many swells are of average length but lengths of over 300 m are not uncommon.

3   Abnormal waves may occur, see 1.18.

Table of swell patterns in the South Indian Ocean
6.21

	0°S to 20°S	20°S to 35°S	35°S to 50°S
25°E to 70°E			
Direction	Predominantly SE	SE through S to SW	Some NW but mainly W to SW
Height	Low or moderate, at times heavy between 10°S and 20°S	Moderate or heavy	Moderate or heavy
70°E to 110°E			
Direction	S to SE	Mainly SW to S	Some NW but mainly W to SW
Height	Mainly moderate	Moderate or heavy	Moderate or heavy, with waves greater than 6 m quite common

Speed reduction in relation to sea conditions
6.22

1   During the north-east monsoon and southern summer period, from about November to March, sea conditions in the Indian Ocean do not call for particular comment except that, S of 40°S, they are such as to cause some ships on W headings to find it necessary to reduce speed for a small proportion of their voyage time. The southern summer is, however, the season of greatest frequency of tropical storms in the South Indian Ocean, see 6.16.

2   At the peak of the south-west monsoon period, in July, speeds of some ships in the Arabian Sea may have to be reduced for a moderate proportion of the time when steaming into, or across, wind and sea and for a smaller proportion of the time in a following sea.

3   In the S hemisphere, seas in winter are higher than in summer and the South-east Trade Winds and the Westerlies are at their strongest. It is apparent that winter storms in the South Indian Ocean have their greatest frequency in about 80°E and lesser concentrations about 60°E and 110°E.

4   During the transitional periods, in April and October, sea conditions, though less severe than in July, may still affect speed on the E–W tracks across the Southern Ocean. Speed reduction may be necessary S of 35°S on these tracks for a small percentage of voyage time.

CURRENTS

North Indian Ocean

General remarks
6.23

1   Currents in the North Indian Ocean are reversed in direction seasonally under the influence of the monsoons. These comprise the currents of the Arabian Sea, the Bay of Bengal and the **Somali Current**. Along the equator, between 2°N and 2°S, an E-going current (the equatorial jet) appears twice a year in the transitions between monsoon seasons.

2   The south-west monsoon circulation is fully established from June to September. April and October are months of transition. In May and November, which are also to some extent transitional, the circulation more resembles that of the four subsequent months.

3   The north-east monsoon circulation occurs from December to March. The currents are therefore described below for four periods:

   December to March
   April and May
   June to September
   October and November

December to March (North-east Monsoon)
6.24

1   In the open waters of the Arabian Sea and Bay of Bengal, the current sets in a general W direction. In the central Indian Ocean, along the equator, the last remnants of the equatorial jet are still running E in December and are replaced by a broad flow to the W in January to March. Round the coasts, the current flows counter-clockwise in December and January, but gradually changes to clockwise in February and March. This is in response to changes in the distribution of wind stress, though the winds are still predominantly from the NE.

2   In the Bay of Bengal, coastal currents are generally clockwise from early February.

3   In the Arabian Sea, the onset is more gradual and is not complete on all parts of the coast until the end of March. In February the current flows SW off the coast of Somalia S of about 8°N, but farther N it sets NE. In March the S limit of the NE flow is near 4°N.

April and May (Transition period)
6.25

1   Currents are weak and variable in the central regions of the Arabian Sea and Bay of Bengal in April and set weakly E in May. The clockwise coastal circulation in the Bay of Bengal is strengthened. Along the equator the E-going equatorial jet appears first in mid-April between 70°E and 90°E. By early May it extends from 55°E to 95°E. On the E African coast, the NE flow is present

everywhere by late April, and the clockwise coastal circulation in the Arabian Sea strengthens during May with the onset of south-west monsoon.

June to September (South-west Monsoon)
6.26

1   The clockwise circulation of the coastal regions of the Arabian Sea and Bay of Bengal persists and strengthens. The Somali Current continues to flow NE, attaining very high rates. Mean values of 3 kn and maximum values of 8 kn have been reported. It should be noted, however, that the synoptic pattern of flow in the Somali Current may differ from the climatological mean. In June the current usually turns away from the coast between 2°N and 6°N and a separate clockwise loop is found between the latitude of that turning point and S of Suquṭrá. The same pattern often persists in July. In late July or August the latitude at which the current turns offshore migrates N to about 10°N. In the open waters of the Arabian Sea and Bay of Bengal currents generally set to the E. Along the equator the equatorial jet decays during June and is replaced at most longitudes by weak W-going currents. In late September the beginnings of the next equatorial jet can be seen.

October and November (Transition period)
6.27

1   Currents in the open waters of the Arabian Sea and Bay of Bengal become weak and variable and gradually increase to the W. The coastal circulation of the Bay of Bengal becomes anti-clockwise in mid-October, increasing in strength through November and continuing along the E and S coasts of Sri Lanka. In the Arabian Sea, the development of an anti-clockwise coastal circulation is more gradual. It is present off the W coast of India and the S coast of Arabia by early November, but off Somalia a NE flow persists at all latitudes until mid-November, which by the end of the month has been replaced by a SW flow between 3°N and 7°N. The E-going equatorial jet increases in rate and length during October and is fully developed by early November.

South Indian Ocean

General remarks
6.28

1   The main surface circulation of the South Indian Ocean is counter-clockwise. The W-going flow of the **South Equatorial Current** of the Indian Ocean lies well S of the equator, thus differing from the South Equatorial Currents of the Atlantic and Pacific Oceans, which extend in latitude a few degrees N of the equator. Its N boundary is about 6°S to 8°S, varying according to longitude and season. To the N of that, in the north-east monsoon the E-going **Equatorial Counter-current** can be seen, at least in the W part of the Indian Ocean, originating at the convergence of the N-going **East African Coast Current** and the SW-going Somali Current.

2   At the beginning and end of the north-east monsoon, the Equatorial Counter-current merges into the equatorial jet, also E-going, and it is difficult to distinguish between them. However, the Counter-current is relatively broad, weak (typically ½ to 1 kn) and variable, whereas the equatorial jet is narrow (within 2° of the equator), of greater rate (typically 1 to 2 kn) and relatively steady.

3   In the south-west monsoon there is little evidence of any E-going flow S of the equator, currents between 2°S and 7°S being mostly weak and variable.

4   The W-going South Equatorial Current splits on reaching the E coast of Madagascar, at 16°S. The N-going branch turns W round the N extremity of Madagascar and continues W to split again off the African coast near Cabo Delgado. Some goes N into the East African Coast Current, the rest goes S into the Moçambique Channel. South of 16°S, the current near the E coast of Madagascar runs SSW. In the open ocean, average rates in the South Equatorial Current are ½ to ¾ kn, but much larger values are found inshore. Within a few miles of the E coast of Madagascar, average rates are 1 to 2 kn and 3 kn or more is sometimes reported.

5   Off the SE coast of South Africa the S-setting **Moçambique Current** is supplemented by the South Equatorial flow setting W to the S of Madagascar. This combined strong SW flow continues along the coast as the **Agulhas Current** reaching average rates of 2 to 3 kn and a maximum of about 5 kn, between 30°S and 34°S. Average rates seem slightly higher during the summer and autumn months than in winter and spring. During the latter two seasons the sparse data available suggests a greater extension of the **Southern Ocean Current** to the N. This in turn restricts the extension of the Agulhas Current to the S in 20°E to 22°E. The low constancy of predominant currents off Cape Town (33°53′S, 18°26′E) in winter also suggests that, at times, the surface flow nearer the coast and into the South Atlantic Ocean may be markedly restricted. In open waters E and SE of the coast between Durban (29°51′S, 31°06′E) and Cape of Good Hope and to the S of Madagascar continuous interaction between the warm waters of the recurving South Equatorial Current and the ENE-going sets of the Southern Ocean Current lead to many eddies and much variability of direction and rate. These ENE-going sets continue across the South Indian Ocean but constancies and rates decrease to the N. Any weak predominance of E-going sets changes to W-going N of about 25°S in summer and near 30°S in winter, at a longitude of 80°E. Approaching the W coast of Australia the circulation is not well defined. During autumn and winter sets off the coast are moderately constant, S-going, turning SE to E off Cape Leeuwin (34°23′S, 115°08′E). In spring and summer coastal eddy and counter-current activity N of 33°S show little marked predominance but W of 113°E there is a tendency for NW or N-going sets to extend in a widening band merging into the South Equatorial Current between 16°S to 20°S and 95°E to 105°E. The more constant part of this current is often referred to as the **West Australian Current** and is evident for much of the year, particularly N of 25°S. South of this latitude during autumn and winter the South Indian Ocean Current extends farther N and E before merging into the S-going coastal flow, with increased rates and constancies off Cape Leeuwin, and entering the Great Australian Bight.

6   Off the NW Australian coast and in the Timor and Arafura Seas historical data is sparse. The general indications are a weak predominance of W-going sets ESE of Timor turning SW for much of the year. During the late summer period a reversal to E-going sets occur in the E part of the Arafura Sea.

ICE

General remarks
6.29

1   The following brief account of ice in the South Indian Ocean should not be taken as complete or in any way all-embracing. More detailed information than can be given here will be found in the following publications, which should be consulted before undertaking passages S of the latitude of Cape Agulhas.

> Appropriate volumes of *Admiralty Sailing Directions.*
> *The Mariner's Handbook.*
> Charts 5126 (1) to 5126 (12) — *Monthly Routeing Charts for the Indian Ocean.*
> Washington, US Navy (HO 705) — *Oceanographic Atlas of the Polar Seas.*

2 A factor always to be borne in mind where ice conditions are concerned is their great variability from year to year. For this reason, and on account of the sparsity of observations in many areas, the charted positions of ice limits should be regarded as approximate.

Pack ice
6.30

1 The long-term average positions of the pack ice (4/8 concentration) is at its greatest extension from August to September, see *Diagram 1.14b World Climatic Chart (July).*

2 It runs from about:

> 55°S, 0°, to
> 58°S, 50°E, thence
> 60°S, 110°E.

Continuing E, the edge lies near 61°S as far as 160°E. For least average extension, see 3.11.

3 None of the normally inhabited places in the South Indian Ocean are affected, but great circle routes between the more S ports of South Africa and Australia are obstructed.

Icebergs
6.31

1 The icebergs that occur in the South Indian Ocean are not, in most cases, calved from glaciers, but consist of portions that have broken away from the great ice shelves which fringe parts of the Antarctic continent. They are consequently flat-topped and may be of immense size.

2 The mean limit of bergs reaches its farthest N between 20°E and 70°E in November and December when it runs from about 44°S in the longitude of Cape Agulhas to about 48°S, 70°E.

3 It is farthest N in February and March E of 70°E, when it runs:

> Between 48°S and 50°S as far as 120°E, thence:
> To about 55°S in the longitude of Tasmania.

4 In May and June the mean limit of bergs is everywhere S of 50°S and between 120°E and the longitude of Tasmania it is S of 55°S.

5 With regard to extreme limits, the season varies considerably from one longitude to another and factors other than climatic may cause abnormalities, so it is best to regard this limit as unrelated to the time of year. Earthquakes, for example, may give rise to an excessive formation of tabular bergs.

6 The extreme limit of icebergs, as indicated on:

> *Indian Ocean Routeing Charts 5126 (1-12)*
> *Diagrams 1.14a and 1.14b World Climatic Charts*
> starts at:
> 35°S near Cape Agulhas, then:
> Gradually recedes to 38°S approaching 100°E, then:
> Advances again to 35°S between 100°E and 110°E, then:
> Recedes through 44°S near 130°E, to:
> About 46°S at 150°E.

NOTES AND CAUTIONS

Oilfields
6.32

Oilfields, with platforms, drilling rigs and associated buoys and pipelines extend nearly 100 miles seaward from Mumbai (Bombay). For recommended routes through them, see *West Coast of India Pilot.*

PASSAGES BETWEEN RED SEA AND PERSIAN GULF

Routes
6.33

Routes through both the Red Sea and the Persian Gulf, and the weather, swell and currents affecting them, are described in *Red Sea and Gulf of Aden Pilot* and *Persian Gulf Pilot.*

Distances
6.33.1

Aden to Suez	1310 miles
Strait of Hormuz to Khawr al Amaya*	495 miles
* Pilot station.	

PASSAGES BETWEEN EAST COAST OF AFRICA, ARABIAN SEA AND BAY OF BENGAL

GENERAL INFORMATION

Coverage
6.34

This section contains details of the following passages:

> From Cape Town to Durban and Moçambique Channel (6.35).
> From Moçambique Channel to Durban or Cape Town (6.36).
> Routes through Moçambique Channel (6.37).
> To and from Moçambique Channel and Aden (6.38).
> To and from African coast and Persian Gulf (6.39).
> From African coast to Karāchi (6.40).
> From Karāchi to Mombasa (6.41).
> From Karāchi to Moçambique Channel, Durban and Cape Town (6.42).
> To and from Aden and Persian Gulf (6.43).
> To and from Aden and Karāchi (6.44).
> From Cape Town and Durban to Mumbai (6.45).
> From Mumbai to Cape Town or Durban (6.46).
> From Cape Town and Durban to Colombo or Bay of Bengal (6.47).
> From Colombo to Durban or Cape Town (6.48).
> From Bay of Bengal to Durban or Cape Town (6.49).
> Routes in Bay of Bengal (6.50).
> From Mombasa to Mumbai (6.51).
> From Mumbai to Mombasa (6.52).
> From Mombasa to Colombo or Dondra Head (6.53).

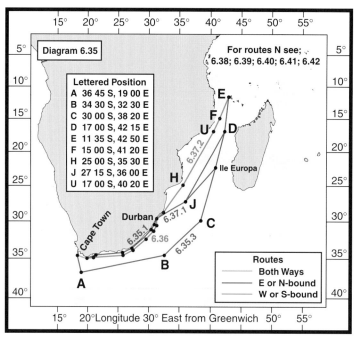

6.35 Cape Town → Durban and Mozambique Channel
6.36 Mozambique Channel → Durban or Cape Town
6.37 Routes through Mozambique Channel

From Colombo or Dondra Head to Mombasa (6.54).
From Aden to Mumbai (6.55).
From Mumbai to Aden (6.56).
From Aden to Colombo or Dondra Head (6.57).
From Colombo and Dondra Head to Aden (6.58).
To and from Strait of Hormuz and Colombo or Dondra Head (6.59).
To and from Karāchi or Mumbai and Colombo or Dondra Head (6.60).
Northern Approaches to Malacca Strait (6.61).
Through Malacca Strait (6.62).

PASSAGES

Cape Town ⇒ Durban and Moçambique Channel

Diagram 6.35
Currents and weather
6.35

1   Dominant factors are the Agulhas Current (6.28) flowing S and W with considerable strength, and the heavy seas and swells generated by S gales.

2   Counter-currents are common near the coast between Cape Agulhas (34°50′S, 20°01′E) and Durban (29°51′S, 31°06′E), though such currents are very narrow to the NE of East London (33°02′S, 27°57′E). Eddies between these counter-currents and the Agulhas Current can create local onshore sets, sometimes strong, which have been the cause of many strandings, particularly to the W of Cape Saint Francis (34°13′S, 24°50′E). See *Africa Pilot, Volume III*.

Cape Town ⇒ Durban
6.35.1

1   From Cape Town (33°53′S, 18°26′E) the route to Durban (29°51′S, 31°06′E) keeps as close to the land as safe navigation permits in order to be out of the strength of the Agulhas Current and to obtain the possible benefit of

counter-currents, while avoiding the heavy and dangerous seas and abnormal waves (1.19) which may be encountered in the vicinity of the 200 m depth contour during S and SW gales, particularly off East London (33°02′S, 27°57′E).

2   At all times great care should be taken to avoid salient points and to be vigilant against indraughts into bays. If uncertain of the position, depths of more than 75 m should be maintained.

3   Tankers navigating off the coast of South Africa are governed by regulations regarding the offing to be kept, see *Africa Pilot, Volume III*.

Traffic Separation Schemes. Two separation zone are established in the vicinity of Alphard Banks (35°03′S, 20°54′E), where there are also several established oilfields. For details, see *Africa Pilot, Volume III*.

For Moçambique Channel the route continues from Durban (29°51′S, 31°06′E) as at 6.37.1.
6.35.2
Distance
Cape Town to Durban (inshore)   800 miles

Cape Town ⇒ Moçambique Channel
6.35.3

1   From Cape Town (33°53′S, 18°26′E) to Moçambique Channel (17°00′S, 41°30′E) passage can also be made by keeping to seaward of the main part of the Agulhas Current and passing through:
36°45′S, 19°00′E **(A)**, thence:
Great circle to 34°30′S, 32°30′E **(B)**, thence:
Rhumb line, with nothing to the W, to 30°00′S, 38°20′E **(C)**, thence:
Through the E part of the Moçambique Channel, passing E of Île Europa (22°20′S, 40°20′E), thence:
Join the N-bound route (6.37.1) in 17°00′S, 42°15′E **(D)**.

6.35.4
Distance

This route is 275 miles longer than the coastal route.

Moçambique Channel ⇒ Durban or Cape Town

Diagram 6.35
Currents and weather
6.36

1 The S-bound route through the Moçambique Channel (6.37.2) is on the W side of the channel, in the Moçambique Current. Thence the Agulhas Current should be held by keeping from 20 to 30 miles from the coast as far as Mossel Bay (34°11′S, 22°09′E).

2 During SW gales off the SE African coast a very dangerous sea, with abnormal waves, will be experienced near the seaward edge of the continental shelf (1.19). There is considerably less sea near the coast and, if a distance of about 3 miles or less is kept off the shore, the reduction in the sea will more than compensate for the loss of favourable current. As directed at 6.35.1, a depth of more than 75 m should be maintained if uncertain of the position.

3 After passing Mossel Bay course should be shaped to round Cape Agulhas (34°50′S, 20°01′E) and Cape of Good Hope (34°21′S, 18°30′E).

4 Tankers navigating off the coast of South Africa are governed by regulations regarding the offing to be kept; see *Africa Pilot, Volume III.*

6.36.1
Distance

Durban to Cape Town   825 miles

Routes through Moçambique Channel

Diagram 6.35
Currents and weather
6.37

1 When choosing between a route through Moçambique Channel or one E of Madagascar, consideration should be given, not only to the navigational hazards presented by the islands and shoals in the N approaches to Moçambique Channel, but to the restrictions they impose on freedom of manoeuvre on the approach of a tropical storm, of which little warning may be expected in these waters.

2 The currents near the W coast of Madagascar are little known. In mid-channel and extending at least halfway towards Madagascar, the predominant flow is NE-going at rates of about ¾ kn, but both direction and rate are highly variable.

3 On the African side of the channel, the Moçambique Current sets strongly in a SSW direction, following the coast. In the region of Moçambique this current is thought to extend about 50 miles off the coast during most of the year, increasing to nearly 100 miles in June, July and August. It is strongest from October to February, rates of 4 kn being attained occasionally. Inshore counter-currents are common near Banco de Sofala (20°30′S, 35°30′E) and in Bahia de Maputo (26°00′S, 32°50′E).

4 The situation in Moçambique Channel, where strong SSW-going currents suddenly give place to moderate or possibly strong currents in the opposite direction, has obvious dangers. The boundaries of the currents vary with season and weather, and their rates may differ by as much as 4 kn from those anticipated.

North-bound
6.37.1

From the vicinity of Durban (29°51′S, 31°06′E), the route opens from the coast to a distance of about 100 miles and passes through:
> 27°15′S, 36°00′E (**J**), thence:
> 17°00′S, 42°15′E (**D**), passing W of Bassas da India (21°28′S, 39°46′E) or E of Île Europa (22°20′S, 40°20′E), thence:
> 11°35′S, 42°50′E (**E**), thence to destination.

The route E of Île Europa will encounter less adverse current but is 30 miles longer.

South-bound
6.37.2

From 11°35′S, 42°50′E (**E**) the route is through:
> 15°00′S, 41°20′E (**F**) in the full strength of the Moçambique Current, thence:
> 17°00′S, 40°20′E (**U**), thence:
> 25°00′S, 35°30′E (**H**), thence to destination.

Moçambique Channel ⇐ ⇒ Aden

Diagram 6.38
Currents and weather
6.38

1 The East African Current (6.28) flows continually N from Cabo Delgado (10°41′S, 40°38′E) coastwise past Mombasa (4°05′S, 39°43′E), giving way in about 2°S to the Somali Current (6.26).

2 The Somali Current is:
> SW-going from December to February at rates of about 2 kn, occasionally reaching 3 or 4 kn.
> In March the SW-going set weakens but continues S of about 4°S; N of that latitude the current turns to the NE.
> Between April and May the S part of the current turns to the NE, so that a NE-going current is established along the whole coast until September, often with rates of from 4 to 5 kn, sometimes as much as 7 kn.
> In October the NE-going current starts to weaken and gives way in November to a SW-going set between 5°N and 10°N, which starts offshore. North of 10°N, the NE-going set continues inshore until December, when it turns to establish the SW-going current along the whole coast.

3 A definite width cannot be assigned to the coastal currents between Cabo Delgado (10°41′S, 40°38′E) and Raas Caseyr (11°50′N, 51°17′E). The NE-going current, which reaches its full strength between June and September, is stronger nearer the coast and decreases rapidly at a distance of over 50 miles offshore. South-bound shipping will therefore benefit by keeping a good offing.

Rounding Raas Caseyr
6.38.1

1 Passage may be made either:
> N of Suquṭrá (12°30′N, 54°00′E), giving that island a berth of at least 40 miles, or:
> Between ‘Abd al Kūrī (12°10′N, 52°15′E) and Ras Caseyr (11°50′N, 51°17′E).

The latter is the more usual route, particularly during the south-west monsoon, when better conditions of wind and sea can be expected S of Suquṭrá.

2 Ships, not fitted with radar or other electronic position fixing systems, and unsure of their position, should take the route N of Suquṭrá where there is ample searoom even if more stormy. See, 6.38.2.

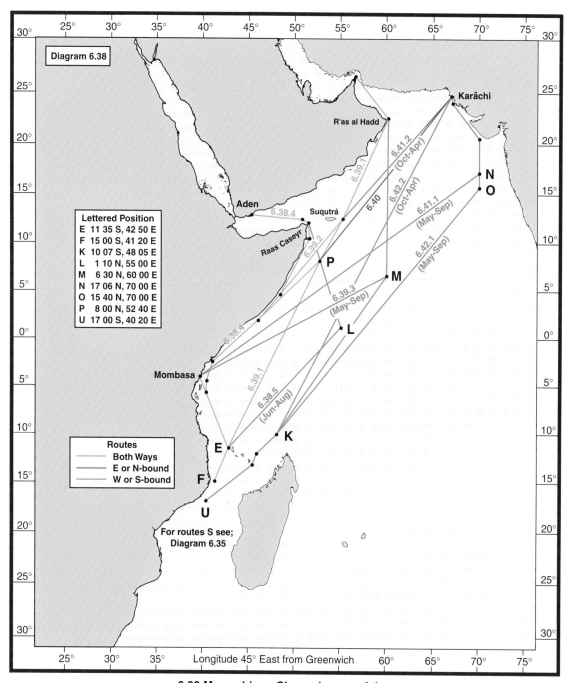

6.38 Mozambique Channel ← → Aden
6.39 African Coast ← → Persian Gulf
6.40 African Coast → Karāchi
6.41 Karāchi → Mombasa
6.42 Karāchi → Mozambique Channel, Durban and Cape Town

Suquṭrá
6.38.2

1   Owing to the imperfect nature of the survey, navigation in the vicinity of Suquṭrá (12°30′N, 54°00′E) must be undertaken with caution.

2   It is dangerous for vessels not fitted with radar, or other electronic position fixing systems, to make Rhiy di-Irīsal (12°35′N, 54°29′E), the E point of Suquṭrá during either monsoon.

> In the north-east monsoon the land may be obscured, about sunset, by heavy rain squalls.

In the south-west monsoon the lower land E of the mountain range is often obscured by haze.

3   Depths off Rhiy di-Irīsal are considerable and sounding gives no warning of the dangers which extend from the shore. Currents in the vicinity are strong and irregular.

Raas Caseyr
6.38.3

1   Raas Caseyr (11°50′N, 51°17′E) and the coastline in its vicinity are charted from old and imperfect surveys and should be approached with caution.

2   Many wrecks have occurred on the coast S of Raas Caseyr and great caution is necessary when steering NW and N towards and past this headland in the south-west monsoon. The weather and sea are then at their worst, the N-going current at its strongest, and the land generally covered by thick haze.

3   A resemblance exists between the profiles of Raas Caseyr and Raas Shannaqiif, 10 miles SSW, but Raas Shannaqiif is 927 m high while Raas Caseyr is only 238 m high. They are separated by a broad, comparatively low, sandy plain. In hazy weather, the steep fall of Raas Shannaqiif may perhaps be dimly seen when it bears less than about 270°. If Raas Caseyr has not been sighted, as often happens if the haze is thicker near sea level and obscures that light-coloured hill, then Raas Shannaqiif may be mistaken for Raas Caseyr with disastrous results.

4   In the south-west monsoon Raas Xaafuun (10°26′N, 51°25′E) should be made before Raas Caseyr.

5   By day there is usually a gradual change in colour of the water from blue to dark green as the land is approached. The sea also becomes smoother and swell tends to come from E of S when N and W of Raas Xaafuun.

6   When making Raas Caluula (11°59′N, 50°47′E) and Raas Caseyr from the Gulf of Aden, allowance must be made for the possibility of a SW or onshore set, particularly during the north-east monsoon.

Routes
6.38.4

1   To summarise, for passages between the Moçambique Channel and Gulf of Aden, the normal route in both directions passes between 'Abd al Kūrī (12°10′N, 52°15′E) and Ras Caseyr (11°50′N, 51°17′E). Between intermediate coastal destinations passages should normally be made coastwise in both directions.

South-bound routes
6.38.5

1   When S-bound, the strongest effects of the south-west monsoon (June to August) and of the NE-going current between Ras Caseyr (11°50′N, 51°17′E) and Comores (12°00′S, 44°30′E) may be avoided by passing:
   Through 8°00′N, 52°40′E (P), thence:
   1°10′N, 55°00′E (L), thence:
   Joining the route through Moçambique Channel in 11°35′S, 42°50′E (E).
This adds 220 miles to the distance between Ras Caseyr and Comores.

Distances
6.38.6

N-bound

Cape Town	Durban		
2620 (6.37.1)†*	1830 (6.37.1)	Mombasa	
4020†*	3230	1620	Aden

   † For Cape Agulhas (15 miles S of) subtract 130 miles
   * For Cape of Good Hope (145 miles S of) add 95 miles

S-bound

Cape Town	Durban		
2580 (6.37.2)†	1770 (6.37.2)	Mombasa	
3970†	3180	1620	Aden

   † For Cape Agulhas (15 miles S of) subtract 130 miles

African Coast ⇐ ⇒ Persian Gulf

Diagram 6.38
Routes
6.39

   The routes described are to and from the S and E coasts of Africa and the Straits of Hormuz.

From ports S of Comores
6.39.1

   Routes are as at 6.35 and 6.37.1:
   To 11°35′S, 42°50′E (E), thence:
   Through 8°00′N, 52°40′E (P), thence:
   As navigation permits to the TSS through Strait of Hormuz, passing at least 50 miles off Suquṭrá (12°30′N, 54°00′E).

From Mombasa
6.39.2

   The route follows the trend of the African coast and joins route 6.39.1 at 8°00′N, 52°40′E (P).

South-bound routes
6.39.3

   During the south-west monsoon (May to September), an alternative route is to cross the monsoon area, with its adverse currents, as quickly as possible.
   From the TSS off Ra's al Hadd (22°30′N, 59°48′E) the route is:
   S along the meridian of 60°E to 6°30′N, 60°00′E (M), thence:
Either
   Direct to Mombasa (4°05′S, 39°43′E), or
   Join the route from Karāchi to Moçambique Channel (6.42) in 10°07′S, 48°05′E (K).

Distances
6.39.4

   Distances from Strait of Hormuz:

	N-bound	S-bound	S-bound (SW Monsoon)
Cape Town	4700†*	4650†	4830†
Durban	3910	3850	4040
Mombasa	2330	2330	2630

   † For Cape Agulhas (15 miles S of) subtract 130 miles
   * For Cape of Good Hope (145 miles S of) add 95 miles

African coast ⇒ Karāchi

Diagram 6.38
Routes
6.40

   From the African coast the routes are as at 6.39.1 and 6.39.2:
   To 8°00′N 52°40′E (P), thence:
   Rhumb line to Karāchi (24°46′N, 66°57′E).

Distances
6.40.1

   Distances from Karāchi:
   Cape Town†*   4730 miles
   Durban   3940 miles
   Mombasa   2360 miles
   † For Cape Agulhas (15 miles S of) subtract 130 miles
   * For Cape of Good Hope (145 miles S of) add 95 miles

Karāchi ⇒ Mombasa

Diagram 6.38
Routes
6.41

Routes are seasonal, as follows:

May to September (SW Monsoon)
6.41.1

Route is as follows:
Parallel to the Indian coast to 70°E, thence:
Due S to 17°06′N, 70°00′E (**N**), thence:
Rhumb line to Mombasa (4°05′S, 39°43′E).
Distance   2720 miles.

October to April (NE Monsoon)
6.41.2

Route is direct by rhumb line.
Distance   2350 miles.

Low-powered vessels (SW Monsoon)
6.41.3

Low-powered, or hampered, vessels may find the route described at 9.145 more suitable for their use.

Karāchi ⇒ Moçambique Channel, Durban and Cape Town

Diagram 6.38
Routes
6.42

Routes are seasonal, as follows:

May to September (SW Monsoon)
6.42.1

Route is as follows:
Parallel to the Indian coast to 70°E, thence:
Due S to 15°40′N, 70°00′E (**O**), thence:
Direct to 10°07′S, 48°05′E (**K**), 20 miles E of Astove Island (10°04′S, 47°43′E), bearing in mind the strong W-going set of the Equatorial Current in that region, thence:
20 miles SE of Île Mayotte (12°50′S, 45°10′E), passing W of Îles Glorieuses (11°35′S, 47°20′E) and Récif du Geyser (12°22′S, 46°25′E), thence:
To 17°00′S 40°20′E (**U**), on the S-bound track through the Moçambique Channel (6.37.2), thence:
As at (6.37.2) and (6.36) to destination.
Caution. The part of the voyage between Astove Island and Île Mayotte should be undertaken by day, if possible, see *South Indian Ocean Pilot*.
Distances.
Durban   4180 miles.
Cape Town†   4970 miles.
† For Cape Agulhas (15 miles S of) subtract 130 miles

October to April (NE Monsoon)
6.42.2

Route is:
Direct to 10°07′S, 48°05′E (**K**) to join route 6.42.1, thence:
As at 6.42.1 to destination.
Distances.
Durban   3920 miles.
Cape Town†   4710 miles.
† For Cape Agulhas (15 miles S of) subtract 130 miles

Low-powered vessels (SW Monsoon)
6.42.3

Low-powered, or hampered, vessels may find the route described at 9.146 more suitable for their use.

Aden ⇐ ⇒ Persian Gulf

Diagram 6.43
Routes
6.43

Routes are given for E-bound and W-bound vessels.

East-bound
6.43.1

1   The route through the Arabian Sea keeps as close as practicable to the Arabian coast, having regard to the variability of the current, and avoiding a close approach to the Kalīj Maşīrah (19°30′N, 58°00′E), see *Red Sea and Gulf of Aden Pilot*.

2   From the TSS off R'as al Hadd (22°30′N, 59°48′E) the route is as direct as navigation permits to the TSS through the Strait of Hormuz (26°27′N, 56°32′E).

3   During the south-west monsoon the weather is generally very hazy along the Arabian coast so that, though the sky may be clear, the land may not be visible until close inshore.

4   Distance   1420 miles.

West-bound
6.43.2

1   After passing Muscaţ (23°37′N, 58°35′E), the route keeps close to the Arabian coast as navigation permits. This is especially advisable during the south-west monsoon, when the full force of the wind, and the NE-going set, will be felt only in the vicinity of Juzur al Ḩalāniyāt (Kuria Muria Islands) (17°30′N, 56°05′E) and of R'as al Kalb (14°02′N, 48°41′E).

Aden ⇐ ⇒ Karāchi

Diagram 6.43
Routes
6.44

In either direction, routes follow those at 6.43 between Aden (12°45′N, 44°57′E) and Juzur al Ḩalāniyāt (Kuria Muria Islands) (17°30′N, 56°05′E), thence direct to and from Karāchi (24°46′N, 66°57′E).
Distance   1460 miles

West-bound
6.44.1

In the south-west monsoon, an alternative route is by keeping parallel to the Indian coast to 70°E, thence:
Due S to 12°50′N, 70°00′E (**E**), thence:
Direct to 13°00′N, 55°00′E (**Q**), thence:
Direct to Aden.
This route is about 800 miles longer, but largely avoids the head wind and heavy seas.

Cape Town and Durban ⇒ Mumbai

Diagram 6.45
Routes
6.45

The routes from Cape Town (33°53′S, 18°26′E) and Durban (29°51′S, 31°06′E) can pass E or W of Madagascar.
Oilfields. For oilfields seaward of Mumbai (18°51′N, 72°50′E), see 6.32.

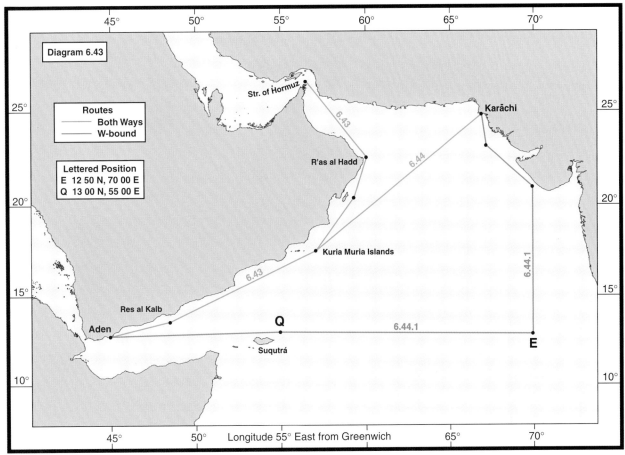

6.43 Aden ← → Persian Gulf
6.44 Aden ← → Karāchi

West of Madagascar
6.45.1
From Cape Town and Durban
Routes are as at 6.35.1; 6.35.3 and 6.37.1 as far as 17°00′S, 42°15′E (**D**) in the Moçambique Channel, thence:

Between Île Anjouan (12°15′S, 44°30′E) and Île Mayotte (12°50′S, 45°10′E), thence:

To 9°30′S, 45°30′E (**L**), 30 miles W of Aldabra Group (9°25′S, 46°20′E), thence:

Direct to Mumbai (18°51′N, 72°50′E).

By day, after passing Île Anjouan, a route may be taken between Assumption Island (9°45′S, 46°30′E) and Cosmoledo Group (9°45′S, 47°35′E) and thence to Mumbai.

When uncertain of the position, the route W of Aldabra Group should always be taken, owing to the strength and variability of the W-going current in the locality.

From Cape of Good Hope
Route is as at 6.35.3 to join 6.45.1 above in 17°00′S, 42°15′E (**D**) in the Moçambique Channel.

Distances to Mumbai.
Cape Town† (coastwise)	4650 miles
Cape of Good Hope (145 miles S of)	4745 miles
Durban	3860 miles

† For Cape Agulhas (15 miles S of) subtract 130 miles

East of Madagascar
6.45.2
From Cape Town (33°53′S, 18°26′E)
The route is as at 6.35.1 as far as Great Fish Point (33°32′S, 27°07′E), thence:

Round the S end of Madagascar at a distance of 60 miles or more offshore, to seaward of the strongest part of the Madagascar Current (6.28), thence:

Direct to Mumbai (18°51′N, 72°50′E), passing W of Saya de Malha Bank (10°00′S, 61°00′E), but giving Agalega Islands (10°26′S, 56°39′E) a wide berth.

From Durban (29°51′S, 31°06′E)
The route runs direct to join that from Cape Town S of Madagascar in 26°45′S, 47°45′E (**M**), keeping the same distance off that island.

From Cape of Good Hope
Route is:

Great circle to 34°30′S, 32°30′E (**B**), thence:

Great circle to join the route from Cape Town S of Madagascar in 26°45′S, 47°45′E (**M**).

Distances to Mumbai.
Cape Town† (coastwise)	4700 miles
Cape of Good Hope (145 miles S of)	4680 miles

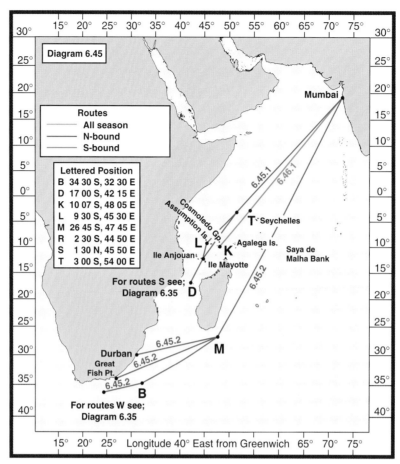

6.45 Cape Town and Durban → Mumbai
6.46 Mumbai → Cape Town and Durban

Durban   3950 miles
† For Cape Agulhas (15 miles S of) subtract 130 miles

Mumbai ⇒ Durban or Cape Town

Diagram 6.45
General information
6.46

The routes to Cape Town (33°53′S, 18°26′E) and Durban (29°51′S, 31°06′E) pass W of Madagascar.

Oilfields. For oilfields seaward of Mumbai, see 6.32.

Routes
6.46.1

Routes are by rhumb line:
> To 10°07′S, 48°05′E (**K**), 20 miles E of Astove Island, thence:
> As at 6.42.1 to destination.

Distances from Mumbai.

Cape Town   4620 miles
For Cape Agulhas (15 miles S of)   4490 miles
Durban   3830 miles

South-west Monsoon (low-powered vessels)
6.46.2

Low-powered, or hampered, vessels, in the full strength of the south-west monsoon, may find the route described at 9.154 more suitable for their use.

Cape Town and Durban ⇒ Colombo or Bay of Bengal

Diagram 6.47 (for Colombo), 6.49 (for Bay of Bengal)
Routes
6.47

The routes from Cape Town (33°53′S, 18°26′E) and Durban (29°51′S, 31°06′E), which are seasonal, can pass W of Madagascar and through the One and Half Degree Channel (1°24′N, 73°20′E), or E of Madagascar.

West of Madagascar to One and Half Degree Channel
6.47.1

Routes are as at 6.35.1, 6.35.3 as far as 17°00′S, 42°15′E (**D**) off Île Juan de Nova in the Moçambique Channel, thence seasonal:

(a) April to October
> Through 8°30′S, 50°40′E (**N**), 30 miles NW of Wizard Reef, passing 30 miles E of Récif du Geyser (12°22′S, 46°25′E) and Îles Glorieuses (11°35′S, 47°20′E), thence:
> To One and Half Degree Channel (1°24′N, 73°20′E) (**J**), taking care to avoid the islands and shoal water at the SE extremity of Seychelles Bank (5°00′S, 56°00′E).

Currents. Attention must be paid to the currents S of 5°S, especially near Wizard Reef (8°50′S, 51°04′E), where they will probably be NW-going and to the W-going current which sets strongly past Geyser Reef and Îles Glorieuses. See *South Indian Ocean Pilot*.

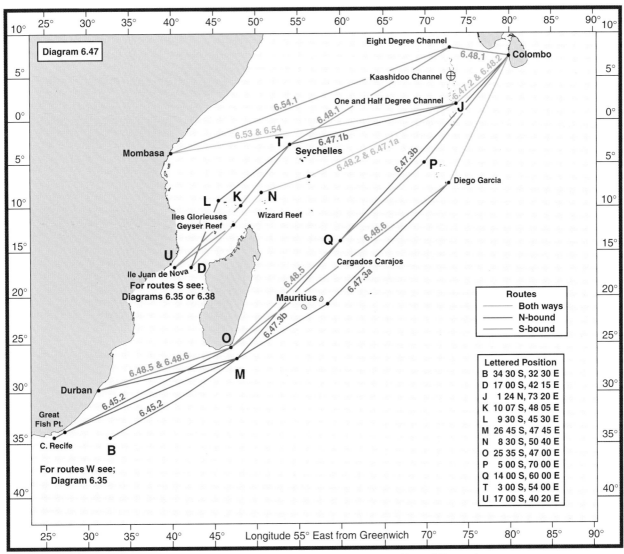

6.47 Cape Town and Durban → Colombo or Bay of Bengal
6.48 Colombo → Cape Town or Durban
6.53 Mombasa → Colombo or Dondra Head
6.54 Colombo or Dondra Head → Mombasa

(b) November to March

Between Île Anjouan (12°15′S, 44°30′E) and Île Mayotte (12°50′S, 45°10′E), thence:

9°30′S, 45°30′E **(L)**, 30 miles W of Aldabra Group, thence:

3°00′S, 54°00′E **(T)**, 50 miles NW of Seychelles Group, thence:

To One and Half Degree Channel (1°24′N, 73°20′E) **(J)**.

Current. Attention must be paid to the W-going current which flows strongly past the N point of Madagascar.

From One and Half Degree Channel
6.47.2

Routes, at all seasons, are as direct as navigation permits to destinations.

To Yangon (Rangoon) (16°46′N, 96°11′E) the shortest route is through Preparis South Channel (14°33′N, 93°27′E) (Diagram 6.49).

East of Madagascar
6.47.3

Routes are as at 6.45.2 as far as 26°45′S, 47°45′E **(M)**

having rounded the S end of Madagascar, at a distance of 60 miles or more, thence seasonal:

(a) May to September

Passing 60 miles SE of Mauritius (20°10′S, 57°30′E), thence:

E of Diego Garcia (7°13′S, 72°23′E), thence:

As navigation permits to destination.

For Yangon (Rangoon) (16°46′N, 96°11′E) the Ten Degree Channel (10°00′N, 92°30′E) should be used (Diagram 6.49).

(b) October to March

Through 14°00′S, 60°00′E **(Q)**, thence:

As navigation permits to destination.

For Yangon (Rangoon) (16°46′N, 96°11′E) the Preparis South Channel (14°33′N, 93°27′E) should be used (Diagram 6.49).

(c) April

In April either route may be used.

(d) From Cape of Good Hope

Route is:

Great circle to 34°30′S, 32°30′E **(B)**, thence:

Great circle to join the routes passing E of Madagascar in 26°45′S, 47°45′E **(M)**.

Distances
6.47.4

West of Madagascar

	April to October		November to March	
	Cape Town*	Durban	Cape Town	Durban
Colombo	4510[a]	3710	4640[a]	3850
Chennai	5030[a]	4240	5160[a]	4370
Paradip	5480[a]	4690	5620[a]	4830
Kolkata (Calcutta) Approach†	5560[a]	4770	5690[a]	4900
Yangon (Rangoon) River Entrance‡	5660[a]	4870	5790[a]	5000

East of Madagascar

	May to September		October to March	
	Cape Town*	Durban	Cape Town	Durban
Colombo	4450[b]	3690	4380[b]	3620
Chennai	4930[b]	4170	4880[b]	4130
Paradip	5360[b]	4600	5310[b]	4560
Kolkata (Calcutta) Approach†	5430[b]	4670	5390[b]	4630
Yangon (Rangoon) River Entrance‡	5520[b]	4770	5470[b]	4740

*   For Cape Agulhas (15 miles S of) subtract 130 miles
†   Kolkata (Calcutta) Approach to Kolkata 125 miles
‡   Yangon (Rangoon) River Entrance to Yangon 40 miles
[a] For Cape of Good Hope (145 miles S of) add 95 miles
[b] For Cape of Good Hope (145 miles S of) subtract 20 miles

Colombo ⇒ Durban or Cape Town

Diagram 6.47 (for Colombo), 6.49 (for Bay of Bengal)
Routes
6.48

The routes to Cape Town (33°53′S, 18°26′E) and Durban (29°51′S, 31°06′E) can pass E or W of Madagascar and some routes are seasonal.

West of Madagascar
6.48.1

In all seasons routes are:
Through Eight Degree Channel (7°24′N, 73°00′E), thence:
To 3°00′S 54°00′E (T), NW of Seychelles Group, thence:
To 10°07′S, 48°05′E (K), 20 miles E of Astove Island, thence:
As at 6.42.1 to destination.

Distances from Colombo
Durban   3820 miles
Cape Town   4610 miles
For Cape Agulhas (15 miles S of)   4480 miles
6.48.2

Alternative route in April and October
In April and October, an alternative route is:
Through One and Half Degree Channel (1°24′N, 73°20′E), thence:
To 8°30′S, 50°40′E (N), 30 miles NW of Wizard Reef, taking care to avoid the islands and shoal water at the SE extremity of Seychelles Bank, thence:
Passing 30 miles E of Îles Glorieuses (11°35′S, 47°20′E) and Récif du Geyser (12°22′S, 46°25′E), to join the S-bound route through Moçambique Channel in 17°00′S, 40°20′E (U), thence:
As at 6.37 and 6.36 to destination.

Distances from Colombo
Durban   3730 miles
Cape Town   4520 miles
For Cape Agulhas (15 miles S of)   4390 miles
6.48.3

Low-powered vessels in south-west monsoon
Low-powered, or hampered, vessels, in the full strength of the south-west monsoon, may find the route described at 9.163 more suitable for their use.

East of Madagascar
6.48.4

Routes are seasonal.
6.48.5

November to March
Route leads:
SW to 5°00′S, 70°00′E (P), passing S of Maldives (4°00′N, 73°00′E) and W of Chagos Archipelago (6°30′S, 72°00′E), thence:
To 14°00′S, 60°00′E (Q), thence:
Passing 20 miles SE of Madagascar (O), to take advantage of the Madagascar Current (6.28), thence:
Either:
Direct to Durban (29°51′S, 31°06′E).
Or:
If bound for ports farther W, to a landfall off Cape Recife (34°02′S, 25°42′E), thence:
As for route 6.36, keeping in the Agulhas Current.
Distances from Colombo.
Durban   3620 miles
Cape Town   4360 miles
For Cape Agulhas (15 miles S of)   4230 miles
6.48.6

April to October
Routes leads:
East of Diego Garcia (7°13′S, 72°23′E), thence:

20 miles SE of Madagascar (O), giving Cargados Carajos Shoals (16°35′S, 59°40′E) a wide berth, thence:

Either:

Direct to Durban (29°51′S, 31°06′E).

Or:

If bound for ports farther W, to a landfall off that port, thence:

Coastwise to destination, in order to avoid the heavy weather prevalent to seaward at this season (6.36).

Distances from Colombo.

Durban   3670 miles

Cape Town   4460 miles

For Cape Agulhas (15 miles S of)   4330 miles

Bay of Bengal ⇒ Durban or Cape Town

Diagram 6.49
Routes
6.49

The routes to Cape Town (33°53′S, 18°26′E) and Durban (29°51′S, 31°06′E) can pass E or W of Madagascar and some routes are seasonal.

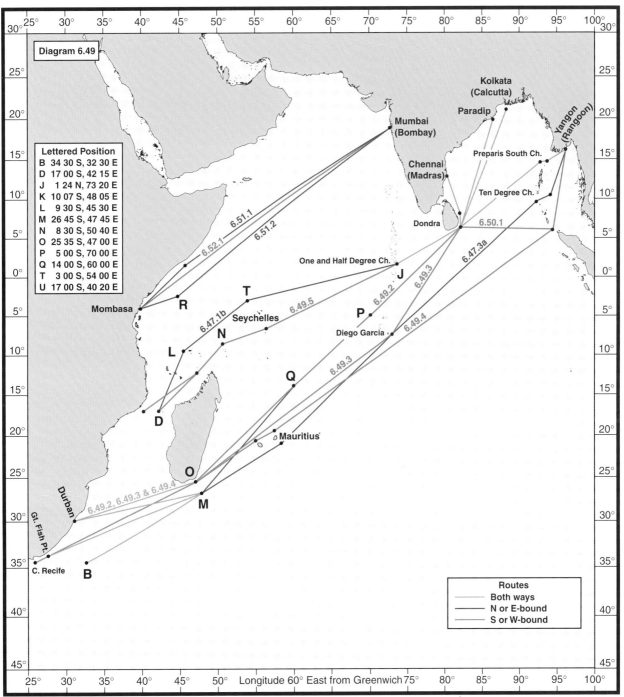

6.49 Bay of Bengal → Durban or Cape Town
6.50 Bay of Bengal
6.51 Mombasa → Mumbai
6.52 Mumbai → Mombasa

East of Madagascar

6.49.1

Routes are seasonal.

6.49.2

November to March

Route leads:

W of Chagos Archipelago (6°30′S, 72°00′E), thence:

Through 5°00′S, 70°00′E (**P**), thence:

As at 6.48.5 to destination.

6.49.3

May to September

Routes leads:

East of Diego Garcia (7°13′S, 72°23′E), thence:

As at 6.48.6 to destination.

6.49.4

June to August

An alternative route from Yangon (Rangoon) (16°46′N, 96°11′E) leads:

S of Great Nicobar Island (7°00′N, 93°50′E), thence:

By rhumb line passing N of Mauritius (20°10′S, 57°30′E) to join the seasonal routes from Colombo (6.48.6), 20 miles SE of Madagascar (**O**).

West of Madagascar

6.49.5

Only in April and October is the route through the Moçambique Channel advised. Routes from Bay of Bengal pass:

Through the TSS off Dondra Head (5°45′N, 80°36′E), thence:

Through One and Half Degree Channel (1°24′N, 73°20′E) (**J**), thence:

As at 6.48.2.

Distances

6.49.6

W of Chagos Archipelago

	Cape Town*	Durban
Chennai	4870	4130
Paradip	5300	4560
Kolkata (Calcutta) Approach†	5380	4630
Yangon (Rangoon) River Entrance‡	5470	4730

E of Chagos Archipelago

	Cape Town*	Durban
Chennai	4950	4160
Paradip	5370	4580
Kolkata (Calcutta) Approach†	5450	4660
Yangon (Rangoon) River Entrance‡	5530	4740

Moçambique Channel

	Cape Town*	Durban
Chennai	5050	4260
Paradip	5480	4690
Kolkata (Calcutta) Approach†	5550	4760
Yangon (Rangoon) River Entrance‡	5650	4860

* For Cape Agulhas (15 miles S of) subtract 130 miles.

† Kolkata (Calcutta) Approach to Kolkata 125 miles.

‡ Yangon (Rangoon) River Entrance to Yangon 40 miles.

Bay of Bengal

Diagram 6.49

Routes

6.50

Routes in the Bay of Bengal offer little opportunity for diversion, even at the cost of distance, to reduce the adverse effects or to take advantage of wind and current.

6.50.1

Yangon (Rangoon) (16°46′N, 96°11′E) to position 5°45′N, 80°36′E, 10 miles S of Dondra Head, is the only route which may be diverted to advantage, where the full strength of the south-west monsoon in June, July and August may be avoided by a route:

From Yangon (Rangoon), thence:

Through 5°50′N, 94°30′E, passing E of Andaman Islands (12°30′N, 93°00′E) and Nicobar Islands (8°00′N, 94°00′E), thence:

Direct to the TSS off Dondra Head.

For details of the route from the Bay of Bengal to Singapore (1°12′N, 103°51′E), via the Malacca Strait, see 6.61, 6.62 and 6.94.2.

6.50.2

Distances

Dondra Head					
550	**Chennai**				
980	570	**Paradip**			
1060	650	110	**Kolkata (Calcutta) Approach†**		
1160	985	680	640	**Yangon (Rangoon) River‡**	
1470	1580	1540	1520	1060	**Singapore**

† Kolkata (Calcutta) Approach to Kolkata 125 miles.

‡ Yangon (Rangoon) River Entrance to Yangon 40 miles.

Mombasa ⇒ Mumbai

Diagram 6.49
Routes
6.51

Routes are seasonal.

Oilfields. For oilfields seaward of Mumbai (18°51′N, 72°50′E), see 6.32.

May to September
6.51.1

The route follows the trend of the African coast, taking advantage of the East African (6.28) and Somali Currents (6.26), both NE-going at that season:

To 1°30′N, 45°50′E, thence:
Rhumb line to Mumbai.
Distance 2400 miles.

October to April
6.51.2

The route is:
Through 2°30′S, 44°50′E **(R)**, thence:
Rhumb line to Mumbai.
Distance 2410 miles.

Mumbai ⇒ Mombasa

Diagram 6.49
Caution
6.52

Oilfields. For oilfields seaward of Mumbai (18°51′N, 72°50′E), see 6.32.

Routes
6.52.1

Route, at all seasons, for fully-powered vessels is direct.
Distance 2390 miles.

Low-powered vessel route
6.52.2

Low-powered, or hampered, vessels, during the full strength of the south-west monsoon, may find the route described at 9.152 more suitable for their use.

Mombasa ⇒ Colombo or Dondra Head

Diagram 6.47
Routes
6.53

At all seasons the route is route is through either One and Half Degree Channel (1°24′N, 73°20′E) or Kaashidoo Channel (5°05′N, 73°24′E). If the position is in doubt, the former is preferable as the W entrance to Kaashidoo Channel is not easily identified. See *West Coast of India Pilot*.

Distances
Colombo	2560 miles
Dondra Head*	2550 miles

  * 10 miles S of.

Colombo or Dondra Head ⇒ Mombasa

Diagram 6.47
Routes
6.54

At all seasons the route is through One and Half Degree Channel (1°24′N, 73°20′E).

Kaashidoo Channel (5°05′N, 73°24′E) can also be used providing Olivelifuri Islet, which marks the N side of the entrance to the channel, can be made between sunrise and noon.

Distances
Colombo	2560 miles
Dondra Head*	2550 miles

  * 10 miles S of.

Alternative route
6.54.1

In October to April an alternative route is through Eight Degree Channel (7°24′N, 73°00′E).

Low-powered vessels
6.54.2

Low-powered, or hampered, vessels may find the route described at 9.161 more suitable for their use.

Aden ⇒ Mumbai

Diagram 6.55
Routes
6.55

Routes are seasonal.

Oilfields. For oilfields seaward of Mumbai (18°51′N, 72°50′E), see 6.32.

October to April
6.55.1

The route is direct.
Distance 1650 miles.

May to September
6.55.2

The route is:
Through 13°00′N, 55°00′E **(Q)**, thence:
Rhumb line to Mumbai.
Distance 1680 miles.

Mumbai ⇒ Aden

Diagram 6.55
Routes
6.56

Routes are seasonal.

Oilfields. For oilfields seaward of Mumbai (18°51′N, 72°50′E), see 6.32.

October to April
6.56.1

The route is direct.
Distance 1650 miles.

May to September
6.56.2

During the south-west monsoon the best route depends on the strength of the monsoon.

In May and September.
The route is through:
19°00′N, 70°00′E **(A)**, thence:
18°30′N, 65°00′E **(B)**, thence:
17°30′N, 60°00′E **(C)**, thence:
As navigation permits to Aden (12°45′N, 44°57′E).
Distance 1660 miles.

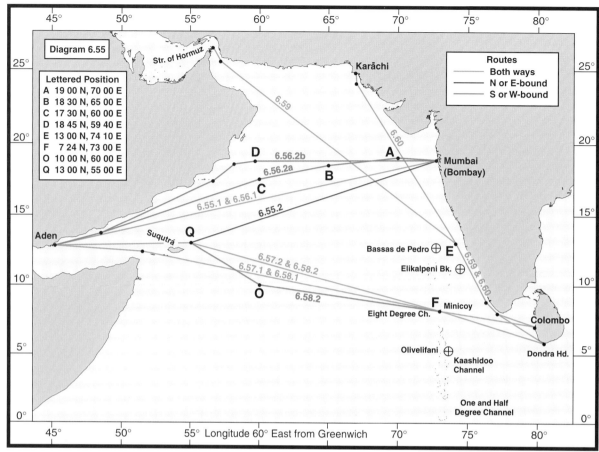

Diagram 6.55

Lettered Position
A 19 00 N, 70 00 E
B 18 30 N, 65 00 E
C 17 30 N, 60 00 E
D 18 45 N, 59 40 E
E 13 00 N, 74 10 E
F 7 24 N, 73 00 E
O 10 00 N, 60 00 E
Q 13 00 N, 55 00 E

6.55 Aden → Mumbai
6.56 Mumbai → Aden
6.57 Aden → Colombo or Dondra Head
6.58 Colombo or Dondra Head → Aden
6.59 Strait of Hormuz ← → Colombo or Dondra Head
6.60 Karachi and Mumbai ← → Colombo or Dondra Head

In June, July and August.

During these months the monsoon is at its strongest, the route follows:

> The parallel of Mumbai to 18°45′N, 59°40′E (**D**), about 100 miles from the Arabian Coast, thence:
> Following the coast as near as navigation permits, as at 6.43.2.

During the south-west monsoon the wind and sea are at their height between 66°E and 60°E. The adverse current may attain a rate of 2 kn in the middle of the Arabian Sea and occasionally 3 kn in the W part of that sea and off the Arabian coast.

Distance   1680 miles.

Low-powered vessel route
6.56.3

Low-powered, or hampered, vessels may find the route described at 9.151 more suitable for their use.

Aden ⇒ Colombo or Dondra Head

Diagram 6.55
Routes
6.57

Routes are seasonal.

In the Gulf of Aden and N of Suquṭrá (12°30′N, 54°00′E), allowance must be made for the possibility of a set towards the S shore.

Mariners using Eight Degree Channel (7°24′N, 73°00′E) should keep nearer to Minicoy Island than to the Maldives; see *West Coast of India Pilot.*

October to April
6.57.1

The route is:
> Between Raas Caseyr (11°50′N, 51°17′E) and ʼAbd al Kūrī (12°10′N, 52°15′E) (6.38.3), thence:
> Through Eight Degree Channel (7°24′N, 73°00′E) (**F**).

Distances
Colombo   2080 miles.
Dondra Head*   2140 miles.
* (10 miles S of).

May to September
6.57.2

To avoid the heavy cross-sea S of Suquṭrá (12°30′N, 54°00′E), caused by the south-west monsoon, the route is:
> Through 13°00′N, 55°00′E (**Q**), passing N of Suquṭrá, thence:

Either:
> Direct to destination.

Or:
> Through Eight Degree Channel (7°24′N, 73°00′E) (**F**), thence to destination.

Distances

Colombo	2110 miles.
Dondra Head*	2170 miles.

* (10 miles S of).

Colombo or Dondra Head ⇒ Aden

Diagram 6.55
Routes
6.58

Routes are seasonal.

October to April
6.58.1

The routes are:
Through Eight Degree Channel (7°24′N, 73°00′E) **(F)**, thence:
(a) Either:
S of Suquṭrá, thence:
Between Raas Caseyr (11°50′N, 51°17′E) and 'Abd al Kŭrī (12°10′N, 52°15′E), bearing in mind the difficulty of identifying the landfall (6.38.3), thence:
Direct to Aden (12°45′N, 44°57′E).
(b) Or:
Through 13°00′N, 55°00′E **(Q)**, thence:
Passing N of Suquṭrá, thence:
Direct to Aden (12°45′N, 44°57′E).

Distances (S of Suquṭrá)

Colombo	2080 miles.
Dondra Head*	2140 miles.

* (10 miles S of).

Distances (N of Suquṭrá)

Colombo	2110 miles.
Dondra Head*	2170 miles.

* (10 miles S of).

May to September
6.58.2

The route is:
Through Eight Degree Channel (7°24′N, 73°00′E) **(F)**, thence:
To 10°00′N, 60°00′E **(O)**, thence:
To 13°00′N, 55°00′E **(Q)**, thence:
N of Suquṭrá to Aden (12°45′N, 44°57′E).

Distances

Colombo	2130 miles.
Dondra Head*	2190 miles.

* (10 miles S of).

Low-powered vessel route
6.58.3

Low-powered, or hampered, vessels may find the routes described at 9.160 more suitable for their use.

Strait of Hormuz ⇐ ⇒ Colombo or Dondra Head

Diagram 6.55
Routes
6.59

The route from the TSS through the Strait of Hormuz (26°27′N, 56°32′E) is:
Direct as navigation permits to 13°00′N, 74°10′E **(E)**, off the Malabar coast to avoid Bassas de Pedro

(13°00′N, 72°25′E) and the shoals E of Lakshadweep, thence:
To destination.

Distances

Colombo	1800 miles.
Dondra Head*	1880 miles.

* (10 miles S of).

Karāchi and Mumbai ⇐ ⇒ Colombo or Dondra Head

Diagram 6.55
Routes
6.60

Routes are direct as navigation permits.
Oilfields. For oilfields seaward of Mumbai (18°51′N, 72°50′E), see 6.32.
Currents. For the possibility of onshore sets, see *West Coast of India Pilot*.
Distances.

	Mumbai	**Colombo**	**Dondra Head***
Karāchi	500	1340	1420
Mumbai	—	885	960

* (10 miles S of).

Northern approaches to Malacca Strait

Principal routes
6.61

1 Malacca Strait may be approached from the Indian Ocean by three routes:
The Great Channel which passes N of Pulau Rondo (6°04′N, 95°07′E).
Between Pulau Rondo and Pulau We (5°50′N, 95°18′E).
Through Selat Benggala between Pulau We and Pulau Breueh (5°42′N, 95°04′E), and continuing through Alur Pelayaran Malaka, but on this route, which was previously recommened, shoal patches have been reported and the Great Channel is preferred.
2 For details of these routes; see *Malacca Strait and West Coast of Sumatera Pilot*.

Malacca Strait

Chart 5502
General notes
6.62

1 Malacca Strait and Singapore Strait together form the main seaway used by vessels from Europe and India bound for Malaysian ports, Singapore and ports farther NE. They also provide the shortest route for ships trading between the Persian Gulf and Japan.
2 The least depth in the fairway is about 25 m, but there are many areas of sandwaves. Depths and the configuration of the channel are liable to change. Deep-draught vessels should therefore take particular note of the latest reports of depths in, or near, the fairway.
3 Tidal streams are strong. Navigational aids are difficult to maintain and may be unreliable.
4 Local fishing craft with nets may be encountered in Malacca Strait.

5   Rules for vessels navigating through Malacca Strait and Singapore Strait are given in *Malacca Strait and West Coast of Sumatera Pilot* and Chart *5502 Mariner's Routeing Guide — Malacca and Singapore Straits* should be consulted.

6   These factors and the density of traffic make navigation through the straits difficult, particularly for deep-draught vessels. The long run of more than 250 miles through the straits demands long periods of considerable vigilance.

Distance.

    Great Channel to Singapore*   615 miles.

PASSAGES TO AND FROM MAURITIUS (PORT LOUIS)

GENERAL INFORMATION

Coverage
6.63

    This section covers details of the following passages:

    To and from Port Louis and Cape Town or Durban (6.64).

    To and from Port Louis and ports in Moçambique Channel (6.65).

    To and from Port Louis and Mombasa (6.66).

    To and from Port Louis and Aden (6.67).

    To and from Port Louis and Karāchi (6.68).

    To and from Port Louis and Mahé Island (6.69).

    To and from Port Louis and Mumbai (6.70).

    To and from Port Louis and Colombo (6.71).

    To and from Port Louis and Singapore (6.72).

    To and from Port Louis and Selat Sunda (6.73).

    To and from Port Louis and Torres Strait or Darwin (6.74).

    To and from Port Louis and Fremantle or Cape Leeuwin (6.75).

Port Louis ⇐ ⇒ Cape Town or Durban

Diagram 6.64

Routes
6.64

    The choice of routes to be followed is influenced by the Agulhas (6.28) and Madagascar Currents (6.28).

West-bound
6.64.1

    From Port Louis (20°08′S, 57°28′E) routes are:

    To a position 20 miles SE of Madagascar (**O**), thence:

    Seasonal as at 6.48.5 and 6.48.6.

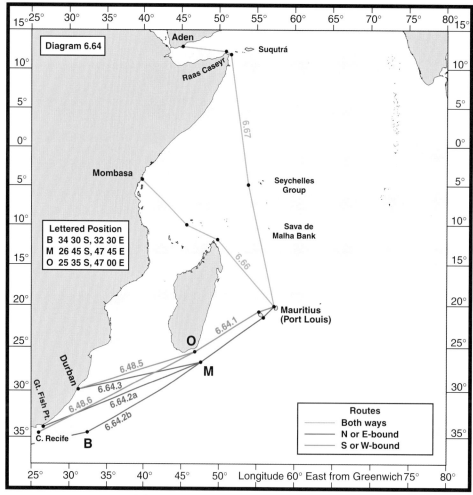

6.64 Port Louis ← → Durban or Cape Town
6.65 Port Louis ← → Mozambique Channel
6.66 Port Louis ← → Mombasa
6.67 Port Louis ← → Aden

East-bound

6.64.2

From Cape Town (33°53′S, 18°26′E) routes are:

(a) Either:

As at 6.35.1 as far as Great Fish Point (33°32′S, 27°07′E), thence:

Rounding the S end of Madagascar at a distance of 60 miles or more offshore **(M)**, to seaward of the strongest part of the Madagascar Current (6.28), thence:

As navigation permits to Port Louis (20°08′S, 57°28′E).

(b) Or, to keep to seaward of the main part of the Agulhas Current:

Through 36°45′S, 19°00′E **(A)**, thence:

Great circle to 34°30′S, 32°30′E **(B)**, thence:

Great circle to pass S of Île de la Réunion (20°55′S, 55°15′E), thence:

To Port Louis (20°08′S, 57°28′E).

6.64.3

From Durban (29°51′S, 31°06′E) the route passes:

S of Madagascar at a distance of 60 miles or more offshore **(M)**, to seaward of the strongest part of the Madagascar Current (6.28), thence:

As navigation permits to Port Louis (20°08′S, 57°28′E).

Distances

6.64.4

Between Port Louis (20°08′S, 57°28′E) and:

	Cape Town	Durban
W-bound	2300† (November to March)	1550
	2350† (April to October)	
E-bound	2300† (Coastwise)	1560
	2450* (Outside Agulhas Current)	

† For Cape Agulhas (15 miles S of) subtract 130 miles
* For Cape of Good Hope (145 miles S of) subtract 180 miles.

Port Louis ⇐ ⇒ Ports in Moçambique Channel

Diagram 6.64

Routes

6.65

(a) Routes between Port Louis (20°08′S, 57°28′E) and the NW coast of Madagascar pass N of that island.

(b) Routes between Port Louis and the W coast of Madagascar, or ports on the African coast S of 18°S, pass S of the island.

Port Louis ⇐ ⇒ Mombasa

Diagram 6.64

Routes

6.66

For fully-powered vessels the route is as direct as navigation permits.

Distance   1430 miles.

Low-powered vessels

6.66.1

Low-powered, or hampered, vessels may find the following routes more suitable for their use:

North-bound. See 9.45 for directions.

South-bound. See 9.53 for directions.

Port Louis ⇐ ⇒ Aden

Diagram 6.64

Routes

6.67

For fully-powered vessels the route is as direct as navigation permits, passing E or W of Seychelles Group (4°30′S, 55°30′E) according to circumstances.

Distance (W of Seychelles Group)   2340 miles.

Low-powered vessels

6.67.1

Low-powered, or hampered, vessels may find the following routes more suitable for their use:

North-bound. See 9.43 for details.

South-bound. See 9.140 for details.

Port Louis ⇐ ⇒ Karāchi

Diagram 6.68

Routes

6.68

The S-bound route is affected by the south-west monsoon.

Fully-powered vessels

6.68.1

The normal route is:

Direct, passing W of Saya de Malha Bank (10°00′S, 61°00′E).

Distance   2740 miles.

South-bound (South-west Monsoon)

6.68.2

During the full strength of the south-west monsoon, the route for fully-powered vessels, from Karāchi (24°46′N, 66°57′E) is:

Parallel to the Indian coast to 70°E, thence:

S to the equator in 70°E **(B)**, thence:

E of Cargados Carajos Shoals (16°35′S, 59°40′E), thence:

To Port Louis (20°08′S, 57°28′E).

Low-powered vessels

6.68.3

Low-powered, or hampered, vessels may find the route from Karāchi to Mauritius, described at 9.147, more suitable for their use.

Port Louis ⇐ ⇒ Mahé Island

Diagram 6.68

Routes

6.69

Weather conditions should not affect fully-powered vessel, the route is as direct as navigation permits.

Distance   950 miles.

Low-powered vessels

6.69.1

Low-powered, or hampered, vessels may find the routes described at 9.168 more suitable for their use.

Port Louis ⇐ ⇒ Mumbai

Diagram 6.68

Routes

6.70

Routes are variable according to direction and season.

Oilfields. For oilfields seaward of Mumbai (18°51′N, 72°50′E), see 6.32.

North-bound
6.70.1

From Port Louis (20°08′S, 57°28′E) the route is:

W of Cargados Carajos Shoals (16°35′S, 59°40′E) and Nazareth Bank (14°30′S, 60°40′E), and

E or W of Saya de Malha Bank (10°00′S, 61°00′E), thence

To Mumbai (18°51′N, 72°50′E).

Distance*   2530 miles.

* W of Saya de Malha Bank.

South-bound
6.70.2

From Mumbai (18°51′N, 72°50′E) the route is:

Direct to the Equator in 66°45′E (**A**), thence:

E of Nazareth Bank (14°30′S, 60°40′E), thence:

E of Cargados Carajos Shoals (16°35′S, 59°40′E), thence:

Direct to Port Louis (20°08′S, 57°28′E).

Distance   2520 miles.

South-west Monsoon
6.70.3

Fully-powered vessels

From Mumbai (18°51′N, 72°50′E) a better route for fully-powered vessels may be:

Along the normal S-bound route (6.70.2) as far as 70°E, thence:

S to the equator in 70°E (**B**), thence:

Rejoining route 6.70.2 E of Cargados Carajos Shoals (16°35′S, 59°40′E).

Low-powered vessels
6.70.4

Low-powered, or hampered, vessels may find the route from Mumbai to Mauritius, described at 9.147, more suitable for their use.

Port Louis ⇐ ⇒ Colombo

Diagram 6.68
Routes
6.71

The route in either direction passes E of Diego Garcia (7°13′S, 72°23′E).

Distance   2140 miles.

Port Louis ⇐ ⇒ Singapore

Diagram 6.68
Routes
6.72

The route from Port Louis (20°08′S, 57°28′E) is direct to the Malacca Strait (6.62), and thence S-bound through the Strait to Singapore (1°12′N, 103°50′E).

For details of the N approaches to the Strait; see 6.61.

Distance   3330 miles.

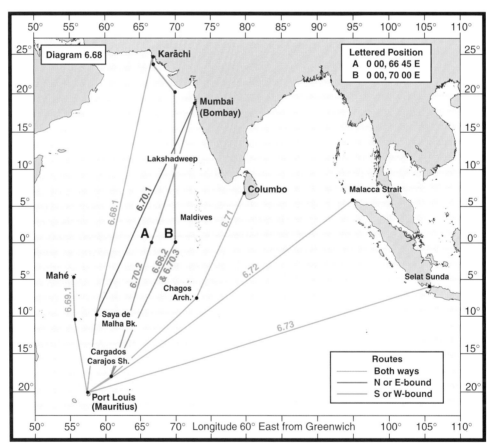

6.68 Port Louis ← → Karāchi
6.69 Port Louis ← → Mahé Island
6.70 Port Louis ← → Mumbai
6.71 Port Louis ← → Colombo
6.72 Port Louis ← → Singapore
6.72 Port Louis ← → Selat Sunda

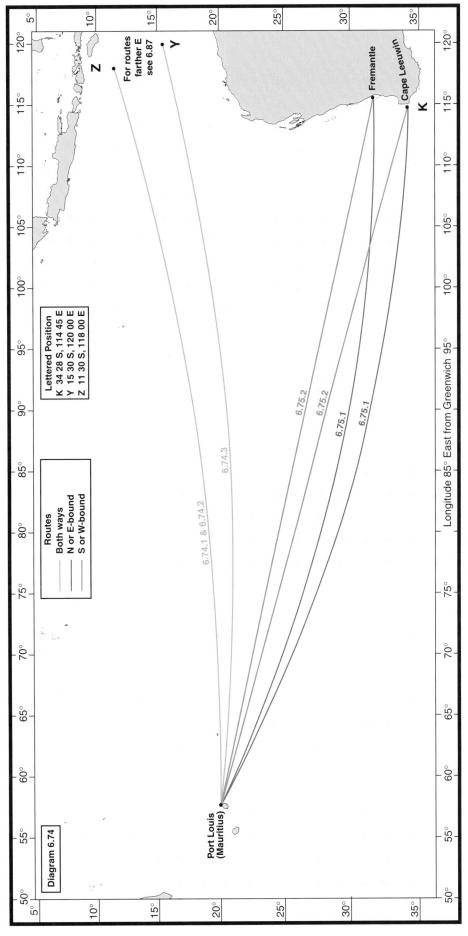

Diagram 6.74

Lettered Position
K 34 28 S, 114 45 E
Y 15 30 S, 120 00 E
Z 11 30 S, 118 00 E

Routes
Both ways
N or E-bound
S or W-bound

For routes
farther E
see 6.87

Z

Y

Fremantle

Cape Leeuwin

K

Port Louis
(Mauritius)

6.74.1 & 6.74.2

6.74.3

6.75.2

6.75.2

6.75.1

6.75.1

Longitude 85° East from Greenwich

6.74 Port Louis ←——→ Torres Strait or Darwin
6.75 Port Louis ←——→ Fremantle or Cape Leeuwin

Port Louis ⇐ ⇒ Selat Sunda

Diagram 6.68
Routes
6.73
Route is direct.
Distance   2960 miles.

Port Louis ⇐ ⇒ Torres Strait or Darwin

Diagram 6.74
Routes
6.74
1   For notes on the approaches to Torres Strait (10°36′S, 141°51′E) and Darwin (12°25′S, 130°47′E), see 6.84-6.87.
2   Routes are seasonal.

October to April
There are two alternative routes from Port Louis (20°08′S, 57°28′E):
6.74.1
    By great circle to 11°30′S, 118°00′E (**Z**), to pass N of the usual track of hurricanes, thence:
    By the ocean route at 6.87.3 to Torres Strait (10°36′S, 141°51′E) or Darwin (12°25′S, 130°47′E).
Distances:
    Darwin   4340 miles
    Torres Strait   4980 miles
6.74.2
    By great circle to 11°30′S, 118°00′E (**Z**), to pass N of the usual track of hurricanes, thence:
    Through 12°40′S, 123°45′E, S of Cartier Island, thence:

Through Osborn Passage (12°40′S, 124°00′E) to join the recommended coastal route to Darwin (12°25′S, 130°47′E) or Torres Strait (10°36′S, 141°51′E).
Distances:
    Darwin   4300 miles
    Torres Strait   4970 miles

May to September
6.74.3
    Routes from Port Louis (20°08′S, 57°28′E) are:
    By great circle to 15°30′S, 120°00′E (**Y**), thence:
    Joining the recommended coastal route (6.85), S of Browse Islet (14°06′S, 123°33′E).
Distances:
    Darwin   4290 miles
    Torres Strait   4970 miles

Port Louis ⇐ ⇒ Fremantle or Cape Leeuwin

Diagram 6.74
Routes
6.75
Routes are direct:
6.75.1
East-bound
    By great circles.
Distances:
    Fremantle   3190 miles
    Cape Leeuwin*   3140 miles
6.75.2
West-bound
    By rhumb line.
Distances:
    Fremantle   3220 miles
    Cape Leeuwin*   3170 miles
* 20 miles WSW of Cape Leeuwin (**K**).

PASSAGES TO AND FROM MAHÉ ISLAND (SEYCHELLES GROUP)

Coverage
6.76
This section covers details of the following routes:
    To and from Mahé and Cape Town or Durban (6.77).
    To and from Mahé and Mombasa (6.78).
    To and from Mahé and Aden (6.79).
    To and from Mahé and Mumbai (6.80).
    To and from Mahé and Colombo (6.81).
    To and from Mahé and Fremantle or Cape Leeuwin (6.82).

Mahé Island ⇐ ⇒ Cape Town or Durban

Diagram 6.77
Routes
6.77
Routes are affected by the Moçambique and Agulhas Currents (6.28).

South-bound
6.77.1
Routes are as navigation permits:
    To 20 miles E of Île Mayotte (12°50′S, 45°10′E), thence:

To 17°00′S, 40°20′E (**U**), thence:
To destination as at 6.37.2 and 6.36.
Distances:
    Cape Town   2920 miles
    Durban   2130 miles

North-bound
6.77.2
Routes are:
    As at 6.35.1 or 6.35.3 and 6.37.1 as far as 17°00′S, 42°15′E (**D**) in the Moçambique Channel, thence:
    As navigation permits to Mahé Island (4°35′S, 55°30′E).
Distances:
    Cape Town   2900 miles
    Durban   2110 miles

Mahé Island ⇐ ⇒ Mombasa

Diagram 6.77
Routes
6.78
For fully-powered vessels the routes are direct in either direction.
    Distance   950 miles.

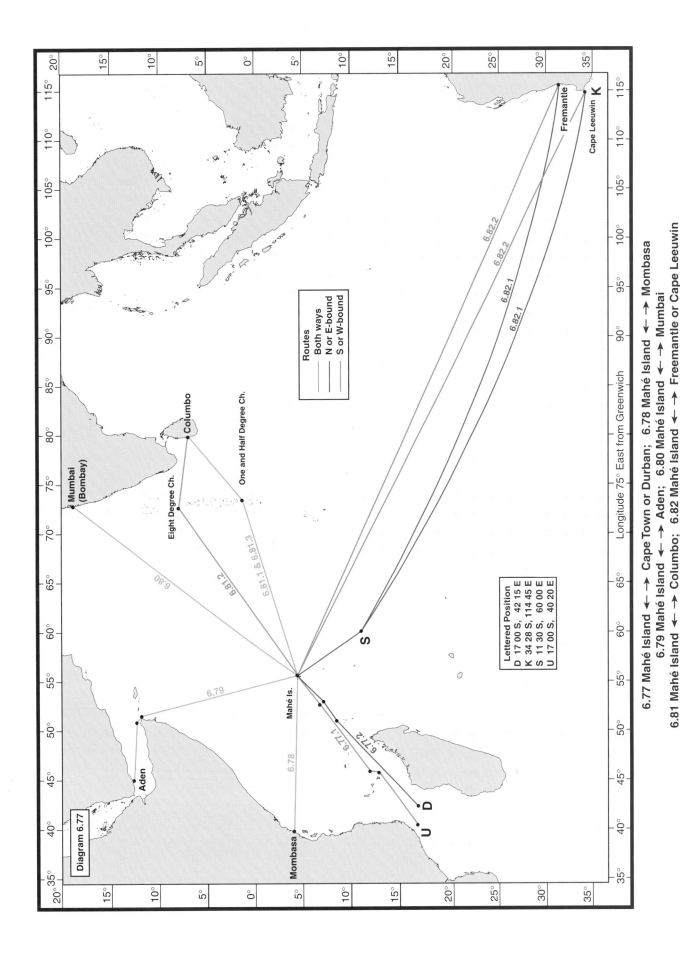

Diagram 6.77

Routes
Both ways
N or E-bound
S or W-bound

Lettered Position
D 17 00 S, 42 15 E
K 34 28 S, 114 45 E
S 11 30 S, 60 00 E
U 17 00 S, 40 20 E

6.77 Mahé Island ←→ Cape Town or Durban; 6.78 Mahé Island ←→ Mombasa
6.79 Mahé Island ←→ Aden; 6.80 Mahé Island ←→ Mumbai
6.81 Mahé Island ←→ Columbo; 6.82 Mahé Island ←→ Freemantle or Cape Leeuwin

Low-powered vessels
6.78.1

Low-powered, or hampered, vessels may find the routes described at 9.169 more suitable for their use.

Mahé Island ⇐ ⇒ Aden

Diagram 6.77
Routes
6.79

The usual route is direct to and from Raas Caseyr (11°50′N, 51°17′E), see 6.38.3.

    Distance   1410 miles.

Low-powered vessels
6.79.1

Low-powered, or hampered, vessels may find the following routes more suitable for their use:

North-bound. See 9.170 for directions.
South-bound. See 9.143 for directions.

Mahé Island ⇐ ⇒ Mumbai

Diagram 6.77
Routes
6.80

For fully-powered vessels the usual routes are direct in either direction.

    Distance   1750 miles.

Oilfields. For oilfields seaward of Mumbai, see 6.32.

Low-powered vessels
6.80.1

Low-powered, or hampered, vessels proceeding from Mumbai to the Seychelles during the south-west monsoon, may find the route described at 9.152 more suitable for their use.

Mahé Island ⇐ ⇒ Colombo

Diagram 6.77
Routes
6.81

Routes are variable according to direction and season.

North-east-bound
6.81.1

The route is through One and Half Degree Channel (1°24′N, 73°20′E).

South-west-bound
6.81.2

October to April:
    The route is through Eight Degree Channel (7°24′N, 73°00′E).
6.81.3

May to September:
    The route is through One and Half Degree Channel (1°24′N, 73°20′E).

Distances:
    via One and Half Degree Channel 1640 miles.
    via Eight Degree Channel   1700 miles.

Low-powered vessels
6.81.4

Low-powered, or hampered, vessels proceeding between Colombo and the Seychelles may find the routes described at 9.171 more suitable for their use.

Mahé Island ⇐ ⇒ Fremantle or Cape Leeuwin

Diagram 6.77
Routes
6.82

Routes vary according to direction.

East-bound
6.82.1

Routes are:
    Through 11°30′S, 60°00′E **(S)**, thence:
    By great circle to destination.
Distances:
    Fremantle   3800 miles.
    Cape Leeuwin*   3780 miles.

West-bound
6.82.2

Routes are:
    Direct by rhumb line to Mahé Island (4°35′S, 55°30′E).
Distances:
    Fremantle   3770 miles.
    Cape Leeuwin*   3760 miles.
  * 20 miles WSW of Cape Leeuwin **(K)**.

WESTERN APPROACHES TO AND PASSAGES OFF THE AUSTRALIAN COAST

General information

Coverage
6.83

This section covers details of the following routes:
    Approaches to Timor and Arafura Seas (6.87).
    Approaches to Fremantle (6.88).
    Approaches to Cape Leeuwin (6.89).
    Routes E of Cape Leeuwin (6.90).
    Routes E of Adelaide (6.91).
    Distances (6.92).

General notes
6.84

The following notes on passages in Australian waters, details of which will be found in Admiralty Sailing Directions, have some bearing on routes in the Indian Ocean which have terminal positions in Australian waters or in the Pacific Ocean.

Recommended tracks
6.85

1   The waters off the N, NW and W coasts of Australia have not been thoroughly surveyed and less water than charted may exist. Many banks in Timor and Arafura Seas, including Sahul Banks, and those lying between them and Melville Island, are unsurveyed and particular attention is necessary in their vicinity.

2   Recommended tracks through these waters, surveyed to a width of 10 miles on either side, are indicated on the charts and described in the appropriate volumes of Admiralty Sailing Directions.

Fishing
6.86

1   Crawfish fishing takes place between the W coast of Australia and the 200 m (110 fm) depth contour between the parallels of 24°S and 34°S, particularly from January to May and in November and December. It is prohibited from 15th August to 14th November.

2   Crawfish and shark fishing fleets operate off the S coast of Australia and the coasts of Tasmania, up to 90 miles from the coast between 37°30′S, 140°00′E and 37°00′S, 149°55′E.

Approaches to Timor and Arafura Seas

Diagram 6.87
Routes
6.87

1   **Archipelagic Sea Lanes (ASL)** (1.29) have been designated through the Indonesian Archipelago. These may affect the choice of route and full details are given at Appendix A.

2   Further details regarding Archipelagic Sea Lanes can be found in *The Mariner's Handbook*.

3   The following Routeing Positions are used for tracks approaching the W coast of Australia.

From South Indian Ocean
6.87.1

1   There are two Routeing Positions for approaching Arafura Sea.

2   The N position, 11°30′S, 118°00′E (**Z**), is used for routes through Timor Sea passing between Timor (9°00′S, 125°00′E) on the N and Sahul Banks (11°30′S, 125°00′E) and the dangers extending NW from Melville Island (11°30′S, 131°00′E) on the S.

3   The S position, 15°30′S, 120°00′E, (**Y**), lies in the approach to the seaward end (14°30′S, 121°30′E) of the recommended coastal track which leads S of Browse Island (14°06′S, 123°33′E) and inshore of Sahul Banks and other off-lying dangers, to Darwin (12°25′S, 130°47′E) and along the N coast of Australia.

4   South African traffic generally uses the S approach, except during the summer season of hurricanes, when the N approach is preferable.

From the Arabian Sea or Bay of Bengal
6.87.2

1   Timor and Arafura Seas may be approached either through Routeing Position 6°25′S, 102°30′E (**O**), to the W of Selat Sunda (6°00′S, 105°52′E) or through Malacca Strait and the Java, Flores and Banda Seas.

2   From the Routeing Position, the principal routes continue, as described in the following paragraphs, to Torres Strait (10°36′S, 141°51′E) and Darwin (12°25′S, 130°47′E).

Ocean Route
6.87.3

1   South of all the islands E of Selat Sunda (6°00′S, 105°52′E) to pass through 11°30′S, 118°00′E (**Z**). Thence, either through 12°40′S, 123°45′E, to the S of Cartier Island and through Osborn Passage (12°40′S, 124°00′E) for Darwin (12°25′S, 130°47′E), or continuing S of Roti (10°45′S, 123°00′E) and through the deep-water trench between Timor and Sahul Banks for Torres Strait (10°36′S, 141°51′E), or to approach Darwin through North Sahul Passage (10°10′S, 127°05′E).

Selat Sumba Route
6.87.4

1   South of Jawa (7°00′S, 110°00′E) and through Selat Sumba (9°00′S, 119°30′E), Selat Ombi (8°35′S, 125°00′E) and Selat Wetar (8°16′S, 127°12′E) for Torres Strait (10°36′S, 141°51′E), or through Selat Sumba, Selat Roti (10°25′S, 123°30′E) and North Sahul Passage (10°10′S, 127°05′E) for Darwin (12°25′S, 130°47′E).

2   These passages are deep and wide and should present no difficulty to navigation.

Java, Flores and Banda Seas Route
6.87.5

1   Through Selat Sunda (6°00′S, 105°52′E), Selat Sapudi (7°00′S, 114°15′E), Flores and Banda Seas to Selat Wetar (8°16′S, 127°12′E), (routes described at 7.105), thence to Torres Strait (10°36′S, 141°51′E) passing S of Duddell and

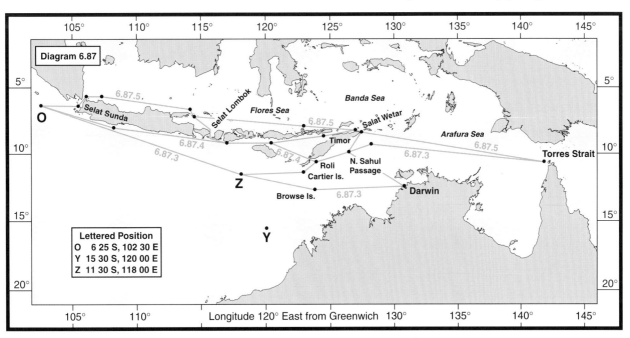

6.87 Approaches to Timor and Arafura Sea

Volsella Shoals, or through North Sahul Passage (10°10′S, 126°50′E) to Darwin (12°25′S, 130°47′E).

2　This route leads between the shoals E of Selat Sunda, through Selat Sapudi and Alur Pelayaran Wetar (8°00′S, 125°30′E) and requires close attention to navigation.

3　**Currents** in Java and Flores Seas are to the advantage of E-bound shipping during the north-west monsoon (November to March), otherwise currents are predominantly W-going on all three routes.

Torres Strait
6.87.6
1　Torres Strait (10°36′S, 141°51′E), which connects Arafura and Coral Seas, is described at 7.42; for further details see the appropriate volume of Admiralty Sailing Directions.

Distances
6.87.7
Ocean Route (6.87.3)*

Torres Strait	2410 miles
Darwin: N Sahul Passage	1770 miles
Osborn Passage	1730 miles

Selat Sumba Route (6.87.4)*

Torres Strait	2380 miles
Darwin	1780 miles

Java, Flores and Banda Seas Route (6.87.5) †

Torres Strait	2190 miles
Darwin	1640 miles

* From Routeing Position 6°25′S, 102°30′E (**O**).

† From Selat Sunda 6°00′S, 105°52′E.

Approaches to Fremantle

General information
6.88

Approaching Fremantle (32°02′S, 115°42′E) from the S, unless certain of the position, the coast between Cape Naturaliste (33°32′S, 115°01′E) and Rottnest Island (32°00′S, 115°30′E) should not be closed to depths of less than 55 m (30 fm) till N of Naturaliste Reefs. Thence, heading N, a vessel may stand into 37 m (20 fm), this depth however can be found within 5 cables of the dangers off the W extremity of Rottnest Island, therefore vessels bound N of Rottnest Island should not go into depths of less than 55 m (30 fm) as the island is approached.

Approaches to Cape Leeuwin

General information
6.89
1　The distance which submerged dangers extend off a long stretch of the coast in the vicinity of Cape Leeuwin

(34°23′S, 115°08′E), and the frequent thick weather that prevails with strong onshore winds and a set towards the coast, render it advisable to give this dangerous cape a wide berth in all but settled weather.

2　From 15 to 20 miles is a good offing to take and the Routeing Position of 34°28′S, 114°45′E (**K**), 20 miles SSW of Cape Leeuwin, has been used, and shown on subsequent routeing diagrams, unless otherwise stated and should suit most passages without appreciable increase in distance.

3　If approaching from the N, vessels passing offshore, and not calling at Fremantle or other ports N of Cape Naturaliste, will generally be outside the 200 m (110 fm) depth contour from Houtman Abrolhos (28°35′S, 113°45′E) to the approaches to the cape. They should keep in depths of more than 55 m (30 fm) until clear of Cape Leeuwin. If running in to make Cape Leeuwin Light, great caution should be exercised. Mist hangs about the land even when it is clear at sea and with any thick weather the light may not be visible as far as Geographe Reef (34°18′S, 114°59′E). Sounding should never be neglected and and vessels should not stand into depths of less than 130 m (70 fm).

Routes East of Cape Leeuwin

Routes
6.90
1　In either direction, routes are direct between:
　　Cape Leeuwin (34°28′S, 114°45′E) (**K**) and Investigator Strait (35°30′S, 137°00°E) (for Adelaide),
　　Cape Leeuwin (34°28′S, 114°45′E) (**K**) and Cape Otway (38°51′S, 143°31′E) (for Bass Strait and Melbourne),
　　Cape Leeuwin (34°28′S, 114°45′E) (**K**) and South West Cape, Tasmania (43°35′S, 146°03′E) (for Hobart).

2　For notes on coastal passages through and E of Bass Strait, see 7.40.

Routes East of Adelaide

General information
6.91
1　Routes between Adelaide (34°38′S, 138°23′E) and ports farther E are through Backstairs Passage (35°45′S, 138°08′E).

2　With SW or W winds, currents setting onto the land at rates of up to 2 kn are sometimes experienced between Cape Willoughby (35°51′S, 138°08′E) and Cape Otway (38°51′S, 143°31′E).

Table of Distances between Torres Strait and Hobart (West-about)
6.92

Torres Strait							
830*	**Darwin**						
1620	920	**Port Hedland**					
2500	1820	980	**Fremantle**				
2630	1940	1110	175	**C. Leeuwin**			
3810	3130	2290	1360	1180	**Adelaide**		
4100	3410	2580	1650	1470	460	**Port Phillip†**	
4280	3590	2760	1820	1650	760		**Hobart**

*　via Cape Van Dieman, 730 miles via Clarence Strait.

†　Port Phillip to Melbourne 40 miles.

For distances E-about, see 7.43.

PASSAGES ON EASTERN SIDE OF INDIAN OCEAN

General information

Coverage
6.93

This section covers details of the following routes:
  To and from Bay of Bengal and N coast of Australia (6.94).
  To and from W side of Bay of Bengal and W coast of Australia (6.95).
  To and from Yangon (Rangoon) and W coast of Australia (6.96).

Bay of Bengal ⇐ ⇒ North coast of Australia

Diagram 6.94
Routes
6.94

1 The choice between an entirely ocean route, W of Sumatera then S of Jawa, and a route passing partly through the Eastern Archipelago, is governed by considerations of draught, weather, distance and season.

2 Archipelagic Sea Lanes (ASL) (1.29) have been designated through the Indonesian Archipelago. These may affect the choice of route and full details are given at Appendix A.

3 Further details regarding Archipelagic Sea Lanes can be found in *The Mariner's Handbook*.

Trans-ocean Routes
6.94.1

Trans-ocean routes may join Ocean Route (6.87.3) or Selat Sumba Route (6.87.4) at position 6°25′S, 102°30′E (O), or the Java, Flores and Banda Sea Route (6.87.5) in Selat Sunda (6°00′S, 105°52′E).

Malacca Strait Route
6.94.2

1 Malacca Strait, which may be entered from the Indian Ocean by three alternative routes (details of which are given at 6.61), leads to Singapore Strait whence a number of alternative routes, given at 7.105, lead through the Eastern Archipelago to Selat Wetar (8°16′S, 127°12′E) and thence into the Arafura Sea.

2 The following route has been used for computation of distances:
  From Singapore Strait, thence:
  Through Selat Karimata (1°43′S, 108°34′E) and Selat Sapudi (7°00′S, 114°15′E), thence:
  Through Selat Sunda (6°00′S, 105°52′E) for Darwin (12°25′S, 130°47′E), or:
  Through Java, Flores and Banda Seas and Selat Wetar (8°16′S, 127°12′E) for Torres Strait (10°36′S, 141°51′E).

3 Bay of Bengal and Malacca Strait, see 6.50 and 6.62.
4 For W approaches to Australian waters, see 6.83.
5 For routes through the Eastern Archipelago, see Chapter 7.

Distances
6.94.3

	Torres Strait	Darwin‡
Ocean Route (6.87.3)		
Chennai	4180	3500
Kolkata Approach†	4280	3600
Yangon River*	3920	3240
Selat Sumba Route (6.87.4)		
Chennai	4150	3550
Kolkata Approach†	4250	3650
Yangon River*	3900	3300
Java, Flores and Banda Seas Route (6.87.5)		
Chennai	4140	3590
Kolkata Approach†	4240	3680
Yangon River*	3880	3320
Malacca Strait Route (6.94.2)		
Chennai	4080	3510
Kolkata Approach†	4020	3460
Yangon River*	3560	2990

† Kolkata (Calcutta) Approach to Kolkata 125 miles.
‡ via Osborne Passage. If via North Sahul Passage, add 45 miles.
* Yangon (Rangoon) River Entrance to Yangon 40 miles.

Western side of Bay of Bengal ⇐ ⇒ West coast of Australia

Diagram 6.94
Routes
6.95

Routes are direct, passing W of Nicobar Islands (8°00′N, 94°00′E) for Paradip (20°15′N, 86°42′E) and Kolkata (Calcutta) (21°00′N, 88°13′E).

Distances (by great circle):
6.95.1

	Chennai	Paradip	Kolkata (Calcutta)†
Port Hedland	3020	3120	3120
Fremantle	3380	3550	3560
C. Leeuwin*	3460	3650	3676

* For rounding Cape Leeuwin, see 6.89.
† Kolkata (Calcutta) approaches to Kolkata 125 miles.

Yangon (Rangoon) ⇐ ⇒ West coast of Australia

Diagram 6.94
Routes
6.96

1 Alternative routes are, either the ocean route S of Nicobar Islands (8°00′N, 94°00′E) and W of Sumatera, or through Malacca Strait (4°00′N, 100°00′E) and Selat Sunda (6°00′S, 105°52′E).

2 For routes from Singapore (1°12′N, 103°51′E) through Eastern Archipelago, see Chapter 7.

Distances from Yangon (Rangoon) River Entrance†:
6.96.1

	Port Hedland	Fremantle	Cape Leeuwin*
Ocean Route	2770	3200	3300
Malacca Strait	2820**	3320	3430

† Yangon (Rangoon) River Entrance to Port of Yangon 40 miles.
* For rounding Cape Leeuwin, see 6.89.
** If via Malacca Strait and Selat Lombok 2780 miles.

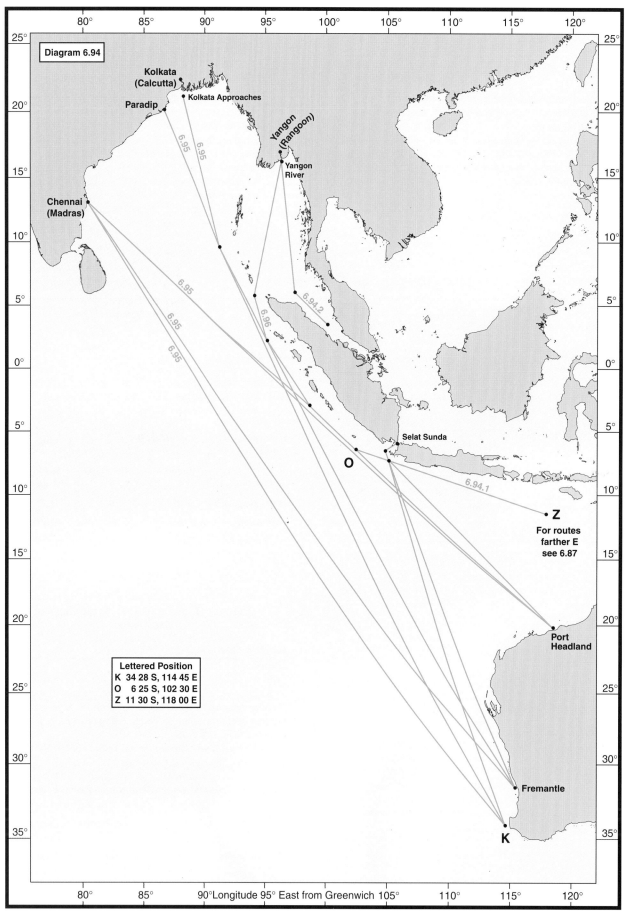

Diagram 6.94

Lettered Position
K 34 28 S, 114 45 E
O   6 25 S, 102 30 E
Z 11 30 S, 118 00 E

For routes
farther E
see 6.87

6.94 Bay of Bengal ← → North coast of Australia
6.95 West side of Bay of Bengal ← → West coast of Australia
6.96 Yangon ← → West coast of Australia

TRANS-OCEAN PASSAGES

GENERAL INFORMATION

Coverage
6.97

This section covers details of the following routes:

1 **Passages between South Africa and Singapore:**
> To and from Cape Town and Selat Sunda (6.98).
> To and from Durban and Selat Sunda (6.99).
> To and from Cape Town or Durban and Singapore (6.100).

2 **Passages between South Africa and Australia or New Zealand:**
> From Cape Town to NW and N coasts of Australia (6.102).
> From Durban to NW and N coasts of Australia (6.103).
> From NW and N coasts of Australia to Durban or Cape Town (6.104).
> From Cape Town to W and S coasts of Australia (6.105).
> From Durban to W and S coasts of Australia (6.106).
> From W and S coasts of Australia to Durban or Cape Town (6.107).
> From Cape Town to New Zealand and Pacific Ocean (6.108).

3 **Passages between Mombasa and Singapore or Australia:**
> To and from Mombasa and Selat Sunda (6.109).
> To and from Mombasa and Singapore (6.110).
> To and from Mombasa and N and W coasts of Australia (6.111).

4 **Passages between Aden and Singapore, Australia or New Zealand:**
> From Aden to Selat Sunda (6.112).
> From Selat Sunda to Aden (6.113).
> To and from Aden and Singapore (6.114).
> From Aden to Darwin or Torres Strait (6.115).
> From Darwin and Torres Strait to Aden (6.116).
> From Aden to Fremantle or S coast of Australia (6.117).
> From S coast of Australia and Fremantle to Aden (6.118).
> From Aden to New Zealand and Pacific Ocean (6.119).

5 **Passages between E side of Indian Ocean and Australia:**
> To and from Strait of Hormuz, Karāchi, Mumbai or Colombo and Darwin or Torres Strait (6.120).
> To and from Strait of Hormuz, Karāchi, Mumbai or Colombo and W and NW coasts of Australia (6.121).
> To and from ports in the Indian Ocean and E coast of Australia (6.122).

PASSAGES BETWEEN SOUTH AFRICA AND SINGAPORE

Cape Town ⇐ ⇒ Selat Sunda

Diagram 6.98
Routes
6.98

The routes vary according to direction and season.

East-bound
6.98.1

At all seasons the route is:
> Across the Agulhas Current to 36°45′S, 19°00′E (**A**), 145 miles S of Cape of Good Hope, thence:
> Great circle to 33°45′S, 36°30′E (**D**), thence:
> Great circle to Selat Sunda (6°00′S, 105°52′E), passing N of Cocos or Keeling Islands (12°05′S, 96°52′E).

Distances

Cape Town	5240 miles.
Cape Agulhas*	5060 miles.

* 15 miles S of.

West-bound
6.98.2

From October to April the route is:
> Great circle to 33°45′S, 36°30′E (**D**), thence:
> To a landfall on the African coast near Great Fish Point (33°32′S, 27°07′E), thence:
> Keeping in the Agulhas Current as at 6.36 as far as Mossel Bay (34°11′S, 22°09′E), thence to Cape Town (33°53′S, 18°26′E).

Distances

Cape Town	5130 miles.
Cape Agulhas*	5000 miles.

* 15 miles S of.

6.98.3

From May to September the route is:
> Great circle to 30°00′S, 56°30′E (**N**), passing N of Cocos or Keeling Islands (12°05′S, 96°52′E), thence:
> Along the parallel of 30°S to a landfall on the African coast near Durban (29°51′S, 31°06′E), thence:
> Coastwise to Cape Town (33°53′S, 18°26′E) to avoid the heavy weather prevalent to seaward at this season (6.36).

Distances

Cape Town	5250 miles.
Cape Agulhas*	5120 miles.

* 15 miles S of.

Durban ⇐ ⇒ Selat Sunda

Diagram 6.98 or 6.99
Routes
6.99

The routes vary according to direction and season.

East-bound
6.99.1

At all seasons the route is:
> Along the parallel of 30°00′S to 56°30′E (**N**), thence:
> Great circle to Selat Sunda (6°00′S, 105°52′E), passing N of Cocos or Keeling Islands (12°05′S, 96°52′E).

Distance   4460 miles.

West-bound
6.99.2

From October to April the route is:
> Direct by great circle.

Distance   4440 miles.

6.99.3

From May to September the route is:
> The E-bound route in reverse.

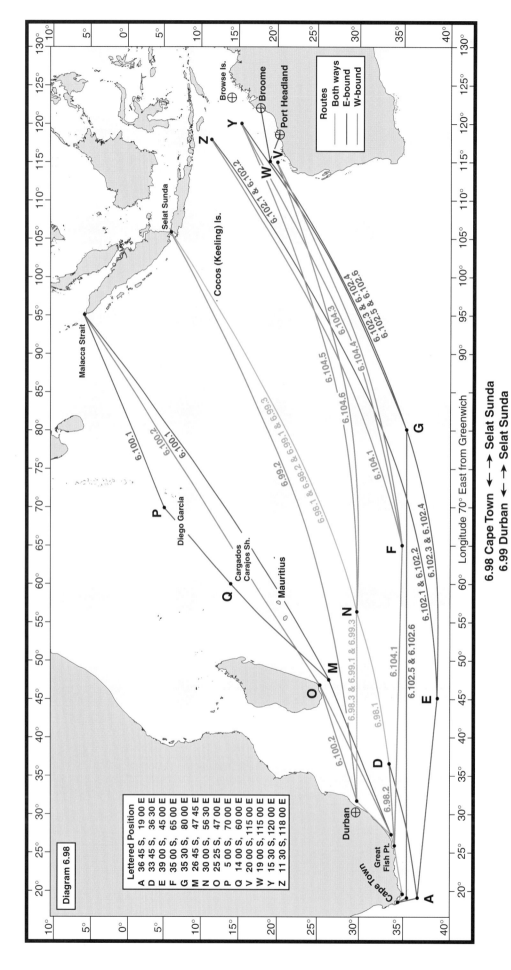

Diagram 6.98

Lettered Position

A 36 45 S, 19 00 E
D 33 45 S, 36 30 E
E 39 00 S, 45 00 E
F 35 00 S, 65 00 E
G 35 30 S, 80 00 E
M 26 45 S, 47 45 E
N 30 00 S, 56 30 E
O 25 25 S, 47 00 E
P 5 00 S, 70 00 E
Q 14 00 S, 60 00 E
V 20 00 S, 115 00 E
W 19 00 S, 115 00 E
Y 15 30 S, 120 00 E
Z 11 30 S, 118 00 E

Routes
Both ways
E-bound
W-bound

Longitude 70° East from Greenwich

6.98 Cape Town ←→ Selat Sunda
6.99 Durban ←→ Selat Sunda
6.100 Cape Town and Durban ←→ Singapore
6.102 Cape Town → North west and north coasts of Australia
6.104 North west and north coasts of Australia → Cape Town

Cape Town and Durban ⇐ ⇒ Singapore

Diagram 6.98 or 6.99
Routes
6.100
The routes are either through Malacca Strait (4°00′N, 100°00′E) or through Selat Sunda (6°00′S, 105°52′E) and vary according to season.

East-bound
6.100.1
For Malacca: routes are as at 6.47.3 (a) and (b) but proceeding to the N approaches to the Malacca Strait, which are described at 6.61 and 6.62.

For Selat Sunda: routes are as at 6.98.1 and 6.99.1.

West-bound
6.100.2
For Malacca Strait: routes are by rhumb line, passing N of Mauritius (20°10′S, 57°30′E), to join the route from Colombo (6.48.4 to 6.48.6) 20 miles SE of Madagascar (**O**).

For Selat Sunda: routes are as at 6.98.2 and 6.99.2, but when W-bound for Durban (29°51′S, 31°06′E) the Malacca Strait route is preferable.

Distances from Singapore
6.100.3

		Cape Town		Durban	
via **Malacca Strait**					
E-bound	Oct–Mar	5710[a]*	Oct–Mar	4950	
	May–Sept	5630[a]*	May–Sept	4890	
W-bound	Nov–Mar	5630*	All seasons	4870	
	Apr–Nov	5680*			
via **Selat Sunda**					
E-bound	All seasons	5810[b]	All seasons	5030	
W-bound	Oct–Apr	5700*	—		
	May–Sept	5820*	—		

* For Cape Agulhas (15 miles S of) subtract 130 miles
[a] For Cape of Good Hope (145 miles S of) subtract 20 miles
[b] For Cape of Good Hope (145 miles S of) subtract 180 miles

PASSAGES BETWEEN SOUTH AFRICA, AUSTRALIA OR NEW ZEALAND

General information
6.101
1 Between the South-east Trade Winds (6.13) and the Roaring Forties (6.15), there is a zone of light variable winds lying with its axis in about 35°S in the S summer and 30°S in the S winter.

2 The Southern Ocean Current, with a mean rate of ½ kn or less, has an indefinite N boundary in the South Indian Ocean, but E-going sets predominate as far N as 30°S, or the S limit of the South-east Trade Wind.

3 The rhumb line between Cape Agulhas (34°50′S, 20°01′E) and Cape Leeuwin (34°23′S, 115°08′E) coincides with the parallel of 35°S and is 4711 miles in length. The corresponding great circle is only 4501 miles in length, but its vertex is in about 45°S, in the storm-ridden waters of the Roaring Forties. Great circle routes to places farther E along the S coast of Australia pass even farther S,

increasing the probability of encountering ice bergs and approaching the limit of pack ice (6.30).

4 It is therefore evident that any attempt to shorten a voyage by great circle sailing between the two continents is likely, except on the most N tracks, to put a vessel at risk of delay due to the weather and in the case of W-bound voyages, due to stronger adverse current, with an additional risk that pack ice may be encountered.

5 Consequently, E-bound routes to the S and W coasts of Australia usually comprise a composite route with a limiting latitude of 40°S in summer and 35°30′S in winter.

6 W-bound routes, to avoid the head winds and adverse currents of the direct routes, keep well to the N. From the S and W coasts of Australia they pass, in all seasons, through or N of 30°S, 100°E. If bound for Cape Agulhas (34°50′S, 20°01′E) in the S summer, the route continues from 30°S, 100°E by great circle to join the parallel of 35°S in 65°E, which it follows to a landfall off Cape Saint Francis (34°13′S, 24°50′E). In the S winter it follows the parallel of 30°S to the African coast off Durban (29°51′S, 31°06′E) before turning S to continue coastwise as at 6.36.

7 Voyages between South Africa and the NW and W coasts of Australia are little affected by the foregoing considerations. However, in view of the possibility of encountering hurricanes off the NW coast of Australia, the routes for Darwin (12°25′S, 130°47′E) and the Arafura Sea, from October to April, keep away from the coast and pass close S of Roti (10°45′S, 123°00′E) and Timor (9°00′S, 125°00′E).

8 For restrictions on tankers navigating off the coast of South Africa, see *Africa Pilot Volumes I and II*.

Cape Town ⇒ North-west and North coasts of Australia

Diagram 6.98
Routes
6.102
Routes are seasonal.

October to April
6.102.1
For Torres Strait (10°36′S, 141°51′E) the route is:
Across the Agulhas Current to 36°45′S, 19°00′E (**A**), 145 miles S of Cape of Good Hope, thence:
Rhumb line to 39°00′S, 45°00′E (**E**), thence:
Great circle to 11°30′S, 118°00′E (**Z**), to avoid the possibility of hurricanes, thence:
Continuing by the Ocean Route (6.87.3) through Timor and Arafura Seas.

6.102.2
For Darwin (12°25′S, 130°47′E) the route is:
As the Torres Strait route (6.102.1) as far as 11°30′S, 118°00′E (**Z**), thence:
Through either Osborn Passage (12°40′S, 124°00′E) or North Sahul Passage (10°10′S, 127°05′E) (6.87.3) to Darwin.

6.102.3
For destinations between Darwin (12°25′S, 130°47′E) **and Yampi Sound** (16°10′S, 123°30′E) the route is:
As the Torres Strait route (6.102.1) as far as 39°00′S, 45°00′E (**E**), thence:
Great circle to 35°30′S, 80°00′E (**G**), thence:
Great circle to 15°30′S, 120°00′E (**Y**), thence:
Joining the recommended track (6.85) S of Browse Islet (14°06′S, 123°33′E), thence:
To destination.

6.102.4

For destinations W of Yampi Sound (16°10′S, 123°30′E) the route is:

As for for destinations between Darwin (12°25′S, 130°47′E) and Yampi Sound (6.102.3) as far as the meridian of 115°E (**W**), thence:

Direct, as navigation permits, to destination.

May to September

6.102.5

For Torres Strait (10°36′S, 141°51′E), **Darwin** (12°25′S, 130°47′E) **and other destinations E of Yampi Sound** (16°10′S, 123°30′E) the route, in spite of the adverse Agulhas Current, is:

Along the parallel of 35°30′S as far as 80°00′E (**G**), thence:

Great circle to 15°30′S, 120°00′E (**Y**), thence:

Joining the recommended track (6.85) S of Browse Islet (14°06′S, 123°33′E).

6.102.6

For destinations W of Yampi Sound (16°10′S, 123°30′E) the route is:

As for destinations E of Yampi Sound (6.102.5) as far as 35°30′S, 80°00′E (**G**), thence:

Great circle to 19°00′S, 115°00′E (**W**), thence:

Direct, as navigation permits, to destination.

Distances

6.102.7

From Cape Town (33°53′S, 18°26′E):

		Oct-Apr	May-Sep
Torres Strait		7050[(b)]	6880[(a)]
Darwin:	N Sahul Passage	6410[(b)]	—
	Coastal route	6370[(b)]*	6200[(a)]**
Port Hedland		5430[(b)]	5400[(a)]

[(a)] For Cape Agulhas (15 miles S of) subtract 130 miles.
[(b)] For Cape of Good Hope (145 miles S of) subtract 180 miles.

   * via Osborn Passage   ** via Browse Islet.

Durban ⇒ North-west and North coasts of Australia

Diagram 6.99
Routes
6.103

Routes are seasonal.

October to April

6.103.1

For Torres Strait (10°36′S, 141°51′E) **and Darwin** (12°25′S, 130°47′E) the routes are:

By great circle to 11°30′S, 118°00′E (**Z**), thence:

As at 6.102.1 or 6.102.2 to destination.

May to September

6.103.2

For Torres Strait (10°36′S, 141°51′E) **and Darwin** (12°25′S, 130°47′E) the route are:

By great circle to 15°30′S, 120°00′ (**Y**), thence:

Joining the recommended track (6.85) S of Browse Islet (14°06′S, 123°33′E).

All seasons

6.103.3

For destinations between Darwin (12°25′S, 130°47′E) **and Yampi Sound** (16°10′S, 123°30′E) the routes are:

By great circle to 15°30′S, 120°00′ (**Y**), thence:

Joining the recommended track (6.85) S of Browse Islet (14°06′S, 123°33′E), thence:

To destination.

6.103.4

For destinations W of Yampi Sound (16°10′S, 123°30′E) the routes are:

By great circle to 19°00′S, 115°00′E (**W**), thence:

As direct as navigation permits to destination.

Distances

6.103.5

From Durban (29°51′S, 31°06′E):

		Oct-Apr	May-Sep
Torres Strait		6350	6250
Darwin:	N Sahul Passage	5720	—
	Coastal route	5670*	5570**
Port Hedland		4770	4770

   * via Osborn Passage   ** via Browse Islet.

North-west and North coasts of Australia ⇒ Durban or Cape Town

Diagrams 6.98 for Cape Town, 6.99 for Durban
Routes
6.104

Routes are seasonal.

October to April

6.104.1

From Torres Strait (10°36′S, 141°51′E) the route:

Follows the Ocean Route (6.87.3) to 11°30′S, 118°00′E (**Z**), thence:

Great circle to Durban (29°51′S, 31°06′E).

Or:

Great circle to 35°00′S, 65°00′E (**F**), thence:

Rhumb line to a landfall on Cape Recife (34°02′S, 25°42′E), thence:

Coastwise as at 6.36 to Cape Town (33°53′S, 18°26′E).

6.104.2

From Darwin (12°25′S, 130°47′E) the route is:

Through North Sahul Passage (10°10′S, 127°05′E) to join the Ocean Route (6.87.3) from Torres Strait SE of Timor (9°00′S, 125°00′E).

6.104.3

From places between Darwin (12°25′S, 130°47′E) **and Yampi Sound** (16°10′S, 123°30′E) routes are:

By the recommended routes (6.85) to 15°30′S, 120°00′E (**Y**) thence:

Great circle to Durban (29°51′S, 31°06′E).

Or:

Great circle to join the route from Torres Strait to Cape Town in 35°00′S, 65°00′E (**F**).

6.104.4

From places W of Yampi Sound (16°10′S, 123°30′E) the route is:

As direct as navigation permits to 20°00′S, 115°00′E (**V**), off Monte Bello Islands, thence:

Great circle to Durban (29°51′S, 31°06′E).

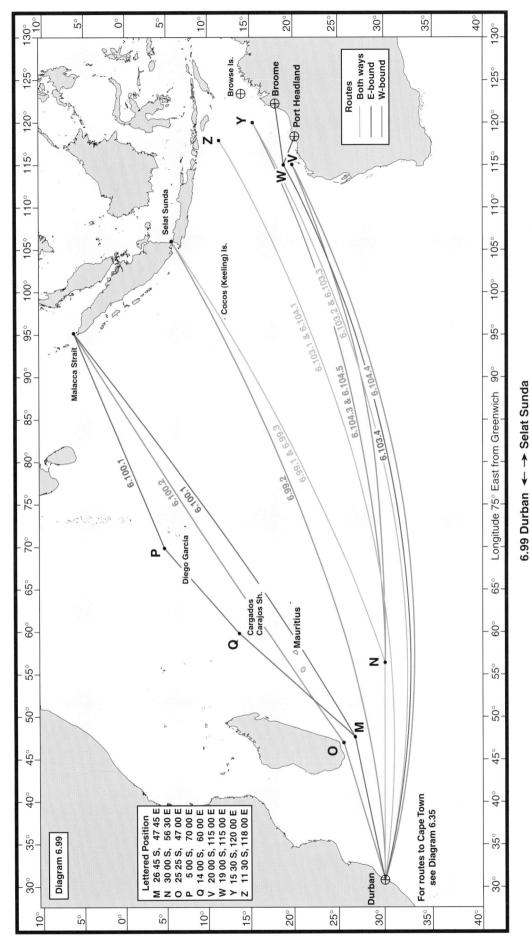

Diagram 6.99

Lettered Position	
M	26 45 S, 47 45 E
N	30 00 S, 56 30 E
O	25 25 S, 47 00 E
P	5 00 S, 70 00 E
Q	14 00 S, 60 00 E
V	20 00 S, 115 00 E
W	19 00 S, 115 00 E
Y	15 30 S, 120 00 E
Z	11 30 S, 118 00 E

Routes
Both ways
E-bound
W-bound

6.99 Durban ←—→ Selat Sunda
6.103 Durban ←—→ North west and north coasts of Australia
6.104 North west and north coasts of Australia —→ Durban

Longitude 75° East from Greenwich

131

Or:

Great circle to join the route from Torres Strait to Cape Town in 35°00'S, 65°00'E (**F**).

May to September
6.104.5

From Yampi Sound (16°10'S, 123°30'E) **and all places farther E** the routes are:

By the recommended routes (6.85) to 15°30'S, 120°00'E (**Y**) thence:

Great circle to 30°00'S, 56°30'E (**N**), thence:

Along the parallel of 30°S to Durban.

Or:

Continuing thence for Cape Town (33°53'S, 18°26'E) as at 6.36.

6.104.6

From places W of Yampi Sound (16°10'S, 123°30'E) the routes are:

Direct as navigation permits to 20°00'S, 115°00'E (**V**), thence:

Great circle to join the route from ports E of Yampi Sound in 30°00'S, 56°30'E (**N**).

Distances
6.104.7

		Durban	Cape Town*
Torres Strait	Oct–Apr	6350	7000
	May–Sept	6260	7050
Darwin:	Oct–Apr	5710	6370
	May–Sept	5580	6370
Port Hedland	Oct–Apr	4730	5360
	May–Sept	4750	5540

* For Cape Agulhas (15 miles S of) subtract 130 miles.

Cape Town ⇒ West and South coasts of Australia

Diagram 6.105
Routes
6.105

Routes are seasonal.

Ice. The possibility of encountering icebergs on these routes, at any time of year, cannot be discounted, see 6.31.

October to April
6.105.1

Routes are:

Across the the Agulhas Current to 36°45'S, 19°00'E (**A**), 145 miles S of Cape of Good Hope, thence:

Rhumb line to 40°00'S, 55°00'E (**M**), thence:

Along the parallel of 40°00'S.

(a) **For Fremantle** (32°02'S, 115°42'S) the route:

Continues by great circle from 40°00'S, 77°00'E (**R**).

(b) **For Adelaide** (34°38'S, 138°23'E) and **Melbourne** (Port Phillip) (38°20'S, 144°34'E) the route:

Continues by great circle from 40°00'S, 100°00'E (**S**).

(c) **For Hobart** (42°58'S, 147°23'E) the route:

Continues by great circle from 40°00'S, 100°00'E (**S**) to 41°30'S, 122°50'E (**U**), on the route to Melbourne, thence:

Great circle to South West Cape, Tasmania (43°35'S, 146°03'E), thence:

As navigation permits to Hobart.

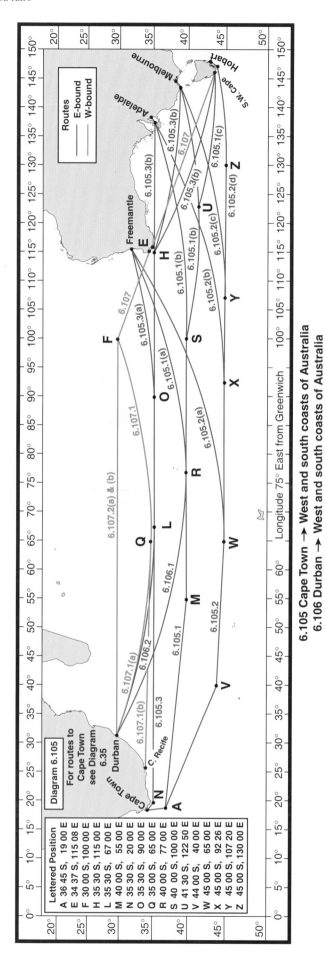

6.105 Cape Town → West and south coasts of Australia
6.106 Durban → West and south coasts of Australia
6.107 West and south coasts of Australia → Durban and Cape Town

Lettered Position	
A	36 45 S, 19 00 E
E	34 37 S, 115 08 E
F	30 00 S, 100 00 E
H	35 30 S, 115 00 E
L	35 30 S, 67 00 E
M	40 00 S, 55 00 E
O	35 30 S, 90 00 E
Q	35 00 S, 65 00 E
R	40 00 S, 77 00 E
S	40 00 S, 100 00 E
U	41 30 S, 122 50 E
V	44 00 S, 40 00 E
W	45 00 S, 65 00 E
X	45 00 S, 92 26 E
Y	45 00 S, 107 20 E
Z	45 00 S, 130 00 E

October to April (alternative route)
6.105.2

Shorter but more boisterous routes are:

Through 36°45′S, 19°00′E (**A**), thence:

Rhumb line to 44°00′S, 40°00′E (**V**), thence:

Rhumb line to 45°00′S, 65°00′E (**W**), thence:

Along the parallel of 45°00′S.

(a) **For Fremantle** (32°02′S, 115°42′S) the route:

Continues by great circle from 45°00′S, 65°00′E (**W**).

(b) **For Adelaide** (34°38′S, 138°23′E) the route:

Continues by great circle from 45°00′S, 92°26′E (**X**) to Investigator Strait.

(c) **For Melbourne (Port Phillip)** (38°20′S, 144°34′E) the route:

Continues by great circle from 45°00′S, 107°20′E (**Y**) to Cape Otway (38°51′S, 143°31′E), thence:

Direct as navigation permits to destination.

(d) **For Hobart** (42°58′S, 147°23′E) the route:

Continues by great circle from 45°00′S, 130°00′E (**Z**) to destination.

May to September
6.105.3

To avoid the probability of foul weather farther S, routes are:

Through 35°30′S, 20°00′E (**N**), thence:

Along the parallel of 35°30′S, which, if strictly followed, passes close under West Cape Howe to Investigator Strait which leads to Adelaide.

(a) **For Fremantle** (32°02′S, 115°42′S) the route is:

By great circle from 35°30′S, 90°00′E (**O**).

(b) **For ports E of Adelaide** (34°38′S, 138°23′E) the routes are:

By great circle from 35°30′S, 115°00′E (**H**), S of Cape Leeuwin.

Distances
6.105.4

From Cape Town (33°53′S, 18°26′E):

	Oct to April		May to Sept
	6.105.1[b]	6.105.2[b]	6.105.3[a]
Fremantle	4840	4790	4870
Adelaide	5820	5660	5950
Port Phillip*	6030	5820	6230
Hobart	6150	5870	6400

* Port Phillip to Melbourne: 40 miles

[a] For Cape Agulhas (15 miles S of) subtract 130 miles.

[b] For Cape of Good Hope (145 miles S of) subtract 180 miles.

Durban ⇒ West and South coasts of Australia

Diagram 6.105
Routes
6.106

Routes are seasonal.

October to April
6.106.1

Routes are:

From Durban (29°51′S, 31°06′E) by great circle to join those from Cape Town (6.105.1) at 40°00′S, 77°00′E (**R**).

May to September
6.106.2

Routes are:

From Durban (29°51′S, 31°06′E) by great circle to join those from Cape Town (6.105.3) at 35°30′S, 67°00′E (**L**).

Distances
6.106.3

From Durban (29°51′S, 31°06′E):

	Oct to April	May to Sept
Fremantle	4250	4270
Adelaide	5230	5350
Port Phillip*	5440	5630
Hobart	5560	5800

* Port Phillip to Melbourne: 40 miles

West and South coasts of Australia ⇒ Durban or Cape Town

Diagram 6.105
Routes
6.107

1 All routes throughout the year pass through, or N of, 30°00′S, 100°00′E (**F**).

2 From the S coast of Australia and Tasmania this position is approached through 34°37′S, 115°08′E (**E**), 15 miles S of Cape Leeuwin and thence by rhumb line.

3 Although a great circle track from Tasmania direct to 30°00′S, 100°00′E (**F**) might appear preferable, it would only save about 20 miles and adverse winds, with head seas, would be more likely.

4 From 30°00′S, 100°00′E (**F**) routes are seasonal.

October to April
6.107.1

Routes are:

(a) **For Durban** (29°51′S, 31°06′E):

Direct by great circle.

(b) **For Cape Town** (33°53′S, 18°26′E):

Great circle to 35°00′S, 65°00′E (**Q**), thence:

Rhumb line to a landfall off Cape Recife (34°02′S, 25°42′E), thence:

Coastwise, as at 6.36 to Cape Town.

May to September
6.107.2

Routes are:

(a) **For Durban** (29°51′S, 31°06′E):

Along the parallel of 30°00′S to Durban.

(b) **For Cape Town** (33°53′S, 18°26′E):

From Durban continuing coastwise as at 6.36.

Distances
6.107.3

	Fremantle	Adelaide	Port Phillip*	Hobart
Durban				
October to April	4350	5510	5800	5980
May to September	4410	5570	5860	6040
Cape Town				
October to April	4970[(a)]	6130[(a)]	6420[(a)]	6600[(a)]
May to September	5200[(a)]	6360[(a)]	6650[(a)]	6830[(a)]

* Port Phillip to Melbourne: 40 miles
[(a)] For Cape Agulhas (15 miles S of) subtract 130 miles.

Cape Town ⇒ New Zealand and Pacific Ocean

Diagram 6.108
Routes
6.108
Routes are seasonal.

October to April
6.108.1
Routes are:
As at 6.105.1 for Hobart as far as 41°30′S, 122°50′E **(U)**, thence:
(a) Either:
Great circle to 47°50′S, 167°50′E **(J)**, ENE of Snares Islands (48°01′S, 166°36′E).
(b) Or:
Great circle passing close S of Tasmania to Cook Strait (41°00′S, 174°30′E).

October to April (alternative route)
6.108.2
If the shorter route (6.105.2) to Hobart is taken it should be left in 45°00′S, 130°00′E **(Z)**, thence:
Great circle to Snares Islands (48°01′S, 166°36′E) or Cook Strait (41°00′S, 174°30′E).

May to September
6.108.3
Routes are:
As at 6.105.3 for ports E of Adelaide as far as 35°30′S, 115°00′E **(H)**, S of Cape Leeuwin, thence:
Great circle to a landfall off South West Cape, Tasmania (43°35′S, 146°03′E), thence:
Great circle to Snares Islands (48°01′S, 166°36′E) or Cook Strait (41°00′S, 174°30′E).

Distances
6.108.4
From Cape Town (33°53′S, 18°26′E):

	Oct to April		May to Sept
	6.108.1[(b)]	6.108.2[(b)]	6.108.3[(a)]
Snares Islands	6970	6650	7240
Cook Strait	7380	7100	7630

[(a)] For Cape Agulhas (15 miles S of) subtract 130 miles.
[(b)] For Cape of Good Hope (145 miles S of) subtract 180 miles.

PASSAGES BETWEEN MOMBASA AND SINGAPORE OR AUSTRALIA

Mombasa ⇐ ⇒ Selat Sunda
Diagram 6.109
Route
6.109
From Mombasa (4°05′S, 39°43′E) the route:
Passes through 3°00′S, 54°00′E **(T)**, 50 miles N of the Seychelles Group, thence:
Through 4°00′S, 73°30′E **(A)**, passing 50 miles N of Chagos Archipelago (6°30′S, 72°00′E), thence:
To Selat Sunda (6°00′S, 105°52′E).
Distance 3980 miles.

Mombasa ⇐ ⇒ Singapore
Diagram 6.109
Route
6.110
From Mombasa (4°05′S, 39°43′E) the route is:
Through One and Half Degree Channel (1°24′N, 73°20′E) **(J)**, thence:
To the N approaches to the Malacca Strait, described at 6.61, thence:
Through Malacca Strait (6.62) to Singapore (1°12′N, 103°51′E).
Distance 3990 miles.

Mombasa ⇐ ⇒ North and West coasts of Australia
Diagram 6.109
Routes
6.111
1 **For Torres Strait** (10°36′S, 141°51′E) and **Darwin** (12°25′S, 130°47′E) the routes are:
(a) Either:
As at 6.109 to Selat Sunda (6°00′S, 105°52′E), thence:
Through Java, Flores and Banda Seas routes of 6.87.5.
(b) Or:
Passing through 3°00′S, 54°00′E **(T)**, 50 miles N of the Seychelles Group, thence:
Through 4°00′S, 73°30′E **(A)**, passing 50 miles N of Chagos Archipelago, thence:
Joining the Ocean route of 6.87.3 in 11°30′S, 118°00′E **(Z)**.
6.111.1
1 **For places between Darwin** (12°25′S, 130°47′E) and **Cape Leeuwin** (34°23′S, 115°08′E) the routes are directional:

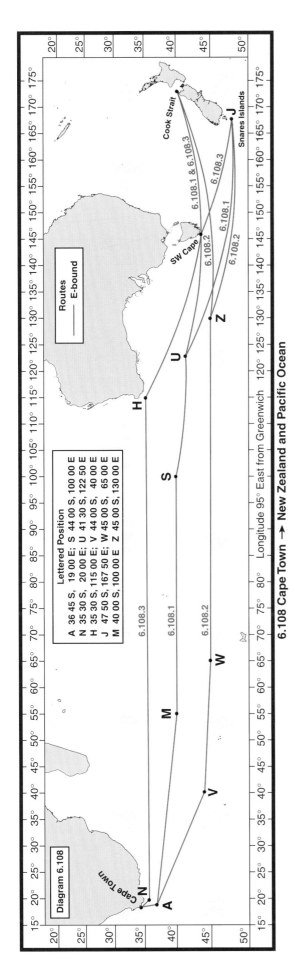

6.108 Cape Town → New Zealand and Pacific Ocean

Longitude 95° East from Greenwich

Diagram 6.108

Routes
—— E-bound

Lettered Position

A 36 45 S, 19 00 E; S 44 00 S, 100 00 E
N 35 30 S, 20 00 E; U 41 30 S, 122 50 E
H 35 30 S, 115 00 E; V 44 00 S, 40 00 E
J 47 50 S, 167 50 E; W 45 00 S, 65 00 E
M 40 00 S, 100 00 E; Z 45 00 S, 130 00 E

(a) East-bound from Mombasa (4°05′S, 39°43′E) the routes are:

Passing through 3°00′S, 54°00′E **(T)**, 50 miles N of the Seychelles Group, thence:

Through 4°00′S, 73°30′E **(A)**, thence:

Through 10°00′S, 80°00′E **(B)**, thence:

As direct as navigation permits, either to the appropriate recommended track (6.85) **(Y)**, or:

To destination.

(b) West-bound from the Australian coast S of Darwin (12°25′S, 130°47′E) are:

By great circle, as near as navigation permits, keeping N of 30°00′S, 100°00′E **(F)**.

Distances
6.111.2

From Mombasa (4°05′S, 39°43′E):

	Java, Flores & Banda Seas Route (6.87.5)	Ocean Route (6.87.3)
Torres Strait	6170	6150
Darwin	5620	5520
	E-bound	**W-bound**
Port Hedland	4790	4720
Fremantle	4840	4570
Cape Leeuwin*	4860	4530

* 20 miles SW 34°28′S, 114°45′E **(K)**.

PASSAGES BETWEEN ADEN AND SINGAPORE, AUSTRALIA OR NEW ZEALAND

Aden ⇒ Selat Sunda

Diagram 6.112
Routes
6.112

Routes are seasonal.

6.112.1

October to April the route is:

Round Raas Caseyr (11°50′N, 51°17′E), thence:

Through One and Half Degree Channel (1°24′N, 73°20′E) **(J)**, thence:

Direct to Selat Sunda (6°00′S, 105°52′E).

6.112.2

May to September the route is:

N of Suquṭrá (12°30′N, 54°00′E) to 13°00′N, 55°00′E **(Q)**, thence:

Through Eight Degree Channel (7°24′N, 73°00′E) **(F)**, thence:

Landfall (5°45′N, 80°36′E), 10 miles S of Dondra Head, thence:

To Selat Sunda (6°00′S, 105°52′E).

Distances
6.112.3

From Aden:

	Oct to April	May to Sept
Selat Sunda	3840	3860
Position (O) 6°25′S, 102°30′E	3650	3670

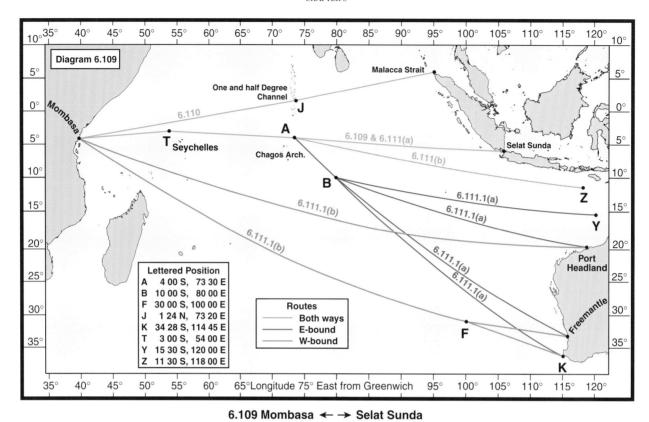

6.109 Mombasa ← → Selat Sunda
6.110 Mombasa ← → Singapore
6.111 Mombasa ← → North and west coast of Australia

Low-powered vessels
6.112.4

Low-powered, or hampered, vessels may find the routes described at 9.137 more suitable for their use.

Selat Sunda ⇒ Aden

Diagram 6.112
Routes
6.113

Routes are seasonal.
6.113.1

October to April the route is:

To a landfall (5°45′N, 80°36′E), 10 miles S of Dondra Head, thence:

Through Eight Degree Channel (7°24′N, 73°00′E) **(F)**, thence:

(a) Either:

S of Suquṭrá (12°30′N, 54°00′E), thence:

Between Raas Caseyr (11°50′N, 51°17′E) and 'Abd al Kŭrī (12°10′N, 52°15′E), bearing mind the difficulty of identifying the landfall (6.38.3), thence:

Direct to Aden (12°45′N, 44°57′E).

(b) Or:

Through 13°00′N, 55°00′E **(Q)**, thence:

Passing N of Suquṭrá (12°30′N, 54°00′E), thence:

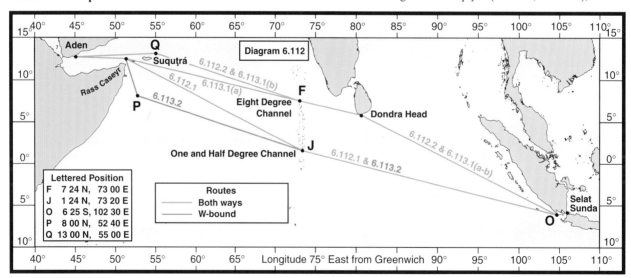

6.112 Aden → Selat Sunda
6.113 Selat Sunda → Aden

Direct to Aden (12°45′N, 44°57′E).

Distance (N or S of Suquṭrá)   3860 miles.

6.113.2

May to September the route is:

Through One and Half Degree Channel (1°24′N, 73°20′E) **(J)**, thence:

To 8°00′N, 52°40′E **(P)**, thence:

Round Raas Caseyr (11°50′N, 51°17′E) to Aden (12°45′N, 44°57′E).

Distance   3940 miles.

Low-powered vessels

6.113.3

Low-powered, or hampered, vessels proceeding between Selat Sunda and Aden may find the routes described at 9.197 more suitable for their use.

Aden ⇐ ⇒ Singapore

Diagram 6.115

Routes

6.114

Routes are directional.

6.114.1

East-bound the routes are:

As for the Malacca Strait route of 6.115.1 and 6.115.2.

Alternative routes E-bound are:

As at 6.112.1 or 6.112.2 from Aden (12°45′N, 44°57′E) to Selat Sunda (6°00′S, 105°52′E), thence:

As at 7.103 to Singapore (1°12′N, 103°51′E).

6.114.2

West-bound the routes are:

As for the Malacca Strait route of 6.116.2 and 6.116.4.

Distances

6.114.3

		Oct to April	May to Sept
E-bound	Malacca Strait	3620 (S)	3650 (N)
	Selat Sunda	4410	4430
W-bound	Malacca Strait	3620 (S)	3780 (S)

(N) or (S) = Through Eight Degree Channel and North or South of Suquṭrá.

Aden ⇒ Darwin or Torres Strait

Diagram 6.115

General information

6.115

1   Across the W part of the Indian Ocean and through the Arabian Sea routes are seasonal.

2   East of Sri Lanka, the choice between an open ocean passage or one through the Eastern Archipelago will be governed by draught, weather, endurance and season.

3   Ocean Route (6.87.3), Selat Sumba Route (6.87.4), Java, Flores and Banda Seas Route (6.87.5) and Malacca Strait Route (6.94.2) are described earlier in the chapter, as indicated.

Distances, see Table at 6.115.3.

Routes

6.115.1

October to April the route is:

Round Raas Caseyr (11°50′N, 51°17′E), thence:

(a) Either:

Through One and Half Degree Channel (1°24′N, 73°20′E) **(J)**, thence:

Direct to Selat Sunda (6°00′S, 105°52′E) as at 6.112.1, and thence:

Via Java, Flores and Banda Seas Route (6.87.5).

(b) Or:

Through One and Half Degree Channel (1°24′N, 73°20′E) **(J)**, thence:

Direct to Routeing Position **(O)** 6°25′S, 102°30′E, and thence:

Via the Ocean (6.87.3) or Selat Sumba (6.87.4) Routes.

(c) Or:

Through Eight Degree Channel (7°24′N, 73°00′E) **(F)**, thence:

Through the TSS off Dondra Head (5°55′N, 80°35′E), thence:

Direct to the N approaches to the Malacca Strait, described at 6.61, thence:

Via Malacca Strait Route (6.94.2).

6.115.2

May to September the route is:

N of Suquṭrá (12°30′N, 54°00′E) to 13°00′N, 55°00′E **(Q)**, thence:

Through Eight Degree Channel (7°24′N, 73°00′E) **(F)**, thence:

(a) Either:

To a landfall (5°45′N, 80°36′E), 10 miles S of Dondra Head, thence:

Direct to Selat Sunda (6°00′S, 105°52′E) and via Java, Flores and Banda Seas Route (6.87.5).

(b) Or:

To a landfall off Dondra Head (5°45′N, 80°36′E), thence:

Direct to Routeing Position **(O)** 6°25′S, 102°30′E and via the Ocean (6.87.3) or Selat Sumba (6.87.4) Routes.

(c) Or:

To a landfall off Dondra Head (5°45′N, 80°36′E), thence:

Direct to the N approaches to the Malacca Strait, described at 6.61, thence:

Via Malacca Strait Route (6.94.2).

Distances

6.115.3

	Torres Strait	Darwin
Ocean Route (6.87.3)		
Oct–Apr	6050	5370†
May–Sept	6080	5400†
Selat Sumba Route (6.87.4)		
Oct–Apr	6020	5390
May–Sept	6050	5420
Java, Flores and Banda Seas Route (6.87.5)		
Oct–Apr	6030	5480
May–Sept	6050	5500
Malacca Strait Route (6.94.2)		
Oct–Apr	6120	5550
May–Sept	6140	5570

†   Through Osborn Passage

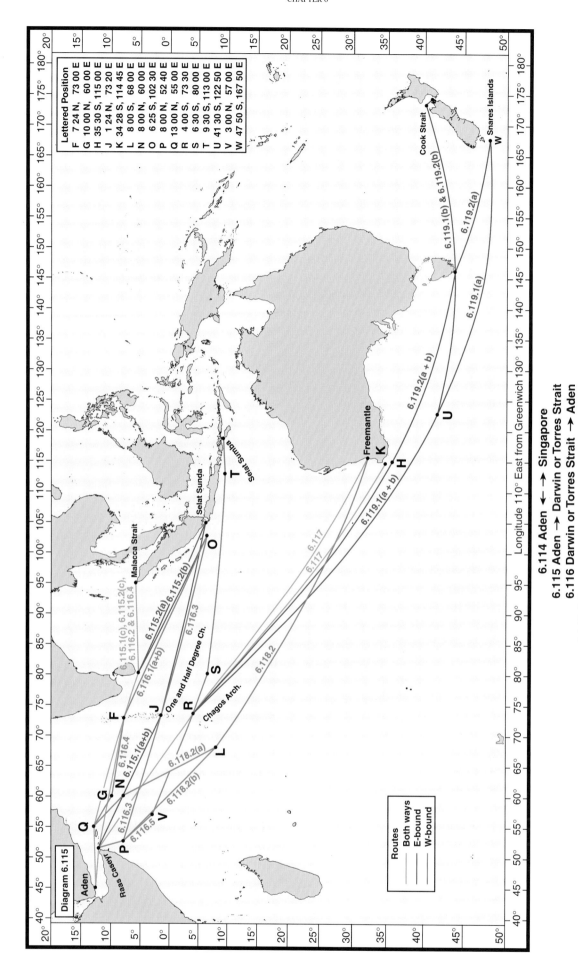

Diagram 6.115

Lettered Position

F	7 24 N, 73 00 E
G	10 00 N, 60 00 E
H	35 30 S, 115 00 E
J	1 24 N, 73 20 E
K	34 28 S, 114 45 E
L	8 00 S, 68 00 E
N	8 00 N, 60 00 E
O	6 25 S, 102 30 E
P	8 00 N, 52 40 E
Q	13 00 N, 55 00 E
R	4 00 S, 73 30 E
S	6 30 S, 80 00 E
T	9 30 S, 113 00 E
U	41 30 S, 122 50 E
V	3 00 N, 57 00 E
W	47 50 S, 167 50 E

6.114 Aden ←→ Singapore
6.115 Aden → Darwin or Torres Strait
6.116 Darwin or Torres Strait → Aden
6.117 Aden → Freemantle or south coast of Australia
6.118 South coast of Australia or Freemantle → Aden
6.119 Aden → New Zealand and Pacific Ocean

138

Darwin or Torres Strait ⇒ Aden

Diagram 6.115
General information
6.116

1 For the part of the voyage E of Sri Lanka the Ocean Route (6.87.3), Selat Sumba Route (6.87.4), Java, Flores and Banda Seas Route (6.87.5) and Malacca Strait Route (6.94.2) as described earlier in the chapter, should be considered.

2 Across the Indian Ocean and Arabian Sea routes are seasonal.

Distances, see Table at 6.116.6.

Routes
October to April the routes are:
6.116.1
From Routeing Position (**O**) 6°25′S, 102°30′E, thence:
To a landfall (5°45′N, 80°36′E), 10 miles S of Dondra Head, thence:
Through Eight Degree Channel (7°24′N, 73°00′E) (**F**), thence:
(a) Either:
S of Suquṭrá (12°30′N, 54°00′E), thence:
Between Raas Caseyr (11°50′N, 51°17′E) and ’Abd al Kŭrī (12°10′N, 52°15′E), bearing mind the difficulty of identifying the landfall (6.38.3), thence:
Direct to Aden (12°45′N, 44°57′E).
(b) Or:
Through 13°00′N, 55°00′E (**Q**), thence:
Passing N of Suquṭrá (12°30′N, 54°00′E), thence:
Direct to Aden (12°45′N, 44°57′E).
Distance (N or S of Suquṭrá)   3860 miles.
6.116.2
From the N approaches to the Malacca Strait, described at 6.61:
Direct to a landfall (5°45′N, 80°36′E), 10 miles S of Dondra Head,, thence:
Through Eight Degree Channel (7°24′N, 73°00′E) (**F**), thence:
As above at 6.116.1 (a) or (b) to Aden (12°45′N, 44°57′E).
May to September the routes are:
6.116.3
From Routeing Position (**O**) 6°25′S, 102°30′E, thence:
Through One and Half Degree Channel (1°24′N, 73°20′E) (**J**), thence:
To 8°00′N, 52°40′E (**P**), thence:
Round Raas Caseyr (11°50′N, 51°17′E) to Aden (12°45′N, 44°57′E).
6.116.4
From the N approaches to the Malacca Strait, described at 6.61:
Direct to a landfall (5°45′N, 80°36′E), 10 miles S of Dondra Head,, thence:
Through Eight Degree Channel (7°24′N, 73°00′E) (**F**), thence:
To 10°00′N, 60°00′E (**G**), thence:
To 13°00′N, 55°00′E (**Q**), thence:
N of Suquṭrá (12°30′N, 54°00′E) to Aden (12°45′N, 44°57′E).
6.116.5
An alternative route from Selat Sumba (9°00′S, 119°30′E) is:
From Selat Sumba to 9°30′S, 113°00′E (**T**), thence:
To 6°30′S, 80°00′E (**S**), thence:
To 4°00′S, 73°30′E (**R**), thence:

N of Chagos Archipelago (6°30′S, 72°00′E) to 3°00′N, 57°00′E (**V**), thence:
To join the route from Selat Sunda (6.113.2) in 8°00′N, 52°40′E (**P**).

Distances
6.116.6

	Torres Strait	Darwin
Ocean Route (6.87.3)		
Oct-Apr	6050*	5370*†
May-Sept	6150	5460†
Selat Sumba Route (6.87.4)		
Oct-Apr	6020*	5390*
May-Sept	6110	5480
May-Sept (Alt)	6220	5600
Java, Flores and Banda Seas Route (6.87.5)		
Oct-Apr	6020*	5470*
May-Sept	6130	5580
Malacca Strait Route (6.94.2)		
Oct-Apr	6120*	5550*
May-Sept	6280	5710

* Passing S of Suquṭrá; if passing N add 30 miles.
† Through Osborn Passage.

Aden ⇒ Fremantle or South coast of Australia

Diagram 6.115
Route
6.117
The route is:
Round Raas Caseyr (11°50′N, 51°17′E), thence:
To 4°00′S, 73°30′E (**R**), thence:
Great circle to Fremantle (32°02′S, 115°42′E) or;
To 34°28′S, 114°45′E (**K**), 20 miles WSW of Cape Leeuwin to join the routes to the E (6.90).
6.117.1
Distances.

Fremantle	4920 miles.
Cape Leeuwin	4950 miles.
Adelaide	6130 miles.
Port Phillip*	6420 miles.

* Port Phillip to Melbourne 40 miles.

South coast of Australia or Fremantle ⇒ Aden

Diagram 6.115
Route
6.118
The routes are seasonal:
6.118.1
October to April the routes are:
As for 6.117 above in reverse.
6.118.2
May to September the routes to Cape Leeuwin from the E are at 6.90. From 34°28′S, 114°45′E (**K**), 20 miles WSW of Cape Leeuwin and Fremantle (32°02′S, 115°42′E) the routes are:
By great circle to 8°00′S, 68°00′E (**L**), thence:
(a) Either:
Through 8°00′N, 60°00′E (**N**), thence:
Through 13°00′N, 55°00′E (**Q**), thence:
N of Suquṭrá (12°30′N, 54°00′E) to Aden (12°45′N, 44°57′E).

(b) Or:

Through 8°00′N, 52°40′E **(P)**, thence:
Round Raas Caseyr (11°50′N, 51°17′E) to Aden
(12°45′N, 44°57′E).

Distances
6.118.3

October to April see 6.117.1.

May to September

	N of Suquṭrá	S of Suquṭrá
Cape Leeuwin	5120	4990
Fremantle	5120	4990
Adelaide	6300	6170
Port Phillip*	6590	6460

*   Port Phillip to Melbourne 40 miles.

Aden ⇒ New Zealand and Pacific Ocean

Diagram 6.115
Route
6.119

The routes are seasonal.

6.119.1

October to April
The routes are:
Round Raas Caseyr (11°50′N, 51°17′E), thence:
Through 4°00′S, 73°30′E **(R)**, thence:
Great circle to 41°30′S, 122°50′E **(U)**, thence:
(a) Either:
Great circle to 47°50′S, 167°50′E **(W)**, to the ENE of
Snares Islands (48°01′S, 166°36′E).
(b) Or:
Great circle, passing close S of Tasmania, to Cook
Strait (41°00′S, 174°30′E).
Distances:
Snares Islands   7400 miles.
Cook Strait   7830 miles.
6.119.2

May to September the routes are:
Round Raas Caseyr (11°50′N, 51°17′E), thence:
Through 4°00′S, 73°30′E **(R)**, thence:
Great circle to 35°30′S, 115°00′E **(H)**, to the S of
Cape Leeuwin, thence:
Great circle to a landfall off South West Cape,
Tasmania (43°35′S, 146°03′E), thence:
(a) Either:
Great circle to 47°50′S, 167°50′E **(W)**, to the ENE of
Snares Islands (48°01′S, 166°36′E).
(b) Or:
Great circle, passing close S of Tasmania, to Cook
Strait (41°00′S, 174°30′E).
Distances:
Snares Islands   7430 miles.
Cook Strait   7820 miles.

PASSAGES BETWEEN EAST SIDE OF INDIAN OCEAN AND AUSTRALIA

Strait of Hormuz, Karāchi, Mumbai and Colombo ⇐ ⇒ Darwin or Torres Strait

Diagram 6.120
Routes
6.120

Between Strait of Hormuz (26°27′N, 56°32′E) and
Dondra Head (5°45′N, 80°36′E), routes are as at 6.59 and
6.60.

East of Sri Lanka, the same considerations apply to the
same routes as those given at 6.115 and 6.116.
6.120.1
Distances

	Strait of Hormuz	Karāchi	Mumbai	Colombo
Ocean Route (6.87.3)				
Torres Strait	5790	5330	4870	4010
Darwin*	5110	4650	4190	3320
Selat Sumba Route (6.87.4)				
Torres Strait	5760	5300	4840	3980
Darwin	5160	4700	4240	3380
Java, Flores and Banda Seas Route (6.87.5)				
Torres Strait	5770	5310	4840	3980
Darwin	5220	4760	4290	3430
Malacca Strait Route (6.94.2)				
Torres Strait	5870	5410	4940	4080
Darwin	5300	4840	4380	3520

*   Through Osborn Passage.

Strait of Hormuz, Karāchi, Mumbai and Colombo ⇐ ⇒ West and North-west coasts of Australia

Diagram 6.120
Routes
6.121

Routes from Strait of Hormuz (26°27′N, 56°32′E),
Karāchi (24°46′N, 66°57′E) and Mumbai (18°51′N,
72°50′E) are:
As at 6.59 and 6.60 as far as the latitude of Cape
Comorin (8°05′N, 77°33′E) if bound for W coast
of Australia, or as far as Pointe de Galle (6°01′N,
80°12′E) if bound for NW coast, thence:
Either:
By great circle to the W coast..
Or:
By rhumb line, and as navigation permits to
destinations on the NW coast.
6.121.1
Distances:

	Cape Leeuwin*	Fremantle	Port Hedland
Strait of Hormuz	4950	4890	4620
Karāchi	4490	4430	4160
Mumbai	4030	3970	3690
Colombo	3180	3120	2830

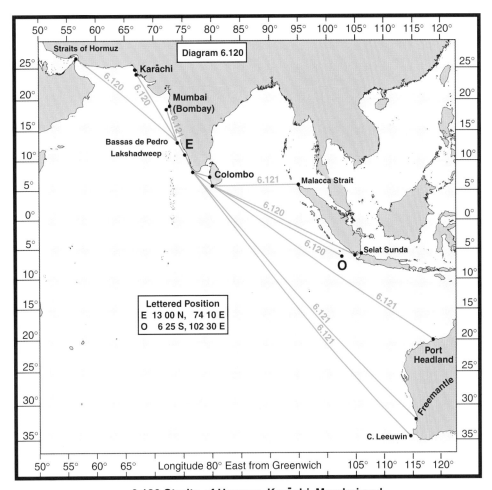

**6.120 Straits of Hormuz, Karāchi, Mumbai and
Colombo ← → Darwin or Torres Strait
6.121 Straits of Hormuz, Karāchi, Mumbai and
Colombo ← → West and North west coasts of Australia**

Ports in the Indian Ocean ⇐ ⇒ East coast of Australia

General information
6.122

1   The choice between a route N or S of Australia depends both on the distance, which may vary with the season, and the climatic conditions on each route.

2   In the following table of distances, Caloundra Head (26°48′S, 153°08′E), 35 miles from the Port of Brisbane, is taken as the central Australian point, and the distances are quoted from it. A comparison for Sydney may be obtained by adding 470 miles to N-about distances and subtracting that amount from S-about distances.

3   Modern bulk carriers, loading in ports N of Brisbane and bound for Europe, may be unable to use the Torres Strait routes due to excessive draught. A route, often used, passes through Jomard Entrance (11°15′S, 152°06′E) (7.151) or Rossel Spit (11°27′S, 154°24′E) (7.151), passing N of New Guinea, through Selat Sagewin (0°55′S, 130°40′E) thence through Selat Lombok (8°47′S, 115°44′E). See 7.41.3 for further details.

4   From Selat Lombok routes to Europe are either via Cape of Good Hope (6.104.1) or the Suez Canal, via Aden (6.116.5).

Distances
6.122.1

Between Caloundra Head* and:		E-bound		W-bound		References
	Season	S-about[a]	N-about[b]	S-about[a]	N-about[b]	7.40 – 7.42
Cape of Good Hope¤	Oct–Apr	6810	8190	—	—	6.102, 6.104, 6.105, 6.107.
Cape Town	Oct–Apr	6990	8370	7390	8320	6.102, 6.104, 6.105, 6.107.
	May–Sept	7190	8200	7620	8370	
Durban	Oct–Apr	6400	7670	6770	7670	6.103, 6.104, 6.106, 6.107.
	May–Sept	6600	7570	6830	7580	
Mombasa		7300	7470*	6970	7470*	6.90, 6.111.

Between Caloundra Head* and:	Season	E-bound		W-bound		References
		S-about[a]	N-about[b]	S-about[a]	N-about[b]	7.40 – 7.42
Aden	Oct–Apr	7390	7360*	7390	7390*	6.90, 6.115–6.118.
	May–Sept	7390	7390*	7430‡	7470*	
Strait of Hormuz		7390	7110*	7390	7110*	6.90, 6.120, 6.121.
Karāchi		6930	6650*	6930	6650*	6.90, 6.120, 6.121.
Mumbai		6470	6190*	6470	6190*	6.90, 6.120, 6.121.
Colombo		5620	5330*	5620	5330*	6.90, 6.120, 6.121.
Chennai		5900	5400†	5900	5400†	6.90, 6.94, 6.95.
Kolkata (Calcutta) Approach§		6110	5340†	6110	5340†	6.90, 6.94, 6.95.
Yangon (Rangoon) River*		5740*	4880†	5740	4880†	6.90, 6.94, 6.96.

¤   Position 145 miles S of Cape of Good Hope.

[a]   All S-about routes via Bass Strait.

[b]   All N-about routes via Torres Strait and Inner Route (7.41.2).

*   Via Ocean Route (6.87.3).

§   Kolkata (Calcutta) approaches to Kolkata 125 miles

‡   S of Suquṭrá.

†   Via Malacca Strait Route (6.94.2)

**   Yangon (Rangoon) River Entrance to Port of Yangon 40 miles.

CHAPTER 7

PACIFIC OCEAN AND ADJACENT SEAS

GENERAL INFORMATION

COVERAGE

Chapter coverage
7.1

This chapter contains details of the following routes:

Admiralty Publications
7.2

Relevant navigational publications should be consulted, in addition to this book, when planning and conducting passages. Details of these, and the areas covered, can be found, as follows:

Admiralty Sailing Directions; see 1.12.1

Admiralty List of Lights and Fog Signals; see 1.12.2

Admiralty Tide Tables; see 1.12.3

Admiralty List of Radio Signals; see 1.12.5

Admiralty Distance Tables; see 1.12.8

7.2.1

See also *The Mariner's Handbook* for notes on electronic products.

NATURAL CONDITIONS

Charts:

5127 (1) to (12) Monthly Routeing Charts for North Pacific Ocean.

5128 (1) to (12) Monthly Routeing Charts for South Pacific Ocean.

5310 World Surface Current Distribution.

WINDS AND WEATHER

General description
7.3

1   The following description of the winds and weather of the Pacific Ocean and adjacent seas amplifies the general statement given in *The Mariner's Handbook*. For more precise information regarding oceanic winds and weather the mariner is referred to Charts *5127 (1) to (12); 5128 (1) to (12);*. Reference should also be made to Chart *5310* and to Diagrams 1.14a *World Climatic Chart (January)* and 1.14b *World Climatic Chart (July)*.

2   Detailed information about specific localities should be sought in the appropriate volume of *Admiralty Sailing Directions*.

3   In the E part of the Pacific Ocean, the winds and weather conform, in the main, with the text book description of oceanic winds and weather published in *The Mariner's Handbook*. In the W part of the ocean, however, the seasonal heating and cooling of the Asiatic land mass results in the establishment here of a monsoonal regime. Conditions are further complicated, in the region between Australia and the Philippines, by numerous islands, many of which are of considerable size and height, causing marked differences in the winds and weather experienced in different localities. These local differences are dealt with in the appropriate volume of *Admiralty Sailing Directions*.

North Pacific Ocean

Intertropical Convergence Zone
7.4

1   The Intertropical Convergence Zone remains permanently N of the equator in longitudes E of about 160°W. To the W of that meridian, it lies in the S hemisphere from about November or December until April or May; in the northern summer it is virtually non-existent W of about 150°E. In the W part of the North Pacific, therefore, the Intertropical Convergence Zone is really only in evidence during the change of the monsoons, from about mid-September to mid-November and from about mid-April to mid-May.

2   The weather of the Intertropical Convergence Zone is that typical of the Zone in other oceans, that is light variable winds, with calm alternating with squalls, heavy showers or thunderstorms, but W of about 130°W the frequency of calms and variable winds is considerably less than in the Intertropical Convergence Zone of other oceans and most winds are from an E point.

3   The mean positions of the Zone are shown, in January on Diagram 1.14a and in July on Diagram 1.14b. The actual position is subject to much variation, as is the width of the Zone, which averages about 150 miles. The worst weather is generally experienced when the Trade Winds of the two hemispheres meet at a wide angle. Visibility is normally good, except in heavy rain.

Seasonal winds of eastern North Pacific Ocean
7.5

1   In summer E of about 120°W and between the Intertropical Convergence Zone and the equator, there is an area covered by prevailing SW winds, see Diagram 1.14b *World Climatic Chart (July)*. These winds are of a monsoonal nature and result from the summer heating of the North American continent, which causes a reduction of pressure over that area and a distortion of the Intertropical Convergence Zone to the N, the South-east Trade Wind of the South Pacific Ocean is drawn across the equator and deflected to the right by the effect of the earth's rotation and is felt as a SW wind in the area under consideration.

2   Over the greater part of the area these winds prevail from about June to October and replace the North-east Trade Wind which prevails there in winter, see 7.6. The duration of the season of these south-westerlies varies with latitude, being longest near the equator, near which S to SW winds are prevalent E of 100°W during most months.

3   Winds are mostly light or moderate, though squalls, in which the wind may at times reach gale force, are rather common. Tropical storms, see 7.12 also produce strong winds and gales at times.

4   The weather is generally cloudy and unsettled and rainfall is considerable. It is in fact, these winds which bring the rainy season to much of Mexico and Central America. Visibility over the open ocean is generally good except in rain.

North-east Trade Wind
7.6

1   The North-east Trade Wind blows on the equatorial side of the large clockwise circulation around the oceanic high pressure area situated in about 30°N. This 'high' lies farther N and is somewhat more intense in summer than in winter and, while in the former season it generally consists of a single cell, in the latter it more often represents the resultant of a succession of anticyclones moving E across the North Pacific from Asia and becoming stationary over the E part of the ocean.

2   In summer the Trade Wind blows in the region E of about 150°E and between the Intertropical Convergence Zone and about 32°N; the limits are not fixed, but fluctuate considerably. To the W of 150°E the Trade Wind gives way to the south-west monsoon of the W part of the North Pacific Ocean, described at 7.8.

3   As with the Trade Winds of other oceans, these winds are remarkable for their persistence and steadiness over large areas. The general direction and steadiness of the wind in different parts of the zone can best be seen from a study of Diagrams 1.14a and 1.14b, which will show that the direction becomes more N (or even NW) near the American coast and mainly E in summer in the SW part of the area covered by these winds.

4   The strength of the North-east Trade Winds averages force 3 to 4, but often freshens to 5 to 6. Winds are likely to reach force 7, or above, on 1 to 3 days per month in the heart of the Trade Wind. In the vicinity of the Mexican coast, N of about 10°N, and between about 90°W and 100°W, the frequency rises to 3 to 6 days per month from November to February. Apart from squalls, winds of this strength are unlikely within about 600 miles of the equator.

5   The typical weather of the Trade Wind zone is fair, with scattered showers and skies half covered by small cumulus cloud. At times the Trade Wind becomes unsteady, being interrupted by a day or two of unsettled showery weather with occasional squalls. In the NE part of the zone, near the American coast, cloud amounts are generally smaller than elsewhere and rain is rare.

6   Visibility over the open ocean is generally good, except in rain, but there is often a light haze which restricts visibility to between 8 and 15 miles; showers, cloud and haze usually increase when the wind freshens. Dust haze is sometimes prevalent off the American coast and is associated with fresh or strong offshore winds.

North-east Monsoon
7.7

1   In the winter of the N hemisphere, the cooling of the Asiatic land mass results in the establishment of an intense area of high pressure over Mongolia and the E part of Siberia. The anticyclonic wind circulation resulting from this pressure distribution gives rise to the establishment, at this season, of NE winds over the W part of the North Pacific Ocean, S of about 30°N, and in the South and East China Seas and Yellow Seas. The N and E limits of the area covered by the Monsoon are not very well defined. On its E side it merges with the North-east Trade Wind of the central and E parts of the North Pacific Ocean, while to the N it gives way to the prevailing westerlies of higher latitudes.

2   The time of the onset of the Monsoon varies with latitude. In the N it begins about September, while towards the equator it does not become established until November. In April it becomes less steady, the prevailing direction becomes more E and winds with a S component are more frequent.

3   The general direction and steadiness of the Monsoon are indicated on Diagram 1.14a *World Climatic Chart (January)*. At the height of the season, in January, winds over the open waters of the South China Sea and E of the Philippines are almost exclusively from between N and E, while in the Yellow Sea the direction becomes more N and over the S part of Japan it is NW. Wind direction become more variable as latitude increases. The strength of the wind over the open sea averages about force 5 in the N part of the monsoon zone, increasing to force 6 in the Taiwan Strait; farther S it decreases to an average of force 5 in the South China Sea and force 4 S of 10°N. It becomes less steady, lighter and more N in direction towards the equator and among the islands of the Sulu Sea and the Celebes Sea.

4   The movement of depressions in an E direction across the area also affects the strength of the wind. As far E as the general longitude of Japan there is often no closed wind circulation round newly formed 'lows'; their passage is marked by a slackening of the monsoon ahead of them and a freshening, often to gale force, in their rear.

5   At the height of the season, in December and January, winds are likely to reach force 7 or above on 6 to 10 days per month over much of the area between Vietnam, Luzon, T'ai-wan and Japan, as indicated on Diagram 1.14a *World Climatic Chart (January)*; the stormiest area is E of Luzon and T'ai-wan where winds of this strength are likely on more than 11 days per month. In the Yellow Sea their frequency is about 3 to 6 days per month, while S of 10°N it decreases to 0 to 3 days per month.

6   To the N of 20°N, overcast skies with periods of light rain or drizzle are typical during this season, especially from January to April, though at times there are periods of more broken skies and in October and November generally fair conditions prevail along the SE coast of China. In Bo Hai (38°30′N, 119°30′E) and Liaodong Wan (40°00′N, 121°00′E), immediately to leeward of the Asiatic land mass, a good deal of fine and settled weather with only small amounts of cloud prevails.. South of about 17°N, over the open seas, skies are only about half covered and there are occasional showers; cloudiness increases again towards the equator and showers become more frequent.

7   The weather in the vicinity of land is greatly affected by the degree of exposure to the prevailing monsoon. When the monsoon blows onshore, and especially when the coast

is backed by high ground, cloud amounts are larger and rainfall is heavier than over the open sea, while to leeward of high ground fairer conditions prevail. Information about specific locations is published in the appropriate volume of *Admiralty Sailing Directions*.

8 Over the open ocean, visibility is good except in rain. Off the coasts of China and Vietnam, poor visibility becomes increasingly frequent after December and mist and fog may may occur on more than 10 days per month in the vicinity of North Vietnam in February and March, and for 8 to 9 days per month off Hong Kong in March and April. In Bo Hai and Liaodong Wan, strong NW winds at times bring bring dust haze from the interior of Mongolia.

South-west Monsoon
7.8

1 In the summer of the N hemisphere, intense heating of the Asiatic land mass results in the formation of an area of low pressure centred approximately over NW India with an extension over the E part of Asia, see Diagram 1.14b *World Climatic Chart (July)*. The South-east Trade Wind of the Pacific and Indian Oceans is drawn across the equator and deflected to the right by the effect of the earth's rotation. This wind, known as the south-west monsoon, is felt in the W part of the North Pacific Ocean and the South and East China Seas and Yellow Sea as a prevailing S to SW wind, and in the Japan Sea as a S to SE wind. The N and E limits of the Monsoon are ill-defined but, W of about 140°E and S of about 40°N, winds are predominantly from between SE and SW at the height of the season, in July. The general direction and steadiness of the winds at this period are indicated on Diagram 1.14b *World Climatic Chart (July)*. The Monsoon is steadiest in the South China Sea, where nearly all winds are from between S and W. Farther N and E they are much more variable in direction and, in the early part of the season, N of about 25°N, travelling depressions may cause winds from any direction. Along the China coast between 20°N and 30°N and in the vicinity of T'ai-wan north-easterlies are still more common than south-westerlies in May.

2 The average strength of the Monsoon over the open sea is about force 3 to 4 in the South China Sea and force 3 elsewhere, but squalls, in which the wind may reach gale force, are fairly common. Apart from these squalls, or in the vicinity of tropical storms (7.12), winds do not often reach force 7 in the monsoon season. Land and sea breezes prevail close to the coast and calms are not uncommon.

3 The weather over the open sea away from the effects of land is mainly fair, with skies about half covered and with occasional showers. Over the coasts, especially if exposed to the Monsoon and backed by high ground, cloudy weather with frequent heavy rain prevails.

4 Visibility over the open ocean is good except when reduced by rain, but along the China coast there is a high frequency of sea fog in certain months, due to the spread of warm moist equatorial air over water previously cooled by the NE winds of the winter Monsoon. The water recovers its normal temperature progressively from S to N and the foggy season reaches its maximum in April off Hong Kong (8 to 9 days per month), in June off Chang Jiang (31°03′N, 122°20′E) (12 days per month) and in July off Shandong Bandao (37°00′N, 122°00′E) (12 days per month). In the Japan Sea, fog occurs on 3 to 4 days per month and on 5 to 7 days per month off N Honshū. After these months the incidence drops sharply to about 2 days per month and fog is rare in the later part of the season.

The Variables
7.9

1 In a belt extending across the central part of the Pacific Ocean and situated in about 25°N to 30°N in winter and 35°N to 40°N in summer, there are variable and mainly light or moderate winds in the vicinity of the oceanic anticyclone. In the E part of this zone winds are mainly N in all seasons and form a N extension of the North-east Trade Wind around the E flank of the oceanic 'high'. In the W part of the zone, in summer, winds become mainly S and merge with the south-west monsoon described at 7.8, while in winter they give way, W of about 150°E, to prevailing NW winds forming part of the circulation of the north-east monsoon.

2 In summer the winds are generally light and are likely to reach force 7 only on rare occasions except in association with tropical storms (7.12), and E of about 140°W where they may be expected to reach this strength on 1 to 4 days per month, the higher figure applying towards the American coast, near which strong N to NW winds are common. At the height of the winter season, in January, winds may be expected to reach force 7, or above, on 1 to 3 days per month E of about 140°W and on 3 to 6 days per month W of that meridian, increasing to 6 to 10 days per month in the area covered by the north-east monsoon W of about 150°E, described at 7.7.

3 The weather in summer is generally fair or fine near the normal position of the oceanic 'high', see Diagram 1.14b *World Climatic Chart (July)*, which at this season usually consists of a single cell, and rain is infrequent. Cloudier conditions prevail E and W of the area of high pressure; rainfall is light on the E side, towards the American coast, but more common to the W. In winter the 'high' shown on Diagram 1.14a *World Climatic Chart (January)* usually consists of a series of E moving anticyclones, near which fair or fine weather prevails, the intervening troughs of relatively low pressure being characterised by cloudy, showery weather.

4 Visibility in winter is mostly good except in rain and over the open ocean fog is not common. In summer fog and poor visibility become increasingly frequent towards the N limit of the zone (40°N at this season). In the W this is due to the N flow of warm moist equatorial air over progressively colder water, aggravated off the E coast of N Honshū by contact with the cold Kamchatka Current (7.26). In the E it is due to a similar cooling by the California Current (7.26). Over much of the zone fog may occur on 3 to 4 days per month at this season, rising to 5 to 7 days per month off the coast of California.

The Westerlies
7.10

1 On the polar side of the oceanic anticyclone, the prevailing winds are from some W point but summer and winter conditions are markedly different and it is convenient to treat the two periods separately.

2 In winter, N of 40°N, the almost continuous passage of depressions from the vicinity of China and Japan in a NE direction towards the Aleutian Islands and S Alaska causes winds to vary greatly in both direction and strength and winds from any direction can be experienced. As can be seen from Diagram 1.14a *World Climatic Chart (January)*, strong winds and gales are frequent. The region of highest gale frequency extends from E of Japan to the area S of the Aleutians and the Alaska peninsular; in this region winds are likely to reach force 7, or above, on 12 to 18 days per month.

3   The main feature of the weather is its great variability, periods of overcast skies and rain or snow alternating with fairer intervals. Fine weather is seldom prolonged and cloud amounts are generally large. Although fog is not common at this season, rain and snow often reduce visibility drastically. Visibility is also often only moderate with winds from a S point, but is generally good (except in precipitation) with N or NW winds.

4   In summer, depressions are less frequent, much less intense, and their tracks are farther N than in winter. Winds, therefore, although they still vary considerably both in direction and strength are much lighter and gales are far less common. Over the greater part of the zone, winds may reach force 7, or above, on 15 days per month S of about 50°N; the quietest month is July, during which winds of this strength are unlikely on more than 1 day on average. North of 50°N observations are scarce, but the frequency of gales is probably the same.

5   Over the greater part of the zone the weather is very cloudy and foggy. West of about 160°W fog occurs on about 5 to 10 days per month in most parts, rising to more than 10 days per month over large areas; see Diagram 1.14b *World Climatic Chart (July)*. This high incidence rate is due to the N flow of warm moist S to SW winds over progressively colder water, in particular over the cold waters of the Kamchatka Current (7.26). East of 160°W the frequency is less but it increases again to 5 to 10 days per month towards the W coast of America over the cold waters of the California Current (7.26). Apart from fog, visibility is generally moderate.

Polar Easterlies
7.11

1   Since, in winter, the tracks of most depressions are S of the Aleutian Islands, the prevailing winds in the Bering Sea at this season are often E. As in the case of the Westerlies, great variations in both strength and direction occur, due to the passage of some depressions close to and across the area. The N part of the zone is not navigable on account of ice while in the S part winds may reach force 7 or more on over 10 days per month.

2   The weather is generally very cloudy and precipitation, usually in the form of snow, is frequent, amounts being greatest in the S. Visibility is often poor because of the snow.

Tropical storms
7.12

1   In the W part of the North Pacific Ocean these storms are known as 'typhoons' and in the E part as 'hurricanes'. They are fully described in *The Mariner's Handbook*, together with their warning signs and advice on avoiding them.

2   The area mainly affected by typhoons is W and N of the Caroline Islands and Marianas Islands and includes the N part of the Philippines, the N half of the South China Sea, the vicinity of the China coast and T'ai-wan, the East China Sea and Japan. Although the typhoons may occur in any month, more than 50% are experienced from July to October and nearly 90% between May and December inclusive. September is the month with the greatest frequency with an average of just over 4 storms. The number experienced in any month varies greatly in different years.

3   Taking the area as a whole, no month is immune from typhoons, but certain parts of it are free from them in certain months, notably the China coast, Taiwan Strait and

the W part of the East China Sea, in which areas it is most likely that they will be encountered from December to April.

4   The area mainly affected by hurricanes is the vicinity of the Pacific coast of America between about 10°N and 30°N; they have, however, been recorded as far W as 130°W to 140°W, generally in the early part of the season.

5   Almost all hurricanes occur in the period from June to October, the month of greatest frequency being September, with an average of 2 storms. They are occasionally recorded in May and November and, very occasionally, in December; they are unknown from January to April. As with all tropical storms the number experienced in different years varies greatly.

6   More detailed information regarding the frequency of typhoons and hurricanes in different localities will be found in the appropriate volume of *Admiralty Sailing Directions*.

South Pacific Ocean
Intertropical Convergence Zone
7.13

1   As stated at 7.4, the Intertropical Convergence Zone remains N of the equator throughout the year in longitudes E of about 160°W. In more W longitudes it lies in the S hemisphere from about November or December to April or May, reaching its extreme S position in February. The seasonal movement of the zone in the W part of the South Pacific is thus large, as also is the day to day variation in its position, especially in the extreme W in the vicinity of N Australia and New Guinea. The width of the zone averages about 150 miles, but it may, at times, be as little as 50 miles and at others over 300 miles.

2   The weather is that typical of the Intertropical Convergence Zone elsewhere, in which calms and light variable winds and fine weather alternate with squalls, heavy rain (most often in the form of showers) and thunderstorms. Conditions are generally more severe in the W part of the South Pacific Ocean than elsewhere in this ocean, due to the wide angle at which the South-east Trade Wind and the north-west monsoon, (see 7.15 and 7.14), meet. Visibility over the open sea is good except in heavy rain.

North-west Monsoon
7.14

1   During the summer of the S hemisphere, pressure is low over the N part of the heated Australian land mass and the Intertropical Convergence Zone is located over that area. The north-east monsoon of the W part of the North Pacific Ocean is drawn across the equator and deflected to the left by the effect of the earth's rotation. It is felt over the South Pacific Ocean, W of about the 180th meridian and between the equator and the Intertropical Convergence Zone as a prevailing NW wind known as the north-west monsoon. The season of this monsoon varies somewhat with latitude; in the vicinity of N Australia it is generally only firmly established in January and February, while farther N in the Java Sea and the Banda Sea it normally blows from December to March.

2   The general wind direction is indicated on Diagram 1.14a *World Climatic Chart (January)*; winds are mainly from N and NE near the equator and back gradually to between NW and W in more S latitudes. Over much of the area the constancy of the Monsoon is not great and winds from other directions are also experienced, though at the height of the season and away from the effects of the land winds from between S and E are uncommon. In the vicinity of the numerous islands local effects may give rise to variation in both the direction and force of the wind.

3   The strength of the Monsoon is generally only light or moderate, but squalls, in which the wind may reach gale force, are rather common. Apart from these, or when in the vicinity of tropical storms (7.18), gale force winds are unlikely.

4   The weather is generally cloudy, and rain, usually in the form of heavy showers, is frequent over most of the area. In the vicinity of land the wind often varies greatly over short distances. Off coasts exposed to the monsoon, especially if backed by high ground, rainfall is often very heavy and cloud amounts are large, while off sheltered coasts fair weather and less cloudy conditions prevail.

5   Visibility over the open ocean is generally good except in heavy rain. Information relating to specific localities will be found in the appropriate volume of *Admiralty Sailing Directions*.

South-east Trade Wind
7.15

1   The South-east Trade Wind blows on the equatorial side of the oceanic high pressure situated in about 30°S. In the E part of the zone the Trade Winds are maintained by the semi-permanent anticyclone situated towards the E side of the ocean and shown on Diagrams 1.14a *World Climatic Chart (January)* and *1.14b World Climatic Chart (July)*, while in the W they are due to migratory anticyclones moving E from the vicinity of Australia.

2   Over the greater part of the ocean the N limit of the Trade Wind is defined by the Intertropical Convergence Zone. In the winter of the S hemisphere, E of about 120°W and W of about 140°E, the N limit is the equator, N of which the Trade Wind recurves to form the South-westerlies and the south-west monsoon of the E and W parts of the North Pacific Ocean respectively; these winds are described at 7.8. The S limit of the Trade Wind is situated in 15°S to 20°S in winter and in 30°S in summer.

3   As with the Trade Winds in the other oceans, those of the South Pacific Ocean are remarkable for their persistence and steadiness. The general direction and constancy of the wind can best be seen by a study of Diagrams 1.14a *World Climatic Chart (January)* and 1.14b *World Climatic Chart (July)*. In the vicinity of the W coast of South America the Trade Wind blows from between S and SE while farther W the direction becomes predominantly E. It becomes SE again, in winter, W of about 160°E and over the seas N of Australia, where it is sometimes known as the south-east monsoon. West of about 140°W, from November until April, the Trade Wind is unsteady over large areas and, though the predominant direction remains from between NE and SE, winds from other directions are rather frequent.

4   The average strength of the Trade Wind is about force 4, but it often freshens to force 5 or 6 over large areas. Over the greater part of the Trade Wind zone winds of force 7, or above, are unlikely on more than 1 or 2 days per month and, apart from short-lived squalls, are rare within 10 degrees of the equator. In an area between the NE coast of Queensland, Nouvelle Calédonie and Vanuatu, however, the frequency rises to 3 to 6 days per month for much of the year.

5   Over the open ocean the characteristic weather of the steady South-east Trade Wind is fair with occasional showers; skies are about half covered with small cumulus clouds and there is a slight haze which reduces visibility to within about 8 to 15 miles. Showers, cloud and haze generally increase when the wind freshens. To the E of about the 180th meridian and between the equator and about 8°S, but varying somewhat with the season, there is a belt in which rainfall and cloud amounts are generally small. This dry belt widens towards the coast of South America to include most of the area covered by the Trade Wind; weather here is cloudier and overcast skies are common.

6   From November to April, W of about 140°W, but excluding the dry belt mentioned above, weather is often unsettled, the Trade Wind becomes unsteady and is followed by a period of cloudy, showery weather before settling in again with increased strength and some squalls from between S and E.

7   Over the seas N of Australia, during the season when the South-east Trade Wind prevails in these regions, namely from April to September or October, cloud amounts and rainfall are small. Extensive dust haze prevails, especially towards the end of the season, due to persistent offshore winds from the increasingly dry interior of the continent. These conditions are most marked in the Timor Sea, but are also prevalent in the Java Sea and Banda Sea and to a lesser extent in the Arafura Sea. Visibility in haze is often less than 5 miles. Fog and mist are rather common towards the coast of South America over the cold waters of the Peru Current (7.27) but rarely occur elsewhere.

The Variables
7.16

1   Between the S limit of the South-east Trade Wind (7.15) and the N limit of the Westerlies (7.17) there is a wide belt of variable winds of mainly moderate strength. The approximate area covered by this belt extends from 25°S to 40°S in summer and from 20°S to 30°S in winter. It does not, however, extend completely across the ocean. To the E of about 85°S winds from S to SE prevail, forming a S extension of the South-east Trade Wind around the E flank of the oceanic 'high'. Except in the E part of the zone referred to above, winds vary considerably in strength as well as direction and, in general, strong winds become more frequent with increasing latitude. Over the greater part of the area winds are likely to reach force 7, or above, on 1 to 3 days per month, rising to 3 to 6 days per month towards the S limits of the zone. This latter frequency is also reached in many months over large areas W of about 160°W.

2   The weather is variable, being mainly governed by the E-moving anticyclones already mentioned. Near the centres of these anticyclones it is fair or fine, while the intervening troughs of low pressure are characterised by cloudy, unsettled weather with rainfall increasing towards the S. To the E of 85°W to 90°W rainfall becomes progressively less towards the N and E and it is very infrequent in the vicinity of the American coast. In this area cloud amounts are often large and overcast skies are common in winter.

3   Visibility is generally good in the N part of the zone, except when reduced by rain, but the frequency of moderate and poor visibility increases with latitude towards the S limits of 40°S in summer and 30°S in winter. Visibility of less than 5 miles is recorded in some 10% to 15% of ships' observations in summer and 5% in winter. It is generally associated with winds from some N point.

4   In the extreme E part of the zone, over the cold waters of the Peru Current (7.27), fog is rather prevalent and off

the W coast of South America it occurs on 3 to 5 days per month towards the S limit of the zone.

The Westerlies
7.17

1 The Westerlies or 'Roaring Forties' predominate S of the belt of high pressure described at 7.15 and 7.16. As in the zone of the Westerlies in other oceans, the almost continuous passage of depressions from W to E causes the wind to vary greatly in both direction and strength. Gales are very common, especially in winter, during which season winds are likely to reach force 7, or above, on 5 to 10 days per month over most of the area between 30°S and 40°S and on more than 12 days per month S of 40°S. One of the stormiest areas is to the W or NW of Cabo de Hornos (56°04′S, 67°15′W), in which region winds of this strength are likely on about 20 days per month from July to September. In summer, gales are somewhat less common and occur farther S. To the E of about 150°W, between 40°S and 45°S, winds are likely to reach force 7, or above, on 5 to 10 days per month and S of 45°S the frequency rises to more than 10 days per month. To the W of 150°W the area of highest gale frequency is farther S, but in few parts of the zone of the Westerlies, namely S of 40°S at this season, is the frequency less than 3 to 5 days per month.

2 Diagrams 1.14a *World Climatic Chart (January)* and 1.14b *World Climatic Chart (July)* give an indication of the distribution of gales in summer and winter respectively.

3 As in the Westerlies of other oceans the weather is very variable, periods of overcast skies and rain or snow associated with the fronts of E-moving depressions alternating with fair weather. Fine weather is seldom prolonged and cloud amounts are generally large at all times.

4 Visibility also varies greatly; with winds from a S point it is generally good, while N winds are often associated with moderate or poor visibility. Fog is rather common in summer and may be expected on 3 to 5 days per month.

Tropical storms
7.18

1 Tropical storms are known as 'hurricanes' or 'cyclones' in the South Pacific Ocean. They are described, and advice on avoiding them is given, in *The Mariner's Handbook*.

2 The area mainly affected is W of about 155°W and S of 8°S to 10°S. Most storms occur from December to April and the season of greatest frequency is from January to March. They are not unknown at other times and the actual number of storms varies from year to year. The movement of storms in this area tends to be erratic and difficult to foretell.

3 More detailed information regarding the frequency of hurricanes in specific localities will be found in the appropriate volume of *Admiralty Sailing Directions*.

SWELL

North Pacific Ocean

East of 160°W
7.19

1 In this region there are large areas devoid of recorded observations of swell, information is therefore confined to certain localities.

2 Off the coast of America between 20°N and 35°N a swell from the NW, mainly low or moderate and rarely heavy, persists throughout the year. North of 35°N the swell is mainly from the NW in summer, but from between NW and N in winter.

3 To the N of 50°N the swell is predominantly from SW to W throughout the year though there is an increased frequency from W to NW in October, December and March. From October to April heavy swell is reported on about 20% of occasions and on rather less than 10% in the remaining months.

4 A swell from NE persists throughout the year SE of Hawaii. It is normally moderate or heavy and may extend as far as 130°W and, in winter, as far as the equator.

5 Off the American coast, S of 20°N, swell is mainly from the N in winter, when it may be heavy, and from the S in summer.

West of 160°W
7.20

1 This region has many islands which frequently interrupt the swell waves, particularly S of 20°N and near the Aleutian Islands. The statements which follow apply to the uninterrupted areas. In the SW part of the North Pacific Ocean the swell is governed by the monsoons.

2 From the equator to 20°N a NE swell predominates from November to March inclusive. It is mainly low or moderate, but is heavy on 10% of occasions.

3 The South China Sea is affected, from June to September inclusive, by a swell from the SW, sometimes moderate but only rarely heavy.

4 To the N of 20°N and W of 140°E, there is no predominant direction, though a swell from the NW is often found. Swell in this region is normally moderate or heavy; the frequency of heavy swells is about 30% in the area close E of Japan.

Length of swell
7.21

1 In the North Pacific Ocean swell is normally average in length, though short and long swells can also be encountered.

South Pacific Ocean

West of 160°W
7.22

1 This region is encumbered by islands which interrupt swell waves. The following statements therefore apply only to areas where there are few islands.

2 From the equator to 20°S swell is predominantly from between NE and SE and is mainly moderate in height.

3 From 20°S to 30°S swell is frequently from between SE and SW but no direction predominates. In this region swell is normally moderate or heavy.

4 From 30°S to 50°S swell is predominantly SW and moderate or heavy. South of 30°S two or even three swells are often present and reports of confused swells are frequent.

5 To the S of 50°S swell comes mainly from between NW and SW, moderate or heavy.

East of 160°W
7.23

1 In this region there are vast areas for which swell data are almost non-exist, particularly between 30°S and 50°S and between 80°W and 120°W. Available data indicate the following.

2 Off the coast of South America, between 10°S and 40°S, a swell from S to SW persists throughout the year, normally moderate but heavy at times.

3   From the equator to 20°S and from 100 miles off the coast of South America to 130°W a SE swell predominates. It is mainly moderate, though occasionally heavy.

4   From 30°S to 50°S and between 130°W and 160°W a moderate to heavy swell from the SW predominates. Swells over 6 m in height are a common feature S of 35°S.

5   To the S of 50°S swell comes mainly from between NW and SW and is either moderate or heavy. As in the W part of this ocean, reports of confused swell are frequent S of 30°S.

Length of swell
7.24

1   In the South Pacific Ocean most swells are short or average in length, but waves of more than 300 m in length occur quite often and it is in this ocean that the longest swells occur.

2   In the South Pacific Ocean abnormal waves may occur, see 1.19.

CURRENTS

General remarks
7.25

1   The Pacific Ocean, like the Atlantic Ocean, is dominated in equatorial regions by the semi-permanent N and S subtropical anticyclones. These generate a wide expanse of W-going sets of moderate constancy between about 25°N and 25°S with a slight movement of the N and S boundaries from summer to winter. Separating the N and S Equatorial Currents is the Equatorial Counter-current which sets E with a low to moderate constancy for much of the year.

2   Between the Poles and the sub-tropical gyres there are E to ESE-going sets in the N hemisphere, with E to ENE-going sets in the S hemisphere which are assisted by the temperate latitude W winds of the Aleutian low pressure in the N and the Roaring Forties of the S hemisphere.

3   The increased frequency of mobile depressions, in higher latitudes leads to a general reduction in both constancy and rate of predominant currents. Often the predominance of a particular direction is only minimal, any other direction being just as likely.

North Pacific Ocean
7.26

1   The combined, generally low constancy, counter-clockwise circulation in the N and the more constant clockwise circulation in the S, show much similarity to those of the North Atlantic Ocean (2.14).

2   **The North Equatorial Current** sets W off the coast of Central America and quickly expands into a broad current spanning 5° to 10° of latitude at its E and W extremities and near 18° in central longitudes near 180°. Its general latitudinal boundaries vary with longitude and season. In middle longitudes some predominance of W-going sets extends to about 25°N in winter and near 30°N in summer. Rates and constancies are higher on the equatorial side and diminish with increasing latitude as the North-east Trade Wind decreases.

3   The North Equatorial Current is strongest during the north-east monsoon (7.7) with an average rate of 1 kn. As it approaches the Philippine Islands in about 12°N it diverges. The S part turns SSW passing through the islands into the Sulu Sea and also rounding the SE coast of Mindanao into the Celebes Sea. Here a counter-clockwise circulation persists throughout the year. Just N of Halmahera in longitude 128°E a combined outflow from the NE part of the Celebes Sea and Molucca Sea merges with the recurving North Equatorial Current as the Equatorial Counter-current (7.28). This sets E, intermittently and with variable constancy, across the ocean so that, at times, its N boundary forms the S limit of the North Equatorial Current (and its S boundary the N limit of the South Equatorial Current).

4   The N part of the North Equatorial Current continues to turn through NNW off the E coast of Luzon and T'ai-wan to N and NE, forming the Japan Current.

5   **The Japan Current**, also known as Kuro Shio, is a warm boundary current similar to the Gulf Stream (2.14) of the North Atlantic Ocean. It passes W of Nansei Shotō (27°00′N, 129°00′E) until their N end where the main flow turns E through the islands and thence NE along the S side of the Japanese islands. A W branch diverges with the weaker part entering the East China Sea and the remainder continuing to the W of Kyūshū and into the Sea of Japan as the **Tsushima Current**.

6   The Japan Current continues NE later turning E and fanning out towards the W coast of N America as the **North Pacific Current**, a broad drift between 35°N and 50°N. To the N the circulation in both the Sea of Okhotsk and the Bering Sea is broadly counter-clockwise. This allows cold water to pass SW along the coast of Kamchatka through Kuril'skiye Ostrova (48°00′N, 153°00′E) towards the NE coast of Japan. This cold current is known as the the **Kamchatka Current** and continues off the Japanese coast to near 35°N where it turns E on the N side of the Japan Current to form the **Aleutian Current**. In a similar manner to that within the Gulf Stream (2.14) eddies are formed along the interaction zone between the cold and warm water. Some of these may last for months, sometimes years, and major displacements of the Japan Current Axis have been observed at such times.

7   The North Pacific Current is joined by the colder water of the Aleutian Current on its N flank as it crosses the Pacific Ocean. On approaching the North American coast in about 135°W and between 45°N and 55°N the North Pacific Current diverges. The N part of the Aleutian Current turns NE then NW past Queen Charlotte Islands. This flow is known as the **Alaska Current** as it continues along the coast of Alaska towards the E end of the Aleutian Chain. Some water passes through the islands in a generally N direction forming the E side of the Bering Sea counter-clockwise circulation. The North Pacific Current and the remainder of the Aleutian Current turns SE then SW as the relatively cool **California Current** which runs parallel with, but clear of, the Californian coast.

8   At certain times of the year, particularly during winter months when the continental high pressure is dominant, a counter-current sets N between the offshore California Current and the coast. Known as the **Davidson Current** this supplements the Alaska Current at times. During the rest of the year the space between the California Current and the coast is filled by irregular eddies.

9   Off the Californian coast the predominant sets for most of the year are to the SE but recurvature into the North

Equatorial Current commences S of 35°N and the resulting divergence to the S enhances local upwelling.

South Pacific Ocean
7.27

1 **The South Equatorial Current** sets WNW to WSW with a moderate to high constancy for much of the year. Its N boundary lies just N of the equator being, on average, near 1°N in the W and E and near 6°N in central longitudes between 150°W and 120°W. Average rates of the predominant W-going sets decrease from near 2 kn in the N to about ¾ kn near 20°S. Constancies decrease to low (less than 50%) S of about 20°S and also generally as far W as about the 180° meridian. This latter effect is less apparent during the S winter when moderate constancies continue as far W as the Solomon Islands. The weaker and less constant W-going sets are sometimes referred to as the **South Sub-tropical Current**.

2 The pattern W of 180° becomes more complex towards Melanesia, Indonesia and Australasia. The South Equatorial Current tends to diverge slightly with the N part, from June to August, continuing WNW along the N coast of New Guinea and then recurving N and NE to join the E-going Equatorial Counter-current. The S part sets WSW through the Fiji Islands towards the E Australian coast. This is the broad pattern for most of the rest of the year, except for a period from December to February, the season of the north-west monsoon with low pressure over the extreme N of Australia. Then the South Equatorial Current does not pass into the Equatorial Counter-current, it recurves to the SW and flows past the N coast of New Guinea in a SE direction. There is thus a complete reversal of current along this coast during the year.

3 In the S part of the Coral Sea, near 20°S, the Equatorial Current diverges on meeting the coast, the N part soon decreasing and becoming variable; the S stream intensifies off the coast as the **East Australian Current**, a typical but limited warm western boundary current. It is weakest during the S winter when the N extension of the Southern Ocean Current increases the formation of eddies E of Tasmania. The East Australian Current mostly degenerates in a marked counter-clockwise eddy in the SW part of the Tasman Sea, but the main flow passes both E and W of New Zealand as it continues ENE across the South Pacific Ocean.

4 A large area of variable current exists within the extensive counter-clockwise gyre of the South Pacific Ocean. Weak predominances of direction depend on the season.

5 During the winter months the approximate latitude at which ENE-going sets are replaced by W-going sets lies near 30°S off the E coast of Australia and in longitude 85°W off Chile. In central longitudes, from 140°W to 170°W, the latitude of changeover is near 20°S.

6 During summer months WSW-going sets extend to about 45°S in these same central longitudes but currents are very variable with no predominances exceeding the low (less than 50%) category. Near 45°S the E-going current splits, the N part recurving through NE to set N and later NW as the **Peru Current**, thus completing the counter-clockwise gyre. The S part continues E turning through SE to S off the coast and is then drawn into the main stream which passes Cabo de Hornos (56°04'S, 67°15'W), through Drake Passage, into the South Atlantic Ocean.

7 As the broad Peru Current moves N, water from its left flank is continuously fed into the South Equatorial Current, so that by about 5°S only a narrow coastal current remains to enter the S part of the Gulf of Panama. This coastal current varies with the season and is subject to much local variability, usually in association with upwelling.

Pacific Equatorial Counter-current
7.28

1 This counter-current sets E between the two W-going Equatorial Currents and is situated entirely N of the equator. The precise mechanism of its formation is not known but many factors probably contribute in different parts at different times.

2 When an average is calculated, using all observations, the current appears as a continuous feature from just N of the Celebes Sea in the W to Gulf of Panama in the E. Its main axis lies slightly convex to the equator being near 2°N to 4°N in the W and E and near 6°N to 8°N between 140°W to 120°W with an average width of about 3° to 4° along its length. However, it appears to expand and contract throughout the year in association with pulsations of the adjacent Equatorial Currents.

3 At times the Equatorial Currents merge and eliminate the surface Counter-current, a process which seems most apparent during March and April between 140°E to 160°E and between 115°W and 95°W. The Counter-current is widest during September and October when it spans up to 6° of latitude for much of its length. Average rates within it, at this time, reach 1 to 1½ kn, but variations do occur, sometimes with complete directional reversals.

4 The Equatorial Currents immediately to the N and S are generally higher both in rate and constancy. Near the boundaries eddies and minor upwellings cause local disturbances of the sea surface with marked areas of rippling.

5 The phenomenon known as '**El Nino**' (see *South America Pilot, Volume III*), seems related to the large scale pattern of currents in the equatorial Pacific Ocean and its interaction with the cool N extensions from the Peru Current (7.27) both in the open ocean and off the coast.

Gulf of Panama
7.29

1 The situation in the general area at the E end of the Equatorial Counter-current (7.28) and Gulf of Panama is very disturbed. Most major features are evident to some degree throughout the year but the interaction between them is constantly changing and each year may be different, though basically similar.

2 Using data averaged over 2-month periods the following broad sequential pattern emerges, commencing in March-April when the Equatorial Counter-current is at a minimum.

3 **March-April**. A narrow current follows the coast of Columbia and circuits Gulf of Panama in a counter-clockwise fashion leaving to the S of Punta Mala (7°28'N, 80°00'W) from whence it continues as a widening current to directions between S and W. On its N flank it is joined by the remnants of the recurving Equatorial Counter-current and on its S flank by the S Equatorial (7.27) flow just N of the equator.

4 **May-June**. The S and W-going sets S of Punta Mala recede N to about 6°N. The Equatorial Counter-current widens and extends to about 2°N turning into the narrow coastal current off Columbia which continues round the Gulf accentuating the counter-clockwise eddy.

5 **July-August**. The main eddy centre is just SE of Punta Mala with the Equatorial Counter-current extending close to the Columbian coast. The outflow from Gulf of Panama

turns W off Punta Mala and continues NW along the Costa Rican coast.

6 **September-October**. The Equatorial Counter-current broadens to the N but appears to split off the Costa Rican coast in longitude 85°W. Part turns NW along the coast and the remainder continues SE then E into the S part of the Gulf eddy.

7 **November-December**. Similar to previous months but the SW flow on the W side of the eddy extends farther S to about 1°N at 81°W. The Equatorial Counter-current has weakened to the W. The N-going flow continues along the Columbian coast.

8 **January-February**. The E end of the Equatorial Counter-current seems to be almost a detached clockwise eddy SW of Coasta Rica with a broad expanse of S to W-going sets emerging from Gulf of Panama and filling the area W of 80°W where they engage the N flank of the South Equatorial Current just N of the equator.

ICE

General remarks
7.30

1 The following brief accounts of ice in the Pacific Ocean should not be taken as complete, or in any way all-embracing. More detailed information than can be given here will be found in the following publications, which should be consulted, as appropriate, before undertaking passages through areas in which ice is likely to be encountered:

Appropriate volumes of *Admiralty Sailing Directions* (see 1.12.1);

Appropriate volumes of *Admiralty List of Radio Signals* (see 1.12.5);

The Mariner's Handbook (see 1.12.11);

Charts 5127 (1) to (12) — *Monthly Routeing Charts for North Pacific Ocean*;

Charts 5128 (1) to (12) — *Monthly Routeing Charts for South Pacific Ocean*;

2 A factor always to be borne in mind where ice conditions are concerned is their great variability from year to year. For this reason, and on account of the sparsity of observations in many areas, the charted positions of the limits should be regarded as approximate.

North Pacific Ocean

Pack ice
7.31

1 Diagrams 1.14a *World Climatic Chart (January)* and 1.14b *World Climatic Chart (July)* indicate the mean limits of 4/8 pack ice in March and September respectively, in which months it attains its greatest and least extent. The Routeing Charts indicate the maximum limit of pack ice in any particular month. An examination of these limits reveals the marked influence of winds and currents; on the W side of the ocean the N winds of winter and the cold Kamchatka Current (7.27) bring the ice to relatively low latitudes; while on the E side, except in the N part of the Bering Sea, open water is maintained by the warm North Pacific Current (7.27).

2 During an average winter, navigation off the E coasts of Asia is impeded as far S as about 45°N.

3 By mid-November coastwise navigation is interrupted as far S as 60°N and is closed N of 62°N. Ice is also present in all coastal waters of the N and W parts of the Sea of Okhotsk, in the N part of the Gulf of Tartary and E of Ostrov Sakhalin N of 50°N.

4 In December, navigation is closed to all ports N of 60°N and ice may be found anywhere in the Gulf of Tartary N of 47°N, as well as along the E coast of Ostrov Sakhalin and along the coasts of the Russian Maritime Province as far S as 43°N.

5 From January to March, the whole of the waters of the Russian Maritime Province, the greater part of the Gulf of Tartary and the coasts of N Hokkaidō and the SW islands of Kuril'skiye Ostrova (48°00′N, 153°00′E) are encumbered with ice in varying degrees, as also is the whole of the Sakhalin area and the greater part of the Sea of Okhotsk, except in the deep central position. Ice is also present in the vicinity of the NE islands of Kuril'skiye Ostrova and along much of the E coast of Kamchatka and the coast farther N.

6 In April the ice edge begins to retreat N and by mid-May, after an average winter, there is little or no ice S of about 52°N.

7 By mid-June ice is confined to the SW part of the Sea of Okhotsk, the N part of Penzhinskiy Zaliv (60°00′N, 158°00′E), Proliv Litke (59°00′N, 163°30′E), Zaliv Olyutorskiy (60°00′N, 169°00′E) and from Anadyrskiy Zaliv (64°00′N, 178°00′E) to the N.

8 By late July vessels can generally pass through Bering Strait.

9 The months during which ports are closed to navigation vary not only with the severity of the season and the prevailing winds, but also with the availability of ice breakers; detailed information should be sought in the appropriate volumes of *Admiralty Sailing Directions*.

10 Ice may also be found in the shallow waters of Bo Hai (38°30′N, 119°30′E) and Liaodong Wan (40°00′N, 121°00′E) between the middle of November and the end of March; the port of Yingkou, at the head of Liaodong Wan, is closed to navigation from December to March.

11 Along the Alaskan coast, in the average winter, ice extends as far as 56°N from December to April; in very severe winters the extreme NE Aleutians may be affected. The ice edge advances S during October and November and retreats N during May and June and ice is not normally found from July to September except near the Bering Strait.

12 During the ice season the N half of the Bering Sea is filled with pack ice, though it is not solidly frozen.

Icebergs
7.32

1 Icebergs are not a feature of the North Pacific Ocean, because the relatively high temperatures of the Gulf of Alaska rapidly melt any bergs which reach the open sea from the glaciers of SE Alaska or the inlets of British Columbia, which are the only breeding grounds. Occasional floebergs may may be expected among the pack ice, particularly in the W part of the Bering Sea.

South Pacific Ocean

Pack ice
7.33

1 Diagrams 1.14a *World Climatic Chart (January)* and 1.14b *World Climatic Chart (July)* indicate the mean limits of 4/8 pack ice in February to March and September to October, in which months it attains its least and greatest extent respectively. None of the normally inhabited places in the South Pacific Ocean is affected, but great circle

sailing between Australian or New Zealand ports and the more S ports of South America is prevented.

Icebergs
7.34

1 The icebergs that occur in the Southern Ocean are not, in most cases, calved from glaciers, but consist of portions that have broken away from the great ice shelves which fringe parts of the Antarctic continent. They are consequently flat-topped and may be of immense size.

2 In November and December, when the mean limit reaches its farthest N, it runs from about 100 miles S of Cabo de Hornos (56°04′S, 67°15′W) along 57°S to 90°W, whence it curves N to 52°S, 120°W. Between 120°W and the 180° meridian it is situated between 50°S and 52°S, whence it continues in a SW direction to about 55°S in the longitude of Tasmania.

3 In May and June the mean limit of icebergs is everywhere S of 55°S. West of 150°W it lies within a degree or two of 60°S.

4 With regard to the extreme limit of icebergs, although information, for the most part, is too scanty for a confident description, it appears to be reached during the latter half of the year, running close to 45°S over most of the region, but locally near 40°S just E of New Zealand and about 50°S to the E of 110°E. Factors other than climatic may be responsible for abnormal numbers or abnormal movements of icebergs. Earthquakes, for example, may increase the number calved. Accordingly it is probably best to regard the extreme limit of icebergs as unrelated to the time of year.

5 Estimates of the extreme limit are indicated on Diagrams 1.14a *World Climatic Chart (January)* and 1.14b *World Climatic Chart (July)* and on Charts *5128 (1) to (12).*

NAVIGATIONAL NOTES

Soundings and dangers
7.35

1 Very little of the Pacific Ocean has been thoroughly surveyed. Certain areas are subject to earthquakes and volcanic activity which could cause shoals to build up even in those areas which have been well surveyed; live coral is continually growing. Many vigias exist, and until disproved must continue to be regarded as potentially dangerous. See 1.9 and *The Mariner's Handbook.*

2 The routes laid down in this book are those considered most likely to lead clear of dangers but, owing to the reasons stated above, the only safeguards are a good look-out and careful sounding.

3 In the interests of all mariners it should be stressed that a sounding over any suspected danger should be obtained and recorded.

Currents among the islands
7.36

1 Particular and constant attention must be paid to the current when navigating among the island groups, for these sometimes deflect and always accelerate it. Again, most of the islands are so low that it is almost impossible to see them at night and ships may be driven onto the barrier or fringing reef, with no warning from sounding, as the reefs generally have very deep water close to.

Flotsam
7.37

1 In navigating the waters of the Eastern Archipelago during the rainy season a sharp look-out must be kept for flotsam. Trees, some of immense size, will be frequently met afloat. They have been found to be especially numerous on the S coast of Luzon; in one instance, near Marinduque Island (13°20′N, 122°00′E), a group of trees was adrift, still upright and resembling an island.

Areas to be avoided
7.38
The following areas of the Pacific Ocean contain hazards to shipping:
7.38.1
Hawaiian Islands. Vessels of more than 1000 grt, carrying cargoes of oil or hazardous material, are cautioned to avoid areas of 50 miles radius centred on certain islands and reefs in the W part of the island chain. For details see *Pacific Islands Pilot, Volume III.*
7.38.2
Java Sea. Considerable exploitation of natural resources takes place in the Java Sea, particularly in the W part. Charts, Notices to Mariners and *Indonesia Pilot, Volumes I and II* should be consulted for latest information available on permanent and moveable structures (not all of which will be charted) and the pipelines between them, together with regulations affecting their localities.

Mariners are advised to avoid these areas whenever possible.
7.38.3
Bass Strait. An Area to be Avoided encloses oil and gas fields extending between 20 miles SE and 45 miles S of Lakes Entrances (37°54′S, 147°59′E) on Ninety Mile Beach. Submarine pipelines are laid between the fields and the shore. For details see *Australia Pilot, Volume II.*

PASSAGES BETWEEN TORRES STRAIT AND EAST COAST OF AUSTRALIA

GENERAL INFORMATION

PASSAGES

Southern coastal passages

Diagram 7.40
General information
7.40

Coverage
7.39
This section contains details of the following passages:
Southern coastal passages (7.40).
Northern coastal passages (7.41).
Torres Strait channels (7.42).

1 Coastwise passages off the S part of the coast of Queensland and the Pacific coast of New South Wales are affected by the East Australian Current (7.27) which sets S at all times off most of this part of the coast. Between

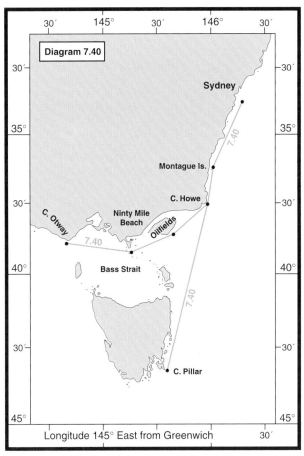

7.40 Southern Coastal Passages

32°S and 34°S the strength and constancy of the current are decreased by reason of the diversion of water in a SE direction towards the open ocean. Between 34°S and Cape Howe (37°30′S, 149°59′E) currents may set in any direction, sometimes with an onshore component; close inshore there may be a predominantly N-going current at times.

2   Cape Pillar (43°14′S, 148°02′E) and Tasman Island (close S of the cape) may be rounded at a distance of 1 mile, but the rest of the E coast of Tasmania should not be closed within 5 miles.

3   North-bound ships navigating off the mainland coast E of Cape Otway (38°51′S, 143°31′E) should keep well inshore as far as Bass Strait (40°00′S, 146°00′E) because of the current. Thence via the appropriate lanes of the TSS through Bass Strait and that passing SE of the area enclosing oil and gas fields extending seaward from Ninety Mile Beach (38°15′S, 147°23′E). Thence the coast may be closed again and passage may be made inside of Montagu Island (36°15′S, 150°14′E).

4   South-bound ships should maintain an offing of about 15 miles, in depths of about 180 m, as far as Cape Howe (37°30′S, 149°59′E).

Northern coastal passages

Diagram 7.41
General information
7.41

1   Between Torres Strait (10°36′S, 141°51′E) and ports S of Caloundra Head (26°48′S, 153°08′E), 35 miles from Brisbane, the N part of the passage may be made by either:

Outer Route (7.41.1) through the Coral Sea, E of Great Barrier Reefs, or:
Inner Route (7.41.2) inshore of the Reefs.

7.41.1

Outer Route is not normally used as numerous large reefs have to be given a wide berth, especially at night, owing to the strong and variable sets which may often be experienced.

From off Sandy Cape (24°42′S, 153°16′E) the most satisfactory track is:
Between Saumarez Reef (21°50′S, 153°40′E) and Frederick Reef (21°00′S, 154°25′E) (both lit), thence:
E of Lihou Reef (17°20′S, 152°00′E), thence:
E of Eastern Fields (10°10′S, 145°40′E), thence:
N of Lagoon Reef (9°27′S, 144°54′E), thence:
Through Bligh Entrance (9°12′S, 144°00′E) to Great North East Channel (9°20′S, 144°00′E) by the recommended track shown on the appropriate charts.

7.41.2

Inner Route is marked by adequate navigational aids and saves considerable distance. It is described in *Australia Pilot, Volume III*, together with the places where pilots may be embarked or disembarked. For the purposes of this book it is assumed that Capricorn Channel (22°45′S, 152°00′E) is used, though Curtis Channel (24°20′S, 152°55′E) may be used to embark or disembark pilots off North Point, Gladstone (23°44′S, 151°21′E), adding little to the distance.

7.41.3

1   **Passages through Great Barrier Reefs**. From ports lying behind Great Barrier Reefs, ships can pass through he reefs, to cross the Coral Sea, by the following deep-water passages:
Capricorn Channel (22°45′S, 152°00′E) passing S and E of the reefs;
Hydrographers Passage;
Palm Passage;
Grafton Passage.

2   For details of these passages see *Australia Pilot, Volume III*.

3   From certain of these ports, ships bound for Singapore, the N part of the Indian Ocean, or Europe via the Suez Canal, and for which the Torres Strait is not suitable, may find a shorter route, with weather and current more favourable, through one of these passages., rather than the alternative route via Bass Strait and the Great Australian Bight.

4   One such route leads from Hydrographers Passage:
Through Diamond Passage (17°30′S, 151°15′E), thence:
Through either Jomard Entrance (11°15′S, 152°06′E) or Rossel Spit (11°27′S, 154°24′E), thence:
Thence proceeding E and N of New Guinea, thence:
Through Selat Sagewin (0°55′S, 130°40′E) and the Eastern Archipelago.
This N route allows advantage to be taken of the Tropical Load Line from April to November.

Torres Strait

Diagram 7.42
Transit channels
7.42

Great North East Channel (9°20′S, 144°00′E) and Prince of Wales Channel (10°30′S, 142°15′E) are well surveyed, as is the area S of the latter channel. The area W and N of

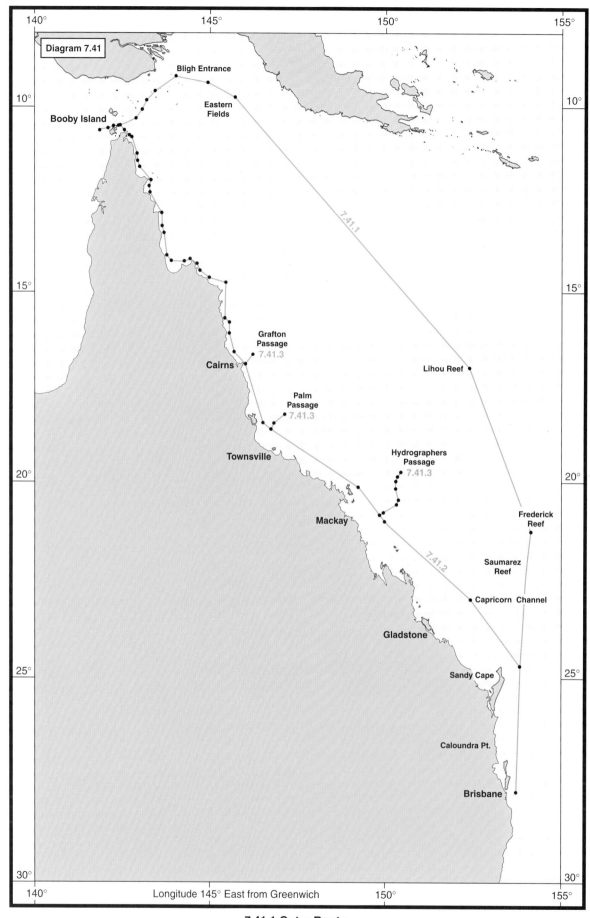

Diagram 7.41

7.41.1 Outer Route
7.41.2 Inner Route
7.41.3 Passages through Great Barrier Reefs

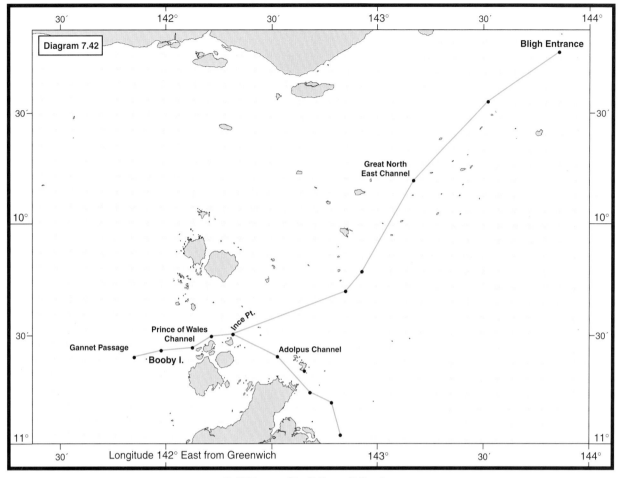

7.42 Torres Strait Transit Routes

of those channels and that off the S coast of New Guinea is unsurveyed.

Prince of Wales Channel, the principal route through the Strait, is approached from:

Coral Sea, through Bligh Entrance (9°12′S, 144°00′E) and Great North East Channel;

Inner Route (7.41.2), through Adolphus Channel (10°40′S, 142°35′E);

Arafura Sea, through Gannet Passage (10°35′S, 141°50′E), within Prince of Wales Channel, where the controlling depth of 12 m will be found.

For latest information on pilotage and limitations of draught, see *Australia Pilot, Volume III*.

Distances

7.43

Distances between the ports on the E and S coasts of Australia and Torres Strait, using Inner Route (7.41.2), are given in the table below):

7.42.1

Distances through Torres Strait channels are as follows:

From 10°36′S, 141°51′E (3½ miles W of Booby Island) to the junction with Inner Route (7.41.2) off Ince Point (10°30′S, 142°19′E): 30 miles.

From junction with Inner Route (7.41.2) off Ince Point to Bligh Entrance (9°12′S, 144°00′E): 130 miles.

From 3½ miles W of Booby Island to Bligh Entrance: 165 miles.

In this book, distances given to or from Torres Strait refer to position 10°36′S, 141°51′E (3½ miles W of Booby Island).

Torres Strait					
1320	**Caloundra Head‡**				
1770	470	**Sydney**			
2300	1000	540	**Port Phillip†**		
2730	1430	965	460	**Adelaide**	
2390	1090	630	*	760	**Hobart**

* Via Bass Strait and E of Furneaux Group   465 miles.
  Via Banks Strait   430 miles.
  W-about   455 miles.
‡ Caloundra Head to Brisbane   35 miles.
† Port Phillip to Melbourne   40 miles.
For distances W-about, see 6.92.

PASSAGES BETWEEN AUSTRALIA, NEW ZEALAND AND ISLANDS IN THE SOUTH PACIFIC OCEAN

GENERAL INFORMATION

Coverage
7.44

This section contains details of the following passages between:

Hobart and Bluff (7.45).
Hobart and Wellington (7.46).
Hobart and Auckland (7.47).
Melbourne and Bluff (7.48).
Melbourne and Wellington (7.49).
Melbourne and Auckland (7.50).
Sydney and Bluff (7.51).
Sydney and Wellington (7.52).
Sydney and Auckland (7.53).
Sydney and Papeete (7.54).
Sydney and Nouméa (7.55).
Sydney and Nuku'alofa (7.56).
Sydney and Suva (7.57).
Sydney and Apia (7.58).
Brisbane and Bluff (7.59).
Brisbane and Wellington (7.60).

Brisbane and Auckland (7.61).
Brisbane and Papeete (7.62).
Brisbane and Nouméa (7.63).
Brisbane and Nuku'alofa (7.64).
Brisbane and Suva (7.65).
Brisbane and Apia (7.66).
Torres Strait and Wellington (7.67).
Torres Strait and Auckland (7.68).
Torres Strait and Papeete (7.69).
Torres Strait and Suva (7.70).
Torres Strait and Apia (7.71).
Wellington and Auckland (7.72).
New Zealand and Papeete (7.73).
New Zealand and Nuku'alofa (7.74).
New Zealand and Apia (7.75).
New Zealand and Nouméa or Suva (7.76).
Suva and Nuku'alofa (7.77).
Suva and Papeete (7.78).
Suva and Apia (7.79).
Nuku'alofa and Apia (7.80).
Nuku'alofa and Papeete (7.81).
Apia and Papeete (7.82).

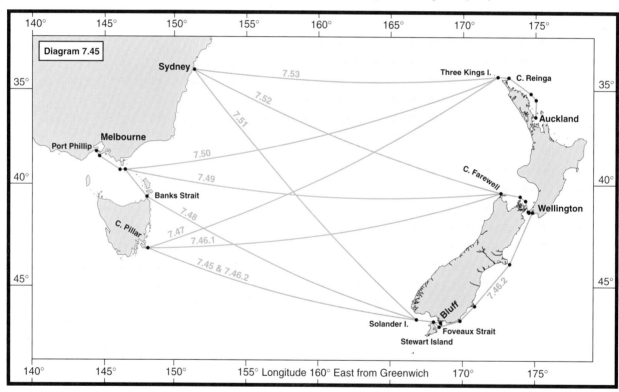

7.45 Hobart ← → Bluff; 7.48 Bluff ← → Wellington; 7.51 Sydney ← → Bluff
7.46 Hobart ← → Wellington; 7.49 Melbourne ← → Wellington; 7.52 Sydney ← → Wellington
7.47 Hobart ← → Auckland; 7.50 Melbourne ← → Auckland; 7.53 Sydney ← → Auckland

PASSAGES

Hobart ⇐ ⇒ Bluff

Diagram 7.45
Route
7.45

From Hobart (42°58′S, 147°23′E) the route is:
By great circle from Cape Pillar (43°14′S, 148°02′E) to Solander Islands (46°34′S, 166°50′E), thence:

As navigation permits to Bluff (46°38′S, 168°21′E).
Distance 930 miles.

Hobart ⇐ ⇒ Wellington

Diagram 7.45

Route
7.46

From Hobart (42°58′S, 147°23′E) there are two alternative routes.

7.46.1

From Cape Pillar (43°14′S, 148°02′E):
By great circle to Cape Farewell (40°30′S, 173°41′E) in the W approach to Cook Strait, thence:
Direct to Wellington (41°22′S, 174°50′E).

7.46.2

From Cape Pillar (43°14′S, 148°02′E):
By great circle to pass SE of South Island, through Foveaux Strait (46°35′S, 168°00′E), thence:
By offshore routes to Wellington (41°22′S, 174°50′E).

7.46.3

Distances:

Via Cook Strait	1290 miles.
Via Foveaux Strait	1370 miles.
Passing S of Stewart Island	1390 miles.

Hobart ⇐ ⇒ Auckland

Diagram 7.45
Route
7.47

From Hobart (42°58′S, 147°23′E) the route is:
By great circle from Cape Pillar (43°14′S, 148°02′E) to a position between Three Kings Islands (34°10′S, 172°00′E) and Cape Reinga (34°25′S, 172°40′E), thence:
Coastwise to Auckland (36°36′S, 174°49′E).
Distance   1510 miles.

Melbourne ⇐ ⇒ Bluff

Diagram 7.45
Route
7.48

From Port Phillip (38°20′S, 144°34′E) the route is:
Through Banks Strait (40°39′S, 148°05′E), thence:
Great circle to Solander Islands (46°34′S, 166°50′E), thence:
As navigation permits to Bluff (46°38′S, 168°21′E).

7.48.1

Distances:

From Port Phillip	1170 miles.
Port Phillip to Melbourne	40 miles.

Melbourne ⇐ ⇒ Wellington

Diagram 7.45
Route
7.49

From Port Phillip (38°20′S, 144°34′E) the route is:
Through the TSS in Bass Strait, thence:
By great circle to Cape Farewell (40°30′S, 173°41′E) in the W approach to Cook Strait, thence:
Direct to Wellington (41°22′S, 174°50′E).

7.49.1

Distances.

From Port Phillip	1450 miles.
Port Phillip to Melbourne	40 miles.

Melbourne ⇐ ⇒ Auckland

Diagram 7.45
Route
7.50

From Port Phillip (38°20′S, 144°34′E) the route is:
Through the TSS in Bass Strait, thence:

By great circle to a position between Three Kings Islands (34°10′S, 172°00′E) and Cape Reinga (34°25′S, 172°40′E), thence:
Coastwise to Auckland (36°36′S, 174°49′E).

7.50.1

Distances:

From Port Phillip	1600 miles.
Port Phillip to Melbourne	40 miles.

Sydney ⇐ ⇒ Bluff

Diagram 7.45
Route
7.51

From Sydney (33°50′S, 151°19′E) the route is:
Great circle to Solander Islands (46°34′S, 166°50′E), thence:
As navigation permits to Bluff (46°38′S, 168°21′E).
Distance   1100 miles.

Sydney ⇐ ⇒ Wellington

Diagram 7.45
Route
7.52

From Sydney (33°50′S, 151°19′E) the route is:
By great circle to Cape Farewell (40°30′S, 173°41′E) in the W approach to Cook Strait, thence:
Direct to Wellington (41°22′S, 174°50′E).
Distance   1230 miles.

Sydney ⇐ ⇒ Auckland

Diagram 7.45
Route
7.53

From Sydney (33°50′S, 151°19′E) the route is:
Direct as navigation permits, passing on either side of Three Kings Islands (34°10′S, 172°00′E), thence:
Coastwise to Auckland (36°36′S, 174°49′E).
Distance   1270 miles.

Sydney ⇐ ⇒ Papeete

Diagram 7.54
Route
7.54

From Sydney (33°50′S, 151°19′E) the route is:
By great circle to 22°30′S, 158°00′W (**A**), passing:
Close S of Wanganella Bank (32°31′S, 167°24′E), thence:
Midway between Raoul Island, thence (29°15′S, 177°54′W) and Macauley Island (30°14′S, 178°26′W) in the Kermadec Islands, thence:
25 miles SSE of both the unnamed reef (Rep 1972) in 24°25′S, 163°39′W and Mangaia (21°53′S, 157°55′W), thence:
By great circle to Papeete (17°30′S, 149°36′W).
Caution. For details of currents near Mangaia see *Pacific Islands Pilot, Volume III.*
Distance   3290 miles.

Sydney ⇐ ⇒ Nouméa

Diagram 7.54
Route
7.55

From Sydney (33°50′S, 151°19′E) the route is:
By great circle, passing 40 miles NW of Middleton Reef (29°28′S, 159°04′E), to Nouméa (22°18′S, 166°25′E).
Distance   1060 miles.

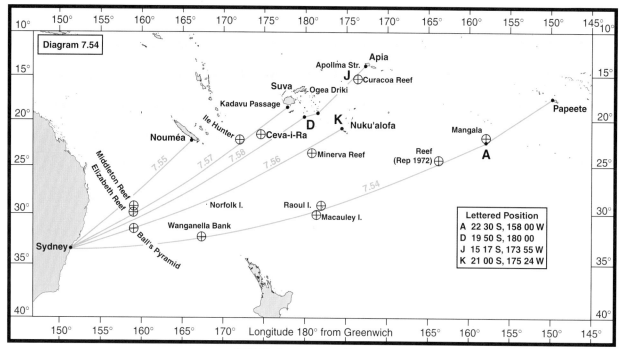

7.54 Sydney ← → Papeete
7.55 Sydney ← → Nouméa
7.56 Sydney ← → Nuku'alofa
7.57 Sydney ← → Suva
7.58 Sydney ← → Apia

Sydney ⇐ ⇒ Nuku'alofa

Diagram 7.54
Route
7.56

From Sydney (33°50′S, 151°19′E) the route is:
By great circle to 21°00′S, 175°24′W (**K**), in the W approach to Ava Lahi, the approach channel to Nuku'alofa (21°00′S, 175°10′W), passing:
Close S of Ball's Pyramid (31°45′S, 159°15′E), thence:
Close N of Norfolk Island (29°02′S, 167°56′E), thence:
30 miles NW of Minerva Reefs (23°45′S, 179°00′W).

Caution. For details of dangers and volcanic activity in the approaches to Tonga Islands and NE of Norfolk Island, see *Pacific Islands Pilot, Volume II* and *Australia Pilot, Volume III*.

Distance   1940 miles.

Sydney ⇐ ⇒ Suva

Diagram 7.54
Route
7.57

From Sydney (33°50′S, 151°19′E) the route is:
By great circle to Kadavu Passage (18°45′S, 178°00′E), passing:
20 miles SE of Elizabeth Reef (29°55′S, 159°02′E), thence:
20 miles SE of Île Hunter (22°24′S, 172°05′E), thence:
30 miles NW of Ceva-i-Ra (21°44′S, 174°38′E), thence:

Through Kadavu Passage to Suva (18°11′S, 178°24′E).

Caution is necessary near Elizabeth Reef owing to the variability of the current, see *Australia Pilot, Volume III*.

Distance   1730 miles.

Sydney ⇐ ⇒ Apia

Diagram 7.54
Route
7.58

From Sydney (33°50′S, 151°19′E) the route is:
By great circle to 19°50′S, 180°00′ (**D**), thence:
10 miles SE of Ogea Driki (19°12′S, 178°24′W), thence:
To 15°17′S, 173°55′W (**J**), 20 miles NW of Curacoa Reef, avoiding Nuku Soge Reef, 3¼ miles ESE of Ogea Driki, thence:
Through Apolima Strait (13°48′S, 172°10′W) to Apia (13°47′S, 171°45′W).

Distance   2360 miles.

Brisbane ⇐ ⇒ Bluff

Diagram 7.59
Route
7.59

From Caloundra Head (26°49′S, 153°10′E) the route is:
By great circle to Solander Islands (46°34′S, 166°50′E), thence:
As navigation permits to Bluff (46°38′S, 168°21′E).

7.59.1
Distances:
From Caloundra Head   1420 miles.
Caloundra Head to Brisbane   35 miles.

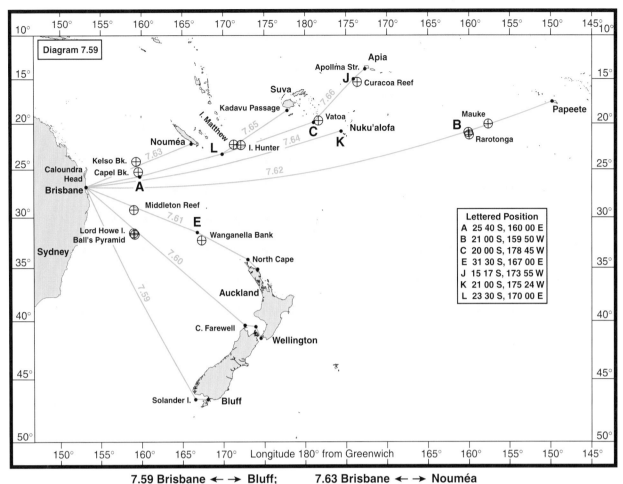

Diagram 7.59

Lettered Position
A 25 40 S, 160 00 E
B 21 00 S, 159 50 W
C 20 00 S, 178 45 W
E 31 30 S, 167 00 E
J 15 17 S, 173 55 W
K 21 00 S, 175 24 W
L 23 30 S, 170 00 E

7.59 Brisbane ←→ Bluff;	7.63 Brisbane ←→ Nouméa
7.60 Brisbane ←→ Wellington;	7.64 Brisbane ←→ Nuku'alofa
7.61 Brisbane ←→ Auckland;	7.65 Brisbane ←→ Suva
7.62 Brisbane ←→ Papeete;	7.66 Brisbane ←→ Apia

Brisbane ⇐ ⇒ Wellington

Diagram 7.59

Route

7.60

From Caloundra Head (26°49′S, 153°10′E) the route is:
By rhumb line to Cape Farewell (40°30′S, 173°41′E) in the W approach to Cook Strait, thence:
Direct to Wellington (41°22′S, 174°50′E).

The rhumb line clears Lord Howe Island (31°32′S, 159°05′E) and Ball's Pyramid (31°45′S, 159°15′E) better than the great circle route.

7.60.1

Distances:

From Caloundra Head	1400 miles.
Caloundra Head to Brisbane	35 miles.

Brisbane ⇐ ⇒ Auckland

Diagram 7.59

Route

7.61

From Caloundra Head (26°49′S, 153°10′E) the route is:
By rhumb line to 31°30′S, 167°00′E (E), passing N of Middleton Reef (29°28′S, 159°04′E), thence:
Rhumb line to North Cape (34°25′S, 173°03′E), passing N of Wanganella Bank (32°31′S, 167°24′E), thence:
Coastwise to Auckland (36°36′S, 174°49′E).

7.61.1

Distances:

From Caloundra Head	1300 miles.
Caloundra Head to Brisbane	35 miles.

Brisbane ⇐ ⇒ Papeete

Diagram 7.59

Route

7.62

From Caloundra Head (26°49′S, 153°10′E) the route is:
By great circle to 21°00′S, 159°50′W (B), 10 miles N of Rarotonga, thence:
To a position 10 miles S of Mauke (20°08′S, 157°23′W), thence:
To Papeete (17°30′S, 149°36′W).

7.62.1

Distances:

From Caloundra Head	3210 miles.
Caloundra Head to Brisbane	35 miles.

Brisbane ⇐ ⇒ Nouméa

Diagram 7.59

Route

7.63

From Caloundra Head (26°49′S, 153°10′E) the route is:
By rhumb line, passing midway between Capel Banc (25°15′S, 159°40′E) and Kelso Banc (24°00′S, 159°20′E), to Nouméa (22°18′S, 166°25′E).

7.63.1

Distances:

From Caloundra Head	790 miles.
Caloundra Head to Brisbane	35 miles.

Brisbane ⇐ ⇒ Nuku'alofa

Diagram 7.59
Route
7.64

From Caloundra Head (26°49′S, 153°10′E) the route is:
By great circle to 21°00′S, 175°24′W (**K**), in the W approach to Ava Lahi, the approach channel to Nuku'alofa (21°00′S, 175°10′W).

Caution. For details of dangers and volcanic activity in the approaches to Tonga Islands, see *Pacific Islands Pilot, Volume II*.

7.64.1

Distances:

From Caloundra Head	1770 miles.
Caloundra Head to Brisbane	35 miles.

Brisbane ⇐ ⇒ Suva

Diagram 7.59
Route
7.65

From Caloundra Head (26°49′S, 153°10′E) the route is:
Through 25°40′S, 160°00′E (**A**), passing S of Capel Banc (25°15′S, 159°40′E), thence:
Through 23°30′S, 170°00′E (**L**) to clear the reported banks and dangers SE of Nouvelle-Calédonie (21°15′S, 165°20′E), thence:
Midway between Île Matthew (22°21′S, 171°21′E) and Île Hunter (22°24′S, 172°05′E), thence:
Through Kadavu Passage (18°45′S, 178°00′E) to Suva (18°11′S, 178°24′E).

7.65.1

Distances:

From Caloundra Head	1510 miles.
Caloundra Head to Brisbane	35 miles.

Brisbane ⇐ ⇒ Apia

Diagram 7.59
Route
7.66

From Caloundra Head (26°49′S, 153°10′E) the route is:
By great circle to 20°00′S, 178°45′W (**C**), about 30 miles WSW of Vatoa, thence:
To 15°17′S, 173°55′W (**J**), 20 miles NW of Curacoa Reef, thence:
Through Apolima Strait (13°48′S, 172°10′W) to Apia (13°47′S, 171°45′W).

7.66.1

Distances:

From Caloundra Head	2150 miles.
Caloundra Head to Brisbane	35 miles.

Torres Strait ⇐ ⇒ Wellington

Diagram 7.67
Route
7.67

From Torres Strait (10°36′S, 141°51′E) the route is:

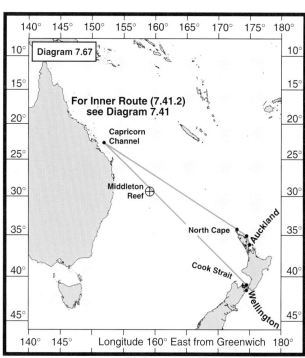

7.67 Torres Strait ← → Wellington
7.68 Torres Strait ← → Auckland

Via Inner Route to Capricorn Channel (22°45′S, 152°00′E) (7.41.2), thence:
Rhumb line to Cook Strait (41°00′S, 174°30′E), passing N of Middleton Reef (29°28′S, 159°04′E), thence:
Direct to Wellington (41°22′S, 174°50′E).

Distance	2660 miles.

Torres Strait ⇐ ⇒ Auckland

Diagram 7.67
Route
7.68

From Torres Strait (10°36′S, 141°51′E) the route is:
Via Inner Route to Capricorn Channel (22°45′S, 152°00′E) (7.41.2), thence:
Rhumb line to North Cape (34°25′S, 173°03′E), thence:
Coastwise to Auckland (36°36′S, 174°49′E).

Distance	2540 miles.

Torres Strait ⇐ ⇒ Papeete

Diagram 7.69
Route
7.69

From Torres Strait (10°36′S, 141°51′E) the route is:
Through Bligh Entrance (9°12′S, 144°00′E) (7.42), thence:
Round the N point of Espiritu Santo, Vanuatu (15°20′S, 166°50′E), thence:
N of Balmoral Reef (15°40′S, 175°50′E) and Fiji Islands to Papeete (17°30′S, 149°36′W), passing:
N of Zephyr Bank (15°55′S, 176°50′W), and:
S of Niua Fo'ou (15°36′S, 175°38′W) and Niuatoputapu (16°00′S, 173°45′W), and:
N of Durham Shoal (16°07′S, 173°49′W).

Distance	4090 miles.

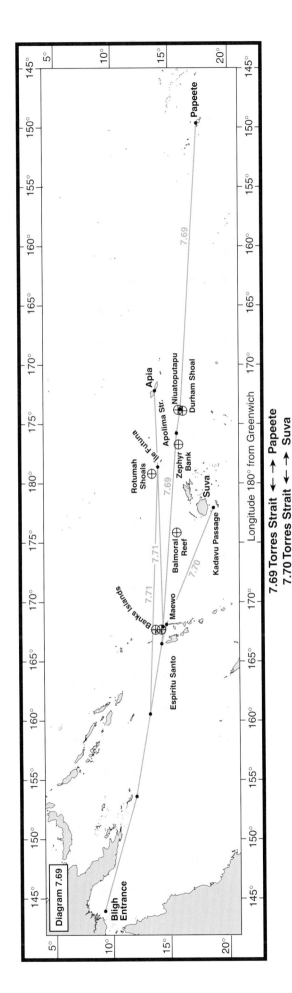

7.69 Torres Strait ← → Papeete
7.70 Torres Strait ← → Suva
7.71 Torres Strait ← → Apia

Torres Strait ⇐ ⇒ Suva

Diagram 7.69
Route
7.70

From Torres Strait (10°36′S, 141°51′E) the route is:
Through Bligh Entrance (9°12′S, 144°00′E) (7.42),
thence:
As navigation permits to pass N of Espiritu Santo
(15°20′S, 166°50′E) and Maéwo Island (15°10′S,
168°05′E), Vanuatu, thence:
Through Kadavu Passage (18°45′S, 178°00′E) to Suva
(18°11′S, 178°24′E).

Distance   2290 miles.

Torres Strait ⇐ ⇒ Apia

Diagram 7.69
Route
7.71

From Torres Strait (10°36′S, 141°51′E) the route is:
Through Bligh Entrance (9°12′S, 144°00′E) (7.42),
thence:
As navigation permits to pass N or S of Banks
Islands (14°00′S, 167°30′E), thence:
Between Rotumah Shoals (13°30′S, 179°12′W) and
Île Futuna (14°20′S, 178°05′W), thence:
Through Apolima Strait (13°48′S, 172°10′W) to Apia
(13°47′S, 171°45′W).

Distances:

Passing N of Banks Islands   2800 miles.
Passing S of Banks Islands   2810 miles.

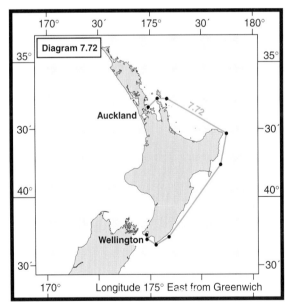

7.72 Wellington ← → Auckland

Wellington ⇐ ⇒ Auckland

Diagram 7.72
Route
7.72

Between Wellington (41°22′S, 174°50′E) and Auckland
(36°36′S, 174°49′E) the routes are coastwise.

Distance   545 miles.

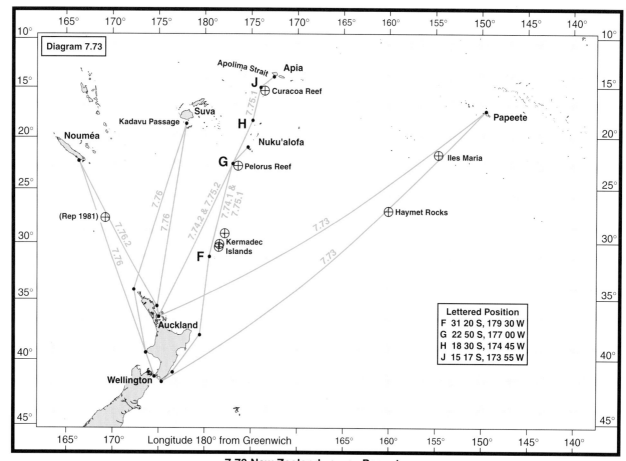

Diagram 7.73

Lettered Position
F 31 20 S, 179 30 W
G 22 50 S, 177 00 W
H 18 30 S, 174 45 W
J 15 17 S, 173 55 W

7.73 New Zealand ← → Papeete
7.74 New Zealand ← → Nuku'alofa
7.75 New Zealand ← → Apia
7.76 New Zealand ← → Numéa or Suva

New Zealand ⇐ ⇒ Papeete

Diagram 7.73
Route
7.73

From Auckland (36°36′S, 174°49′E) and ports on the SE coast of New Zealand the routes are:

By great circle, as navigation permits, to Papeete (17°30′S, 149°36′W).

From Wellington (41°22′S, 174°50′E) the route is:

By great circle, which passes near to Haymet Rocks (27°11′S, 160°13′W), whose existence has been confirmed though the position is doubtful, thence:

About 30 miles SE of Îles Maria (21°48′S, 154°42′W), thence:

To Papeete (17°30′S, 149°36′W).

7.73.1
Distances:

Auckland	2210 miles.
Wellington	2340 miles.

New Zealand ⇐ ⇒ Nuku'alofa

Diagram 7.73
Route
7.74

The sea bed between Kermadec Islands (30°30′S, 178°30′W) and Fiji and Tonga Islands is very uneven and liable to volcanic activity and the whole region must be regarded with suspicion, see *New Zealand Pilot* and *Pacific Islands Pilot, Volume II.*

7.74.1

From Wellington (41°22′S, 174°50′E) and South Island ports the routes are:

E of North Island, thence:

Through 31°20′S, 179°30′W (**F**), thence:

Through 22°50′S, 177°00′W (**G**), passing W of Kermadec Islands, Pelorus Reef (22°51′S, 176°26′W) and other charted dangers, thence:

As navigation permits to Nuku'alofa (21°00′S, 175°10′W).

7.74.2

From Auckland (36°36′S, 174°49′E) the route is direct to join the route from Wellington (7.74.1) W of Pelorus Reef in 22°50′S, 177°00′W (**G**).

7.74.3
Distances:

Auckland	1100 miles.
Wellington	1400 miles.

New Zealand ⇐ ⇒ Apia

Diagram 7.73
Route
7.75

For remarks on the region S of Tonga Islands, see 7.74.

7.75.1

From Wellington (41°22′S, 174°50′E) and South Island ports the routes are:

E of North Island, thence:

Through 31°20′S, 179°30′W (**F**), thence:
Through 22°50′S, 177°00′W (**G**), thence:
Through 18°30′S, 174°45′W (**H**), thence:
Through 15°17′S, 173°55′W (**J**), thence:
Through Apolima Strait (13°48′S, 172°10′W) to Apia (13°47′S, 171°45′W).

The track passes W of Kermadec Islands (30°30′S, 178°30′W), Pelorus Reef (22°51′S, 176°26′W) and dangers reported W of Tonga Islands, and Curacoa Reef (15°29′S, 173°37′W).

7.75.2

From Auckland (36°36′S, 174°49′E) the route is direct to join the route from Wellington (7.75.1) W of Pelorus Reef in 22°50′S, 177°00′W (**G**).

7.75.3

Distances:

Auckland	1580 miles.
Wellington	1890 miles.

New Zealand ⇐ ⇒ Nouméa or Suva

Diagram 7.73
Route
7.76

Routes from Wellington (41°22′S, 174°50′E) and Auckland (36°36′S, 174°49′E) are direct:
To Nouméa (22°18′S, 166°25′E), or:
To Kadavu Passage (18°45′S, 178°00′E) for Suva (18°11′S, 178°24′E).

7.76.1

From Wellington and South Island ports the routes pass W of North Island.

7.76.2

From Auckland the route to Nouméa passes E of the submarine volcano (reported 1981) in 27°45′S, 169°09′E.

7.76.3

Distances:

From Auckland		
	To Nouméa	980 miles.
	To Suva	1140 miles.
From Wellington		
	To Nouméa	1230 miles.
	To Suva	1470 miles.

Suva ⇐ ⇒ Nuku'alofa

Diagram 7.77
Route
7.77

Between Suva (18°11′S, 178°24′E) and Nuku'alofa (21°00′S, 175°10′W) the routes are as direct as navigation permits.

Distance	410 miles.

Suva ⇐ ⇒ Papeete

Diagram 7.77
Route
7.78

Between Suva (18°11′S, 178°24′E) and Papeete (17°30′S, 149°36′W) the routes are as direct as navigation permits, either through Nanuku Passage (16°40′S, 179°10′W) or through Lakeba Passage (17°50′S, 178°30′W).

Although the distance by Lakeba Passage is some 30 miles shorter, this passage is only recommended for use by vessels fitted with modern navigational equipment or in fine weather with extreme visibility.

Distance (via Nanuka Passage)   1880 miles.

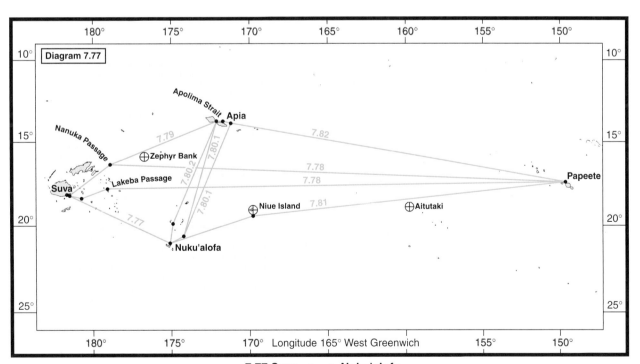

7.77 Suva ← → Nuku'alofa
7.78 Suva ← → Papeete
7.79 Suva ← → Apia
7.80 Nuku'alofa ← → Apia
7.81 Nuku'alofa ← → Papeete
7.82 Apia ← → Papeete

Suva ⇐ ⇒ Apia

Diagram 7.77
Route
7.79
From Suva (18°11'S, 178°24'E) the route is direct:
Through Nanuku Passage (16°40'S, 179°10'W), and:
N of Zephyr Bank (15°55'S, 176°50'W), thence,
Through Apolima Strait (13°48'S, 172°10'W) to Apia
(13°47'S, 171°45'W).
Distance   640 miles.

Nuku'alofa ⇐ ⇒ Apia

Diagram 7.77
Routes
7.80
From Nuku'alofa (21°00'S, 175°10'W) routes pass either
E or W of Tonga Islands (20°00'S, 174°30'W).
7.80.1
East of Tonga Islands.
Either through Apolima Strait (13°48'S, 172°10'W) to
Apia (13°47'S, 171°45'W), or:
E of Upolu Island (13°55'S, 171°45'W) to Apia.
Distances:
Through Apolima Strait   510 miles.
E of Upolu Island   525 miles.

7.80.2
West of Tonga Islands.
Through Apolima Strait
Distance   520 miles.

Nuku'alofa ⇐ ⇒ Papeete

Diagram 7.77
Routes
7.81
From Nuku'alofa (21°00'S, 175°10'W) the route passes:
20 miles S of Niue Island (19°03'S, 169°51'E),
thence:
N of Aitutaki (18°54'S, 159°47'W), thence:
To Papeete (17°30'S, 149°36'W).
Distance   1470 miles.

Apia ⇐ ⇒ Papeete

Diagram 7.77
Routes
7.82
Between Apia (13°47'S, 171°45'W) and Papeete
(17°30'S, 149°36'W) the routes are as direct as navigation
permits.
Distance   1300 miles.

PASSAGES THROUGH EASTERN ARCHIPELAGO, SOUTH CHINA SEA AND SOLOMON SEA

GENERAL INFORMATION
Coverage
7.83
This section contains details of the following passages:
Straits and Passages in Eastern Archipelago (7.84).
To and from Selat Sunda, Jakarta or Java Sea and
South China Sea (7.85).
North-South routes through South China Sea (7.86).
To and from Singapore and Krung Thep (Bangkok) or
Ho Chi Minh City (7.87).
From Singapore to Hong Kong (7.88).
From Hong Kong to Singapore (7.89).
To and from Hong Kong and Shanghai or more N
ports (7.90).
From Singapore to Shanghai (7.91).
From Shanghai to Singapore (7.92).
From Singapore to Nagasaki (7.93).
From Nagasaki to Singapore (7.94).
From Singapore to Yokohama (7.95).
From Yokohama to Singapore (7.96).
To and from Singapore and Manila or Verde Island
Passage (7.97).
To and from Manila or Verde Island Passage and
Guam (7.98).
To and from Singapore and San Bernadino Strait,
Surigao Strait or Guam (7.99).
To and from Singapore and Yap (7.100).
To and from Singapore and Sulu Sea or Basilan Strait
(7.101).
To and from Singapore and Palawan Passage (7.102).
To and from Singapore and Selat Sunda or Jakarta
(7.103).
To and from Singapore and Selat Lombok (7.104).
To and from Singapore and Selat Wetar (7.105).

To and from Singapore and Torres Strait (7.106).
To and from Selat Sunda or Jakarta and Selat Wetar
(7.107).
To and from Selat Sunda or Jakarta and Selat
Lombok (7.108).
To and from Singapore and Ambon (7.109).
To and from Singapore and Ujungpandang (7.110).
To and from Singapore and Surabaya (7.111).
To and from Singapore and Balikpapan (7.112).
To and from Singapore and Tarakan (7.113).
To and from Krung Thep (Bangkok), Ho Chi Minh
City or Hong Kong and Manila, Verde Island
Passage or Mindoro Strait (7.114).
To and from Krung Thep (Bangkok) or Ho Chi Minh
City and Selat Sunda, Jakarta or Surabaya (7.115).
To and from Hong Kong and Selat Sunda, Jakarta or
Surabaya (7.116).
To and from Hong Kong and Sandakan (7.117).
To and from Hong Kong and Tarakan, Balikpapan or
Ujungpandang (7.118).
To and from Hong Kong and Ambon (7.119).
To and from Hong Kong and Iloilo or Cebu (7.120).
To and from Hong Kong and Guam or Yap (7.121).
To and from Manila or Verde Island Passage and
Selat Sunda or Jakarta (7.122).
To and from Manila, Verde Island Passage or
Mindoro Strait and Tarakan, Balikpapan,
Ujungpandang or Surabaya (7.123).
To and from Manila, Verde Island Passage or
Mindoro Strait and Sandakan (7.124).
To and from Manila, Verde Island Passage or
Mindoro Strait and Iloilo or Cebu (7.125).
To and from Torres Strait and Balikpapan (7.126).
To and from Torres Strait and Tarakan (7.127).

Straits and Passages in Eastern Archipelago
7.84

The following notes are for use when planning passages between the Indian Ocean and North Pacific Ocean, in addition the relevant volume of *Admiralty Sailing Directions* should be consulted.

Strait or Passage	Position	Remarks
Alas, Selat	8°40′S, 116°40′E	On route from the Indian Ocean to Selat Makasar and Selat Sapudi, has no dangers and may be used as an alternative to Selat Lombok if anchorage is desired.
Alor, Selat	8°15′S, 123°55′E	Savu Sea to Banda Sea, deep water.
Bali, Selat	8°10′S, 114°25′E	Seldom used except by local traffic.
Bangka, Selat (Java Sea)	2°06′S, 105°03′E	Between the coast of Sumatera and Banka. The shortest route between Selat Sunda and Singapore.
Bangka, Selat (Celebes Sea)	1°45′N, 125°05′E	Molucca Sea to Celebes Sea. The shortest route round NE end of Sulawesi, but is not lighted.
Bangka Passage (Celebes Sea)	2°00′N, 125°15′E	Molucca Sea to Celebes Sea. Separates islands NE of Sulawesi from Pulau-pulau Sanihe; wide and deep.
Basilan Strait	6°50′N, 122°00′E	Between Sulu Sea and Celebes Sea; the shortest route SW of Mindanao.
Berhala Selat	0°50′S, 104°15′E	Between coast of Sumatera and Singkep on the inner route between Singapore and Selat Sunda; is lighted.
Boleng, Selat and Selat Lamakera	8°25′S, 123°20′E	Savu Sea to Banda Sea; deep but somewhat exposed with strong tidal streams.
Butung, Alur Pelayaran	5°20′S, 123°15′E	Deep, wide, clear and lighted.
Butung, Selat	4°56′S, 122°47′E	Coastal route easy to navigate in daylight, but has no routeing advantage over Alur Pelayaran Buton. There is a depth of 18 m in South Narrows.
Dampier, Selat	0°40′S, 130°30′E	Connects Pacific Ocean with Halmahera Sea and thence to Ceram Sea NW of New Guinea.
Durian, Selat	0°40′N, 103°39′E	Entrance to Singapore Strait from the inner route from Selat Sunda, there is a controlling depth of 15 m.
Flores, Selat	8°20′S, 123°00′E	Savu Sea to Banda Sea. Deep and clear except for Narrows at N end. Strong tidal streams in parts calling for good reserves of power.
Gelasa, Selat	2°53′S, 107°18′E	Frequently in use between Java Sea and South China Sea as alternative to Selat Bangka.
Hinatuan Passage	9°52′N, 125°28′E	Connects Pacific Ocean with Surigao Strait, N of Mindanao.
Jailolo, Selat	0°06′N, 129°10′E	Between Pacific Ocean and Halmahera Sea leading to Ceram Sea; deep water.
Karimata, Selat*	1°43′S, 108°34′E	Wide passage connecting South China Sea with E part of Java Sea.
Lamakera, Selat	8°30′S, 123°10′E	See Boleng, Selat.
Linapacan Strait	11°37′N, 119°57′E	Between Sulu Sea and South China Sea. Deep and clear but has dangers in E approaches.
Lombok, Selat*	8°47′S, 115°44′E	The most important passage between Java Sea or Selat Makasar and the Indian Ocean. Wide and easy to navigate.
Makasar, Selat*	2°00′S, 118°00′E	About 400 miles in length, connecting Celebes Sea with Java Sea and Flores Sea.
Malacca Strait	4°00′N, 100°00′E	About 250 miles long in its narrower part connecting Andaman Sea with Singapore Strait and Selat Durian. Depths are irregular and apt to vary. See *Malacca Strait and West Coast of Sumatera Pilot* also *Chart 5502, Mariner's Routeing Guide - Malacca and Singapore Straits.*
Manipa, Selat	3°20′S, 127°22′E	Connects Ceram Sea with Banda Sea; a wide and deep passage.
Mindoro Strait	13°00′N, 120°00′E	Connects South China Sea with Sulu Sea. In frequent use between Manila and islands to the S; a wide and deep strait.
Obi, Selat	1°00′S, 127°00′E	Connects Molucca Sea with Halmahera Sea and Selat Jailolo, wide and deep.
Ombi, Selat	8°35′S, 125°00′E	Connects Savu Sea with Banda Sea and E to Arafura Sea through Selat Wetar and Timor Sea; wide and deep.
Pantar, Selat	8°20′S, 124°20′E	Connects between Banda Sea and Selat Ombi; used by local traffic.
Riau, Selat	0°55′N, 104°20′E	Approach to Singapore Strait from S. Well lighted and buoyed with controlling depth of about 18 m.
Roti, Selat	10°25′S, 123°30′E	Connects between Savu Sea and Timor Sea, SW of Timor; deep water.
Sagewin, Selat	0°55′S, 130°40′E	Connects Pacific Ocean with Ceram Sea, NW of New Guinea.

Strait or Passage	Position	Remarks
Salayar, Selat	5°43′S, 120°30′E	Regularly used on route between Java Sea and Molukka Archipelago; deep water.
San Bernadino Strait	12°33′N, 124°12′E	Important passage on Pacific Ocean routes; wide and deep.
Sape, Selat	8°30′S, 119°20′E	Connects between Selat Sumba and Flores Sea.
Sapudi, Selat	7°00′S, 114°15′E	Regularly used on route between Java Sea and Selat Lombok or Flores Sea; lighted.
Sele, Selat	1°00′S, 131°00′E	Connects Pacific Ocean with Ceram Sea, NW of New Guinea.
Singapore Strait	1°10′N, 103°50′E	Connects between Malacca Strait and South China Sea; well lighted and buoyed with deep water.
Sumba, Selat	9°00′S, 119°30′E	Connects Savu Sea and Indian Ocean, N of Sumba; a wide and deep passage.
Sunda, Selat*	6°00′S, 105°52′E	Principal connection between Indian Ocean and Java Sea, but limited for deep-draught vessels by lack of deep water towards the NE.
Surigao Strait	9°53′N, 125°21′E	Connects between Pacific Ocean and Bohol Sea to Sulu Sea; safe and deep.
Wetar, Selat	8°16′S, 127°12′E	Connects between Banda Sea and Timor Sea through Alur Pelayaran Wetar; used for main routes between Singapore and Australia.

* Straits or passages incorporated in Indonesian Archipelagic Sea Lanes (7.130).

PASSAGES

Selat Sunda, Jakarta and Java Sea ⇐ ⇒ South China Sea

Diagram 7.85
Routes
7.85

Selat Sunda and Jakarta.
From Selat Sunda (6°00′S, 105°52′E), Jakarta (6°03′S, 106°53′E) and the E part of the Java Sea the routes are via ASL I (see 7.130 for details) which leads:

From Selat Sunda, passing between numerous oilfields and S of Pulau Jagautara (5°12′S, 106°27′E), thence:
Through Selat Karimata (1°43′S, 108°34′E), thence:
Between Natuna Besar (4°00′N, 108°00′E) and Subi Kecil (3°03′N, 108°51′E) into the South China Sea.

Caution. Areas of the Java Sea are being exploited for natural resources, see 7.38.2.

North-South routes through South China Sea

Diagram 7.85
Caution
7.86

A large area of dangerous ground, incompletely surveyed and encumbered with many coral reefs and shoals, lies in the SE part of the South China Sea. It lies between the parallels of 7°30′N and 12°00′N, separated on its SE side from Palawan and adjacent islands by Palawan Passage, and separated on its NW side from the coast of Vietnam by a wider and less encumbered part of the sea.

Routes
7.86.1

Main route.
This route is recommended for full-powered vessel, N-bound or S-bound at all seasons.
From Singapore (1°12′N, 103°51′E) the route passes:
Midway between Pulau Aur (2°27′N, 104°31′E) and Pulau-pulau Anambas (3°00′N, 106°00′E), thence:

Through 10°00′N, 110°05′E **(A)**, passing 25 miles SE of Charlotte Bank (7°08′N, 107°35′E), thence:
Midway between Macclesfield Bank (15°50′N, 114°30′E) and Bombay Reef (16°02′N, 112°30′E), thence:
Either side of Pratas Reef (20°40′N, 116°45′E), thence:
To Taiwan Strait, passing E of Taiwan Banks (23°00′N, 118°30′E) and through P'eng-hu Kang-tao (23°30′N, 119°53′E).

For directions when passing Pratas Reef, and cautions regarding currents and Macclesfield Bank, see *China Sea Pilot, Volume I.*
7.86.2

Low-powered vessels N-bound during the north-east monsoon may find the directions given at 10.65 more suitable.

Singapore ⇐ ⇒ Krung Thep or Ho Chi Minh City

Diagram 7.85
Routes
7.87

From Singapore (1°12′N, 103°51′E) routes to Krung Thep (Bangkok)) (13°23′N, 100°35′E) or to Song Sai Gon (10°18′N, 107°04′E) for Ho Chi Minh City, are as direct as navigation permits.

Distances.

Krung Thep	825 miles.
Ho Chi Minh City	600 miles.

7.87.1

In July during the strength of the south-west monsoon, or in December or January during the north-east monsoon, low-powered, or hampered, vessels may find the routes, described at 10.70, 10.72, 10.75 or 10.78 more advantageous.

Singapore ⇒ Hong Kong

Diagram 7.85
Routes
7.88

From Singapore (1°12′N, 103°51′E) the Main route (7.86.1) is the usual route for all seasons until midway

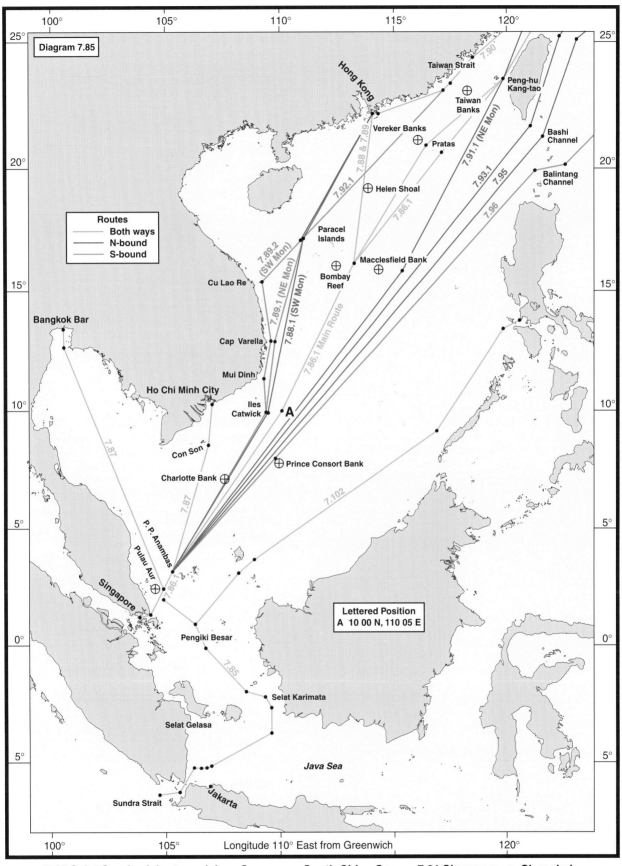

Diagram 7.85

Routes
— Both ways
— N-bound
— S-bound

Lettered Position
A 10 00 N, 110 05 E

Longitude 110° East from Greenwich

7.85 Selat Sunda, Jakarta and Java Sea ← → South China Sea;
7.86 North - South routes through South China Sea;
7.87 Singapore ← → Bangkok or Ho Chi Minh City;
7.88 Singapore → Hong Kong;
7.89 Hong Kong → Singapore;
7.90 Hong Kong ← → Shanghai and northern ports;

7.91 Singapore → Shanghai
7.92 Shanghai → Singapore
7.93 Singapore → Nagasaki
7.94 Nagasaki → Singapore
7.95 Singapore → Yokohama
7.96 Yokohama → Singapore

between Macclesfield Bank (15°50′N, 114°30′E) and Bombay Reef (16°02′N, 112°30′E), thence the route to Hong Kong (22°17′N, 114°10′E) passes 15 miles W of Helen Shoal (19°12′N, 113°53′E).

Distance 1460 miles.

7.88.1

During the strength of the south-west monsoon, smoother water will be found by keeping closer to the coast of Vietnam and passing W of the Paracel Islands (16°30′N, 112°00′E); this route is 20 miles shorter than the Main route.

7.88.2

Low-powered vessels, during the north-east monsoon, may prefer the route through Palawan Passage, described at 10.65.

Distance via Palawan Passage 1460 miles.

Hong Kong ⇒ Singapore

Diagram 7.85
Routes
7.89

From Hong Kong (22°17′N, 114°10′E) routes are either by Main route (7.88) in reverse, or are seasonal.

7.89.1

North-east Monsoon (December and January).

To take advantage of the predominantly S-flowing current of this monsoon, the route passes:

30 miles W of Paracel Islands (16°30′N, 112°00′E), thence:

15 to 20 miles E of Cap Varella (12°54′N, 109°28′E), thence:

E of Îles Catwick (10°00′N, 109°00′E), thence:

To Singapore Strait (1°10′N, 103°50′E).

7.89.2

South-west Monsoon (July).

During this monsoon the route passes:

30 miles W of Paracel Islands (16°30′N, 112°00′E), thence:

To a landfall off Cu Lao Ré (15°23′N, 109°07′E) if the monsoon is strong; or off Cap Varella (12°54′N, 109°28′E) if the monsoon is light, thence in either case:

Coastwise to Mui Dinh (11°22′N, 109°01′E), keeping 10 miles offshore, thence:

To Pulau Aur (2°27′N, 104°31′E), keeping E of Îles Catwick (10°00′N, 109°00′E), unless the weather is clear or the ship's position certain, thence:

To Singapore Strait (1°10′N, 103°50′E).

Hong Kong ⇐ ⇒ Shanghai and northern ports

Diagram 7.85 and 7.90
Cautions
7.90

1 **Tidal streams** are very strong in places along this section of the coast and care is necessary at all times.

2 **Soundings.** There have been many strandings, on outlying islands off the coast of China, between Hong Kong (22°17′N, 114°10′E) and the entrance to Chang Jiang (31°03′N, 122°20′E). In many cases the stranding would not have occurred if attention had been paid to the necessity for constantly sounding in thick or misty weather. Many lighthouses on the islands are of considerable

elevation and often the upper parts of the islands, and the lights, are obscured by fog. As a general rule, when unsure of the ship's position, course should be adjusted to keep in safe depths and continued sounding should be maintained until the position is ascertained.

3 **Fishing.** Large fleets of fishing junks, many not carrying lights, may be met off the coast of China.

Routes
7.90.1

Routes between Hong Kong (22°17′N, 114°10′E) and Shanghai (31°03′N, 122°20′E) keep about 5 to 10 miles E of the outer islands. For details see *China Sea Pilot, Volumes I and II.*

7.90.2

Low-powered, hampered or small vessels proceeding along this coast, during the north-east monsoon, may prefer an alternative route described at 10.90.

7.90.3

Distances:

Shanghai 825 miles.
Nakhodka (42°48′N, 132°57′E) 1630 miles.

Singapore ⇒ Shanghai

Diagram 7.85 and 7.90
Route
7.91

The route from Singapore Strait (1°10′N, 103°50′E) is via the Main Route (7.86.1) as far as Taiwan Strait, thence as in *China Sea Pilot, Volume II* to Shanghai (31°03′N, 122°20′E).

Distance* 2200 miles.
* Through Xiaoban Men (30°12′N, 122°36′E).

7.91.1

North-east Monsoon.

During a strong north-east monsoon, an alternative route is:

W of Pulau-pulau Anambas (3°00′N, 106°00′E), thence:

W of Prince Consort Bank (7°53′N, 110°00′E), thence:

E of Macclesfield Bank (15°50′N, 114°30′E), thence:

Through P'eng-hu Kang-tao (23°30′N, 119°53′E), thence:

N along the W coast of T'ai-wan, thence:

Through Xiaoban Men (30°12′N, 122°36′E), as in *China Sea Pilot, Volume II* to Shanghai (31°03′N, 122°20′E).

In addition to getting smoother water and a favourable current, a great advantage in using P'eng-hu Kang-tao is the absence of big fleets of fishing junks, which are encountered along the China coast. The channel is well lighted.

Distance 2230 miles.

Shanghai ⇒ Singapore

Diagram 7.85 and 7.90
Route
7.92

From Shanghai (31°03′N, 122°20′E) the usual route is through Xiaoban Men (30°12′N, 122°36′E) to Taiwan Strait, as described in *China Sea Pilot, Volume II*, thence

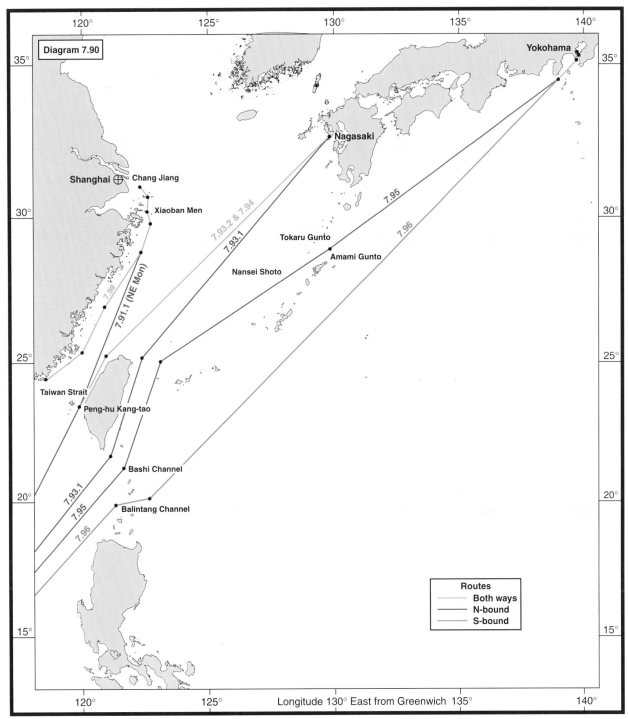

7.90 Hong Kong ← → Shanghai and northern ports;
7.91 Singapore → Shanghai;
7.92 Shanghai → Singapore;
7.93 Singapore → Nagasaki

7.94 Nagasaki → Singapore
7.95 Singapore → Yokohama
7.96 Yokhama → Singapore

by the Main Route (7.86.1) through South China Sea to Singapore Strait (1°10′N, 103°50′E).

Distance 2200 miles.

7.92.1

South-west Monsoon.

From May to August if the south-west monsoon is strong, an alternative route, after passing through Taiwan Strait, is:

NW of Vereker Banks (21°05′N, 116°00′E), Helen Shoal (19°12′N, 113°53′E) and Paracel Islands (16°30′N, 112°00′E), to make landfall on Cu Lao Ré (15°23′N, 109°07′E), thence:

Via the south-west monsoon Route from Hong Kong (7.89.2) to Singapore Strait (1°10′N, 103°50′E).

Distance 2240 miles.

Singapore ⇒ Nagasaki

Diagram 7.85 and *7.90*
Routes
7.93

From Singapore Strait (1°10′N, 103°50′E) there are two alternative routes:
7.93.1

    W of Pulau-pulau Anambas (3°00′N, 106°00′E), thence:

    NW of Prince Consort Bank (7°53′N, 110°00′E), thence:

    South of T'ai-wan, thence:

    N along the E coast of T'ai-wan, in the main stream of the Japan Current (7.26), thence:

    As navigation permits to Nagasaki (32°42′N, 129°49′E).

Distance   2430 miles.

7.93.2

    Via the Main Route (7.86.1) through South China Sea to Taiwan Strait, passing through P'eng-hu Kang-tao (23°30′N, 119°53′E) thence:

    As navigation permits to Nagasaki (32°42′N, 129°49′E).

Distance   2430 miles.

Nagasaki ⇒ Singapore

Diagram 7.85 and *7.90*
Route
7.94

From Nagasaki (32°42′N, 129°49′E), to Singapore Strait (1°10′N, 103°50′E) the route is:

    As navigation permits to Taiwan Strait, passing through P'eng-hu Kang-tao (23°30′N, 119°53′E), thence:

    Joining the Main Route (7.86.1) to Singapore.

An alternative is to joining the south-west monsoon route from Shanghai to Singapore (7.92.1).

The Japan Current (7.26) flows NE-ward along the E coast of T'ai-wan throughout the year; in the East China Sea and the Taiwan Strait currents change direction according to the monsoon season.
7.94.1

Distances:

Via Main Route	2430 miles.
Via south-west monsoon Route	2480 miles.

Singapore ⇒ Yokohama

Diagram 7.85 and *7.90*
Routes
7.95

From Singapore Strait (1°10′N, 103°50′E) the route is:

    W of Pulau-pulau Anambas (3°00′N, 106°00′E), thence:

    NW of Prince Consort Bank (7°53′N, 110°00′E), thence:

    Through Bashi Channel (21°20′N, 121°00′E), thence:

    As at 7.163.1 to Yokohama (35°26′N, 139°43′E).

Distance   2970 miles.

Alternatively the route at 7.96, in reverse, can be used although this is less favourable as regards current.

Yokohama ⇒ Singapore

Diagram 7.85 and *7.90*
Route
7.96

From Yokohama (35°26′N, 139°43′E) the route is:

    Direct to Balintang Channel, as at 7.163.2, thence:

    NW of Prince Consort Bank (7°53′N, 110°00′E), thence:

    W of Pulau-pulau Anambas (3°00′N, 106°00′E), thence:

    As navigation permits to to Singapore Strait (1°10′N, 103°50′E).

Distance   2890 miles.

Singapore ⇐ ⇒ Manila or Verde Island Passage

Diagram 7.97
Routes
7.97

From Singapore (1°12′N, 103°51′E) the route is:

    W of Pulau-pulau Anambas (3°00′N, 106°00′E), thence:

    NW of Prince Consort Bank (7°53′N, 110°00′E), thence:

    30 miles NW of North Danger Reef (11°25′N, 114°21′E), thence:

    Direct to Manila (14°32′N, 120°56′E) or Verde Island Passage (13°36′N, 121°00′E).

7.97.1

During the north-east monsoon, an alternative route is through Palawan Passage, for details see 7.102 for full-powered vessels and 10.65 for low-powered vessels..
7.97.2

Distances:

	Manila	Verde Island Passage
W of P.-P. Anambas	1340	1320
Via Palawan Passage	1370	1330

Manila or Verde Island Passage ⇐ ⇒ Guam

Diagram 7.97
Routes
7.98

The routes from Manila (14°32′N, 120°56′E) and Verde Island Passage (13°36′N, 121°00′E) to San Bernadino Strait (12°33′N, 124°12′E) are described in *China Sea Pilot, Volume II* and *Philippine Islands Pilot*.

From San Bernadino Strait they continue direct to Guam (13°27′N, 144°35′E).
7.98.1

Distances:

Manila	1510 miles.
Verde Island Passage	1430 miles.

Singapore ⇐ ⇒ San Bernadino Strait, Surigao Strait or Guam

Diagram 7.97
Routes
7.99

From Singapore (1°12′N, 103°51′E) the route:

    Follows that at 7.97 to Verde Island Passage (13°36′N, 121°00′E), thence:

    As directed in *Philippine Islands Pilot* to San Bernadino Strait (12°33′N, 124°12′E), thence:

    Direct to Guam (13°27′N, 144°35′E).

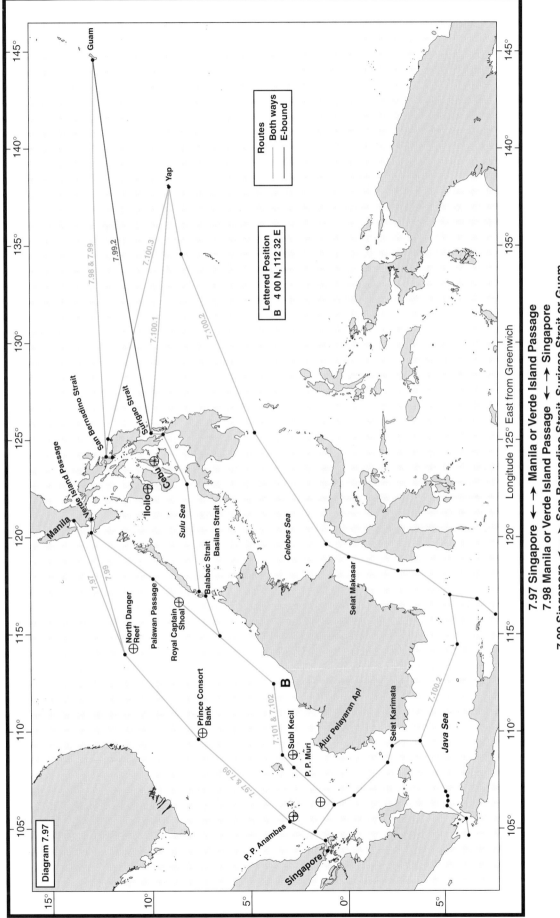

Diagram 7.97

7.97 Singapore ←→ Manila or Verde Island Passage
7.98 Manila or Verde Island Passage ←→ Singapore
7.99 Singapore ←→ San Bernadino Strait, Surigao Strait or Guam
7.100 Singapore ←→ Yap
7.101 Singapore ←→ Sulu Sea and Basilan Strait
7.102 Singapore ←→ Palawan Passage

7.99.1

Distances:

San Bernadino Strait	1530 miles.
Guam	2750 miles.

7.99.2

Alternatively, particularly E-bound during the north-east monsoon, the following route from Singapore may be taken:

As at 7.102.1 to the S entrance of Palawan Passage (10°20′N, 118°00′E), thence:

Through Balabac Strait (7°34′N, 116°55′E), thence:

Through Sulu Sea to Surigao Strait (9°53′N, 125°21′E), thence:

Direct to Guam (13°27′N, 144°35′E).

Distance   2640 miles.

Singapore ⇐ ⇒ Yap

Diagram 7.97

Routes

7.100

From Singapore (1°12′N, 103°51′E) there are three principal routes:

7.100.1

The first route is:

As at 7.102.1 to the S entrance of Palawan Passage (10°20′N, 118°00′E), thence:

Through Balabac Strait (7°34′N, 116°55′E), thence:

Through Sulu Sea to Surigao Strait (9°53′N, 125°21′E), thence:

Direct to Yap (9°28′N, 138°09′E).

Distance   2230 miles.

1   Surigao Strait is safe and deep throughout its length, and the islands bordering it are steep-to. It is the only passage for large vessels from the Pacific Ocean to the interior waters of the Philippine Islands, with the exception of San Bernadino Strait (12°33′N, 124°12′E). It is of particular use for ships bound for S Philippine Islands or the Sulu Sea.

2   During the north-east monsoon, Hinatuan Passage (9°52′N, 125°28′E) between the NE end of Mindanao and the off-lying islands may afford some protection against the weather, but it is not recommended for large or low-powered vessels without a pilot due to strong tidal streams. For details see *Philippine Islands Pilot*.

7.100.2

The second route passes:

S via ASL IA and ASL I (see 7.130 for details), through Selat Karimata (1°43′S 108°34′E) and E across the Java Sea to join ASL II which leads N through Selat Makasar (2°00′S, 118°00′E), thence:

Across the Celebes Sea to pass E of Mindanao, thence:

Direct to Yap (9°28′N, 138°09′E).

Distance   2520 miles.

7.100.3

The third route is:

By the route from Singapore to Guam (7.99) as far as San Bernadino Strait (12°33′N, 124°12′E), thence:

Direct to Yap (9°28′N, 138°09′E).

Distance   2410 miles.

Singapore ⇐ ⇒ Sulu Sea and Basilan Strait

Diagram 7.97

Routes

7.101

From Singapore (1°12′N, 103°51′E) the route is:

As at 7.102.1 to the S entrance of Palawan Passage (10°20′N, 118°00′E), thence:

Through Balabac Strait (7°34′N, 116°55′E). For details of channels and cautions on dangers and currents see *Philippine Islands Pilot*, thence:

As navigation permits to destination.

7.101.1

Distances:

Iloilo (10°42′N, 122°36′E)	1290 miles.
Cebu (10°18′N, 123°55′E)	1380 miles.
Basilan Strait (6°50′N, 122°00′E)	1210 miles.
S point of Mindanao (5°00′N, 125°30′E) for Central Pacific Route (7.300.1)	1480 miles.
Sandakan (5°49′N, 118°07′E) via Banggi South Channel	1040 miles.

Singapore ⇐ ⇒ Palawan Passage

Diagram 7.97

Routes

7.102

From Singapore (1°12′N, 103°51′E) the route passes:

5 miles N of Subi Kecil (3°03′N, 108°51′E), thence:

Through 4°00′N, 112°32′E (**B**), thence:

As navigation permits to Palawan Passage (10°20′N, 118°00′E).

Distance   1020 miles.

7.102.1

1   The narrowest and most dangerous part of Palawan Passage, where it is only 29 miles wide between dangers, lies abreast Royal Captain Shoal (9°03′N, 116°41′E) off the S part of Palawan.

2   For cautions on shoals near the track in 3°55′N, 112°15′E, position fixing, currents and directions for Palawan Passage, see *China Sea Pilot, Volume II*.

Singapore ⇐ ⇒ Selat Sunda or Jakarta

Diagram 7.103

Routes

7.103

From Singapore (1°12′N, 103°51′E) to Selat Sunda (6°00′S, 105°52′E) or Jakarta (6°03′S, 106°53′E) the route is:

Via ASL IA and ASL I (see 7.130 for details), passing through Selat Karimata (1°43′S, 108°34′E).

Distance   820 miles.

Caution. Areas in the N approaches to Selat Sunda are being exploited for natural resources, see 7.38.2.

Singapore ⇐ ⇒ Selat Lombok

Diagram 7.103

Routes

7.104

From Singapore (1°12′N, 103°51′E) the route is:

Via ASL IA and ASL I (see 7.130 for details), passing through Selat Karimata (1°43′S, 108°34′E), thence:

E across the Java Sea passing either N or S of Pulau Bawean (5°50′S, 112°40′E), thence:

Either through Selat Sapudi (7°00′S, 114°15′E), or, by day, S of Pulau-pulau Kangean (6°50′S, 115°25′E), thence:

To Selat Lombok (8°47′S, 115°44′E).

Distance (by either route)   980 miles.

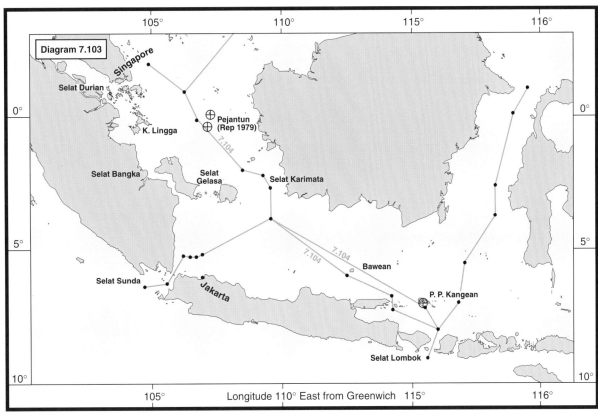

7.103 Singapore ← → Selat Sunda or Jakarta
7.104 Singapore ← → Selat Lombok

Singapore ⇐ ⇒ Selat Wetar

Diagram 7.105
Routes
7.105

From Singapore (1°12′N, 103°51′E) the route is:
South bound via ASL IA and ASL I (see 7.130 for details), passing through Selat Karimata (1°43′S, 108°34′E), thence:

From Selat Karimata the passage continues by one of the following four routes:

7.105.1

The first route passes:
Either N or S of Pulau Bawean (5°50′S, 112°40′E), thence:
Through Selat Sapudi (7°00′S, 114°15′E), thence:
Joining the parallel of 8°S, to the S of Gosong Sakunci (7°50′S, 117°15′), thence:
Proceeding along that parallel to Alur Pelayaran Wetar (8°00′S, 125°30′E), thence:
Into Selat Wetar (8°16′S, 127°12′E).
Distance to Alur Pelayaran Wetar   1510 miles.

7.105.2

The second route, by day, passes:
N of Pulau Bawean (5°50′S, 112°40′E) thence:
S of Pulau-pulau Kangean (6°50′S, 115°25′E), thence:
Joining the parallel of 8°S, to the S of Gosong Sakunci (7°50′S, 117°15′), thence:
Proceeding along that parallel to Alur Pelayaran Wetar (8°00′S, 125°30′E), thence:
Into Selat Wetar (8°16′S, 127°12′E).
Distance to Alur Pelayaran Wetar   1490 miles.

7.105.3

The third route passes:

N of Pulau Masalembo Kecil (5°25′S, 114°25′E), thence:
Between Gosong Takarewataya (6°04°S, 118°55′E) and Pulau-pulau Sabalana (7°00′S, 118°30′E), thence:
S of Pulau Kalao (7°17′S, 120°55′E), thence:
Joining the parallel of 8°S, to the S of Pasir Layaran (7°47′S, 122°18′E), thence:
Proceeding along that parallel to Alur Pelayaran Wetar (8°00′S, 125°30′E), thence:
Into Selat Wetar (8°16′S, 127°12′E).
Distance to Alur Pelayaran Wetar   1470 miles.

7.105.4

The fourth route passes:
N of Pulau Masalembo Kecil (5°25′S, 114°25′E), thence:
S of Gosong Takarewataya (6°04°S, 118°55′E), thence:
Through Selat Salayar (5°43′S, 120°30′E), thence:
Through Alur Pelayaran Wetar (8°00′S, 125°30′E), thence:
Into Selat Wetar (8°16′S, 127°12′E).
Distance to Alur Pelayaran Wetar   1480 miles.

7.105.5

In choosing a route it should be noted that an E-going current is more often experienced on the S sides of the Java and Flores Seas, rather than on the N sides.

Singapore ⇐ ⇒ Torres Strait

Diagram 7.105
Routes
7.106

Currents, on the routes given below, are predominantly W-going, except in the Java and Flores Seas, where the

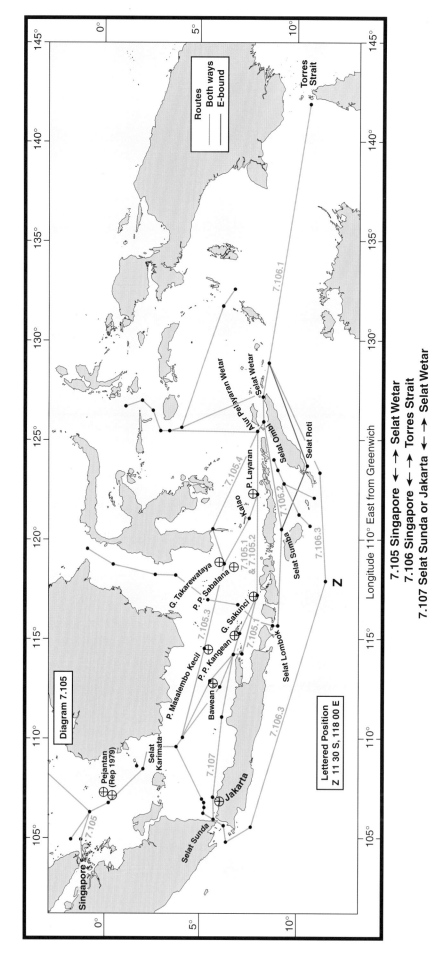

Diagram 7.105

7.105 Singapore ←→ Selat Wetar
7.106 Singapore ←→ Torres Strait
7.107 Selat Sunda or Jakarta ←→ Selat Wetar

Longitude 110° East from Greenwich

Lettered Position
Z 11 30 S, 118 00 E

Routes
Both ways
E-bound

current is E-going during the north-west monsoon (November to March).

7.106.1

East-bound (April to October), or:

West-bound (November to March).

From Singapore (1°12′N, 103°51′E) the routes are:

Through Selat Karimata to Selat Wetar (as described at 7.105), thence:

Direct to Torres Strait (10°36′S, 141°51′E).

Distances:

Singapore to Alur Palayaran Wetar, depending on route used, see 7.105.

Alur Palayaran Wetar to Torres Strait   980 miles.

7.106.2

West-bound (April to October).

East-bound (November to March).

The routes are:

If E-bound, passing S of Timor (9°00′S, 125°00′E), keeping in the deep-water trench S of Timor and Pulau-pulau Tanimbar (7°30′S, 131°30′E) to avoid the shoal area in the Timor Sea, and through Selat Roti (10°25′S, 123°30′E) or:

If W-bound, passing N of Timor (9°00′S, 125°00′E), and through Selat Ombi (8°35′S, 125°00′E).

Thence, in either case passing:

Through Selat Sumba (9°00′S, 119°30′E), thence:

Through either Selat Lombok (8°47′S, 115°44′E), or Selat Alas (8°40′S, 116°40′E) thence:

Through Java Sea and via ASL I and ASL IA, through Selat Karimata (1°43′S, 108°34′E), to Singapore (1°12′N, 103°51′E) as described at 7.104.

The route N of Timor avoids most of the dangerous areas in the Timor Sea and makes use of the bold shores of the strait as aids to navigation.

Distances via Selat Lombok:

N of Timor	2560 miles.
S of Timor	2570 miles.

7.106.3

East-bound or West-bound (at all seasons).

From Singapore (1°12′N, 103°51′E), an alternative route is:

Through Selat Sunda (6°00′S, 105°52′E), as described at 7.103, thence:

Through 11°30′S, 118°00′E (**Z**), passing S of all the islands E of Selat Sunda, thence:

As navigation permits to Torres Strait (10°36′S, 141°51′E).

The distance is greater and the only advantage over the other routes is ease of navigation over much of the route.

Distance*   2880 miles.

* For distances from Singapore to Australian Ports, see 7.142.

Selat Sunda or Jakarta ⇐ ⇒ Selat Wetar

Diagram 7.105
Routes
7.107

From Selat Sunda (6°00′S, 105°52′E) and Jakarta (6°03′S, 106°53′E) the route is:

Coastwise along the N coast of Jawa and Madura, thence:

As at 7.105.1 or 7.105.2 to Selat Wetar (8°16′S, 127°12′E).

Distances via Selat Sapudi (7.105.1) to Alur Pelayaran Wetar (8°00′S, 125°30′E) in the NW approaches to Selat Wetar:

Selat Sunda	1200 miles.
Jakarta	1140 miles.

Selat Sunda or Jakarta ⇐ ⇒ Selat Lombok

Diagram 7.108
Routes
7.108

From Selat Sunda (6°00′S, 105°52′E) and Jakarta (6°03′S, 106°53′E) the route is:

Coastwise along the N coast of Jawa, passing to seaward of a charted restricted entry area, thence:

S of Pulau-Pulau Karimunjawa (5°54′S, 110°15′E) and N of Pulau Madura (7°00′S, 113°30′E), thence:

Through Selat Sapudi (7°00′S, 114°15′E), thence:

To Selat Lombok (8°47′S, 115°44′E).

Distances:

Selat Sunda	660 miles.
Jakarta	615 miles.

Singapore ⇐ ⇒ Ambon

Diagram 7.108
Routes
7.109

From Singapore (1°12′N, 103°51′E) the route is:

South bound via ASL IA and ASL I (see 7.130 for details), passing through Selat Karimata (1°43′S, 108°34′E), thence:

S of Pulau Bawean (5°50′S, 112°40′E), thence:

S of Gosong Takarewataya (6°04′S, 118°55′E), thence:

Through Selat Salayar (5°43′S, 120°30′E), thence:

Through Selat Butung (Buton) (4°56′S, 122°47′E), thence:

To Ambon (3°41′S, 128°10′E).

This is probably a better route to the Molucca Sea, in either Monsoon, than passage N of Borneo.

Distance   1690 miles.

Singapore ⇐ ⇒ Ujungpandang

Diagram 7.108
Routes
7.110

From Singapore (1°12′N, 103°51′E) the route is:

South bound via ASL IA and ASL I (see 7.130 for details), passing through Selat Karimata (1°43′S, 108°34′E), thence:

From Selat Karimata the passage continues by one of the following two routes:

7.110.1

The first route passes:

S of Pulau Laut (3°40′S, 116°15′E), thence:

N of Batu Butunga (4°40′S, 117°00′E), thence:

Through the swept channel leading to Ujungpandang (5°08′S, 119°22′E).

Distance   1110 miles.

7.110.2

The second route passes:

S of Pulau-pulau Laurot (4°50′S, 115°45′E), thence:

Through Pulau-pulau Lima (5°05′S, 117°04′E), thence:

Through the swept channel leading to Ujungpandang (5°08′S, 119°22′E).

For directions see *Indonesia Pilot, Volume II.*

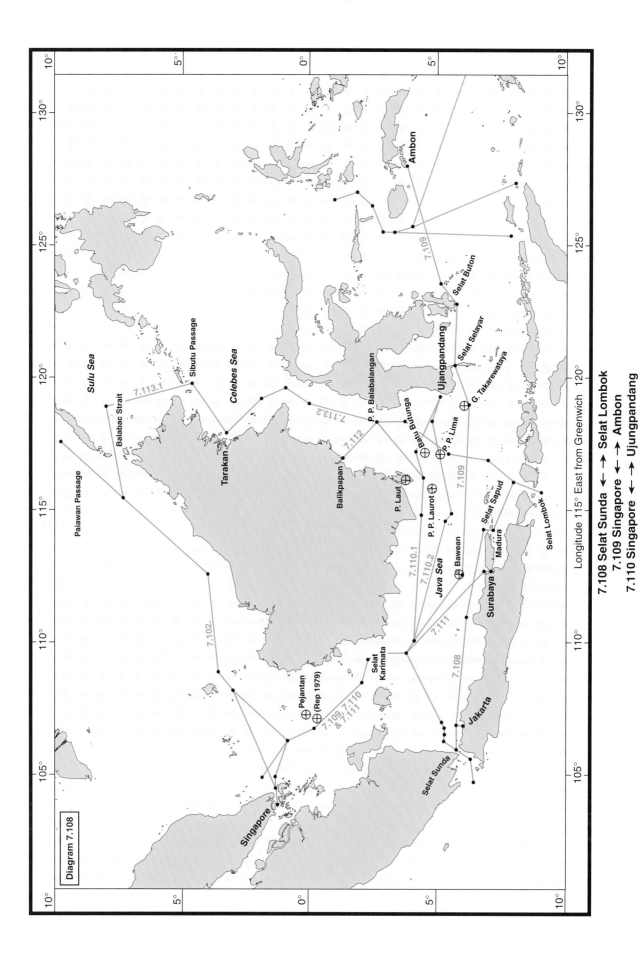

Diagram 7.108

7.108 Selat Sunda ←→ Selat Lombok
7.109 Singapore ←→ Ambon
7.110 Singapore ←→ Ujungpandang
7.111 Singapore ←→ Surabaya
7.112 Singapore ←→ Balikpapan
7.113 Singapore ←→ Tarakan

Longitude 115° East from Greenwich 120°

Singapore ⇐ ⇒ Surabaya

Diagram 7.108
Routes
7.111

From Singapore (1°12′N, 103°51′E) the route is:

South bound via ASL IA and ASL I (see 7.130 for details), passing through Selat Karimata (1°43′S, 108°34′E), thence:

Direct to Surabaya (7°12′S, 112°44′E).

Distance 770 miles.

Singapore ⇐ ⇒ Balikpapan

Diagram 7.108
Routes
7.112

From Singapore (1°12′N, 103°51′E) the route is:

South bound via ASL IA and ASL I (see 7.130 for details), passing through Selat Karimata (1°43′S, 108°34′E), thence:

S and E of Pulau Laut (3°40′S, 116°15′E), thence:

W of Pulau-pulau Balabalangan (2°15′S, 117°30′E), thence:

To Balikpapan (1°21′S, 116°56′E).

Distance 1080 miles.

Singapore ⇐ ⇒ Tarakan

Diagram 7.108
Routes
7.113

From Singapore (1°12′N, 103°51′E) there are two alternative routes.

7.113.1

The first route, N of Borneo, passes:

As at 7.102 to the S entrance of Palawan Passage (10°20′N, 118°00′E), thence:

Through Balabac Strait (7°34′N, 116°55′E), thence:

Through the Sulu Sea to Sibutu Passage (4°55′N, 119°37′E); for directions see *Philippine Islands Pilot*. Thence:

To Tarakan (3°15′N, 117°54′E).

7.113.2

The second route, S of Borneo, passes:

South bound via ASL IA and ASL I (see 7.130 for details), passing through Selat Karimata (1°43′S, 108°34′E), thence:

To a position W of Pulau-pulau Balabalangan (2°15′S, 117°30′E), as described at 7.112, thence:

Joining ASL II (see 7.130 for details), passing through Selat Makasar (2°00′S, 118°00′E), thence:

As navigation permits through Celebes Sea to Tarakan (3°15′N, 117°54′E).

Distances:

N of Borneo	1290 miles.
S of Borneo	1440 miles.

Krung Thep, Ho Chi Minh City or Hong Kong ⇐ ⇒ Manila, Verde Island Passage or Mindoro Strait

Diagram 7.114
Routes
7.114

From Krung Thep (13°23′N, 100°35′E), Ho Chi Minh City (10°18′N, 107°04′E) and Hong Kong (22°17′N, 114°10′E) routes to Manila (14°32′N, 120°56′E), Verde Island Passage (13°36′N, 121°00′E) or Mindoro Strait (13°00′N, 120°00′E) are as direct as safe navigation permits.

7.114.1

Distances:

	Manila	Verde Island Passage	Mindoro Strait
Krung Thep	1450	1430	1390
Ho Chi Minh City	880	865	820
Hong Kong	630	660	635

Krung Thep or Ho Chi Minh City ⇐ ⇒ Selat Sunda, Jakarta or Surabaya

Diagram 7.114
Routes
7.115

From Krung Thep (13°23′N, 100°35′E) or Ho Chi Minh City (10°18′N, 107°04′E) routes are as direct as safe navigation permits:

To join ASL IA and ASL I (see 7.130 for details), passing through Selat Karimata (1°43′S, 108°34′E), thence:

Direct via ASL I to Selat Sunda (6°00′S, 105°52′E) and Jakarta (6°03′S, 106°53′E), or:

Leaving ASL I when clear of Selat Karimata (1°43′S, 108°34′E), thence:

As navigation permits for Surabaya (7°12′S, 112°44′E).

7.115.1

Distances:

	Selat Sunda	Jakarta	Surabaya
Krung Thep	1280	1260	1450
Ho Chi Minh City	1010	990	1160

7.115.2

For low-powered vessels it may be preferable to follow the routes described at 10.75 and 10.78 to the E entrance to Singapore Strait (1°10′N, 103°50′E).

7.115.3

Caution. Details regarding Areas to be Avoided, in the Java Sea, can be found at 7.38.2.

Hong Kong ⇐ ⇒ Selat Sunda, Jakarta or Surabaya

Diagram 7.114
Routes
7.116

North-bound or South-bound the normal route from Hong Kong (22°17′N, 114°10′E) is:

15 miles W of Helen Shoal (19°12′N, 113°53′E), thence:

Midway between Macclesfield Bank (15°50′N, 114°30′E) and Bombay Reef (16°02′N, 112°30′E), thence:

To 10°00′N, 110°05′E (**A**), thence:

W of Vanguard Bank (7°28′N, 109°37′E), thence:

To join ASL I (see 7.130 for details), passing through Selat Karimata (1°43′S, 108°34′E), thence:

Direct via ASL I to Selat Sunda (6°00′S, 105°52′E) and Jakarta (6°03′S, 106°53′E), or:

Leaving ASL I when clear of Selat Karimata (1°43′S, 108°34′E), thence:

As navigation permits for Surabaya (7°12′S, 112°44′E) or the E part of the Java Sea.

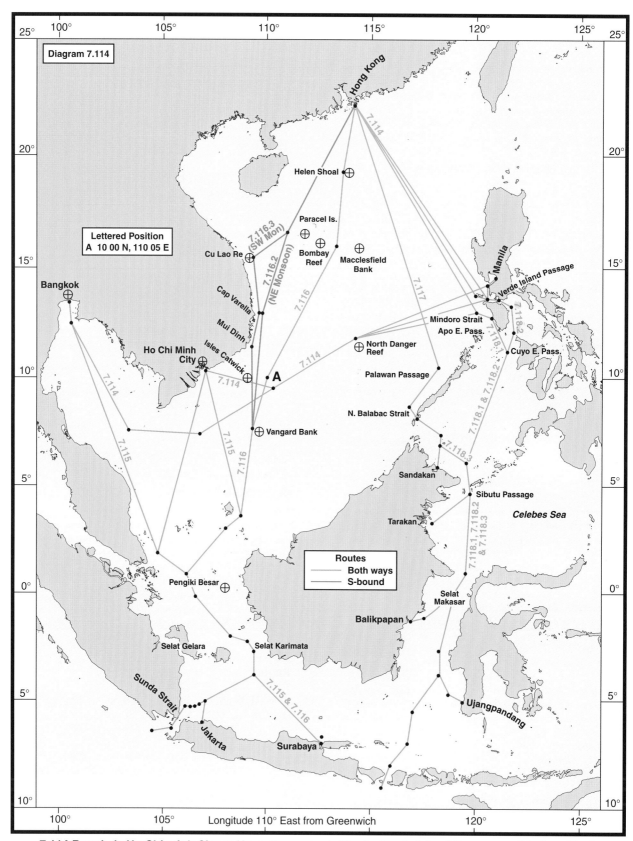

Diagram 7.114

Lettered Position
A 10 00 N, 110 05 E

Routes
— Both ways
— S-bound

Longitude 110° East from Greenwich

7.114 Bangkok, Ho Chi minh City or Hong Kong ← → Manila, Verde Island Passage or Mindoro Strait
7.115 Bangkok or Ho Chi Minh City ← → Selat Sunda, Jakarta or Surabaya
7.116 Hong Kong ← → Selat Sunda, Jakarta or Surabaya
7.117 Hong Kong ← → Sandakan
7.118 Hong Kong ← → Tarakan, Balikpapan or Ujungpandang

Distances:

Selat Sunda	1800 miles.
Jakarta	1780 miles.
Surabaya	1940 miles.

7.116.1
South-bound there are alternative seasonal routes.
7.116.2
North-east Monsoon. To take advantage of the predominant current of this monsoon, setting S in the W part of the South China Sea, the route is:

30 miles W of Paracel Islands (16°30′N, 112°00′E), thence:

15 to 20 miles E of Cap Varella (12°54′N, 109°28′E), thence:

E of Îles Catwick (10°00′N, 109°00′E), thence:

Joining the normal route (7.116) W of Vanguard Bank.

7.116.3
South-west Monsoon. For the most sheltered water during this monsoon, the route is:

30 miles W of Paracel Islands (16°30′N, 112°00′E), thence:

If the monsoon is strong, to a landfall off Cu Lao Ré (15°23′N, 109°07′E), or if the monsoon is light, off Cap Varella (12°54′N, 109°28′E), thence:

Coastwise to Mui Dinh (11°22′N, 109°01′E), keeping 10 miles offshore, thence:

E of Îles Catwick (10°00′N, 109°00′E), thence:

Joining the normal route (7.116) W of Vanguard Bank.

Hong Kong ⇐ ⇒ Sandakan

Diagram 7.114
Routes
7.117
From Hong Kong (22°17′N, 114°10′E) the route is:

Through Palawan Passage (10°20′N, 118°00′E), thence:

Through North Balabac Strait (8°15′N, 117°00′E), thence:

To Sandakan, (5°49′N 118°07′E).

Distance via North Balabac Strait   1100 miles

7.117.1
Cautions on dangers and currents in Palawan Passage and North Balabac Strait are described in *China Sea Pilot, Volume II* and *Philippine Islands Pilot*.

Hong Kong ⇐ ⇒ Tarakan, Balikpapan or Ujungpandang

Diagram 7.114
Routes
7.118
From Hong Kong (22°17′N, 114°10′E) there are three alternative routes to Celebes Sea and Selat Makasar and the ports of Tarakan (3°15′N, 117°54′E), Balikpapan (1°21′S, 116°56′E) and Ujungpandang (5°08′S, 119°22′E).
7.118.1
The first route passes:

Through Mindoro Strait (13°00′N, 120°00′E), thence:
Through Apo East Pass (12°40′N, 120°45′E), thence:
To Cuyo East Pass (10°30′N, 121°30′E), thence:
Across the Sulu Sea to Sibutu Passage (4°55′N, 119°37′E), as described in *Philippine Islands Pilot*, thence:

Through Sibutu Passage into Celebes Sea and joining ASL II (see 7.130 for details) to pass through Selat Makasar, thence:
To destination.

7.118.2
The second route passes:

Through Verde Island Passage (13°36′N, 121°00′E), thence:
To Cuyo East Pass (10°30′N, 121°30′E) where it joins the Mindoro Strait route (7.118.1).

This route offers shelter from the strength of either monsoon but is 20 miles longer than the Mindoro Strait route.
7.118.3
The third route passes:

Through Palawan Passage (10°20′N, 118°00′E), thence:
Through North Balabac Strait (8°15′N, 117°00′E), into the Sulu Sea thence:
Across the Sulu Sea to Sibutu Passage (4°55′N, 119°37′E) where it joins routes 7.118.1 and 7.118.2.

7.118.4
Distances via Mindoro Strait*:

Tarakan	1350 miles.
Balikpapan	1650 miles.
Ujungpandang	1830 miles.

* If via North Balabac Strait, subtract 20 miles.

Hong Kong ⇐ ⇒ Ambon

Diagram 7.119
Routes
7.119
From Hong Kong (22°17′N, 114°10′E) there are two alternative routes, either:
7.119.1
Through Mindoro Strait (13°00′N, 120°00′E), thence:
Through Apo East Pass (12°40′N, 120°45′E), thence:
Through Basilan Strait (6°50′N, 122°00′E). Or:
7.119.2
Through Verde Island Passage (13°36′N, 121°00′E), thence:
To Cuyo East Pass (10°30′N, 121°30′E), thence:
Through Basilan Strait (6°50′N, 122°00′E).
7.119.3
From the Basilan Strait routes lead:

Across the Celebes Sea to join ASL IIIE and ASLIIIA (Part1) (see 7.130 for details), thence:
To a position W of Buru (3°30′S, 126°30′E), thence:
As navigation permits, passing S of Buru, to Ambon (3°41′S, 128°10′E).

7.119.4
Distance via Mindoro Strait   1830 miles.

Hong Kong ⇐ ⇒ Iloilo or Cebu

Diagram 7.119
Routes
7.120
From Hong Kong (22°17′N, 114°10′E) the routes to Iloilo (10°42′N, 122°36′E) or Cebu (10°18′N, 123°55′E) are:

Through Verde Island Passage (13°36′N, 121°00′E) (see 7.114), thence:
As described in *Philippine Islands Pilot*, to destination.

Distances:

Verde Island Passage   660 miles.
Iloilo (passing W of Panay or through Jintotlo Channel)   930 miles.
Cebu   960 miles.

Hong Kong ⇐ ⇒ Guam or Yap

Diagram 7.119
Routes
7.121

From Hong Kong (22°17′N, 114°10′E) the routes are:
As navigation permits to Balintang Channel (20°00′N, 122°20′E), taking care to avoid the vicinity of Pratas Reef (20°40′N, 116°45′E) and Vereker

Banks (21°05′N, 116°00′E), (details in *China Sea Pilot, Volume I*), thence:
By great circle to Guam (13°27′N, 144°35′E) or Yap (9°28′N, 138°09′E).

Distances:

Guam   1820 miles.
Yap   1600 miles.

Manila or Verde Island Passage ⇐ ⇒ Selat Sunda or Jakarta

Diagram 7.122
Routes
7.122

From Manila (14°32′N, 120°56′E) and Verde Island Passage (13°36′N, 121°00′E) the route is:

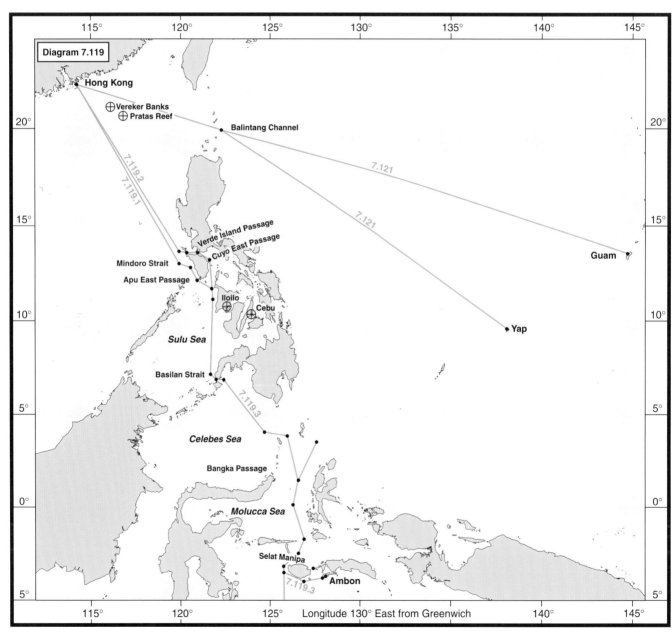

7.119 Hong Kong ← → Ambon
7.120 Hong Kong ← → Verde Island Passage, Iloilo or Cebu
7.121 Hong Kong ← → Guam or Yap

Through Palawan Passage (10°20′N, 118°00′E), thence:

Joining ASL I (see 7.130 for details) passing through Selat Karimata (1°43′S, 108°34′E), thence:

To Selat Sunda (6°00′S, 105°52′E) or Jakarta (6°03′S, 106°53′E).

7.122.1

Distances:

	Manila	Verde Island Passage
Selat Sunda	1580	1550
Jakarta	1560	1530

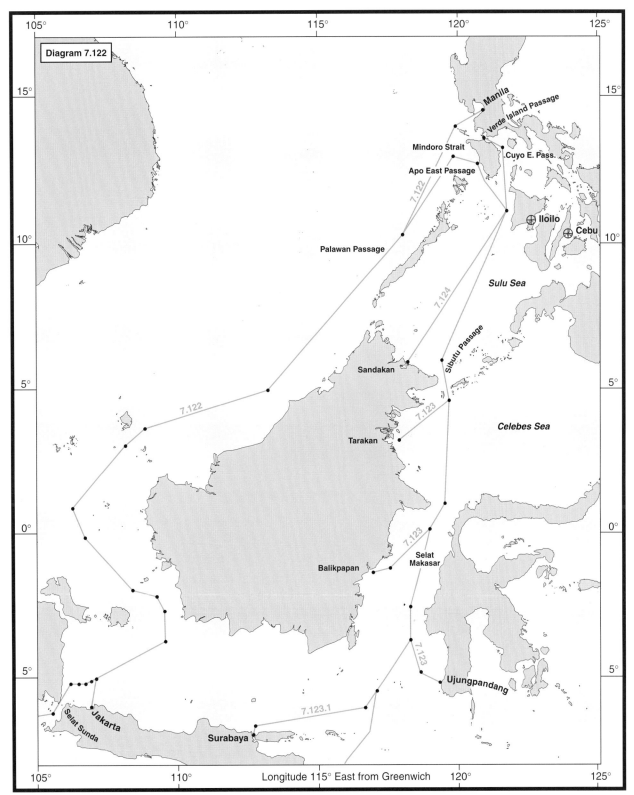

Diagram 7.122

7.122 Manila or Verde Island Passage ← → Selat Sunda or Jakarta
7.123 Manila, Verde Island Passage or Mindoro Strait ← → Torakan, Balikpapan, Ujungpandang or Surabaya
7.124 Manila, Verde Island Passage or Mindoro Strait ← → Sandakan
7.125 Manila, Verde Island Passage or Mindoro Strait ← → Iloilo or Cebu

7.122.2

Cautions on dangers and currents in Palawan Passage are described in *China Sea Pilot, Volume II*.

Manila, Verde Island Passage or Mindoro Strait ⇐ ⇒ Tarakan, Balikpapan, Ujungpandang or Surabaya

Diagram 7.122
Routes
7.123

From Manila (14°32′N, 120°56′E) the routes to Tarakan (3°15′N, 117°54′E), Balikpapan (1°21′S, 116°56′E) and Ujungpandang (5°08′S, 119°22′E) should join those from Hong Kong in Mindoro Strait (7.118.1) or Verde Island Passage (7.118.2), which is slightly shorter, thence:
> Across the Sulu Sea to Sibutu Passage (4°55′N, 119°37′E), as described in *Philippine Islands Pilot*, thence:
> Through Sibutu Passage into Celebes Sea and direct to Tarakan, or;
> Joining ASL II (see 7.130 for details) to pass through Selat Makasar (2°00′S, 118°00′E) proceeding, as appropriate, to Balikpapan or Ujungpandang, or;
> For Surabaya (7°12′S, 112°44′E), the route continues through Selat Makasar and across the Java Sea.

7.123.1
Distances:

	Manila*	Verde Island Passage	Mindoro Strait
Tarakan	810	710	720
Balikpapan	1100	1010	1010
Ujungpandang	1290	1190	1200
Surabaya	1570	1470	1480

* Via Mindoro Strait.

Manila, Verde Island Passage or Mindoro Strait ⇐ ⇒ Sandakan

Diagram 7.122
Routes
7.124

From Manila (14°32′N, 120°56′E) the routes are either:
> Through Mindoro Strait (13°00′N, 120°00′E) and Apo East Pass (12°40′N, 120°45′E) to Cuyo East Pass (10°30′N, 121°30′E), or:
> Through Verde Island Passage (13°36′N, 121°00′E) to Cuyo East Pass, this route is sheltered from the strength of the monsoons and slightly shorter.

From Cuyo East Pass the route leads across the Sulu Sea to Sandakan (5°49′N 118°07′E).

7.124.1
> Distances via Cuyo East Pass:
> Manila (via Mindoro Strait)   645 miles.
> Manila (via Verde Island Passage)   635 miles.
> Verde Island Passage   550 miles.
> Mindoro Strait (Cape Calavite)   555 miles.

Manila, Verde Island Passage and Mindoro Strait ⇐ ⇒ Iloilo or Cebu

Diagram 7.122
Routes
7.125

From Manila (14°32′N, 120°56′E) the routes are through Mindoro Strait (13°00′N, 120°00′E) or Verde Island Passage (13°36′N, 121°00′E), thence routes to Iloilo (10°42′N, 122°36′E) or Cebu (10°18′N, 123°55′E) are those described in *Philippine Islands Pilot*.

7.125.1
Distances:

	Manila	Verde Island Passage	Mindoro Strait
Iloilo:			
via Jintotlo Channel	360*	275	
Passing W of Panay	350*	265	270
Cebu:			
via Jintotlo Channel	390*	310	360

* via Verde Island Passage.

Torres Strait ⇐ ⇒ Balikpapan

Diagram 7.126
Routes
7.126

There are two alternative routes from Torres Strait (10°36′S, 141°51′E) which lead:
> Through Selat Wetar (8°16′S, 127°12′E) and Alur Pelayaran Wetar (8°00′S, 125°30′E) into Banda Sea, thence either:

7.126.1
> Along the N coast of Flores to 8°00′S, 121°00′E (**A**), thence:
> E of Pulau-pulau Sabalana (7°00′S, 118°30′E), thence:
> E of Karang Takarewataya (6°04′S, 118°55′E), thence:
> Joining ASL II (see 7.130 for details) to pass N through Selat Makasar (2°00′S, 118°00′E) proceeding, as appropriate, to Balikpapan (1°21′S, 116°56′E). Or:

7.126.2
> Through Selat Salayar (5°43′S, 120°30′E), thence:
> Round the SW end of Sulawesi, thence:
> Joining ASL II (see 7.130 for details) to pass N through Selat Makasar (2°00′S, 118°00′E) proceeding, as appropriate, to Balikpapan (1°21′S, 116°56′E).

7.126.3
Distances:
> Via N coast of Flores   1750 miles.
> Via Selat Selayar   1720 miles.

Torres Strait ⇐ ⇒ Tarakan

Diagram 7.126
Routes
7.127

There are alternative seasonal routes from Torres Strait (10°36′S, 141°51′E).

7.127.1
> The general route, for all seasons, is:
> To follow one of the routes described at 7.126 to join ASL II (see 7.130 for details) to pass N through Selat Makasar (2°00′S, 118°00′E), thence:
> As navigation permits to Tarakan (3°15′N, 117°54′E).

Distance   1980 mile.

7.127.2
North-bound during the south-east monsoon (May to September) one of the following routes may be preferred, either:

By the Torres Strait Main Route, as described at 7.141.1, passing:
> Through Arafura and Banda Seas to join ASL IIIC (see 7.130 for details), thence:

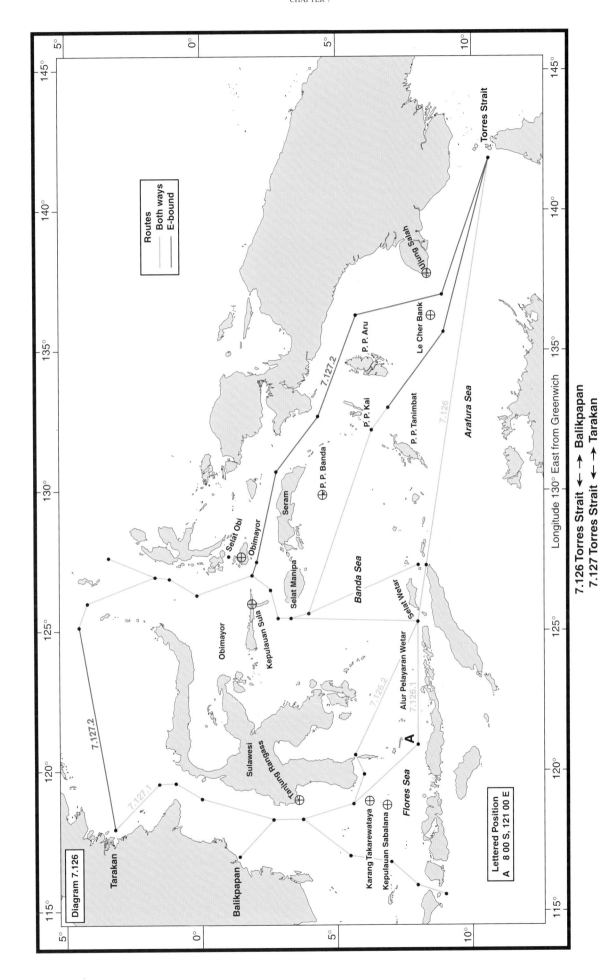

Diagram 7.126

Routes
Both ways
E-bound

Tarakan

Balikpapan

Sulawesi

Tanjung Pangass

Obimayor

Kepulauan Sula

Selat Obi

Obimayor

Seram

Selat Manipa

P. P. Banda

P. P. Kai

P. P. Aru

P. P. Tanimbat

Le Cher Bank

Ujung Salah

Torres Strait

Karang Takarewataya

Kepulauan Sabalana

Flores Sea

Alur Pelayaran Wetar

Selat Wetar

Banda Sea

Arafura Sea

7.127.2

7.127.1

7.127.2

7.126.2

7.126.1

7.126

7.127.2

A

Lettered Position
A 8 00 S, 121 00 E

Longitude 130° East from Greenwich

7.126 Torres Strait ←—→ Balikpapan
7.127 Torres Strait ←—→ Tarakan

Following ASL IIIC; ASL IIIA (Part 1) and ASL IIIE, thence:

Direct to Tarakan (3°15′N, 117°54′E).

Or by an alternative route, described at 7.141.3, passing N of Seram, to a position between Obi Mayor

(1°25′S, 127°25′E) and Kepulauan Sula (1°50′S, 125°00′E), thence:

Joining ASL IIIA (Part 1) and ASL IIIE, thence: Direct to Tarakan (3°15′N, 117°54′E).

Distance (via either route)   1780 miles.

PASSAGES THROUGH SEAS OF THE WESTERN PACIFIC OCEAN

GENERAL INFORMATION

Coverage
7.128

This section contains details of the following passages:

Passages through Eastern Archipelago:
    Principal Routes (7.130).
    Western Route (7.131).
    Central Route (7.132).
    Central Branch Routes (7.133).
    Eastern Route (7.134).

Passages connecting Australia with principal passages of Eastern Archipelago:
    East coast of Australia (7.136).
    South coast of Australia (7.137).
    West coast of Australia (7.138).
    Port Hedland (7.139).
    Darwin (7.140).
    Torres Strait (7.141).

Ocean Passages between Australia and Philippine Islands, South and East China Seas and North-west part of the Pacific Ocean:
    To and from Melbourne, Sydney or Brisbane and Yap (7.144).
    To and from Melbourne, Sydney or Brisbane and San Bernadino Strait, Verde Island Passage or Manila (7.145).
    To and from Melbourne, Sydney or Brisbane and Balintang Channel or Hong Kong (7.146).
    To and from Melbourne, Sydney or Brisbane and Guam (7.147).
    To and from Melbourne, Sydney or Brisbane and Shanghai (7.148).
    To and from Melbourne, Sydney or Brisbane and ports N of Shanghai in the East China Sea and Japan (7.149).
    To and from Melbourne, Sydney or Brisbane and ports NE of Japan (7.150).
    To and from Torres Strait and Solomon Sea (7.151).
    To and from Torres Strait and Yap, San Bernadino Strait, Verde Island Passage, Manila, Balintang Channel or Hong Kong (7.152).
    To and from Torres Strait and Guam (7.153).
    To and from Torres Strait and Shanghai (7.154).
    To and from Torres Strait and ports N of Shanghai in the East China Sea, in Japan or NE of Japan (7.155).

Ocean Passages between New Zealand and Philippine Islands, South and East China Seas and North-west part of the Pacific Ocean:
    Passages to and from New Zealand through Eastern Archipelago (7.158).
    Ocean passages to and from New Zealand (7.159).

Passage planning
7.129

1   The complicated pattern of N-S routes on the W side of the Pacific Ocean gives a variety of choice when planning a passage. In selecting the most direct route, the circumnavigation of Australia, the comparative merits of the various routes through the Eastern Archipelago or Solomon Sea, and the depths required to suit the vessel's draught, may be important factors.

2   The seasonal variations of winds, currents and weather in an area extending 45° or more, N and S of the equator, must also play a large part in determining a route agreeable to the characteristics of the vessel and the object of the voyage.

3   For descriptions of seasonal winds and currents, see 7.4 - 7.8; 7.13 - 7.15 and 7.26.

PASSAGES THROUGH EASTERN ARCHIPELAGO

Principal routes

Diagram 7.130
Details
7.130

1   **Archipelagic Sea Lanes (ASL)** (1.30) have been designated through the Indonesian Archipelago. Full details of these are given at Appendix A, together with a diagram which the axis lines of these sea lanes. It should be noted that the axis of the archipelagic sea lane does not indicate the deepest water, nor any recommended route or track.

2   The Principal Routes through the Eastern Archipelago, described in this book, follow the axis lines of these ASLs where appropriate.. Further details regarding Archipelagic Sea Lanes can be found in *The Mariner's Handbook*.

3   **Principal Routes**. There are three principal N-S routes, and their branches, through Eastern Archipelago:
    Western Route (7.131), consisting of ASL I and ASL IA.
    Central Route (7.132), includes ASL II.
    Central Branch routes (7.133).
    Eastern Route (7.134), includes ASL IIIA (parts 1 and 2); ASL IIIB; ASL IIIC; ASL IIID and ASL IIIE.

4   They are used for both N-bound and S-bound passages between Australia or New Zealand and South China Sea or the NW Pacific Ocean.

5   For deep-draught vessels, the most direct routes between Australia and South China Sea, East China Sea, Japan or ports farther N, are through the Central Route, or the Central branch routes to Selat Karimata, Balabac Strait or the Pacific Ocean.

6   Notes on straits and passages in the Eastern Archipelago are given at 7.84.

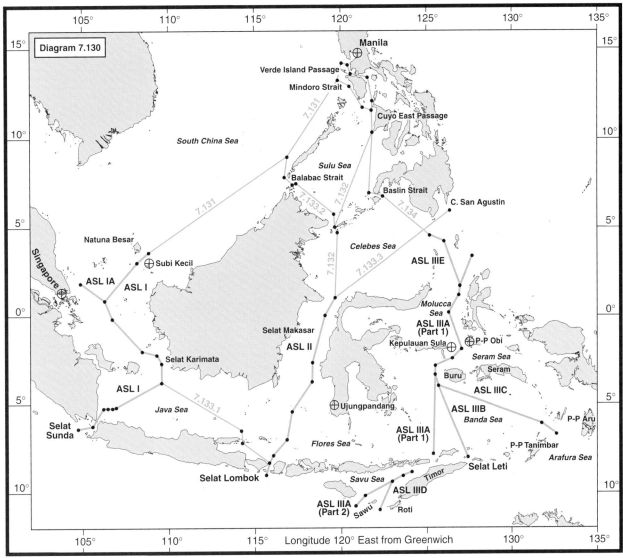

7.130 Principal Routes through Eastern Archipelago
7.131 Western Route
7.132 Central Route
7.133 Central Branch Routes
7.134 Eastern Route

Western Route

Diagram 7.130
Details
7.131

This route follows ASL I from Selat Sunda (6°00′S, 105°52′E) leading:

> Between numerous oilfields and S of Pulau Jagautara (5°12′S, 106°27′E), thence:
> Through Selat Karimata (1°43′S, 108°34′E), thence:
> Between Natuna Besar (4°00′N, 108°00′E) and Subi Kecil (3°03′N, 108°51′E) into the South China Sea.

For Singapore Strait follow ASL I and ASL IA.

For details of Archipelagic Sea Lanes I and IA, see *Indonesian Pilot, Volume I.*

There is a least charted depth of 20 m in the NE approach to Selat Sunda.

Caution. Areas of the Java Sea are being exploited for natural resources, see 7.38.2.

Central Route

Diagram 7.130
Details
7.132

This route follows ASL II from Selat Lombok (8°47′S, 115°44′E) leading:

> Through Selat Makasar (2°00′S, 118°00′E), thence:

The Central Route continues through the Celebes Sea passing:

> Through Sibutu Passage (4°55′N, 119°37′E) and through the Sulu Sea, thence:
> Through Cuyo East Pass (10°30′N, 121°30′E), thence:
> Through Mindoro Strait (13°00′N, 120°00′E) to South China Sea.

For details of Archipelagic Sea Lane II, see *Indonesian Pilot, Volume II.*

7.132.1

The route from Cuyo East Pass through Tablas Strait (13°00′N, 121°45′E) and Verde Island Passage (13°36′N, 121°00′E) and between Lubang Islands (13°45′N, 120°10′E) and the W coast of Luzon, offers an alternative to the passage through Mindoro Strait affording shelter from the full strength of either monsoon. The route is clear, though care must be taken near Baco Islets, SE of Verde Island (13°33′N, 121°05′E).

Central Branch Routes

Diagram 7.130
Details
7.133

The South China Sea can also be reached from Selat Lombok (8°47′S, 115°44′E) by two alternative routes:
7.133.1

The first route leads W through the Java Sea to join ASL I leading:

> Through Selat Karimata (1°43′S, 108°34′E), thence:
> Between Natuna Besar (4°00′N, 108°00′E) and Subi Kecil (3°03′N, 108°51′E) into the South China Sea.

7.133.2

The second route, passing E and N of Borneo:

> Follows the Central Route (7.132) as far as Sibutu Passage (4°55′N, 119°37′E), thence:
> Through Balabac Strait (7°34′N, 116°55′E) into the South China Sea.

7.133.3

The Philippine Sea may be reached, to enable a course to the N to be set passing E of Philippine Islands, by:

> Leaving the Central Route (7.132) after passing through Selat Makasar (2°00′S, 118°00′E) to cross the Celebes Sea, thence:
> Passing 18 miles S of Cape San Agustin (6°15′N, 126°30′E) into the Philippine Sea.

Eastern Route

Diagram 7.130
Details
7.134

This route follows ASL IIIA (part 1) which leads through the Banda, Ceram and Molucca Seas.

> It is joined:
> > From W of Timor (9°00′S, 125°00′E), via ASL IIIA (part 2) or ASL IIID, or:
> > From E of Timor, via ASL IIIB, or:
> > From the Arafura Sea, via ASL IIIC.
> ASL IIIA (part 1) then leads:
> > W of Buru Island (3°30′S, 126°30′E), thence
> > Between Pulau-pulau Obi (1°30′S, 127°40′E) and Kepulauan Sula (1°50′S, 125°00′E), thence:
> > W and N of Halmahera (0°30′N, 128°00′E) into the North Pacific Ocean, or:
> > Via ASL IIIE into the Celebes Sea.
> From ASL IIIE the Eastern Route then continues:
> > Through the Celebes Sea, thence:
> > Through Basilan Strait (6°50′N, 122°00′E), thence:
> > Through Cuyo East Pass (10°30′N, 121°30′E), thence:
> > Through Mindoro Strait (13°00′N, 120°00′E) or Verde Island Passage (13°36′N, 121°00′E) to South China Sea.

For details of Archipelagic Sea Lanes IIIA (parts 1 and 2); IIIB; IIIC; IIID and IIIE, see *Indonesian Pilot, Volumes II* and *III*.

Distances through Principal Routes

Diagram 7.130
7.135

	Western Route	Central Route	Eastern Route (entered)			
	via ASL I	via ASL II	via ASL IIIA	via ASL IIIB	via ASL IIIC	via ASL IIID
Selat Karimata	410	580				
Sibutu Passage		880				
Balabac Strait		1150				
Basilan Strait			1400	1100	1280	1350
Cape San Augustin		1160				
Verde Island Passage		1470	1900	1600	1780	1850
Mindoro Strait:						
Via Sulu Sea		1470	1840	1540	1720	1790
Via Selat Karimata	1840*	2000*				
Via Balabac Strait		1570*				
Manila:						
Via Sulu Sea		1520	1960	1660	1840	1910
Via Selat Karimata	1950*	2090*				
Via Balabac Strait		1660*				

  * Via Palawan Passage

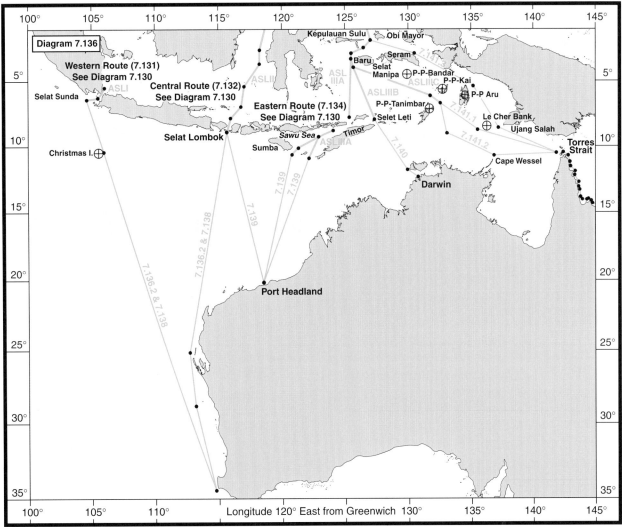

| Diagram 7.136 |

Western Route (7.131)
See Diagram 7.130

Central Route (7.132)
See Diagram 7.130

Eastern Route (7.134)
See Diagram 7.130

7.136 East coast of Australia to Principal Routes
7.137 South coast of Australia to Principal Routes
7.138 West coast of Australia to Principal Routes

7.139 Port Headland to Principal Routes
7.140 Darwin to Principal Routes
7.141 Torres Strait to Principal Routes

ROUTES CONNECTING AUSTRALIA WITH PRINCIPAL PASSAGES OF EASTERN ARCHIPELAGO

East coast of Australia

Diagram 7.136
Details
7.136

From E coast of Australia there are two alternative routes:

7.136.1

North-about:

To the South China Sea, or the W coast of the Philippines, the route is:

By the Inner Route (7.41.2) to the Torres Strait (10°36′S, 141°51′E) described at 7.42, thence:

By one the Torres Strait routes (7.141) to join ASL IIIC, ASL IIIA, ASL IIIE and the Eastern Route through the Celebes Sea, as described at 7.134.

To the Pacific Ocean the route is:

By one of the Torres Strait routes (7.141) to join ASL IIIC and ASL IIIA, thence:

W and N of Halmahera (0°30′N, 128°00′E) into the North Pacific Ocean.

The controlling depth by these routes is in Torres Strait; for details, see *Australia Pilot, Volume III*.

7.136.2

South-about:

The routes are:

For Selat Sunda (6°00′S, 105°52′E):

To Cape Otway (38°51′S, 143°31′E) as described at 7.40, thence:

Direct to Cape Leeuwin (34°23′S, 115°08′E), for details see 6.89 and 6.90, thence:

20 miles E of Christmas Island (10°24′S, 105°43′E), thence:

To Selat Sunda to join ASL I and the Western Route (7.131).

For Selat Lombok (8°47′S, 115°44′E):

From Cape Leeuwin as navigation permits to Selat Lombok to join ASL II and the Central Route (7.132).

If proceeding S-about, the advantage of the S-going East Australian Current (7.27) must be weighed against the frequency of W and NW gales S of Australia.

7.136.3

Distances from Sydney (33°50′S, 151°19′E) to Singapore (1°12′N, 103°51′E) are roughly the same via either Torres Strait or Cape Leeuwin.

South coast of Australia

Diagram 7.136
Details
7.137

From S coast of Australia there are two alternative routes:

7.137.1

East-about:

The routes are:

Through Bass Strait (40°00′S, 146°00′E), details at 6.90 and 6.91, thence:

By coastwise routes, as described at 7.40, to join the N-about routes (7.136.1) from the East coast.

7.137.2

West-about:

The routes are:

Direct to Cape Leeuwin (34°23′S, 115°08′E), for details see 6.89 and 6.90, thence:

Joining the S-about routes (7.136.2) from the East coast.

7.137.3

From ports E of Adelaide (34°48′S, 138°23′E) the shortest distance to the W end of the Torres Strait is E-about.

From Melbourne (Port Phillip) (38°20′S, 144°34′E) distances to Hong Kong are roughly the same via either the Torres Strait or Cape Leeuwin.

West coast of Australia

Diagram 7.136
Details
7.138

From W coast of Australia routes are as navigation permits to:

Either Selat Sunda (6°00′S, 105°52′E) to join ASL I and the Western Route (7.131), (if proceeding from Cape Leeuwin (34°23′S, 115°08′E) see 7.136.2), or:

Selat Lombok (8°47′S, 115°44′E) to join ASL II and the Central Route (7.132).

See *Australia Pilot, Volume V* for recommended routes.

Port Hedland

Diagram 7.136
Details
7.139

From Port Hedland (20°12′S, 118°32′E) routes are:

Either to Selat Lombok (8°47′S, 115°44′E) for ASL II and the Central Route (7.132), or:

W of Timor (9°00′S, 125°00′E), thence:

Through Selat Ombi (8°35′S, 125°00′E) via ASL IIIA (part 2) or ASL IIID to join ASL IIIA (part 1) and the Eastern Route, as described at 7.134.

Darwin

Diagram 7.136
Details
7.140

From Darwin (12°25′S, 130°47′E) the route is:

E of Timor (9°00′S, 125°00′E) to join ASL IIIB, thence:

Joining ASL IIIA (part 1) W of Buru Island (3°30′S, 126°30′E) and the Eastern Route, as described at 7.134.

Torres Strait

Diagram 7.136
Details
7.141

From Torres Strait (10°36′S, 141°51′E) there are three alternative routes:

7.141.1

Main Route.

From Torres Strait the Main Route, for all seasons, is:

S of Le Cher Bank (8°29′S, 136°17′E) and the unexamined shoals W of it, thence:

Joining ASL IIIC between Pulau-pulau Tanimbar (7°30′S, 131°30′E) and Pulau-pulau Kai (5°40′S, 132°50′E), giving the S end of Pulau-pulau Aru (6°00′S, 134°30′E) a wide berth, thence:

Joining ASL IIIA (part 1) W of Buru Island (3°30′S, 126°30′E) and the Eastern Route, as described at 7.134.

For details of the route see, *Australia Pilot, Volume V* and *Indonesia Pilot, Volume III*.

7.141.2

The second route is:

By the recommended tracks leading from the Torres Strait to Cape Wessel (11°00′S, 136°45′E), thence:

To 9°00′S, 133°05′E whence course can be set to join the Main Route and ASL IIIC (7.141.1) between Pulau-pulau Tanimbar and Pulau-pulau Kai.

This route passes well clear of Duddell Shoal (9°57′S, 136°00′E) and the dangers W of Le Cher Bank, but is 50 miles longer than the Main Route.

7.141.3

Alternative Route.

The third route, which provides some protection from the seas in the Banda Sea during the north-east monsoon (December to March) and the south-east monsoon (May to September), is less regularly used.

The route passes:

W of Ujung Salah (8°25′S, 137°39′E), giving that cape a wide berth, thence:

N of Pulau-pulau Aru (6°00′S, 134°30′E), thence:

N of Seram (3°00′S, 129°00′E), thence:

Joining ASL IIIA (part 1) between Obi Mayor (1°30′S, 127°40′E) and Kepulauan Sula (1°50′S, 125°00′E) and the Eastern Route, as described at 7.134.

For further details, see *Indonesia Pilot, Volume III*.

Australia ⇐ ⇒ South and East China Seas and the Pacific Ocean

Distances through Eastern Archipelago
7.142
All routes follow the Indonesian Archipelagic Sea Lanes (7.130) where appropriate.

	Route Notes	West-about							East-about		
		Port Phillip*	Adelaide	Cape Leeuwin	Fremantle	Port Hedland	Darwin	Torres Strait	Caloundra Head**	Sydney	Port Phillip*
To Principal Routes (7.130)											
Western Route [1]	W	3240	2950	1770	1650						
Central Route [2]	C	3020	2730	1550	1430	690					
Eastern Route [3]	E					580[3a] 590[3b]	270[3c]	605[3d]	1940[3d]	2380[3d]	2910[3d]
Through Eastern Archipelago											
Mindoro Strait	W[A]	5110	4820	3640	3520						
	C[B]	4460	4170	2990	2870	2130					
	E[C]					2420[3a] 2380[3b]	1810[3c]	2330[3d]	3670[3d]	4110[3d]	4640[3d]
Verde Island Passage	C[B]	4470	4180	3000	2880	2140					
	E[C]					2480[3a] 2440[3b]	1870[3c]	2390[3d]	3720[3d]	4160[3d]	4690[3d]
Manila	W[A]	5180	4890	3710	3590						
	C[B]	4530	4240	3060	2940	2190					
	E[C]					2540[3a] 2500[3b]	1930[3c]	2450[3d]	3780[3d]	4220[3d]	4750[3d]
Hong Kong	W	5360	5070	3890	3770						
	C[D]	5120	4830	3650	3530	2790					
	E[C]					3050[3a] 3010[3b]	2440[3c]	2960[3d]	4290[3d]	4730[3d]	5250[3d]
Shanghai	W[F]	6120	5830	4650	4530						
	C[G]	5570	5280	4100	3980	3240					
	E[H]					3530[3a] 3490[3b]	2920[3c]	3440[3d]	4770[3d]	5210[3d]	5740[3d]
Nagasaki	W[J]		6120	4940	4820						
	C[K]		5470	4290	4170	3430					
	E[P]					3450[3a] 3410[3b]	2840[3c]	3360[3d]	4700[3d]	5140[3d]	5670[3d]
Yokohama	W[L]		6510	5460	5340						
	C[M]		5810	4630	4510	3770					
	E[P]		5850[3a] 5840[3b]	4670[3a] 4660[3b]	4560[3a] 4540[3b]	3700[3a] 3660[3b]	3090[3c]	3610[3d]	4940[3d]	5380[3d]	5910[3d]
Tsugara Kaikyō	W[N]		6930	5750	5630						
	C[O]		6120	4940	4820	4080					
	E[E]		6280[3a] 6260[3b]	5100[3a] 5080[3b]	4990[3a] 4960[3b]	4120[3a] 4080[3b]	3510[3c]	4030[3d]	5370[3d]	5810[3d]	6340[3d]
Guam	C[Q]			3880	3760	3020					
	E[P]					2840[3a] 2800[3b]	2230[3c]	2740[3d]	4080[3d]	4520[3d]	5050[3d]
Yap	C[Q]			3450	3330	2590					
	E[P]					2390[3a] 2350[3b]	1780[3c]	2300[3d]	3630[3d]	4070[3d]	4600[3d]
Singapore											
		4070	3780	2600	2480	1690[R]	1960[R]	2590[S]	3920	4360	4890
		Via Western Route (ASL I & IA)				Via Selat Lombok			Via Torres Strait/ Selat Lombok		

7.142.1

Routes are indicated in the second column of the table, for notes affecting them, see 7.142.3.

W = Western Route (7.131).

C = Central Route (7.132).

E = Eastern Route (7.134).

7.142.2

Distances to the Principal Routes are given as follows:

[1] Western Route commences at the S point of ASL I (6°25′S, 104°41′E).

[2] Central Route commences at S point of ASL II (9°01′S, 115°36′E).

[3] Eastern Route has four approaches which commence, as indicated, at:

[3a] S point of ASL IIIA (part 2) (10°45′S, 120°46′E).

[3b] S point of ASL IIID (10°58′S, 122°11′E).

[3c] S point of ASL IIIB (8°03′S, 127°21′E).

[3d] S point of ASL IIIC (6°44′S, 132°35′E).

General information:

* Port Phillip to Melbourne 40 miles.

** Caloundra Head to Brisbane 35 miles.

Distances are in miles and, except where indicated in the notes below, are by the Principal Routes (7.130) indicated in the second column of the tables.

Distances from Torres Strait by the Eastern Route (7.134) are via the Torres Strait Main Route (7.141.1).

Distances E-about around Australia are by the Inner Route, via Capricorn Channel (7.41.2).

7.142.3

Notes affecting the routes are indicated, in superscript, next to the appropriate route.

[A] Via ASL I and Palawan Passage.

[B] If through Selat Lombok thence ASL I and Palawan Passage, add 600 miles.

[C] If from Torres Strait by Alternative Route (7.141.3), subtract 170 miles.

[D] If through Selat Lombok thence ASL I and South China Sea, add 175 miles.

[E] Through ASL IIIA thence S and E of Honshū.

[F] Via South China Sea Main Route (7.86.1) and through Taiwan Strait and Xiaoban Men.

[G] E or W of T'ai-wan and through Xiaoban Men (7.162.1 or 7.162.2). If through ASL II, S of Mindanao and E of Philippine Islands and T'ai-wan, add 150 miles.

[H] E or W of T'ai-wan and through Xiaoban Men (7.162.1 or 7.162.2).

[J] E or W of T'ai-wan.

[K] Either Central Route (7.132), thence Mindoro Strait and E of T'ai-wan; if through ASL II and S of Mindanao, add 66 miles.

[L] Through Palawan Passage and Balintang Channel.

[M] Through ASL II and S of Mindanao.

[N] Through Taiwan Strait and Korea Strait.

[O] Through ASL II and S of Mindanao and Honshū; if E of T'ai-wan and through Korea Strait add 80 miles.

[P] Through ASL IIIA and N of Halmahera.

[Q] Through ASL II and S of Mindanao.

[R] Through Selat Lombok, thence joining ASL I at Selat Karimata.

[S] Through Selat Wetar, Selat Ombi, Selat Sumba, Selat Lombok, Java Sea (7.106.2) and ASL I and IA. If through Selat Roti, Selat Sumba, Selat Lombok, Java Sea (7.106.2) and ASL I and IA, add 10 miles.

OCEAN PASSAGES BETWEEN AUSTRALIA AND PHILIPPINE ISLANDS, SOUTH AND EAST CHINA SEAS AND NORTH-WEST PART OF THE PACIFIC OCEAN

Details

7.143

From the E coast of Australia, Ocean Routes to the NW Pacific Ocean pass from Coral Sea, through Solomon Sea, and continue NW and N through Caroline Islands.

For distances see Table 7.156.

Melbourne, Sydney or Brisbane ⇐ ⇒ Yap

Diagram 7.144

Routes

7.144

Routes from N of Brisbane (Caloundra Head) (26°49′S, 153°10′E) are:

Either 20 miles E of Frederick Reef (21°00′S, 154°25′E), thence:

Midway between Rossel Spit (11°27′S, 154°24′E) and Pocklington Reef (10°50′S, 155°45′E), thence:

Between New Ireland (3°20′S, 152°00′E) and Bougainville Island (6°00′S, 155°15′E), thence:

Direct to Yap (9°28′N, 138°09′E). Or:

Through Jomard Entrance (11°15′S, 152°06′E) (details at 7.151), thence:

By the routes described in *Pacific Islands Pilot, Volume I* to Bomatu Point (8°23′S, 151°07′E), thence:

Between New Ireland (3°20′S, 152°00′E) and Bougainville Island (6°00′S, 155°15′E), thence:

Direct to Yap (9°28′N, 138°09′E).

7.144.1

Alternative routes have been used:

From Bomatu Point (8°23′S, 151°07′E) through Saint George's Channel (4°20′S, 152°32′E) or Vitiaz Strait (5°51′S, 147°30′E) and Isumrud Strait (4°45′S, 145°50′E), or:

From Rossel Spit (11°27′S, 154°24′E), passing N of Laughlan Islands (9°17′S, 153°41′E), thence through Saint George's Channel or Vitiaz Strait and Isumrud Strait.

They offer a considerable reduction in distance to Yap and other destinations.

Melbourne, Sydney or Brisbane ⇐ ⇒ San Bernadino Strait, Verde Island Passage or Manila

Diagram 7.145

Routes

7.145

Routes from N of Brisbane (Caloundra Head) (26°49′S, 153°10′E) are:

As described at 7.144, as far as the strait between New Ireland (3°20′S, 152°00′E) and Bougainville Island (6°00′S, 155°15′E) or, if using one of the alternative routes, described at 7.144.1, then as far as George's Channel (4°20′S, 152°32′E) or Isumrud Strait (4°45′S, 145°50′E), thence:

These routes continue to San Bernadino Strait (12°33′N, 124°12′E), passing S of Palau Islands (7°30′N, 134°30′E), thence:

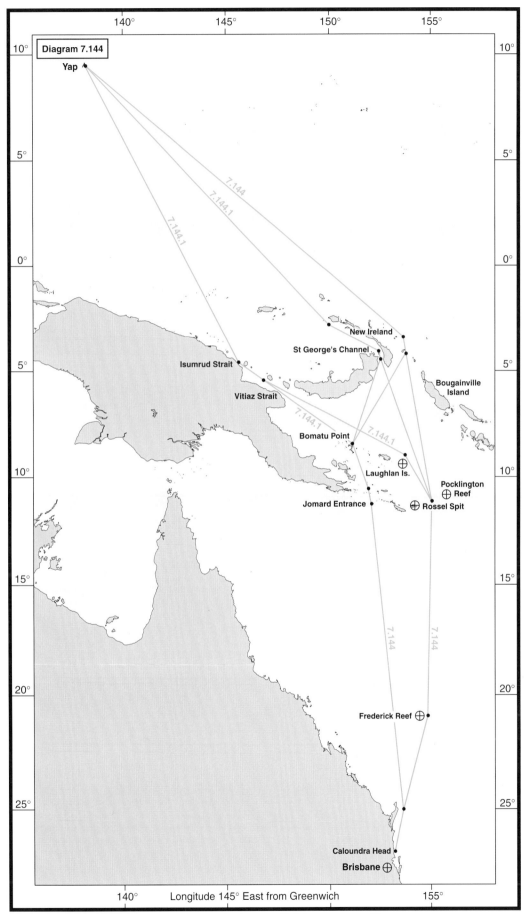

Diagram 7.144

Yap

140°　145°　150°　155°

10°　10°

5°　5°

0°　0°

7.144

7.144.1

7.144.1

New Ireland

St George's Channel

Isumrud Strait

Bougainville
Island

Vitiaz Strait

7.144.1

Bomatu Point

7.144.1

Laughlan Is.

Pocklington
Reef

Jomard Entrance

Rossel Spit

5°　5°

10°　10°

15°　15°

7.144

7.144

Frederick Reef

20°　20°

25°　25°

Caloundra Head

Brisbane

140°　Longitude 145° East from Greenwich　155°

7.144 Melbourne, Sydney or Brisbane ← → Yap

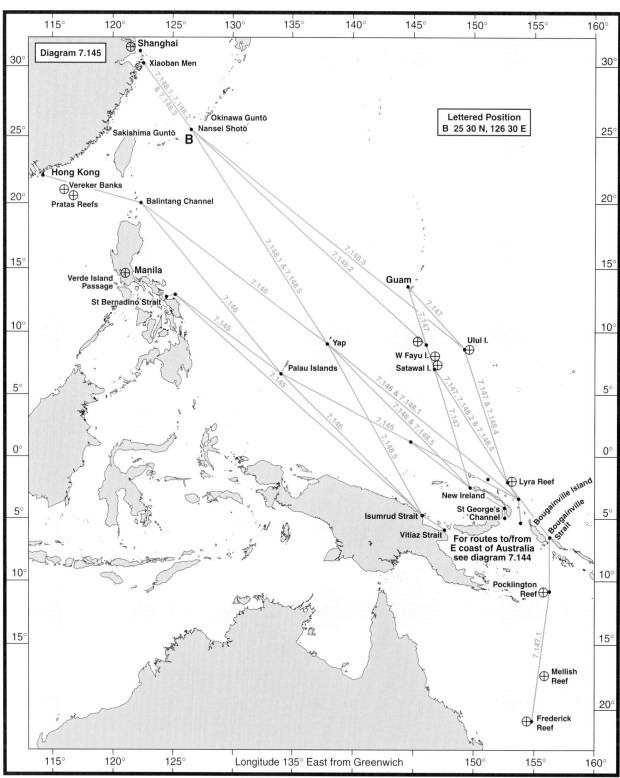

7.145 Melbourne, Sydney or Brisbane ← → San Bernadino Strait, Verde Island Passage or Manila
7.146 Melbourne, Sydney or Brisbane ← → Balintang Channel or Hong Kong
7.147 Melbourne, Sydney or Brisbane ← → Guam
7.148 Melbourne, Sydney or Brisbane ← → Shanghai

Routes continue to Verde Island Passage (13°36′N, 121°00′E) or Manila (14°32′N, 120°56′E) as described in *Philippine Islands Pilot*.

Melbourne, Sydney or Brisbane ⇐ ⇒ Balintang Channel or Hong Kong

Diagram 7.145
Routes
7.146

Routes from N of Brisbane (Caloundra Head) (26°49′S, 153°10′E) are:

As described at 7.144, as far as the strait between New Ireland (3°20′S, 152°00′E) and Bougainville Island (6°00′S, 155°15′E) or, if using one of the alternative routes, described at 7.144.1, then as far as Saint George's Channel (4°20′S, 152°32′E) or Isumrud Strait (4°45′S, 145°50′E), thence:

From the strait between New Ireland and Bougainville Island or from Saint George's Channel the route leads to Balintang Channel (20°00′N, 122°20′E), passing close S of Yap (9°28′N, 138°09′E), or from Isumrud Strait the route leads to Balintang Channel, passing close S of Palau Islands (7°30′N, 134°30′E), thence:

As navigation permits to Hong Kong (22°17′N, 114°10′E), taking care to avoid the vicinity of Pratas Reef (20°40′N, 116°45′E) and Vereker Banks (21°05′N, 116°00′E), (details in *China Sea Pilot, Volume I*).

Melbourne, Sydney or Brisbane ⇐ ⇒ Guam

Diagram 7.145
Routes
7.147

Routes from N of Brisbane (Caloundra Head) (26°49′S, 153°10′E) are:

As described at 7.144, as far as the strait between New Ireland (3°20′S, 152°00′E) and Bougainville Island (6°00′S, 155°15′E) or, if using one of the alternative routes, described at 7.144.1, then as far as Saint George's Channel (4°20′S, 152°32′E), thence:

From the strait between New Ireland and Bougainville Island the route continues W of Lyra Reef (1°50′S, 153°25′E), thence:

Either W of Satawal Island (7°20′N, 147°02′E) and West Fayu Island (8°04′N, 146°42′E), or; 20 miles W of Ulul Island (8°35′N, 149°40′E) to Guam (13°27′N, 144°35′E).

From Saint George's Channel the route continues W of Satawal Island and West Fayu Island to Guam.

7.147.1

An alternative route passes:

20 miles E of Frederick Reef (21°00′S, 154°25′E), thence:

30 miles E of Pocklington Reef (10°50′S, 155°45′E), taking care to avoid Mellish Reef (17°25′S, 155°51′E), thence:

Through Bougainville Strait (6°40′S, 156°15′E), thence:

Either W of Satawal Island (7°20′N, 147°02′E) and West Fayu Island (8°04′N, 146°42′E), or; 20 miles W of Ulul Island (8°35′N, 149°40′E) to Guam (13°27′N, 144°35′E).

Melbourne, Sydney or Brisbane ⇐ ⇒ Shanghai

Diagram 7.145
Routes
7.148

Routes from N of Brisbane (Caloundra Head) (26°49′S, 153°10′E) are:

As described at 7.144, as far as the strait between New Ireland (3°20′S, 152°00′E) and Bougainville Island (6°00′S, 155°15′E) or as at 7.147.1 to Bougainville Strait (6°40′S, 156°15′E).

7.148.1

From the strait between New Ireland and Bougainville Island the route passes:

Close S of Yap (9°28′N, 138°09′E), thence:

Through 25°30′N, 126°30′E **(B)**, to pass through Nansei Shotō, between Okinawa Guntō (26°30′N, 128°00′E) and Sakishima Guntō (24°40′N, 124°45′E), thence:

Through Xiaoban Men (30°12′N, 122°36′E), as in *China Sea Pilot, Volume II* to Shanghai (31°03′N, 122°20′E).

7.148.2

Another route is, as described at 7.147, to pass W of Satawal Island (7°20′N, 147°02′E) and West Fayu Island (8°04′N, 146°42′E), thence:

Passing N of Gaferut Island (9°14′N, 145°36′E) to 25°30′N, 126°30′E **(B)** to join the route (7.148.1) which passes S of Yap.

7.148.3

Another route is, as described at 7.147, to pass 20 miles W of Ulul Island (8°35′N, 149°40′E), thence:

Through 25°30′N, 126°30′E **(B)** to join the route (7.148.1) which passes S of Yap.

7.148.4

From Bougainville Strait the routes join those from the strait between New Ireland and Bougainville Island (7.148.1) W of either Satawal Island (7°20′N, 147°02′E) or Ulul Island (8°35′N, 149°40′E).

7.148.5

If using the Alternative Route (7.144.1) through Saint George's Channel or Vitiaz Strait the routes join those from the strait between New Ireland and Bougainville Island (7.148.1) close S of Yap (9°28′N, 138°09′E).

Melbourne, Sydney or Brisbane ⇐ ⇒ Ports North of Shanghai in the East China Sea and Japan

Diagram 7.149
Routes
7.149

Routes from N of Brisbane (Caloundra Head) (26°49′S, 153°10′E) are:

As at 7.147 or 7.147.1 either through the strait between New Ireland and Bougainville Island or through Bougainville Strait as far as West Fayu Island (8°04′N, 146°42′E) or Ulul Island (8°35′N, 149°40′E), thence:

As navigation permits to destination.

For Yokohama and ports farther E, routes pass at least 20 miles E of Mariana Islands (17°00′N, 146°00′E) and Ogasawara Guntō (27°00′N, 142°00′E), to destination.

Caution. For details of volcanic activity NW of Mariana Islands, see *Pacific Islands Pilot, Volume I.*

Melbourne, Sydney or Brisbane ⇐ ⇒ Ports North-east of Japan

Diagram 7.149
Routes
7.150

Routes from N of Brisbane (Caloundra Head) (26°49′S, 153°10′E) are:

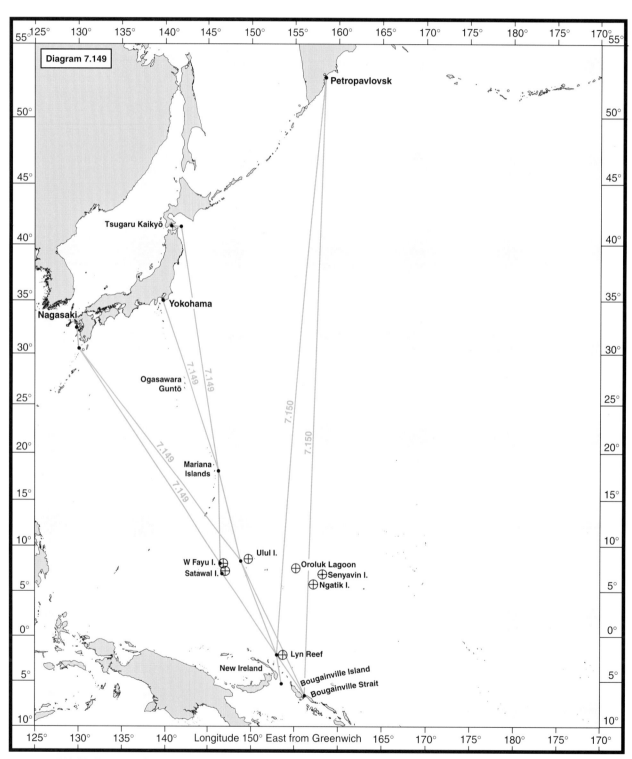

7.149 Melbourne, Sydney or Brisbane ← → Ports North of Shanghai in East China Sea and Japan
7.150 Melbourne, Sydney or Brisbane ← → Ports North East of Japan

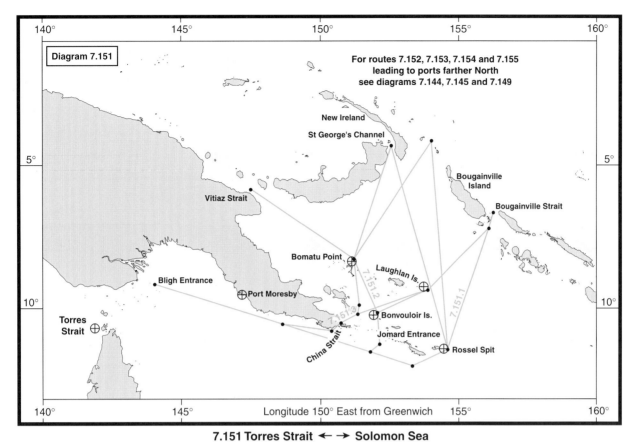

Diagram 7.151

140° 145° 150° 155° 160°

For routes 7.152, 7.153, 7.154 and 7.155
leading to ports farther North
see diagrams 7.144, 7.145 and 7.149

New Ireland
St George's Channel
Bougainville Island
Bougainville Strait
Vitiaz Strait
Bomatu Point
Laughlan Is.
Bligh Entrance
Port Moresby
Bonvouloir Is.
Torres Strait
Jomard Entrance
China Strait
Rossel Spit

Longitude 150° East from Greenwich

7.151 Torres Strait ← → Solomon Sea
7.152 Torres Strait ← → Yap, San Bernadino Strait, Verde Island Passage, Manila, Balintang Channel or Hong Kong
7.53 Torres Strait ← → Guam
7.154 Torres Strait ← → Shanghai
7.155 Torres Strait ← → Ports North of Shanghai in the East China Sea, in Japan, or North East of Japan

As at 7.147 or 7.147.1 either through the strait between New Ireland and Bougainville Island or Bougainville Strait, thence:

W of Ngatik Islands (5°50′N, 157°11′E), thence:

Between Senyavin Islands (6°55′N, 158°10′E) and Oroluk Lagoon (7°30′N, 155°20′E), thence:

To destination.

Torres Strait ⇐ ⇒ Solomon Sea

Diagram 7.151

Routes
7.151

From Torres Strait (10°36′S, 141°51′E) the following routes are used to enter and cross the Solomon Sea from Bligh Entrance (9°12′S, 144°00′E):

East of Rossel Spit (7.151.1), with no navigational hazards.

Jomard Entrance (7.151.2), three miles wide with strong tidal streams.

China Strait (7.151.3), although the shortest route the streams are strong and navigation intricate.

In this volume, distances for routes via the strait between New Ireland and Bougainville Island or Bougainville Strait are given for the routes passing E of Rossel Spit.

7.151.1

East of Rossel Spit (11°27′S, 154°24′E). The route from Bligh Entrance (9°12′S, 144°00′E) is:

Along the S coast Of New Guinea from Port Moresby (9°25′S, 147°05′E) to its E end, thence:

Parallel to the S side of the Louisiade Archipelago until Rossel Spit can be rounded, thence:

To Saint George's Channel (4°20′S, 152°32′E), or the strait between New Ireland (3°20′S, 152°00′E) and Bougainville Island (6°00′S, 155°15′E), or Bougainville Strait (6°40′S, 156°15′E).

7.151.2

Jomard Entrance (11°15′S, 152°06′E). The route from Bligh Entrance is:

Along the S coast of New Guinea from Port Moresby (9°25′S, 147°05′E) to its E end, thence:

Parallel to the S side of the Louisiade Archipelago until Jomard Entrance is reached, thence:

Through Jomard Entrance as described in *Pacific Islands Pilot, Volume I* to Bomatu Point (8°23′S, 151°07′E), thence:

Close N of Bomatu Point, thence:

Through the strait between New Ireland (3°20′S, 152°00′E) and Bougainville Island (6°00′S, 155°15′E), or if using the alternative routes

(7.144.1) to Saint George's Channel (4°20'S, 152°32'E) or Vitiaz Strait (5°51'S, 147°30'E).

Alternatively the route may be left when N of Bonvouloir Islands (10°21'S, 151°52'E) to pass E of Laughlan Islands (9°17'S, 153°41'E) to Bougainville Strait (6°40'S, 156°15'E).

7.151.3

China Strait (10°35'S, 150°40'E). The route from Bligh Entrance is:

Along the S coast of New Guinea from Port Moresby (9°25'S, 147°05'E) to the S entrance to China Strait, thence:

Joining route 7.151.2, either off Bomatu Point (8°23'S, 151°07'E), for the strait between New Ireland (3°20'S, 152°00'E) and Bougainville Island (6°00'S, 155°15'E), or Vitiaz Strait (5°51'S, 147°30'E); or Saint George's Channel (4°20'S, 152°32'E); or SE of Laughlan Islands (9°17'S, 153°41'E) for Bougainville Strait (6°40'S, 156°15'E).

7.151.4

Distances from Torres Strait through Solomon Sea

	China Strait	Jomard Entrance	E of Rossel Spit
Vitiaz Strait*	1020	1130	1400
Saint George's Channel**	1010	1120	1270
Strait between New Ireland and Bougainville Island (Pioneer Channel) †	1020	1140	1230
Bougainville Strait†	1020	1110	1130

To connect to *Admiralty Distance Tables — Pacific Ocean*:

* To places W of Vitiaz Strait:

Subtract 135 miles from the distances given from Lae in the Distance Tables to obtain distances from Vitiaz Strait.

** To places W of Saint George's Channel:

Add 10 miles to distances given from Rabaul in the Distance Tables to obtain distances from Saint George's Channel.

† Stations of *Admiralty Distance Tables — Pacific Ocean*.

Torres Strait ⇐ ⇒ Yap, San Bernadino Strait, Verde Island Passage, Manila, Balintang Channel or Hong Kong

Diagram 7.151
Routes
7.152

From Torres Strait (10°36'S, 141°51'E) the routes join those routes (7.144 to 7.146) from Australian ports, either:

E of Rossel Spit (11°27'S, 154°24'E), or:

In Jomard Entrance (11°15'S, 152°06'E), or:

If through China Strait (10°35'S, 150°40'E), as described at 7.151.3, off Bomatu Point (8°23'S, 151°07'E).

Thence to:

To the strait between New Ireland (3°20'S, 152°00'E) and Bougainville Island (6°00'S, 155°15'E), or:

The alternative routes (7.144.1) through Saint George's Channel (4°20'S, 152°32'E) or Vitiaz Strait (5°51'S, 147°30'E):

Torres Strait ⇐ ⇒ Guam

Diagram 7.151
Routes
7.153

From Torres Strait (10°36'S, 141°51'E) the routes join those routes (7.147 and 7.147.1) from Australian ports in either:

The strait between New Ireland (3°20'S, 152°00'E) and Bougainville Island (6°00'S, 155°15'E), or:

Bougainville Strait (6°40'S, 156°15'E), or:

Saint George's Channel (4°20'S, 152°32'E).

Torres Strait ⇐ ⇒ Shanghai

Diagram 7.151
Routes
7.154

From Torres Strait (10°36'S, 141°51'E) the routes join those routes (7.148) from Australian ports either:

E of Rossel Spit (11°27'S, 154°24'E), or:

In Jomard Entrance (11°15'S, 152°06'E), or:

In Bougainville Strait (6°40'S, 156°15'E), or:

If through China Strait (10°35'S, 150°40'E), as described at 7.151.3, off Bomatu Point (8°23'S, 151°07'E).

Thence to:

The strait between New Ireland (3°20'S, 152°00'E) and Bougainville Island (6°00'S, 155°15'E), or:

The alternative routes (7.144.1) through Saint George's Channel (4°20'S, 152°32'E) or Vitiaz Strait (5°51'S, 147°30'E):

Torres Strait ⇐ ⇒ Ports North of Shanghai in the East China Sea, in Japan or North-east of Japan

Diagram 7.151
Routes
7.155

From Torres Strait (10°36'S, 141°51'E) the routes pass either:

E of Rossel Spit (11°27'S, 154°24'E), or:

Through Jomard Entrance (11°15'S, 152°06'E), or:

Through China Strait (10°35'S, 150°40'E), as described at 7.151.3.

Thence joining those routes (7.149 and 7.150) from Australian ports either in:

The strait between New Ireland (3°20'S, 152°00'E) and Bougainville Island (6°00'S, 155°15'E), or:

Bougainville Strait (6°40'S, 156°15'E).

Australia ⇐ ⇒ Philippine Islands, South and East China Seas and North-west part of the Pacific Ocean

Distances by Ocean Routes
7.156

	Torres Strait				Caloundra Head**		Sydney		Port Phillip*	
	Vitiaz Strait	Saint Georges Channel	Strait between New Ireland and Bougainville Island	Bougainville Strait	Saint Georges Channel	Strait btween New Ireland and Bougainville Island	Saint Georges Channel	Strait between New Ireland and Bougainville Island	Saint Georges Channel	Strait between New Ireland and Bougainville Island
San Bernadino Strait										
E of Rossel Spit	3210	3260	3330		3380[N]	3440	3820[N]	3890	4350[N]	4420
Jomard Entrance	2940[A]	3120[A]	3240[A]		3410[O]		3850[O]		4380[O]	
Verde Island Passage										
E of Rossel Spit	3420	3470	3540		3590[N]	3650	4030[N]	4100	4560[N]	4630
Jomard Entrance	3150[A]	3330[A]	3450[A]		3620[O]		4060[O]		4590[O]	
Manila										
E of Rossel Spit	3500	3550	3620		3670[N]	3740	4120[N]	4180	4650[N]	4710
Jomard Entrance	3240[A]	3410[A]	3530[A]		3700[O]		4140[O]		4670[O]	
Yap										
E of Rossel Spit	2480	2460	2510		2580[P]	2620	3030[P]	3060	3560[P]	3590
Jomard Entrance	2210[A]	2320[A]	2410[A]		2610[Q]		3050[Q]		3580[Q]	
Balintang Channel										
E of Rossel Spit	3550	3570	3630		3690[R]	3750	4140[R]	4190	4670[R]	4720
Jomard Entrance	3290[A]	3430[A]	3530[A]		3720[S]		4160[S]		4690[S]	
Hong Kong										
E of Rossel Spit	4030	4050	4110		4170[R]	4230	4620[R]	4670	5150[R]	5200
Jomard Entrance	3760[A]	3910[A]	4000[A]		4200[S]		4640[S]		5170[S]	
Guam										
E of Rossel Spit		2740	2460	2550[C]	2590	2570[T]	3030	3020[T]	3560	3550[T]
Jomard Entrance		2320[A]	2370[B]	2520[D]	2610		3060		3590	
Shanghai										
E of Rossel Spit	4120	4100	4130[E]		4220[P]	4240[U]	4660[P]	4690[U]	5190[P]	5220[U]
Jomard Entrance	3850[A]	3960[A]	4040[F]		4250[Q]		4690[Q]		5220[Q]	
Nagasaki										
E of Rossel Spit		3900	3890[G]		4020	4000[V]	4470	4450[V]	5000	4980[V]
Jomard Entrance		3760[A]	3800[H]		4050	4090[G]	4490	4530[G]	5020	5060[G]
Yokohama										
E of Rossel Spit			3770[I]	3830[K]		3890[W]		4330[W]		4860[W]
Jomard Entrance			3680[J]	3810[L]		3970[I]		4410[I]		4940[I]
Tsugara Kaikyō										
E of Rossel Spit			4130	4180		4240[X]		4680[X]		5210[X]
Jomard Entrance			4030[A]	4160[M]		4320		4760		5300
Petropavlovsk										
E of Rossel Spit			4690	4700		4810[Y]		5250[Y]		5780[Y]
Jomard Entrance			4600[A]	4680[M]		4890		5330		5860

**   Caloundra Head to Brisbane 35 miles.

*   Port Phillip to Melbourne 40 miles.

7.156.1

Notes affecting the routes are indicated, in superscript, next to the appropriate distance.

^A If via China Strait, subtract 110 miles.

^B Passing W of Satawal Island and West Fayu Island. If passing W of Ulul Island, add 30 miles. If via China Strait and W of Satawal Island and West Fayu Island, subtract 110 miles.

^C Passing W of either Satawal Island and West Fayu Island or Ulul Island.

^D Passing W of either Satawal Island and West Fayu Island or Ulul Island. If via China Strait, subtract 90 miles.

^E Passing W of Satawal Island and West Fayu Island. If passing W of Ulul Island, add 30 miles. If passing S of Yap, add 15 miles.

^F Passing W of Satawal Island and West Fayu Island. If passing W of Ulul Island, add 30 miles. If passing S of Yap, add 15 miles. If via China Strait, subtract 115 miles.

^G Passing W of Satawal Island and West Fayu Island. If passing W of Ulul Island, add 30 miles.

^H Passing W of Satawal Island and West Fayu Island. If passing W of Ulul Island, add 30 miles. If via China Strait, subtract 110 miles.

^I Passing W of Ulul Island. If passing W of Satawal Island and West Fayu Island, add 50 miles.

^J Passing W of Ulul Island. If passing W of Satawal Island and West Fayu Island, add 50 miles. If via China Strait, subtract 110 miles.

^K Passing W of Ulul Island. If passing W of Satawal Island and West Fayu Island, add 80 miles.

^L Passing W of Ulul Island. If passing W of Satawal Island and West Fayu Island, add 80 miles. If via China Strait, subtract 90 miles.

^M If via China Strait, subtract 90 miles.

^N If via Vitiaz Strait, subtract 50 miles.

^O If via Vitiaz Strait, subtract 170 miles.

^P If via Vitiaz Strait, add 20 miles.

^Q If via Vitiaz Strait, subtract 100 miles.

^R If via Vitiaz Strait, subtract 20 miles.

^S If via Vitiaz Strait, subtract 140 miles.

^T Passing W of Satawal Island and West Fayu Island. If passing W of Ulul Island, add 25 miles. If via Bougainville Strait and passing W of either Satawal Island and West Fayu Island or Ulul Island, add 80 miles.

^U Passing W of Satawal Island and West Fayu Island. If passing W of Ulul Island, add 60 miles. If passing S of Yap, add 15 miles. If via Bougainville Strait and passing W of Satawal Island and West Fayu Island, add 80 miles. If passing W of Ulul Island, add 100 miles.

^V Passing W of Satawal Island and West Fayu Island. If passing W of Ulul Island, add 30 miles. If via Bougainville Strait and passing W of either Satawal Island and West Fayu Island or Ulul Island, add 70 miles.

^W Passing W of Ulul Island. If passing W of Satawal Island and West Fayu Island, add 50 miles. If via Bougainville Strait and passing W of Ulul Island, add 50 miles. If passing W of Satawal Island and West Fayu Island, add 120 miles.

^X If via Bougainville Strait, add 40 miles.

^Y If via Bougainville Strait, subtract 10 miles.

OCEAN PASSAGES BETWEEN NEW ZEALAND AND PHILIPPINE ISLANDS, SOUTH AND EAST CHINA SEAS AND NORTH-WEST PART OF THE PACIFIC OCEAN

General information

7.157

1   From New Zealand routes are either through Torres Strait (10°36′S, 141°51′E) and Eastern Archipelago, or by the Ocean Passages through Coral and Solomon Seas and Caroline Islands.

2   For general considerations affecting such passages, see 7.129.

3   For distances by these passages, see Table 7.160.

Passages through Eastern Archipelago

Diagram 7.158

Routes

7.158

From New Zealand routes are:

Through Inner Route (7.67 or 7.68) to Torres Strait (10°36′S, 141°51′E), thence:

By one of the Torres Strait routes (7.141—7.141.3) to join ASL IIIC and ASL IIIA (part 1), thence:

W and N of Halmahera (0°30′N, 128°00′E) into the North Pacific Ocean.

As navigation permits to destination.

For Shanghai, as described at 7.166.

For Nagasaki pass E of T'ai-wan, thence as described at 7.93.1.

For Yokohama, as described at 7.163.

For Singapore from Torres Strait see details at 7.105 and 7.106.

Ocean Passages

Diagram 7.158

Routes

7.159

From New Zealand routes are:

Through 21°00′S, 157°30′E (**A**), avoiding Kelso Bank (24°00′S, 159°40′E) and Bellona Reefs (21°30′S, 158°50′E), thence:

Joining the routes from Australian ports (7.143 to 7.150) midway between Rossel Spit (11°27′S, 154°24′E) and Pocklington Reef (10°50′S, 155°45′E), or in Jomard Entrance (11°15′S, 152°06′E), taking care to avoid Mellish Reef (17°25′S, 155°51′E), or 30 miles E of Pocklington Reef for Bougainville Strait (6°40′S, 156°15′E).

7.159.1

For ports E of Japan, shorter routes are either:

Through the passage W of Guadalcanal Island (9°40′S, 160°00′E), thence:

Through Indispensable Strait (9°00′S, 160°30′E), or:

Through the passage E of both Guadalcanal and Malaita Island (9°00′S, 161°00′E).

These passages appear deep and safe.

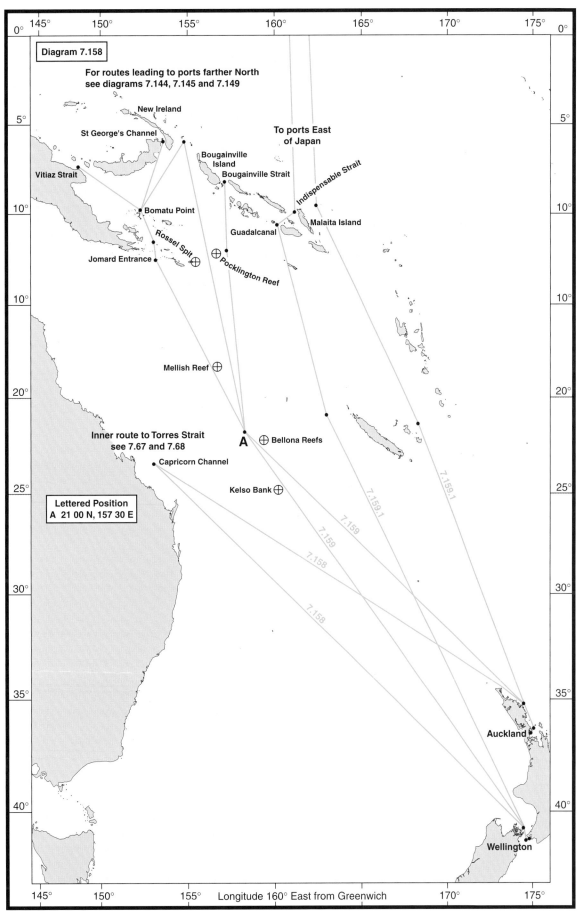

Diagram 7.158

For routes leading to ports farther North
see diagrams 7.144, 7.145 and 7.149

New Ireland

St George's Channel

To ports East
of Japan

Bougainville
Island
Bougainville Strait

Vitiaz Strait

Indispensable Strait

Bomatu Point

Malaita Island

Rossel Spit

Guadalcanal

Jomard Entrance

Pocklington Reef

Mellish Reef

Inner route to Torres Strait
see 7.67 and 7.68

A

Bellona Reefs

Capricorn Channel

7.159.1

7.159.1

Lettered Position
A 21 00 N, 157 30 E

Kelso Bank

7.159

7.159

7.158

7.158

7.158

Auckland

Wellington

Longitude 160° East from Greenwich

7.158 Passages through Eastern Archipelago from New Zealand
7.159 Ocean Passages from New Zealand

Distances through Eastern Archipelago and by Ocean Routes
7.160

	Auckland via			Wellington via		
	Solomon Sea		Eastern Archipelago	Solomon Sea		Eastern Archipelago
	Saint George's Channel	Strait between New Ireland and Bougainville Island		Saint George's Channel	Strait between New Ireland and Bougainville Island	
San Bernadino Strait						
E of Rossel Spit	4360[A]	4430		4550[A]	4610	
Jomard Entrance	4430[B]			4610[B]		
Verde Island Passage						
E of Rossel Spit	4570[A]	4640	4660[N]	4760[A]	4820	4780[N]
Jomard Entrance	4630[B]			4820[B]		
Manila						
E of Rossel Spit	4660[A]	4720	4750[N]	4840[A]	4910	4870[N]
Jomard Entrance	4720[B]			4910[B]		
Yap						
E of Rossel Spit	3560[C]	3610		3750[C]	3790	
Jomard Entrance	3630[D]			3810[D]		
Balintang Channel						
E of Rossel Spit	4860[E]	4730		4860[E]	4920	
Jomard Entrance	4740[F]			4930[F]		
Hong Kong						
E of Rossel Spit	5160[E]	5120	5290[N]	5340[E]	5400	5410[N]
Jomard Entrance	5220[F]			5410[F]		
Guam						
E of Rossel Spit	3570	3560[G]		3760	3740[G]	
Jomard Entrance	3630			3820		
Shanghai						
E of Rossel Spit	5200[C]	5230[H]	5810[O]	5390[C]	5420[H]	5930[O]
Jomard Entrance	5270[D]			5450[D]		
Nagasaki						
E of Rossel Spit	5010	4990[I]		5190	5170[I]	
Jomard Entrance	5070	5110[J]		5250	5290[J]	
Yokohama						
E of Rossel Spit		4870[K]			5060[K]	
Jomard Entrance		4990[L]			5170[L]	
Tsugaru Kaikyō						
E of Rossel Spit		5220[M]			5410[M]	
Jomard Entrance		5300			5490	

1 Distances through Eastern Archipelago are by:
  Inner Route (7.67 or 7.68) to Torres Strait (10°36′S, 141°51′E), thence:
  Main Route (7.141.1) to join ASL IIIC, thence:
  Eastern Route (7.134) to South China Sea.
2 Distances through Solomon Sea are by:
  Appropriate routes passing E of Rossel Spit (11°27′S, 154°24′E) or through Jomard Entrance (11°15′S, 152°06′E), thence either:
  Through the strait between New Ireland (3°20′S, 152°00′E) and Bougainville Island (6°00′S, 155°15′E), or:
  By the alternative routes (7.144.1) through Saint George's Channel (4°20′S, 152°32′E) or Vitiaz (5°51′S, 147°30′E) and Isumrud Strait (4°45′S, 145°50′E) Straits.
3 Distances are also given for routes through Bougainville Strait (6°40′S, 156°15′E) where appropriate.

7.160.1

Notes affecting the routes are indicated, in superscript, next to the appropriate distance.

A   If via Vitiaz Strait, subtract 50 miles.

B   If via Vitiaz Strait, subtract 170 miles.

C   If via Vitiaz Strait, add 20 miles.

D   If via Vitiaz Strait, subtract 100 miles.

E   If via Vitiaz Strait, subtract 20 miles.

F   If via Vitiaz Strait, subtract 140 miles.

G   Passing W of Satawal Island and West Fayu Island. If passing W of Ulul Island, add 25 miles. If via Bougainville Strait and passing W of either Satawal Island and West Fayu Island or Ulul Island, add 60 miles.

H   Passing W of Satawal Island and West Fayu Island. If passing W of Ulul Island, add 30 miles. If passing S of Yap, add 15 miles. If via Bougainville Strait and passing W of Satawal Island and West Fayu Island, add 70 miles. If passing W of Ulul Island, add 90 miles.

I   Passing W of Satawal Island and West Fayu Island. If passing W of Ulul Island, add 30 miles. If via Bougainville Strait and passing W of either Satawal Island and West Fayu Island or Ulul Island, add 50 miles.

J   Passing W of Satawal Island and West Fayu Island. If passing W of Ulul Island, add 30 miles.

K   Passing W of Ulul Island. If passing W of Satawal Island and West Fayu Island, add 50 miles. If via Bougainville Strait and passing W of Ulul Island, add 30 miles; or passing W of Satawal Island and West Fayu Island, add 100 miles.

L   Passing W of Ulul Island. If passing W of Satawal Island and West Fayu Island, add 50 miles.

M   If via Bougainville Strait, add 20 miles.

N   If via Alternative Route (7.141.3) from Torres Strait, add 20 miles.

O   Continuing from South China Sea, E or W of T'ai-wan, through Xiaoban Men to Shanghai. If through Verde Island Passage and W of T'ai-wan, add 25 miles. If via Alternative Route (7.141.3) from Torres Strait to Selat Jailolo, thence E of Philippine Islands and T'ai-wan (7.166.2), subtract 180 miles.

PASSAGES ON THE WESTERN SIDE OF THE PACIFIC OCEAN

GENERAL INFORMATION

Coverage
7.161

This section contains details of the following passages:

To and from Manila, Verde Island Passage or Mindoro Strait and Shanghai (7.162).

To and from Manila, Verde Island Passage or Mindoro Strait and Yokohama (7.163).

To and from Hong Kong and Japan (7.164).

To and from Shanghai and Yokohama (7.165).

To and from Shanghai and Torres Strait (7.166).

To and from Yokohama or Tsugaru Kaikyō and Petropavlovsk (7.167).

To and from Apia or Suva and Yap, Verde Island Passage, Manila or Hong Kong (7.168).

To and from Yokohama and Guam or Yap (7.169).

To and from Apia and Yokohama (7.170).

To and from Suva and Yokohama (7.171).

To and from Apia and Guam or Shanghai (7.172).

To and from Suva and Guam or Shanghai (7.173).

PASSAGES

Manila, Verde Island Passage or Mindoro Strait ⇐ ⇒ Shanghai

Diagram 7.162
Routes
7.162

Routes from Manila (14°32′N, 120°56′E), Verde Island Passage (13°36′N, 121°00′E) or Mindoro Strait (13°00′N, 120°00′E) are seasonal and are:

7.162.1

South-west Monsoon (May to September):

E of T'ai-wan, thence:

15 to 20 miles E of P'eng-chia Yü (25°38′N, 122°04′E), thence:

Through Xiaoban Men (30°12′N, 122°36′E), as in *China Sea Pilot, Volume II*, thence:

To Shanghai (31°03′N, 122°20′E).

The influence of the NE-going Japan Current will be felt during the greater part of this passage.

7.162.2

North-east Monsoon (December to March):

W of T'ai-wan, either through P'eng-hu Kang-tao (23°30′N, 119°53′E) or W of P'eng-hu Ch'un-tao (23°30′N, 119°30′E), thence:

Through Xiaoban Men (30°12′N, 122°36′E), as in *China Sea Pilot, Volume II*, thence:

To Shanghai (31°03′N, 122°20′E).

For cautions on approaching Taiwan Banks (23°00′N, 118°30′E), see *China Sea Pilot, Volume II*.

7.162.3

Distances:

	Manila	Verde Island Passage	Mindoro Strait
E of T'ai-wan	1130	1170	1150
W of T'ai-wan	1140	1170	1160

Manila, Verde Island Passage or Mindoro Strait ⇐ ⇒ Yokohama

Diagram 7.162
Routes
7.163

Routes are directional and are:

7.163.1

North-bound, from Manila (14°32′N, 120°56′E), Verde Island Passage (13°36′N, 121°00′E) or Mindoro Strait (13°00′N, 120°00′E), taking full advantage of the Japan Current:

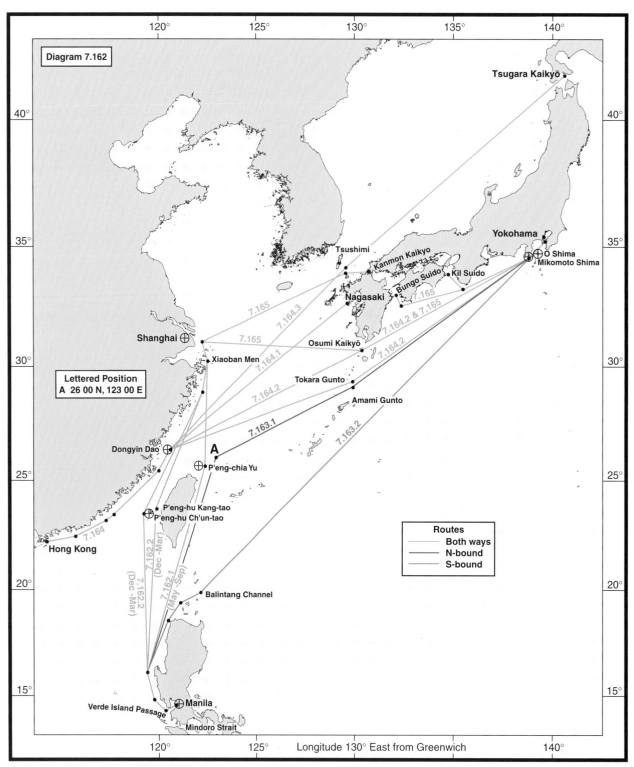

Diagram 7.162

40°

Tsugara Kaikyō

Yokohama

35°

Tsushimi
Kanmon Kaikyo
Kil Suido
Bungo Suido
Nagasaki
Osumi Kaikyō

O Shima
Mikomoto Shima

7.165
7.164.3
7.165
7.164.1
7.164.2 & 7.165
7.164.2
7.165

Shanghai

Xiaoban Men

30°

Lettered Position
A 26 00 N, 123 00 E

Tokara Gunto

Amami Gunto

7.164.2

7.163.1

7.163.2

Dongyin Dao

A

P'eng-chia Yu

25°

P'eng-hu Kang-tao
P'eng-hu Ch'un-tao

Routes
Both ways
N-bound
S-bound

7.164

Hong Kong

7.162.2 (Dec.-Mar)
7.162.1 (May-Sep)
7.162.2 (Dec.-Mar)

20°

Balintang Channel

15°

Verde Island Passage

Manila

Mindoro Strait

Longitude 130° East from Greenwich

7.162 Manila, Verde Island Passage or Mindoro Strait ← → Shanghai
7.163 Manila, Verde Island Passage or Mindoro Strait ← → Yokohama
7.164 Hong Kong ← → Japan
7.165 Shanghai ← → Yokohama

E of T'ai-wan to 26°00′N, 123°00′E (**A**), thence:
Through Nansei Shotō, passing between Amami
Guntō (28°00′N, 129°05′E) and Tokara Guntō
(29°20′N, 129°30′E), thence:
To Yokohama (35°26′N, 139°43′E).

7.163.2

South-bound from Yokohama (35°26′N, 139°43′E), or
alternatively N-bound though less favourable as regards
current, the route is:
By rhumb line to Balintang Channel (20°00′N,
122°20′E), thence:
Coastwise to destination.

7.163.3
Distances:

	Manila	Verde Island Passage	Mindoro Strait
N-bound	1830	1860	1850
S-bound	1760	1790	1780

Hong Kong ⇐ ⇒ Japan

Diagram 7.162
Routes
7.164

Routes are through Taiwan Strait.

When N-bound from Hong Kong (22°17′N, 114°10′E) they keep as close as prudent to the coast of China, (see *China Sea Pilot, Volume II* for details), during the north-east monsoon until abreast of Dongyin Dao (26°22′N, 120°30′E).

Thence:

7.164.1

For **Nagasaki** (32°42′N, 129°49′E) the route is direct.

7.164.2

For **Yokohama** (35°26′N, 139°43′E) the route is either:
Through Ōsumi Kaikyō (30°55′N, 130°40′E), or:
Between Amami Guntō (28°00′N, 129°05′E) and Tokara Guntō (29°20′N, 129°30′E) in about 29°20′N.

Distances differ little by either route and the Japan Current sets strongly NE-ward on both.

South-bound from Yokohama the route keeps as close to the S coast of Japan as safety permits to avoid the strength of the Japan Current, thence through Ōsumi Kaikyō and Taiwan Strait.

With E winds there is often a strong indraught into the deep bays, especially between Ō Shima (34°45′N, 139°22′E) and Mikomoto Shima (34°35′N, 138°57′E). During typhoon months, the currents in this locality are, at times, very irregular.

7.164.3

For **Tsugaru Kaikyō** (41°39′N, 140°48′E) the route is through Korea Strait either side of Tsushima (34°25′N, 129°20′E).

7.164.4

Distances from Hong Kong:

Nagasaki	1070 miles.
Yokohama	1590 miles.
Tsugaru Kaikyō	1820 miles.

Shanghai ⇐ ⇒ Yokohama

Diagram 7.162
Routes
7.165

Routes from Shanghai (31°03′N, 122°20′E) are either:
Through Ōsumi Kaikyō (30°55′N, 130°40′E), thence:
As navigation permits to Yokohama (35°26′N, 139°43′E), or:
Through Kanmon Kaikyō (33°58′N, 130°52′E) to enter Seto Naikai (The Inland Sea of Japan), thence:
Routes are through Bungo Suidō (33°00′N, 132°15′E) or Kii Suidō (34°00′N, 134°50′E) to Yokohama.
For details of routes through Seto Naikai, see *Japan Pilot, Volume II*.

7.165.1
Distances:

Through Ōsumi Kaikyō		1030 miles.
Through Seto Naikai		
	via Bungo Suidō	1080 miles.
	via Kii Suidō	1120 miles.

Shanghai ⇐ ⇒ Torres Strait

Diagram 7.166
Routes
7.166

Routes are directional and are:

7.166.1

South-bound, from Shanghai (31°03′N, 122°20′E) the route is:
Through Taiwan Strait as described at 7.162.2, thence:
Through Mindoro Strait (13°00′N, 120°00′E) to join the Eastern Route (7.134), thence:
Through Basilan Strait (6°50′N, 122°00′E) to join ASL IIIE and ASL IIIA (part 1), thence:
Joining ASL IIIC in a position W of Buru Island (3°30′S, 126°30′E), thence:
As described at 7.141.1 and 7.141.2, to Torres Strait (10°36′S, 141°51′E) or:
Leaving ASL IIIA (part 1) between Pulau-pulau Obi (1°30′S, 127°40′E) and Kepulauan Sula (1°50′S, 125°00′E) to proceed as described at 7.141.3 to Torres Strait.

This S-bound route avoids the full strength of the Japan Current.

7.166.2

North-bound, from Torres Strait (10°36′S, 141°51′E) the route is either:
As described at 7.141.1 and 7.141.2, to join ASL IIIC, thence:
Joining ASL IIIA (part 1) in a position W of Buru Island (3°30′S, 126°30′E), thence:
W and N of Halmahera (0°30′N, 128°00′E) into the North Pacific Ocean.
As navigation permits to Shanghai (31°03′N, 122°20′E).

7.166.3
Distances:

South-bound	3440 miles.
North-bound	3170 miles.

Yokohama or Tsugaru Kaikyō ⇐ ⇒ Petropavlovsk

Diagram 7.167
Routes
7.167

Routes from Yokohama (35°26′N, 139°43′E) or Tsugaru Kaikyō (41°39′N, 140°48′E) to Petropavlovsk (53°00′N, 158°38′E) are as direct as navigation permits.

7.167.1

General information:
Pack ice may be found off the SE coast of Hokkaidō during February, March and April.
The effect of the SW-going Kamchatka Current (7.26) may possibly be reduced by keeping 60 miles or more off Kuril'skiye Ostrova.

7.167.2

Distances:

Yokohama	1420 miles.
Tsugaru Kaikyō	1060 miles.

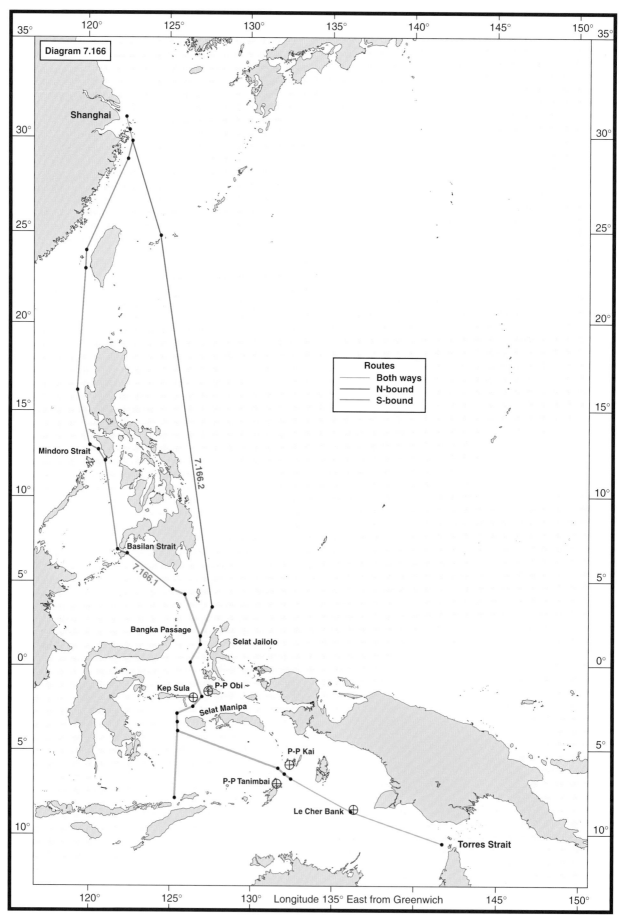

Diagram 7.166

35°

30° Shanghai

25°

20°

15°

Mindoro Strait

10°

7.166.2

5° Basilan Strait

7.166.1

Bangka Passage

Selat Jailolo

0°

Kep Sula   P-P Obi

Selat Manipa

5° P-P Kai

P-P Tanimbai

Le Cher Bank

10° Torres Strait

Routes
Both ways
N-bound
S-bound

120° 125° 130° Longitude 135° East from Greenwich 145° 150°

7.166 Shanghai ← → Torres Strait

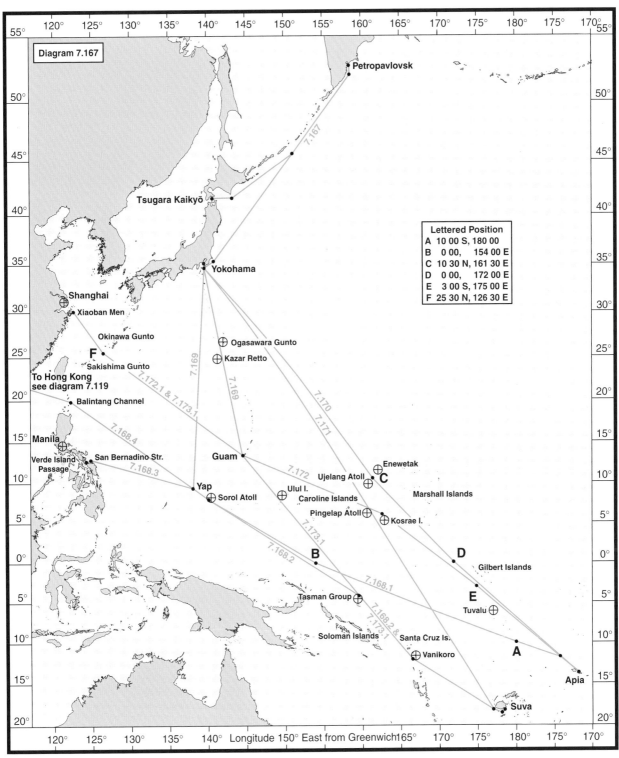

Diagram 7.167

Lettered Position
A 10 00 S, 180 00
B   0 00,   154 00 E
C 10 30 N, 161 30 E
D   0 00,   172 00 E
E   3 00 S, 175 00 E
F 25 30 N, 126 30 E

7.167 Yokohama or Tsugaru Kaikyō ← → Petropavlovsk
7.168 Apia or Suva ← → Yap, Verde Island Passage, Manila or Hong Kong
7.169 Yokohama ← → Guam or Yap
7.170 Apia ← → Yokohama
7.171 Suva ← → Yokohama
7.172 Apia ← → Guam or Shanghai
7.173 Suva ← → Guam or Shanghai

Apia or Suva ⇐ ⇒ Yap, Verde Island Passage, Manila or Hong Kong

Diagram 7.167
Routes
7.168

All routes pass close S of Yap (9°28′N, 138°09′E).
7.168.1

From Apia (13°47′S, 171°45′W) the route is:
Through 10°00′S, 180°00′ **(A)**, thence:
Across the equator in 154°00′E **(B)**, thence:
20 miles SW of Sorol Atoll (8°10′N, 140°25′E), thence:
To Yap.
7.168.2

From Suva (18°11′S, 178°24′E) the route is:
Through Kadavu Passage (18°45′S, 178°00′E), thence:
S of Vanikoro (11°40′S, 166°50′E), thence:
Between Santa Cruz Islands (11°00′S, 166°15′E) and Solomon Islands (8°00′S, 158°00′E), thence:
N of Tasman Islands (4°35′S, 159°25′E), thence:
20 miles SW of Sorol Atoll (8°10′N, 140°25′E), thence:
To Yap.
7.168.3

For Verde Island Passage or **Manila** the route from close S of Yap (9°28′N, 138°09′E) is:
To San Bernadino Strait (12°33′N, 124°12′E), as described at 7.100.3, thence:
As described in *Philippine Islands Pilot* to Verde Island Passage (13°36′N, 121°00′E), thence:
To Manila (14°32′N, 120°56′E).
7.168.4

For Hong Kong the route from close S of Yap (9°28′N, 138°09′E) is as described at 7.121:
Through Balintang Channel (20°00′N, 122°20′E), thence:
To Hong Kong (22°17′N, 114°10′E).
7.168.5

Distances:

	Apia	Suva
Yap	3320	2970
Verde Island Passage	4390	4040
	Apia	Suva
Manila	4480	4130
Hong Kong	4920	4570

Yokohama ⇐ ⇒ Guam or Yap

Diagram 7.167
Routes
7.169

Routes from Yokohama (35°26′N, 139°43′E) are by rhumb line to Guam (13°27′N, 144°35′E) and Yap (9°28′N, 138°09′E).

The route for Guam passes through Nanpō Shotō between Ogasawara Guntō (27°00′N, 142°00′E) and Kazan Rettō (25°00′N, 141°20′E). For cautions on volcanic activity in Nanpō Shotō and NW of the Mariana Islands (17°00′N, 146°00′E), see *Japan Pilot, Volume II* and *Pacific Islands Pilot, Volume I*.
7.169.1

Distances:

Guam	1350 miles.
Yap	1570 miles.

Apia ⇐ ⇒ Yokohama

Diagram 7.167
Routes
7.170

The great circle track between Yokohama (35°26′N, 139°43′E) and Apia (13°47′S, 171°45′W) passes through the Marshall Islands (10°00′N, 170°00′E) and the Gilbert Group (0°00′, 174°00′E). This part of the track, though navigable, is normally best avoided owing to the incompleteness of the surveys and uncertainty of the currents.

From Yokohama the preferable track is:
By great circle to 10°30′N, 161°30′E **(C)** between Enewetak (11°30′N, 162°15′E) and Ujelang (9°50′N, 160°54′E) Atolls, thence:
Rhumb line to the equator in 172° 00′E **(D)**, passing W of Marshall Islands and Gilbert Group, thence:
Rhumb line to Apia, passing between Gilbert Group and Tuvalu (8°00′S, 179°00′E).
7.170.1

Distances:

By great circle	4050 miles.
By rhumb line	4080 miles.

Suva ⇐ ⇒ Yokohama

Diagram 7.167
Routes
7.171

The route from Suva (18°11′S, 178°24′E) is direct by great circle between Kadavu Passage (18°45′S, 178°00′E) and Yokohama (35°26′N, 139°43′E), passing between Kosrae Island (5°20′N, 163°00′E) and Pingelap Atoll (6°10′N, 160°55′E).
7.171.1

Distance	3950 miles.

Apia ⇐ ⇒ Guam or Shanghai

Diagram 7.167
Routes
7.172

The route from Apia (13°47′S, 171°45′W) is:
Through 3°00′S, 175°00′E **(E)**, passing between Gilbert Group (0°00′, 174°00′E) and Tuvalu (8°00′S, 179°00′E), thence:
N of Kosrae Island (5°20′N, 163°00′E) and the other Caroline Islands to Guam (13°27′N, 144°35′E).
7.172.1

For Shanghai the route continues from Guam passing:
Through 25°30′N, 126°30′E **(F)**, to pass through Nansei Shotō, between Okinawa Guntō (26°30′N, 128°00′E) and Sakishima Guntō (24°40′N, 124°45′E), thence:
Through Xiaoban Men (30°12′N, 122°36′E), as in *China Sea Pilot, Volume II*, to Shanghai (31°03′N, 122°20′E).
7.172.2

Distances:

Guam	3090 miles.
Shanghai	4830 miles.

Suva ⇐ ⇒ Guam or Shanghai

Diagram 7.167
Routes
7.173

The route from Suva (18°11′S, 178°24′E) is:
Through Kadavu Passage (18°45′S, 178°00′E), thence:

S of Vanikoro (11°40′S, 166°50′E), thence:
Between Santa Cruz Islands (11°00′S, 166°15′E) and
    Solomon Islands (8°00′S, 158°00′E), thence:
20 miles W of Ulul Island (8°35′N, 149°40′E),
    thence:
To Guam (13°27′N, 144°35′E).

7.173.1
    For Shanghai (31°03′N, 122°20′E) the route continues
from Guam as described at 7.172.1.
7.173.2
    Distances:

Guam	2820 miles.
Shanghai	4560 miles.

PASSAGES ON THE EASTERN SIDE OF THE PACIFIC OCEAN

GENERAL INFORMATION

Coverage
7.174
    This section contains details of the following passages:
To and from Dutch Harbour and North and Central
    America (7.177).
To and from San Francisco or San Diego and Callao
    or Iquique (7.178).
To and from San Francisco or San Diego and
    Valparaíso (7.179).
To and from San Francisco and Estrecho de
    Magallanes (7.180).
To and from Panama and the Pacific coast of South
    America (7.181).

PASSAGES

Passages along the Pacific coasts of North and Central America

Diagram 7.175
Passage planning
7.175
1   On many passages N of Juan de Fuca Strait (48°30′N, 124°47′W) the choice may be made between an ocean route and a passage inshore of the islands fringing the coast. The inshore passages are described in *British Columbia Pilot, Volumes I and II* and *South-east Alaska Pilot*. They afford smooth water, suitable anchorages at moderate distances apart, and protection against the oceanic weather. Navigation is, however, intricate in many parts and it should be borne in mind that many of the minor passages may have only been partially examined.
2   Navigation along the Pacific coast of the United States requires due caution. Courses between salient points are, in general, long, and must be traversed during frequent periods of thick weather, with the vessel subject to the action of currents, the rate and direction of which are uncertain.
3   Studies of investigations into the causes of strandings on this coast, have found that many were due to lack of ordinary precautions essential to safe navigation, for example soundings, knowledge of compass errors, etc.

United States Coastal Route
7.176
1   The recommended track from Juan de Fuca Strait (48°30′N, 124°47′W) to Panama is described in *Pacific Coasts of Central America and United States Pilot*. This inshore route is as direct as possible and enables best use to be made of navigational aids and soundings and avoids the heavy seas in the offing.
2   Offshore the California Current (7.26) flows SE, but from November to January or February, the Davidson Current (7.26) flows N close inshore between the California

Current (7.26) and the coast N of Point Conception (34°27′N, 120°28′W), or sometimes farther S.
3   For distances, see table at 7.182.

Dutch Harbour ⇐ ⇒ North and Central America

Diagram 7.175
Routes
7.177
    From Dutch Harbour (53°56′N, 166°29′W) routes are as follows:
7.177.1
    For San Diego (32°38′N, 117°15′W) and destinations farther N as direct by great circle as navigation permits.
    Prince Rupert (54°19′N, 130°20′W) is approached through Dixon Entrance (54°30′N, 132°30′W).
7.177.2
    For destinations S of San Diego by great circle to join US Coastal Route (7.176) in 28°00′N, 116°00′W (**A**).
7.177.3
    Distances in miles:

Prince Rupert via Dixon Entrance	1210
Juan de Fuca Strait (48°30′N, 124°47′W)	1600
San Francisco (37°45′N, 122°40′W)	2060
San Diego (32°38′N, 117°15′W)	2470
Panama (Balboa) (8°53′N, 79°30′W)	5270

San Francisco or San Diego ⇐ ⇒ Callao or Iquique

Diagram 7.178
Routes
7.178
    From San Francisco (37°45′N, 122°40′W) and San Diego (32°38′N, 117°15′W) the routes are:
US Coastal Route (7.176) to 28°00′N, 116°00′W (**A**),
    thence:
Rhumb line to 26°40′N, 115°00′W (**B**), thence:
Rhumb line to Callao (12°02′S, 77°14′W), passing E
    of Archipiélago de Colón (0°00′, 90°00′W).
For Iquique (20°12′S, 70°10′W) the route follows that
    for Callao as far as the equator, thence as
    navigation permits.
For distances, see table at 7.182.

San Francisco or San Diego ⇐ ⇒ Valparaíso

Diagram 7.178
Routes
7.179
    From San Francisco (37°45′N, 122°40′W) and San Diego (32°38′N, 117°15′W) the routes are:
US Coastal Route (7.176) to 28°00′N, 116°00′W (**A**),
    thence:
Rhumb line to 26°40′N, 115°00′W (**B**), thence:
Rhumb line to 7°00′S, 90°00′W (**C**), thence:

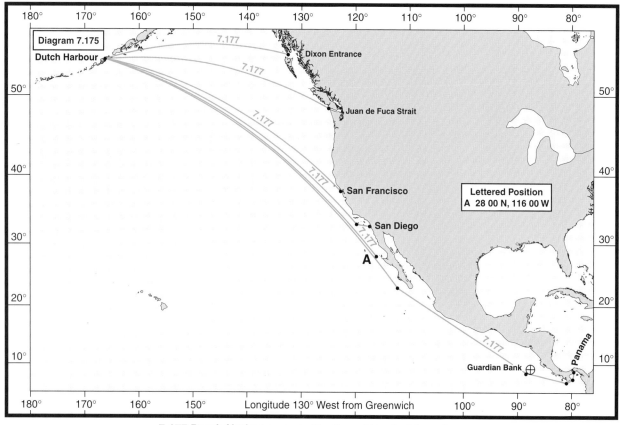

7.177 Dutch Harbour ← → North and Central America

Great circle to Valparaíso (33°02'S, 71°37'W). For distances, see table at 7.182.

San Francisco ⇐ ⇒ Estrecho de Magallanes

Diagram 7.178
Routes
7.180

From San Francisco (37°45'N, 122°40'W) the route is:
  20 miles W of Isla de Guadalupe (29°00'N, 118°17'W), thence:
  Great circle to the equator in 106°30'W (**D**), thence:
  Great circle to position 52°38'S, 74°46'W, 5 miles NNW of Cabo Pilar, the W entrance to Estrecho de Magallanes.

The track passes through Islas Revilla Gigedo (19°25'N, 110°30'W) between Roca Partida (19°00'N, 122°04'W) and Isla Clarion (18°20'N, 114°45'W). It also passes 70 miles

W of Île Clipperton (10°18'N, 109°13'W) and about 45 miles W of Germaine Bank (5°09'N, 107°53'W).
  For distances, see table at 7.182.

Panama ⇐ ⇒ Pacific coast of South America

Diagram 7.178
Routes
7.181

In all cases routes to and from Panama (Balboa) (8°53'N, 79°30'W) are as direct as navigation permits.

Off this coast the Peru Current (7.27) flows predominantly N, particularly near the land.

Fog is most frequent off the coast of Peru and least frequent in the parts N of 6°S and, except in April and May, between 15°S and 30°S. The highest and lowest frequencies of fog, over the region as a whole, occur in April and October respectively.

For distances, see table at 7.182.

Distances along Eastern shore of the Pacific Ocean

7.182

Juan de Fuca Strait							
680	San Francisco						
1110	435	San Diego					
3920	3230	2840	Panama				
4650	3970	3570	1340	Callao			
5280	4600	4210	1980	650	Iquique		
5800	5120	4720	2600	1800	780	Valparaiso	
6640	5970	5590	3730	2460	2000	124	Estrecho de Magallanes*

* 5 miles NNW of Cabo Pilar (52°43'S, 74°41'W)

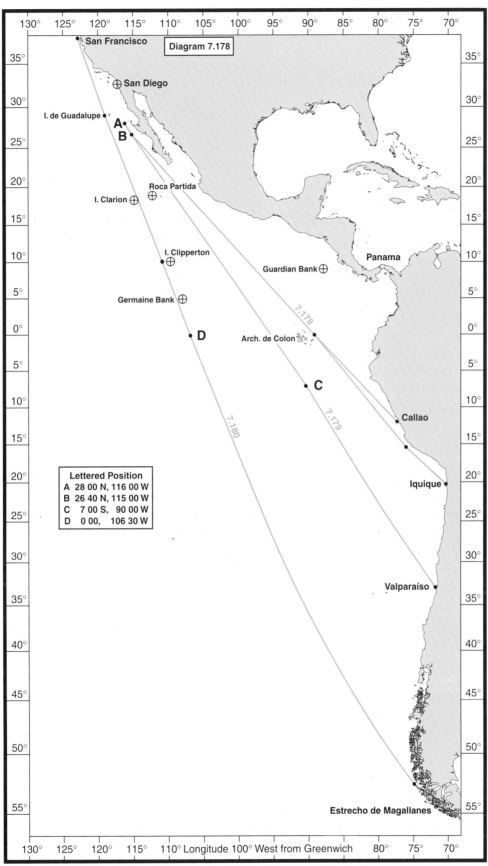

Diagram 7.178

Lettered Position
A 28 00 N, 116 00 W
B 26 40 N, 115 00 W
C   7 00 S, 90 00 W
D   0 00,   106 30 W

7.178 San Francisco and San Diego ← → Callao or Iquique
7.179 San Francisco and San Diego ← → Valparaíso
7.180 San Francisco and San Diego ← → Estrecho de Magallanes
7.181 Panama ← → Pacific coast of South America

PASSAGES TO AND FROM HONOLULU

GENERAL INFORMATION

Coverage
7.183

This section contains details of the following passages to and from:

Honolulu and Sydney or Brisbane (7.184).
Honolulu and Torres Strait (7.185).
Honolulu and New Zealand (7.186).
Honolulu and Apia (7.187).
Honolulu and Suva (7.188).
Honolulu and Nuku'alofa (7.189).
Honolulu and Guam or Yap (7.190).
Honolulu and Singapore (7.191).
Honolulu and Verde Island Passage or Manila (7.192).
Honolulu and Hong Kong (7.193).
Honolulu and Shanghai (7.194).
Honolulu and Yokohama (7.195).
Honolulu and Tsugaru Kaikyō (7.196).
Honolulu and Dutch Harbour (7.197).
Honolulu and Prince Rupert (7.198).
Honolulu and Juan de Fuca Strait, San Francisco or San Diego (7.199).
Honolulu and Panama (7.200).
Honolulu and Papeete (7.201).
Honolulu and South America (7.202).

PASSAGES

Honolulu ⇐ ⇒ Sydney or Brisbane

Diagram 7.184
Routes
7.184

The routes are:
7.184.1

From Sydney (33°50'S, 151°19'E):
Coastwise to Sugarloaf Point (32°27'S, 152°32'E), thence:
30 miles E of Cato Island (23°15'S, 155°32'E), thence:
30 miles NW of Bampton Reefs (19°00'S, 158°40'E), thence:
Through 10°00'S, 170°00'E **(F)**, passing midway between Torres Island (13°15'S, 166°35'E) and Vanikoro (11°40'S, 166°50'E), thence:
Across the equator in 178°50'W **(G)**, passing between Gilbert Group (0°00', 174°00'E) and Tuvalu (8°00'S, 179°00'E), thence:
Great circle to Honolulu (21°17'N, 157°53'W).
7.184.2

From Brisbane (Caloundra Head) (26°49'S, 153°10'E):
Through 21°30'S, 156°05'E **(H)**, passing between Cato Island and Wreck Reefs (22°10'S, 155°20'E), thence:
Joining Route 7.184.1 in a position 30 miles NW of Bampton Reefs.
7.184.3

Distances:

From Sydney	4490 miles.
From Caloundra Head*	4090 miles.

* Brisbane to Caloundra Head 35 miles.

Honolulu ⇐ ⇒ Torres Strait

Diagram 7.184
Routes
7.185

From Torres Strait (10°36'S, 141°51'E), via Bligh Entrance (9°12'S, 144°00'E), the routes are:
7.185.1

East of Rossel Spit (7.151.1), or through Jomard Entrance (7.151.2), or through China Strait (7.151.3). Thence:
Through the Solomon Sea to Bougainville Strait (6°40'S, 156°15'E), thence:
NW of Nukumanu Islands (Tasman Islands) (4°35'S, 159°25'E), thence:
Between Marshall Islands and Gilbert Group (0°00', 174°00'E), thence:
To Honolulu (21°17'N, 157°53'W).
For clarity on diagram 7.184 only the route through Jomard Entrance is shown.
7.185.2

Passing 35 miles S of Rossel Spit (11°27'S, 154°24'E), thence:
Between Guadalcanal (9°40'S, 160°00'E) and San Cristóbal Islands (10°30'S, 161°45'E), thence:
N of Ulawa Island (9°45'S, 162°00'E), thence:
Between Gilbert Group (0°00', 174°00'E) and Tuvalu Group (8°00'S, 179°00'E), thence:
Great circle to Honolulu (21°17'N, 157°53'W).
7.185.3

South of Indispensable Reefs (12°30'S, 160°25'E), thence:
S of Vanikoro (11°40'S, 166°50'E), thence:
Between Gilbert Group (0°00', 174°00'E) and Tuvalu Group (8°00'S, 179°00'E), thence:
Great circle to Honolulu (21°17'N, 157°53'W).
7.185.4

Distances:

E of Rossel Spit	4310 miles.
Via Jomard Entrance	4290 miles.
Via China Strait	4200 miles.
S of Rossel Spit	4270 miles.
S of Indispensable Reefs	4430 miles.

Honolulu ⇐ ⇒ New Zealand

Diagram 7.184
Routes
7.186

The routes are:
7.186.1

From Auckland (36°36'S, 174°49'E):
Through 18°30'S, 174°45'W **(J)**, thence:
20 miles W of Curacoa Reef (15°29'S, 173°37'W), thence:
40 miles NW of Savai'i (13°35'S, 172°30'W), thence:
Rhumb line to 10°30'S, 171°00'W **(K)**, between Tokelau (9°00'S, 172°00'W) and Swains Island (11°00'S, 171°05'W), thence:
To Honolulu (21°17'N, 157°53'W).
7.186.2

From Wellington (41°22'S, 174°50'E) and South Island ports:
E of North Island, thence:
Through 31°20'S, 179°30'W **(L)**, thence:
Joining the route from Auckland (7.186.1) in 22°50'S, 177°00'W **(M)**, passing W of Kermadec Islands

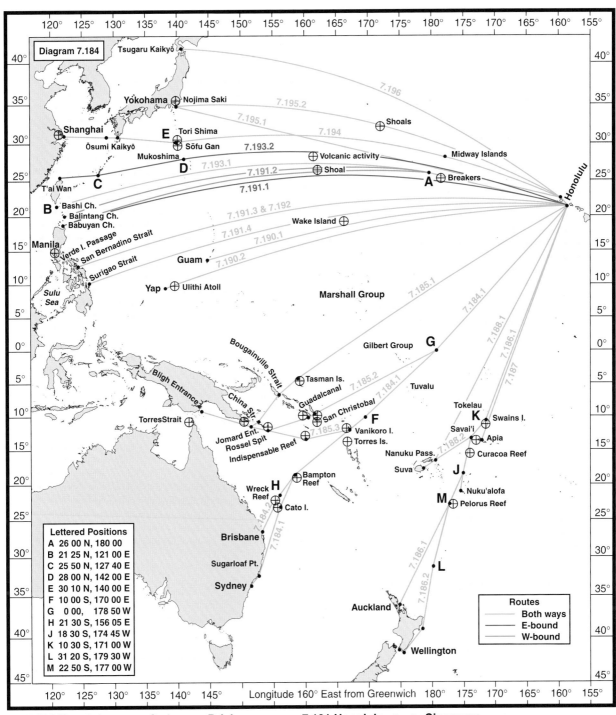

Diagram 7.184

Lettered Positions
A 26 00 N, 180 00
B 21 25 N, 121 00 E
C 25 50 N, 127 40 E
D 28 00 N, 142 00 E
E 30 10 N, 140 00 E
F 10 00 S, 170 00 E
G 0 00,   178 50 W
H 21 30 S, 156 05 E
J 18 30 S, 174 45 W
K 10 30 S, 171 00 W
L 31 20 S, 179 30 W
M 22 50 S, 177 00 W

Routes
— Both ways
— E-bound
— W-bound

Longitude 160° East from Greenwich

7.184 Honolulu ← → Sydney or Brisbane;
7.185 Honolulu ← → Torres Strait;
7.186 Honolulu ← → New Zealand;
7.187 Honolulu ← → Apia;
7.188 Honolulu ← → Suva;
7.189 Honolulu ← → Nuku'alofa;
7.190 Honolulu ← → Guam or Yap

7.191 Honolulu ← → Singapore
7.192 Honolulu ← → Verde Island Passage or Manilla
7.193 Honolulu ← → Hong Kong
7.194 Honolulu ← → Shanghai
7.195 Honolulu ← → Yokohama
7.196 Honolulu ← → Tsugaru Kaikyō

(30°30′S, 178°30′W), Pelorus Reef (22°51′S, 176°26′W) and other charted dangers.

For dangers between New Zealand and Tonga Islands, see 7.74.

7.186.3

Distances:

From Auckland	3800 miles.
From Wellington	4120 miles.

Honolulu ⇐ ⇒ Apia

Diagram 7.184

Routes

7.187

The route between Honolulu (21°17′N, 157°53′W) and Apia (13°47′S, 171°45′W) is by great circle.

Distance 2250 miles.

Honolulu ⇐ ⇒ Suva

Diagram 7.184
Routes
7.188

From Suva (18°11'S, 178°24'E) the routes are:
7.188.1

Through Nanuku Passage (16°40'S, 179°10'W), thence:

Great circle to Honolulu (21°17'N, 157°53'W).

This route entails passage through the islands and dangers NE of Fiji Islands (18°00'S, 180°00') and through Phoenix Group (4°00'S, 173°00'W) which are all low and not easily sighted.
7.188.2

An alternative route passes E and clear of the dangers noted at 7.188.1, passing:

40 miles NW of Savai'i (13°35'S, 172°30'W), thence
As described at 7.186.1.
7.188.3

Distances:

By great circle	2760 miles.
NW of Savai'i	2840 miles.

Honolulu ⇐ ⇒ Nuku'alofa

Diagram 7.184
Routes
7.189

From Nuku'alofa (21°00'S, 175°10'W) the route is:

As navigation permits to pass 20 miles W Curacoa Reef (15°29'S, 173°37'W), thence:

As described at 7.186.1
Distance   2730 miles.

Honolulu ⇐ ⇒ Guam or Yap

Diagram 7.184
Routes
7.190

In each case the route from Honolulu (21°17'N, 157°53'W) is by great circle:
7.190.1

For Guam (13°27'N, 144°35'E) the track passes about 30 miles S of Wake Island (19°17'N, 166°39'E) which, although only 6 m high, is a good radar target.
7.190.2

For Yap (9°28'N, 138°09'E) the track requires a diversion to avoid Ulithi Atoll (10°00'N, 139°40'E).
7.190.3

Distances:

Guam	3320 miles.
Yap	3750 miles.

Honolulu ⇐ ⇒ Singapore

Diagram 7.184
Routes
7.191

There are several alternative routes:
7.191.1

East-bound from Singapore (1°12'N, 103°51'E) the route, as described at 7.97, to pass:

NW of Prince Consort Bank (7°53'N, 110°00'E), thence:

Through Babuyan Channel (18°40'N, 122°00'E), thence:

Great circle to Honolulu (21°17'N, 157°53'W).
Distance   6010 miles.

7.191.2

West-bound from Honolulu (21°17'N, 157°53'W) the route is:

Rhumb line to 26°00'N, 180°00' **(A)**, thence:

Great circle to either Babuyan Channel (18°40'N, 122°00'E) or Balingtang Channel (20°00'N, 122°20'E), thence:

As described at 7.191.1, in reverse.

The rhumb line passes close N of the breakers reported in 25°17'N, 178°03'W, while the great circle tracks pass close to a shoal reported in 26°34'N 162°19'E.
Distance   6010 miles.
7.191.3

An alternative route, in either direction, leads from Singapore through the South China Sea, passing:

W of Pulau-pulau Anambas (3°00'N, 106°00'E), thence:

NW of Prince Consort Bank (7°53'N, 110°00'E), thence:

30 miles NW of North Danger Reef (11°25'N, 114°21'E), thence:

Through Verde Island Passage (13°36'N, 121°00'E), thence:

To San Bernadino Strait (12°33'N, 124°12'E) as described in *Philippine Islands Pilot*, thence:

Great circle to Honolulu (21°17'N, 157°53'W).
Distance   6000 miles.
7.191.4

A shorter route leads:

As at 7.102 or 7.102.1 to the S entrance of Palawan Passage (10°20'N, 118°00'E), thence:

Through Balabac Strait (7°34'N, 116°55'E), as described at 7.99.2, thence:

Through Sulu Sea to Surigao Strait (9°53'N, 125°21'E), thence:

Great circle to Honolulu (21°17'N, 157°53'W).
Distance   5910 miles.

Honolulu ⇐ ⇒ Verde Island Passage or Manila

Diagram 7.184
Routes
7.192

From Manila (14°32'N, 120°56'E) the route is:

Through Verde Island Passage (13°36'N, 121°00'E), thence:

To San Bernadino Strait (12°33'N, 124°12'E) as described in *Philippine Islands Pilot*, thence:

Great circle to Honolulu (21°17'N, 157°53'W).
7.192.1

Distances:

Verde Island Passage	4690 miles.
Manila	4780 miles.

Honolulu ⇐ ⇒ Hong Kong

Diagram 7.184
Routes
7.193

The routes are:
7.193.1

In either direction the route from Hong Kong (22°17'N, 114°10'E) is:

Through 21°25'N, 121°00'E **(B)**, in Bashi Channel, thence:

Great circle to 26°00'N, 180°00' **(A)**, thence:

Rhumb line to Honolulu (21°17'N, 157°53'W), keeping S of Hawaiian Islands.

For areas to be avoided near Hawaiian Islands, see 7.38.1 and *Pacific Islands Pilot, Volume III*.

Distance   4880 miles.

7.193.2

Eastbound, an alternative route from Hong Kong (22°17′N, 114°10′E), leads:

Through Taiwan Strait, thence:

Round the N end of T'ai-wan, thence:

Rhumb line to 25°50′N, 127°40′E **(C)**, thence:

Through 28°00′N, 142°00′E **(D)**, N of Mukoshima Rettō, thence:

Great circle to Honolulu (21°17′N, 157°53′W), passing N of volcanic activity in 28°14′N, 161°44′E (see *Pacific Islands Pilot, Volume III* for details) and S of Hawaiian Islands.

Distance   4870 miles.

Honolulu ⇐ ⇒ Shanghai

Diagram 7.184
Routes
7.194

From Shanghai (31°03′N, 122°20′E) the route is:

As navigation permits to pass through Ōsumi Kaikyō (30°55′N, 130°40′E); for details see *Japan Pilot, Volume II*. Thence:

To 30°10′N, 140°00′E **(E)**, between Tori Shima and Sōfu Gan, thence:

Great circle to Midway Islands (28°13′N, 177°21′W), thence:

As navigation permits to Honolulu (21°17′N, 157°53′W), passing N of Hawaiian Islands.

For areas to be avoided near Hawaiian Islands, see 7.38.1 and *Pacific Islands Pilot, Volume III*.

Distance   4350 miles.

Honolulu ⇐ ⇒ Yokohama

Diagram 7.184
Routes
7.195

Routes are either by rhumb line or great circle:

7.195.1

Good weather is usually experienced on the rhumb line route which passes 20 miles S of the Hawaiian Islands.

For areas to be avoided near Hawaiian Islands, see 7.38.1 and *Pacific Islands Pilot, Volume III*.

7.195.2

The great circle route leads from Kauai Channel (21°50′N, 158°45′W) to a landfall off Nojima Saki (34°54′N, 139°53′E) and passes 50 miles N of Hawaiian Islands, but close to reported dangers in approximately 32°15′N, 172°20′E.

7.195.3

   Distances:

   Rhumb line   3440 miles.
   Great circle   3400 miles.

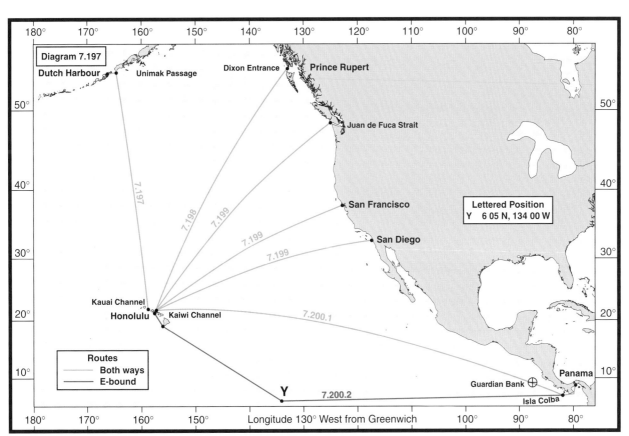

7.197 Honolulu ← → Dutch Harbour
7.198 Honolulu ← → Prince Rupert
7.199 Honolulu ← → Juan de Fuca Strait, San Francisco or San diego
7.200 Honolulu ← → Panama

Honolulu ⇐ ⇒ Tsugaru Kaikyō

Diagram 7.184
Routes
7.196

Route is by great circle between Kauai Channel (21°50′N, 158°45′W) and the E approach to Tsugaru Kaikyō (41°39′N, 140°48′E).

For areas to be avoided near Hawaiian Islands, see 7.38.1 and *Pacific Islands Pilot, Volume III.*

Distance 3300 miles.

Honolulu ⇐ ⇒ Dutch Harbour

Diagram 7.197
Routes
7.197

The route is by great circle between Kauai Channel (21°50′N, 158°45′W) and Unimak Pass (54°15′N, 164°30′W), thence as navigation permits to Dutch Harbour (53°56′N, 166°29′W).

Distance 2100 miles.

Honolulu ⇐ ⇒ Prince Rupert

Diagram 7.197
Routes
7.198

The route is by great circle between Kaiwi Channel (21°12′N, 157°40′W) and Dixon Entrance (54°30′N, 132°30′W), thence as navigation permits to Prince Rupert (54°19′N, 130°20′W).

Distance 2400 miles.

Honolulu ⇐ ⇒ Juan de Fuca Strait, San Francisco or San Diego

Diagram 7.197
Routes
7.199

Routes are by great circle between Kaiwi Channel (21°12′N, 157°40′W) and Juan de Fuca Strait (48°30′N, 124°47′W), San Francisco (37°45′N, 122°40′W) or San Diego (32°38′N, 117°15′W).

7.199.1
Distances:

Juan de Fuca Strait	2290 miles.
San Francisco	2080 miles.
San Diego	2270 miles.

Honolulu ⇐ ⇒ Panama

Diagram 7.197
Routes
7.200

There are alternative routes:

7.200.1

In either direction the route is by great circle between Kaiwi Channel (21°12′N, 157°40′W) and a landfall off Isla Coiba (7°25′N, 81°45′W), avoiding Guardian Bank (9°30′N, 87°30′W).

7.200.2

East-bound, an alternative route is as navigation permits to join the Central Route (7.232) in 6°05′N, 134°00′W **(Y)**, thence to Panama (Balboa) (8°53′N, 79°30′W).

7.200.3
Distances:

Great circle	4690 miles.
Central Route	5030 miles.

Honolulu ⇐ ⇒ Papeete

Diagram 7.201
Routes
7.201

The direct route between Honolulu (21°17′N, 157°53′W) and Papeete (17°30′S, 149°36′W) passes between Caroline Island (9°57′S, 150°13′E) and Vostok Island (10°06′S, 152°23′W) and close to the position of the breakers (reported 1926) in 5°48′S, 152°16′W and about 30 miles SW of Filippo Reef (5°31′S, 151°40′W).

Distance 2370 miles.

Honolulu ⇐ ⇒ South America

Diagram 7.201
Routes
7.202

Routes are by great circle, departure being taken from the NE side of the Hawaiian Islands for destinations N of about 35°S.

The great circle track for a vessel intending to round Cabo de Hornos (56°04′S, 67°15′W) runs to 55°00′S, 80°00′W **(B)**, passing about 40 miles E of Îles Marquises (9°00′S, 140°00′W) and Henderson Island (24°22′S, 128°19′W), thence by rhumb line to round Cabo de Hornos.

The great circle track to Iquique requires a diversion to avoid the position 6°40′S, 99°40′W where breakers were reported in 1906.

7.202.1
Distances in miles:

Callao (12°02′S, 77°14′W)	5160
Iquique (20°12′S, 70°10′W)	5720
Valparaíso (33°02′S, 71°37′W)	5920
Estrecho de Magallanes (52°38′S, 74°46′W)*	6170
Cabo de Hornos (56°04′S, 67°15′W)**	6470

Positions indicated are:
*   5 miles NNW of Cabo Pilar
** 5 miles S of Cabo de Hornos

PASSAGES TO AND FROM PAPEETE

GENERAL INFORMATION

Coverage
7.203

This section contains details of the following passages to and from:

Papeete and Guam (7.204).

Papeete and Yap, Verde Island Passage, Manila or Hong Kong (7.205).

Papeete and Shanghai (7.206).

Papeete and Yokohama (7.207).

Papeete and Canada, United States and Mexico between Prince Rupert and Gulf of Caifornia (7.208).

Papeete and Panama (7.209).
Papeete and Callao (7.210).
Papeete and Iquique (7.211).
Papeete and Valparaíso (7.212).
Papeete and Estrecho de Magallanes (7.213).
Papeete and Cabo de Hornos (7.214).

PASSAGES

Papeete ⇐ ⇒ Guam

Diagram 7.204
Routes
7.204

From Papeete (17°30′S, 149°36′W) the route is:
S of the W islands of Îles de la Société and
Suwarrow (13°15′S, 163°05′W), thence:

Between Tokelau (9°00′S, 172°00′W) and Swains
Island (11°00′S, 171°05′W), thence:
S of Tamana Island (2°30′S, 176°00′E) and Arorae
Island (2°39′S, 176°54′E) of the Gilbert Group,
thence:
Across the equator in 171°30′E (**B**), thence:
N of Kosrae (5°20′N, 163°00′E) and the other
Caroline Islands, thence:
To Guam (13°27′N, 144°35′E).
Landfall should not be attempted on Manuae (16°30′S,
154°40′W) at night or in thick weather.
Distance   4360 miles.

Papeete ⇐ ⇒ Yap, Verde Island Passage, Manila or Hong Kong

Diagram 7.204
Routes
7.205

From Papeete (17°30′S, 149°36′W) the routes are:

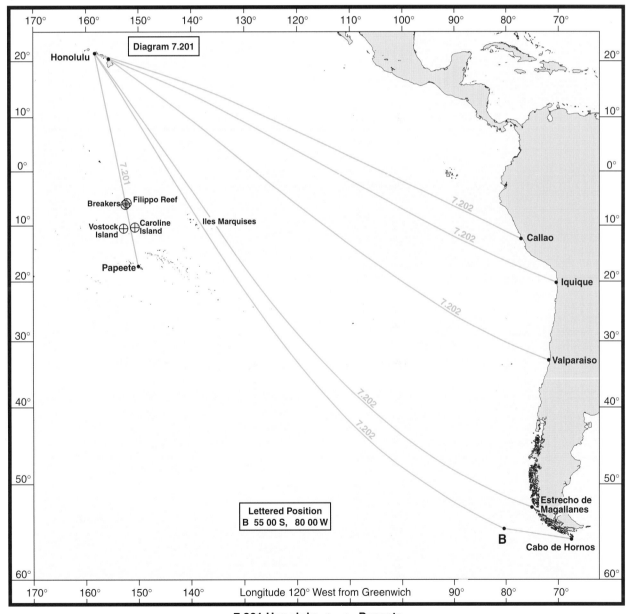

7.201 Honolulu ← → Papeete
7.202 Honolulu ← → South America

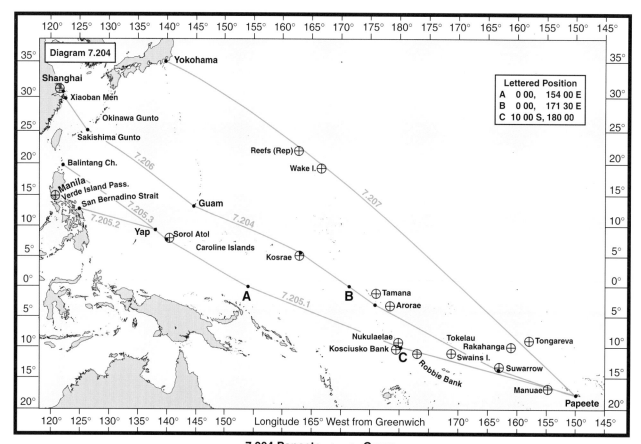

7.204 Papeete ← → Guam
7.205 Papeete ← → Yap, Verde Island Passage, Manila or Hong Kong
7.206 Papeete ← → Shanghai
7.207 Papeete ← → Yokohama

7.205.1

For Yap:
S of the W islands of Îles de la Société, thence:
Rhumb line to 10°00′S, 180°00′ (**C**), which passes close N of Robbie Bank (11°03′S, 176°57′W), thence:
Rhumb line to the equator in 154°00′E (**A**), passing between Nukulaelae Atoll (9°22′S, 179°52′E) and Kosciusko Bank (10°26′S, 179°30′E), thence:
20 miles SW of Sorol Atoll (8°10′N, 140°25′E), thence:
To Yap (9°28′N, 138°09′E).

7.205.2

For Verde Island Passage or Manila:
As at 7.205.1 but passing close S of Yap, thence:
To San Bernadino Strait (12°33′N, 124°12′E), thence:
As described in *Philippine Islands Pilot*, to Verde Island Passage (13°36′N, 121°00′E) or Manila (14°32′N, 120°56′E).

7.205.3

For Hong Kong:
As at 7.205.1 but passing close S of Yap, thence:
Through Balintang Channel (20°00′N, 122°20′E), thence:
As at 7.121 to Hong Kong (22°17′N, 114°10′E).

7.205.4

Distances:

	Papeete	Yap
Yap	4580	
Verde Island Passage	5640	1060
Manila	5730	1150
Hong Kong	6180	1600

Papeete ⇐ ⇒ Shanghai

Diagram 7.204

Routes

7.206

From Papeete (17°30′S, 149°36′W) the route is:
As at 7.204 to Guam (13°27′N, 144°35′E), thence:
As at 7.172.1 to Shanghai (31°03′N, 122°20′E).
Distance 6100 miles.

Papeete ⇐ ⇒ Yokohama

Diagram 7.204

Routes

7.207

From Yokohama (35°26′N, 139°43′E) the route is by great circle, which passes:
30 miles NE of Wake Island (19°17′N, 166°39′E) and close to a number of reefs reported to lie 300 miles NW of that island, thence:

About 80 miles SW of Tongareva (9°00′S, 158°03′W) and NE of Rakahanga (10°03′S, 161°06′W), thence:

To Papeete (17°30′S, 149°36′W).

Distance   5130 miles.

Papeete ⇐ ⇒ Canada, United States and Mexico between Prince Rupert and Gulf of California

Diagram 7.208

Routes

7.208

From Papeete (17°30′S, 149°36′W) the routes are by great circle from 14°45′S, 148°55′W (**D**), NW of Matahiva, passing NW of Îles Marquises.

Prince Rupert can be approached through either Dixon Entrance (54°30′N, 132°30′W) or Hecate Strait (53°00′N, 131°00′W).

7.208.1

Distances in miles:

Prince Rupert via Dixon Entrance	4500
Juan de Fuca Strait (48°30′N, 124°47′W)	4170
San Francisco (37°45′N, 122°40′W)	3650
San Diego (32°38′N, 117°15′W)	3500

Papeete ⇐ ⇒ Panama

Diagram 7.208

Routes

7.209

From Papeete (17°30′S, 149°36′W) the routes are:

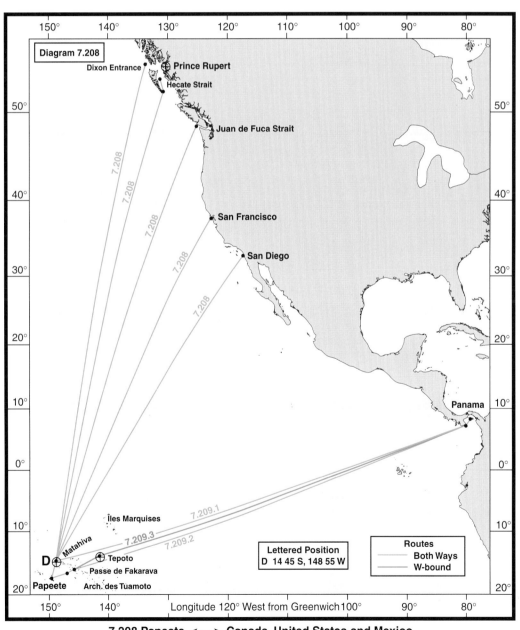

7.208 Papeete ← → Canada, United States and Mexico
7.209 Papeete ← → Panama

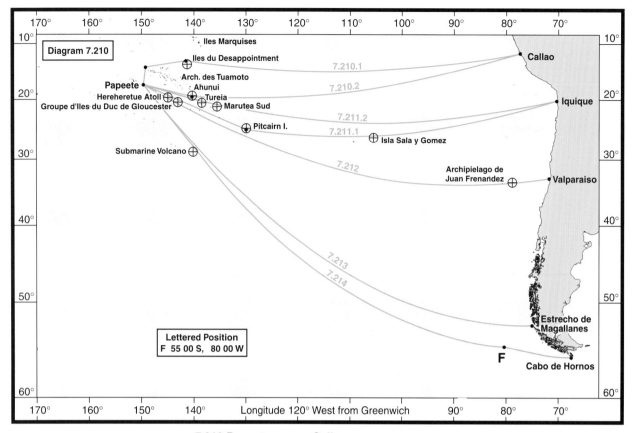

Diagram 7.210

Lettered Position
F 55 00 S, 80 00 W

7.210 Papeete ← → Callao;
7.211 Papeete ← → Iquique;
7.212 Papeete ← → Valparaíso
7.213 Papeete ← → Estrecho de Magallanes
7.214 Papeete ← → Cabo de Hornos

7.209.1

W and N of Archipel des Tuamoto (18°00′S, 141°00′W), thence:

Great circle to Gulf of Panama (8°00′N, 79°00′W).

7.209.2

An alternative route from Papeete leads:

Through Passe de Fakarava (16°00′S, 145°50′W), thence:

Great circle to Gulf of Panama (8°00′N, 79°00′W).

The pass is not lighted and good electronic aids are essential. For details, see *Pacific Islands Pilot, Volume III.*

7.209.3

A route West-bound from Gulf of Panama is:

Great circle to a landfall on Tepoto (14°06′S, 141°27′W), one of Îles du Désappointment, thence:

Through Passe de Fakarava.

7.209.4

Distances:

W and N of Archipel des Tuamoto	4610 miles.
Through Passe de Fakarava	4500 miles.

Papeete ⇐ ⇒ Callao

Diagram 7.210

Routes

7.210

From Papeete (17°30′S, 149°36′W) the alternative routes are:

7.210.1

W and N of Archipel des Tuamoto (18°00′S, 141°00′W), to position N of Îles du Désappointment (14°10′S, 141°20′W), thence:

Great circle to Callao (12°02′S, 77°14′W).

7.210.2

Through Archipel des Tuamoto (18°00′S, 141°00′W) to a position S of Ahunui (19°40′S, 140°25′W), thence:

Great circle to Callao (12°02′S, 77°14′W).

7.210.3

Distances:

W and N of Archipel des Tuamoto	4370 miles.
Through Archipel des Tuamoto	4210 miles.

Papeete ⇐ ⇒ Iquique

Diagram 7.210

Routes

7.211

From Papeete (17°30′S, 149°36′W) the alternative routes are:

7.211.1

S of Archipel des Tuamoto (18°00′S, 141°00′W) to Pitcairn Island (25°04′S, 130°05′W), thence:

Great circle, passing close N of Isla Sala y Gomez (26°28′S, 105°28′W), to Iquique (20°12′S, 70°10′W).

7.211.2

Great circle through Archipel des Tuamoto (18°00′S, 141°00′W), S of Ahunui (19°40′S, 140°25′W) and N of Tureia (20°50′S, 138°35′W) and Marutea Sud

218

(21°30′S, 135°35′W), to Iquique (20°12′S, 70°10′W).

7.211.3

Distances:

S of Archipel des Tuamoto	4510 miles.
Through Archipel des Tuamoto	4480 miles.

Papeete ⇐ ⇒ Valparaíso

Diagram 7.210

Routes

7.212

From Papeete (17°30′S, 149°36′W) the route is:

S of Hereheretue Atoll (19°52′S, 145°00′W) and Groupe d'Îles du Duc de Gloucester (20°38′S, 143°17′W), thence:

Great circle, passing close S of Archipiélago de Juan Fernandez (33°40′S, 78°50′W), to Valparaíso (33°02′S, 71°37′W).

Distance 4260 miles.

Papeete ⇐ ⇒ Estrecho de Magallanes

Diagram 7.210

Routes

7.213

From Papeete (17°30′S, 149°36′W) the route is by great circle to position 52°38′S, 74°46′W, 5 miles NNW of Cabo Pilar, at the W entrance to Estrecho de Magallanes.

Distance 4040 miles.

Papeete ⇐ ⇒ Cabo de Hornos

Diagram 7.210

Routes

7.214

From Papeete (17°30′S, 149°36′W) the route is:

Great circle to 55°00′S, 80°00′W **(F)**, on the Southern Route (7.216), passing close N of the submarine volcano (29°00′S, 140°15′W) reported in 1981, thence:

Rhumb line to a position 56°04′S, 67°15′W, 5 miles S of Cabo de Hornos.

Distance 4300 miles.

SOUTH PACIFIC TRANS-OCEAN PASSAGES

GENERAL INFORMATION

Coverage

7.215

This section contains details of the following passages:

Main E-bound routes:

Southern Route (7.216).

Southern Route connections to other routes (7.218).

From Torres Strait to South America (7.221).

From Hobart to Panama (7.222).

From Wellington to Panama (7.223).

From Auckland to Panama (7.224).

Main W-bound routes:

West-bound Route (7.225).

From Panama to New Zealand (7.226).

Chile and Perú to East coast of Australia and New Zealand (7.227).

Other main routes:

To and from Apia and South America (7.228).

To and from Suva and South America (7.229)

MAIN EAST-BOUND ROUTES ACROSS SOUTHERN PACIFIC OCEAN

Southern Route

Diagram 7.216

Details

7.216

The most S route usually adopted, referred to as the Southern Route, passes E-bound through the following positions:

48°30′S, 165°00′W	**(A)**
49°30′S, 150°00′W	**(B)**
50°00′S, 140°00′W	**(C)**
51°30′S, 120°00′W	**(D)**
52°45′S, 100°00′W	**(E)**
55°00′S, 80°00′W	**(F)**

Joining Southern Route

7.217

When great circle passages, between terminal points, pass S of the Southern Route it is best to steer, by great circle, to join this route at a convenient position.

Similarly it is best to leave the Southern Route at a position to enable the destination to be reached by great circle, if possible, without passing S of the above route.

Passages for which the Southern Route, or part of it, are appropriate can best be seen from *Chart 5098 Gnomic Chart - South Pacific and Southern Oceans.*

Southern Route connections to other routes

7.218

The following are the best joining and leaving positions:

From	Join in
Hobart or Snares Island	48°30′S, 165°00′W **(A)**
Cook Strait	49°30′S, 150°00′W **(B)**
Auckland	50°00′S, 140°00′W **(C)**

For	Leave in
Callao	48°30′S, 165°00′W **(A)**
Iquique	49°30′S, 150°00′W **(B)**
Valparaíso	50°00′S, 140°00′W **(C)**
Estrecho de Magallanes	52°45′S, 100°00′W **(E)**
Cabo de Hornos	55°00′S, 80°00′W **(F)**

From Sydney or Brisbane for Callao and destinations farther S, the route is through Cook Strait.

Alternatively from Sydney for Valparaíso and destinations farther S, the route S of New Zealand is practicable and only slightly longer.

Ice

7.219

Icebergs may be encountered on the Southern Route in all seasons, for details see 7.34.

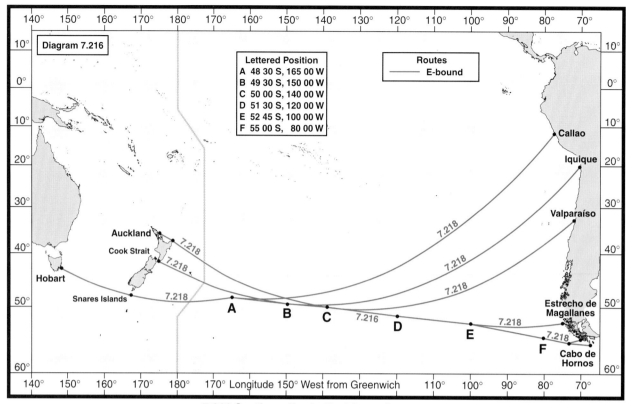

7.216 Southern Route East-bound
7.218 Southern Route connections to other routes.

Distances
7.220

Distances are given for E-bound routes incorporating the Southern Route (7.216).

	Callao	Iquique	Valparaíso	Estrecho de Magallanes [a]	Cabo de Hornos
Hobart	6780	6700	6120	5430	5670
Port Phillip [b]	7040	6960	6370	5680	5930
Sydney [c]	6940 [f]	6880	6290	5600	5850
Sydney [d]	—	—	6330	5640	5880
Caloundra Head [e]	7110	7050	6460	5770	6020
Wellington	5720 [f]	5660	5070	4380	4630
Auckland	5840 [g]	5820 [g]	5250	4560	4810

[a] 5 miles NNW of Cabo Pilar.	[e] Brisbane to Caloundra Head 35 miles. Routes via Cook Strait.
[b] Melbourne to Port Phillip 40 miles.	
[c] Via Cook Strait.	[f] Direct after passing N of Chatham Island.
[d] Via Snares Islands.	[g] Direct.

Torres Strait ⇒ South America

Diagram 7.221

Routes
7.221

Routes are:

Through Bligh Entrance (9°12′S, 144°00′E), thence:
Through 28°30′S, 170°00′E **(J)**, passing S of Bellona Reefs (21°30′S, 158°50′E) and the submarine volcano (27°45′S, 169°09′E), thence as follows:

7.221.1

For Cabo de Hornos (56°04′S, 67°15′W) or Estrecho de Magallanes (52°43′S, 74°41′W) by great circle to join Southern Route at 50°00′S, 140°00′E **(C)**.

7.221.2

For Valparaíso (33°02′S, 71°37′W) or Iquique (20°12′S, 70°10′W) by great circle to destination. The route to Iquique passes close S of Isla San Ambrosio (26°20′S, 79°52′W).

7.221.3

For Callao (12°02′S, 77°14′W) by great circle to 38°00′S, 150°00′W **(H)**, thence great circle to Callao.

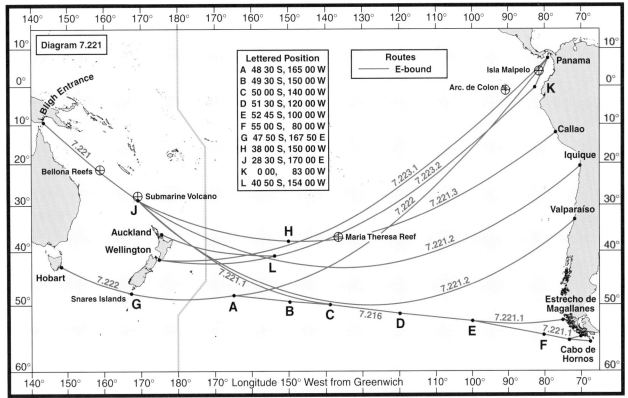

7.221 Torres Strait → South America
7.222 Hobart → Panama
7.223 Wellington → Panama
7.224 Auckland → Panama

Maria Theresa Reef (36°50′S, 136°39′W), the existence of which is doubtful, lies close N of the track.

7.221.4

Distances:

Cabo de Hornos	7360 miles.
Estrecho de Magallanes*	7110 miles.
Valparaíso	7800 miles.
Iquique	8340 miles.
Callao	8300 miles.

*   5 miles NNW of Cabo Pilar.

Hobart ⇒ Panama

Diagram 7.221

Routes

7.222

From Hobart (42°58′S, 147°23′E) the route is by great circle to 47°50′S, 167°50′E (**G**), ENE of Snares Islands (48°01′S, 166°36′E), thence by great circle to Gulf of Panama (8°00′N, 79°00′W).

Distance   7640 miles.

Wellington ⇒ Panama

Diagram 7.221

Routes

7.223

From Wellington (41°22′S, 174°50′E) there are two alternative routes by great circle:

7.223.1

To Gulf of Panama (8°00′N, 79°00′W), passing S of Archipiélago de Colón (0°00′, 90°00′W).

Distance   6490 miles.

7.223.2

To the equator in 83°00′W (**K**), thence as navigation permits to Gulf of Panama (8°00′N, 79°00′W), passing E of Isla Malpelo (4°00′N, 81°35′W).

This route is slightly longer than 7.223.1, but more favourable winds and currents may be experienced when approaching the coast of South America.

Distance   6540 miles.

Auckland ⇒ Panama

Diagram 7.221

Routes

7.224

The route from Auckland (36°36′S, 174°49′E) is by great circle to join route 7.223.2 from Wellington in 40°50′S, 154°00′W (**L**).

Distance   6610 miles.

MAIN WEST-BOUND ROUTES ACROSS SOUTHERN PACIFIC OCEAN

Route

Diagram 7.225

West-bound Route

7.225

West-bound routes across the South Pacific Ocean lie far N of the Southern Route, following the parallel of 30°S for various distances between the meridians of 120°W and 150°W and through the following positions:

30°00′S, 150°00′W (**P**);
30°00′S, 140°00′W (**Q**);
30°00′S, 139°00′W (**R**);
30°00′S, 120°00′W (**S**).

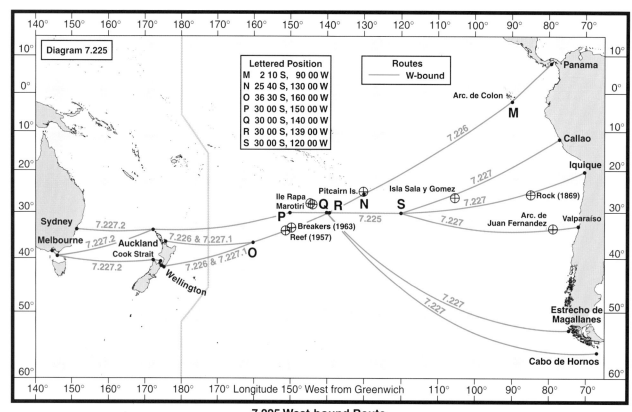

7.225 West-bound Route
7.226 Panama → New Zealand
7.227 Chile and Perú → East coasts of Australia and New Zealand

Panama ⇒ New Zealand

Diagram 7.225
Routes
7.226

From the Gulf of Panama (8°00′N, 79°00′W) the route is:

As navigation permits to 2°10′S, 90°00′W (**M**), 50 miles S of Archipiélago de Colón, thence:

Great circle to 25°40′S, 130°00′W (**N**), 30 miles S of Pitcairn Island, thence:

Great circle to 36°30′S, 160°00′W (**O**), with due regard to the breakers reported (1963) in 33°22′S, 149°38′W and the reef reported (1957) in 34°00′S, 151°00′W; for details, see *Pacific Islands Pilot, Volume III*. Thence:

Great circle to Auckland (36°36′S, 174°49′E) or Wellington (41°22′S, 174°50′E).

7.226.1
Distances:

To Auckland	6530 miles.
To Wellington	6530 miles.

Chile and Perú ⇒ East coasts of Australia and New Zealand

Diagram 7.225
Routes
7.227

The parallel of 30°S forms part of all routes from these coasts, for details see 7.225. It is reached by great circle from the departure points and joined in the following longitudes:

140°W (**Q**) from:
Cabo de Hornos (56°04′S, 67°15′W);

Estrecho de Magallanes (52°43′S, 74°41′W).

120°W (**S**) from:
Valparaíso (33°02′S, 71°37′W), passing close S of Archipiélago de Juan Fernandez (33°40′S, 78°50′W);

Iquique (20°12′S, 70°10′W), passing N of the rock reported (1869) in 25°40′S, 85°00′W, the existence of which is doubtful;

Callao (12°02′S, 77°14′W).

7.227.1

For New Zealand the routes continue from 30°00′S, 140°00′W (**Q**) by great circle to join route 7.226 from Panama in 36°30′S, 160°00′W (**O**), with due regard to the dangers reported, to Auckland (36°36′S, 174°49′E), Wellington (41°22′S, 174°50′E) or other destinations.

7.227.2

For Australian ports the routes continue:
Either through Cook Strait (41°00′S, 174°30′E), or:
Continue along the parallel of 30° to 150°W (**P**) passing N of New Zealand, thence:
To Sydney (33°50′S, 151°19′E), Melbourne (Port Phillip) (38°20′S, 144°34′E) or other destinations.

7.227.3

Distances in miles:

	Auckland/ Wellington	Sydney
Cabo de Hornos	5670	6870
Est. de Magallanes	5410	6610
Valparaíso	5740	6940
Iquique	6090	7290
Callao	5950	7150

* Port Phillip to Melbourne 40 miles.

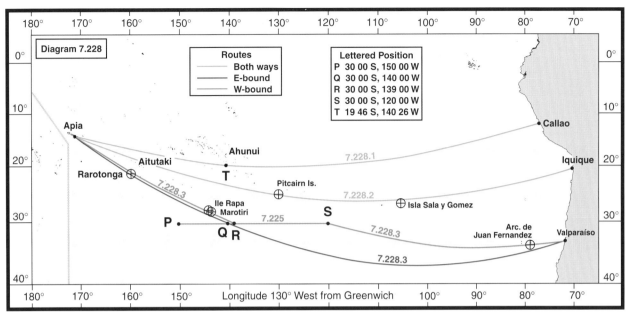

7.228 Apia ← → South America

OTHER MAIN ROUTES ACROSS SOUTHERN PACIFIC OCEAN

Apia ⇐ ⇒ South America

Diagram 7.228

Routes

7.228

From Apia (13°47′S, 171°45′W) the routes are as follows:

7.228.1

To Callao the route is:

By great circle to 19°46′S, 140°26′W (**T**), S of Ahunui (19°40′S, 140°25′W), thence:

Great circle to Callao (12°02′S, 77°14′W).

Distance 5500 miles.

7.228.2

To Iquique (20°12′S, 70°10′W) the route is by great circle, passing close to Pitcairn Island (25°04′S, 130°05′W) and Isla Sala y Gomez (26°28′S, 105°28′W).

Distance 5760 miles.

7.228.3

To Valparaíso (33°02′S, 71°37′W) the routes are directional.

East-bound the route is by great circle. It passes close to Rarotonga (21°10′S, 159°47′W) and the dangers SE of Îles Australes (23°00′S, 151°00′W) and Île Rapa (27°37′S, 144°17′W) and Marotiri Islands (27°55′S, 143°30′E).

West-bound the route from Valparaíso (33°02′S, 71°37′W) is:

By great circle to 30°00′S, 120°00′W (**S**), as at 7.227, passing close S of Archipiélago de Juan Fernandez (33°40′S, 78°50′W), thence:

Along the parallel of 30°00′S to 139°00′W (**R**), thence:

By great circle to Apia (13°47′S, 171°45′W).

Distances:

E-bound 5460 miles.
W-bound 5510 miles.

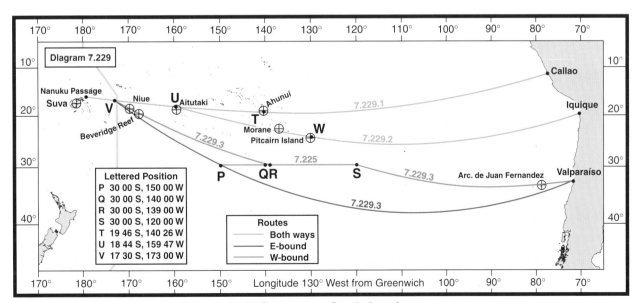

7.229 Suva ← → South America

Suva ⇐ ⇒ South America

Diagram 7.229
Routes
7.229

From Suva (18°11′S, 178°24′E) the routes are as follows:
7.229.1

To Callao the route is:

Through Nanuku Passage (16°40′S, 179°10′W), thence:

Rhumb line to 18°44′S, 159°47′W (U), N of Aitutaki (18°54′S, 159°47′W), thence:

Rhumb line to 19°46′S, 140°26′W (T), SE of Ahunui (19°40′S, 140°25′W), thence:

Great circle to Callao (12°02′S, 77°14′W).
Distance   6060 miles.
7.229.2

To Iquique the route is:

Through Nanuku Passage (16°40′S, 179°10′W), thence:

Rhumb line to 18°44′S, 159°47′W (U), thence:

Rhumb line to 24°55′S, 130°10′W (W), N of Pitcairn Island (25°04′S, 130°05′W), passing close S of

Morane (23°10′S, 137°08′W), thence:

Great circle to Iquique (20°12′S, 70°10′W).
Distance   6300 miles.
7.229.3

To Valparaíso (33°02′S, 71°37′W) the routes are directional.

East-bound the route is:

Through Nanuku Passage (16°40′S, 179°10′W), thence:

Rhumb line to 17°30′S, 173°00′W (V), thence:

By great circle to Valparaíso.

West-bound the route from Valparaíso (33°02′S, 71°37′W) is:

Great circle to 30°00′S, 120°00′W (S), passing close S of Archipiélago de Juan Fernandez (33°40′S, 78°50′W), thence:

Along the parallel of 30°S to 139°W (R), thence:

Great circle to 17°30′S, 173°00′W (V), which passes close to Beveridge Reef (20°00′S, 167°47′W) and Niue (19°03′S, 169°51′E), thence:

Through Nanuku Passage (16°40′S, 179°10′W) to Suva (18°11′S, 178°24′E).

Distances:

E-bound	5920 miles.
W-bound	5990 miles.

MID-PACIFIC TRANS-OCEAN PASSAGES

GENERAL INFORMATION

Coverage
7.230

This section contains details of the following passages:

Central Route (7.231).

To and from Suva and Panama (7.234).

To and from Apia and Panama (7.235).

From Melbourne to Panama (7.236).

From Sydney to Panama (7.237).

From Panama to Sydney (7.238).

From Brisbane to Panama (7.239).

From Panama to Brisbane (7.240).

To and from Torres Strait and Panama (7.241).

From Yap to Panama (7.242).

From Guam to Panama (7.243).

From Panama to Guam (7.244).

From Basilan Strait to Panama (7.245).

From San Bernadino Strait to Panama (7.246).

To and from Balintang Channel and Panama (7.247).

From Singapore to Panama (7.248).

From Panama to Verde Island Passage and Manila (7.249).

From Panama to Singapore (7.250).

To and from Apia and Prince Rupert (7.251).

To and from Apia and Juan de Fuca Strait (7.252).

To and from Apia and San Francisco (7.253).

To and from Apia and San Diego (7.254).

To and from Suva and Prince Rupert (7.255).

To and from Suva and Juan de Fuca Strait (7.256).

To and from Suva and San Francisco (7.257).

To and from Suva and San Diego (7.258).

To and from Auckland and Prince Rupert (7.259).

To and from Auckland and Juan de Fuca Strait (7.260).

To and from Auckland and San Francisco (7.261).

To and from Auckland and San Diego (7.262).

To and from Wellington and Prince Rupert (7.263).

To and from Wellington and Juan de Fuca Strait (7.264).

To and from Wellington and San Francisco (7.265).

To and from Wellington and San Diego (7.266).

To and from Sydney or Brisbane and San Diego or San Francisco (7.267).

To and from Sydney or Brisbane and Juan de Fuca Strait or Prince Rupert (7.268).

To and from Torres Strait and San Diego or San Francisco (7.269).

To and from Torres Strait and Juan de Fuca Strait or Prince Rupert (7.270).

To and from Guam or Yap and North America (7.271).

PASSAGES

Central Route

Diagram 7.231
General information
7.231

1   The constant W flow of water in the equatorial part of the Pacific Ocean, between roughly the latitude of Hawaii (19°40′N, 155°30′W) in the N and Îles de la Société (16°00′S, 152°00′W) in the S, together with the North-east and South-east Trade Winds (7.6 and 7.15) which blow on either side of the Intertropical Convergence Zone (7.4 and 7.13), tend to lengthen voyage times and to increase fuel and maintenance costs on ships E-bound through these waters.

2   Passages from ports between Hong Kong (22°17′N, 114°10′E) and Sydney (33°50′S, 151°19′E) to the coasts of Central America and equatorial South America may merit diversion to take advantage of the Equatorial Counter-current (7.28). For much of the year, this current flows from W to E in a narrow belt a few degrees N of the equator, between the North and South Equatorial Currents (7.26 and 7.27) setting in the opposite direction.

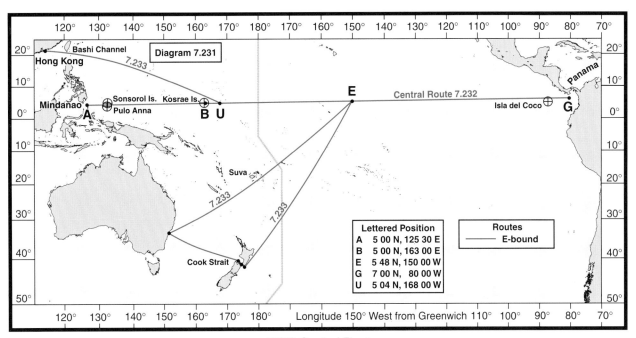

7.232 Central Route
7.233 Central Route connections

Furthermore, the central and E part of the ocean is favoured by the light weather of the Intertropical Convergence Zone.

3 For further details consult the appropriate volumes of *Admiralty Sailing Directions*.

Route description
7.232

1 The Central Route, used in this book, follows the average Equatorial Counter-current from Celebes Sea to Gulf of Panama (8°00′N, 79°00′W).

2 From 5°00′N, 125°30′E (**A**), S of Mindanao (8°00′N, 125°00′E), it runs:

Along the parallel of 5°N, passing between Sonsorol Islands (5°20′N, 132°13′E) and Pulo Anna (4°40′N, 131°58′E), thence:

To 5°00′N, 163°00′E (**B**), S of Kosrae Island (5°20′N, 163°00′E), thence:

Rhumb line, on a course of 089°, to 7°00′N, 80°00′W (**G**).

The overall length of this route is 9250 miles.

Central Route connections
7.233

1 Positions for joining or leaving the route depend on local as well as climatic considerations.

2 With this in view:

Panama-bound ships from Hong Kong might join in 168°E (**U**), between the Marshall and Caroline Islands, at a cost of 820 miles over the shortest navigable distance of 9270 miles.

Panama-bound ships from Sydney, joining in 150°W, (**E**) would accept an extra distance of about 700 miles compared with the 7700 miles of the route via Cook Strait (7.51 and 7.223.1).

3 Several joining routes are described in the routes for Panama in the following sections.

Suva ⇐ ⇒ Panama

Diagram 7.234
Route
7.234

There are several alternative routes:

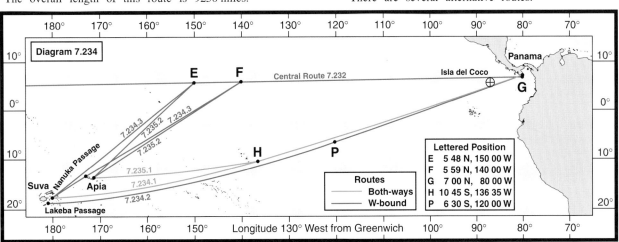

7.234 Suva ← → Panama
7.235 Apia ← → Panama

7.234.1

In either direction the route is:

From Suva (18°11′S, 178°24′E), through Nanuku Passage (16°40′S, 179°10′W), thence:

Great circle to 10°45′S, 136°35′W (**H**), thence:

Great circle to Gulf of Panama (8°00′N, 79°00′W), passing S of Isla del Coco (5°32′N, 87°04′W).

Distance 6350 miles.

7.234.2

An East-bound route leads:

Through Lakeba Passage (17°50′S, 178°30′W), for cautions in this passage, see 7.78. Thence:

Joining the route, from Papeete to Panama (7.209), in 6°30′S, 120°00′W (**P**).

Distance 6330 miles.

7.234.3

An alternative East-bound route leads:

Through Nanuku Passage (16°40′S, 179°10′W), thence:

NW of Savai'i (13°35′S, 172°30′W) to join Central Route (7.232) in either 5°48′N, 150°00′W (**E**) or 5°59′N, 140°00′W (**F**), thence:

Via the Central Route to Gulf of Panama (8°00′N, 79°00′W).

Distance 6660 miles.

Note. If joining Central Route in 5°59′N, 140°00′W (**F**), rather than 5°48′N, 150°00′W (**E**), the distance over which the chance of a favourable Equatorial Counter-current may be expected will be reduced by about 600 miles, while only reducing the total distance by about 110 miles.

Apia ⇐ ⇒ Panama

Diagram 7.234

Route

7.235

There are several alternative routes:

7.235.1

In either direction from Apia (13°47′S, 171°45′W) the route is:

Great circle to 10°45′S, 136°35′W (**H**), passing close to Suwarrow (13°15′S, 163°05′W) thence:

Great circle to Gulf of Panama (8°00′N, 79°00′W), passing S of Isla del Coco (5°32′N, 87°04′W).

Distance 5740 miles.

7.235.2

An East-bound route leads from Apia (13°47′S, 171°45′W) by great circle to join the route from Suva (7.234.3) in either 5°48′N, 150°00′W (**E**) or 5°59′N, 140°00′W (**F**).

The great circle to 5°48′N, 150°00′W (**E**) passes close to the position of the breakers reported in 5°12′S, 162°18′W.

Distance 6040 miles (See also Note at 7.234.3).

Melbourne ⇒ Panama

Diagram 7.236

Route

7.236

The route from Port Phillip (38°20′S, 144°34′E) is:

Through the appropriate TSS through Bass Strait (40°00′S, 146°00′E), thence:

Great circle across the Tasman Sea to pass N of New Zealand, N or S of Three Kings Islands (34°10′S, 172°00′E), with due regard to local tidal streams, for details, see *New Zealand Pilot*, thence:

Great circle to 30°00′S, 150°00′W (**J**), thence:

Rhumb line to 25°40′S, 130°00′W (**K**), S of Pitcairn Island, thence:

Great circle to 2°10′S, 90°00′W (**L**), S of Archipiélago de Colón, thence:

Great circle to Punta Mala (7°28′N, 80°00′W), thence:

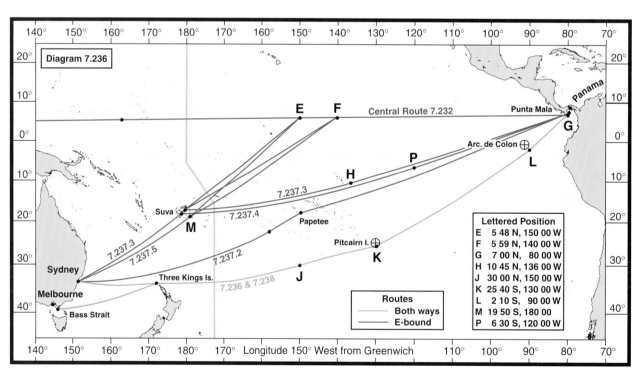

7.236 Melbourne → Panama
7.237 Sydney → Panama
7.238 Panama → Sydney

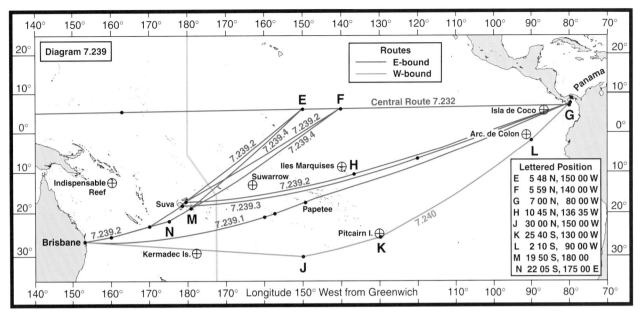

7.239 Brisbane → Panama
7.240 Panama → Brisbane

As navigation permits to Panama (Balboa) (8°53′N, 79°30′W).

Distance   8050 miles.
(Melbourne to Port Phillip 40 miles).

Sydney ⇒ Panama

Diagram 7.236
Route
7.237

From Sydney (33°50′S, 151°19′E) there is a choice of routes:

7.237.1

Great circle across the Tasman Sea to pass N of New Zealand, N or S of Three Kings Islands (34°10′S, 172°00′E), thence:

As described at 7.236 to Panama (Balboa) (8°53′N, 79°30′W).

Distance   7700 miles.

7.237.2

Via Papeete (17°30′S, 149°36′W), as described at 7.54, thence:

To Panama (Balboa) (8°53′N, 79°30′W), as described at 7.209.

Distance   7900 miles.

7.237.3

Via Suva, as described at 7.57, and through Nanuku Passage (16°40′S, 179°10′W), thence:

To Panama (Balboa) (8°53′N, 79°30′W), as described at 7.234.1 or 7.234.3.

Distance   8080 miles.

7.237.4

Via Suva, as described at 7.57, and Lakeba Passage (17°50′S, 178°30′W), for cautions in this passage, see 7.78. Thence:

To Panama (Balboa) (8°53′N, 79°30′W), as described at 7.234.2.

Distance   8060 miles.

7.237.5

Great circle to 19°50′S, 180°00′W (**M**), thence:

10 miles S of Ogea Driki (19°12′S, 178°24′W), thence:

NW of Savai'i (13°35′S, 172°30′W), thence:
Joining Central Route (7.232) in either 5°48′N, 150°00′W (**E**) or 5°59′N, 140°00′W (**F**), thence:
Via Central Route to Gulf of Panama (8°00′N, 79°00′W).

Distance   8390 miles (See also Note at 7.234.3).

Panama ⇒ Sydney

Diagram 7.236
Route
7.238

From Panama (Balboa) (8°53′N, 79°30′W) the route is as described at 7.236 in reverse to a position 5 miles N of Three Kings Islands (34°10′S, 172°00′E), thence by great circle to Sydney (33°50′S, 151°19′E).

Distance   7700 miles.

Brisbane ⇒ Panama

Diagram 7.239
Route
7.239

From Brisbane (Caloundra Head) (26°49′S, 153°10′E): there is a choice of routes:

7.239.1

Via Papeete (17°30′S, 149°36′W), as described at 7.62, thence:

To Panama (Balboa) (8°53′N, 79°30′W), as described at 7.209.

Distance   7820 miles.

7.239.2

Via Suva, as described at 7.65, and through Nanuku Passage (16°40′S, 179°10′W), thence:

To Panama (Balboa) (8°53′N, 79°30′W), as described at 7.234.1 or 7.234.3.

Distance   7860 miles.

7.239.3

Via Suva, as described at 7.65, and Lakeba Passage (17°50′S, 178°30′W); for cautions in this passage, see 7.78. Thence:

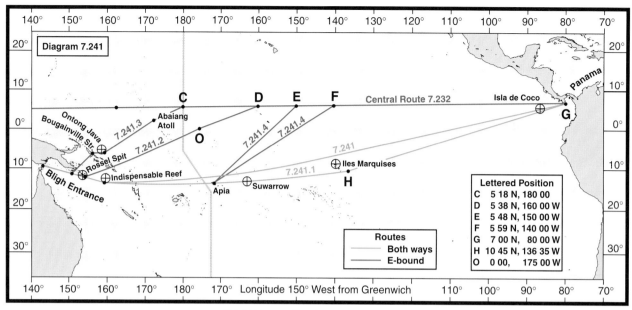

Diagram 7.241

7.241 Torres Strait ⟵ ⟶ Panama

To Panama (Balboa) (8°53′N, 79°30′W), as described at 7.234.2.

Distance 7840 miles.

7.239.4

Great circle to 22°05′S, 175°00′E (**N**), 30 miles SE of Ceva-i-Ra, thence:

Rhumb line to 19°50′S, 180°00′ (**M**), thence:

Joining Central Route (7.232) in either 5°48′N, 150°00′W (**E**) or 5°59′N, 140°00′W (**F**), thence:

Via Central Route to Gulf of Panama (8°00′N, 79°00′W).

Distance 8180 miles (See also Note at 7.234.3).

Panama ⇒ Brisbane

Diagram 7.239
Route
7.240

From Panama (Balboa) (8°53′N, 79°30′W) the route is:

As described at 7.236 in reverse as far as 30°00′S, 150°00′W (**J**), thence:

By rhumb line to Caloundra Head (26°49′S, 153°10′E), passing N of Kermadec Islands (30°30′S, 178°30′W).

Distance 7740 miles.

Torres Strait ⇐ ⇒ Panama

Diagram 7.241
Route
7.241

From Bligh Entrance (9°12′S, 144°00′E,), at the E end of the Torres Strait there are several routes to Panama. The great circle track between 13°10′S, 160°00′E and Gulf of Panama (8°00′N, 79°00′W) is encumbered with dangers between Indispensable Reefs (12°30′S, 160°25′E) and Îles Marquises (9°00′S, 140°00′W). The distance by this route, neglecting navigational diversions, is 8520 miles.

7.241.1

In either direction the recommended route is via Apia (13°47′S, 171°45′W) as described at 7.71 and 7.235.1. This route is comparatively free of navigational hazards.

Distance 8540 miles.

7.241.2

An East-bound route leads:

As navigation permits to 35 miles S of Rossel Spit (11°27′S, 154°24′E), thence:

Between Guadalcanal (9°40′S, 160°00′E) and San Cristóbal Islands (10°30′S, 161°45′E), thence:

Across the equator in 175°00′W (**O**), passing between Gilbert Group and Tuvalu Group (8°00′S, 179°00′E), thence:

Rhumb line to join Central Route (7.232) in 5°38′N, 160°00′W (**D**), thence:

By Central Route to Gulf of Panama (8°00′N, 79°00′W).

Distance 8620 miles.

7.241.3

An alternative East-bound route leads:

Through Jomard Entrance (11°15′S, 152′06′E) or China Strait (10°35′S, 150°40′E), as described at 7.151.2 and 7.151.3, thence:

Through Bougainville Strait (6°40′S, 156°15′E), thence:

S of Ontong Java Group (5°30′S, 159°30′E), thence:

N of Abaiang Atoll (1°58′N, 172°50′E), thence:

Joining Central Route (7.232) in 5°18′N, 180°00′ (**C**), thence:

By Central Route to Gulf of Panama (8°00′N, 79°00′W).

This route (using Jomard Entrance), though 90 miles longer than that at 7.241.2, will allow favourable weather and currents to be carried for an additional 1200 miles.

Distances:

Via China Strait 8620 miles.
Via Jomard Entrance 8710 miles.

7.241.4

Another alternative East-bound route leads via Apia (13°47′S, 171°45′W) as described at 7.71 and 7.235.2.

Yap ⇒ Panama

Diagram 7.242
Route
7.242

From Yap (9°28′N, 138°09′E) there are two alternative routes.

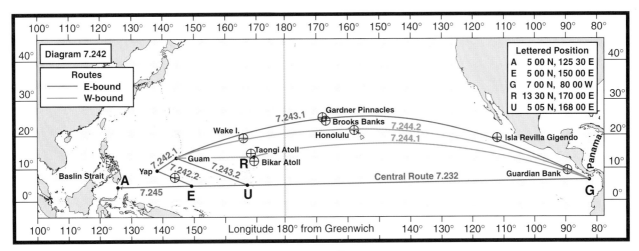

7.242 Yap → Panama
7.243 Guam → Panama
7.244 Panama → Guam

7.242.1

As navigation permits to Guam (13°27′N, 144°35′E), thence as detailed at 7.243.1 to Gulf of Panama (8°00′N, 79°00′W).

Distance   8450 miles.

7.242.2

Rhumb line to join the Central Route (7.232) in 5°00′N, 150°00′E (**E**), passing S of Woleai Islands (7°20′N, 143°50′E), thence to Gulf of Panama (8°00′N, 79°00′W).

Distance   8670 miles.

Guam ⇒ Panama

Diagram 7.242
Route
7.243

From Guam (13°27′N, 144°35′E) there are two alternative routes.

7.243.1

Great circle to Gulf of Panama (8°00′N, 79°00′W), passing:

Through Hawaiian Islands, between Gardner Pinnacles (25°00′N, 168°00′W) and Brooks Banks (24°10′N, 167°00′W), and:

Between Islas Revilla Gigendo (19°25′N, 110°30′W), and:

Avoiding Guardian Bank (9°30′N, 87°30′W).

See 7.38.1 for details of areas to be avoided near Hawaiian Islands.

Distance   8000 miles.

7.243.2

After rounding the S point of Guam by rhumb line to join the Central Route (7.232) in 5°05′N, 168°00′E (**U**), passing between Namorik Atoll (5°35′N, 168°08′E) and Ebon Atoll (4°38′N, 168°43′E).

Distance   8300 miles.

Panama ⇒ Guam

Diagram 7.242
Route
7.244

From Gulf of Panama (8°00′N, 79°00′W) there are two alternative routes.

7.244.1

After clearing Punta Mala (7°28′N, 80°00′W) the route is:

Great circle to 13°30′N, 170°00′E (**R**), thence:

To Guam (13°27′N, 144°35′E), passing between Bikar Atoll (12°15′N, 170°06′E) and Taongi Atoll (14°38′N, 168°59′E), paying due attention to the shoal reported midway between them.

On this route the North Equatorial Current (7.26) will be of advantage for most of the ocean crossing.

Distance   8090 miles.

7.244.2

An alternative route, or if refuelling is required, may be made via Honolulu (21°17′N, 157°53′W) as described at 7.200.1 and 7.190.1. This route is shorter, but slightly less favourable conditions may be encountered.

Distance   8040 miles.

Basilan Strait ⇒ Panama

Diagram 7.245
Route
7.245

From Basilan Strait (6°50′N, 122°00′E) routes are across the Celebes Sea to 5°00′N, 125°30′E (**A**), S of Mindanao, thence following Central Route (7.232) to Gulf of Panama (8°00′N, 79°00′W).

Distances:

From Basilan Strait	9580 miles.
From S of Mindanao	9360 miles.

San Bernadino Strait ⇒ Panama

Diagram 7.245
Route
7.246

From San Bernadino Strait (12°33′N, 124°12′E) there are alternative routes.

7.246.1

Great circle to join US Coastal Route (7.176) off Manzanillo in 20°00′N, 107°45′W (**S**), thence to Gulf of Panama (8°00′N, 79°00′W).

Distance   9090 miles.

7.246.2

Rhumb line to pass S of Palau Islands (7°30′N, 134°30′E), thence rhumb line to join Central Route (7.232)

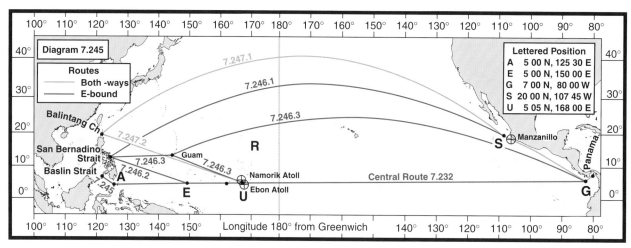

7.245 Basilan Strait → Panama
7.246 San Bernadino Strait → Panama
7.247 Balintang Channel → Panama

in 5°00′N, 150°00′E (**E**), thence to Gulf of Panama (8°00′N, 79°00′W).

Distance 9530 miles.

7.246.3

Rhumb line to Guam (13°27′N, 144°35′E), thence as described at 7.243.1 to Gulf of Panama (8°00′N, 79°00′W).

Distance 9200 miles.

Balintang Channel ⇐ ⇒ Panama

Diagram 7.245

Route

7.247

From Balintang Channel (20°00′N, 122°20′E) there are alternative routes.

7.247.1

Great circle to join US Coastal Route (7.176) off Manzanillo in 20°00′N, 107°45′W (**S**), passing NE of Rocas Alijos (24°57′N, 115°45′W), thence to Gulf of Panama (8°00′N, 79°00′W).

Distance 8960 miles.

7.247.2

Rhumb line to join Central Route (7.232) in 5°05′N, 168°00′E (**U**), passing close S of Guam (13°27′N,

144°35′E) and between Namorik Atoll (5°35′N, 168°08′E) and Ebon Atoll (4°38′N, 168°43′E), thence to Gulf of Panama (8°00′N, 79°00′W).

Distance 9620 miles.

Singapore ⇒ Panama

Diagram 7.248

Route

7.248

From Singapore (1°12′N, 103°51′E) there is a choice of several routes, each with different characteristics of depths, navigational hazards, shelter, weather, currents and bunkering facilities.

The principal routes are as follows (distances between ports include fuelling stops at the ports mentioned).

7.248.1

To Yokohama (35°26′N, 139°43′E), as described at 7.95, distance 2970 miles, thence:

To Gulf of Panama (8°00′N, 79°00′W), as described at 7.283, distance 7700 miles.

Total distance 10 670 miles.

7.248.2

To Yokohama (35°26′N, 139°43′E), as described at 7.95, distance 2970 miles, thence:

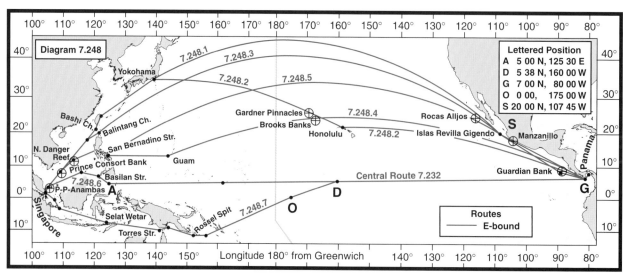

7.248 Singapore → Panama

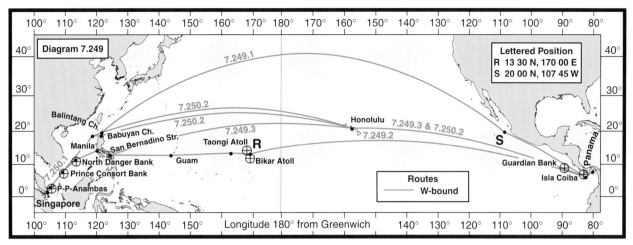

7.249 Panama → Verde Island Passage and Manila
7.250 Panama → Singapore

To Honolulu (21°17′N, 157°53′W), as described at 7.195, distance 3440 miles, thence:

To Gulf of Panama (8°00′N, 79°00′W), as described at 7.200.1, distance 4690 miles.

Total distance   11 100 miles.

7.248.3

To Balintang Channel (20°00′N, 122°20′E), as described at 7.96 in reverse, distance 1550 miles, thence:

To Gulf of Panama (8°00′N, 79°00′W), as described at 7.247.1, distance 8960 miles.

Total distance   10 510 miles.

7.248.4

Through South China Sea, Verde Island Passage and San Bernadino Strait, as described at 7.97 and 7.98, thence to Guam (13°27′N, 144°35′E), as described at 7.98, distance 2750 miles, thence:

To Gulf of Panama (8°00′N, 79°00′W), as described at 7.243.1, distance 8000 miles.

Total distance   10 750 miles.

7.248.5

To San Bernadino Strait (12°33′N, 124°12′E), as described at 7.97 and 7.98, distance 1530 miles, thence:

Great circle to join US Coastal Route (7.176) off Manzanillo in 20°00′N, 107°45′W **(S)**, thence to Gulf of Panama (8°00′N, 79°00′W), distance 9090 miles.

Total distance   10 630 miles.

7.248.6

To Balabac Strait (7°34′N, 116°55′E) and Basilan Strait (6°50′N, 122°00′E), as described at 7.101, distance 1210 miles, thence:

Across the Celebes Sea to 5°00′N, 125°30′E **(A)**, S of Mindanao, thence:

Following Central Route (7.232) to Gulf of Panama (8°00′N, 79°00′W), distance 9580 miles.

Total distance   10 790 miles.

7.248.7

Through Eastern Archipelago, via Selat Wetar (8°16′S, 127°12′E) to Torres Strait (10°36′S, 141°51′E), as described at 7.106, thence:

S of Rossel Spit (11°27′S, 154°24′E) and via Central Route (7.232) to Gulf of Panama (8°00′N, 79°00′W), as described at 7.241.2.

Total distance   11 170 miles.

Panama ⇒ Verde Island Passage and Manila

Diagram 7.249
Route
7.249

From Gulf of Panama (8°00′N, 79°00′W) there are alternative routes.

7.249.1

From Gulf of Panama the reverse of the route described at 7.247.1 to Balintang Channel (20°00′N, 122°20′E) offers the shortest distance to both Manila (9380 miles) and Singapore (10 150 miles), but that part of the route which follows a great circle from the US Coast Route off Manzanillo, from 20°00′N, 107°45′W **(S)**, reaches the parallel of 40°N in about 170°W. Adverse winds and currents can be expected along most of this part of the route throughout the year.

However a route farther S, but keeping N of the E-bound Central Route (7.232), will enable the North Equatorial Current (7.26) to be of advantage, probably with favourable weather, for most of the ocean crossing. If fuel is required it can be obtained at Guam (13°27′N, 144°35′E).

7.249.2

The recommended route therefore from Punta Mala (7°28′N, 80°00′W) is as described at 7.244, to pass close S of Guam, thence to San Bernadino Strait (12°33′N, 124°12′E), as described at 7.98. The routes from San Bernadino Strait to Manila (14°32′N, 120°56′E) and Verde Island Passage (13°36′N, 121°00′E) are described in *Philippine Islands Pilot*.

7.249.3

A shorter route, but on which slightly less favourable conditions may be expected, is one calling at Honolulu (21°17′N, 157°53′W), as described at 7.200.1 and 7.192.

7.249.4
Distance:

	Verde Island Passage	Manila
Recommended Route	9490	9580
Via Honolulu	9400	9490

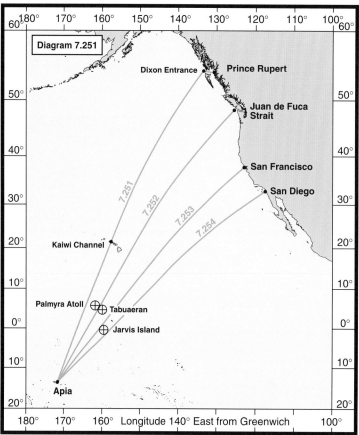

7.251 Apia ← → Prince Rupert
7.252 Apia ← → Juan de Fuca Strait
7.253 Apia ← → San Francisco
7.254 Apia ← → San Diego

Panama ⇒ Singapore

Diagram 7.249
Route
7.250

There are two alternative routes.

7.250.1

As described at 7.249 as far as Verde Island Passage (13°36′N, 121°00′E), continuing thence through the South China Sea, as described at 7.97, to Singapore (1°12′N, 103°51′E).

Distance   10 820 miles.

7.250.2

Via Honolulu, as described at 7.200.1 and 7.191.2.

Distance   10 720 miles.

Apia ⇐ ⇒ Prince Rupert

Diagram 7.251
Routes
7.251

From Apia (13°47′S, 171°45′W) the route is by great circle which passes through Kaiwi Channel (21°12′N, 157°40′W) to Dixon Entrance (54°30′N, 132°30′W), thence as navigation permits to Prince Rupert (54°19′N, 130°20′W).

Distance   4640 miles.

Apia ⇐ ⇒ Juan de Fuca Strait

Diagram 7.251
Routes
7.252

From Apia (13°47′S, 171°45′W) to Juan de Fuca Strait (48°30′N, 124°47′W) the route is by great circle passing between Palmyra Atoll (5°53′N, 162°05′W) and Tabuaeran (4°43′N, 160°25′W).

Distance   4490 miles.

Apia ⇐ ⇒ San Francisco

Diagram 7.251
Routes
7.253

From Apia (13°47′S, 171°45′W) to San Francisco (37°45′N, 122°40′W) the route is by great circle passing about 30 miles NW of Jarvis Island (0°23′S, 160°01′W).

Distance   4140 miles.

Apia ⇐ ⇒ San Diego

Diagram 7.251
Routes
7.254

From Apia (13°47′S, 171°45′W) to San Diego (32°38′N, 117°15′W) the route is by great circle.

Distance   4180 miles.

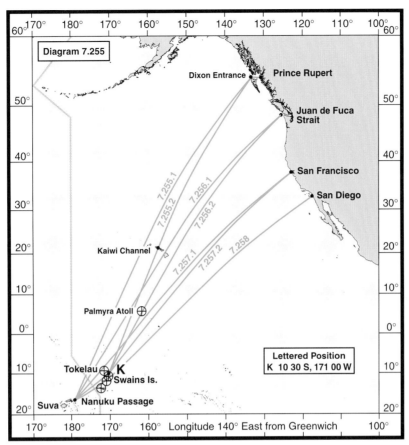

Diagram 7.255

Lettered Position
K 10 30 S, 171 00 W

7.255 Suva ← → Prince Rupert
7.256 Suva ← → Juan de Fuca Strait
7.257 Suva ← → San Francisco
7.258 Suva ← → San Diego

Suva ⇐ ⇒ Prince Rupert

Diagram 7.255
Routes
7.255

From Suva (18°11′S, 178°24′E) the routes are:
7.255.1

To all North American ports through Nanuku Passage
(16°40′S, 179°10′W), thence by great circle to destination.

These routes however pass through the islands and
dangers NE of the Fiji Islands (18°00′S, 180°00′) and
through Phoenix Group (4°00′S, 173°00′W) or Tokelau
(9°00′S, 172°00′W), which are all low and not easily
sighted.
7.255.2

An alternative route passing E of these islands leads:
    Through Nanuku Passage (16°40′S, 179°10′W),
      thence:
    40 miles NW of Savai'i (13°35′S, 172°30′W), thence:
    Through 10°30′S, 171°00′W (**K**), between Tokelau
      (9°00′S, 172°00′W) and Swains Island (11°00′S,
      171°05′W), thence:
    Great circle through Kaiwi Channel (21°12′N,
      157°40′W) to Dixon Entrance (54°30′N,
      132°30′W), thence:
    As navigation permits to Prince Rupert (54°19′N,
      130°20′W).
7.255.3
   Distances:

By great circle	5140 miles.
Alternative route	5230 miles.

Suva ⇐ ⇒ Juan de Fuca Strait

Diagram 7.255
Routes
7.256

From Suva (18°11′S, 178°24′E) to Juan de Fuca Strait
(48°30′N, 124°47′W) the routes are:
7.256.1

By great circle from Nanuku Passage (16°40′S,
179°10′W), but see 7.255.1.
7.256.2

An alternative route is:
    As at 7.255.2 to 10°30′S, 171°00′W (**K**), thence:
    By great circle, passing 30 miles SE of Palmyra Atoll
      (5°53′N, 162°05′W), to Juan de Fuca Strait.
7.256.3
   Distances:

By great circle	5030 miles.
Alternative route	5080 miles.

Suva ⇐ ⇒ San Francisco

Diagram 7.255
Routes
7.257

From Suva (18°11′S, 178°24′E) to San Francisco
(37°45′N, 122°40′W) the routes are:
7.257.1

By great circle from Nanuku Passage (16°40′S,
179°10′W), but see 7.255.1.

7.257.2

An alternative route is:

As at 7.255.2 to 10°30′S, 171°00′W **(K)**, thence:

By great circle, passing 10 miles S of Teraina (3°51′N, 159°22′W), to San Francisco.

7.257.3

Distances:

By great circle	4730 miles.
Alternative route	4750 miles.

Suva ⇐ ⇒ San Diego

Diagram 7.255

Routes

7.258

From Suva (18°11′S, 178°24′E) the route is:

Through Nanuku Passage (16°40′S, 179°10′W), thence:

As navigation permits to pass 40 miles S of Swains Island (11°00′S, 171°05′W), thence:

Great circle to San Diego (32°38′N, 117°15′W).

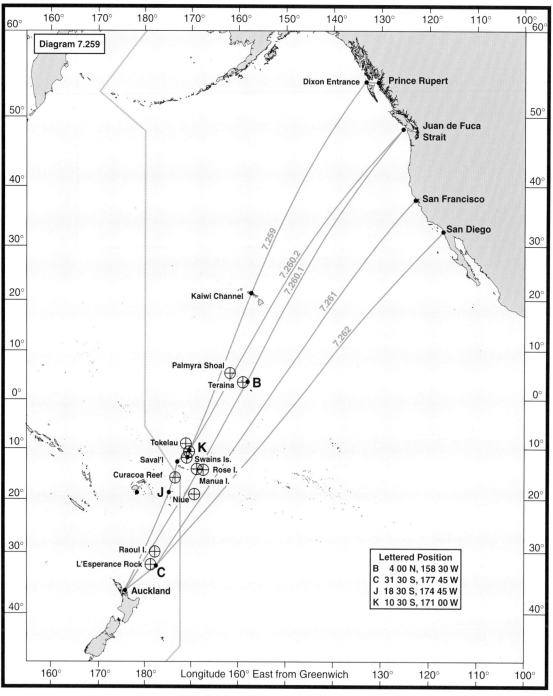

Diagram 7.259

Lettered Position
B   4 00 N, 158 30 W
C   31 30 S, 177 45 W
J   18 30 S, 174 45 W
K   10 30 S, 171 00 W

7.259 Auckland ← → Prince Rupert
7.260 Auckland ← → Juan de Fuca Strait
7.261 Auckland ← → San Francisco
7.262 Auckland ← → San Diego

7.258.1
Distance   4800 miles.

Auckland ⇐ ⇒ Prince Rupert

Diagram 7.259
Routes
7.259
    From Auckland (36°36′S, 174°49′E) the route is:
      Towards Honolulu (21°17′N, 157°53′W), as described
          at 7.186.1, as far as 10°30′S, 171°00′W **(K)**,
          thence:
      As described at 7.255.2 to Prince Rupert (54°19′N,
          130°20′W).
Distance   6190 miles.

Auckland ⇐ ⇒ Juan de Fuca Strait

Diagram 7.259
Routes
7.260
    From Auckland (36°36′S, 174°49′E) the alternative
routes are:
7.260.1
    Great circle to 4°00′N, 158°30′W **(B)**, passing:
      Thirty miles W of Raoul Island (29°15′S, 177°54′W),
          thence:
      Between Manua Islands (14°13′S, 169°35′E) and
          Rose Island (14°33′S, 168°09′W), thence:
    30 miles SE of Teraina (3°51′N, 159°22′W), thence:
    Great circle to Juan de Fuca Strait (48°30′N, 124°47′W).
    Distance   6040 miles.
7.260.2
      Towards Honolulu (21°17′N, 157°53′W), as described
          at 7.186.1, as far as 10°30′S, 171°00′W **(K)**,
          thence:
      As described at 7.256.2 to Juan de Fuca Strait
          (48°30′N, 124°47′W)
Distance   6050 miles.

Auckland ⇐ ⇒ San Francisco

Diagram 7.259
Routes
7.261
    From Auckland (36°36′S, 174°49′E) the route is:
      Great circle to 31°30′S, 177°45′W **(C)** (60 miles ESE
          of L'Esperance Rock), thence:
      Great circle to a position 80 miles SE of Niue
          (19°03′S, 169°51′E), thence:
      Great circle to San Francisco (37°45′N, 122°40′W).
Distance   5660 miles.

Auckland ⇐ ⇒ San Diego

Diagram 7.259
Routes
7.262
    From Auckland (36°36′S, 174°49′E) the route is by
great circle to San Diego (32°38′N, 117°15′W), taking care
to avoid Palmerston Island (18°05′S, 163°15′W) which lies
within 10 miles of the track.
    Distance   5650 miles.

Wellington ⇐ ⇒ Prince Rupert

Diagram 7.263
Routes
7.263
    From Wellington (41°22′S, 174°50′E) the routes pass E
of North Island and are:
7.263.1
      Towards Honolulu (21°17′N, 157°53′W), as described
          at 7.186.1, as far as 10°30′S, 171°00′W **(K)**,
          thence:
      As described at 7.255.2 to Prince Rupert (54°19′N,
          130°20′W).
7.263.2
      Coastwise until off Mahia Peninsula (39°10′S,
          177°55′E), thence:
      Rhumb line to the equator in 157°00′W **(D)**, passing
          about 30 miles SE of Antiope Reef (18°14′S,
          168°20′W), thence:
      Great circle to Dixon Entrance (54°30′N, 132°30′W),
          passing 40 miles E of Kiritimati Atoll (1°55′N,
          157°20′W), thence:
      As navigation permits to Prince Rupert (54°19′N,
          130°20′W).
Distance by either route   6520 miles.

Wellington ⇐ ⇒ Juan de Fuca Strait

Diagram 7.263
Routes
7.264
    From Wellington (41°22′S, 174°50′E) the routes pass E
of North Island and are as described at 7.263.2 as far as
crossing the equator in 157°00′W **(D)**, thence by great
circle to Juan de Fuca Strait (48°30′N, 124°47′W).
    Distance   6300 miles.

Wellington ⇐ ⇒ San Francisco

Diagram 7.263
Routes
7.265
    From Wellington (41°22′S, 174°50′E) the route to San
Francisco (37°45′N, 122°40′W) is by great circle passing E
of North Island.
    The track passes about 40 miles E of Tongareva (9°00′S,
158°03′W), Starbuck Island (5°37′S, 155°55′W) and
Malden Island (4°00′S, 154°55′W); for cautions on currents
near these islands, see *Pacific Islands Pilot, Volume III.*
    Distance   5870 miles.

Wellington ⇐ ⇒ San Diego

Diagram 7.263
Routes
7.266
    From Wellington (41°22′S, 174°50′E) the route to San
Diego (32°38′N, 117°15′W) is by great circle passing E of
North Island.
    The track passes:
      Within 10 miles of the reef reported in 24°25′S,
          163°39′W, and:
      Between Aitutaki (18°54′S, 159°47′W) and Manuae
          Islands (19°15′S, 158°55′W), and:
      20 miles SE of Vostok Island (10°06′S, 152°23′W).
Distance   5830 miles.

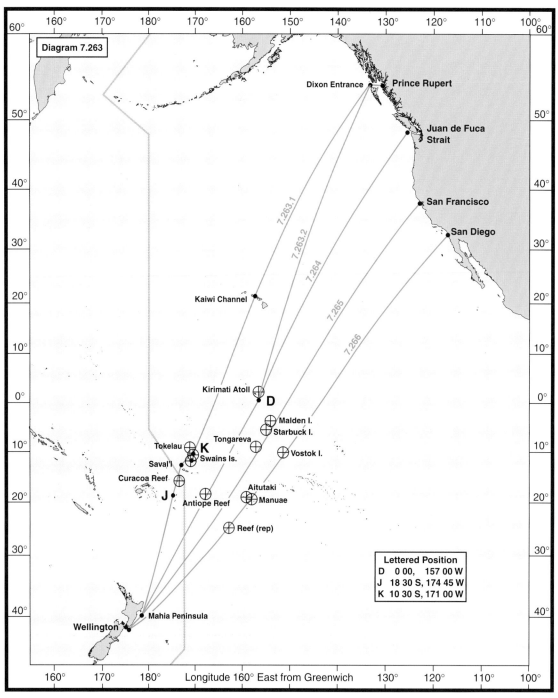

Diagram 7.263

7.263 Wellington ← → Prince Rupert
7.264 Wellington ← → Juan de Fuca Strait
7.265 Wellington ← → San Francisco
7.266 Wellington ← → San Diego

Sydney or Brisbane ⇐ ⇒ San Diego or San Francisco

Diagram 7.267

Routes

7.267

From these Australian ports the alternative routes are:

7.267.1

To Suva (18°11′S, 178°24′E), as described at 7.57 (from Sydney 33°50′S, 151°19′E) and 7.65 (from Brisbane 26°49′S, 153°10′E).

Thence from Suva to destination, as described at 7.257 (for San Francisco 37°45′N, 122°40′W) and 7.258 (for San Diego 32°38′N, 117°15′W).

7.267.2

To cross the equator at 178°50′W **(E)**, as described at 7.184.1 (from Sydney) and 7.184.2 (from Brisbane), thence by great circle to destination.

For San Francisco a diversion from the great circle must be made to pass through Alenhuihaha Channel (20°20′N, 156°00′W) between Hawaii and Maui.

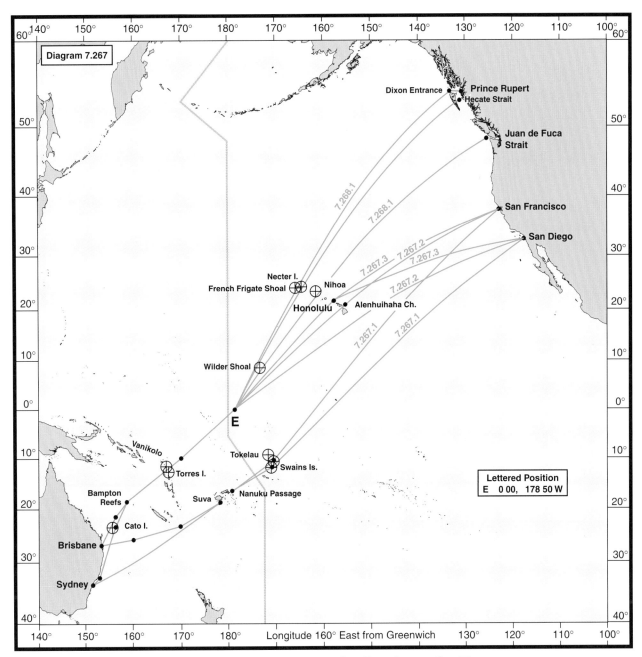

Diagram 7.267

7.267 Sydney or Brisbane ← → San Francisco or San Diego
7.268 Sydney or Brisbane ← → Juan de Fuca Strait or Prince Rupert

7.267.3

To Honolulu (21°17′N, 157°53′W) as described at 7.184.1 (from Sydney) and 7.184.2 (from Brisbane).

Thence from Honolulu to destination, as described at 7.199.

7.267.4

Distances:

		Sydney	Brisbane*
San Francisco	via 7.267.1	6460	6240
	via 7.267.2	6550	6150
	via 7.267.3	6570	6170
San Diego	via 7.267.1	6520	6300
	via 7.267.2	6700	6300
	via 7.267.3	6760	6360

* Distances are to Caloundra Head (26°49′S, 153°10′E).
Brisbane to Caloundra Head 35 miles.

Sydney or Brisbane ⇐ ⇒ Juan de Fuca Strait or Prince Rupert

Diagram 7.267

Routes

7.268

From these Australian ports the routes are:

7.268.1

Towards Honolulu (21°17′N, 157°53′W) as described at 7.184.1 (from Sydney) and 7.184.2 (from Brisbane), as far as the equator at 178°50′W (**E**), thence by great circle to destination. For areas to be avoided near the Hawaiian Islands, see 7.38.1.

For Juan de Fuca Strait (48°30′N, 124°47′W), the great circle passes close SE of the reported position of Wilder Shoal (8°16′N, 173°26′W) and through the Hawaiian Islands E of Nihoa (23°00′N, 162°00′W).

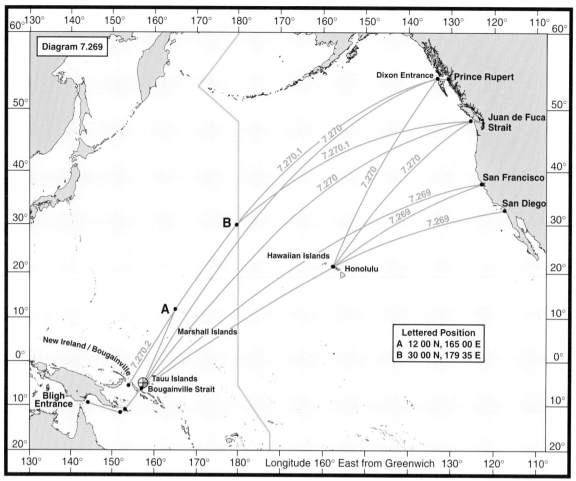

7.269 Torres Strait ← → San Francisco or San Diego
7.270 Torres Strait ← → Juan de Fuca Strait or Prince Rupert

For Hecate Strait (53°00′N, 131°00′W), the great circle passes E of Necker Island (23°35′N, 164°45′W).

For Dixon Entrance (54°30′N, 132°30′W), the great circle passes E of French Frigate Shoals (23°45′N, 166°10′W).

7.268.2

Refuelling calls, if required, can be made at:

Honolulu:

From Sydney (7.184.1) or Brisbane (7.184.2):

To Juan de Fuca Strait (7.199) or Prince Rupert (7.198).

Suva:

From Sydney (7.57) or Brisbane (7.65):

To Juan de Fuca Strait (7.256) or Prince Rupert (7.255).

7.268.3

Distances:

		Sydney	Brisbane*
Juan de Fuca Strait	Direct	6740	6340
	via Honolulu	6780	6380
	via Suva	6810	6590
Prince Rupert**	Direct	6780	6380
	via Honolulu	6890	6490
	via Suva	6960	6740

* Distances are to Caloundra Head (26°49′S, 153°10′E).

Brisbane to Caloundra Head 35 miles.

** Via Dixon Entrance

Torres Strait ⇐ ⇒ San Diego or San Francisco

Diagram 7.269

Routes

7.269

From Torres Strait (10°36′S, 141°51′E), via Bligh Entrance (9°12′S, 144°00′E), alternative routes lead to Bougainville Strait (6°40′S, 156°15′E), as described at 7.185.

From Bougainville Strait a direct great circle to San Diego (32°38′N, 117°15′W) passes close S of Honolulu (21°17′N, 157°53′W); it is clear of all charted dangers. Great circles from Bougainville Strait to ports N of San Diego pass through Marshall Islands and Hawaiian Islands.

Routes via Honolulu, as described at 7.185 and 7.199, are therefore usually preferred as being clear of dangers and inappreciably longer than the direct great circles.

7.269.1

Distances are from Torres Strait by routes passing E of Rossel Spit and through Bougainville Strait:

	Via Honolulu	Direct
San Diego	6580	6575
San Francisco	6390	6360

Torres Strait ⇐ ⇒ Juan de Fuca Strait or Prince Rupert

Diagram 7.269
Routes
7.270

From Torres Strait (10°36′S, 141°51′E), via Bligh Entrance (9°12′S, 144°00′E), alternative routes lead to Bougainville Strait (6°40′S, 156°15′E), as described at 7.185.

From Bougainville Strait direct great circles to Juan de Fuca Strait (48°30′N, 124°47′W) and ports farther N on the coast of Canada pass through Marshall Islands and Hawaiian Islands.

Therefore routes from Bougainville Strait by great circle to Honolulu (21°17′N, 157°53′W) as described at 7.185.1, which pass clear of charted dangers, thence by great circle to destination, are often used, particularly when W-bound.
7.270.1

Alternative routes passing NW of Marshall Islands and Hawaiian Islands are shorter than the routes via Honolulu and for an E-bound passage may encounter more favourable conditions.

From Bougainville Strait (6°40′S, 156°15′E) these routes are:

> By rhumb line to 20 miles E of Tauu Islands (5°25′S, 154°20′E), thence:
>
> Great circle to 12°00′N, 165°00′E (**A**), 20 miles NW of Bikini Atoll, thence:
>
> For Prince Rupert (54°19′N, 130°20′W) by great circle to Dixon Entrance (54°30′N, 132°30′W) thence as navigation permits.
>
> For Juan de Fuca Strait (48°30′N, 124°47′W) by the great circle route to Dixon Entrance, as above, as far as 30°00′N, 179°35′E (**B**), thence great circle to destination.

7.270.2

Alternatively passage can be made through the channel between New Ireland (3°20′S, 152°00′E) and Bougainville Island (6°00′S, 155°15′E), thence joining the route from Bougainville Strait in 12°00′N, 165°00′E (**A**), by great circle which passes about 20 miles SE of Pingelap Atoll (6°10′N, 160°55′E).
7.270.3

Distances are from Torres Strait by routes passing E of Rossel Spit and through Bougainville Strait or the channel between New Ireland and Bougainville Island:

	Juan de Fuca Strait	Prince Rupert
Via Honolulu	6580	6710
Via Bougainville Strait	6460	6320
Via New Ireland/Bougainville Channel	6530	6390

Guam or Yap ⇐ ⇒ North America

Diagram 7.271
Routes
7.271

From Guam (13°27′N, 144°35′E) or Yap (9°28′N, 138°09′E) routes pass through the Mariana Islands (17°00′N, 146°00′E) keeping as near as practical to the great circle tracks.
7.271.1
Distances:

	Guam	Yap
San Diego (32°38′N, 117°15′W)	5380	5830
San Francisco (37°45′N, 122°40′W)	5050	5500
Juan de Fuca Strait (48°30′N, 124°47′W)	4830	5260
Prince Rupert (54°19′N, 130°20′W)	4600	5010

7.271.2

For low-powered, or hampered vessels, W-bound in winter, the routes described at 10.132 may be preferred to avoid the worst of the weather and adverse current.

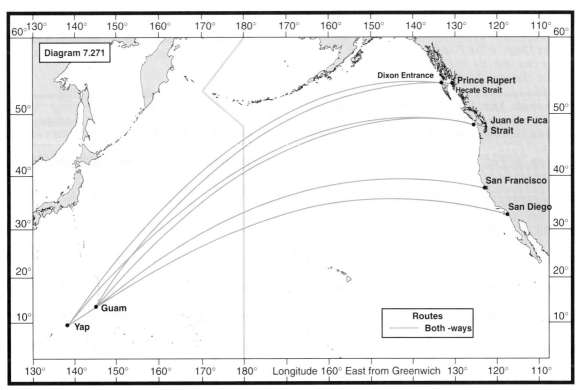

7.271 Guam or Yap ← → North America

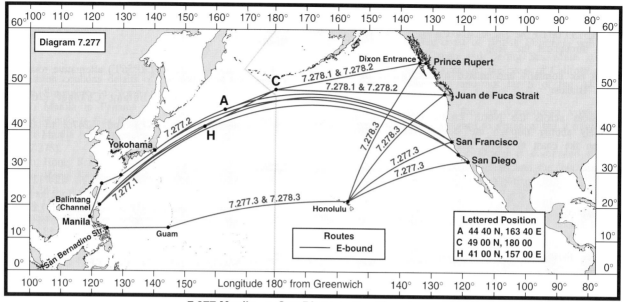

7.277 Manila → San Diego or San Francisco
7.278 Manila → Juan de Fuca Strait or Prince Rupert

Great circle to San Diego (32°38′N, 117°15′W) or San Francisco (37°45′N, 122°40′W).

Distances:

San Diego	6620 miles.
San Francisco	6240 miles.

7.277.2

To Yokohama (35°26′N, 139°43′E), as described at 7.163.1, thence:

As described at 7.284 to San Diego (32°38′N, 117°15′W) or San Francisco (37°45′N, 122°40′W).

Distances:

San Diego	6760 miles.
San Francisco	6360 miles.

7.277.3

To Guam (13°27′N, 144°35′E), as described at 7.98, thence:

To Honolulu (21°17′N, 157°53′W), as described at 7.190.1, thence:

By great circle to San Diego (32°38′N, 117°15′W) or San Francisco (37°45′N, 122°40′W).

Distances:

San Diego	7100 miles.
San Francisco	6910 miles.

Manila ⇒ Juan de Fuca Strait or Prince Rupert

Diagram 7.277

Routes

7.278

From Manila (14°32′N, 120°56′E) alternative routes are:

7.278.1

W of Luzon to Balintang Channel (20°00′N, 122°20′E), thence:

Great circle to 41°00′N, 157°00′E (**H**), thence:
Rhumb line to 49°00′N, 180°00′ (**C**), thence:
Rhumb line to Juan de Fuca Strait (48°30′N, 124°47′W) or Dixon Entrance (54°30′N, 132°30′W) for Prince Rupert (54°19′N, 130°20′W).

7.278.2

To Yokohama (35°26′N, 139°43′E), as described at 7.163.1, thence:

Great circle to 44°40′N, 163°40′E (**A**), thence:
Rhumb line to 49°00′N, 180°00′ (**C**), thence:
Rhumb line to Juan de Fuca Strait (48°30′N, 124°47′W) or Dixon Entrance (54°30′N, 132°30′W) for Prince Rupert (54°19′N, 130°20′W).

7.278.3

To Guam (13°27′N, 144°35′E), as described at 7.98, thence:

To Honolulu (21°17′N, 157°53′W), as described at 7.190.1, thence:

By great circle to Juan de Fuca Strait (48°30′N, 124°47′W) or Dixon Entrance (54°30′N, 132°30′W) for Prince Rupert (54°19′N, 130°20′W).

7.278.4

Distances:

	Juan de Fuca Strait	Prince Rupert
Via Balintang Channel	5910	5600
Via Yokohama	6040	5730
Via Guam and Honolulu	7120	7230

Hong Kong ⇒ Panama

Diagram 7.279

Routes

7.279

From Hong Kong (22°17′N, 114°10′E) alternative routes are:

7.279.1

As described at 7.164.2 to 34°40′N, 140°00′E (**K**), off Tōkyō Wan, thence:

By great circle to join the US Coastal Route (7.176) in 20°00′N, 107°45′W (**S**), thence:

To Gulf of Panama (8°00′N, 79°00′W).

Distance 9210 miles.

7.279.2

As navigation permits to Balintang Channel (20°00′N, 122°20′E) as described at 7.121, thence:

242

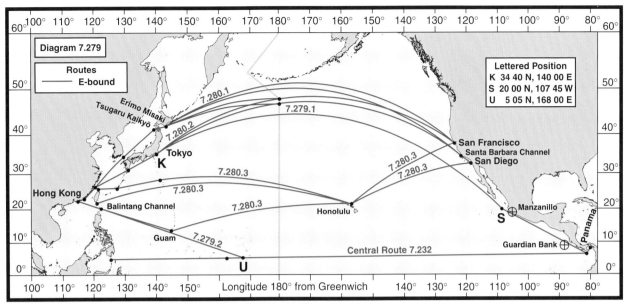

7.279 Hong Kong → Panama
7.280 Hong Kong → San Diego or San Francisco

Rhumb line to join Central Route (7.232) in 5°05′N, 168°00′E **(U)**, as described at 7.247.2, thence:
To Gulf of Panama (8°00′N, 79°00′W).
Distance 10 090 miles.

Hong Kong ⇒ San Diego or San Francisco

Diagram 7.279
Routes
7.280
From Hong Kong (22°17′N, 114°10′E) alternative routes are:
7.280.1
To Tsugaru Kaikyō (41°39′N, 140°48′E) passing through Korea Strait, as described at 7.164.3, thence:
As described at 7.288.1 or 7.288.2 to San Diego (32°38′N, 117°15′W) or San Francisco (37°45′N, 122°40′W).
Caution, These routes pass near to Russian Regulated Areas described at 7.274.7.

7.280.2
As described at 7.164.2 to 34°40′N, 140°00′E **(K)**, off Tōkyō Wan, thence:
By great circle to San Diego (32°38′N, 117°15′W) or San Francisco (37°45′N, 122°40′W).
The highest latitude on the route to San Francisco occurs at 47°30′N, 170°00′W.
7.280.3
A favourable current may be expected over most parts of the above routes, but the vessel will be exposed to the weather of the North Pacific Ocean.
Better weather but less favourable currents are likely to be experienced on the longer routes via Honolulu (7.193 and 7.199) or via Guam and Honolulu (7.121, 7.190 and 7.199).

7.280.4
Distances:

	San Diego	San Francisco
Via Tsugaru Kaikyō	6450	6050
Via 34°40′N, 140°00′E	6460	6060
Via Honolulu	7140	6950
Via Guam and Honolulu	7430	7240

Hong Kong ⇒ Juan de Fuca Strait or Prince Rupert

Diagram 7.281
Routes
7.281
From Hong Kong (22°17′N, 114°10′E) alternative routes are:
7.281.1
To Tsugaru Kaikyō (41°39′N, 140°48′E) passing through Korea Strait, as described at 7.164.3, thence:
To Erimo Misaki (41°56′N, 143°14′E). thence:
By rhumb line to Juan de Fuca Strait (48°30′N, 124°47′W) or Dixon Entrance (54°30′N, 132°30′W) for Prince Rupert (54°19′N, 130°20′W).
Caution, These routes pass near to Russian Regulated Areas described at 7.274.7.
7.281.2
As described at 7.164.2 to 34°40′N, 140°00′E **(K)**, off Tōkyō Wan, thence:
Great circle to 44°40′N, 163°40′E **(A)**, thence:
Rhumb line to 49°00′N, 180°00′ **(C)**, thence:
By rhumb line to Juan de Fuca Strait (48°30′N, 124°47′W) or Dixon Entrance (54°30′N, 132°30′W) for Prince Rupert (54°19′N, 130°20′W).

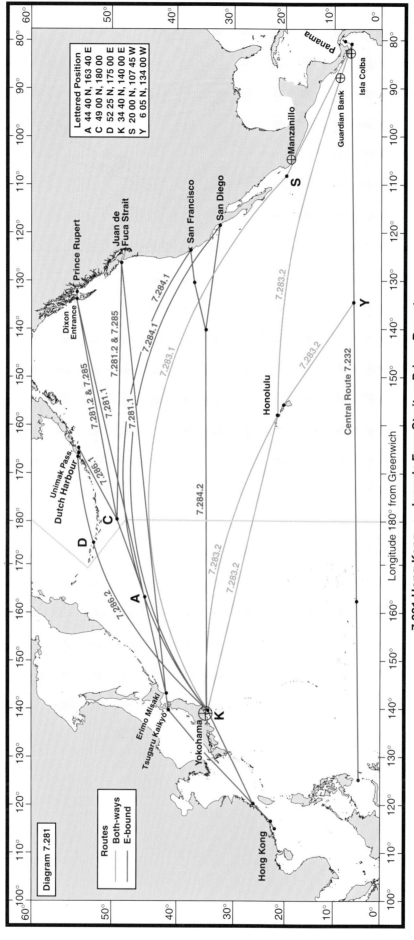

Diagram 7.281

Routes
Both-ways
E-bound

Lettered Position
A 44 40 N, 163 40 E
C 49 00 N, 180 00
D 52 25 N, 175 00 E
K 34 40 N, 140 00 E
S 20 00 N, 107 45 W
Y 6 05 N, 134 00 W

7.281 Hong Kong → Juan de Fuca Strait or Prince Rupert
7.283 Yokohama → Panama
7.284 Yokohama → San Diego or San Francisco
7.285 Yokohama → Juan de Fuca Strait or Prince Rupert
7.286 Yokohama → Dutch Harbour

Torres Strait ⇐ ⇒ Juan de Fuca Strait or Prince Rupert

Diagram 7.269
Routes
7.270

From Torres Strait (10°36'S, 141°51'E), via Bligh Entrance (9°12'S, 144°00'E), alternative routes lead to Bougainville Strait (6°40'S, 156°15'E), as described at 7.185.

From Bougainville Strait direct great circles to Juan de Fuca Strait (48°30'N, 124°47'W) and ports farther N on the coast of Canada pass through Marshall Islands and Hawaiian Islands.

Therefore routes from Bougainville Strait by great circle to Honolulu (21°17'N, 157°53'W) as described at 7.185.1, which pass clear of charted dangers, thence by great circle to destination, are often used, particularly when W-bound.
7.270.1

Alternative routes passing NW of Marshall Islands and Hawaiian Islands are shorter than the routes via Honolulu and for an E-bound passage may encounter more favourable conditions.

From Bougainville Strait (6°40'S, 156°15'E) these routes are:

By rhumb line to 20 miles E of Tauu Islands (5°25'S, 154°20'E), thence:

Great circle to 12°00'N, 165°00'E (**A**), 20 miles NW of Bikini Atoll, thence:

For Prince Rupert (54°19'N, 130°20'W) by great circle to Dixon Entrance (54°30'N, 132°30'W) thence as navigation permits.

For Juan de Fuca Strait (48°30'N, 124°47'W) by the great circle route to Dixon Entrance, as above, as far as 30°00'N, 179°35'E (**B**), thence great circle to destination.
7.270.2

Alternatively passage can be made through the channel between New Ireland (3°20'S, 152°00'E) and Bougainville Island (6°00'S, 155°15'E), thence joining the route from Bougainville Strait in 12°00'N, 165°00'E (**A**), by great circle which passes about 20 miles SE of Pingelap Atoll (6°10'N, 160°55'E).
7.270.3

Distances are from Torres Strait by routes passing E of Rossel Spit and through Bougainville Strait or the channel between New Ireland and Bougainville Island:

	Juan de Fuca Strait	Prince Rupert
Via Honolulu	6580	6710
Via Bougainville Strait	6460	6320
Via New Ireland/Bougainville Channel	6530	6390

Guam or Yap ⇐ ⇒ North America

Diagram 7.271
Routes
7.271

From Guam (13°27'N, 144°35'E) or Yap (9°28'N, 138°09'E) routes pass through the Mariana Islands (17°00'N, 146°00'E) keeping as near as practical to the great circle tracks.
7.271.1
Distances:

	Guam	Yap
San Diego (32°38'N, 117°15'W)	5380	5830
San Francisco (37°45'N, 122°40'W)	5050	5500
Juan de Fuca Strait (48°30'N, 124°47'W)	4830	5260
Prince Rupert (54°19'N, 130°20'W)	4600	5010

7.271.2

For low-powered, or hampered vessels, W-bound in winter, the routes described at 10.132 may be preferred to avoid the worst of the weather and adverse current.

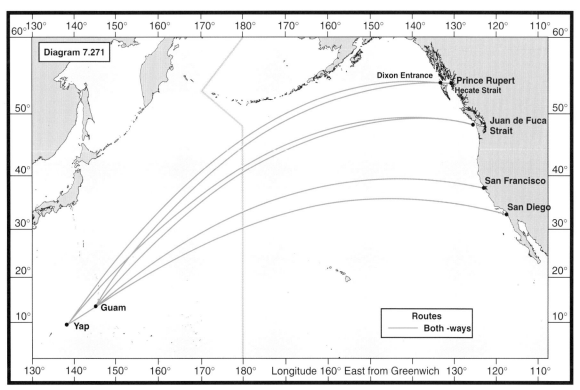

7.271 Guam or Yap ← → North America

NORTH PACIFIC TRANS-OCEAN PASSAGES

GENERAL INFORMATION

Coverage
7.272

This section contains details of the following passages:

From Singapore to North America (7.275).

From Manila to Panama (7.276).

From Manila to San Diego or San Francisco (7.277).

From Manila to Juan de Fuca Strait or Prince Rupert (7.278).

From Hong Kong to Panama (7.279).

From Hong Kong to San Diego or San Francisco (7.280).

From Hong Kong to Juan de Fuca Strait or Prince Rupert (7.281).

From Shanghai to North America (7.282).

To and from Yokohama and Panama (7.283).

From Yokohama to San Diego or San Francisco (7.284).

From Yokohama to Juan de Fuca Strait or Prince Rupert (7.285).

From Yokohama to Dutch Harbour (7.286).

From Tsugaru Kaikyō to Panama (7.287).

From Tsugaru Kaikyō to San Diego or San Francisco (7.288).

From Tsugaru Kaikyō to Juan de Fuca Strait or Prince Rupert (7.289).

From Tsugaru Kaikyō to Dutch Harbour (7.290).

From Nakhodka to Dutch Harbour and North America (7.291).

From Panama to Hong Kong or Shanghai (7.292).

From San Diego or San Francisco to Manila (7.293).

From San Diego or San Francisco to Singapore (7.294).

From San Diego or San Francisco to Ports on the coast of China S of Fuzhou (7.295).

From San Diego or San Francisco to Ports on the coast of China N of Fuzhou (7.296).

From San Diego or San Francisco to Yokohama (7.297).

From San Diego or San Francisco to Ports in the East China Sea and Bo Hai (7.298).

From San Diego or San Francisco to Tsugaru Kaikyō (7.299).

From Juan de Fuca Strait to Manila or Singapore (7.300).

From Juan de Fuca Strait to Ports on the coast of China (7.301).

From Juan de Fuca Strait to Yokohama (7.302).

From Juan de Fuca Strait to Tsugaru Kaikyō (7.303).

From Juan de Fuca Strait to Nakhodka (7.304).

From Prince Rupert to Manila or Singapore (7.305).

From Prince Rupert to Ports on the coast of China (7.306).

From Prince Rupert to Yokohama (7.307).

From Prince Rupert to Tsugaru Kaikyō (7.308).

From Prince Rupert to Nakhodka (7.309).

From Dutch Harbour to Yokohama or Tsugaru Kaikyō (7.310).

To and from Dutch Harbour and Australian ports or Torres Strait (7.311).

Route selection
7.273

1 Broadly speaking, the trend of the coastline bordering the North Pacific basin follows the arc of a great circle. In fact a great circle drawn between a position in Luzon Strait and a position on the coast of British Columbia will pass through the Sea of Japan and Bering Sea while a great circle between Luzon Strait and the coast of California will pass close to Yokohama and not far S of the Aleutian Islands.

2 A high-latitude route for the trans-ocean voyage is therefore attractive with regard to distance, but it may have disadvantages in weather and currents.

3 Weather in the N Pacific Ocean is dominated by the high pressure system over the ocean and the low pressure system which moves along the Aleutian chain of islands.

4 In summer, calm and clear conditions prevail over the E part of the N Pacific Ocean N of 40°N. Depressions, however, less frequent than in winter, move across the N part of the ocean bringing extensive fog for much of the time along the W part of the northern routes.

5 In winter, the Aleutian low pressure system intensifies and moves W from the vicinity of Bristol Bay (57°30′N, 160°30′W) to the W part of the Aleutian Islands near the date line. Violent storms sweep from China and Japanese waters towards the centre of the depression and then into the Gulf of Alaska, while storms from the central Pacific move NE towards the Gulf of Alaska. These storms bring rain, sleet, snow and violent winds to many of the northern routes.

6 In spring, the E coast of Japan is fully exposed to the strong E gales then prevalent.

7 General notes on winds, weather, currents and ice can be found at 7.3 – 7.11; 7.25 – 7.26; 7.30 – 7.32. For details of swell conditions, see 7.19 – 7.21.

8 East-bound the choice of route depends mainly on currents likely to be encountered and navigational requirements.

9 West-bound it may be preferable to take a route N of the Aleutian Islands or, alternatively, one well S of the northern routes, based on the parallel of 35°N, or even farther S, compromising between extra distance and the disadvantages of adverse winds and currents.

10 If using the S route, calls can be made if necessary for refuelling at Honolulu or Guam, without adding greatly to the distance.

Routes north of the Aleutian Islands
7.274

Vessels following routes N of the Aleutian Islands will be affected by the following conditions:

7.274.1

Bering Sea is N of the usual track of storms that sweep across the North Pacific Ocean. A W-bound vessel N of the islands will therefore be in the favourable semi-circle of most of these storms and so experience following winds and sea.

Local storms are frequent, particularly in Autumn, but the weather is characterised by persistently overcast skies, rapid change and instability, rather than by the violence of the winds.

7.274.2

Fog is prevalent in spring, summer and early autumn particularly off Kuril'skiye Ostrova (48°00′N, 153°00′E). Close to the Aleutian Islands conditions are generally more favourable on their N side rather than on their S side.

7.274.3

Currents are weak N of the Aleutian Islands, being strongest in summer, when they are E-going. Along the S side of the islands the Alaska Current (7.26) flows to the

W throughout the year. Off the E side of Poluostrov Kamchatskiy (56°30′N, 159°00′E) and Kuril'skiye Ostrova (48°00′N, 153°00′E) the Kamchatka Current (7.26) sets to the SW throughout the year.

7.274.4

Ice is not normally encountered in the vicinity of the Aleutian Islands.

7.274.5

Distances across the North Pacific Ocean are only appreciably shorter through the Bering Sea if between places on the coast of North America and places N of Japan.

7.274.6

Channels separating the Aleutian Islands should be transited with caution; for descriptions of these channels see *Bering Sea and Strait Pilot*.

7.274.7

Russian Regulated Areas exist in this region; for details see *Bering Sea and Strait Pilot*.

PASSAGES

Singapore ⇒ North America

Routes

7.275

From Singapore (1°12′N, 103°51′E) to Yokohama (35°26′N, 139°43′E), as described at 7.95, which is near the great circle joining Singapore with positions on the North American coast and a convenient fuelling point.

| | Distance | 2970 miles. |

7.275.1

From Yokohama the onward routes are as follows:

	Route	Distances*
Dutch Harbour	7.286	5740 (via Unimak Pass)
		5530 (N of Aleutians)
Prince Rupert	7.285	6860
Juan de Fuca Strait	7.285	7170
San Francisco	7.284	7500
San Diego	7.284	7900

* Distances in miles from Singapore

Manila ⇒ Panama

Diagram 7.276

Routes

7.276

From Manila (14°32′N, 120°56′E) alternative routes are:

7.276.1

W of Luzon to Balintang Channel (20°00′N, 122°20′E), thence:

As described at 7.247.1 and 7.247.2 to Gulf of Panama (8°00′N, 79°00′W).

| Distance | 9380 miles. |

7.276.2

To San Bernadino Strait (12°33′N, 124°12′E), as described in *Philippine Islands Pilot*, thence:

Via Central Route (7.232) as described at 7.246.2 to Gulf of Panama (8°00′N, 79°00′W).

| Distance | 9820 miles. |

7.276.3

To Yokohama (35°26′N, 139°43′E), as described at 7.163.1, thence:

By great circle to join the US Coastal Route (7.176) in 20°00′N, 107°45′W (**S**), thence:

To Gulf of Panama (8°00′N, 79°00′W).

| Distance | 9530 miles. |

7.276.4

To Guam (13°27′N, 144°35′E), as described at 7.98, thence:

To Honolulu (21°17′N, 157°53′W), as described at 7.190.1, thence:

To Gulf of Panama (8°00′N, 79°00′W), as described at 7.200.1 or 7.200.2.

| Distance | 9520 miles. |

Manila ⇒ San Diego or San Francisco

Diagram 7.277

Routes

7.277

From Manila (14°32′N, 120°56′E) alternative routes are:

7.277.1

W of Luzon to Balintang Channel (20°00′N, 122°20′E), thence:

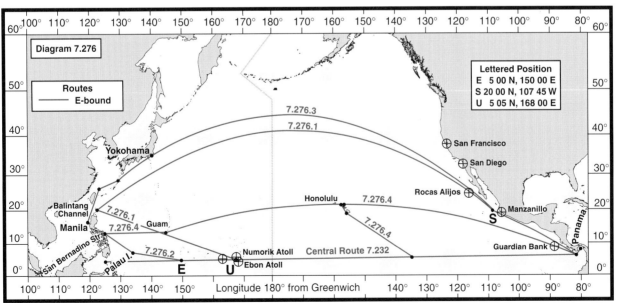

7.276 Manila → Panama

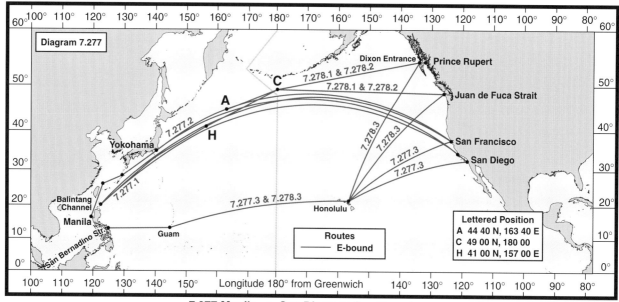

7.277 Manila → San Diego or San Francisco
7.278 Manila → Juan de Fuca Strait or Prince Rupert

Great circle to San Diego (32°38′N, 117°15′W) or San Francisco (37°45′N, 122°40′W).

Distances:

San Diego	6620 miles.
San Francisco	6240 miles.

7.277.2

To Yokohama (35°26′N, 139°43′E), as described at 7.163.1, thence:

As described at 7.284 to San Diego (32°38′N, 117°15′W) or San Francisco (37°45′N, 122°40′W).

Distances:

San Diego	6760 miles.
San Francisco	6360 miles.

7.277.3

To Guam (13°27′N, 144°35′E), as described at 7.98, thence:

To Honolulu (21°17′N, 157°53′W), as described at 7.190.1, thence:

By great circle to San Diego (32°38′N, 117°15′W) or San Francisco (37°45′N, 122°40′W).

Distances:

San Diego	7100 miles.
San Francisco	6910 miles.

Manila ⇒ Juan de Fuca Strait or Prince Rupert

Diagram 7.277

Routes

7.278

From Manila (14°32′N, 120°56′E) alternative routes are:

7.278.1

W of Luzon to Balintang Channel (20°00′N, 122°20′E), thence:

Great circle to 41°00′N, 157°00′E (**H**), thence:

Rhumb line to 49°00′N, 180°00′ (**C**), thence:

Rhumb line to Juan de Fuca Strait (48°30′N, 124°47′W) or Dixon Entrance (54°30′N, 132°30′W) for Prince Rupert (54°19′N, 130°20′W).

7.278.2

To Yokohama (35°26′N, 139°43′E), as described at 7.163.1, thence:

Great circle to 44°40′N, 163°40′E (**A**), thence:

Rhumb line to 49°00′N, 180°00′ (**C**), thence:

Rhumb line to Juan de Fuca Strait (48°30′N, 124°47′W) or Dixon Entrance (54°30′N, 132°30′W) for Prince Rupert (54°19′N, 130°20′W).

7.278.3

To Guam (13°27′N, 144°35′E), as described at 7.98, thence:

To Honolulu (21°17′N, 157°53′W), as described at 7.190.1, thence:

By great circle to Juan de Fuca Strait (48°30′N, 124°47′W) or Dixon Entrance (54°30′N, 132°30′W) for Prince Rupert (54°19′N, 130°20′W).

7.278.4

Distances:

	Juan de Fuca Strait	Prince Rupert
Via Balintang Channel	5910	5600
Via Yokohama	6040	5730
Via Guam and Honolulu	7120	7230

Hong Kong ⇒ Panama

Diagram 7.279

Routes

7.279

From Hong Kong (22°17′N, 114°10′E) alternative routes are:

7.279.1

As described at 7.164.2 to 34°40′N, 140°00′E (**K**), off Tōkyō Wan, thence:

By great circle to join the US Coastal Route (7.176) in 20°00′N, 107°45′W (**S**), thence:

To Gulf of Panama (8°00′N, 79°00′W).

Distance   9210 miles.

7.279.2

As navigation permits to Balintang Channel (20°00′N, 122°20′E) as described at 7.121, thence:

7.281.3
Distances:

	Juan de Fuca Strait	Prince Rupert
Via Tsugaru Kaikyō	5720	5410
Via 34°40′N, 140°00′E	5730	5420

Shanghai ⇒ North America

Routes
7.282
From Shanghai (31°03′N, 122°20′E) the shortest route is through the Korea Strait to Tsugaru Kaikyō (41°39′N, 140°48′E) thence by one of the following routes:

	Route	Distances
Prince Rupert	7.289	4770
Juan de Fuca Strait	7.289	5080
San Francisco	7.288.1 or 7.288.2	5400
San Diego	7.288.1 or 7.288.2	5810
Panama	7.287	8590

Yokohama ⇐ ⇒ Panama

Diagram 7.281
Routes
7.283
From Yokohama (35°26′N, 139°43′E) alternative routes are:
7.283.1
Great circle to to join the US Coastal Route (7.176), in 20°00′N, 107°45′W **(S)**, thence:
To Gulf of Panama (8°00′N, 79°00′W).
Distance   7700 miles.
7.283.2
Via Honolulu (21°17′N, 157°53′W), as described at 7.195 and 7.200, particularly if W-bound.
This avoids the generally E-going current which can be expected on the great circle route, carries the probability of better weather and affords the opportunity of fuelling midway.
Distance   8090 miles.

Yokohama ⇒ San Diego or San Francisco

Diagram 7.281
Routes
7.284
Routes from Yokohama (35°26′N, 139°43′E) are:
7.284.1
Direct by great circle to San Diego (32°38′N, 117°15′W) or San Francisco (37°45′N, 122°40′W), with the highest latitude, on the route to San Francisco, of 47°30′N, 170°00′W.
Distances:
San Diego   4930 miles.
San Francisco   4530 miles.
7.284.2
To avoid bad weather, an alternative route is as described at 7.297.2 in reverse.
Distances:
San Diego   5160 miles.
San Francisco   4850 miles.

Yokohama ⇒ Juan de Fuca Strait or Prince Rupert

Diagram 7.281
Routes
7.285
Routes from Yokohama (35°26′N, 139°43′E) are:
By great circle to 44°40′N, 163°40′E **(A)**, thence:
Rhumb line to 49°00′N, 180°00′ **(C)**, thence:
Rumb line to Juan de Fuca Strait (48°30′N, 124°47′W) or Dixon Entrance (54°30′N, 132°30′W) for Prince Rupert (54°19′N, 130°20′W).
Distances:
Juan de Fuca Strait   4200 miles.
Prince Rupert   3890 miles.

Yokohama ⇒ Dutch Harbour

Diagram 7.281
Routes
7.286
From Yokohama (35°26′N, 139°43′E) alternative routes are S or N of the Aleutian Islands:
7.286.1
South of the Aleutian Islands:
By great circle to 44°40′N, 163°40′E **(A)**, thence:
Rhumb line to 49°00′N, 180°00′ **(C)**, thence:
Through Unimak Pass (54°15′N, 164°30′W), or if conditions are suitable through Akutan Pass (54°05′N, 166°00′W), to Dutch Harbour (53°56′N, 166°29′W).
For descriptions of channels between the Aleutian Islands, see *Bering Sea and Strait Pilot*.
Distance (via Unimak Pass)   2770 miles.
7.286.2
North of the Aleutian Islands:
By great circle to 52°25′N, 175°00′E **(D)**, between Near Islands and Rat Islands, thence:
Great circle to Dutch Harbour (53°56′N, 166°29′W).
Caution, These routes pass near to Russian Regulated Areas described at 7.274.7.
Distance   2560 miles.

Tsugaru Kaikyō ⇒ Panama

Diagram 7.287
Routes
7.287
From Tsugaru Kaikyō (41°39′N, 140°48′E), after clearing Erimo Misaki (41°56′N, 143°14′E), the route is:
By great circle to 28°40′N, 118°20′W **(E)**, to the S of Isla da Guadalupe (29°00′N, 118°17′W), thence:
Rhumb line to join the US Coastal Route (7.176), in 20°00′N, 107°45′W **(S)**, thence:
To Gulf of Panama (8°00′N, 79°00′W).
Distance   7420 miles.

Tsugaru Kaikyō ⇒ San Diego or San Francisco

Diagram 7.287
Routes
7.288
From Tsugaru Kaikyō (41°39′N, 140°48′E) passages may be made either by direct great circles, or by tracks farther S where a favourable current is more likely.
7.288.1
Direct routes.
After clearing Erimo Misaki (41°56′N, 143°14′E), the routes are:

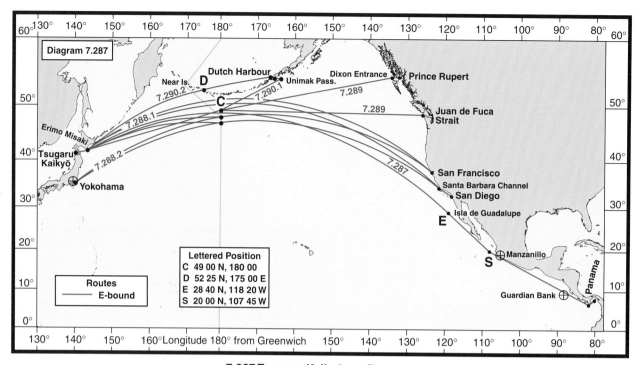

7.287 Tsugaru Kaikyō ➞ Panama
7.288 Tsugaru Kaikyō ➞ San Diego or San Francisco
7.289 Tsugaru Kaikyō ➞ Juan de Fuca Strait or Prince Rupert
7.290 Tsugaru Kaikyō ➞ Dutch Harbour

For San Diego (32°38′N, 117°15′W):

By great circle to the entrance to Santa Barbara Channel (34°15′N, 120°00′W), passing about 70 miles S of the Aleutian Islands.

For San Francisco (37°45′N, 122°40′W):

By great circle, passing about 30 miles S of the Aleutian Islands.

Caution, These routes pass near to Russian Regulated Areas described at 7.274.7.

Distances:

San Diego	4640 miles.
San Francisco	4230 miles.

7.288.2

Routes farther South.

From Erimo Misaki (41°56′N, 143°14′E), the routes are by either rhumb line or great circle to join the direct great circle routes from Yokohama (7.284.1):

For San Diego at 46°50′N, 180°00′:

For San Francisco at 47°50′N, 180°00′.

Distances:

San Diego	4660 miles.
San Francisco	4250 miles.

Tsugaru Kaikyō ⇒ Juan de Fuca Strait or Prince Rupert

Diagram 7.287

Routes

7.289

From Tsugaru Kaikyō (41°39′N, 140°48′E), after clearing Erimo Misaki (41°56′N, 143°14′E), the routes are:

By rhumb line to 49°00′N, 180°00′ (**C**), thence:

By rhumb line to Juan de Fuca Strait (48°30′N, 124°47′W) or Dixon Entrance (54°30′N, 132°30′W) for Prince Rupert (54°19′N, 130°20′W).

Caution, These routes pass near to Russian Regulated Areas described at 7.274.7.

Distances:

Juan de Fuca Strait	3910 miles.
San Francisco	3600 miles.

Tsugaru Kaikyō ⇒ Dutch Harbour

Diagram 7.287

Routes

7.290

From Tsugaru Kaikyō (41°39′N, 140°48′E), after clearing Erimo Misaki (41°56′N, 143°14′E), alternative routes are S or N of the Aleutian Islands.

Caution, These routes pass near to Russian Regulated Areas described at 7.274.7.

7.290.1

South of the Aleutian Islands:

Either as direct as navigation permits to Dutch Harbour (53°56′N, 166°29′W).

Distance   2440 miles.

An alternative route, to hold a favourable current as long as possible, is:

Rhumb line to 49°00′N, 180°00′ (**C**), thence:

Through Unimak Pass (54°15′N, 164°30′W), or if conditions are suitable through Akutan Pass (54°05′N, 166°00′W), to Dutch Harbour (53°56′N, 166°29′W).

For descriptions of channels between the Aleutian Islands, see *Bering Sea and Strait Pilot*.

Distance (via Unimak Pass)   2480 miles.

7.290.2

North of the Aleutian Islands:

By great circle to 52°25′N, 175°00′E (**D**), between Near Islands and Rat Islands, thence:

Great circle to Dutch Harbour (53°56′N, 166°29′W).

Distance   2220 miles.

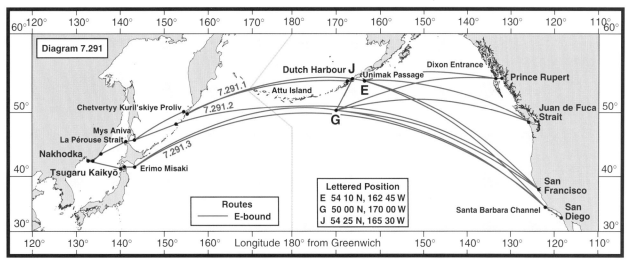

7.291 Nakhodka → Dutch Harbour and North America

Nakhodka ⇒ Dutch Harbour and North America

Diagram 7.291
Routes
7.291

From Nakhodka (42°48′N, 132°57′E) routes are either through La Pérouse Strait (45°45′N, 142°00′E) and then N or S of the Aleutian Islands, or through Tsugaru Kaikyō (41°39′N, 140°48′E).

For general considerations on routes N of Aleutian Islands, see 7.274.

7.291.1

North of the Aleutian Islands:

As navigation permits to the TSS through La Pérouse Strait and off Mys Aniva (46°02′N, 143°25′E), thence:

Through the TSS through Chetvertyy Kuril'skiye Proliv (50°00′N, 155°00′E), thence:

By great circle, passing N of Attu Island (52°55′N, 173°00′E), to Dutch Harbour (53°56′N, 166°29′W).

If bound for ports on the W coast of America:

Great circle to 54°25′N, 165°30′W (**J**), thence:

Through Unimak Pass (54°15′N, 164°30′W) to

54°10′N, 162°45′W (**E**), thence:

Great circle to destination.

7.291.2

South of the Aleutian Islands:

As navigation permits to the TSS through La Pérouse Strait and off Mys Aniva (46°02′N, 143°25′E), thence:

Great circle to Proliv Golovnina (48°12′N, 153°15′E), thence:

Great circle to 50°00′N, 173°00′W, thence:

Rhumb line to 50°00′N, 170°00′W (**G**), thence:

As navigation permits to Dutch Harbour (53°56′N, 166°29′W).

If bound for ports on the W coast of America:

By great circle to destination.

Caution, These routes pass near to Russian Regulated Areas described at 7.274.7.

7.291.3

Through Tsugaru Kaikyō:

If proceeding to San Francisco (37°45′N, 122°40′W) or ports farther S:

As navigation permits to Tsugaru Kaikyō (41°39′N, 140°48′E), thence:

As described at 7.288.1.

7.291.4

Distances:

	N of Aleutian Islands	S of Aleutian Islands	Through Tsugaru Kaikyō
Dutch Harbour (53°56′N, 166°29′W)	2520		
Prince Rupert (54°19′N, 130°20′W)	3780	3850	
Juan de Fuca Strait (48°30′N, 124°47′W)	4090	4120	
San Francisco (37°45′N, 122°40′W)	4550	4510	4610
San Diego (32°38′N, 117°15′W)	4960	4910	5020

Panama ⇒ Hong Kong or Shanghai

Diagram 7.292
Routes
7.292

From Panama (8°53′N, 79°30′W) the passage recommended is:

Via Guam (13°27′N, 144°35′E) as described at 7.244.1, via 13°30′N, 170°00′E (**R**); or:

Via Honolulu (21°17′N, 157°53′W) as described at 7.200.1 and 7.190.1 to Guam, thence:

To Hong Kong (22°17′N, 114°10′E), via Balintang Channel (20°00′N, 122°20′E) as described at 7.121.

To Shanghai (31°03′N, 122°20′E), through Nansei Shotō (27°00′N, 129°00′E), as described at 7.172.1.

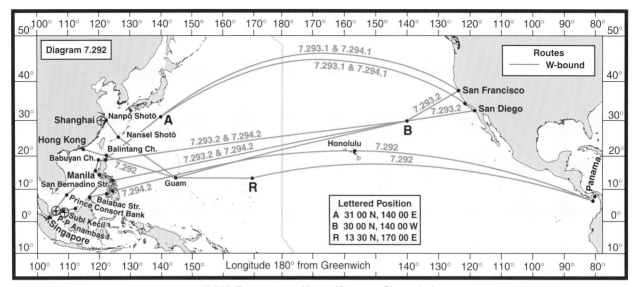

7.292 Panama → Hong Kong or Shanghai
7.293 San Diego or San Francisco → Manila
7.294 San Diego or San Francisco → Singapore

Distances (via Guam (7.244.1):
Hong Kong   9910 miles.
Shanghai   9830 miles.

Distances:

	San Diego	San Francisco
Via Babuyan Channel	7000	6800
Via San Bernadino Strait	6910	6710

San Diego or San Francisco ⇒ Manila

Diagram 7.292
Routes
7.293

From San Diego (32°38′N, 117°15′W) or San Francisco (37°45′N, 122°40′W) the following seasonal routes are recommended:

7.293.1

June to September:
Great circle to 31°00′N, 140°00′E (**A**), thence:
Rhumb line to Babuyan Channel (18°40′N, 122°00′E), thence:
Along the W coast of Luzon to Manila (14°32′N, 120°56′E).

This route, keeping S of the direct great circle route to Luzon Strait, reduces the effect of the Japan Current (7.26) and takes advantage of the clear passage through Nanpō Shotō in 31°00′N, 140°00′E.

Caution. For details of a volcanic area in Nanpō Shotō, see *Japan Pilot, Volume II*.

Distances:
San Diego   6650 miles.
San Francisco   6250 miles.

7.293.2

October to May:
Rhumb line to 30°00′N, 140°00′W (**B**), thence either:
Rhumb line to Babuyan Channel (18°40′N, 122°00′E), thence:
Along the W coast of Luzon to Manila (14°32′N, 120°56′E), or:
Rhumb line to San Bernadino Strait (12°33′N, 124°12′E), thence:
As described in *Philippine Islands Pilot* to Manila.

San Diego or San Francisco ⇒ Singapore

Diagram 7.292
Routes
7.294

From San Diego (32°38′N, 117°15′W) or San Francisco (37°45′N, 122°40′W) the following seasonal routes are recommended:

7.294.1

June to September:
Great circle to 31°00′N, 140°00′E (**A**), thence:
Rhumb line to Babuyan Channel (18°40′N, 122°00′E), thence:
NW of Prince Consort Bank (7°53′N, 110°00′E) to join route 7.96 from Yokohama to Singapore (1°12′N, 103°51′E).

7.294.2

October to May:
Rhumb line to 30°00′N, 140°00′W (**B**), thence by one of the following routes:
Via Babuyan Channel:
Rhumb line to Babuyan Channel (18°40′N, 122°00′E), thence:
NW of Prince Consort Bank (7°53′N, 110°00′E), thence:
W of Pulau-pulau Anambas (3°00′N, 106°00′E), thence:
To Singapore (1°12′N, 103°51′E).
Via San Bernadino Strait:
Rhumb line to San Bernadino Strait (12°33′N, 124°12′E), thence:
As described in *Philippine Islands Pilot* to Verde Island Passage (13°36′N, 121°00′E), thence:
NW of Prince Consort Bank (7°53′N, 110°00′E), thence:
W of Pulau-pulau Anambas (3°00′N, 106°00′E), thence:

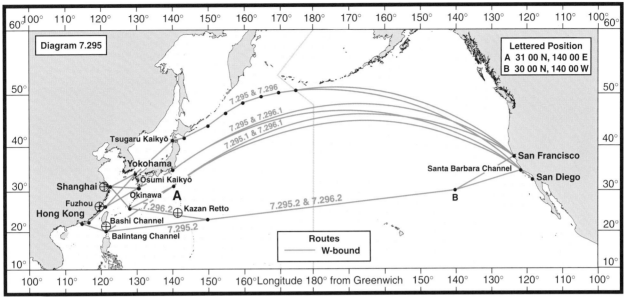

7.295 San Diego or San Francisco → Ports on the coast of China south of Fuzhou
7.296 San Diego or San Francisco → Ports on the coast of China north of Fuzhou

To Singapore (1°12′N, 103°51′E).
Via Surigao Strait:

Rhumb line to Surigao Strait (9°53′N, 125°21′E), thence:

Through Sulu Sea and through Balabac Strait (7°34′N, 116°55′E) to the S entrance of Palawan Passage (10°20′N, 118°00′E), thence:

Following routes 7.102 or 7.102.1 to Singapore (1°12′N, 103°51′E).

These routes offer favourable currents and protection from the north-east monsoon (December to March).

For details of channels and cautions on dangers and currents in the Sulu Sea see *Philippine Islands Pilot*.

For information on areas to be avoided near the Hawaiian Islands, see 7.38.1.

7.294.3
Distances:

	San Diego	San Francisco
Via Babuyan Channel		
June–September	7770	7380
October–May	8120	7920
Via San Bernadino Strait	8160	7960
Via Surigao Strait	8050	7850

San Diego or San Francisco ⇒ Ports on the coast of China south of Fuzhou

Diagram 7.295

Routes
7.295

From San Diego (32°38′N, 117°15′W) or San Francisco (37°45′N, 122°40′W) the shortest routes are either through Tsugaru Kaikyō and the Sea of Japan (7.299.1 or 7.299.2 and 7.164.3) or through Ōsumi Kaikyō (7.296.1 and 7.164.2).

If not calling at an intermediate port the following routes are recommended to avoid the unfavourable conditions of weather and sea (described at 7.273) which can often be expected on the N routes.

7.295.1
June to September:

Great circle to 31°00′N, 140°00′E (**A**), thence:

Rhumb line to Balintang Channel (20°00′N, 122°20′E) or Bashi Channel (21°20′N, 121°00′E), thence:

As navigation permits to destination.

This route, keeping S of the direct great circle route to Luzon Strait, reduces the effect of the Japan Current (7.26) and takes advantage of the clear passage through Nanpō Shotō in 31°00′N, 140°00′E.

Caution. For details of a volcanic area in Nanpō Shotō, see *Japan Pilot, Volume II*.

7.295.2
October to May:

Rhumb line to 30°00′N, 140°00′W (**B**), thence:

Rhumb line to Balintang Channel (20°00′N, 122°20′E) or Bashi Channel (21°20′N, 121°00′E), thence:

As navigation permits to destination.

For information on areas to be avoided near the Hawaiian Islands, see 7.38.1.

7.295.3

Distances to Hong Kong (22°17′N, 114°10′E):

	San Diego	San Francisco
Via Tsugaru Kaikyō	6450	6040
Via Ōsumi Kaikyō	6440	6050
Via Balingtang Channel		
June–September	6660	6820
October–May	7020	6820

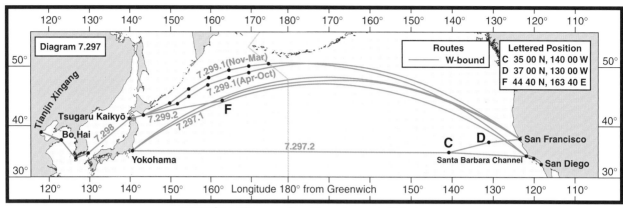

7.297 San Diego or San Francisco → Yokohama
7.298 San Diego or San Francisco → East China Sea and Bo Hai
7.299 San Diego or San Francisco → Tsugaru Kaikyō

San Diego or San Francisco ⇒ Ports on the coast of China north of Fuzhou

Diagram 7.295
Routes
7.296

From San Diego (32°38′N, 117°15′W) or San Francisco (37°45′N, 122°40′W) the shortest routes are either through Tsugaru Kaikyō and the Sea of Japan (7.299.1 or 7.299.2).

Distances to Shanghai (31°03′N, 122°20′E):

San Diego	5810 miles.
San Francisco	5400 miles.

If not calling at an intermediate port the following routes are recommended to avoid the unfavourable conditions of weather and sea (described at 7.273) which can often be expected on the N routes.

7.296.1

June to September:
Great circle to 31°00′N, 140°00′E **(A)**, thence:
S of Okinawa Guntō (26°30′N, 128°00′E), thence:
As navigation permits to destination.

Alternatively:
By great circle to the approaches to Yokohama (35°26′N, 139°43′E), thence:
Through Ōsumi Kaikyō (30°55′N, 130°40′E) or Seto Naikai (34°10′N, 133°30′E), thence:
As navigation permits to destination.

Distances to Shanghai (31°03′N, 122°20′E):

	San Diego	San Francisco
Via 31°00′N, 140°00′E **(A)**	5980	5590
Via Ōsumi Kaikyō/Seto Naikai	5900	5500

7.296.2

October to May:
Rhumb line to 30°00′N, 140°00′W **(B)**, thence:
Rhumb line to Bashi Channel (21°20′N, 121°00′E) as far as 150°00′E, thence:
S of Kazan Rettō (25°00′N, 141°20′E), thence:
S of Okinawa Guntō (26°30′N, 128°00′E), thence:
As navigation permits to destination.

Distances to Shanghai (31°03′N, 122°20′E):

San Diego	6630 miles.
San Francisco	6440 miles.

San Diego or San Francisco ⇒ Yokohama

Diagram 7.297
Routes
7.297

From San Diego (32°38′N, 117°15′W) or San Francisco (37°45′N, 122°40′W) the routes are seasonal and are as follows:

7.297.1

June to August:
Great circle to Yokohama (35°26′N, 139°43′E). A contrary current is likely throughout the voyage.

Distances:

San Diego	4930 miles.
San Francisco	4530 miles.

7.297.2

September to May:
From San Diego (32°38′N, 117°15′W):
Rhumb line to 35°00′N, 140°00′W **(C)**, thence:
To Yokohama (35°26′N, 139°43′E).
From San Francisco (37°45′N, 122°40′W):
Rhumb line to 37°00′N, 130°00W **(D)**, thence:
To 35°00′N, 140°00′W **(C)**, thence:
To Yokohama (35°26′N, 139°43′E).

Bad weather is unlikely on these routes and the strength of the contrary current should not be felt until approaching Japan.

Distances:

San Diego	5160 miles.
San Francisco	4850 miles.

San Diego or San Francisco ⇒ Ports in the East China Sea and Bo Hai

Diagram 7.297
Routes
7.298

From San Diego (32°38′N, 117°15′W) or San Francisco (37°45′N, 122°40′W) the routes are:
To Tsugaru Kaikyō (41°39′N, 140°48′E), as described at 7.299, thence:
Through Korea Strait, thence:
As navigation permits to destination.

Distances to Tianjin Xingang (38°56′N, 117°59′E):

San Diego	6010 miles.
San Francisco	5600 miles.

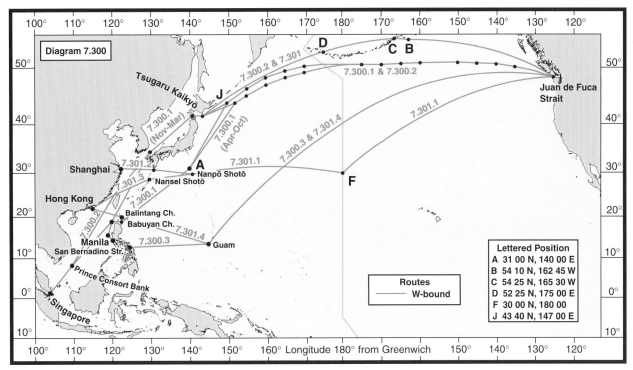

7.300 Juan de Fuca Strait ➤ Manila or Singapore
7.301 Juan de Fuca Strait ➤ Ports on the coast of China

San Diego or San Francisco ⇒ Tsugaru Kaikyō

Diagram 7.297
Routes
7.299

From San Diego (32°38′N, 117°15′W) or San Francisco (37°45′N, 122°40′W) alternative routes are:

7.299.1

Great circle to 50°30′N, 180°00′, thence:

By the appropriate seasonal route as described at 7.302 to 44°00′N, thence:

As navigation permits to Tsugaru Kaikyō (41°39′N, 140°48′E).

7.299.2

Great circle to 44°40′N, 163°40′E (**F**), with a highest latitude on the route at 47°30′N, 170°00′W, thence:

Rhumb line to Tsugaru Kaikyō (41°39′N, 140°48′E).

On these routes a contrary current may be expected throughout the voyage.

A call at Dutch Harbour (53°56′N, 166°29′W) as described at 7.177 and 7.310, entails a slightly longer passage, but with a favourable current in parts.

Passage N of the Aleutian Islands, see details at 7.274, may avoid heavy seas and head winds which can be expected, particularly in winter S of the islands.

Distances to Shanghai (31°03′N, 122°20′E):

	San Diego	San Francisco
7.299.1 (either seasonal route)	4650	4240
7.299.1	4670	4260
Dutch Harbour and N of Aleutians	4690	4280

Juan de Fuca Strait ⇒ Manila or Singapore

Diagram 7.300
Routes
7.300

From Juan de Fuca Strait (48°30′N, 124°47′W) a number of routes, as described below, merit consideration depending on the season and the prevailing weather.

7.300.1

By following routes 7.302 to Yokohama (35°26′N, 139°43′E) as far as 44°00′N on the appropriate seasonal route, thence:

Rhumb line to 31°00′N, 140°00′E (**A**), thence:

Rhumb line to the Babuyan Channel (18°40′N, 122°00′E), thence:

As navigation permits to Manila (14°32′N, 120°56′E), or NW of Prince Consort Bank (7°53′N, 110°00′E) to join route 7.96 from Yokohama to Singapore (1°12′N, 103°51′E).

This route, through 31°00′N, 140°00′E (**A**), passes well clear of the SE part of Honshū to reduce the effect of the Japan Current (7.26) and to take advantage of the clear passage through Nanpō Shotō (30°00′N, 141°00′E).

Caution. For details of the a volcanic area in Nanpō Shotō, see *Japan Pilot, Volume II.*

7.300.2

By following one of the routes described at 7.303.1 or 7.303.2 to Tsugaru Kaikyō (41°39′N, 140°48′E), either N or S of the Aleutian Islands, thence:

Through the Sea of Japan to Korea Strait, thence:

E of T'ai-wan (24°00′N, 121°00′E) and along the W coast of Luzon to Manila (14°32′N, 120°56′E), or through Taiwan Strait and South China Sea to Singapore (1°12′N, 103°51′E) by the Main Route described at 7.86.1.

Caution, These routes pass near to Russian Regulated Areas described at 7.274.7.

7.300.3

To avoid the worst of the adverse currents and winter weather of high latitudes a route passing farther S leads:

By great circle to Guam (13°27′N, 144°35′E), as described at 7.271, thence:

Through San Bernadino Strait (12°33′N, 124°12′E), thence:

As described in *Philippine Islands Pilot* to Manila (14°32′N, 120°56′E), or as described at 7.97 to Singapore (1°12′N, 103°51′E).

From Guam a favourable current can be expected, except between June and August, when it will be adverse in the South China Sea.

7.300.4

For low-powered, or hampered, vessels one of the alternative routes, described at 10.133, may be preferred to the great circle routes described above.

7.300.5

Distances:

	Manila	Singapore
7.300.1 (Via Babuyan Channel)		
November – March	5950	7080
April – October	5920	7050
7.300.2 (Via Taiwan Strait)		
N of Aleutian Islands	5910	6960
S of Aleutian Islands	5960	7010
7.300.3 (Via Guam)	6340	7580

Juan de Fuca Strait ⇒ Ports on the coast of China

Diagram 7.300

Routes

7.301

From Juan de Fuca Strait (48°30′N, 124°47′W) the shortest route in every case is through Tsugaru Kaikyō (41°39′N, 140°48′E) and the Sea of Japan, as described at 7.303.1 or 7.303.2 and either N or S of the Aleutian Islands.

Distances:

	Shanghai	Hong Kong	Tianjin Xingang
N of Aleutian Islands	4960	5610	5160
S of Aleutian Islands	5010	5600	5210

To avoid the worst of the adverse current and the winter weather of high latitudes the following alternative routes are recommended.

7.301.1

Great circle to 30°00′N, 180°00′ (**F**), thence:

Through Nanpō Shotō (30°00′N, 141°00′E), between Tori Shima and Sōfu Gan, thence:

7.301.2

For Shanghai (31°03′N, 122°20′E) and ports farther N:

Through Ōsumi Kaikyō (30°55′N, 130°40′E), thence:

As navigation permits to destination.

The Japan Current (7.26) sets NE on the W part of this route.

Distance to Shanghai   5780 miles.

7.301.3

For ports S of Shanghai (31°03′N, 122°20′E):

Through Nansei Shotō (27°00′N, 129°00′E), between Tokara Guntō and Amami Guntō, thence:

As navigation permits to destination.

The Japan Current (7.26) sets NE on the W part of this route.

Distance to Hong Kong   6320 miles.

7.301.4

An alternative route for Hong Kong (22°17′N, 114°10′E) and ports farther S leads:

To Guam (13°27′N, 144°35′E), as described at 7.271, thence:

To destination as described at 7.121.

From Guam a favourable current can be expected, except between June and August, when it will be adverse in the South China Sea.

Distance to Hong Kong   6650 miles.

7.301.5

For low-powered, or hampered, vessels one of the alternative routes, described at 10.134, may be preferred to the great circle routes described above.

Juan de Fuca Strait ⇒ Yokohama

Diagram 7.302

Routes

7.302

From Juan de Fuca Strait (48°30′N, 124°47′W) the route is to the 180°00′ meridian, by rhumb line, through the following positions:

49°30′N, 130°00′W	50°50′N, 160°00′W
50°10′N, 135°00′W	50°40′N, 165°00′W
50°35′N, 140°00′W	50°30′N, 170°00′W
50°45′N, 145°00′W	50°30′N, 175°00′W
50°50′N, 150°00′W	50°30′N, 180°00′

From 50°30′N, 180°00′ the routes are seasonal, by rhumb lines, through the following positions:

November – March	April – October
50°30′N, 175°00′E	50°00′N, 175°00′E
50°10′N, 170°00′E	49°15′N, 170°00′E
49°30′N, 165°00′E	48°20′N, 165°00′E
48°20′N, 160°00′E	47°10′N, 160°00′E
46°30′N, 155°00′E	45°20′N, 155°00′E
44°00′N, 150°00′E	44°00′N, 152°00′E

Then in each case by rhumb line to make a landfall on Inubo Saki (35°42′N, 140°52′E) or Kinkasan Tō (38°18′N, 141°35′E), thence as navigation permits to Yokohama (35°26′N, 139°43′E).

These routes, leading close S of the Aleutian Islands, are usually N of the W winds and favoured by the W-going current throughout.

Caution, These routes pass near to Russian Regulated Areas described at 7.274.7.

Distances:

November to March	4180 miles.
April to October	4160 miles.

Juan de Fuca Strait ⇒ Tsugaru Kaikyō

Diagram 7.302

Routes

7.303

From Juan de Fuca Strait (48°30′N, 124°47′W) the routes are either N or S of the Aleutian Islands, for details of conditions affecting these routes see 7.274.

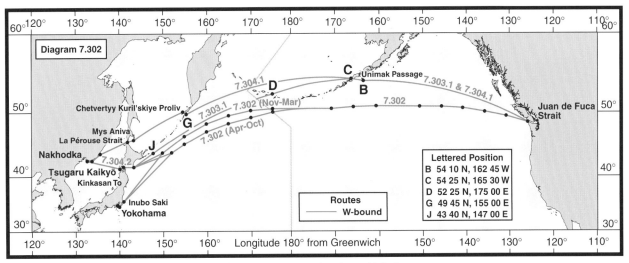

7.302 Juan de Fuca Strait → Yokohama
7.303 Juan de Fuca Strait → Tsugaru Kaikyō
7.304 Juan de Fuca Strait → Nakhodka

7.303.1

North of the Aleutian Islands the route is:
By great circle to 54°10′N, 162°45′W (**B**), thence:
Through Unimak Pass to 54°25′N, 165°30′W (**C**), thence:
Great circle to 52°25′N, 175°00′E (**D**), thence:
Great circle to 43°40′N, 147°00′E (**J**), thence:
As navigation permits to Tsugaru Kaikyō (41°39′N, 140°48′E).

7.303.2

South of the Aleutian Islands the route is:
As for Yokohama, described at 7.302, by the appropriate seasonal route as far as 44°00′N, thence:
As navigation permits to Tsugaru Kaikyō (41°39′N, 140°48′E).

7.303.3

Caution, These routes pass near to Russian Regulated Areas described at 7.274.7.

Distances:

North of Aleutians	3800 miles.
South of Aleutians (All seasons)	3850 miles.

Juan de Fuca Strait ⇒ Nakhodka

Diagram 7.302
Routes
7.304

From Juan de Fuca Strait (48°30′N, 124°47′W) the routes are either N or S of the Aleutian Islands, for details of conditions affecting these routes see 7.274.

7.304.1

North of the Aleutian Islands the route is:
By great circle to 54°10′N, 162°45′W (**B**), thence:
Through Unimak Pass to 54°25′N, 165°30′W (**C**), thence:
Great circle, passing N of Attu Island (52°55′N, 173°00′E), to 49°45′N, 155°00′E (**G**), thence:
Through Kuril'skiye Ostrova (48°00′N, 153°00′E) via the TSS through Chetvertyy Kuril'skiye Proliv (50°00′N, 155°00′E), thence:
Great circle to the TSS off Mys Aniva (46°02′N, 143°25′E), thence:

To the TSS through La Pérouse Strait (45°45′N, 142°00′E), thence:
As navigation permits to Nakhodka (42°48′N, 132°57′E).

7.304.2

South of the Aleutian Islands the route is:
As for Yokohama, described at 7.302, by the appropriate seasonal route as far as 44°00′N, thence:
As navigation permits to Tsugaru Kaikyō (41°39′N, 140°48′E), thence:.
As navigation permits to Nakhodka (42°48′N, 132°57′E).

7.304.3

Distances:

North of Aleutians	4050 miles.
South of Aleutians	4230 miles.

Prince Rupert ⇒ Manila or Singapore

Diagram 7.305
Routes
7.305

From Prince Rupert (54°19′N, 130°20′W) a number of routes, as described below, merit consideration depending on the season and the prevailing weather.

7.305.1

From Dixon Entrance (54°30′N, 132°30′W), thence following route 7.307.2 to Yokohama (35°26′N, 139°43′E), passing S of the Aleutian Islands, as far as 44°00′N on the appropriate seasonal route, thence:
Via 31°00′N, 140°00′E (**A**) and Babuyan Channel (18°40′N, 122°00′E), as described at 7.300.1, to Manila (14°32′N, 120°56′E) or Singapore (1°12′N, 103°51′E).

7.305.2

From Dixon Entrance (54°30′N, 132°30′W), thence following route 7.307.1 to Yokohama (35°26′N, 139°43′E), passing N of the Aleutian Islands, as far as 52°25′N, 175°00′E (**D**), thence:
By great circle to join route 7.300.1 in 31°00′N, 140°00′E (**A**), thence:
As described at 7.300.1 to Manila (14°32′N, 120°56′E) or Singapore (1°12′N, 103°51′E).

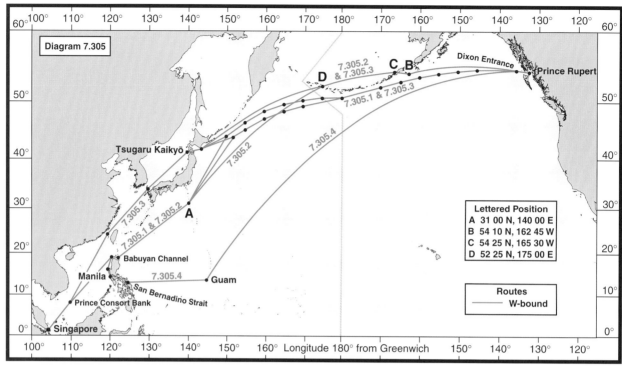

Diagram 7.305

Lettered Position
A 31 00 N, 140 00 E
B 54 10 N, 162 45 W
C 54 25 N, 165 30 W
D 52 25 N, 175 00 E

Routes
——— W-bound

7.305 Prince Rupert → Manila or Singapore

7.305.3

> From Dixon Entrance (54°30′N, 132°30′W), thence following one of the routes 7.308.1 or 7.308.2, passing N or S of the Aleutian Islands, to Tsugaru Kaikyō (41°39′N, 140°48′E), thence:
> Via the Taiwan Strait as described at 7.300.2 to Manila (14°32′N, 120°56′E) or Singapore (1°12′N, 103°51′E).

7.305.4

To avoid the worst of the adverse currents and winter weather of high latitudes a route passing farther S leads:

> By great circle to Guam (13°27′N, 144°35′E), as described at 7.271, thence:
> Through San Bernadino Strait (12°33′N, 124°12′E), thence:
> As described in *Philippine Islands Pilot* to Manila (14°32′N, 120°56′E), or as described at 7.97 to Singapore (1°12′N, 103°51′E).

From Guam a favourable current can be expected, except between June and August, when it will be adverse in the South China Sea.

7.305.5
Distances:

	Manila	Singapore
7.305.1 (S of Aleutian Islands)		
November – March	5630	6770
April – October	5600	6740
7.305.2 (N of Aleutian Islands)	5570	6700
7.305.3 (Via Taiwan Strait)		
N of Aleutian Islands	5600	6650
S of Aleutian Islands	5640	6690
7.305.4 (Via Guam)	6070	7310

7.305.6

For low-powered, or hampered, vessels one of the alternative routes, described at 10.135, may be preferred to the great circle routes described above.

Prince Rupert ⇒ Ports on the coast of China

Diagram 7.306
Routes
7.306

From Prince Rupert (54°19′N, 130°20′W) the shortest route in every case is through Tsugaru Kaikyō (41°39′N, 140°48′E) and the Sea of Japan, as described at 7.308.1 or 7.308.2 and either N or S of the Aleutian Islands.
Distances:

	Shanghai	Hong Kong	Tianjin Xingang
N of Aleutian Islands	4650	5300	4850
S of Aleutian Islands	4700	5340	4900

To avoid the worst of the adverse current and the winter weather of high latitudes the following alternative routes, from Dixon Entrance (54°30′N, 132°30′W) are recommended.

7.306.1

> Great circle to 30°00′N, 180°00′ (**F**), thence:
> Through Nanpō Shotō (30°00′N, 141°00′E), between Tori Shima and Sōfu Gan, thence:

7.306.2

> For Shanghai (31°03′N, 122°20′E) and ports farther N:
> Through Ōsumi Kaikyō (30°55′N, 130°40′E), thence:
> As navigation permits to destination.

The Japan Current (7.26) sets NE on the W part of this route.

Distance to Shanghai   5610 miles.

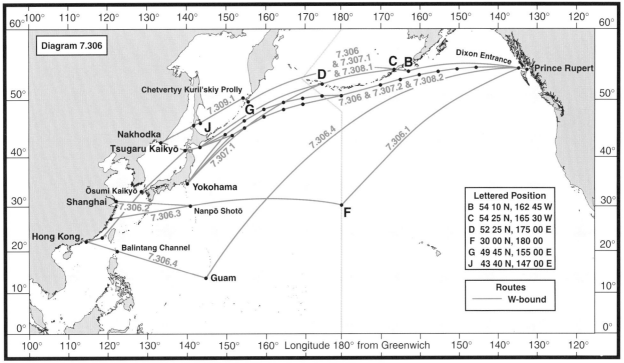

7.306 Prince Rupert → Ports on the coast of China
7.307 Prince Rupert → Yokohama
7.308 Prince Rupert → Tsugaru Kaikyō
7.309 Prince Rupert → Nakhodka

7.306.3

For ports S of Shanghai (31°03′N, 122°20′E):

Through Nansei Shotō (27°00′N, 129°00′E), between Tokara Guntō and Amami Guntō, thence:

As navigation permits to destination.

The Japan Current (7.26) sets NE on the W part of this route.

Distance to Hong Kong   6150 miles.

7.306.4

An alternative route for Hong Kong (22°17′N, 114°10′E) and ports farther S leads:

To Guam (13°27′N, 144°35′E), as described at 7.271, thence:

To destination as described at 7.121.

From Guam a favourable current can be expected, except between June and August, when it will be adverse in the South China Sea.

Distance to Hong Kong   6420 miles.

7.306.5

For low-powered, or hampered, vessels one of the alternative routes, described at 10.136, may be preferred to the great circle routes described above.

Prince Rupert ⇒ Yokohama

Diagram 7.306
Routes
7.307

From Prince Rupert (54°19′N, 130°20′W) the routes are either N or S of the Aleutian Islands, for details of conditions affecting these routes see 7.274.

7.307.1

North of the Aleutian Islands the route from Dixon Entrance (54°30′N, 132°30′W) is:

By great circle to 54°10′N, 162°45′W (**B**), thence:

Through Unimak Pass (54°15′N, 164°30′W) to 54°25′N, 165°30′W (**C**), thence:

Great circle to 52°25′N, 175°00′E (**D**), thence:

Great circle to 34°49′N, 140°00′E, SE of Nojima Saki, thence:

As navigation permits to Yokohama (35°26′N, 139°43′E).

Distance   3820 miles.

Caution, These routes pass near to Russian Regulated Areas described at 7.274.7.

7.307.2

South of the Aleutian Islands the route from Dixon Entrance (54°30′N, 132°30′W) is by rhumb line through the following positions:

54°40′N, 135°00′W	53°40′N, 160°00′W
54°50′N, 140°00′W	53°00′N, 165°00′W
54°50′N, 145°00′W	52°15′N, 170°00′W
54°30′N, 150°00′W	51°30′N, 175°00′W
54°10′N, 155°00′W	50°30′N, 180°00′

This part of the route passes about 30 miles S of the Aleutian Islands and has the benefit of the W-going Alaska Current (7.26).

At 50°30′N, 180°00′ the route joins the appropriate seasonal route from Juan de Fuca Strait to Yokohama, described at 7.302.

7.307.3

Caution, These routes pass near to Russian Regulated Areas described at 7.274.7.

Distances:

November to March   3860 miles.
April to October   3850 miles.

Prince Rupert ⇒ Tsugaru Kaikyō

Diagram 7.306
Routes
7.308

From Prince Rupert (54°19′N, 130°20′W) the routes are either N or S of the Aleutian Islands, for details of conditions affecting these routes see 7.274.

7.308.1

North of the Aleutian Islands the route from Dixon Entrance (54°30′N, 132°30′W) is:

By great circle to 54°10′N, 162°45′W **(B)**, thence:

Through Unimak ·Pass (54°15′N, 164°30′W) to 54°25′N, 165°30′W **(C)**, thence:

Great circle to 52°25′N, 175°00′E **(D)**, thence:

Great circle to 43°40′N, 147°00′E **(J)**, thence:

As navigation permits to Tsugaru Kaikyō (41°39′N, 140°48′E).

7.308.2

South of the Aleutian Islands the route, from Dixon Entrance (54°30′N, 132°30′W), is:

By rhumb line through the positions indicated at route 7.307.2 as far as 50°30′N, 180°00′, thence:

By the appropriate seasonal route from Juan de Fuca Strait to Yokohama, described at 7.302, as far as 44°00′N, thence:

As navigation permits to Tsugaru Kaikyō (41°39′N, 140°48′E).

7.308.3

Caution, These routes pass near to Russian Regulated Areas described at 7.274.7.

Distances:

North of Aleutians	3490 miles.
South of Aleutians (All seasons)	3530 miles.

Prince Rupert ⇒ Nakhodka

Diagram 7.306
Routes
7.309

From Prince Rupert (54°19′N, 130°20′W) the routes are either N or S of the Aleutian Islands, for details of conditions affecting these routes see 7.274.

7.309.1

North of the Aleutian Islands the route from Dixon Entrance (54°30′N, 132°30′W) is:

By great circle to 54°10′N, 162°45′W **(B)**, thence:

Through Unimak Pass (54°15′N, 164°30′W) to 54°25′N, 165°30′W **(C)**, thence:

Great circle, passing N of Attu Island (52°55′N, 173°00′E), to 49°45′N, 155°00′E **(G)**, thence:

Through Kuril'skiye Ostrova (48°00′N, 153°00′E) via the TSS through Chetvertyy Kuril'skiye Proliv (50°00′N, 155°00′E), thence:

Great circle to the TSS off Mys Aniva (46°02′N, 143°25′E), thence:

To the TSS through La Pérouse Strait (45°45′N, 142°00′E), thence:

As navigation permits to Nakhodka (42°48′N, 132°57′E).

7.309.2

South of the Aleutian Islands the route from Dixon Entrance (54°30′N, 132°30′W) is:

By rhumb line through the positions indicated at route 7.307.2 as far as 50°30′N, 180°00′, thence:

By the appropriate seasonal route from Juan de Fuca Strait to Yokohama, described at 7.302, as far as 44°00′N, thence:

As navigation permits to Tsugaru Kaikyō (41°39′N, 140°48′E), thence:

As navigation permits to Nakhodka (42°48′N, 132°57′E).

7.309.3

Caution, These routes pass near to Russian Regulated Areas described at 7.274.7.

Distances:

North of Aleutians	3740 miles.
South of Aleutians (All seasons)	3910 miles.

Dutch Harbour ⇒ Yokohama or Tsugaru Kaikyō

Diagram 7.310
Routes
7.310

From Dutch Harbour (53°56′N, 166°29′W) the routes are either N or S of the Aleutian Islands, for details of conditions affecting these routes see 7.274.

7.310.1

North of the Aleutian Islands the route is:

Great circle to 52°25′N, 175°00′E **(D)**, thence:

As described at 7.307.1 to Yokohama (35°26′N, 139°43′E), or:.

As described at 7.303.1 to Tsugaru Kaikyō (41°39′N, 140°48′E).

Distances:

Yokohama	2560 miles.
Tsugaru Kaikyō	2220 miles.

7.310.2

South of the Aleutian Islands the route is:

As navigation permits to 50°30′N, 180°00′, thence:

By the appropriate seasonal route, described at 7.302, as far as 44°00′N, thence:

Continuing as at 7.302 to Yokohama (35°26′N, 139°43′E), or:

As navigation permits to Tsugaru Kaikyō (41°39′N, 140°48′E).

Distances:

Yokohama	
November to March	2770 miles.
April to October	2760 miles.
Tsugaru Kaikyō (All seasons)	2440 miles.

7.310.3

Caution, These routes pass near to Russian Regulated Areas described at 7.274.7.

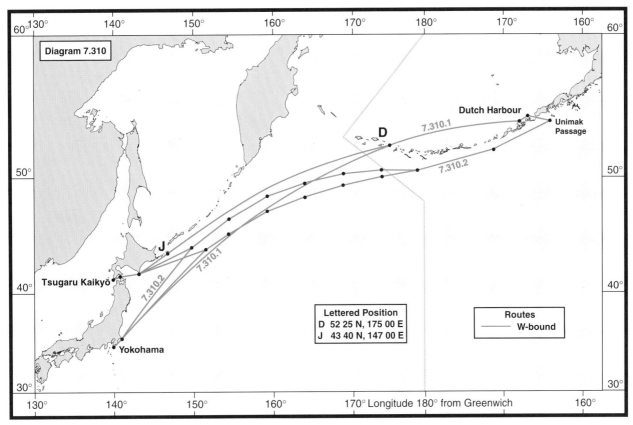

7.310 Dutch Harbour → Yokohama or Tsugaru Kaikyō

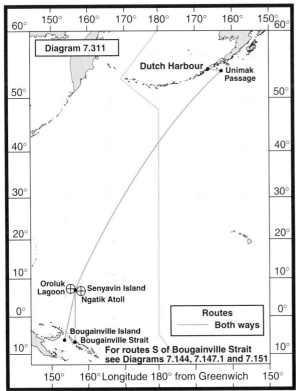

7.311 Dutch Harbour ← → Australian Ports and Torres Strait

Dutch Harbour ⇐ ⇒ Australian ports or Torres Strait

Diagram 7.311

Routes

7.311

From Dutch Harbour (53°56′N, 166°29′W) the routes lead:

> Through Unimak Pass (54°15′N, 164°30′W), thence:
> Great circle to pass between Senyavin Islands (6°55′N, 158°10′E) and Oroluk Lagoon (7°30′N, 155°20′E), thence:
> W of Ngatik Atoll (5°50′N, 157°11′E), thence:
> Through the strait between New Ireland (3°20′S, 152°00′E) and Bougainville Island (6°00′S, 155°15′E) or through Bougainville Strait (6°40′S, 156°15′E) into the Solomon Sea, thence:
> As at 7.144, 7.147.1 or 7.152 to destination.

7.311.1

Distances:

Torres Strait (10°36′S, 141°51′E)   5370 miles.
Caloundra Head (26°48′S, 153°08′E)*   5530 miles.
Sydney (33°50′S, 151°19′E)   5970 miles.
Port Phillip (38°20′S, 144°34′E)**   6540 miles.

* Caloundra Head to Brisbane   35 miles.
** Port Phillip to Melbourne   40 miles.

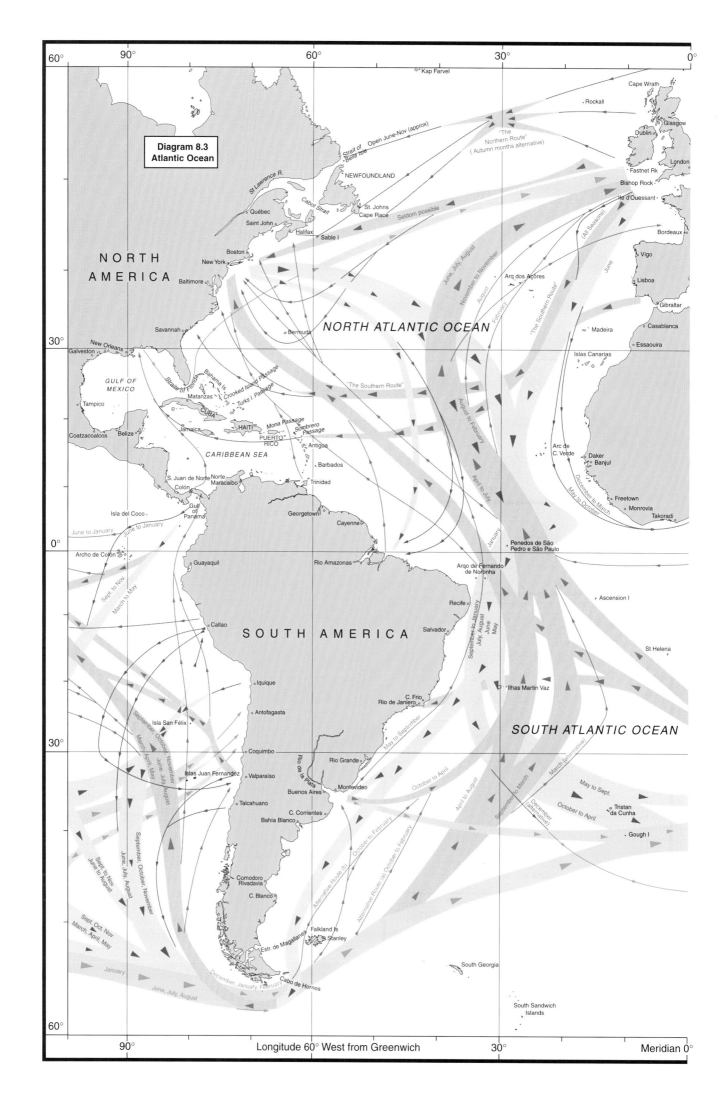

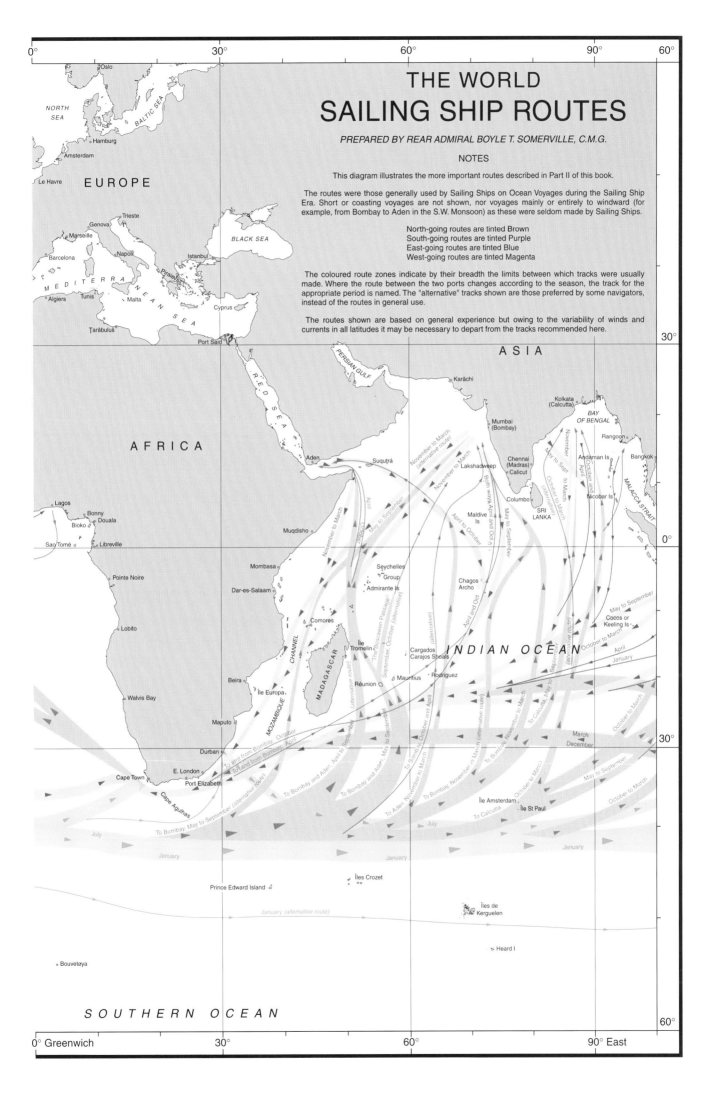

THE WORLD
SAILING SHIP ROUTES

PREPARED BY REAR ADMIRAL BOYLE T. SOMERVILLE, C.M.G.

NOTES

This diagram illustrates the more important routes described in Part II of this book.

The routes were those generally used by Sailing Ships on Ocean Voyages during the Sailing Ship Era. Short or coasting voyages are not shown, nor voyages mainly or entirely to windward (for example, from Bombay to Aden in the S.W. Monsoon) as these were seldom made by Sailing Ships.

North-going routes are tinted Brown
South-going routes are tinted Purple
East-going routes are tinted Blue
West-going routes are tinted Magenta

The coloured route zones indicate by their breadth the limits between which tracks were usually made. Where the route between the two ports changes according to the season, the track for the appropriate period is named. The "alternative" tracks shown are those preferred by some navigators, instead of the routes in general use.

The routes shown are based on general experience but owing to the variability of winds and currents in all latitudes it may be necessary to depart from the tracks recommended here.

NOTES

CHAPTER 8

SAILING PASSAGES FOR ATLANTIC OCEAN AND MEDITERRANEAN SEA

GENERAL INFORMATION

COVERAGE

Chapter coverage

8.1

This chapter contains details of the following passages for sailing vessels and includes recommended routes for low-powered, or hampered, vessels:

Passages from ports on the eastern side of the Atlantic Ocean (8.6).

Passages in the Mediterranean Sea (8.34).

Passages from ports on the west coast of Africa and from Atlantic islands (8.40).

Passages from ports on the western side of the Atlantic Ocean (8.65).

Admiralty Publications

8.2

Relevant navigational publications should be consulted, in addition to this book, when planning and conducting passages. Details of these, and the areas covered, can be found, as follows:

Admiralty Sailing Directions; see 1.12.1

Admiralty List of Lights and Fog Signals; see 1.12.2

Admiralty Tide Tables; see 1.12.3

Admiralty Tidal Stream Atlases; see 1.12.4

Admiralty List of Radio Signals; see 1.12.5

Admiralty Distance Tables; see 1.12.8

8.2.1

See also *The Mariner's Handbook* for notes on electronic products.

INTRODUCTORY REMARKS

General information

8.3

1   It can be argued that a standard work on Ocean Passages should be confined to the needs of contemporary seamen operating full-powered modern vessels (1.2) and that directions for sailing ships, or other hampered vessels, are not required.

2   Chapters two to seven, covering ocean passages for the world, are written entirely to this criteria and details of alternative routes for low-powered, or hampered, vessels are not covered in these chapters.

3   Chapters eight to ten have, therefore, been modified, from the previous edition of this book, for the benefit of masters of low-powered, hampered or ocean-going sailing vessels which are susceptible to the main wind and current circulation of the ocean.

4   A selection of the routes described are shown on Diagrams 8.3, 9.3 and 10.3; taken from *Chart 5308 The World — Sailing ship routes*, at the start of these chapters. Also *Chart 5309 — Tracks followed by Sailing and Auxiliary Powered Vessels* (1.11) shows many routes additional to those contained in this volume.

Restrictions

8.4

The routes described have not been modified to take account of the many Offshore installations (1.27), Traffic Separation Schemes (1.28) or Areas to be Avoided (1.30) located in in different parts of the world, details of which can be found in Admiralty Notices to Mariners, charts and the appropriate volumes of *Admiralty Sailing Directions*.

Duration of passages

8.5

1   Regarding distances traversed it has been considered more useful, based on historical data, to express them as the average number of days taken in ordinary weather by a well-found vessel of about 2000 tons which, in good conditions, could log speeds of 10 to 12 kn but generally averaged 100 to 150 miles per day.

2   The following list, compiled from the historical records of operators of commercial sailing vessels, gives the average duration of a number of voyages:

Passage	Number of days
English Channel to New York (Winter)	35–40
English Channel to New York (Summer)	40–50
English Channel to New Orleans	45–55
English Channel to Rio de Janiero or Santos	45–60
English Channel to Río de La Plata	55–65
English Channel to Valparaíso (round Cabo de Hornos)	90–100
English Channel to Callao (round Cabo de Hornos)	95–120
English Channel to San Francisco (round Cabo de Hornos)	125–150
English Channel to Cape Town	50–60
English Channel to Durban	60–65
English Channel to Bombay	100–110
English Channel to Calcutta	100–120
English Channel to Rangoon	100–120
English Channel to Selat Sunda	90–100
English Channel to Hong Kong (south-west monsoon)	100–120
English Channel to Adelaide	80–90
English Channel to Melbourne	80–90
English Channel to Sydney (or Newcastle)	85–100
New York to English Channel	25–30
New York to Cape Town	65–70
New York to Río de La Plata	60–65
New York to Melbourne	100–120

Passage	Number of days
New York to Selat Sunda	100–110
Cape Town to Melbourne	35–40
Cape Town to Wellington	40–45
Cape Town to Río de La Plata (across the Atlantic)	45
Cape Town to Río de La Plata (round Cabo de Hornos)	110
Cape Town to Calcutta	40–50
Cape Town to Shanghai (via Selat Sunda, south-west monsoon)	60
Calcutta to Sydney	60
Calcutta to Cape Town	45
Calcutta to English Channel	90–100
Hong Kong to English Channel (north-east monsoon)	110–120
Hong Kong to San Francisco	40
Hong Kong to Sydney	50–60
Melbourne to Valparaíso	40–50
Melbourne to San Francisco	60–70

Passage	Number of days
Melbourne to Río de La Plata (round Cabo de Hornos)	70–80
Melbourne to English Channel (round Cabo de Hornos)	80–100
Melbourne to Hampton Roads	80–95
Wellington to San Francisco	60–70
Wellington to Valparaíso	30–35
Wellington to Río de La Plata	55–60
Wellington to English Channel	80–100
Valparaíso to English Channel (round Cabo de Hornos)	80–90
Valparaíso to New York (round Cabo de Hornos)	75–85
Valparaíso to Cape Town	65
Río de La Plata to English Channel	70–80
Río de La Plata to New York	60–70
Río de La Plata to Cape Town	20
Río de La Plata to Melbourne	50–55
New Orleans to English Channel	45–50

PASSAGES FROM PORTS ON EASTERN SIDE OF ATLANTIC OCEAN

Charts:

> *5124 (1) to (12) Monthly Routeing Charts for North Atlantic Ocean.*
> *5125 (1) to (12) Monthly Routeing Charts for South Atlantic Ocean.*
> *5309 Tracks followed by Sailing and Auxiliary Powered Vessels.*
> *5310 World Surface Current Distribution.*
> *Diagram 8.3 World Sailing Ship Routes.*

From Norwegian and Baltic ports ⇒ Eastern Canadian and United States ports

Routes

8.6

There are two main routes, Northern and Southern and a Direct route.

8.6.1

Northern route should only be taken in autumn, when it is clear of ice, vessels pass N of Orkney Islands (59°00′N, 3°00′W), or of Shetland Islands (60°30′N, 1°00′W) if weather so dictates. Thence they should stand W to cross 30°W in about 55°N and continue by the Northern route from the English Channel (8.12).

8.6.2

Southern route vessels should pass N of Orkney Islands, or of Shetland Islands if necessary and stand W far enough to ensure weathering the British Isles. When clear they should stand S to join the Southern route from the English Channel (8.13).

8.6.3

Direct route is seldom taken as it it almost directly against the prevailing winds and the North Atlantic Current (2.14). To follow it, round the Orkney Islands or Shetland Islands, as above and make W to at least 10°W, thence SW to join the Direct route from the English Channel (8.14) in 47°N, 40°W.

From Norwegian and Baltic ports ⇒ other Atlantic ports, Cape of Good Hope or Cabo de Hornos

Routes

8.7

Use the Southern route (8.6.2) and join the appropriate route from the English Channel (8.15 to 8.21 and 8.24 to 8.27.2) in about 40°00′N.

From North Sea ports ⇒ Eastern Canadian and United States ports

Routes

8.8

1   There are two main routes, Northern and Southern and a direct route, as at 8.6.

2   On the Northern route, with W winds in summer, a vessel will probably do better going N-about round the British Isles than by beating down the English Channel.

3   On the Southern and Direct routes the latter is preferable.

From North Sea ports ⇒ other Atlantic ports, Cape of Good Hope or Cabo de Hornos

Routes

8.9

Proceed via the English Channel and the appropriate route from (8.15 to 8.21 and 8.24 to 8.27.2).

From Irish Sea and River Clyde ⇒ Atlantic Ocean ports

Routes
8.10

1   If taking the Northern (8.6.1) or Direct (8.6.3) routes to Newfoundland, Canada or the United States when bound from Liverpool (53°25′N, 3°00′W) or the River Clyde (55°45′N, 4°58′W), it is better to pass N of Ireland with W winds in summer.

2   On the Southern route (8.6.2), and on the routes to other ports, the weather at the time of sailing will determine the most advantageous course to join the routes from the English Channel described below.

From English Channel ⇒ Canada and United States

Routes
8.11

There are two main routes, Northern and Southern and a Direct route. The Northern route should, as a rule, only be taken in autumn when it is free from ice.

Northern route
8.12

1   Although heavy weather is frequently experienced, the winds are generally more favourable and the currents from the Arctic assist in the latter part of the voyage.

2   When clear of the British Isles, stand W and cross the meridian of 30°W in about 55°N, then steer according to destination for the Straits of Belle Isle (51°44′N, 56°00′W), St John's Harbour, Newfoundland (47°34′N, 52°38′W) or for other Canadian and United States ports.

8.12.1

1   For Cabot Strait (47°32′N, 59°30′W) or Halifax (44°31′N, 63°30′W) either try to make Cape Race (46°39′N, 53°04′W) by passing N of Virgin Rocks (46°29′N, 50°46′W) or, in order to avoid the ice, cross the banks on the parallel of 40°N and haul up on the proper course on reaching 55°W, heavy ice being seldom met with W of that meridian. Make Cape Race if the weather is clear and thence steer for a position S of Île Saint-Pierre (46°47′N, 56°10′W). While on the Grand Banks during fog or when there is uncertainty regarding the position, soundings should be obtained frequently and an indraught towards the S coast of Newfoundland must be guarded against.

8.12.2

1   **Cautions.** For details of hazards on the Grand Banks of Newfoundland, see 2.43.

2   The S coast of Newfoundland, E of Cape Ray (47°37′N, 59°18′W), is broken, rocky and dangerous and the tidal streams are influenced by the winds. Winds from the SE, and often also SW, bring a thick fog which is most dense near the lee shore. This coast, therefore, should be approached except with a decidedly N wind and clear weather.

3   Sable Island (43°55′N, 60°00′W) should be given a wide berth as it is a very dangerous locality owing to the prevalent fog and variable currents near it. Soundings should never be neglected when crossing the banks and should be continuous whether bound for a Nova Scotia or a United States port. In thick weather the thermometer is also a useful guide when approaching the banks off Newfoundland as the temperature of the water falls on nearing them.

4   With SW winds, while foggy E of the meridian of Flint Island (46°11′N, 59°46′W), it is frequently clear for some miles off the land W of it.

5   Between Île Saint-Pierre (46°47′N, 56°10′W) and Cape Breton Island (46°00′N, 60°30′W), when feeling the way by sounding in foggy weather, the edges of the deep water channel running through the banks into Cabot Strait are especially good guides. Cape Pine (46°37′N, 53°32′W) should not be approached within depths of 70 m, nor Cape St Mary's (46°49′N, 54°12′W) within depths of 90 m, in fog. There is deep water, of 180 m to 260 m, in the approach to Placentia Bay (47°00′, 54°30′W).

Southern route
8.13

1   This is the best route to follow for the whole of the year except autumn, on account of the better weather likely to be encountered, the certainty of the wind and the avoidance of both fog and ice off Newfoundland banks during spring and the early part of the summer.

2   By this route, leaving the English Channel with a favourable wind, steer a direct course as long as it lasts and, at least, ensure a sufficient westing to avoid the danger of being set into the Bay of Biscay. When the favourable wind fails, take the Madeira (32°40′N, 17°00′W) route (8.18), and if the wind permits pass midway between that island and the Arquipélago dos Açores (38°00′N, 27°00′W), into the North-east Trade Wind (2.5). If the wind is not favourable then the Trade Wind will usually be gained sooner by passing nearer to Madeira. In that neighbourhood it is usually found, in summer, between 32°N and 31°N; in winter a degree or so farther S.

8.13.1

1   For Halifax (44°31′N, 63°30′W) or other Canadian ports, when well within the Trade Wind limits, run W until in about 48°00′W and thence edge off to the NW passing about 200 miles E of Bermuda (32°23′N, 64°38′W) and direct for Halifax, allowing for the Gulf Stream (2.14) setting ENE across the track.

8.13.2

1   For New York (40°28′N, 73°50′W) or other United States ports, when well into the Trade Wind limits, run W keeping S of 25°N until in about 65°W, then steer NW for any United States port, hauling out earlier for ports on the N part of this coast. The Gulf Stream (2.14) will have to be crossed in the latter part of this route.

Direct route
8.14

1   This route across the Atlantic Ocean, from the English Channel to New York, is about 1000 miles shorter than the Southern route (8.13.2), but can seldom be taken on account of the prevailing W winds and of the North Atlantic Current and Gulf Stream (2.14) combined, running contrary to the prescribed track. It is, however, recommended by some navigators, making as directly as possible from the English Channel to cross 50°W at 45°N and thence to the desired port.

From English Channel ⇒ Bermuda

Routes
8.15

1   There are two routes, the Direct route and the Southern route.

2   **Caution.** When approaching the islands of Bermuda every opportunity should be taken to verify the vessel's position and should it be at all doubtful, or the weather too unfavourable for seeing the lights, then the parallel of the islands should not be crossed during the night as the 180 m depth contour is too close to the reefs for soundings to give warning. See also 2.44 for Areas to be Avoided.

8.15.1

1   **Direct route**. This route generally follows that detailed at 8.14.

8.15.2

1   **Southern route**. Proceed as directed at 8.18 as far as Madeira (32°40′N, 17°00′W) and thence steer SW until within the limits of the North-east Trade Wind (2.5), which is usually found, in summer, between 32°N and 31°N; in winter a degree or so farther S, when course should be altered gradually towards the W, while keeping within the limits of the Trade Winds. Cross 25°N in 40°W, which parallel should be preserved until crossing 60°W, whence a course for Bermuda (32°23′N, 64°38′W) should be steered.

From English Channel ⇒ West Indies, Gulf of Mexico and North coast of South America

Sailing ship routes
8.16

1   Proceed as directed at 8.18 as far as Madeira (32°40′N, 17°00′W). After passing Madeira try to cross the parallel of 25°N between 25°W and 30°W, the object being to reach the North-east Trade Wind (2.5) as soon as possible. The season must be taken into consideration as to how far S it will be necessary to go in order to hold the Trade Wind.

8.16.1

1   For Cuba, if bound for Habana (23°10′N, 82°22′W) or Puerto de Matanzas (23°03′N, 81°34′W) pass through North-West Providence Channel (26°10′N, 78°35′W) close along the W edge of Great Bahama Bank (23°30′N, 78°00′W), round North Elbow Cay (23°56′N, 80°29′W) of Double-headed Shot Cays (23°57′N, 80°28′W) on Cay Sal Bank (23°40′N, 80°00′W) and towards Punta Guanos (23°09′N, 81°39′W), on the W side of Matanzas, out of the stream. Old Bahama Channel (21°30′N, 75°45′W) may also be used, or, if approaching from the W, Cabo San Antonio (21°52′N, 84°57′W) may be rounded.

2   Bound for any port on the S side of Cuba, it is better to pass N of Puerto Rico (18°15′N, 66°30′W) and Hispaniola (19°00′N, 71°00′W) during periods of S winds, which is the rainy season, and S of these islands when N winds are prevalent.

8.16.2

1   For Leeward Islands, Jamaica, Belize or Gulf of Mexico, cross 18°N in about 40°W and thence steer direct to pass between Antigua (17°05′N, 61°45′W) and Guadeloupe (16°15′N, 61°30′W); thence pass close S of Hispaniola (19°00′N, 71°00′W) and Jamaica (18°00′N, 77°30′W) and thence continue nearly direct to destination.

2   The channel between Antigua and Guadeloupe is 30 miles wide and there is generally much less current here than farther N or S. It will be better, however, in using this channel to keep close to the Antigua shore and to sight the island on the parallel of 17°N. Vessels sometimes pass between Antigua and Barbuda (17°40′N, 61°45′W), this may be done without much risk by day but not at night as soundings are so irregular that in running down it would be difficult to tell whether to haul N or S.

3   To ports in the Gulf of Mexico, North-West Providence Channel (26°10′N, 78°35′W) is used by a great number of vessels, keeping on the edges of the banks, to avoid the strength of the current. Old Bahama Channel (21°30′N, 75°45′W) is also used but less commonly.

8.16.3

1   For Venezuelan, Colombian and Caribbean ports, as far W as San Juan del Norte (10°56′N, 83°43′W), cross the meridian of 40°W in about 13°N, thence steer direct to the NE point of Trinidad and thence W to the required destination, keeping in the strength of the prevailing W-going current.

8.16.4

1   For islands N of Trinidad. The season must be considered as to how far S it will be necessary to go to ensure holding the Trade Wind. In making for any of the Windward Islands get in the parallel of the island about 100 miles E of it.

8.16.5

1   For Cayenne (5°02′N, 52°19′W). Cross the meridian of 40°W in 9°N, thence steering to make the parallel of Cayenne from 100 to 200 miles to windward to allow for the strong W-going current which prevails at all seasons, thence gradually closing the shore in depths of from 13 m to 18 m.

8.16.6

1   For Paramaribo (5°50′N, 55°10′W) and Georgetown (6°50′N, 58°10′W). Cross the meridian of 40°W between the parallels of 11°N and 12°N and thence steer to make the land to windward as for Cayenne.

Low-powered vessel routes
8.17

1   For low-powered vessels, proceeding from Bishop Rock (49°47′N, 6°27′W) to North-east Providence Channel (25°50′N, 77°00′W), an alternative to the full-powered route, given at 2.83, may be preferred:

By great circle to Position 36°00′N, 35°00′W, thence:
By great circle to destination (25°50′N, 77°00′W).
Distance   3730 miles.

8.17.1

1   For low-powered vessels proceeding from Bermuda (32°23′N, 64°38′W) to Habana (23°10′N, 82°22′W) an alternative route to the full-powered route, given at 4.9, leads:

2   From the W end of North-West Providence Channel (26°10′N, 78°35′W):

Either:
Across Straits of Florida to Fowey Rocks (25°35′N, 80°06′W), thence:
Close off Florida Reefs, as directed in *West Indies Pilot, Volume I* to Sand Key, thence:
Across the Florida Current again to Habana (23°10′N, 82°22′W).
Or, preferably by day:
Along the W edge of Great Bahama Bank (23°30′N, 78°00′W) to Habana. See *West Indies Pilot, Volume I.*

From English Channel ⇒ South America
English Channel to Arquipélago de Cabo Verde
8.18

1   On leaving the English Channel at once make westing, as the prevailing winds are from that direction. With a favourable wind from Lizard Point (49°58′N, 5°12′W) steer a course to the WSW to gain an offing in 10°W to 12°W.

2   If the wind should be from W keep on the tack which enables most westing to be made to get a good offing and keep clear of the Bay of Biscay, even standing NW until well able to weather Cabo Finisterre (42°53′N, 9°16′W) on the starboard tack. By making a long board to the W nothing is lost, as the wind will generally be found to veer, so that a change of wind will be favourable and even permit a vessel to pursue a course with a free wind; whereas if embayed in the Bay of Biscay any change of wind would necessitate beating to windward against the current.

3   It must be borne in mind that the prevailing winds and currents have a tendency to set towards Île d'Ouessant (48°28′N, 5°05′W) and into the Bay of Biscay when S of it. To get well to the W is therefore of the greatest importance. Île d'Ouessant should in no case be sighted.

4   From 10°W to 12°W shape course to pass Madeira (32°40′N, 17°00′W) at any convenient distance, giving a wide berth to Cabo Finisterre, in passing it, as the current from the Atlantic Ocean usually sets right onshore there. In winter it is preferable to pass W of Madeira for the strong W gales, which occur from November to January, produce eddy winds and heavy squalls E of the island. From Madeira the best track is to pass W of, but just in sight of Arquipélago de Cabo Verde (17°00′N, 24°00′W), as the winds are stronger and steadier W than E of them.

Arquipélago de Cabo Verde ⇒ N coast of Brazil
8.19

1   No particular crossings of the equator are necessary (see 8.20), as the E coast of South America does not have to be weathered. From abreast Arquipélago de Cabo Verde (17°00′N, 24°00′W) steer a direct course, taking care to make the coast E of the destination, and thence steering along the coast in depths of from 18 m to 27 m.

Arquipélago de Cabo Verde ⇒ the equator
8.20

1   In considering where to cross the equator it is necessary to bear in mind that if a vessel crosses far to the W there will be a less interval of doldrums to cross, but it may be necessary to tack in order to weather the coast of South America and these crossings vary during the year as the direction of the South-east Trade Wind is more S when the sun is N of the equator than when S of it.

2   After passing Arquipélago de Cabo Verde stand S between the meridians of 26°W and 29°W, being nearer to 26°W from May to October and nearer 29°W from November to April.

3   The equator should be crossed at points varying according to season, as follows:

     January to April, when the North-east Trade Wind is well to the S, continue on a S course and cross the parallel of 5°N between 25°W and 28°W and the equator between 28°W and 31°W.

     May and June, when the S winds will be met with between 5°N and 10°N. On meeting them stand on the starboard tack so as to cross the parallel of 5°N between 18°W and 20°W. Between 4°N and 5°N go round on the port tack and cross the equator between 23°W and 25°W.

     July to September, when the S winds will be met with between 10°N and 12°N. On meeting them steer on the starboard tack so as to cross the parallel of 5°N between 17°W and 19°W. Go round then on the port tack and cross the equator between 23°W and 25°W.

     October to December, when the S winds will be met with between 6°N and 8°N. On meeting them steer so as to cross the parallel of 5°N between 20°W and 23°W, then take the tack that gives the most southing and cross the equator between 24°W and 29°W.

8.20.1

1   **Caution**. The South Equatorial Current (3.9) is not so strong in the winter of the N hemisphere as in summer and autumn, but the mariner must remember that the strength of the current increases as it advances towards the American coast.

From the equator southward
8.21

1   Having crossed the equator, as recommended, stand across the South-east Trade Wind (3.4) on the port tack, even if the vessel should fall off to about 260°, for the wind will draw more to the E as the vessel advances and finally to due E at the S limits of the Trade Wind. When in the vicinity of Penedos de São Pedro e São Paolo (0°55′N, 29°21′W), the vessels position should be frequently checked, the current observed and allowed for and a good look-out kept, as these rocks are steep-to and, even on a clear day, can only be seen from a distance of about 8 miles. The same precautions are necessary if passing W of Ilha de Fernando de Noronha (3°50′S, 32°37′W), when approaching Atol das Rocas (3°52′S, 33°49′W).

2   On approaching the Brazilian coast:

     March to September, when the wind is from the SE and the current near the coast sets N, it will be better to keep from 120 to 150 miles off the land until well S and steer to be to windward of the port of destination.

     October to January, when the NE winds prevail and the current sets SW, the coast may be approached with prudence and a vessel may steer according to circumstances for her intended port.

8.21.1

1   For Rio de Janeiro (22°59′S, 43°10′W), from October to March make Cabo Frio (23°01′S, 42°00′W) and give the coast a prudent berth, as a constant and sometimes heavy swell sets in. The islands at the entrance to the harbour should not be approached until the sea breeze is well set in, as a vessel may run into a calm and be exposed to the swell and current.

8.21.2

1   For Montevideo (34°54′S, 56°15′W) or Río de La Plata (35°10′S, 56°15′W), stand direct through the South-east Trade Wind, passing about 200 miles E of Rio de Janeiro (22°59′S, 43°10′W).

8.21.3

1   For Bahía Blanca (39°10′S, 61°45′W) and ports southward. When bound for Bahía Blanca, or if bound farther S from Montevideo or Río de La Plata, keep well in with the coast. This can be done with safety, as the winds are almost always from W, and an E gale never comes on without without ample warning. Pass Cabo Corrientes (38°01′S, 57°32′W), at a distance of 40 to 50 miles, and make the land S of Cabo Blanco (47°12′S, 65°45′W) and afterwards keep it topping on the horizon until the entrance to Estracho de Magallanes (52°27′S, 68°26′W) has been passed.

2   This coastal route is more advantageous than one farther offshore, and a vessel would do well to make a tack inshore, even with an apparent loss of ground, in order to maintain it. As long as the wind does not back to the E of S the water will be smooth and more sail can be carried than if farther out. Should the wind come from SE, unless just off Cabo Blanco, the land recedes so much as to afford plenty of sea room.

Atlantic Ocean ⇒ Pacific Ocean

Estrecho de Magallanes
8.22

1   Estrecho de Magallanes (52°27′S, 68°26′W), where there are violent squalls and a lack of searoom, is not advised as a route for sailing vessels. Some historical passages took 80 days, or more, between Puerto del Hambre (53°40′S, 70°55′W) and Cabo Pilar (52°43′S, 74°41′W).

Cabo de Hornos
8.23

1   West-bound round Cabo de Hornos (56°04′S, 67°15′W) the usual track is to take as direct a route as possible from a position 200 miles E of Rio de Janeiro (22°59′S, 43°10′W) to about 45°S, 60°W and from thence to pass 30 or 40 miles E of Isla de los Estados (54°50′S, 64°10′W). This track lies between 120 and 200 miles E of the Patagonian coast and is the most direct route for a large and well-found ship. The old navigators, however, recommended that sailing ships should keep within 100 miles of the coast in order to avoid the heavy sea that is raised by the W gales and to profit by the variableness of the inshore winds when from a W direction.

2   Near the coast from April to September, when the sun has a N declination, the winds prevail more from WNW to NNW than from any other quarter. Gales from the E are of very rare occurrence and even when they do blow, the direction being oblique upon the coast, it is not hazardous to keep the land aboard.

3   From October to March, when the sun has a S declination, though the winds shift to the S of W and frequently blow hard, yet as it is a weather shore, the sea goes down immediately after a gale. The winds at this time are certainly against making quick progress, yet as they seldom remain fixed in one point, and frequently back or veer 6 or 8 points in as many hours, advantage may be taken of the changes so as to keep close in with the coast.

4   When passing Isla de los Estados the usual course is E of the island, but there is, off its E extremity, a heavy tide-rip which extends for a distance of 5 or 6 miles, or even more to seaward. When the wind is strong, and opposed to the tidal stream, the overfalls are overwhelming and very dangerous, even to a large and well-found vessel. Every precaution should be used to avoid this perilous area.

5   Estrecho de Le Maire (54°50′S, 65°00′W) provides the shortest route round Cabo de Hornos with a valuable saving, when the difficulty of making westing is considered, of some 60 miles. Furthermore, a vessel is to some extent protected from W gales and heavy seas when between Estrecho de Le Maire and Cabo de Hornos, and she will avoid the NE-going current which is encountered between the E extremity of Isla de los Estados and Cabo de Hornos.

6   However, conditions must be suitable for passage of Estrecho de Le Maire, which is best attempted during daylight, with a favourable wind and tide; the best time for beginning the passage being one hour after high water. A vessel should, if necessary, heave-to off the entrance to the strait until that time. Under these conditions, even should the wind fail or come adverse, a vessel would probably drive through rapidly, for the tidal streams are strong. With a S wind, it would not be advisable to attempt the strait, for, with a weather-going tide, the sea is very turbulent and might severely endanger the safety of a small vessel and do much damage to a large one. In calm weather it would be still more imprudent, unless the W side of the strait can be reached, where a vessel might anchor, on account of the tidal streams which set towards Isla de los Estados where, if it becomes necessary to anchor, it would be very deep water and close to the land.

7   Should the wind fail, and the tidal stream not be sufficiently strong to carry a vessel through, there is a convenient anchorage in Bahía Buen Suceso (54°49′S, 65°14′W).

8   Winds from N and NE are often accompanied by thick, misty weather; vessel approaching the strait are thus often compelled to lie-to for a time.

9   June and July are the best months for making a W-bound passage round Cabo de Hornos as the wind is often then in the E quarter. The days are short, however, and the weather is cold. August and September are bad months with heavy gales, snow and ice occurring at about the time of the equinox. From October to March, in summer, the winds are almost invariably W.

10   In April and May the winds are slightly more favourable. The passage from E to W round Cabo de Hornos should usually be made in about 57°S, or at about 100 miles S of the cape, but if, after passing Isla de los Estados, the wind is W the vessel should be kept on the starboard tack until in 60°S, unless the wind veers from S to SSW, and thence on the tack that offers the most westing. On this parallel the wind is thought by some to prevail more from the E than any other quarter.

11   It would usually be necessary to stand S, in this manner, from August to March; but from April fair passages have been made by keeping nearer the land and sighting Islas Diego Ramirez (56°30′S, 68°43′W). There is no advantage to be gained by attempting, even with a favourable wind, to go close to Cabo de Hornos, for the E-going Southern Ocean Current (3.9) sets close past the cape and appears to flow with greater velocity under the land than farther seaward on the route from Cabo San Juan (54°43′S, 63°49′W).

From English Channel south-bound ⇒ European ports

General directions
8.24

1   For all destinations, at once make westing as the prevailing winds are from that direction. With a favourable wind from Lizard Point (49°58′N, 5°12′W) steer a course to the WSW to gain an offing in 10°W to 12°W.

2   **Traffic Separation Schemes** are established in the W approaches to the English Channel, off the W coasts of Spain and Portugal and in the Strait of Gibraltar.

8.24.1

1   For Western French ports. The WSW course (8.24) should be modified, according to weather, in order to reach the destination more directly. It must be borne in mind that the prevailing winds and currents have a tendency to set a vessel towards Île d'Ouessant (48°28′N, 5°23′W) and the many surrounding dangers. If circumstances require it, shelter may be obtained in one of the French anchorages, until weather improves, but Pointe de Penmarc'h (47°48′N, 4°23′W) should never be made.

8.24.2

1   For Lisboa (38°36′N, 9°24′W). Having gained an offing in 10°W to 12°W (8.24) and with the wind from the W, haul to the wind on the tack that will best enable the approach to the proper course to be made without being drawn into the Bay of Biscay, which is especially to be avoided. During and after SW gales the indraft of the bay is strongest and is to be guarded against.

2   Should S or SE gales have been experienced the vessel will have been driven W and if this is the case the aim should be to progress S. On the other hand, if W gales have prevailed, and the vessel has become embayed, it may be found difficult to weather Cabo Finisterre (42°53′N, 9°16′W) or even Cabo Ortegal (43°46′N, 7°52′W); in these circumstances refuge may be found in El Ferrol (43°28′N, 8°20′W), La Coruña (43°22′N, 8°24′W), Ría del Barquero (43°45′N, 7°38′W) or Ría de Vivero (43°43′N, 7°35′W) and, in extreme cases, in the ports and roadsteads of France from La Gironde (45°40′N, 1°28′W) to Brest (48°22′N, 4°30′W).

3   Rather than run any risk of becoming embayed in this manner, it will be better to make to the W and, since W winds generally veer, once a good offing has been made a more SE course can afterwards be resumed, making due allowance for a SE-going set.

4   Proceeding S of Cabo Finisterre, shape course to clear Os Farilhões (39°29′N, 9°33′W) and Isla Berlenga (39°25′N, 9°30′W), which should be given a wide berth in thick weather. With SW winds it is better to keep off the land to avoid the N-going current that sets along the coast with those winds also to be in a position to profit by any change of wind to the W and NW. In short it is better to run to the S, at some distance from the coast of Portugal, as W winds will make it a lee shore. In winter these gales are frequent, blow with great strength and continue for several days.

5   Bound for Lisboa, when abreast of Isla Berlenga, steer for a position off Cabo da Roca (38°46′N, 9°30′W).

8.24.3

1   For Strait of Gibraltar (36°00′N, 5°21′W). Take the Lisboa route (8.24.2) to clear Os Farilhões (39°29′N, 9°33′W) and Isla Berlenga (39°25′N, 9°30′W) and then continue down the coast as far as Cabo de São Vicente (37°01′N, 9°00′W), thence shape a course for the Strait of Gibraltar.

2   Cabo de São Vicente should be sighted and, after rounding it and proceeding SE, the state of the wind and weather and the indraft and current of the Strait of Gibraltar must be considered and allowed for. With the wind from NW through N to NE make Cabo Trafalgar (36°11′N, 6°02′W); with it from W through S to E make Cabo Espartel (35°47′N, 5°56′W).

3   In thick weather the safety of the vessel may be assured by making the bank which extends about 20 miles from the coast abreast Cabo Trafalgar, but care must be taken on nearing Isla de Tarifa (36°00′N, 5°37′W), to avoid Los Cabezos (36°01′N, 5°43′W).

4   Cabo Espartel is safe to approach and can be seen from a long distance. To the S of the cape the land falls and has been mistaken for the mouth of the strait; so that, at night when the light is not seen, caution is necessary. If an E wind is met, and it is too strong to beat against, shelter will be found under Cabo Espartel, the vessel either keeping under way or anchoring off Playa de Jeremias, about 3 miles S of the cape.

5   When working through the Strait of Gibraltar, against an E wind, keep in mid-channel to have the advantage of the current while the W-going tidal stream is running, but, with the E-going stream either shore may be approached with a chance of meeting favourable slants of wind. When Isla de Tarifa is passed the force of the wind lessens.

6   When the E wind inclines to the N it is advisable to keep on the Spanish coast, avoiding La Perla (36°03′N, 5°26′W), but when it inclines to the S the African coast is preferable.

From English Channel ⇒ Cape of Good Hope

Routes
8.25

1   Follow the directions given at 8.18 and 8.20 to cross the equator between 23°W and 31°W, according to season, passing W of Arquipélago de Cabo Verde (17°00′N, 24°00′W). Having crossed the equator head across the South-east Trade Wind on the port tack, even if the vessel cannot make a better course than WSW, for the wind will draw more to the E as the vessel advances and finally to E at the S limit of the Trade Wind. When in the vicinity of Penedos de São Pedro e São Paolo (0°55′N, 29°21′W) and Ilha de Fernando de Noronha (3°50′S, 32°37′W) precautions should be taken as described at 8.21. During the greater part of the year the South-east Trade Wind fails on a line drawn from Cape of Good Hope (34°21′S, 18°30′E) to Ilha da Trinidade (20°30′S, 29°20′W) and Ilhas Martin Vaz (20°31′S, 28°51′W). This limit varies according to season.

2   When S of the South-east Trade Wind fresh winds, variable in direction, will be met. Those from NE through N to NW, if accompanied by cloudy weather, often shift suddenly to SW or S, but sometimes the wind steadies between W and WSW. From Ilha da Trinidade shape course to the SE to cross the parallel of 30°S in about 22°W and the meridian of Greenwich in about 35°S to 37°S, whence to Cape of Good Hope winds from W and S usually prevail. If E-bound round the Cape of Good Hope, cross the meridian of Greenwich in about 40°S.

3   After passing the meridian of Greenwich a strong N-going current will frequently be experienced and, on nearing the land when bound to Table Bay (33°50′S, 18°30′E), great caution is required, as there it will be found almost constantly running strongly to the N and, if it is disregarded, a vessel may well have difficulty and lose time in reaching the bay. If bound for Simons Bay (34°11′S, 18°26′E) during the S summer, it will be better to make the land about Cape Hangklip (34°28′S, 18°50′E), as a strong current sets at that period across the entrance to Valsbaai (False Bay) (34°15′S, 18°40′E) towards Cape Point (34°21′S, 18°30′E).

4   If near the coast at night, and land is not visible, keep to the SW until the position is ascertained. In any circumstances, at night, there is great difficulty in judging the distance of lights situated under high land. Therefore, the prudent course for a stranger to pursue when making Table Bay is to keep off and on until daylight, far enough W of Green Point (33°54′S, 18°24′E) to prevent being becalmed near the land and set in upon the coast by the heave of the sea.

5   For continuation to the Indian Ocean, see 8.95.

8.25.1

1   **Note**. As the wind seldom, if ever, blows from the E or NE (ie directly off the peninsula), sailing vessels bound either for Table Bay or round Cape of Good Hope should ensure a weatherly position to the N or S according to season. Those for Simons Bay have been detained many days by south-easters off Lion's Head (33°56′S, 18°23′E) and Hout Bay (34°04′S, 18°21′E), in consequence of their making land too far to the N during the summer season. The same winds would have been favourable for them had they been 30 miles farther S. On the other hand a vessel bound for Table Bay in the winter season will find it difficult to make her port from a position off Cape Point, during the continuance of N and NW winds,

notwithstanding the general prevalence of a NNW-going current.

From English Channel ⇒ West African ports

Routes
8.26

For ports in West Africa, follow the directions given at 8.18 as far as Madeira (32°40′N, 17°00′W). Thence steer so as to pass 60 to 100 miles W of Islas Canarias (29°00′N, 15°00′W) and from abreast these islands take one of the following routes.

8.26.1

1   For Banjul (Bathurst) (13°32′N, 16°54′W) and Freetown (8°30′N, 13°19′W), according to season.

2   November to April:
Steer due S to about 20°N, and then edge over to the African coast and steer as directly as navigation permits to destination.

3   May to October:
Keep more to the W so as to sight Arquipélago de Cabo Verde (17°00′N, 24°00′W); and, after picking up the South-east Trade Wind in about 10°N, stand on the starboard tack direct to destination.

Note. Some navigators recommend standing to the W of Arquipélago de Cabo Verde from abreast Islas Canarias; the North-east Trade Wind being sometimes held longer by doing so and turning E after passing them.

8.26.2

1   For Lagos (6°23′N, 3°24′E) or Calabar River (4°24′N, 8°18′E), according to season.

2   November to April:
After edging towards the coast, as directed in 8.26.1, keep it at about 60 miles distance until abreast of the port of destination.

3   May to October:
Keep about 200 miles off the African coast during the south-west monsoon (2.4). Turn to the E in about 6° or 7°N, 15°W and, closing the land to keep in the Guinea Current (2.15), steer to destination.

English Channel ⇒ Saint Helena

Routes
8.27

There are alternative routes, depending on season.

8.27.1

1   **The usual route** is as for Cape of Good Hope, described at 8.25, to beyond the South-east Trade Wind, then making enough easting to be able to enter the Trade Wind again and weather Saint Helena (15°57′S, 5°43′W), which should be approached from SE. As a rule, avoid going on the starboard tack, or decreasing latitude, until Saint Helena bears about 035°.

8.27.2

1   **Northern Route**. This route, is deemed by some authorities, to be preferable from January to April. It passes E of Arquipélago de Cabo Verde (17°00′N, 24°00′W) and along the African coast until past Cape Palmas (4°22′N, 7°42′W), thence, keeping in the Guinea Current (2.15), to pass close to São Tomé (0°12′N, 6°37′E). In March, try to reach about 7°or 8°S, 4°or 5°E from whence Saint Helena will generally be fetched on the port tack; but in June and early July it will probably be sufficient to get as far as 4°or 5°S in the same longitude, as the wind is then generally more E.

Bay of Biscay and west coasts of Spain and Portugal ⇒ Atlantic Ocean and English Channel

Sailing ship route
8.28

1   On the usual route, whether N-bound or S-bound, the importance of making to the W as quickly as possible to join the routes to or from the Channel, cannot be too strongly stressed. From the Bay of Biscay it may even be advisable to postpone sailing until a favourable wind enables a vessel to avoid all risk of being embayed. The indraft of the Bay is strongest after SW gales.

2   See 8.11 to 8.18 and 8.24 to 8.27 for routes from English Channel and 8.30 for route from Gibraltar.

Low-powered vessel route
8.29

1   For low-powered vessels west-bound from Vigo (42°13′N, 8°50′W), Lisboa (38°36′N, 9°24′W) and Strait of Gibraltar (36°00′N, 5°21′W) for Boston (42°20′N, 70°46′W) Chesapeake Bay (36°56′N, 75°58′W) and places between them, alternatives to the full-powered route, given at 2.71, may be preferred the routes are seasonal.

2   May to September:
S of Arquipélago dos Açores (38°00′N, 27°00′W), thence:
Along the parallel of 36°N to 60°W, thence:
As navigation permits to destination.

3   October to April:
Direct to 33°15′N, 20°00′W, thence:
Along the parallel of 33°15′N to 65°W, thence:
As navigation permits to destination.

Gibraltar ⇒ English Channel

General directions
8.30

1   **Traffic Separation Schemes** are established off the W coasts of Spain and Portugal and in the Strait of Gibraltar.

2   The W-bound passage through the Strait of Gibraltar (36°00′N, 5°21′W) against the general E-going current is, even with a favourable wind (especially during neap tides), somewhat difficult for sailing vessels, but with W winds, which increase the strength of the current, it is, for a large ship, almost impossible.

3   From Europa Point (36°06′N, 5°21′W) work along the coast of Spain during the W-going tidal stream until reaching Isla de Tarifa (36°00′N, 5°37′W) and, if necessary, anchor there to await the next favourable stream. If from Algeciras (36°08′N, 5°27′W) get under way at half ebb and so reach Punta del Acebuche (36°02′N, 5°27′W) by the commencement of the W-going stream.

4   If successful in rounding Isla de Tarifa, by keeping to the Spanish coast, continue working up Playa de los Lances (36°01′N, 5°37′W) while the tidal stream remains favourable. After gaining Punta de la Peña tower (36°03′N, 5°40′W), if not proceeding inshore of Los Cabezos (36°01′N, 5°43′W), cross to the African coast and work up under that, as directed below. If the wind is from SW, with moderate weather, keep to the Spanish coast, as by crossing to the African coast, where the wind will probably be less, a vessel will be set to leeward. Should the wind shift to WNW or NW, the Spanish coast should still be kept.

5   If unable to fetch Tangier (35°47′N, 5°47′W), by following these directions, cross to the African coast and work up that coast with the favouring stream, anchoring when necessary, until Tangier Bay is reached; but Isla de Tarifa should be reached before standing across, otherwise there will be no certainty of weathering Punta Cires (35°55′N, 5°29′W) and should a vessel fall to leeward of it then it would be difficult even to regain Gibraltar Bay.

6   Having weathered Punta Cires, work within the counter-current and near the shore to take advantage of any slant of wind that may occur, then doubling Punta Malabata (35°49′N, 5°45′W), gain Tangier Bay, whence it will be easy to regain the Spanish coast. When the meridian of Tangier is passed there is less current and a more manageable wind than in the narrows.

7   With W winds, if a small vessel makes Peninsula de la Almina, Ceuta (35°54′N, 5°17′W), instead of Europa Point, she may work up on the African coast, within the limits of the tidal streams, anchoring during the E-going stream.

8   From S of Ceuta to Gibraltar work up as far as Punta Cires, then taking advantage of the W-going stream cross the strait sailing a point free. If the wind is SW this is more easily done, with the favourable slants of wind met with on the African coast.

9   In light winds, preserve a good offing when in the vicinity of Cabo de São Vicente (37°01′N, 9°00′W), as the currents generally set strongly along the land and have a tendency towards the cape. Ripples are occasionally seen about 3 miles SW of and off the cape. After passing Cabo de São Vicente stand out to the NW on the prevailing N winds until a favourable wind is met. Get an offing of at least 100 to 150 miles to avoid the S and SE-going current near the coast of Portugal.

10   If S winds should be met with, stand to the N keeping sufficiently to the W to be able to weather Île d'Ouessant (48°28′N, 5°23′W) easily and do not steer E until N of the parallel of that island.

Gibraltar ⇒ Halifax or New York

Sailing ship route
8.31
Having cleared Strait of Gibraltar (36°00′N, 5°21′W) as described at 8.29, stand to the SW into the North-east Trade Wind. Thence as directed at 8.13 to 8.13.2

Low-powered vessel route
8.32
For low-powered vessels W-bound from Strait of Gibraltar (36°00′N, 5°21′W) bound for Halifax (44°31′N, 63°30′W), an alternative to the full-powered route, given at 2.71, may be preferred this route is by rhumb line, S of Arquipélago dos Açores, to 36°00′N, 45°00′W, thence as navigation permits.

Gibraltar ⇒ West Indies, South America, Cabo de Hornos or Cape of Good Hope

General directions
8.33
Having cleared Strait of Gibraltar (36°00′N, 5°21′W) as described at 8.30, stand to the SW to join the appropriate route 8.16 to 8.21.3 and 8.24 to 8.25.1.

PASSAGES IN MEDITERRANEAN SEA

Charts:
> *5124 (1) to (12) Monthly Routeing Charts for North Atlantic Ocean.*
> *5309 Tracks followed by Sailing and Auxiliary Powered Vessels.*
> *5310 World Surface Current Distribution.*
> *Diagram 8.3 World Sailing Ship Routes.*

Gibraltar ⇒ Gulf of Lions and Genova

Routes
8.34
1   After leaving the Strait of Gibraltar (36°00′N, 5°21′W), keep in mid-channel whether the wind is from E or W then follow the appropriate seasonal route described below.

2   **Traffic Separation Schemes** are established in the Strait of Gibraltar, off Cap Bon and Îles Cani.

8.34.1
1   **Summer route**. Pass between Islas Baleares (39°45′N, 3°00′E) and Spain. A vessel bound for Marseille (43°18′N, 5°20′E) should sight Cabo de San Sebastian (41°53′N, 3°12′E) or Cabo Creus (42°19′N, 3°19′E) before crossing the Gulf of Lions; but if bound for Gulf of Genoa she should make land about Îles d'Hyères (43°00′N, 6°20′E). In most cases, when bound for Genova (44°23′N, 8°53′E) or Livorno (43°33′N, 10°19′E), the sooner the coast of Provence is made, the more secure the voyage, unless the wind should be settled from SE to SW.

8.34.2
1   **Winter route**. Keep along the coast of Spain up to Cabo Creus, where shelter may be obtained in Bahía de Rosas (42°15′N, 3°09′E) in the case of a N gale or bad weather and thence, if bound for Marseille stand across the Gulf of Lions and pass well W of Île de Planier (43°12′N,

5°14′E), but in case of a SE wind try to make easting as quickly as possible as far as 5°E. If bound for Gulf of Genoa make Îles d'Hyères.
8.34.3
1   **Cautions**. In winter, sailing vessels rounding Cap Corse (43°01′N, 9°22′E), the N end of Corse, should give it a berth of 6 to 8 miles, as within that distance dangerous whirlpools and squalls come off from the cape.

2   When approaching the N shore of the Gulf of Lions, with S winds, the greatest caution is necessary as the currents with these winds set strongly N and NW and many vessels have been wrecked.

Gibraltar ⇒ Sardegna, Sicilia or Napoli

Routes
8.35
1   After leaving the Strait of Gibraltar (36°00′N, 5°21′W) the routes are seasonal as described below.
8.35.1
1   **Summer route**. With a favourable wind pass between Isla de Alborán (35°56′N, 3°02′W) and the coast of Spain and midway between Islas Baleares (39°45′N, 3°00′E) and the coast of Africa, along the S coast of Sardegna and N or S of Sicilia, according to destination.

2   With an E wind work to windward in mid-channel and then between Islas Baleares and the coast of Africa, keeping nearer the African coast when the wind is to the S of E and nearer to the islands if the wind is N of E.
8.35.2
1   **Winter route**. Keep along the coast of Spain as far as Cabo de Palos (37°38′N, 0°41′W), thence make for the S end of Sardegna and pass N or S of Sicilia, according to destination.

Gibraltar ⇒ Malta

Routes
8.36

1   After leaving the Strait of Gibraltar (36°00′N, 5°21′W) the routes are seasonal as described below.

2   **Traffic Separation Schemes** are established in the Strait of Gibraltar, off Cap Bon and Îles Cani.

8.36.1

1   **Summer route**. From May to September, steer midway between Spain and Africa until abreast of Cabo de Gata (36°43′N, 2°11′W), thence keeping to the African coast as far as Cap Bon (37°05′N, 11°03′E), to profit from the E-going current, passing N of Îles de la Galite (37°30′N, 8°55′E). Thence proceed direct for Malta (35°55′N, 14°33′E), passing N or S of Isola di Pantelleria (36°47′N, 12°00′E) and the Maltese Islands, according to circumstances.

8.36.2

1   **Winter route**. From October to April, W winds (SW to NW) principally prevail, making it desirable to keep along the coast of Spain as far as Cabo de Palos (37°38′N, 0°41′W), to steer for the S coast of Sardegna. In all circumstances the African coast should be avoided in winter, as the N gales make it a dangerous lee shore. From S of Sardegna make for Cap Bon and pass N of Isola di Pantelleria and Għawdex (36°03′N, 14°15′E). With a strong SW wind, however, the African coast may be kept as far as Cap Bon (37°05′N, 11°03′E).

8.36.3

1   **Note**. If leaving Gibraltar with an E wind, work to windward in mid-channel as far as Cabo de Palos, thence make for the S end of Sardegna. Thence make for Cap Bon and pass N or S of Isola di Pantelleria and the Maltese Islands, according to circumstances.

Malta ⇒ Gibraltar

Routes
8.37

1   After leaving Malta (35°55′N, 14°33′E) there are alternative routes as described below.

2   **Traffic Separation Schemes** are established in the Strait of Gibraltar, off Cap Bon and Îles Cani.

8.37.1

1   **Usual route**. With a favourable wind, after passing Îles Cani (37°21′N, 10°07′E), keep well off the African coast to avoid the E-going current, and make the Spanish coast about Cabo de Palos (37°38′N, 0°41′W) afterwards keeping along it to Strait of Gibraltar (36°00′N, 5°21′W).

2   Great care is needed in making Strait of Gibraltar in the thick weather which usually accompanies E winds, as vessels, mistaking the Rock of Gibraltar for Jbel Musa (35°54′N, 5°25′W) and supposing they were passing through the strait, have been wrecked in Bahía Mala (36°10′N, 5°20′W) and Ensenada de Tétouan (35°35′N, 5°15′W), where the land is low.

3   With NW winds work along the coast of Sicilia to Isola Marettimo (37°58′N, 12°03′E) and then work across to the S coast of Sardegna and thence the S coast of Spain. The difficulty of getting to windward with W winds increases as the Strait of Gibraltar is approached, vessels being frequently required to remain some days at anchor off the coast. Short tacks should be made along the Spanish coast to avoid the E-going current in mid-channel.

4   If a NW gale is encountered between Malta and Isola di Pantelleria (36°47′N, 12°00′E), it is better to put back to Malta rather than risk damage in the heavy sea which will be met in that channel.

8.37.2

1   **Alternative route**. A route, recommended as a better one, is on leaving Malta to stand on the starboard tack towards the coast of Africa and work along it up to Cap Bon (37°05′N, 11°03′E), subsequently keeping well off the coast of Africa.

Napoli, Sicilia or Sardegna ⇒ Gibraltar

Routes
8.38

   At all times of the year pass along the S coast of Sardegna and Islas Baleares (39°45′N, 3°00′E) and keep along the S coast of Spain, from Cabo de Palos (37°38′N, 0°41′W), noting the remarks in 8.37.1.

Genova and Gulf of Lions ⇒ Gibraltar

Routes
8.39

   At all times of the year make for Cabo de San Antonio (38°48′N, 0°12′E) and then keep along the S coast of Spain, noting the remarks in 8.37.1.

PASSAGES FROM PORTS ON WEST COAST OF AFRICA AND FROM ATLANTIC ISLANDS

Charts:

     5124 (1) to (12) Monthly Routeing Charts for North Atlantic Ocean.

     5125 (1) to (12) Monthly Routeing Charts for South Atlantic Ocean.

     5309 Tracks followed by Sailing and Auxiliary Powered Vessels.

     5310 World Surface Current Distribution.

     Diagram 8.3 World Sailing Ship Routes.

Freetown or Arquipélago de Cabo Verde ⇒ English Channel

Routes
8.40

1   From Freetown (8°30′N, 13°19′W) head to the NW in the North-east Trade Wind. Run through the Trade Wind, passing W of Arquipélago de Cabo Verde (17°00′N,

24°00′W), then follow the route recommended for sailing vessels from Capetown to English Channel described at 8.60.

Freetown ⇒ Ascension Island

Routes
8.41

1 From Freetown (8°30′N, 13°19′W), when clear of Saint Ann Shoals (8°00′N, 13°30′W), run along the coast within 50 miles of the land until past Cape Palmas (4°22′N, 7°42′W), when an endeavour should be made to cross the equator between the meridians of 3°W and 8°W, whence, without making a tack, Ascension Island (7°55′S, 14°25′W) will be fetched.

2 During November, long continued calms and a strong NW-going current are experienced in the vicinity of Saint Ann Shoals.

Ghana, Nigeria or Bight of Biafra ⇒ Freetown or intermediate ports

Routes
8.42

1 For Freetown, head S from the N part of the Bight of Biafra (3°00′N, 7°00′E) and, if possible, pass W of Bioko (3°30′N, 8°45′E) and cross the equator as soon as possible, unless the vessel can point as high as WNW. When S of the equator head W in the South Equatorial Current (3.9) and, as westing is made, the wind will be found to shift gradually round to the SE. When in about 10°W recross the equator and shape course for Freetown (8°30′N, 13°19′W).

2 From any place in the Gulf of Guinea, E of Cape Palmas (4°22′N, 7°42′W), head S into the South Equatorial Current (3.9) and then proceed as above.

3 For intermediate ports, in working to windward in the Bight of Benin (5°00′N, 3°00′E), it is advisable to head off on the starboard tack during the day and inshore on the port tack during the night, tacking if the wind should veer. If going some distance along the Guinea coast it is advisable to head across the equator and make westing in the South Equatorial Current (3.9).

4 In the Harmattan season (November to February) the Guinea Current (2.15) near the land in this bight is checked and inshore a W-going set is felt.

Ghana, Nigeria or Bight of Biafra ⇒ English Channel

Routes
8.43

1 Head S of the equator in the South Equatorial Current (3.9) and then make westing as from the Bight of Biafra to Freetown, described at 8.42. Recross the equator in about 20°W and then as from the Cape of Good Hope (8.60) run through the North-east Trade Wind (2.5) and shape a course for the English Channel.

Ghana, Nigeria or Bight of Biafra ⇒ South America

Routes
8.44

There are two alternative routes, either via Saint Helena or via Ascension Island.

8.44.1

1 Via Saint Helena (15°57′S, 5°43′W), keep in the Guinea Current (2.15) until in the Bight of Biafra (3°00′N, 7°00′E)

and then work along the coast as far as 6°S, whence there will be little difficulty in reaching Saint Helena by keeping on the port tack. From Cape Palmas (4°22′N, 7°42′W) a vessel on the starboard tack will generally reach Cap Lopez (0°37′S, 8°43′E) and often S of Isla de Pagalu (1°26′S, 5°37′E).

2 From Saint Helena keep in the South-east Trade Wind (3.4), at about 20°S, until leaving Ilha da Trinidade (20°30′S, 29°20′W), whence edge off to Rio de Janiero (22°59′S, 43°10′W) or directly to the required destination.

8.44.2

1 Via Ascension Island (7°55′S, 14°25′W), head S on the starboard tack generally weathering Isla de São Tomé (0°12′N, 6°37′E), as far as the equator, then head W, taking care to stay in the South Equatorial Current (3.9). Progress will be slow at first, but as westing is made the South-east Trade Wind (3.4) will be felt.

2 From Cape Coast Castle (5°06′N, 1°14′W) head across the equator on the starboard tack, then as above.

3 For vessels from the coast S of the equator the winds are always favourable, gradually backing from SW to SE as the island is approached.

4 From Ascension Island, head through the South-east Trade Wind (3.4), to pick up the route from the English Channel to South America, described from 8.18 to 8.21, proceeding by it to the required destination.

Ghana and Bight of Biafra ⇒ Cape Town and Cape of Good Hope

To Cap Lopez
8.45

1 Along the whole shore of the Bight of Biafra (3°00′N, 7°00′E) work to windward as far Cap Lopez (0°37′S, 8°43′E) with the land and sea breezes, anchoring when necessary to avoid being set N by the current, especially during April and May, the season of calms and tornadoes.

Cap Lopez to River Congo
8.46

1 From Cap Lopez to River Congo (6°00′S, 12°30′E) maintain a good offing, only approaching the shore to take advantage of the land breezes, which begin to blow at, or a few hours before, sunrise. In February, and sometimes in October, the sea breeze extends so well to the W as to enable vessels to head along the coast on either tack, but during May the wind blows steadily along the coast from a S direction, night and day, with a N-going current of 1 kn.

2 To cross the stream from the River Congo (for details, see *Africa Pilot, Volume II*) either keep 200 miles off the coast or keep in anchoring ground, the latter is preferable. The usual course is to beat alongshore as far as Ponta Vermelha (5°40′S, 12°10′E), keeping on the bank of soundings in order to anchor if the wind falls light, crossing the stream when the sea breeze has well set in.

River Congo to Luanda
8.47

1 From River Congo to Luanda (8°47′S, 13°16′E) anchor every night when the sea breeze falls light, weigh with the first of the land breeze and continue on the port tack until about 1300 hours; then tack and by the time the sea breeze fails good progress will have been made to the S.

Luanda to Baía Namibe
8.48

1 From Luanda to Baía Namibe (15°10′S, 12°07′E) the currents set N with considerable force in the neighbourhood of Ponta das Palmeirinhas (9°06′S, 13°00′E) and a good tack off the land for 50 or 60 miles will enable the vessel to weather the point; it is seldom advisable to work inshore. Do not get more than 50 or 60 miles from the land as beyond these limits the sea breeze declines in force and draws more to the S, which would necessarily cause a loss of ground on the inshore tack; besides which the alternative land and sea breezes which are almost invariably experienced closer inshore would be lost.

South of Baía Namibe
8.49

1   To the S of Cabo Negro (15°40′S, 11°55′E) there is no difficulty in working S if advantage is taken of the variations of wind and the tacks are arranged accordingly. As rollers are frequent the shore must be given a good berth. To the S of Cape Frio (18°26′S, 12°00′E) N winds may be expected from May to August.

For Cape of Good Hope
8.50

1   For Cape of Good Hope (34°21′S, 18°30′E) head off and run through the Trade Wind and approach the Cape as described at 8.25.

Ascension Island ⇒ English Channel

Route
8.51

1   Ascension Island (7°55′S, 14°25′W) lies on the direct route from Cape Town (33°53′S, 18°26′E) to the English Channel. Follow the relevant part of the directions described in 8.60.

Ascension Island ⇒ South America

Route
8.52

1   From Ascension Island (7°55′S, 14°25′W) follow the general directions given at 8.44.2 and 8.21.

Ascension Island ⇒ Saint Helena

Route
8.53

1   From Ascension Island (7°55′S, 14°25′W) proceed S on the port tack and when beyond the limit of the Trade Wind make easting and re-enter the Trade Wind far enough to windward to ensure weathering Saint Helena (15°57′S, 5°43′W). Avoid going on the starboard tack, or decreasing the latitude, until Saint Helena bears about 010°.

Ascension Island ⇒ Cape Town or Cape of Good Hope

Route
8.54

1   From Ascension Island (7°55′S, 14°25′W) run to the S on the port tack, through and then out of, the South-east Trade Wind. Then head SE with the object of crossing the Greenwich meridian between 35°S and 37°S; the parallel of 30°S will be crossed probably not farther W of about 14°W.

2   Continue between the parallels of 35°S and 37°S and make Cape of Good Hope (34°21′S, 18°30′E) from SW.

   If bound for the Indian Ocean without calling at an intermediate port, proceed as directed at 8.95.

Ascension Island ⇒ Equatorial and South-west coasts of Africa

Notes and precautions
8.55

1   Leaving Ascension Island (7°55′S, 14°25′W) on the starboard tack a vessel will fetch the coast of Africa, according to the season, at some point between Cap Lopez (0°37′S, 8°43′E) and Luanda (8°47′S, 13°16′E) or even farther S. In May, however, the wind tends to be more E and a vessel may not weather Isla de Pagalu (1°26′S, 5°37′E); on the other hand a good vessel sailing well may make landfall S of River Congo (6°00′S, 12°30′E). Two precautions, however, are necessary:

   The first is not to get N of the parallels of 3°N or 4°N; and:

   The second is not to bring the port of destination to bear more than 160°.

2   An occasional short tack, as the wind shifts a little, may therefore be necessary, but the whole passage may sometimes be made with a free wind.

Saint Helena ⇒ South America

Route
8.56

1   From Saint Helena (15°57′S, 5°43′W) proceed as directed at 8.44.1 and if bound for Montevideo (34°54′S, 56°15′W) pass about 100 miles S of Ilha da Trinidade (20°30′S, 29°20′W), thence steer as directly as possible to destination.

Saint Helena ⇒ Ascension Island and English Channel

Route
8.57

1   Both Saint Helena (15°57′S, 5°43′W) and Ascension Island (7°55′S, 14°25′W) lie on the direct route from Cape Town (33°53′S, 18°26′E) to the English Channel. Follow the relevant part of the directions described in 8.60.

Saint Helena ⇒ West coast of Africa

Route
8.58

1   From Saint Helena (15°57′S, 5°43′W) vessels will generally fetch as far S as Benguela (12°35′S, 13°24′E), except in May, when the South-east Trade Wind has more easting in it and the lee current is strong. To all destinations N of Benguela, therefore, the winds are favourable. They veer from SE to S and SW as the coast is approached.

Saint Helena ⇒ Cape Town or Cape of Good Hope

Route
8.59

1   From Saint Helena (15°57′S, 5°43′W) run S on the port tack, through and out of, the Trade Wind and then head SE with the object of crossing the Greenwich meridian between 35°S and 37°S (probably not getting W of 10°W to 14°W). Continue between the parallels of 35°S and 37°S, as in the passages from the English Channel (8.25) and make Cape of Good Hope (34°21′S, 18°30′E) from SW.

2   If bound for the Indian Ocean without calling at an intermediate port, proceed as directed at 8.95.

Cape Town or Cape of Good Hope ⇒ Saint Helena, Ascension Island, English Channel or Bordeaux

Route
8.60

1   For the English Channel, from Cape of Good Hope (34°21′S, 18°30′E), first obtain a good offing to the NW as squalls from NW and WNW are not infrequent near the coast and have been experienced in both seasons. Then shape course for Saint Helena (15°57′S, 5°43′W).

2   From Saint Helena steer a direct course for Ascension Island (7°55′S, 14°25′W), passing it on either side, and crossing the equator between 25°W and 30°W (in July between 20°W and 25°W, to ensure better winds). Then make a course to the N to reach the North-east Trade Wind (2.5) as soon as possible (in July and August crossing the parallel of 10°N to the W of 30°W) and run through it. The North-east Trade Wind will probably be lost in about 26°N to 28°N and 38°W to 40°W, when W winds may be expected and, on reaching these, shape course for the English Channel.

3   It is seldom advisable to pass E of Arquipélago dos Açores (38°00′N, 27°00′W) but should the wind draw to the NW when near them the most convenient channel through them may be taken. If E winds are experienced after passing Arquipélago dos Açores the vessel should still be kept on the starboard tack, as W winds will probably be sooner found.

4   From November to February a vessel should pass about 50 miles W of Flores (39°25′N, 31°15′W) and Corvo (39°40′N, 31°05′W); but from June to August at about 250 miles W of these islands. At other times of the year at intermediate positions.

5   For Bordeaux, proceed as for the English Channel but begin to make easting on reaching the parallel of 30°N, passing between Terceira (38°48′N, 27°15′W) and São Miguel (37°45′N, 25°30′W) and rounding the NW point of Spain at a distance of 60 to 80 miles.

Cape Town or Cape of Good Hope ⇒ North and Central America and West Indies

Route
8.61

1   To cross the equator, from Cape of Good Hope (34°21′S, 18°30′E) follow the route to the English Channel (8.60) as far as 5°E and then steer with a favourable Trade Wind towards Ilha de Fernando de Noronha (3°50′S, 32°37′W). On reaching 10°S, at about 30°W, head more to the N so as to cross the equator between 31°W and 34°W and, as soon as the North-east Trade Wind (2.5) has been picked up, steer through it and thence as follows.

8.61.1

1   For New York (40°28′N, 73°50′W), try to reach 30°N, 70°W and thence steer as directly as possible to destination.

8.61.2

1   For north coast of South America, Trinidad, Guyana, Surinam and Guyane Française, proceed towards New York as described above in 8.61.1, leaving that route when clear of the Equatorial Counter-current (2.15) and proceed to destination as described from 8.16.3 to 8.16.6.

8.61.3

1   For Leeward Islands (16°00′N, 62°00′W), Jamaica (18°00′N, 77°30′W), Belize (17°20′N, 88°01′W) or ports in the Gulf of Mexico, proceed as described above in 8.61.2, making W as described at 8.16.2.

8.61.4

For Cuba (22°30′N, 80°00′W), proceed as described above in 8.61.3 but making W as described at 8.16.1.

Cape Town ⇒ South American ports

Sailing ship route
8.62

1   From Cape of Good Hope (34°21′S, 18°30′E) follow the directions at 8.61 as far as 20°S then run along this parallel with a favourable South-east Trade Wind (3.4) as far as 30°W, thence steer direct to Rio de Janeiro (22°59′S, 43°10′W), or if bound for ports farther S pick up, at 35°W, the appropriate outward route from the English Channel, described from 8.21.1 to 8.21.3.

Low-powered vessel routes
8.63

The following routes are recommended for low-powered vessels.

8.63.1

1   Cape Town or Indian Ocean to Río de La Plata (35°10′S, 56°15′W). An alternative route to that given at 3.43 follows the direct rhumb line for Rio de Janiero (3.41) as far as 20°W and continues thence by rhumb line to Río de La Plata (35°10′S, 56°15′W).

2   Although this route increases the distances by about 250 miles, lighter winds and more favourable currents are experienced.

8.63.2

1   Cape Town or Indian Ocean to Estrecho de Magallanes (52°27′S, 68°26′W), Falkland Islands (51°40′S, 57°49′W) or Cabo de Hornos (56°04′S, 67°15′W). Alternative routes to those given at 3.47 are:

By the rhumb line tracks to Rio de Janiero (3.41) as far as 27°00′S, 20°00′W, thence:
Rhumb line to 35°00′S, 40°00′W, thence:
Rhumb line to destination.

2   Although this route is about 150 miles longer, the advantages of lighter winds and more favourable currents should more than compensate.

8.63.3

1   **Cape Town or Indian Ocean ⇒ Rio de Janiero** (22°59′S, 43°10′W) An alternative route, to that given at 3.41, which takes advantage of better weather and more favourable currents, leads through:

29°50′S, 10°00′E, thence:
25°50′S, 0°00′, thence:
22°50′S, 10°00′W, thence:
21°10′S, 20°00′W, thence:
21°10′S, 30°00′W, passing S of Ilha de Trinidade (20°30′S, 29°20′W) and Ilhas Martin Vaz (20°31′S, 28°51′W), thence:
To Rio de Janiero (22°59′S, 43°10′W).

Cape Town ⇒ West coast of Africa

Route
8.64

1   From Cape of Good Hope (34°21′S, 18°30′E) obtain a good offing to the NW, as squalls from NW and WNW are not infrequent near the coast and have been experienced in all seasons. Steer to the N in the South-east Trade Wind (3.4), taking advantage of the Benguela Current (3.9).

2   If bound for ports E of Cape Palmas (4°22′N, 7°42′W), proceed as directly as navigation permits after first obtaining the offing described above.

PASSAGES FROM PORTS ON THE WESTERN SIDE OF THE ATLANTIC OCEAN

Charts:

> *5124 (1) to (12) Monthly Routeing Charts for North Atlantic Ocean.*
> *5125 (1) to (12) Monthly Routeing Charts for South Atlantic Ocean.*
> *5309 Tracks followed by Sailing and Auxiliary Powered Vessels.*
> *5310 World Surface Current Distribution.*
> *Diagram 8.3 World Sailing Ship Routes.*

From east coasts of Canada and United States

Routes
8.65

Routes are as follows:

8.65.1

1   For the English Channel, owing to the prevailing favourable winds and favourable currents, great circle or rhumb line courses may be steered as desired, provided care is taken to avoid ice.

8.65.2

1   For Capetown (33°53′S, 18°26′E) or Cape of Good Hope (34°21′S, 18°30′E), make for 35°N, 45°W. It is better to be about 60 miles N of this position in mid-summer and the same amount S in mid-winter. From this position there are two main routes, according to the time of year, offering the quickest passage through the Doldrums (2.3).

2   May to September, steer from about 36°N, 45°W to 25°N, 30°W and thence through the North-east Trade Wind (2.5) until meeting the South-east Trade Wind (3.4) between the parallels of 1°N and 5°N. Then proceed on the starboard tack, crossing the parallel of 5°N between 17°W and 20°W (the more E longitude in July and August). Then put the vessel about, so as to cross the equator between 23°W and 25°W. Thence head S through the South-east Trade Wind and begin to make easting from 25°S, 30°W, running due E along the parallel of 35°S as soon as it is reached, direct to destination.

3   October to April, from about 34°N, 35′W take a direct track through the North-east Trade Wind (2.5) so as to cross the parallel of 5°N between 20°W and 23°W. The S winds will be met in about 7°N, and on doing so put the ship on whichever tack gives the most southing and cross the equator between 20°W and 24°W. Directions for crossing the equator are given at 8.20 and after crossing it head through the South-east Trade Wind (3.4), and when it is lost, steer SE so as to cross the parallel of 30°S in about 30°W and thence so as to cross the meridian of Greenwich in 40°S. From this point steer direct for Cape Town, taking care not to be set N by the Southern Ocean and Benguela Currents (3.9), which make NE across the track.

4   If bound for the Indian Ocean without calling at an intermediate port, proceed as directed at 8.95.

8.65.3

1   For South American ports, proceed according to the directions for Cape Town, described at 8.65.2, as far as 5°S and then follow the appropriate outward route from the English Channel, described from 8.21 to 8.21.3, to required destination.

8.65.4

1   For Rio Amazonas (1°16′N, 49°30′W), head E to about 33°N, 50°W. Then turn S and make as directly as possible to destination, but nothing to the W of 43°W until reaching 5°N, on account of the strong W-going North Equatorial Current (2.14).

2   In July and August it will be advisable to make for 20°N, 37°W and then head S until the South-east Trade Wind (3.4) is picked up between 5°N and 10°N, thus approaching Rio Amazonas from well to the E.

8.65.5

1   For Caribbean Sea and Gulf of Mexico. Bound for Barbados (13°06′N, 59°39′W) or Trinidad (10°38′N, 61°34′W), make good easting passing either side of Bermuda (32°23′N, 64°38′W), but steering so as to cross the meridian of 60°W, or even 56°W, according to the season before entering the tropics and steering to the S, always allowing for the current to leeward.

2   **Caution**. For details of dangers in the vicinity of Bermuda, see 2.44.

3   If bound for Antigua (17°05′N, 61°45′W) or Leeward Islands (16°00′N, 62°00′W) it will not be necessary to go as far E as 60°W. For Mona Passage (18°20′N, 67°50′W) 66°W will give enough easting.

4   If bound for Jamaica (18°00′N, 77°30′W) or Colón (9°23′N, 79°55′W), make a good easting, as for Barbados, and then take Turks Island Passage (21°48′N, 71°16′W) and Windward Passage (20°00′N, 74°00′W), which is the shortest route.

5   If bound for Puerto La Guaira (10°36′N, 66°56′W), or ports to the E, make good the easting, as for Trinidad, and use Mona Passage, if required. A vessel making the South American coast W of her port will have considerable difficulty and lose much time working to windward to gain it.

6   If bound for the Gulf of Mexico, proceed as above by Turks Island Passage and Windward passage to pass S of Cuba and through Yucatan Channel (21°30′N, 85°40′W).

7   Further to the directions for approaching the Mississippi River (28°59′N, 89°07′W), given in *East coasts of Central America and Gulf of Mexico Pilot*, it may be said that currents near the mouth of the river are uncertain, with fog and haze prevalent, especially in summer and autumn. The mud banks are low and the wind is generally from the E; soundings should therefore be obtained well to windward. If approaching from S or SW great attention should be paid to checking the latitude, for the bank is steep-to.

8   If bound for the S shores of the Gulf of Mexico, a vessel should cross the E edge of Banco de Campeche (22°30′N, 90°00′W) between the parallels of 22°00′N and 22°30′N, and a knowledge of the exact point made is of great importance to check the longitude, especially during the rainy season (March to September) when observations can seldom be obtained. Soundings, therefore, must be used early and constantly. In this season it is best to take the inshore track across the bank as regular land and sea breezes then prevail. If bound for Vera Cruz (19°12′N, 96°08′W) in the "Norther" season (2.5) it is best to pursue the outer track, which runs between Arrecife Sisal (21°20′N, 90°10′W) and the outer cays and into the open between Triangulo Oeste (20°58′N, 92°18′W) and Bajo Nuevo (21°50′N, 92°05′W).

Mississippi River ⇒ East coast of North America or English Channel

Routes
8.66

1   From the mouth of the Mississippi River (28°59′N, 89°07′W) pass through the Straits of Florida (27°00′N, 79°49′W), taking full advantage of the Gulf Stream (2.14) and proceeding N-bound in it up the coast of the United States.

2 If bound for the English Channel head NE to 40°N, 60°W, and thence continue as directly as possible, with a favourable current and prevailing W winds, to destination.

Mississippi River ⇒ Colón or ports to Cabo Gracias á Dias

Routes
8.67

1 From the mouth of the Mississippi River (28°59′N, 89°07′W) pass between 5 and 10 miles off Cabo San Antonio (21°52′N, 84°57′W) to a position 25 miles ENE of Farrall Rocks (15°50′N, 82°19′W), and thence pass between 5 and 10 miles W of Isla de Providencia (13°23′N, 81°22′W) and direct to Colón (9°23′N, 79°55′W).

2 For ports to Cabo Gracias á Dias (15°00′N, 83°10′W), after passing Cabo San Antonio, steer to pass W of Islas Santanilla (17°25′N, 83°52′W) and Cayos Vivorillo (15°51′N, 83°18′W) and thence through Miskito Channel (14°20′N, 83°10′W).

South-west part of Gulf of Mexico ⇒ Atlantic Ocean

Routes
8.68

1 From the SW ports in the Gulf of Mexico take the passage inshore along the coast of Yucatan, where the adverse current is weak. Bound E, pass over Banco de Campeche (22°30′N, 90°00′W) within the shoals, but the passage between Arrecife Sisal (21°20′N, 90°10′W) and the coast should only be taken in daylight. In passing through the Straits of Florida (27°00′N, 79°49′W) and in proceeding N off the Atlantic coast of the United States, take all possible advantage of the Gulf Stream (2.14).

2 If bound for the English Channel head NE to 40°N, 60°W, and thence continue as directly as possible, with a favourable current and prevailing W winds, to destination.

Belize ⇒ English Channel or North American ports

Routes
8.69

1 From Belize (17°20′N, 88°01′W) the routes are as follows.

8.69.1

1 For the English Channel, proceed via Yucatan Channel (21°30′N, 85°40′W), thence through Straits of Florida (27°00′N, 79°49′W) with the Gulf Stream (2.14) to a position midway between Bermuda (32°23′N, 64°38′W) and Halifax (44°31′N, 63°30′W), thence after crossing the meridian of 40°W in about 45°N, continue direct to destination.

8.69.2

1 For east coast of North America, proceed as directed at 8.66.

8.69.3

1 For north shore of Gulf of Mexico, pass 35 miles E of Isla Mujeres (21°12′N, 86°43′W) and continue as directly as possible to destination.

Colón or Colombian ports ⇒ English Channel, New York or Mississippi River

Routes
8.70

1 From Colón (9°23′N, 79°55′W) and Colombian ports the routes are as follows.

8.70.1

1 For the English Channel, steer to pass through the Windward Passage (20°00′N, 74°00′W), between Haiti and Cuba, thence making northing on the starboard tack. When in the Westerlies (2.8), steer to cross the meridian of 40°W in about 44°N in summer, and about 40°N in winter. Thence continue as directly as possible.

8.70.2

1 For east coast of United States, pass through Windward Passage as described above, and having cleared Turks Islands (21°31′N, 71°08′W), head NW in the Antilles Current, until picking up the Gulf Stream (2.14) N of the Bahamas (25°00′N, 75°00′W) and thence proceed as directly as possible along the Atlantic coast of the United States.

8.70.3

1 For north shore of Gulf of Mexico, take the reverse of the New Orleans to Colón route (8.67) as far as the position off Farrall Rocks (15°50′N, 82°19′W) after which a course can be shaped to pass 30 and 40 miles W of Cabo San Antonio (21°52′N, 84°57′W).

From the S shores of the Caribbean Sea northward

Routes
8.71

1 From any of the ports along the Venezuelan coast and along the S shores of the Caribbean Sea, work E coastwise in the eddies or counter-current, until able to fetch the desired port on the starboard tack. At times, however, off Venezuela the W-going current is so far inshore that vessels need to cross it and work up to the N, as described in the route to Curaçao at 8.73. In winter, when the wind is well NE, it is necessary to make more easting than in summer, when the wind is more E of S.

2 From any of the Venezuelan ports, Mona Passage (18°20′N, 67°50′W) gives the best route for sailing vessels bound to any United States Atlantic port, or to the English Channel.

Jamaica ⇒ New York, Halifax or English Channel

Route
8.72

1 April to September, run to leeward round the W end of Cuba and then through the Straits of Florida (27°00′N, 79°49′W), thus getting full benefit from the Gulf Stream (2.14). Then proceed as described at 8.66.

2 October to March, N winds prevail in the Straits of Florida and the Windward Passage (20°00′N, 74°00′W) should be preferred, although ships are frequently opposed there by contrary winds and currents. These may to some extent be overcome by keeping nearer the coast of Haiti, as there a windward current is frequently found. When through the Windward Passage, use either Crooked Island (23°15′N, 74°25′W), Mayaguana (22°30′N, 73°20′W) or Caicos (22°15′N, 72°20′W) Passages, depending as to which the wind favours. From there proceed direct to New York (40°28′N, 73°50′W) or Halifax (44°31′N, 63°30′W). For English Channel see route 8.70.1 onward from Windward Passage.

Jamaica ⇒ Curaçao and the Southern shores of the Caribbean Sea

Routes

8.73

1   Work to windward along the S coast of Haiti; where at spring tides and also near the time of autumnal equinox, there is a counter, or even E-going, set, until on or to windward of the meridian of Curaçao (12°06′N, 68°56′W); then head across for that island when certain of fetching well to windward to allow for the prevailing W-going current. In summer more easting is necessary than in winter, as the wind has more southing in it and the current in summer is stronger.

2   A vessel which makes a landfall to leeward of her port will usually find a counter or E-going set near the shore, in which she can work up; failing to do this, if E of 70°W, she may possibly have to recross the prevailing W-going set and work up to the N of 14°N or 15°N.

West Indies ⇒ ports on east coasts of Canada and United States or English Channel

Routes

8.74

1   The great object for sailing vessels is to get N into the W winds as speedily as possible, Bermuda (32°23′N, 64°38′W) lies in, or near, the best track for this purpose, though a course E or W of it may be taken according to the direction of the wind met with and the season. A more N route is followed in summer than in winter.

2   **Caution**. For details of dangers in the vicinity of Bermuda, see 2.44.

3   From Barbados (13°06′N, 59°39′W), fetch to windward of all the islands, but from the other Windward Islands pass close to the leeward of Antigua (17°05′N, 61°45′W), taking care not come within a depth of 20 m.

4   Having cleared the other islands, and when steering directly for Bermuda, vessels sometimes fall to the E of the course and find it very difficult to make Bermuda when W winds prevail; in this case take advantage of the Trade Wind to reach the meridians of 68°W or 70°W before going N of the parallel of 25°N.

5   When bound for the English Channel, or to western Europe, it is seldom advisable to pass E of Arquipélago dos Açores (38°00′N, 27°00′W), although a passage between Flores (39°25′N, 31°15′W) and Corvo (39°40′N, 31°05′W) and the other islands of the archipelago is recommended by some navigators. If E winds are met with after passing Arquipélago dos Açores, still keep on the starboard tack, as by doing so W winds will probably be found sooner.

Barbados ⇒ North-east coast of South America

Routes

8.75

1   From Barbados (13°06′N, 59°39′W), work SE until abreast of the destination before attempting to cross the prevailing W-going current, particularly during and near the months of August and September, when the current is strong and the wind well to the S of E. It has been recommended, at this season, that vessels should not come S of 8°N until they are certain of fetching their destination on the port tack.

Rio Amazonas ⇒ Recife

Routes

8.76

1   From Rio Amazonas (1°16′N, 49°30′W) there are two alternative routes, inshore and offshore.

8.76.1

1   The usual route, is close inshore out of the influence of the W-going current, and by taking advantage of the current, tidal stream and every slant of wind, a sailing vessel will generally complete the voyage to landfall off Recife (8°00′S, 34°40′W) in about 30 days. During the prevalence of ENE and NE winds a current sets ESE along and near the coast; this fact is well known to the masters of local coasting craft and is taken advantage of by them.

2   When the weather will permit, a vessel may anchor off any part of the coast without danger. In working inshore, the dry season (July to December) is considered preferable, as the winds are then fresh and steady. Head offshore during the day and in towards the land at night, so as to be near the coast in the morning to take advantage of the land breeze, by which a good sailing vessel will make from 40 to 50 miles a day.

3   In the rainy season (January to June), working to windward is more tedious as calms, light variable winds, squalls and rain prevail. In this case head on the tack that is most favourable and, as a general rule, do not go outside a depth of 55 m. If the wind is steady, tack as in the dry season but do not lose sight of the coast.

8.76.2

1   An alternative route, leads farther offshore. Head directly N across the equator in about 10°N and then tack. This will save wear of sails and rigging and will probably take no longer than the inshore route.

Rio Amazonas ⇒ New York or English Channel

Routes

8.77

1   From Rio Amazonas (1°16′N, 49°30′W), after getting a good offing, head N so as to pick up the main Atlantic routes to the N:

        That to New York (40°28′N, 73°50′W) (8.61.1) in about 10°N to 15°N, and:

        That to English Channel (8.60) between 25°N and 30°N, passing W of Arquipélago dos Açores (38°00′N, 27°00′W).

Recife and North-east coast of South America ⇒ English Channel or New York

Routes

8.78

1   From landfall off Recife (8°00′S, 34°40′W) or other parts of the NE coast of South America the routes are as follows.

8.78.1

1   For the English Channel, after obtaining a good offing, head N and, after crossing the Doldrums (2.3), head through the North-east Trade Wind (2.5) into the Westerlies (2.8), passing W of Arquipélago dos Açores (38°00′N, 27°00′W) as directed at 8.60.

8.78.2

1   For New York (40°28′N, 73°50′W) and ports to the N, proceed as above (8.78.1) and, when the Doldrums are crossed and the North-east Trade Wind is reached, head direct for the required destination by the main route from Cape of Good Hope (8.61.1), from about 10°N.

Porto de Salvador ⇒ Europe or North America

Routes
8.79

1   In leaving Porto de Salvador (13°00′S, 38°32′W), and other ports immediately S of Recife (8°00′S, 34°40′W), for Europe, the NE winds sometimes compel sailing vessels to keep on the port tack for 10 to 15 days and to head SSE or even SE to the parallels of 28°S or 32°S and as far E as the meridian of Ilha da Trinidade (20°30′S, 29°20′W). Then on the starboard tack it should be possible to weather the E part of the coast and also Ilha de Fernando de Noronha (3°50′S, 32°37′W). As northing is made the wind will veer from E to SE and the equator should be crossed in 27°W to 29°W.

Rio de Janeiro ⇒ Porto de Salvador or Recife

Routes
8.80

1   From Rio de Janeiro (22°59′S, 43°10′W), N bound along the E coast of Brazil, it is preferable first to make a stretch to the SE. Working along this coast, bordered by reefs and subject to currents and light winds at night, is not recommended.

2   The effects of seasonal heating of the South American continent can produce very variable and sometimes strong adverse wind conditions along the Brazil coast, particularly from September to March. These effects decrease with distance from the coast and are of minor significance when more than 120 or 150 miles out to sea. Beyond this limit the Trade Wind is found, generally blowing from between SE and E.

3   From November to February, while fresh NE winds and a S-going current of 1 or 1½ kn extends along the coast, especially in the vicinity of Cabo de São Tomé (22°02′S, 41°03′W), the wind being also more N than in the offing, it is necessary to head for 450 to 600 miles to the ESE before tacking. This season, particularly December and January, is the most unfavourable time of year for the N-bound passage. In October and March, do not head farther E than actually necessary for weathering Arquipélago dos Abrolhos (18°00′S, 38°40′W), as to the N of this latitude the winds will be about E or E by S.

4   From March to September, close the coast as near as possible, taking advantage of the land and sea breezes, and making short tacks to the E on meeting the fresh NE winds which are common off Cabo Frio (23°01′S, 42°00′W) and Cabo de São Tomé. Then continue along the coast at distances of from 30 to 90 miles. A route more to the E is generally used, but if bound for Porto de Salvador (13°00′S, 38°32′W) it does not appear advantageous to head too far off the land.

Rio de Janeiro ⇒ Europe and North America

Routes
8.81

1   From Rio de Janeiro (22°59′S, 43°10′W), make a stretch to the SE to about 35°W and then head N in the South-east Trade Wind (3.4), crossing the equator between 27°W and 32°W; after passing through the Doldrums (2.3) steer direct for the American ports, or to the NW and W of Arquipélago dos Açores (38°00′N, 27°00′W), as directed at 8.60, if bound for European ports.

Rio de Janeiro ⇒ Cape of Good Hope

Routes
8.82

1   From Rio de Janeiro (22°59′S, 43°10′W), head to the SE to about 32°S, 30°W, thence through:
>   35°30′S, 20°00′W, thence:
>   37°00′S, 10°00′W, thence:
>   37°30′S, 0°00′, thence:
>   37°00′S, 10°00′E, thence:
>   Making Cape of Good Hope (34°21′S, 18°30′E) from the SW.

2   If bound to the Indian Ocean, without calling at intermediate ports, cross the meridian of Greenwich in about 40°S and run E on that parallel; from November to March, on 45°S, as directed at 8.95.

Río de La Plata ⇒ Europe and North America

Routes
8.83

1   From Río de La Plata (35°10′S, 56°15′W), from May to September, proceed direct to Cabo Frio (23°01′S, 42°00′W) and thence across the equator in 27°W to 29°W, thence as for the route from Rio de Janeiro described at 8.81.

2   From October to April head E to beyond 30°W, thence N into and through the South-east Trade Wind (3.4) as for the route from Rio de Janeiro described at 8.81.

Río de La Plata ⇒ Cape of Good Hope

Sailing ship route
8.84

1   From Río de La Plata (35°10′S, 56°15′W), pick up the parallel of 40°S in 30°W. Thence keep along that parallel as far as the Greenwich meridian thence steer directly for Cape of Good Hope (34°21′S, 18°30′E).

2   If bound to the Indian Ocean, without calling at intermediate ports, continue E along 40°S, or from November to March on 45°S, as directed at 8.95.

Low-powered vessel route
8.85

1   Low-powered vessels may usually avoid ice by passing:
>   Through 36°00′S, 40°00′W, thence:
>   Rhumb line to 36°00′S, 25°00′W, thence:
>   Great circle to Cape Town or Cape Agulhas, passing close S of Tristan da Cunha Group (37°00′S, 12°20′W).

2   If entering the Indian Ocean via Cape of Good Hope (145 miles S of) the farthest S point will be 38°30′S, 1°00′W.

3   This route is an alternative to that for full-powered vessels given at 3.42.

Río de La Plata ⇒ Falkland Islands

Route
8.86

1   From Río de La Plata (35°10′S, 56°15′W), keep well W of the direct track until nearing the Falkland Islands (51°40′S, 57°49′W).

Río de La Plata ⇒ round Cabo de Hornos

Sailing ship routes
8.87

1   From Río de La Plata (35°10′S, 56°15′W) to Cabo de Hornos (56°04′S, 67°15′W), two routes are recommended, either:

To steer SE and pick up the route from Rio de Janeiro, described at 8.23, or:

To sail coastwise as described at 8.21.3.

2   In either case Estrecho de Le Maire (54°50′S, 65°00′W) offers the alternative to the passage E of Isla de los Estados (54°50′S, 64°10′W) described at 8.23.

Low-powered vessel route
8.87.1

1   Low-powered vessels south-bound for Estrecho de Magallanes (52°27′S, 68°26′W) or intermediate ports, are advised to keep inshore to avoid the strength of the Falkland Current, which at a distance of 50 miles offshore, has been known to set N at a rate of 50 miles a day. South of Cabo Corrientes (38°01′S, 57°32′W), only the tidal streams are felt within 20 miles of the land. With W winds, better weather is experienced close inshore than in the offing.

Cabo de Hornos ⇒ English Channel

Rounding Cabo de Hornos
8.88

1   Rounding Cabo de Hornos (56°04′S, 67°15′W) from W to E is a comparatively easy matter, for the prevailing winds are favourable and the current near the cape sets strongly E. The passage is usually made between 56°00′S and 57°30′S, to the N of the W-bound route. December and January are the most favourable months; June and July, when E winds are not unusual, are the least favourable. Heavy W gales, with snow and hail, may be expected in August and September. From June to September a track about 80 miles S of Cabo de Hornos is recommended.

2   The best landfall after rounding Cabo de Hornos is W of Estrecho de Le Maire (54°50′S, 65°00′W), where the coast is free of outlying dangers. The islands make a lee during SW and W winds. Keep in mid-channel through Estrecho de Le Maire, avoiding the overfalls off Cabo San Diego (54°39′S, 65°07′W).

3   A vessel in difficulty should run boldly through the the strait and round up under the land if necessary.

Cabo de Hornos to the equator
8.89

1   The usual routes is about 80 miles S of the Falkland Islands (51°40′S, 57°49′W) and to a position in about 35°S, 30′W, making W of that position between April and August and E of it from September to March.

2   Continue according to season:
From April to August, heading N as far as 10°S, 25°W, keeping as much as possible to the W of 25°W throughout and cross the equator between 25°W and 28°W. It might even be possible to pass W of Ilha da Trinidade (20°30′S, 29°20′W) at this time of year.
From September to March, head NNE from 35°S, 30′W to about 25°S, 20°W and then run N with the South-east Trade Wind (3.4) to cross the equator between 22°W and 25°W.

8.89.1

1   **Alternative routes to the equator**. If ice is prevalent, particularly between October and February, steer so as to cross 50°S in about 51°W, and 40°S in about 45°W. Then make northing until the South-east Trade Wind is met, joining the route from Río de La Plata (8.83) in about 35°S.

2   Alternatively some navigators recommend passage W of the Falkland Islands (51°40′S, 57°49′W), from October to

February, on account of the greater freedom from ice in that region. Then a course to NE should be steered to join the other alternative route in about 35°S, 41°W. If unable to pass W of the Falkland Islands, pass as close E as the wind will allow.

3   **Caution**. If meeting with a foul wind while S of 40°S, it would be better to head NW than to the E, as ice is likely not far E of the Falkland Islands.

From the equator to the English Channel
8.90

1   Join the route from Cape of Good Hope (8.60) on meeting the North-east Trade Wind (2.5).

Cabo de Hornos ⇒ East coast of North America

Route
8.91

1   From Cabo de Hornos (56°04′S, 67°15′W) the recommended routes are seasonal.

2   From April to August, follow the directions, given in 8.89, to 10°S, 25°W where the route meets the track from Cape of Good Hope to North America (8.61); follow this route to destination.

3   From September to March, follow the directions in 8.89, for these months, to 15°S, 20°W where the route meets the track from Cape of Good Hope to North America (8.61); follow this route to destination.

Cabo de Hornos ⇒ East coast of South America

Sailing ship route
8.92

1   At all times of the year, after rounding Cabo de Hornos (56°04′S, 67°15′W), head N with the Falklands Current (3.9) between Falkland Islands (51°40′S, 57°49′W) and Tierra del Fuego (54°30′S, 68°00′W) and carry it up the coast, with the prevailing W winds to Bahia Blanca (39°10′S, 61°45′W) or Río de La Plata (35°10′S, 56°15′W).

2   From Río de La Plata onwards to Cabo Frio (23°01′S, 42°00′W) or Rio de Janeiro (22°59′S, 43°10′W) follow directions given at 8.83; and from Rio de Janeiro to Porto de Salvador (13°00′S, 38°32′W) or Recife (8°00′S, 34°40′W) follow directions given at 8.80.

Low-powered vessel routes
8.92.1

Low-powered vessels north-bound between Estrecho de Magallanes (52°27′S, 68°26′W) and Río de la Plata (35°10′S, 56°15′W), are advised to keep between 20 and 50 miles to seaward of the rhumb line track to obtain full benefit from the Falklands Current; this increases the overall distance by about 40 miles.

Cabo de Hornos ⇒ Cape of Good Hope

Sailing ship route
8.93

1   From Cabo de Hornos (56°04′S, 67°15′W), follow the directions given in 8.89 and 8.89.1, taking particular notice of the remarks on ice, to pass S of the Falkland Islands (51°40′S, 57°49′W) to a position about 50°S, 45°W. From whence, at all seasons, steer a direct course with the prevailing W wind and a favourable current to 40°S on the Greenwich meridian, thence steer in a NE direction towards the Cape of Good Hope (34°21′S, 18°30′E). See the relevant parts of the directions given in 8.25, the portions

dealing with the voyage after passing the Greenwich meridian are equally applicable to this route as regards winds, currents when approaching Table Bay (33°50'S, 18°30'E).

Low-powered vessel routes
8.94

1   For low-powered vessels from Cabo de Hornos (56°04'S, 67°15'W), the route is:

      Rhumb line to 36°00'S, 40°00'W, thence:
      Rhumb line to 36°00'S, 25°00'W, thence:
      Great circle to destination, passing S of Tristan da Cunha Group (37°00'S, 12°20'W).

2   For low-powered vessels from Estrecho de Magallanes (52°27'S, 68°26'W), the route is:

      Rhumb line to 36°00'S, 40°00'W, thence:
      Rhumb line to 36°00'S, 25°00'W, thence:

      Great circle to destination, passing close S of Tristan da Cunha Group.

3   These routes offer an alternative to those for full-powered vessels given at 3.46, 3.48 and 3.49.

Cabo de Hornos eastwards ⇒ Indian Ocean and Australian ports

Routes
8.95

1   From Cabo de Hornos (56°04'S, 67°15'W), follow the directions given in 8.93 as far as 40°S on the Greenwich meridian and from this position continue due E along that parallel. From November to March a quicker passage will probably be made in about 45°S, though better weather will be found in 40°S.

2   For continuation onward through the Indian Ocean, see the appropriate routes from Cape of Good Hope at 9.4 to 9.31.

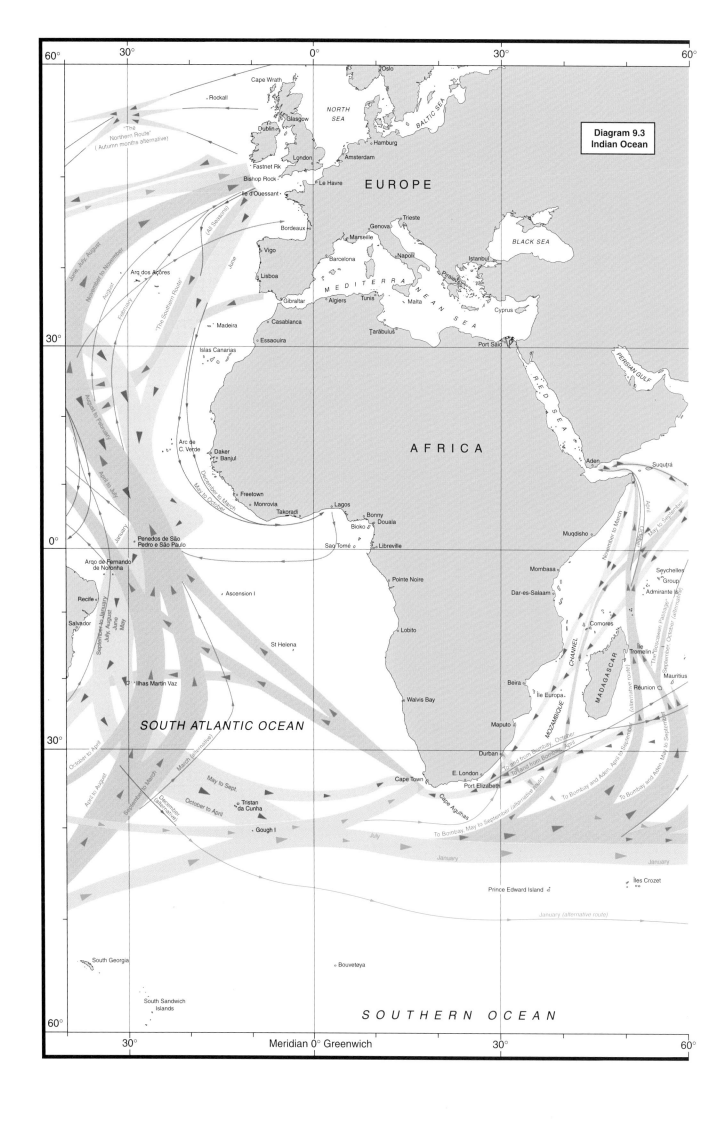

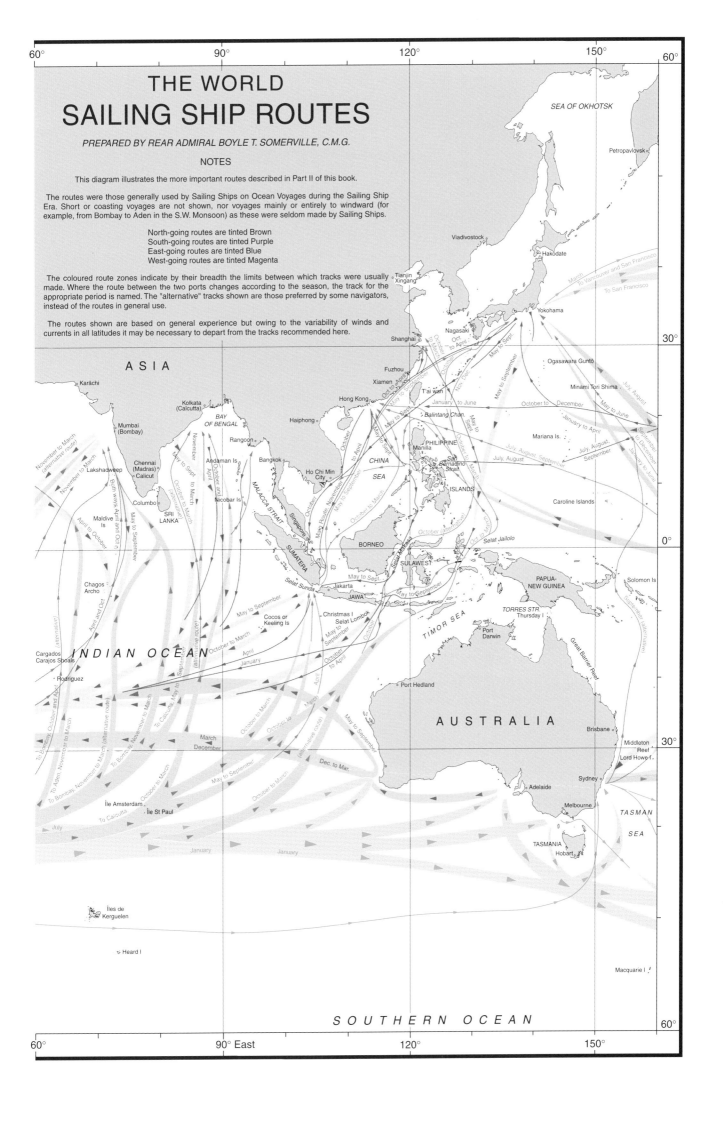

THE WORLD
SAILING SHIP ROUTES

PREPARED BY REAR ADMIRAL BOYLE T. SOMERVILLE, C.M.G.

NOTES

This diagram illustrates the more important routes described in Part II of this book.

The routes were those generally used by Sailing Ships on Ocean Voyages during the Sailing Ship Era. Short or coasting voyages are not shown, nor voyages mainly or entirely to windward (for example, from Bombay to Aden in the S.W. Monsoon) as these were seldom made by Sailing Ships.

North-going routes are tinted Brown
South-going routes are tinted Purple
East-going routes are tinted Blue
West-going routes are tinted Magenta

The coloured route zones indicate by their breadth the limits between which tracks were usually made. Where the route between the two ports changes according to the season, the track for the appropriate period is named. The "alternative" tracks shown are those preferred by some navigators, instead of the routes in general use.

The routes shown are based on general experience but owing to the variability of winds and currents in all latitudes it may be necessary to depart from the tracks recommended here.

NOTES

CHAPTER 9

SAILING PASSAGES FOR INDIAN OCEAN,
RED SEA AND EASTERN ARCHIPELAGO

GENERAL INFORMATION

COVERAGE

Chapter coverage
9.1

This chapter contains details of the following passages for sailing vessels and includes recommended routes for low-powered, or hampered, vessels:

Passages from Cape Town or Cape of Good Hope (9.3).

Passages from east coast of Africa and Mauritius (9.32).

Passages through the Eastern Archipelago (9.56)
North-bound (9.67).
South-bound (9.113).

Passages through the Red Sea (9.131).

Passages from Aden (9.134).

Passages from west coast of India and Sri Lanka (9.144).

Passages to and from Seychelles (9.168).

Passages from ports in Bay of Bengal (9.172).

Passages from ports in Burma (Myanmar) (9.182).

Passages S-bound or W-bound from Singapore or Eastern Archipelago (9.188).

Passages from northern Australia to Sydney, Indian Ocean and South China Sea (9.200).

Passages from south-west and south Australia (9.207).

Passages from Sydney to ports in the Indian Ocean (9.220).

Admiralty Publications
9.2

Relevant navigational publications should be consulted, in addition to this book, when planning and conducting passages. Details of these, and the areas covered, can be found, as follows:

Admiralty Sailing Directions; see 1.12.1
Admiralty List of Lights and Fog Signals; see 1.12.2
Admiralty Tide Tables; see 1.12.3
Admiralty List of Radio Signals; see 1.12.5
Admiralty Distance Tables; see 1.12.8

9.2.1

See also *The Mariner's Handbook* for notes on electronic products.

PASSAGES FROM CAPE TOWN OR CAPE OF GOOD HOPE

Charts:
5126 (1) to (12) Monthly Routeing Charts for Indian Ocean.
5309 Tracks followed by Sailing and Auxiliary Powered Vessels.
5310 World Surface Current Distribution.
Diagram 9.3 World Sailing Ship Routes.

Cape Town or Cape of Good Hope

Ice
9.3

Icebergs are most numerous SE of the Cape of Good Hope (34°21′S, 18°30′E) and midway between Îles Kerguelen (49°00′S, 69°30′E) and the meridian of Cape Leeuwin (34°23′S, 115°08′E). The periods of frequency vary greatly. It may happen that while vessels are encountering ice in lower latitudes, others, in higher latitudes find the ocean free of it. For details of icebergs in this region, see 6.31.

Rounding Cape of Good Hope
9.4

1   From Cape Town (33°53′S, 18°26′E), vessels are recommended to pick up the E-bound track from Cabo de Hornos (8.95) at the point that it is met by the track from the North Atlantic (8.25) bound for the Indian Ocean, namely in about 40°S, 20°E. There is little difficulty in passing the Cape of Good Hope (34°21′S, 18°30′E), E-bound at any time of the year, though a greater proportion of gales will be met from April to September.

2   From October to April, E winds prevail as far S as the tail of the Agulhas Bank (about 17°S), with variable but chiefly W winds beyond it. In May and September, at the tail of the bank, E and W winds are in equal proportion, but between these months W winds prevail, extending sometimes close in to the coast. Should a SE wind be blowing on leaving Table Bay (33°50′S, 18°30′E) or Simons Bay (34°11′S, 18°26′E), stand boldly to the SW until the W winds are reached or the wind changes to a more favourable direction. In all cases when making for the parallel of 40°S from S of the Cape of Good Hope steer nothing E of S, so as to avoid the area SE of the tail of Agulhas Bank, where gales are frequent, and heavy and dangerous breaking cross seas prevail.

Crossing the Indian Ocean
9.5

1   Having crossed to the S of the W-going Agulhas Current (6.28) and picked up the W winds, the best latitude in which to cross the ocean must, to some extent, depend on circumstances.

2   Vessels bound for Australian ports would make the passage at about the parallels of 39°S or 40°S, but those bound to Tasmania or New Zealand would do so between 42°S or 43°S, especially from October to March. Between 39°S and 43°S the winds generally blow from a W direction and seldom with more strength than will admit the carrying of sail. In a higher latitude the weather is frequently more boisterous and stormy, sudden changes of wind with squally wet weather are almost constantly to be expected, especially in winter. Île Amsterdam (37°50′S, 77°35′E) may be seen from a distance of 60 miles in clear weather.

3   In summer many vessels take a route farther S, some going as far S as 52°S, but the steadiness and comparatively moderate strength of the winds, smoother seas and more genial climate N of 40°S compensate by comfort and security for the time saved by taking the shorter S route. Severe gales, sudden violent shifts of wind, accompanied by hail and snow, and terrific and irregular seas are often encountered in the higher latitudes. The islands in the higher latitudes are frequently shrouded in fog.

Approaching Bass Strait
9.6

1   When approaching Bass Strait (40°00′S, 146°00′E), passage N of King Island (39°50′S, 144°00′E) is recommended. In this approach, when making the land at Moonlight Head (38°46′S, 143°14′E) or Cape Otway (38°51′S, 143°31′E), the currents must be carefully watched, particularly during SW or W winds, vessels have been wrecked on King Island by not steering for Cape Otway. In normal weather it is desirable to round Cape Otway at a distance of not less than 3 to 4 miles. When approaching Bass Strait in thick weather, or when uncertain of the vessel's position, do not reduce the soundings below 70 m. Soundings of 110 m to 130 m will be found 25 to 30 miles W of King Island. Outside this limit the soundings deepen rapidly to over 180 m.

2   **Caution**. In approaching King Island from the W, especially during thick or hazy weather, caution is required on account of the variable strength of the current, which sets SE at rates which vary between ½ kn to 2½ kn, according to the strength and duration of the W winds, and sounding is recommended.

3   The entrance to Bass Strait between King Island and Hunter Group is not recommended on account of Bell Reef (40°23′S, 144°05′E) and Reid Rocks (40°15′S, 144°10′E) which lie in it. If, from necessity or choice, entering Bass Strait by this passage keep S of Reid Rocks and Bell Reef, the latter being passed at a distance of 2½ miles S of it by steering for Black Pyramid (40°29′S, 144°20′E) on a bearing of 098°. With a commanding breeze the passage between King Island and Reid Rocks may be taken without danger by paying attention to the tidal streams, which set somewhat across the channel at times. From Black Pyramid pass about 1 mile N of Albatross Islet (40°23′S, 144°39′E), whence, if bound for Port Dalrymple (41°02′S, 146°45′E), round Mermaid Rock (40°23′S, 144°57′E), off Three Hummock Island (40°26′S, 144°55′E) and then make a direct course.

Passage ⇒ Fremantle
9.7

1   For Fremantle (32°02′S, 115°42′E), leave the trans-ocean route (9.5) in about 90°E and then make direct for Fremantle. Some navigators, however, recommend continuing E as far as 100°E before turning NE.

2   From October to March, to avoid being set to the N of Rottnest Island (32°00′S, 115°30′E), it is advisable to make the land about Cape Naturaliste (33°32′S, 115°01′E), see 6.88 for details.

Passage ⇒ Adelaide
9.8

1   For Adelaide (34°48′S, 138°23′E), leave the trans-ocean route (9.5) on the meridian of Cape Leeuwin (34°23′S, 115°08′E), or about 115°E, thence direct to Cape Borda (35°45′S, 136°35′E).

Passage ⇒ Melbourne
9.9

1   For Melbourne (37°50′S, 144°54′E), leave the trans-ocean route (9.5) in 135°E and proceed directly to Bass Strait (40°00′S, 146°00′E), as described at 9.6. If Cape Otway (38°51′S, 143°31′E) is rounded early in the evening with a fresh S wind, beware of over running the distance, as a strong current after a prevalence of S gales, often sets NE along the land; bearings of Split Point Light (38°28′S, 144°06′E) give a good check. When abeam of Split Point, if there is not enough daylight to get into pilotage waters, stand off and on shore until daylight, keeping in more than 35 m of water. Do not heave to.

Passage ⇒ Sydney
9.10

1   For Sydney (33°50′S, 151°19′E), in summer, leave the trans-ocean route (9.5) in about 120°E and steer to pass S of Tasmania.

2   After rounding South East Cape (43°39′S, 146°50′E), give a berth of 20 to 30 miles to Cape Pillar (43°14′S, 148°02′E) and the E coast of Tasmania, to escape the baffling winds and calms, which frequently occur inshore, while a steady breeze is blowing in the offing. This is more desirable from December to March, when E winds prevail, and a current is said to be experienced off the SE coast, at 20 to 60 miles from the shore, running N at a rate of ¾ kn, while inshore it is running in the opposite direction at at nearly double that rate. From a position about 30 miles E of Cape Pillar, proceed on a course of about 012° for about 350 miles to a position 15 miles E of Cape Howe (37°30′S, 149°59′E), whence continue as directly as possible to make Sydney, but keeping at first at a distance from the coast, in order to lessen the strength of the S-going East Australian Current (7.27), not closing the land until N of South Head, Port Jackson (33°50′S, 15°17′E).

3   Some navigators prefer to stand E as far as 155°E before turning N for Port Jackson and thus escape, almost altogether, the S-going set.

4   In winter, follow the route for Melbourne (9.9) as far as Cape Otway (38°51′S, 143°31′E) and then steer to pass through Bass Strait about 2 miles S of Anser Group (39°18′S, 146°30′E), 3 miles N of Rodondo Island (39°14′S, 146°23′E) and 2 miles S of South East Point, Wilson's Promontory (39°08′S, 146°26′E). Then steer to pass about 5 miles SE Rame Head (37°47′S, 149°30′E) and Gabo Island (37°34′S, 149°55′E). Occasionally, and especially during and after E gales, the current sets strongly towards the land; in thick weather sounding should not be neglected. See Navigational Notes on Bass Strait in 9.6 and 9.220.

5   From Rame Head stand on to the E to about 154°E, before turning to the N in order to escape from the S-going current along the coast of New South Wales and approach Port Jackson from a point slightly to the N. During E gales (June to August), an offing may be maintained by watching the shifts of wind and keeping on the starboard tack as long as prudent, thus bringing the prevailing S-going current on the lee bow.

6   **Traffic Separation Schemes** have been established in the Bass Strait and SE of the Area to be Avoided (7.38.3) SE of Ninety Mile Beach. See also 8.4.

Passage ⇒ Hobart
9.11

1   For Hobart (42°58′S, 147°23′E), leave the trans-ocean route (9.5) in about 120°E and steer for a position 10 miles

S of South West Cape, Tasmania (43°35′S, 146°03′E), or in any case far enough S to ensure avoiding the rocky W coast at night through any error in the reckoning, or being caught on a lee shore by a SW gale. In fine weather, from W of South West Cape, pass between Maatsuyker Island (43°36′S, 146°20′E) and Mewstone (43°45′S, 146°22′E), then steer to pass 3 miles S of South East Cape (43°39′S, 146°50′E). When blowing heavily from SW or S, especially if unable to obtain observations before making the land, it is desirable to keep more to the S, passing S of Mewstone and on either side of Pedra Branca (43°52′S, 146°58′E) and Eddystone (43°52′S, 146°59′E) and taking care to avoid Sidmouth Rock (43°51′S, 147°01′E). Proceed to Hobart through Storm Bay (43°15′S, 147°35′E).

Passage West of Tasmania and ⇒ New Zealand ports
9.12

1   It is often necessary, and in heavy W weather desirable, to make the passage down the W coast of Tasmania at from 120 to 250 miles from the coast and often at the same distance round the S end of the island.

2   From the Indian Ocean, for New Zealand ports, it is normal in summer to leave the trans-ocean route (9.5) in about 110°E and to proceed S of Tasmania in 45°S to 47°S, whence the main route is also taken across the Pacific Ocean (see 10.8 or 10.9 for details). In winter take the winter route for Sydney (described at 9.10) through Bass Strait.

3   Both in summer and winter, if bound for Auckland (36°36′S, 174°49′E) proceed round the N point of New Zealand; if for Wellington (41°22′S, 174°50′E) through Cook Strait (41°00′S, 174°30′E) and for Otago (45°45′S, 170°44′E) or Lyttleton (43°36′S, 172°49′E), S of New Zealand and through Foveaux Strait (46°35′S, 168°00′E) or S of Stewart Island (47°00′S, 167°45′E).

4   See also 10.13 and 10.14 for routes to South and North America.

Cape Town or Cape of Good Hope ⇒ Singapore or South China Sea

General directions
9.13

1   Although this voyage takes a vessel out of Indian Ocean waters and into the Eastern Archipelago and South China Sea (which is covered in Chapter 10), it is more convenient to treat this passage as a continuous voyage and describe it as such here. The voyage is complicated by a choice of routes due to the monsoons, and also by a choice of alternative routes through the islands of the Eastern Archipelago.

2   The routes below give, in detail, the passage from the Cape of Good Hope (34°21′S, 18°30′E) as far as the S entrance to the Eastern Archipelago.

3   Then follows a summary of the various routes through the Indonesian islands to Singapore (1°12′N, 103°51′E) or the South China Sea.

4   **Archipelagic Sea Lanes (ASLs)** through the Indonesian Islands have been established by the Indonesian Authorities, details of these are given at 7.130, 9.57 and at Appendix A.

5   These sea lanes, however, are not always practical for use by sailing vessels.

6   The routes described for sailing vessels in this area, therefore, follow the traditional routes used by sailing vessels in the past. The passages through the Eastern Archipelago are given from 9.56 to 9.130.

Weather considerations
9.14

1   The monsoon periods, on which these routes depend, are:

   May to September, when a south-east or east monsoon prevails in the Eastern Archipelago and a south-west monsoon, not usually strong, in the South China Sea.

   October to March, when a north-west or west monsoon prevails in the Eastern Archipelago and a north-east monsoon, usually strong, in the South China Sea.

2   The object of a vessel, bound for the South China Sea, being to get as far to windward as possible in the Indian Ocean before arriving in the monsoon area.

   During the south-west monsoon period of the South China Sea (May to September) make for the W end of the island chain, and:

   During the north-east monsoon period of the South China Sea (October to March) make for the more E passages of the chain.

3   An alternative to the South China Sea, suitable only in October and November, passes through the central part of the island chain.

4   The following variations to the routes above are sometimes taken but do not appear to possess any particular advantage:

   Though the October to March route is usually made via the passages farther E through the islands it is possible to make it by passing through Selat Sunda (6°00′S, 105°52′E) and proceeding N through the South China Sea, along the N coast of Borneo, via the Palawan Passage (described at 10.64.1 and 10.64.2).

   Another route is to transit Selat Sunda, then proceed E through the Java Sea to join the E passages.

5   Vessels bound for Singapore only should use Selat Sunda at all time.

6   Vessels bound for ports on the E coast of Borneo, or in Selat Makasar should use Selat Sunda (6°00′S, 105°52′E), Selat Bali (8°10′S, 114°25′E), Selat Lombok (8°47′S, 115°44′E) or Selat Alas (8°40′S, 116°40′E) as appropriate. These straits should also be used during the local north-west monsoon period, instead of passage through the islands farther E, then stand E to pick up the regular Eastern route.

For Singapore
9.15

From Cape of Good Hope (34°21′S, 18°30′E) the routes are seasonal.
9.15.1

May to September take route across the Indian Ocean, as directed at 9.5, along the parallel of 39°S or 40°S as far as 75°E. Thence edge away to the NE crossing 30°S in about 100°E and 20°S in 105°E, passing close W of Christmas Island (10°24′S, 105°43′E) and up to Tanjung Gedeh (6°46′S, 105°13′E), the E entrance point of Selat Sunda. Care must be taken to keep well to the E, especially in June, July and August when the south-east monsoon and the W-going current are at their strongest, or the vessel may fall to leeward of Tanjung Gedeh and find great difficulty in recovering it against wind and current.
9.15.2

October to March take the route across the Indian Ocean, as above, to 75°E; thence steer to pass through 25°S, 98°E. Thence directly N for Selat Sunda, passing midway between Christmas Island and Cocos or Keeling Islands (12°05'S, 96°52'E), steering for Tanjung Cukubalimbing (5°56'S, 104°33'E) on the W side of Selat Sunda, as in this season the E-going current is strong and W winds blow at time with considerable strength. If contrary winds are met with after passing Île Saint-Paul (38°43'S, 77°32'E), stand N through the South-east Trade Wind, into the north-west monsoon, thence direct to Selat Sunda.

9.15.3

March/April and September/October. During the changes of monsoon it is advisable to make to the E until due S of the entrance to Selat Sunda and then steer directly for it.

9.15.4

1 **Selat Sunda to Singapore Strait** provides three alternative routes.

2 The usual route is through Selat Bangka (2°06'S, 105°03'E) and Selat Riau (0°55'N, 104°20'E). Directions are from 9.71 to 9.76.

3 The first alternative route is through Selat Gelasa (2°53'S, 107°18'E) and then either by Selat Riau to Singapore; or else from Selat Gelasa continue N to the E of Bintan (1°00'N, 104°30'E) and through Singapore Strait from the E entrance. Directions are at 9.82 and from 9.84 to 9.92.

4 The second alternative route, known as the Inner Route, should be taken between October and March only, when the north-east monsoon is blowing strong in the South China Sea, and a vessel having passed through Selat Bangka or Selat Gelasa is confronted by a head wind, heavy seas and an adverse current. It is described from 9.71 to 9.74 and from 9.77 to 9.80.

For South China Sea

9.16

From Cape of Good Hope (34°21'S, 18°30'E) the routes are seasonal.

9.16.1

1 **May to September** Proceed to Selat Sunda as directed at 9.15.1 and thence by one of the three alternative routes that follow.

2 The most direct route is to pass through Selat Bangka (2°06'S, 105°03'E) or Selat Gelasa (2°53'S, 107°18'E). Selat Gelasa offers a more direct route than Selat Bangka, but should not be approached in thick weather whereas Selat Bangka can be navigated without risk. Thence pass between Pulau-Pulau Anambas (3°00'N, 106°00'E) and Pulau-Pulau Natuna (4°00'N, 108°00'E) into the South China Sea.

3 Thence for Hong Kong (22°17'N, 114°10'E) and Taiwan Strait (24°00'N, 119°00'E), pass between the Paracel Islands (16°30'N, 112°00'E) and Macclesfield Bank (15°50'N, 114°30'E), thence as navigation permits to destination.

4 An alternative route should be taken if the north-east monsoon is likely to begin before reaching Hong Kong. This leads from Selat Gelasa, through the Palawan Passage (10°20'N, 118°00'E), thence along the coast of Luzon as far as Cape Bolinao (16°19'N, 119°47'E), thence to destination.

5 The second alternative passes through Selat Karimata (1°43'S, 108°34'E) into the South China Sea direct. Selat Karimata is not much frequented, except by sailing vessels returning from China or by vessels making E through the Java Sea, as it is difficult to get through W-bound due to the effects of winds and currents. Its great breadth, in comparison to the other straits, is of advantage to vessels working to windward; but this is partly counterbalanced by several shoals lying in or near the fairway and out of sight of land, also there are irregular currents.

6 If having passed through Selat Sunda into the Java Sea, and the north-east monsoon has already started to blow in the South China Sea, do not attempt to proceed farther N, but instead turn E and pass through the Java Sea and Flores Sea, S of Borneo to Selat Salayar (5°43'S, 120°30'E), thence by passage between Butung (4°56'S, 122°47'E) and the islands S of it, into the Banda Sea and thence through the Ceram Sea into the open Pacific Ocean either through Selat Jailolo (0°06'N, 129°10'E) or Selat Dampier (0°40'S, 130°30'E). When in the Pacific Ocean pass E of the Palau Islands (7°30'N, 134°30'E) and enter the South China Sea through one of the passages between Luzon and T'ai-wan.

9.16.2

1 **September.** Leaving Cape of Good Hope, in September, for the South China Sea proceed as directed at 9.10 for Sydney and pass:

S and E of Australia or Tasmania, thence:

W of Nouvelle-Calédonie (21°15'S, 165°20'E), thence:

Through the Pioneer Channel (4°40'S, 153°45'E), thence:

Cross the equator at about 156°E, thence:

Direct to destination.

2 This is known as the 'Great Eastern Route'.

9.16.3

1 **October to March** the route is taken via Selat Ombi (8°35'S, 125°00'E) and the Second Eastern Passage, as directed below.

2 Cross the Indian Ocean as described at 9.5 as far as 75°E and then steer for 20°S, 100°E. Some navigators recommend making farther E, to 90°E, before turning N to make for 20°S, 100°E. From 20°S course may be directed to Selat Ombi.

3 The usual, and best recommended passage through the islands, after crossing the Indian Ocean, passing between the NW point of Timor (9°00'S, 125°00'E) and Pulau-Pulau Alor (8°15'S, 124°30'E), thence either W of Buru (3°30'S, 126°30'E) or through Selat Manipa (3°20'S, 127°22'E) into the Ceram Sea. Thence through Selat Jailolo (0°06'N, 129°10'E) or Selat Dampier (0°40'S, 130°30'E) into the open Pacific Ocean. This route is known as the Second Eastern Passage and directions are given at 9.101

4 When in the Pacific Ocean make to the E, between 1°30'N and 3°00'E, until able to pass E of Palau Islands (7°30'N, 134°30'E), however after February pass W of this group. Having passed Palau Islands make to the NW to pass through the Surigao Strait (9°53'N, 125°21'E) to cross the Sulu Sea and into South China Sea via Mindoro Strait (13°00'N, 120°00'E) or Verde Island Passage (13°36'N, 121°00'E) and on to Manila (14°32'N, 120°56'E) or Hong Kong (22°17'N, 114°10'E).

5 A more usual route, however, after passing Palau Islands is to proceed NNW, keeping E of the Philippine Islands, and passing N of Luzon, through the Balintang Channel (20°00'N, 122°20'E) to Hong Kong; or if bound for Shanghai (31°03'N, 122°20'E) to continue N to pass between Okinawa Guntō (26°30'N, 128°00'E) and Sakishima Guntō (24°40'N, 124°45'E) towards the mouth of the Chang Jiang; or to proceed NNE, in the full strength of the NE-going Japan Current (7.26) to Yokohama (35°26'N, 139°43'E) or other Japanese ports.

6 For directions for these routes, see from 9.103 to 9.106.

9.16.4

1   **October and November**. During these months only, the passage from the Indian Ocean to the South China Seas may be made through the central passages of the Eastern Archipelago by a route known as the First Eastern Passage, as follows.

2   From 20°S, 110°E make Selat Bali (8°10′S, 114°25′E) Selat Lombok (8°47′S, 115°44′E) or Selat Alas (8°40′S, 116°40′E). Then pass through Selat Makasar (2°00′S, 118°00′E) into the Celebes Sea, thence through Basilan Strait (6°50′N, 122°00′E) into the SE part of the Sulu Sea. Pass along the W coasts of Mindanao, Negros and Panay and enter the South China Sea via Mindoro Strait (13°00′N, 120°00′E) or Verde Island Passage (13°36′N, 121°00′E). Then work along the coast of Luzon to Cape Bojeador (18°30′N, 120°34′E), before crossing the South China Sea to Hong Kong.

3   This route to China, previously known as the First Eastern Passage, though often used in former days, has little to recommend it, on account of the adverse current setting S through Selat Makasar, often strongly, at all seasons. The winds are boisterous and uncertain at the S end of Selat Makasar and light and variable at the N end, while the navigation causes concern throughout most of the voyage to the open South China Sea.

4   The First Eastern Passage, described in detail at 9.107, is more suitable for S-bound traffic, but in the case of vessels from Cape of Good Hope bound for ports in Selat Makasar, it is mentioned here.

Cape Town or Cape of Good Hope ⇒ Bay of Bengal

Route selection
9.17

1   From the Cape of Good Hope (34°21′S, 18°30′E) there are three principal routes, two of which are appropriate to the south-west monsoon and one to the north-east monsoon.

2   The choice of route rests, not so much on the month in which departure is made from the Cape of Good Hope, as on the month in which a vessel may be expected to arrive in the region affected by the monsoons (6.4 to 6.7). These regions comprise of the Arabian Sea, the Bay of Bengal and the Indian Ocean N of the equator. The references to months, given for the three routes below, refer, therefore, to the months of a vessels anticipated arrival in Indian waters.

Inner Route
9.18

1   From May to September this route, via the Moçambique Channel (17°00′S, 41°30′E), is the most direct for vessels to any port in India during the south-west monsoon, but it must not be taken unless there is a certainty of reaching the port before the close of the monsoon.

2   On leaving the Cape of Good Hope (34°21′S, 18°30′E):
    With SE winds stand S and run down to the E in 39°S to 40°S as far as about 30°E, or:
    With W winds run along the coast, guarding against any indraughts, keeping S of 35°S until in 37°E.

3   Steer for the Moçambique Channel, passing E of Île Europa (22°20′S, 40°20′E) and on either side of Île Juan de Nova (17°03′S, 42°42′E); then pass through the Comores (12°00′S, 44°30′E) and cross the equator in about 54°E. Thence steer direct for the Eight Degree Channel (7°24′N, 73°00′E) and pass S of Sri Lanka into the Bay of Bengal. For further details, see 9.21 and 9.26.1.

First Outer Route
9.19

1   From May to September this route is more usually preferred to the Inner Route (9.18) on account of ease of navigation, but it must only be used if certain of reaching port before the close of the south-west monsoon.

2   From Cape of Good Hope (34°21′S, 18°30′E) cross the Indian Ocean, as directed in 9.5, between the parallels of 39°S and 40°S as far as about 60°E. From this position proceed NE to cross 30°S in about 80°E and then stand N as directly as possible for destination as detailed below:
    For **Chennai (Madras)** (13°07′N, 80°20′E), or adjacent ports, cross the equator in 82°E.
    For **Kolkata (Calcutta)** (22°30′N, 88°15′E), cross the equator in 88°E.
    For **Yangon (Rangoon)** (16°09′N, 96°17′E), cross the equator as for Kolkata, but leave that route at 10°N and steer for Yangon, passing N of Andaman Islands (12°30′N, 93°00′E) in about 15°N.

3   For farther details of passage through the Bay of Bengal see 9.21.

4   **Note** on the voyage N from 30°S, it is advisable to gain distance to the E to counteract the W-going current and to be prepared for the wind shifting to N, for in the South-east Trade Wind it often happens, particularly in April and May, that the wind is more from E and ENE than SE. The South-east Trade Wind at this season extends to the equator; and, between May and September, from 1°N to 2°N, the south-west monsoon is a fair wind to Kolkata or any part of the Bay of Bengal.

Second Outer Route
9.20

1   From October to March this route is taken when it is likely that the Bay of Bengal will not be reached before the south-west monsoon is over (September) or when expecting to arrive in the Bay of Bengal in November when the the north-east monsoon has set in.

2   From Cape of Good Hope (34°21′S, 18°30′E) cross the Indian Ocean, as directed in 9.5, between the parallels of 39°S and 40°S as far as about 70°E. From this position steer ENE so as to cross 35°S in about 82°E and thence proceed NNE, through the Trade Winds, to cross the equator in 92°E.

3   From this position, steer to pass 150 miles W of the NW extremity of Sumatera and about 60 miles W of Nicobar Islands (8°00′N, 94°00′E) and Andaman Islands (12°30′N, 93°00′E) and thence as directed below.

4   If the wind is W give the islands a good berth, but if NW steer up the Bay of Bengal close-hauled. In about 16°N to 17°N the wind often shifts to the N, when favourable tracks may be made to the E.

5   For Kolkata (22°30′N, 88°15′E), do not approach either shore, but work to windward in the middle of the Bay of Bengal, where there is smooth water and moderate winds; from close W of the Nicobar Islands, the entrance to the Hugli River (21°00′N, 88°13′E) has often been reached without tacking. If the equator is crossed late in February, or in March, keep well to the W side of the Bay of Bengal.

6   For Chennai (13°07′N, 80°20′E), shape direct course from the position off the Nicobar Islands.

7   For Yangon (16°09′N, 96°17′E), leave the route at about 3°N and steer to the NE, passing between Pulau We (5°50′N, 95°18′E) and the Nicobar Islands as directly as possible to destination, keeping midway between the Malaysian coast and the Andaman Islands.

General directions for the Bay of Bengal
9.21

1   From mid-January to end of May, when N-bound keep to the W side of the Bay; when S-bound keep to the E side.

2   In June, July and August, when N-bound keep in the middle of the Bay; when S-bound in the middle or E of the Andaman Islands.

3   In September, October or November, when N-bound take the E side of the Bay; when S-bound the W side.

4   From early December to mid-January, when N or S-bound keep in the middle of the Bay.

General remarks on storms in the Bay of Bengal
9.22

1   When in the .Bay of Bengal a strong SW wind, occasional squalls and rain and a slowly falling barometer, indicates bad weather prevailing somewhere to the N.

2   Between early June and the middle of September the storm centre is probably N of 16°N, and in July or August still farther to the N, and a sailing vessel should steer to the E to take advantage of the S and SE winds on the E side of the storm as it moves NW. But should the weather rapidly worsen and the barometer continue to fall, then heave-to and determine the position with regard to the movement of the storm before proceeding.

3   In May, October or November the storm travels in some direction from W, through N, to NE and its course should be definitely ascertained before any attempt is made to round its E side as, if it is moving NE, such a procedure would be dangerous.

4   See also 6.16 and 9.172 for further details of weather in the Bay of Bengal.

Cape Town or Cape of Good Hope ⇒ Colombo

Inner Route
9.23

1   From Cape of Good Hope (34°21′S, 18°30′E), from May to September, proceed as directed for the Bay of Bengal, described at 9.18, and after passing through the Eight Degree Channel (7°24′N, 73°00′E) steer as directly as possible for Colombo (6°58′N, 79°50′E).

2   Notes on the Inner Route through the Moçambique Channel are given at 9.26.1.

Passage East of Madagascar
9.24

1   From Cape of Good Hope (34°21′S, 18°30′E), routes are seasonal.

2   From May to September, proceed as directed for Mumbai described at 9.26.2 until across the equator in about 62°E, after which steer directly for Colombo (6°58′N, 79°58′E).

3   In April and October only, proceed as directed for Mumbai described at 9.27 until the equator is crossed, thence steer directly for Colombo.

4   From November to March, proceed as directed for Mumbai described at 9.28 until the equator is crossed, thence steer directly for Colombo. Alternatively take the Second Outer Route for the Bay of Bengal, described at 9.20, as far as 20°S thence steer directly for Colombo.

Cape Town or Cape of Good Hope ⇒ Mumbai

Route selection
9.25

1   From Cape of Good Hope (34°21′S, 18°30′E), there are six routes; three are available during the south-west monsoon, one during the periods between monsoons and two during the north-east monsoon. The month references given below refer to the expected date of arrival in Indian waters and not to the date of departure from the Cape of Good Hope.

South-west monsoon (May to September)
9.26

1   The following alternative routes are available if reaching India during the south-west monsoon, however, these routes must not be taken unless there is a certainty of reaching India before the close of the monsoon (September)
9.26.1

1   **Through Moçambique Channel**. From May to September, the Inner Route through Moçambique Channel (9.18) is the most direct route for vessels bound for any destination in India.

2   On leaving the Cape of Good Hope (34°21′S, 18°30′E):
    With SE winds stand S and run down to the E in 39°S to 40°S as far as about 30°E, or:
    With W winds run along the coast, guarding against any indraughts, keeping S of 35°S until 37°E.

3   In either case steer for the Moçambique Channel, passing E of Île Europa (22°20′S, 40°20′E) and on either side of Île Juan de Nova (17°03′S, 42°42′E); then pass through the Comores (12°00′S, 44°30′E) and cross the equator in 53°E or 54°E. Thence steer direct for the Mumbai (18°51′N, 72°50′E). At the height of the south-west monsoon (June to August), when the weather is thick and heavy and observations are very uncertain, vessels should sound frequently when making land.

4   By using this track through the Moçambique Channel vessels will avoid the strongest part of the SW-going current and will be nearly sure of a fair wind until about half-way through the Channel, when adverse winds may be expected; should such occur it is better to make to the E on the port tack rather than the W; thus avoiding the African coast with its prevailing S-going current. The passage on the E side of Île Europa is recommended, but vessels should not approach that island nor Bassas da India (21°28′S, 39°46′E) at night, the currents in their vicinity being very strong and uncertain.

5   The winds in Moçambique Channel do not blow with the same regularity that is found farther N, and are generally stronger in the middle of the Channel. The north-east monsoon sets in between mid-September and mid-October and the change is usually accompanied by squally weather. When near the Madagascar coast, advantage may be taken of the alternating land and sea breezes.
9.26.2

1   **East of Madagascar**. From May to September passage E of Madagascar is often preferred to the Moçambique Channel route, as it is less dangerous and the winds are more steady, particularly in August and September, when light and variable winds are found in the Moçambique Channel.

2   On leaving Cape of Good Hope, make to the E in 39°S to 40°S until in about 45°E and then stand to the NE, crossing 30°S in about 53°E. From thence run through the South-east Trade Wind, passing W of Île de la Réunion (20°55′S, 55°15′E), Farquhar Group (10°10′S, 51°10′E) and

Amirante Isles (5°30′N, 53°20′E); then cross the equator in 53°E or 54°E and steer direct for Mumbai (18°51′N, 72°50′E).

3   If arrival at Mumbai before the start of the north-east monsoon is uncertain, make to the E in 39°S to 40°S but stand NE on reaching 40°E. Cross 30°S in about 59°E and then run N, passing between Mauritius (20°10′S, 57°30′E) and Île de la Réunion. After passing these islands there are two courses; one being to join with the route from Moçambique Channel (9.26.1) at about 15°S and to continue on it to destination; the other is to stand on directly N, through the Trade Wind, passing W of Saya de Malha Bank (10°00′S, 61°00′E). Cross the equator in about 62°E and then steer direct for Mumbai.

4   In case the north-east monsoon has started, keep towards the coast of India after passing the Maldives (4°00′N, 73°00′E).

Inter-monsoon period (April and October)
9.27

1   During these periods a W wind is often experienced on leaving the Cape of Good Hope. If this happens, during these months, run along the coast as directed for a W wind in 9.26.1, but stand on ENE past Madagascar and into the Indian Ocean, making directly for a position in about 15°S, 70°E, passing either, between the islands of Mauritius (20°10′S, 57°30′E) and Rodriguez (19°38′S, 63°25′E), or E of the latter. From this point steer to cross the equator in 75°E, passing E of the Chagos Archipelago (6°30′S, 72°00′E) and then N to the E of the Maldives (4°00′N, 73°00′E) and Lakshadweep (12°00′N, 72°30′E), parallel with the coast of India and working land and sea breezes, to Mumbai (18°51′N, 72°50′E).

North-east monsoon (November to March)
9.28

1   From November to March two routes are available, both making use of the W portion of the route for Australia described in 9.5, making to the E between the parallels of 39°S and 40°S.

9.28.1

1   The more W of the two routes leaves that latitude in 60°E and the vessel should then stand NE to 35°S, 70°E and thence stand N through the South-east Trade Winds to 10°S. Course should then be made NNE so as to cross the equator in 80°E, and making northing into the north-east monsoon, standing for Cape Comorin (8°05′N, 77°33′E) and finally working land and sea breezes along the Malabar coast to destination.

9.28.2

1   Alternatively, for a more E route, proceed as above to 35°S, 70°E, after which some authorities consider it more prudent to make farther to the E, so as to be well to windward on reaching the north-east monsoon, and make first for a position in about 25°S, 80°E. From this position turn N, run through the Trade Wind and the north-west monsoon, cross the equator in 82°E to 85°E and the onward track is as above making northing into the north-east monsoon, standing for Cape Comorin (8°05′N, 77°33′E) and finally working land and sea breezes along the Malabar coast to destination.

Cape Town or Cape of Good Hope ⇒ Mauritius
Routes
9.29

1   From Cape of Good Hope (34°21′S, 18°30′E), at all times of the year, make to the E between 39°S and 40°S as far as 40°E. Then stand NE, crossing 30°S in about 59°E, as far as 25°S thence steering direct for Port Louis, Mauritius (20°08′S, 57°28′E) in the South-east Trade Wind.

2   **Note**, vessels from this direction bound for Port Louis should pass E of Mauritius and round its N end, in order to avoid the calms caused by high land near the SW extremity of the island.

3   It may sometimes be possible to follow the route given in 9.27, leaving it when abreast of Mauritius, which it passes about 100 miles to the S.

Cape Town or Cape of Good Hope ⇒ Aden
Routes
9.30

1   From Cape of Good Hope (34°21′S, 18°30′E), there are three routes available from available from April to September; and one from November to March. In October either route may be taken.

9.30.1

1   From April to September or October, take the Inner Route for Mumbai, as described at 9.26.1, as far as the equator, crossing it in 53°E; thence continue to Aden (12°45′N, 44°57′E), passing between Raas Caseyr (11°50′N, 51°17′E) and Suquṭrá (12°30′N, 54°00′E). Work along the African coast as far as Jasiired Maydh (11°14′N, 47°13′E) before standing across the Gulf of Aden.

2   Alternatively, pass E of Madagascar as directed at 9.26.2, standing direct for Raas Caseyr after passing Amirante Isles (5°30′N, 53°20′E) and Seychelles Group (4°30′S, 55°30′E). Round Raas Caseyr closely and then proceed as directly as possible to destination.

3   Another alternative route passes between Mauritius (20°10′S, 57°30′E) and Île de la Réunion (20°55′S, 55°15′E) to join the above route in about 15°S.

9.30.2

1   From October or November to March, follow the directions given in 9.4 and 9.5 from Cape of Good Hope, continuing the route, between the parallels of 39°S and 40°S as far as 60°E. From this position stand N to pass about 200 miles E of Rodriguez Island (19°38′S, 63°25′E) and to cross the equator in 68°E. At this point, turn to the NW, steering so as to cross the meridian 60°E in 10°N and thence N of Suquṭrá (12°30′N, 54°00′E), which should be given a berth of 40 to 60 miles, to Aden (12°45′N, 44°57′E).

Cape Town or Cape of Good Hope ⇒ Mombasa and adjacent ports
Routes
9.31

1   From Cape of Good Hope (34°21′S, 18°30′E), the shortest route is through the Moçambique Channel as directed in 9.26.1, but passing W of the Comores (12°00′S, 44°30′E) and steering as directly as possible to Mombasa (4°05′S, 39°43′E).

2   The preferable route, for all seasons, is E of Madagascar. Follow the directions given in 9.26.2, as far as the N end of Madagascar, from which point both wind and current are favourable for Mombasa and adjacent ports.

PASSAGES FROM EAST COAST OF AFRICA AND MAURITIUS

Charts:

> *5126 (1) to (12) Monthly Routeing Charts for Indian Ocean.*
> *5309 Tracks followed by Sailing and Auxiliary Powered Vessels.*
> *5310 World Surface Current Distribution.*
> *Diagram 9.3 World Sailing Ship Routes.*

Durban ⇒ Australia, New Zealand, India, Singapore and South China Sea

Durban to Australia and New Zealand
9.32

1　From Durban (29°51′S, 31°06′E), stand SE and pick up the main route across the Indian Ocean (9.5) in 50°E and from that position follow the directions given from 9.5 to 9.12.

Durban to India, Singapore and South China Sea
9.33

1　From Durban (29°51′S, 31°06′E), stand SE and make to the E in about 35°S until picking up the route to pass through the Eastern Archipelago, to the Bay of Bengal, or to Mumbai, according to the season, as directed from 9.13 to 9.28.

Durban ⇒ Mauritius, East Africa and Aden

Durban to Mauritius
9.34

1　From Durban (29°51′S, 31°06′E), stand SE and make to the E in about 35°S. From about 50°E, keep gradually more to the N, crossing the parallel of 30°S in 58°E or 59°E, then steer direct through the Trade Winds to Mauritius (20°10′S, 57°30′E).

Durban to Mombasa and adjacent ports
9.35

1　From Durban (29°51′S, 31°06′E), the shortest route is by the Moçambique Channel, steering first to the SE across the Moçambique Current (6.28) until picking up the route from Cape Town to Mumbai (9.26.1), and passing W of the Comores (12°00′S, 44°30′E); but the preferable route is to proceed as directed in 9.34 to Mauritius and taking the route either W or E of Île de la Réunion (20°55′S, 55°15′E), round the N end of Madagascar to Mombasa (4°05′S, 39°43′E) with a favourable wind and current.

Durban to Aden
9.36

1　From Durban (29°51′S, 31°06′E), the routes are the same as from Cape Town (9.30), according to season, making to the E to pick up those routes that pass E of Madagascar on about the parallel of 35°S.

Durban ⇒ Cape of Good Hope

Route
9.37

1　From Durban (29°51′S, 31°06′E), proceed as directly as possible, at 20 miles or more from the coast. A favourable current will be carried throughout the passage to Cape of Good Hope (34°21′S, 18°30′E).

2　See also note under 9.47 and relevant part of 9.153.

3　**Caution**. Abnormal waves have occurred off the SE coast of South Africa, see 1.19 and 6.36 for details.

Mauritius ⇒ Australia and New Zealand

Mauritius to Fremantle, southern Australia and New Zealand
9.38

1　From Mauritius (20°10′S, 57°30′E), make to the S to pick up the main track across the Indian Ocean (9.5). Thence follow the appropriate routes 9.6 to 9.12 to destination.

Mauritius to northern Australia
9.39

1　From April to October, during the south-east monsoon on the N coast of Australia, stand S from Mauritius (20°10′S, 57°30′E), as in route 9.38, and proceed to Bass Strait (40°00′S, 146°00′E), thence to destination via the E coast of Australia and the Torres Strait (10°36′S, 141°51′E). For directions from Sydney (33°50′S, 151°19′E) to Torres Strait, see 10.37.

2　From November to April, during the north-west monsoon, stand N into that monsoon and then proceed as directly as possible.

Mauritius ⇒ Singapore or South China Sea

Routes
9.40

1　From Mauritius (20°10′S, 57°30′E), stand SE or E to pick up the route from Cape Town (9.13 to 9.16) according to destination and season.

Mauritius ⇒ Indian ports

Routes
9.41

1　**From April to October**.

2　The route described at 9.26.2 passes Mauritius (20°10′S, 57°30′E) closely, and may be followed to Mumbai (18°51′N, 72°50′E).

3　For the Bay of Bengal, leave the route in 5°N and pass through Eight Degree Channel (7°24′N, 73°00′E) and round the S end of Sri Lanka.

4　For Colombo (6°58′N, 79°50′E) steer direct from Eight Degree Channel.

5　**From November to March**.

6　From Mauritius, stand E or SE to pick up the routes given in:

> 9.20 for Bay of Bengal:
> 9.24 for Colombo:
> 9.28 for Mumbai.

Mauritius ⇒ Aden

Sailing ship routes
9.42

1　**From April to October**, from Mauritius (20°10′S, 57°30′E) join the route from Cape Town (9.30.1), which passes E of Madagascar, in about 15°S.

2　**From November to March**, run N through the South-east Trade Wind and the north-west monsoon, to pick up the route described at 9.30.2, crossing the equator in about 68°E.

Low-powered vessel routes
9.43

1　From Mauritius (20°10′S, 57°30′E) the route for low-powered vessels is as direct as navigation permits to Aden (12°45′N, 44°57′E), as described at 6.67 for

Mauritius ⇒ Mombasa and adjacent ports

Sailing ship routes
9.44
1   From Mauritius (20°10′S, 57°30′E) proceed as directly as possible, passing N of Madagascar, to Mombasa (4°05′S, 39°43′E).

Low-powered vessel routes
9.45
1   From Mauritius (20°10′S, 57°30′E) the route for low-powered vessels is as direct as navigation permits to Mombasa (4°05′S, 39°43′E), as described at 6.66 for full-powered vessels. The low-powered route from Mombasa to Mauritius is given at 9.53.

Mauritius ⇒ Durban or Cape of Good Hope

Mauritius to Durban
9.46
1   From Mauritius (20°10′S, 57°30′E) proceed as directly as possible, passing about 10 miles S of Madagascar, and making the African coast well N of Durban (29°51′S, 31°06′E).

Mauritius to Cape of Good Hope
9.47
1   From Mauritius (20°10′S, 57°30′E) pass about 200 miles S of Madagascar, and make the African coast about 200 miles SW of Durban (29°51′S, 31°06′E), afterwards keeping in the strength of the Agulhas Current (6.28) until abreast of Mossel Bay (34°11′S, 22°09′E), thence steer direct to round Cape Agulhas (34°50′S, 20°01′E) at a prudent distance.
2   **Note**. When nearing Cape of Good Hope (34°21′S, 18°30′E) with strong W winds, keep on Agulhas Bank (35°30′S, 21°00′E) not more than 40 or 50 miles from the coast, where smoother water will be found.
3   **Caution**. Abnormal waves have occurred off the SE coast of South Africa, see 1.19 and 6.36 for details.

Mombasa and adjacent ports ⇒ Aden

Routes
9.48
1   Routes from Mombasa (4°05′S, 39°43′E) to Aden (12°45′N, 44°57′E) are seasonal.
9.48.1
1   **April to October**, keep coastwise in the strength of the current and pick up the route from Cape of Good Hope to Aden, described at 9.30.1, in about 10°N.
9.48.2
1   **November to March**, work to the E into the north-west monsoon keeping as far N as the wind will permit until that monsoon is reached; then run E, edging to the N at the latter part, as far as about 68°E, then stand N into the north-east monsoon, and from thence direct for the Gulf of Aden.
2   The same route may be taken from the Seychelles Group (4°30′S, 55°30′E).
3   Suquṭrá (12°30′N, 54°00′E) should be weathered if possible. If efforts are made to pass S of it, in a fresh monsoon, there is a great chance of being swept to

windward of Raas Caseyr (11°50′N, 51°17′E). If leaving Mombasa in March, do not go E of the Seychelles Group before standing N, as S winds might be expected before reaching Raas Caseyr.

Mombasa and adjacent ports ⇒ Mumbai

Routes
9.49
1   Routes from Mombasa (4°05′S, 39°43′E) to Mumbai (18°51′N, 72°50′E) are seasonal.
9.49.1
1   **April to October**, keep coastwise in the strength of the current to about 5°N and then steer directly as possible to Mumbai, making a landfall on the parallel of Kanhoji Angre (Khānderi) Island (18°42′N, 72°49′E) if the weather is thick.
9.49.2
1   **November to March**, work E into the north-west monsoon, then run E on about the parallel of 5°S until 82°E or 84°E, when stand N across the equator into the north-east monsoon and make the S end of Sri Lanka, and then Cape Comorin (8°05′N, 77°33′E); thence work up the Malabar coast with the land and sea breezes.
2   In March it would perhaps be better to go direct, when the north-west monsoon is met with, as NW winds are prevalent in the Arabian Sea at the end of the north-east monsoon.

Mombasa and adjacent ports ⇒ Colombo or Kolkata

Routes
9.50
1   Routes from Mombasa (4°05′S, 39°43′E) to Colombo (6°58′N, 79°58′E) or Kolkata (22°30′N, 88°15′E) are seasonal.
9.50.1
1   **May to September**, stand E on the starboard tack and make for the Eight Degree Channel (7°24′N, 73°00′E), if bound for Colombo, or pass through the more direct route offered by the Kardiva Channel (5°05′N, 73°24′E), but not at night, unless entrance has been made before dark or the vessel's position is accurately known.
2   If bound for Kolkata pass S of Sri Lanka and pick up the Bay of Bengal route from the Cape of Good Hope (described at 9.19). See also 9.21 for transit of Bay of Bengal.
9.50.2
1   **October to April**, the route passes close S of the Seychelles Group (4°30′S, 55°30′E). On leaving Mombasa keep N of the direct route to Seychelles Group, while working to the E until the north-west monsoon is picked up, which may be expected after passing 45°E, although this is very uncertain. Light winds and calms render this a generally tedious passage.
2   After passing the Seychelles Group, and if bound for Colombo, run E in about 5°S, cross the equator in 82°E to 84°E, and stand N into the north-east monsoon, then making for the SW end of Sri Lanka; then work up the coast, taking advantage of land and sea breezes.
3   If bound for Kolkata, continue to make farther E until the route from the Cape of Good Hope (described at 9.20) is picked up in about 92°E and follow it to destination. See also 9.21 for notes on transit of the Bay of Bengal.

The low-powered route from Aden to Mauritius is given at 9.140.

Mombasa and adjacent ports ⇒ Mauritius and Australia

General information
9.51

1 In all seasons the route to Australia from Mombasa (4°05′S, 39°43′E) is taken via, or passing close to, Mauritius (20°10′S, 57°30′E).

Sailing ship routes to Mauritius
9.52

1 From Mombasa (4°05′S, 39°43′E) the following seasonal routes are advised.

9.52.1

1 **April to October**, stand E, regardless of crossing the equator in so doing, until E of Chagos Archipelago (6°30′S, 72°00′E), then head farther S into the Trade Wind and then a direct course steered for Mauritius.

9.52.2

1 **November to March**, the recommended route is to make to the E with the north-east and north-west monsoons and cross 10°S in about 70°E and then steer direct through the Trade Wind for Mauritius. Vessels should keep N of a line drawn from Zanzibar (6°10′S, 39°18′E) to the Seychelles Group (4°30′S, 55°30′E) until in the north-west monsoon.

2 An alternative route from November to March is to stand down through Moçambique Channel, taking advantage of the current on the African coast. Then from the S end of the Channel, stand SE into the W winds and make to the E to the S of 35°S. Recross 30°S in about 58°E and 59°E, then make direct for Mauritius through the Trade Wind.

3 **Caution.** The cyclone season is from November to March and the first route is therefore the safer as the path of these cyclones is then more easily avoided.

Low-powered vessel routes to Mauritius
9.53

1 From Mombasa (4°05′S, 39°43′E) the following seasonal routes for low-powered vessels are advised:

2 **November to March**:
  N of Seychelles Group (4°30′S, 55°30′E), thence:
  E of Saya de Malha Bank (10°00′S, 61°00′E), thence:
  S into the South-east Trade Winds, thence:
  To Port Louis (20°08′S, 57°28′E).

3 **April to October**:
  Similar to that for November to March but the E course should be held to about 70°E before:
  Turning S and making into the South-east Trade Winds, thence:
  To about 12°S, 70°E, thence:
  To Port Louis (20°08′S, 57°28′E).

4 The low-powered route from Mauritius to Mombasa is given at 9.45.

Sailing ship routes to Australia
9.54

1 From Mombasa (4°05′S, 39°43′E) follow the directions given in 9.38 or 9.39 after calling at, or passing close to Mauritius (20°10′S, 57°30′E) as described at 9.52.2.

Mombasa and adjacent ports ⇒ Durban or Cape of Good Hope

Route
9.55

1 From Mombasa (4°05′S, 39°43′E), at all seasons, proceed as directly as possible keeping in the strength of the Moçambique and Agulhas Currents (6.28). See also 9.153 for further details.

2 **Caution.** Abnormal waves have occurred off the SE coast of South Africa, see 1.19 and 6.36 for details.

PASSAGES THROUGH THE EASTERN ARCHIPELAGO

Charts:
  5126 (1) to (12) Monthly Routeing Charts for Indian Ocean.
  5309 Tracks followed by Sailing and Auxiliary Powered Vessels.
  5310 World Surface Current Distribution.
  Diagram 9.3 World Sailing Ship Routes.

GENERAL NOTES ON PRESENTATION

General information
9.56

1 This section contains directions for the various routes through the Eastern Archipelago, these constitute an important link between the Indian Ocean and the Pacific Ocean.

2 Each particular route passes through a number of straits or channels and a list of the seas and straits, within the archipelago, are listed 9.58.1, where the directions are indexed for ease of reference.

Archipelagic Sea Lanes

Diagram 9.57
Archipelagic Sea Lanes
9.57

1 **Archipelagic Sea Lanes (ASL)** (1.30) have been designated, by the Indonesian Authorities through the Eastern Archipelago. Appendix A gives full details of these sea lanes. Further details regarding Archipelagic Sea Lanes can be found in *The Mariner's Handbook*.

2 These sea lanes, however, are not practical for use by sailing vessels, during certain periods of the year, due to contrary winds and currents. Also they tend to be in deep water, making it difficult for sailing vessels to find anchorage in calms, light variable winds, or when set back by winds or currents.

3 The routes described for sailing vessels in the Eastern Archipelago, therefore, describe the traditional routes used by sailing vessels in the past. As these routes do not adhere to any of the Archipelagic Sea Lanes, mariners are strongly advised to confirm their intended routes with the relevant authority. The passages through the Eastern Archipelago are given from 9.67 to 9.130.

4 Reference to the appropriate ASL is given where any of the routes described cross or transit part of it.

Summary of routes through Eastern Archipelago

Route selection
9.58

1 The following paragraphs are intended as a guide for the selection of the best route through the Eastern Archipelago and the season for which it is recommended. An index giving the relevant paragraphs applicable to the various straits and sea areas is given below.

9.58.1

Index of all routes through Eastern Archipelago:

Name	Directions N-bound	Directions S-bound
Abang, Selat (0°32′N, 104°15′E)	9.75; 9.77.	
Alas, Selat (8°40′S, 116°40′E)	9.101; 9.107; 9.108; 9.110.	9.113; 9.123; 9.126; 9.127; 9.130.
Bali, Selat (8°10′S, 114°25′E)	9.101; 9.107; 9.108.	9.113; 9.123; 9.126; 9.127; 9.130.
Balintang, Channel (20°00′N, 122°20′E)	9.101 9.106.	9.123; 9.124.
Banda Sea	9.99; 9.100.	9.123; 9.126; 9.127.
Bangka, Selat (2°06′S, 105°03′E)	9.71; 9.72; 9.74; 9.75; 9.90, 9.91	9.114; 9.116; 9.117; 9.121;
Bashi Channel (21°20′N, 121°00′E)	9.101 9.106.	9.123; 9.124.
Basilan Strait (6°50′N, 122°00′E)	9.107; 9.112.	9.127; 9.129.
Baur, Selat (3°00′S, 107°18′E)	9.84; 9.86.	9.117; 9.119.
Berhala, Selat (0°50′S, 104°15′E)	9.75; 9.77; 9.78; 9.79.	9.114.
Celebes Sea	9.107; 9.111; 9.112.	9.127; 9.129.
Ceram Sea	9.99; 9.100; 9.101; 9.102; 9.103.	9.123; 9.125; 9.126; 9.127.
Dampier, Selat (0°40′S, 130°30′E)	9.101; 9.103.	9.123; 9.125.
Durian, Selat (0°40′N, 103°40′E)	9.75; 9.77; 9.79; 9.80.	9.114.
Gelasa, Selat (2°53′S, 107°18′E)	9.84; 9.85; 9.86; 9.88; 9.90.	9.114; 9.117; 9.118; 9.121.
Jailolo, Selat (0°06′N, 129°10′E)	9.101; 9.103.	9.113; 9.123; 9.125.
Java Sea	9.71; 9.85; 9.94; 9.99; 9.100; 9.107; 9.111.	9.121; 9.122; 9.127; 9.130.
Karimata, Selat (1°43′S, 108°34′E)*	9.94; 9.96.	9.114; 9.120; 9.121.
Leplia, Selat (2°53′S, 106°58′E)	9.84; 9.85.	9.117; 9.118.
Limende, Selat (3°00′S, 107°12′E)	9.84; 9.87.	9.119.
Lombok, Selat (8°47′S, 115°44′E)*	9.101; 9.107; 9.108; 9.109.	9.113; 9.123; 9.126; 9.127; 9.130.
Makasar, Selat (2°00′S, 118°00′E)*	9.99; 9.107; 9.111.	9.113; 9.127; 9.129; 9.130.
Manipa, Selat (3°20′S, 127°22′E)	9.100; 9.101; 9.102.	9.123; 9.126; 9.127.
Mindoro Strait (13°00′N, 120°00′E)	9.101; 9.104; 9.107; 9.112.	9.127; 9.128; 9.129.

Name	Directions N-bound	Directions S-bound
Molucca Sea	9.101; 9.103.	9.123; 9.125; 9.127.
Ombi, Selat (8°35′S, 125°00′E)*	9.101; 9.102.	9.113; 9.123; 9.126;
Pengelap, Selat (0°30′N, 104°20′E)	9.75; 9.77.	—
Riau, Selat (0°55′N, 104°20′E)	9.75; 9.76; 9.81; 9.82; 9.88.	9.114; 9.115.
San Bernadino Strait (12°33′N, 124°12′E)	9.101; 9.104; 9.105;	—
Sapudi, Selat (7°00′S, 114°15′E)	9.100.	9.123; 9.127; 9.130.
Sibutu Passage (4°55′N, 119°37′E)	—	9.127; 9.129.
Singapore Strait (1°10′N, 103°50′E)	9.75; 9.76; 9.77; 9.80; 9.81; 9.82; 9.83; 9.91; 9.92.	9.114.
Sunda, Selat (6°00′S, 105°52′E)*	9.67; 9.69; 9.70; 9.75; 9.84; 9.85; 9.94; 9.99; 9.101.	9.113; 9.114; 9.121; 9.122; 9.127; 9.130.
Sulu Sea	9.104; 9.107; 9.112.	9.127; 9.128; 9.129.
Surigao Strait (9°53′N, 125°21′E)	9.101; 9.104.	—
Verde Island Passage (13°36′N, 121°00′E)	9.101; 9.105; 9.107; 9.112.	—

* Waterways which form part of Indonesian Archipelagic Sea Lanes

Selat Sunda ⇒ Singapore

9.59

1   From Selat Sunda (6°00′S, 105°52′E) there are six possible routes. The usual route, though principally for the period May to September, is via Selat Bangka (2°06′S, 105°03′E) and Selat Riau (0°55′N, 104°20′E). References for this passage are 9.67 - 9.76, 9.81.

2   From October to March, a route known as the Inner Route, via Selat Bangka, Selat Berhala (0°50′S, 104°15′E) and Selat Durian (0°40′N, 103°40′), may be used. References are 9.67 - 9.74 and 9.77 - 9.81.

3   In December, January and February, a route via Selat Bangka, Selat Berhala, Selat Pengelap (0°30′N, 104°20′E) or Selat Abang (0°32′N, 104°15′E) and Selat Riau, is recommended. References are 9.67 - 9.76, 9.81.

4   As an alternative, from May to September, proceed via Selat Bangka and E of Pulau Bintan (1°00′N, 104°30′E). References are 9.67 - 9.74, 9.81 - 9.91.

5   A second alternative, from May to September, is via Selat Gelasa (2°53′S, 107°18′E) and Selat Riau. A vessel which, having chosen this route, finds that the north-east monsoon is blowing strongly in the South China Sea, should steer for Selat Berhala and continue through Selat Durian. References are 9.67 - 9.70.2, 9.76, 9.84, 9.88.

6   A third alternative, from May to September, is via Selat Gelasa and E of Pulau Bintan. References are 9.67 - 9.70.2, 9.84, 9.92.

Selat Sunda ⇒ South China Sea
9.60

1   From May to September, proceed either via Selat Bangka (2°06'S, 105°03'E) (9.67 - 9.71, 9.93.1); or via Selat Gelasa (2°53'S, 107°18'E) (9.67 - 9.70.2, 9.84, 9.93.2); or via Selat Karimata (1°43'S, 108°34'E) (9.67 - 9.70.2, 9.95).

2   From November to February, or if on entering the Java Sea it is found that the north-west monsoon of the Java Sea, or the north-east monsoon of the South China Sea, have begun proceed E through the Java Sea and join the Second Eastern Passage (9.67 - 9.70.2, 9.99, 9.101).

3   Alternatively join the First Eastern Passage (9.67 - 9.70.2, 9.99, 9.107) in Selat Makasar (2°00'S, 118°00'E).

Selat Ombi ⇒ South China Sea
9.61

1   This route, known as the Second Eastern Passage, is usable from October to March. Main references for the route are at 9.101.

2   The Second Eastern Passage passes through Selat Ombi (8°35'S, 125°00'E), Banda Sea and Selat Manipa (3°20'S, 127°22'E) (9.102), Ceram Sea, Selat Jailolo (0°06'N, 129°10'E), Selat Dampier (0°40'S, 130°30'E) or the Molucca Sea to the Pacific Ocean (9.103); and to the South China Sea via Surigao Strait (9°53'N, 125°21'E) (9.104) or by San Bernadino Strait (12°33'N, 124°12'E) and Verde Island Passage (13°36'N, 121°00'E) (9.105). Alternatively E of the Philippine Islands and through the Balintang Channel (20°00'N, 122°20'E) or Bashi Channel (21°20'N, 121°00'E) (9.106) to South China Sea.

Selat Bali, Selat Lombok or Selat Alas ⇒ South China Sea
9.62

1   This route known as the First Eastern Passage is usable in October and November only. From the approach strait (9.108) it passes through the Java Sea and Selat Makasar (2°00'S, 118°00'E) to the Celebes Sea (9.111). It continues through Basilan Strait (6°50'N, 122°00'E), the Sulu Sea and Mindoro Strait (13°00'N, 120°00'E) (9.112) to the South China sea.

North Australia ⇒ Singapore
9.63

1   Either N of Timor (9°00'S, 125°00'E) and through the Java Sea or S of all the islands and through Selat Sunda (6°00'S, 105°52'E), see 9.205.

Singapore ⇒ Selat Sunda
9.64

1   From November to April, proceed via Selat Riau (0°55'N, 104°20'E) and Selat Bangka (2°06'S, 105°03'E) (9.115, 9.116, 9.121, 9.122), or, from October to April, Selat Gelasa (2°53'S, 107°18'E) (9.117) may be used instead of Selat Bangka.

2   Also from October to April, passage may be made through Selat Durian (0°40'N, 103°40'E), Selat Berhala (0°50'S, 104°15'E) and Selat Bangka. This is known as the Inner Route (9.83, 9.80, 9.79, 9.78 reversed, and 9.116, 9.121, 9.122).

3   From May to September a route known as the Outer Route should be taken, passing E of Pulau Bintan (1°00'N, 104°30'E) and through either Selat Karimata (1°43'S, 108°34'E) or Selat Gelasa (9.82 reversed, 9.120 or 9.117, 9.121, 9.122).

South China Sea ⇒ Selat Sunda
9.65

1   This route between the South China Sea and the Indian Ocean is known as the Western Route (10.79.1). From October to April, ships having used the north-east monsoon route through the South China Sea, which passes between Pulau-Pulau Anambas (3°00'N, 106°00'E) and Pulau-Pulau Natuna (4°00'N, 108°00'E), should use Selat Gelasa (2°53'S, 107°18'E) (9.117) or, from November to April, Selat Bangka (2°06'S, 105°03'E) (9.116).

2   From May to September, Palawan Passage (10°20'N, 118°00'E) and the coastwise route off Borneo are used in the South China Sea and, either Selat Gelasa or Selat Karimata (1°43'S, 108°34'E) should be taken in continuing for Selat Sunda (6°00'S, 105°52'E), see 9.114.

South China Sea ⇒ Indian Ocean
9.66

1   From mid-May till the end of July the Indian Ocean should be approached through Selat Ombi (8°35'S, 125°00'E), or by either Selat Alas (8°40'S, 116°40'E), Selat Lombok (8°47'S, 115°44'E) or Selat Bali (8°10'S, 114°25'E). The Eastern Route from the South China Sea is used, passing through either the Balintang Channel (20°00'N, 122°20'E) or Bashi Channel (21°20'N, 121°00'E) into the Pacific Ocean and thence through the archipelago via Selat Jailolo (0°06'N, 129°10'E), Selat Dampier (0°40'S, 130°30'E) or the Molucca Sea to the Ceram Sea and thence through Selat Manipa (3°20'S, 127°22'E) and the chosen entrance channel to the Indian Ocean, see 9.123.

2   In May only the Central Route from the South China Sea may be used. It enters the Sulu Sea via Mindoro Strait (13°00'N, 120°00'E) or Verde Island Passage (13°36'N, 121°00'E) and leads thence through either Basilan Strait (6°50'N, 122°00'E) or Sibutu Passage (4°55'N, 119°37'E) to the Celebes Sea and Selat Makasar (2°00'S, 118°00'E). Either Selat Alas, Selat Lombok or Selat Bali are then used in the approach to the Indian Ocean, or a route through the Java Sea and Selat Sunda (6°00'S, 105°52'E) may be taken, see 9.127.

PASSAGES NORTH-BOUND THROUGH EASTERN ARCHIPELAGO

Charts:
> 5126 (1) to (12) Monthly Routeing Charts for Indian Ocean.
> 5309 Tracks followed by Sailing and Auxiliary Powered Vessels.
> 5310 World Surface Current Distribution.
> Diagram 9.3 World Sailing Ship Routes.

Approaches ⇒ and passage through Selat Sunda

Landfall
9.67

1   Selat Sunda (6°00'S, 105°52'E) and its approaches are described in *Indonesia Pilot, Volume I*.

2   Coming from the S in the south-east monsoon, keep well to the E, especially in June, July and August, when the monsoon and the W-going current are at their strongest or the vessel may fall to leeward of Tanjung Gedeh (6°46'S, 105°13'E) and find great difficulty in recovering it against wind and current.

3   In December, January and February, considerable swell rolls into the strait and the sea is heaviest when the tidal stream, combining with the prevailing SW-going current, runs contrary to the wind. The sea is said to be calmest in March, July and November.

4   Having made a landfall, shape course to pass between Pulau Rakata (6°09′S, 105°26′E) and the Jawa shore, or between Pulau Sebesi (5°57′S, 105°29′E) and Pulau Sebuku (5°53′S, 105°31′E). The former is recommended except for those with local knowledge and then only in daylight.

Selat Panaitan
9.68

1   Selat Panaitan (6°40′S, 105°15′E), between the NW side of Ujung Kulon and Pulau Panaitan (6°35′S, 105°13′E), possesses the great advantage of affording anchorage to sailing vessels becalmed, which the channel N of the island does not. Light baffling winds and calms are experienced about the entrance to Selat Sunda (6°00′S, 105°52′E), occurring even in the strength of the east monsoon, and sailing vessels unable to anchor are liable to be set back by adverse currents.

2   Selat Panaitan is entirely clear, but Pulau Panaitan must not be approached within 1 mile on account of Pulau-Pulau Karangjajar (6°41′S, 105°11′E) and the coastal reef which extends from the S side of Tanjung Semadang (6°38′S, 105°14′E), these dangers are always marked by surf.

3   Working through the passage keep nearer the Jawa coastline than Pulau Panaitan, especially in the south-west monsoon.

4   **Conservation Areas** surround both Pulau Panaitan and Ujung Kulon, see *Indonesia Pilot, Volume I.*

Channel north of Pulau Panaitan
9.69

1   This channel, sometimes known as 'Great Channel', although the widest into Selat Sunda, and much frequented being considered free from dangers, has the disadvantage of being too deep for anchoring if becalmed; in which case a vessel may drift out of the strait and into the W-going stream.

2   Entering Selat Sunda by this channel, keep nearer Pulau Panaitan and when farther in, keep towards the same or Jawa side.

3   The channel is recommended for the later part of the west monsoon period and for the transition period.

Passage through Selat Sunda
9.70

1   It is presumed that a sailing vessel will make her way through Selat Sunda on the Jawa side, whether she makes landfall on Tanjung Gedeh (6°46′S, 105°13′E) or Tanjung Cukubalimbing (5°56′S, 104°33′E), or halfway between them, in the two latter cases a vessel is presumed to have passed S of Pulau Rakata (6°09′S, 105°26′E).
9.70.1
Passage with a favourable wind.

1   There is not much difficulty in proceeding through the strait during the north-west monsoon period; the Jawa side of the strait and the channel S of Pulau Sangian (5°57′S, 105°51′E) being recommended. Pass about 2 miles off Tanjung Cikoneng (6°04′S, 105°53′E) and between Pulau Tempurung (5°54′S, 105°56′E) and Pulau Merak Besar (5°56′S, 105°59′E). If becalmed, anchorage is available, while the channel N of Sangian is not favourable for this purpose due to the deeper water. Two dangerous reefs, Terumbu Koliot (5°55′S, 105° 49′E) and Terumbu Gosal (5°53′S, 105°54′E) lie in or near the most N or W channel.

9.70.2
Passage during south-east monsoon.

1   During the south-east monsoon the winds may be E and variable and sometimes strong from the NE towards midday. This combined with an adverse current, possibly from 2 to 3 kn in mid-stream, renders the passage more tedious and it may become necessary to anchor to avoid losing ground.

2   Therefore the coast of Jawa should be kept, where anchorage may be had in many places and where the current is much weaker, and at times nil, when the S-going tidal stream is at its strongest.

3   A vessel having to work up may stand into Teluk Miskam (6°30′S, 105°45′E), when N of Pasir Gundul (6°27′S, 105°42′E), to a depth of about 15 m but when near Pulau Popole (6°24′S, 105°48′E) into not less than 18 m to avoid Gosong Panjang (6°25′S, 105°47′E) and into not less than 27 m or 2 miles off Caringin (6°21′S, 105°49′E) to avoid Karang Kebua (6°22′S, 105°48′E). To the N of it the shore may be approached closer, by sounding. Approaching Pasangtenang (6°09′S, 105°51′E), stand into not less than 22 m, or 1 mile from the shore. Tanjung Cikoneng (6°04′S, 105°53′E) and the coast E is fringed by reef to a distance of 1 to 2 cables, but is steep-to. There is good anchorage S of Tanjung Cikoneng in about 11 m and also off Anyer Lor (6°03′S, 105°55′E) to the E of the point, but it is not so good off Tanjung Cikoneng itself.

4   Passing between Pulau Sangian and Jawa it is advisable to keep outside a depth of 36 m unless seeking anchorage.

Selat Sunda ⇒ Selat Bangka

Directions
9.71

1   **Cautions**. Areas of the Java Sea are being exploited for natural resources, see 7.38.2.

2   This route passes close to oilfields, with extensive areas of prohibited anchorage, it would be prudent to avoid these areas altogether wherever possible.

3   From a position to the N of Pulau Tempurung (5°54′S, 105°56′E) and with a favourable wind, steer direct for Pulau Segama (5°10′S, 106°06′E), keeping the S islet bearing less than 010° to lead E of Layang-layang (5°18′S, 106°04′E); the islets can be passed on either side, during daylight, W of Karang Basa (5°12′S, 106°12′E) and Gosong-gosong Serdang (5°05′S, 106°15′E). Gosong Syahbandar (5°07′S, 105°56′E) must not be approached in depths of less than 13 m.

4   Working to the N, it will be prudent to keep on the Sumatera side and when N of Karang Sybrandi (5°41′S, 105°51′E) and standing towards the shore, to tack when in a depth of between 11 m and 15 m; the directions for clearing Layang Layang, described above, must be noted.

5   After passing Segama, a safe guide is to keep in depths of about 18 m; approaching the coast when the depths increase to between 22 m and 24 m and holding out when they decrease to 17 m. Vessels working up must give the Sumatera coast a wide berth when N of Segama.

6   While powered vessels can pass E of Five Fathom Banks (3°48′S, 106°29′E), some authorities consider that a sailing vessel, working up the coast, may pass between those banks and Gosong Menjangan (3°45′S, 106°12′E) and, after passing the latter, may stand on the inshore tack to a depth of 9 m. It should be noted, however, that the 5 m depth contour is some 14 to 15 miles from the coast in some places. If making for Alur Pelayaran Stanton (3°00′S,

106°15′E), a vessel should always pass E of Five Fathom Banks.

Approaches ⇒ and north-bound passage through Selat Bangka

Natural conditions
9.72

1 **Winds** in Selat Bangka (2°06′S, 105°03′E) follow the direction of the coast, though with slight variations from the influence of the land and sea breezes; fresh breezes may always be expected when working against the monsoon. During the later part of the south-east monsoon, it frequently blows hard from the SW. Land breezes occur at night.

2 **Tidal streams**. A full description is given in *Indonesia Pilot, Volume I*, but it can be noted here that, due to the variations in the predominant streams in the two monsoonal seasons, it is preferable to work N on the Bangka side of the strait during the north-west monsoon and on the Sumatera side during the south-east monsoon.

Directions for Alur Pelayaran Maspari and Alur Pelayaran Stanton

9.73

1 The navigation of Alur Pelayaran Stanton (3°00′S, 106°15′E) is difficult for sailing vessels working N at night; there are not enough marks for fixing the vessel's position, soundings are not a trustworthy guide and the usually strong tidal streams make the position uncertain.

2 Alur Pelayaran Maspari (3°08′S, 106°06′E) should never be attempted at night, except in clear weather with local knowledge and it is not possible with adverse winds. If Pulau Maspari (3°13′S, 106°13′E) is not visible at a distance of 4 miles it is advisable to anchor.

3 Sailing vessels working N through Alur Pelayaran Maspari by day can safely approach the bank extending from the Sumatera coast by sounding, but they must not stand into a depth of less than 11 m when 5 miles S of Tanjung Jati (2°58′S, 106°03′E) and must keep in depths of 18 m, or more, when off that point.

4 Pulau Maspari can be approached to within 3½ miles on the S side, about 4 miles on the SE side and to within 1 mile on the W side. Soundings generally give enough warning when standing towards the banks on the E side of that channel.

5 In Alur Pelayaran Stanton, sailing vessels with a fair wind can follow the directions given for power vessels in *Indonesia Pilot, Volume I*. Working N through Alur Pelayaran Stanton, by day, they can approach Pulau Dapur (3°08′S, 106°31′E) within about 7 cables. The summit of Pegunungan Permisan (2°36′S, 105°57′E) bearing about 323° and open NE of Pulau Besar light-structure (2°53′S, 106°08′E) leads NE of Gosong Melvill (3°02′S, 106°15′E). As soon as Tanjung Labu (2°58′S, 106°20′E) bears more than 035°, vessels will remain clear of the banks on either side of the channel by keeping in depths of not less than 20 m.

Directions north-bound in Selat Bangka
9.74

1 After rounding Tanjung Lelari (2°49′S, 105°57′E), work up under the Bangka side of the strait; the landmarks here are more conspicuous and vessels can derive more advantage from the land winds, which are somewhat

stronger and more regular than those in the middle of the strait.

2 Some sailing vessels, and even power vessels, make use of the narrow channel between Karang Tembaga (2°41′S, 105°53′E) and the Bangka coast, when this is feasible, as the tidal streams are more favourable there. Farther N, the coastal bank extending from Bangka is fairly steep-to, and nearing Pulau Nangka Besar (2°25′S, 105°48′E) vessels must keep in depths of not less than 13 m in order to clear the bank which surrounds these islands.

3 Standing over to the Sumatera side, the bights in the coast may be approached by sounding, but the points must never be approached in depths of less than 20 m, as within this the depths decrease very suddenly. From about 5 miles E to 6 miles W of Tanjung Katimabongko (2°20′S, 105°14′E), the coastal bank is steep-to and it is very hard W of this point. Farther W the depths decrease regularly towards the shore and vessels can approach it into depths of 9 m.

4 The passage between the Sumatera coast and Karang Ular (1°58′S, 104°57′E) can be easily negotiated by sounding; the coast there can be approached into depths of 8 m, but vessels should tack away from the E side of the passage immediately the depths increase to more than 16 m. If taking the channel E of Karang Ular, vessels must not stand to far over towards the Bangka shore, on account of the reefs lying as much as 2½ miles offshore between Tanjung Kelian (2°05′S, 105°08′E) and Tanjung Ular (1°57′S, 105°08′E).

Selat Bangka ⇒ Selat Riau

Routes
9.75

1 From Selat Bangka (2°06′S, 105°03′E) to Selat Riau (0°55′N, 104°20′E), there are two routes, one direct and the other via Selat Berhala (0°50′S, 104°15′E) and Selat Pengelap (0°30′N, 104°20′E) or Selat Abang (0°32′N, 104°15′E). The latter route is recommended for the months of December, January and February (see also 9.77, Selat Bangka to Singapore Strait via Selat Berhala and Selat Durian).

2 The route from Selat Bangka to Singapore, E of Pulau-Pulau Lingga (0°00′, 104°30′E) and through Selat Riau, is the one commonly adopted by vessels proceeding either way between Selat Sunda and Singapore, as being safe, sheltered and easily navigable. The route E of Pulau Bintan (1°00′N, 104°30′E), however, is exposed in both monsoons, and the fairway is encumbered with many dangers, which render it necessary for vessels to keep a considerable distance from the land. Selat Riau is suitable for all classes of vessel, both by day and night. The swept channels and their depths are detailed in *Indonesia Pilot, Volume I*.

9.75.1

1 **The direct route**, ordinarily used by sailing vessels N-bound from Selat Bangka, is between Pulau-Pulau Tuju (1°15′S, 105°15′E) and Pulau Saya (0°47′S, 104°56′E); they may, however, pass either side of Saya, which being high and bold, is very convenient to make particularly in thick weather or at night.

2 Soundings may be very useful in detecting the drift, caused by cross-currents, between Pulau-Pulau Tuju and the coast of Sumatera, particularly in thick weather or at night. The depth decreases generally towards Sumatera and increases towards the islands, but care should be taken in approaching them, as irregularities of the currents have brought many vessels into danger. The bottom near Sumatera is mud, mixed with sand, while that near the islands is mud only.

3   From Pulau Saya keep NE to a position some 12 to 13 miles NE of Tanjung Jang (0°18′S, 105°00′E), the SE extremity of Pulau-Pulau Lingga (0°00′, 104°30′E) and from thence as follows:

By day, with a fair wind, steer directly for the fairway into Selat Riau, taking particular note of the tidal stream, especially when setting strongly to the SE.

At night, steer a little more to the N to give more clearance to the shoals and head in for Selat Riau when the bank extending NE from Pulau Mesanak (0°25′N, 104°32′E) has been crossed.

4   When working N, it is seldom necessary to work along near the islands from Mesanak to Korek Rapat (0°41′N, 104°20′E); it is generally found advantageous to stand to the N, in case of meeting a NW wind. But it may occasionally happen that advantage will be derived by standing towards them; in which case, when standing towards the N side of Mesanak, keep the summit of Pulau Benan (0°28′N, 104°27′E) bearing less than 275°, which will lead N of the extensive shoal (least depth 4·5 m), NE of the E point of Mesanak. To clear Karang Leman (0°28′N, 104°28′E) keep the E extreme of Mesanak bearing more than 133°, and the N extreme of Pulau Katang Lingga (0°30′N, 104°25′E) less than 285°.

5   If working in towards Selat Dempo (0°36′N, 104°15′E), do not get S of the line joining the N extreme of Katang Lingga, Pulau-Pulau Selanga (0°30′N, 104°21′E) and Pulau Udiep (0°32′N, 104°18′E).

6   When standing to the W towards Pulau Galangbaru (0°40′N, 104°15′E), tack before Karas Kecil (0°44′N, 104°22′E) is shut in by Korek Rapat; or farther N, to clear the shoal water between Tanjung Cakang (0°37′N, 104°17′E) and Korekrapat, when reaching a depth of 18 m; Pulau Dempo (0°36′N, 104°18′E) bearing 214° is a safe turning mark. Between Korekrapat and Karas Kecil it is possible to stand into depths of 15 m before tacking, but care should be taken to give Karang Seguci (0°43′N, 104°22′E) a good berth. Karas Kecil and Karas Besar (0°45′N, 104°20′E) should be given a berth of 5 cables.

7   From this point continue as directed at 9.76.

8   When standing to the E towards Pulau Telang Kecil (0°42′N, 104°36′E), at the S entrance to Selat Riau, be careful to give the SE side of that island a berth of 2 miles and to keep the prominent hill on Tanjung Punggung (0°45′N, 104°31′E), the SW extremity of Mantang, well open of Telang Kecil, bearing 304°, to avoid Karang Sendara (0°41′N, 104°37′E). Tanjung Punggung and Pulau Ranggas (0°45′N, 104°29′E) may be approached to a prudent distance.

9   On-going routes:

Continuation through Selat Riau, at 9.76;

Passage to Singapore Strait from Selat Bangka, passing E of Bintan, at 9.91;

Passage from Selat Bangka to South China Sea, at 9.93.

9.75.2

1   **The alternative route**, via Selat Berhala (0°50′S, 104°15′E) and Selat Pengelap (0°30′N, 104°20′E) or Selat Abang (0°32′N, 104°15′E) is in fact an alternative to the 'Inner Route' from Selat Bangka (2°06′S, 105°03′E) to Singapore Strait, via Selat Berhala and Selat Durian (0°40′N, 103°40′), described at 9.77 - 9.80, but is also an alternative route to Selat Riau. The passage through Selat

Berhala is usually taken in December, January and February, when strong N winds prevail; there is then smooth water, good anchorage and little tidal stream.

2   To either Selat Pengelap or Selat Abang follow the directions given in 9.79 towards Selat Durian, as far as necessary.

3   Selat Pengelap is the wider of the two straits. Owing to the uneven nature of the bottom the tidal streams, near spring tides, cause whirls and overfalls which can be alarming to strangers; the strait is, however, clear except for a sand patch and rock on the W side of the fairway and is easy to navigate. Approaching the strait, steer to pass about 1 mile, or less, off Pulau-Pulau Alor (0°28′N, 104°18′E), preferably N of Batubelayar (0°25′N, 104°16′E). Batubelayar bearing 224° and well open SE of Pulau-Pulau Alor, leads through Selat Pengelap.

4   Selat Abang, between Dedap and Pengelap on the SE side and Abang Kecil on the NW, is reduced to a breadth of ¾ mile by the reefs on either side, but is clear and deep in the fairway.

5   Having passed through either of the above straits, the directions for approaching Selat Riau are the same as those given in the relevant part of 9.75.1.

Passage through Selat Riau ⇒ Singapore Strait

Directions
9.76

1   When working through Selat Riau (0°55′N, 104°20′E) from S, with a fair wind, the strait offers few difficulties and vessels should keep in the main fairway.

2   Continuing the directions given in 9.75.1 for standing to the W towards Galang Baru, when between Karas Besar (0°45′N, 104°20′E) and Mabut Laut (0°49′N, 104°18′E), stand in to a depth of 15 m; Karas Kecil (0°44′N, 104°22′E) well open of Karas Besar is a good turning point to avoid the bank off the latter; Lobam Kecil (0°59′N, 104°13′E) to the N, open E of Mabut Laut, leads E of the bank extending S of the latter island.

3   The main channel passes E of a 7 m patch, marked by a light-buoy (E cardinal), lying 2¼ miles E of Tanjung Sembulang (0°52′N, 104°16′E) and when working N keep the E extreme of Mabut Laut bearing more than 163° in order that the bank, which extends NW from the island, may be cleared. If intending to pass W of the 7 m patch do not cross this W limit until Tanjung Sembulang bears 287°.

4   When N of Tanjung Sembulang, stand farther W, but keep the W extremity of Mabut Laut well open of Tanjung Sembulang to clear Gosong Cemara (0°49′N, 104°14′E). Give Pulau Tanjuk (0°57′N, 104°12′E) a berth of about 5 cables and, when to the N, keep Tanjung Sembulang well open of it, to avoid the bank E of Pelanduk Subang Mas (0°57′N, 104°10′E) and the reef E of Pulau Pencaras (0°58′N, 104°10′E).

5   To avoid the reef about 7 cables E of the S extremity of Pulau Nginang (1°01′N, 104°10′E), keep Pulau Tanjungsau light-structure (1°03′N, 104°11′E) well open of Pulau Nginang. To clear the reef fringing that island keep the E point of Pulau Sau well open of the point under the light-structure. After passing Sau, in standing to the W, keep Tanjungsau light-structure open of Pulau Sau, this will clear Pulau Tubu (1°05′N, 104°10′E) and shoals as well as

the 7 m patch about 7 cables SE of Karang Passo (1°08′N, 104°10′E).

6   **Leading lights** on Pulau Tanjuk indicate the fairway through this section of Selat Riau.

7   Vessel are recommended to pass out of Selat Riau into Singapore Strait E of Karang Galang (1°09′N, 104°11′E). If, however, it is decided to pass W of this reef then keep Karang Passo bearing more than 180° until able to pass between Terumbu Betata (1°11′N, 104°09′E) and Karang Galang.

8   **Leading lights** on the E coast of Pulau Batam (1°07′N, 104°07′E) indicate the fairway passing E of Karang Galang.

9   Continuing the directions given in 9.75.1 for standing to the E towards Pulau Telang Kecil (0°42′N, 104°36′E), when nearing Pulau-Pulau Tapai (0°05′N, 104°29′E), the hill on Tanjung Punggung (0°45′N, 104°31′E) kept open of the S point of Pulau Ranggas (0°45′N, 104°29′E), bearing 098°, leads S of them in a depth of about 8 m and of Karang Kata (0°46′N, 104°25′E) in about 7 m. The S extremity of Pulau Pangkil (0°50′N, 104°22′E) kept bearing more than 325° leads W of Karang Kata and the other shoals SE of Pulau Pangkil. To clear the 7 m bank extending 2½ miles S of Pulau-Pulau Tapai, keep the prominent hill on Pulau Siulung (0°47′N, 104°36′E) open of the hill on Tanjung Punggung bearing less than 077°, until the summit of Lobam (0°59′N, 104°15′E) is open W of Pangkil, bearing 327°, or more.

10   The SW end of Pulau Pangkil should not be approached nearer than ½ mile, as its reef is steep-to; the W side may be approached to a depth of 13 m, but off its N end keep Karas Kecil (0°44′N, 104°22′E) light-structure open of Pulau Pangkil bearing 167°, or less, to avoid the fringing reefs.

11   Between Pulau Pangkil and Gosong Tula (0°58′N, 104°15′E), stand to the E into depths of 15 m, or until Pulau Terkulai (0°57′N, 104°20′E) light-structure bears 000°, but do not bring the NE extremity of Pulau Pangkil to bear more than 158°, or Pulau Terkulai light-structure less than 355°, to avoid Karang Soreh (0°53′N, 104°22′E). The Pulau Tanjungsau (1°03′N, 104°11′E) light-structure in line with the W side of Lobam Kecil (0°59′N, 104°13′E) bearing 329° leads W of Gosong Tula. The SE extreme of the Lobam group, bearing less than 090° until Pulau Tanjungsau light-structure bears more than 338°, leads W of Karang Lolo (0°59′N, 104°13′E). When N of Karang Lolo do not bring the W extremes of Lobam Kecil to bear more than 160° until Tanjung Taloh (1°01′N, 104°14′E) bears 090°, which will avoid the dangers near Karang Plasit (1°01′N, 104°13′E). Tanjung Taloh is steep-to and both it and Pulau Buau (1°03′N, 104°13′E) may be approached to about 3 cables, except near the extremes of that island. Tanjung Uban (1°04′N, 104°12′E) is bold, but do not approach the shore N of it, to Malang Jarum (1°06′N, 104°13′E), nearer than 5 cables. The above-water rocks on the edge of the shore reef are useful guides.

12   To the N of Malang Jarum there are depths of 7 m close to the edge of the shallow bank which fronts this part of the coast to a distance of nearly 1 mile. This bank, as well as Netscher Shoal (1°09′N, 104°15′E) and Crocodile Shoal (1°11′N, 104°16′E), and the shoal between them, will be avoided by keeping Pulau Tanjungsau light-structure bearing less than 205°. If the weather is hazy and the light-structure cannot be made out at this distance, Malang Jarum, which will be seen well clear of the extreme of the land as Netscher Shoal is neared, must be kept bearing less than 200° until Karang Galang (1°09′N, 104°11′E)

light-structure bears 248° or the N extremity of Tanjung Tondang (1°14′N, 104°19′E) bears 095°, a vessel will then be N of those dangers and in the Singapore Strait.

Inner Route from Selat Bangka ⇒ Singapore Strait

Detail
9.77

1   The Inner Route is suitable between October and March.

2   Selat Berhala (0°50′S, 104°15′E) forms the S part of the Inner Route to Singapore and Selat Durian (0°40′N, 103°40′) the N part. The intermediate part, between the W side of Pulau-Pulau Lingga (0°00′, 104°30′E) and the E side of Sumatera, has no specific denomination. The total distance from Berhala to Singapore is about 120 miles.

3   The Inner Route is lighted and buoyed and is suitable for all classes of vessels. The least depth in the fairway, from 10 m to 11 m, is in the S part SW of Pulau Muci (0°32′S, 104°02′E).

4   Sailing vessels, bound from Selat Berhala to Singapore, during the strength of the north-east monsoon, frequently adopt this inner route. During the prevalence of strong N winds in December, January and February, they will save much time by doing so, for these straits have smooth water, good anchorage and little tidal stream, whereas on the E side of Pulau-Pulau Lingga, at this season, there is generally a heavy sea and a S-going current sometimes running at a rate of 3 kn. In Selat Berhala sailing vessels will also be greatly assisted by squalls from the Sumatera coast.

5   In order to avoid the difficulty and delay sometimes experienced in getting from the N part of Selat Durian to the Singapore Strait, many sailing vessels have preferred the alternative of passing from the Inner Route by Selat Abang or Selat Pengelap into Selat Riau. It seems probable that the best passages might be made in this way, for the great depth of water in the W part of the Singapore Strait can be a disadvantage in light winds, as there is no suitable anchorage ground on which to bring up in case of the wind failing. For details of this alternative route, see 9.75.2.

Selat Bangka ⇒ Selat Berhala
9.78

1   For Selat Bangka follow the directions given at 9.74. Having passed Karang Ular (1°58′S, 104°57′E) shape course for the light-structure on Pulau Berhala (0°52′S, 104°24′E), distant about 74 miles, avoiding the shoal water extending SE from Tanjung Jabung (1°01′S, 104°22′E). The bank along the Sumatera coast being shelving, soundings will be the best guide, and the rule is to keep in depths of from 10 m to 13 m. In working, the coast may be approached, with care, to a depth of 9 m, observing that the bank with less than 9 m extends nearly 13 miles SE of Tanjung Jabung.

2   Pass through Selat Berhala, using the passage S of Berhala; the channel between Berhala and Pulau-Pulau Singkeplaut (0°42′S, 104°28′E) is not safe as there are several rocks in it and uncharted dangers may exist. In Selat Berhala keep in depths of from 18 m to 22 m, in order to be well clear of the bank projecting from the shore W of Tanjung Jabung; thence, in working along the coast to the W the bank is steep-to and may only be approached occasionally, with care, to a least depth of 13 m.

Selat Berhala ⇒ Selat Durian
9.79

1   From abeam of Pulau Berhala (0°52′S, 104°24′E) with a favourable wind shape course to pass 2 to 3 miles W of the

light-structure on Pulau Muci (0°32′S, 104°02′E). With a working breeze, the Sumatera coast may be approached to a depth of 11 m to 13 m, but the vessel's position must be fixed frequently, as the tidal streams are very irregular off Kuala Niur (1°00′S, 103°48′E). The mud-bank W of Tanjung Jabung (1°01′S, 104°22′E), for a distance of 14 to 15 miles, is nearly dry at low water, spring tides, and extends 4 to 5 miles seaward.

2   There is no difficulty in standing E in the vicinity of Karang Speke (0°37′S, 104°06′E) and Pulau Muci, both of which are lighted, but, when nearing Pulau Muci, tack when it bears 000°, to avoid Karang Atkin (0°33′S, 104°02′E); it is best however to pass Pulau Muci at a distance of about 2 miles, as mentioned above.

3   With a fair wind, having passed Pulau Muci, steer for Pulau Rukan Selatan (0°33′N, 103°46′E), passing either side of it, but preferably to the E which is the main channel. Bukit Jora, the summit of Pulau Durian Besar (0°43′N, 103°43′E), which is visible from from a considerable distance, bearing 344° is a good mark for making towards Pulau Rukan Selatan.

4   In working, be careful not to stand nearer than 2 miles to Tanjung Bakau (0°21′S, 103°47′E) or Tanjung Dato (0°00′, 103°48′E), the entrance points of Teluk Kualacenaku, and when between them off that bay, remember that the bank, which extends beyond a line joining those points, is steep-to and soundings will give no warning. Excepting abreast the S part of Pulau Kateman (0°16′N, 103°40′E), about 12 miles N of Tanjung Dato, the depths decrease more regularly towards the bank, which may from thence be approached by sounding into depths of 13 m towards Pulau Durai (0°32′N, 103°36′E) and the other nearby islands. In standing to the E, when abreast Tanjung Dato, do not deepen above 35 m, for the ground on that side is foul and unsuitable for anchorage.

5   Batu Kameleon (0°31′N, 104°07′E) is out of the fairway track, but if standing so far to the E, the summit of Pulau Petong (0°37′N, 104°05′E) bearing 350°, or more, leads well W of it.

Selat Durian ⇒ Singapore Strait
9.80

1   The initial part of this passage may be taken either E or W of Pulau-Pulau Rukan (0°35′N, 103°45′E). If taking the E side, having passed E of Pulau Rukan Tengah (0°35′N, 103°46′E) and standing towards Gosong-gosong Timur (0°40′S, 103°50′E), tack while Bukit Jora, the summit of Pulau Durian Besar still bears more than 308°, in order to avoid the banks. Having passed Pulau Rukan Utara steer to pass between Pulau Perasi Besar (0°43′N, 103°39′E) and Pulau Pelangkat (0°45′N, 103°35′E); in working by keeping Pulau Perasi Kecil (0°46′N, 103°38′E) open W of Pulau Durian Kecil (0°44′N, 103°40′E) bearing more than 318°, Karang Genting, 1 mile S of Pulau Durian Kecil, will be avoided.

2   If taking the route W of Pulau-Pulau Rukan, pass about 1½ miles W of these islands in depths of 18 m to 25 m, but do not enter Selat Durian (0°40′N, 103°40′E) until Pulau Perasi Kecil is well open E of Pulau Perasi Besar, bearing 322°, to avoid Karang Richardson (0°37′N, 103°43′E). When in the strait, steer to pass between Perasi Besar and Pelangkat as above.

3   Continuing N, the peak of Pulau Sanglang Besar (0°37′N, 103°41′E), astern, in line with the apex of Pulau Perasi Kecil, bearing 159°, leads between Karang Melvill (0°52′N, 103°37′E) and Karang Tengah (0°51′N, 103°34′E). Thence steer to pass through Selat Phillip if bound to

Singapore or to the N and NW if bound for the Malacca Strait.

4   If the channel W of Karang Tengah is taken, the water will be found to shoal gradually towards the W shore over a bottom of soft mud, suitable for anchorage. The E point of Pulau Degong (0°47′N, 103°32′E) bearing 180° leads E of the dangers extending off Pulau Buru (0°52′N, 103°30′E) and the islands N of it. Pulau Karimun Kecil (1°09′N, 103°23′E) bearing less than 325° also leads E of the dangers which project 3 miles from Pulau Karimun Besar (1°05′N, 103°20′E).

5   When working to the N after passing Perasi Kecil and standing E, keep W of the alignment of Bukit Manilang the summit of Sanglang Besar, with the NW extremity of Durian Kecil bearing about 168°; this will avoid the reef about 1¼ miles W of the S extremity of Pulau Belukar (0°51′N, 103°39′E).

Singapore Strait

Details
9.81

1   **Traffic Separation Schemes** and **Prohibited Areas** have been extensively established throughout the Singapore Strait.

2   Heavy rain squalls, during which visibility is moderate or poor are frequently experienced in the Singapore Strait.

3   In the following directions the Strait is considered in two parts:

> The eastern part for vessels coming from, or going to, the South China Sea, or the Eastern Archipelago, via Selat Riau or E of Pulau Bintan;

> The western part for vessels coming from, or going to, the Malacca Strait or Selat Durian.

4   Of the three channels into which the E entrance to the Singapore Strait is divided:

> Middle Channel is recommended.

> North Channel has no advantage except perhaps for vessels bound N along the coast and should only be used with local knowledge.

> South Channel is not recommended for deep-draught vessels, the bottom is generally rocky and uneven and the channel is encumbered with shoals.

Passage westward through the eastern part of the Singapore Strait.
9.82

If approaching from E of Pulau Bintan (1°00′N, 104°30′E), South Channel may be used, but in view of the remarks in 9.81, vessels are recommended to stand on and pass through the Middle Channel.

There is no difficulty in identifying Singapore Strait when coming from the E in clear weather; both Gunung Bintan Besar (1°05′N, 104°27′E) and Bukit Pelali (1°25′N, 104°12′E) are good marks and Horsburgh Light (1°20′N, 104°24′E) marks the S side of Middle Channel.

Sailing vessels will experience no difficulty in working either direction through Middle Channel and the E part of the strait. The best plan is to keep towards the N shore, in case of having to anchor, as the depths are more convenient on that side. The shore may be approached to depths of 20 m; Pulau Mungging (1°22′N, 104°18′E) kept open of Tanjung Ayam (1°21′N, 104°12′E), bearing 075°, leads S of Johor Shoal (1°19′N, 104°04′E) and, when standing towards this danger, if these objects cannot be seen, preserve the depths mentioned for the shoal is steep-to.

When E of Tanjung Ayam keep Tanjung Setapa (1°20′N, 104°08′E) in line with, or open of, Tanjung Ayam, bearing 274°, until Pulau Mungging bears less than 360°. When standing towards Falloden Hall Shoal (1°21′N, 104°19′E), keeping Tanjung Ayam bearing more than 266°. When standing towards Congalton Skar (1°22′N, 104°19′E), and the shoals N of it, keep Tanjung Punggai (1°26′N, 104°18′E) bearing less than 337°. A vessel may stand towards Remunia Shoals (1°27′N, 103°39′E) until the extremity of Pulau Mungging bears 256°.

There are no dangers on the S side of the strait excepting those fronting the coast of Pulau Bintan and Crocodile Shoal (1°11′N, 104°16′E), Terumbu Betata (1°11′N, 104°09′E) and Karang Galang (1°09′N, 104°11′E), in the entrance to Selat Riau. Do not stand so far over as to get near these dangers, for no advantage will be gained by doing so, and the depths there are not convenient for anchorage.

Small vessels bound for Singapore Road from E will have no difficulty as they merely have to proceed to a convenient anchorage.

Passage eastward through the Main Strait.
9.83

1 **Traffic Separation Schemes** and **Prohibited Areas** have been extensively established throughout Singapore Strait.

2 Owing to the strong tidal streams in the W part of Singapore Strait, sailing vessels are frequently obliged to anchor, for which purpose the N side of the channel is to be preferred. The S side is unsuitable for anchorage as the water is deep, the bottom rocky and violent squalls are common. Due to the extensive new developments, reclamation and establishment of TSSs and Prohibited Areas, masters of sailing vessels are advised to contact local Authorities before anchoring.

3 There is a great density of powered vessel traffic in both directions through the Straits, including many very long vessels of very deep draught.

4 Continuing E towards the South China Sea, follow generally the directions at 9.82, in reverse.

Selat Sunda to, and through, Selat Gelasa
Details
9.84

1 Of the three principal passages through Selat Gelasa (2°53′S, 107°18′E), namely from W to E, Selat Leplia (2°53′S, 106°58′E), Selat Limende (3°00′S, 107°12′E) and Selat Baur (3°00′S, 107°18′E), the latter being preferable for sailing vessels N-bound with a fair wind, being the broadest and having no dangers in the fairway. Sailing vessels working through and vessels of low-power should use Selat Leplia during the north-west monsoon (but see notes in 9.85) and Selat Baur during the south-east monsoon, the currents being less favourable. Selat Limende is seldom used.

Selat Sunda to, and through, Selat Leplia
9.85

1 **Cautions**. Areas of the Java Sea are being exploited for natural resources, see 7.38.2. This route passes close to oilfields, with extensive areas of prohibited anchorage, it would be prudent to avoid these areas altogether wherever possible.

2 After leaving Selat Sunda, pass between Layang-Layang (5°18′S, 106°04′E) and Batu Karang Permatan (5°24′S, 106°16′E). Then, with a fair wind, make for a position about 4 miles W of Karang-karang Suji (3°34′S, 106°55′E),

avoiding the oilfields wherever possible. Then steer to make good 005°, passing W of Karang Kait (3°27′S, 107°00′E) and Karang Medang (3°22′S, 106°56′E), when land will soon be sighted; Pulau Simedang (3°19′S, 107°12′E), however, should not be approached close enough to be sighted by day.

3 When past Karang Medang, steer to pass midway between Karang-karang Baginda (3°07′S, 107°05′E) and the 4 m shoal, lying 11 miles farther W, taking care to avoid the shoals lying about 6 miles S of the latter, and thence N between Batu-Batu Discovery (2°53′S, 106°56′E) and Pulau Celaka (2°52′S, 107°01′E). Give Pulau Celaka a berth of about 2 miles, taking care to avoid the 8·8 m patch, lying 2¾ miles E of Tanjung Labu (2°57′S, 106°55′E), and the other dangers in this vicinity.

4 When the N extremity of Pulau Liat (2°52′S, 107°03′E) bears 090° steer to pass E of Pulau Gelasa (2°25′S, 107°04′E) or E of Tanjung Berikat (2°34′S, 106°51′E), according to destination.

5 **Note**, in thick weather it is advisable to anchor on the bank, round Karang-karang Suji, in depths of from 12 m to 18 m, and await more favourable conditions. Vessels coming from the Java Sea and uncertain of their position, can approach the coast of Sumatera to a depth of 17 m.

6 Vessels proceeding through Selat Leplia at night should should take care that they sight Tanjung Murung (3°02′S, 106°54′E) during daylight, if coming from the S; if approaching from the N then Tanjung Berikat should be sighted during daylight.

7 Working through Selat Leplia from the S, it is almost impossible to work through Selat Gelasa during the strength of the north-west monsoon; even in the latter part of the monsoon, about March, when winds are light, sailing vessels are often obliged to anchor on account of the strength of the S-going current. In the south-east monsoon also, vessels will often meet with light variable winds, rendering it impossible for them to preserve a direct course.

8 The approach to Selat Leplia does not afford convenient clearing marks, but the following directions, as far as can be judged, are the best for the purpose. As, however, the some of the objects are at a considerable distance from the dangers, navigators should not depend too implicitly on being able to recognise them. Particular attention should be paid to the set of the tidal streams, currents and to soundings.

9 Coming from Selat Sunda a sailing vessel is advised to work up the coast of Sumatera, as described at 9.71. Approaching Selat Leplia proceed as follows.

10 If proceeding E, to the N of Karang Suji, head towards Karang Pasir (3°29′S, 107°10E), which dries, until it is about 4 miles distant, or within half a mile of Karang Haaien (3°29′S, 107°07′E), giving Karang Medang and Karang Kait a wide berth. Pulau Simedang bearing 028° leads 1 mile W of Karang Ombak (3°25′S, 107°10′E). Simedang should not be approached nearer then 3 miles, on account of the dangers lying W of it.

11 Tanjung Murung, kept bearing more than 318°, leads SW of Karang-Karang Baginda and Pulau Kalangbahu (3°02′S, 107°10′E), bearing 054°, leads NW of the S and central portions of these reefs; Bukit Keladi (2°54′S, 107°03′E), on Pulau Liat, which is not easily recognised, kept bearing more than 005°, leads W of the most W shoal, which is awash at low water, and the N end of Pulau Aur (2°27′N, 104°31′E), open N of Pulau Bakau (3°02′S, 107°09′E), bearing 064°, leads N of the reefs.

12 When N of Karang-Karang Baginda, keep Pulau Bakau bearing less than 108°, and Pulau Selemar (2°59′S,

107°06′E) more than 000°, to avoid the shoals between them. To clear the reef extending 3 miles S of Pulau Liat keep Pulau Kueel (2°59′S, 107°08′E) bearing less than 108° with Bukit Keladi (2°54′S, 107°03′E) bearing less than 349°.

13   To clear the reefs and shoals lying SW and W of Pulau Liat, Pulau Bakau must be kept bearing less than 134° until Pulau Celaka bears 090°. To clear Batu-Batu Discovery keep Tanjung Labu (2°57′S, 106°55′E) bearing more than 220° until Pulau Celaka bears 090°, which also clears the rocks to the N. To clear the reefs extending off the NW side of Pulau Liat, keep Tanjung Labu bearing less than 215° until the S extreme of Pulau Kelapan (2°51′S, 106°50′E) bears 247°.

14   When standing to the W, to clear the banks between Karang Medang and Tanjung Murung keep Bakung, a hill 2 miles W of Tanjung Labu, open E of Tanjung Murung, bearing 344° until Baginda, a hill (167 m) about 1 mile NW of Tanjung Baginda (3°05′S, 106°44′E), bears 276°; after which it is possible to head W until Tanjung Murung bears 017°.

15   To clear the shoals, extending up to 3 miles E of Tanjung Labu, Tanjung Murung must be kept bearing more than 219° until the NE extreme of Pulau Kelapan bears 308°. To clear Batu-Batu Discovery, see above.

16   To avoid Gosong Raya (2°40′S, 106°53′E), Bakung must be kept bearing more than 186°, or Tanjung Labu more than 180°, until Tanjung Berikat bears less than 322°.

North-bound through Selat Baur
9.86
1   For sailing vessels with a fair wind, Selat Baur (3°00′S, 107°18′E) is preferable to the others, and should present no difficulties. The land is, in the fine weather of the north-west monsoon, visible from the outer dangers. The greater breadth of Selat Baur enables sailing vessels to make longer tacks and, as most of the islands can be seen at night, the vessel's position is more easily fixed.

2   The shoals that lie within the strait appear to form the only drawback to the adoption of this channel, however, in clear weather good hill peaks, with which to fix the vessel's position, are visible on all sides, up to 35 miles distance.

3   With a fair wind, making for Selat Baur from the S, once clear of the restricted oilfield areas in the vicinity of Pulau Segama and Pulau Jagautara, shape course for Karang Larabe (3°32′S, 107°10′E) during the north-west monsoon and for Karang Genting (3°34′S, 107°41′E) during the south-east monsoon. In clear weather the mountains in the SW part of Pulau Belitung (3°00′S, 108°00′E) will be sighted some distance S of these dangers. Bukit Ludai (3°09′S, 107°44′E), 3½ miles E of Bukit Beluru, which may be visible from about 12 miles S of Karang Genting, first comes into sight and shortly afterwards Bukit Beluru will be sighted. When near Karang Larabe other mountains on Pulau Belitung, as well as Pulau Simedang (3°19′S, 107°12′E), should be sighted, so that in clear weather there is no difficulty in making the strait. If a vessel is far to the E of the track, Pulau Kebatu (3°48′S, 108°04′E) will be a useful mark for fixing the position.

4   When landfall has been made, steer a N course, passing about 6 miles E of Pulau Simedang, midway between Pulau Kasenga (3°03′S, 107°21′E) and Pulau Geresik (3°00′S, 107°16′E) and not less than 2 miles W of Tanjung Ayerlancur (2°53′S, 107°20′E).

5   With bad visibility, or in thick weather, sounding must be depended on entirely; in such cases it is advisable to make the S edge of the bank, with depths of from 13 m to 18 m, clay with sand, which extends about 25 miles S from Simedang, then immediately steer E until in depths of more than 18 m, then steer N keeping in depths of more than 18 m and, when passing E of Pulau Simedang keeping in depths of not less than 29 m. If, however, depths of over 36 m have been obtained when making for the S entrance, it may be presumed that the vessel is well over on the E side of the channel and a NW course may then be steered, taking care to remain within those depths. In unfavourable conditions, or if in any doubt as to which side of the strait the vessel may be, it is advisable to anchor as bad visibility does not usually last for any length of time.

6   At night Selat Baur can be approached from the S without danger, in clear weather, as the light on Pulau Simedang is visible up to 3 miles S of Karang Hancock (3°34′S, 107°05′E), the S danger on the W side of the approach. When this light is sighted steer to pass about 2 miles E of Pulau Simedang, and thence proceed N until in the arc of visibility of the light on Tanjung Ayerlancur, which must be kept between the bearings of 003° and 022°. When Pulau Geresik (3°00′S, 107°16′E) is sighted, the position can be fixed by bearings of this island and Tanjung Ayerlancur Light and course may then be shaped to pass E or W of Beting Akbar (2°39′S, 107°15′E), according to destination. Vessels passing E of this shoal have the advantage of being able to fix its position by bearings of Langkuas Light (2°32′S, 107°37′E) in addition to Tanjung Ayerlancur Light. Vessels passing W make for Pulau Gelasa (2°25′S, 107°04′E).

7   Working through Selat Baur, and heading E towards Karang Genting keep Bukit Belaru (360 m), 6½ miles NE of Tanjung Genting, bearing more than 011°; and to clear Karang Naga (3°27′S, 107°37′E), Gosong Awal (3°24′S, 107°36′E) and Karang Cooper (3°22′S, 107°35′E) keep Tanjung Marangbolo (3°15′S, 107°31′E) bearing more than 350°. To avoid the dangers N of Karang Cooper keep farther W with Marangbolo bearing more than 010°.

8   Pass Batu Malang (3°15′S, 107°28′E) at a distance of at least 1 mile, the approach from the S being on a bearing of more than 001°, but to the N it should it should not bear more than 112° until the N point of Pulau Seliu bears 073° to clear Karang Tiga (3°14′S, 107°27′E). After passing Karang Tiga then Batu Malang must not bear more than 146°, or Tanjung Marangbolo more than 124°, until Karang Nyera (3°12′S, 107°E) and the 11 m patch NW of it are cleared.

9   After passing these dangers, Tanjung Marangbolo bearing less than 132° will lead S and SW, and Tanjung Ayerlancur bearing more than 003° to the W, of all dangers until N of Pulau-Pulau Lima (3°03′S, 107°24′E). After Pulau Geresik bears 270° stand a little farther to the E but keeping Tanjung Ayerlancur bearing more than 355° until within about 2 miles.

10   Give the light-structure on Tanjung Ayerlancur a berth of about 1½ miles and keep it bearing less than 158° until Pulau Langir (2°48′S, 107°22′E) bears 046° to clear the reef round Karang Kembung (2°51′S, 107°20′E); to the N, Karang Kembung in line with the light-structure bearing about 180° leads a full mile W of Malang Wankang (2°48′S, 107°21′E) and will clear all the reefs between Karang Kembung and Pulau Langir.

11   Give the coast between Pulau Mendanau (2°52′S, 107°25′E) and Pulau Langkuas (2°32′S, 107°37′E) a berth of 6 to 7 miles, keeping Pulau Langir bearing less than 214° and Pulau Langkuas bearing more than 046°.

12   When standing to the W, Pulau Simedang bearing less than 000°, will lead E of all the shoals to the S of it, and soundings will also give good warning when standing towards them, as they lie some 6 or 7 miles within the charted 20 m depth contour. Pulau Simedang and Pulau Simedang Kecil must be approached with caution as soundings do not give much warning when nearing their outlying reefs; they should on no account be approached closer than 2½ miles.

13   Pulau Simedang Kecil bearing 183°, astern, leads E of Karang Blis (3°16′S, 107°12′E), between which and Pulau Aur (3°00′S, 107°13′E) a vessel may stand to the W until the summit of Aur bears 023°, which will lead E of Karang-Karang Baginda (3°07′S, 107°05′E). Pulau Kalangbahu (3°02′S, 107°10′E), bearing 265°, leads S of the dangers extending from Pulau Aur and Pulau Geresik (3°00′S, 107°16′E). The E side of Pulau Geresik may be approached to a distance of 1 mile, but the E side of Pulau Kelemar (2°58′S, 107°13′E) has a rock lying 1 mile off, which will be avoided if Pulau Geresik bears more than 160°. The summit of Pulau Aur, in line with the E extreme of Pulau Kelemar bearing 180°, leads 1½ miles E of Karang Pandan (2°53′S, 107°12′E).

14   Having passed Karang Pandan, head farther W towards Pulau Lait, but the SE extreme of that island must bear more than 200° to clear the reefs off its NE side. When N of all the reefs off the N side of Pulau Liat, indicated when Tanjung Ayerlancur Light (by night) or Bukit Sagoweel, 2 miles SE, (by day) bear more than 111°, head W towards the coast of Pulau Bangka.

Selat Limende
9.87

1   This strait (3°00′S, 107°12′E), passing E of Pulau Liat (2°52′S, 107°03′E), is narrower and more encumbered with dangers than either Selat Leplia or Selat Baur between which it lies. It is easily navigable by sailing vessels with a favourable wind, during daylight. No vessel, however, would from choice attempt to beat to windward through Selat Limende as the other two straits are much better adapted for that purpose; but it is possible that a vessel embarrassed by light baffling winds, may find it convenient to proceed through some part of it. The numerous islets afford every opportunity for fixing a vessel's position.

Selat Gelasa ⇒ Selat Riau
9.88

1   Most vessels, N-bound from Selat Gelasa (2°53′S, 107°18′E), prefer passing E of Pulau Gelasa (2°25′S, 107°04′E), which is the safer route; but some, especially when bound for Singapore by Selat Riau (0°55′N, 104°20′E), prefer the less safe but more direct route between the shoals W of Pulau Gelasa; an alternative is to pass between Tanjung Berikat (2°34′S, 106°51′E) and Pulau Berikat (2°28′S, 106°58′E). These routes W of Pulau Berikat are not recommended as reefs and dangers extend more than 30 miles off the NE coast of Pulau Bangka and the area has not been thoroughly surveyed. During the strength of the north-west monsoon, N winds will be met along the coast of Pulau Bangka and the adverse current off and W of Tanjung Berikat will make it difficult to beat to windward.

Passage east of Pulau Gelasa
9.89

1   With a fair wind pass about 3 miles E of Pulau Gelasa (2°25′S, 107°04′E). Continue N keeping Pulau Gelasa bearing more than 180° to clear Karang Belvedere (2°12′S,

107°02′E), Karang Magdalena (2°02′S, 107°00′E) and Karang Lanrick (1°53′S, 106°57′E). As the summit of Pulau Gelasa is visible from a distance of 30 miles in clear weather, a vessel should be abreast of Karang Lanrick before losing it.

2   After clearing Karang Lanrick pass E of Karang Severn (1°37′S, 106°31′E) and between that reef and the group of reefs about 27 miles to the NNE. Soundings give no warning of the approach to any of the above, as they are steep-to, but in the vicinity of Karang Severn, in fine weather, the highest hill on Tanjung Tuing (1°37′S, 106°03′E) and Raja, a hill close W of Tanjung Raja (1°54′S, 106°11′E) are visible.

3   If the wind should prevent a direct course being steered from abreast of Pulau Liat, then Pulau Gelasa should be kept bearing more than 338° until the vessel is N of Beting Akbar (2°39′S, 107°15′E). After passing about 3 miles E of Pulau Gelasa proceed as above.

4   Having passed Karang Severn steer E of Pulau Toty (0°55′S, 105°46′E) and continue NW to join the route, described in 9.75.1, NE of Tanjung Jang.

Selat Bangka or Selat Gelasa ⇒ Singapore passing E of Pulau Bintan

Detail
9.90

1   These routes are alternatives to those described from 9.75 to 9.89. They are not recommended for use during the seasons of N and NW winds, from November to March.

Selat Bangka ⇒ Singapore Strait
9.91

1   Having passed E of Pulau Saya (0°47′S, 104°56′E), as described at 9.75.1, a vessel should steer to the N so as to pass E of Karang Heluputan (0°38′N, 105°08′E), crossing the equator in depths of about 37 m. At night it is advisable to keep in depths of not less than 43 m when between the parallels of 0°30′N, and 0°50′N; Pulau Merapas (0°56′N, 104°55′E), bearing 315° or less, leads NE of Karang Helupatan and Gosong Ara (0°48′N, 104°57′E). Having rounded the NE point of Pulau Bintan (1°00′N, 104°30′E), proceed as directed in 9.82, preferably using Middle Channel.

Selat Gelasa ⇒ Singapore Strait
9.92

1   First proceed as directed in 9.89 as far as Karang Lanrick (1°53′S, 106°57′E). Then continue on a course to make 000° until Pulau Pejantan (0°07′N, 107°12′E) is sighted, thence shape course for Singapore Strait, as described at 9.82.

Selat Bangka or Selat Gelasa ⇒ South China Sea (May to September)
9.93

1   In either case the route into the South China Sea is as direct as possible. When the north-east monsoon is likely to develop before Hong Kong is reached, pass through Alur Pelayaran Api (2°00′N, 109°08′E) and Palawan Passage (10°20′N, 118°00′E), for details see 10.64.1 and 10.64.2.
9.93.1

1   From Selat Bangka, having passed E of Pulau Saya (0°47′S, 104°56′E), as described at 9.75.1, a vessel should steer to pass between Pulau-Pulau Anambas (3°00′N, 106°00′E) and Pulau-Pulau Natuna (4°00′N, 108°00′E), see 10.63 to 10.64.2 and 10.73 for further details.

9.93.2

1   From Selat Gelasa, either proceed N as directed in 9.92 until Pulau Pejantan (0°07′N, 107°12′E) is sighted, and thence E of Pulau-Pulau Tambelan (1°00′N, 107°34′E) and between those islands and the coast of Borneo; or steer directly to pass W of Pulau Pengibu (1°35′N, 106°19′E). In either case continue N to pass between Pulau-Pulau Anambas (3°00′N, 106°00′E) and Pulau-Pulau Natuna (4°00′N, 108°00′E), see 10.63 to 10.64.2 and 10.73 for further details.

Selat Sunda ⇒ Selat Karimata and South China Sea

Details
9.94

1   **Archipelagic Sea Lane I** passes through Selat Karimata (1°43′S, 108°34′E), for details regarding Archipelagic Sea Lanes, see Appendix A.

2   **Cautions**. Areas of the Java Sea are being exploited for natural resources, see 7.38.2. This route passes close to oilfields, with extensive areas of prohibited anchorage, it would be prudent to avoid these areas altogether wherever possible.

3   Selat Karimata is the passage which leads between Pulau Belitung (3°00′S, 108°00′E) and Pulau-Pulau Monparang (2°35′S, 108°45′E) on the W side and Pulau-Pulau Karimata (1°43′S, 108°54′E) and the coast of Borneo on the E side. It is the customary route taken by vessels bound for the South China Sea or Singapore from the E part of the Java Sea. Such vessels pass well E of all dangers lying off the E side of Pulau Belitung and, hardly ever sight either Pulau Belitung or the Borneo coast, the direct route to Pontianak (0°01′S, 109°20′E) from Java Sea is E of Pulau-Pulau Karimata.

4   The main route lies E of Gosong Mampango (3°35′S, 109°11′E) and Terumbu Manggar (2°54′S, 108°57′E). A line joining Gosong Mampango, a position 20 miles E of Terumbu Manggar, Karang Tenang (2°31′S, 108°55′E) and Karang Ontario (2°00′S, 108°39′E) must be considered as the W limit of safe navigation for large vessels passing through Selat Karimata.

5   Besides the main channel, there are several other channels between the numerous islands lying E and NE of Pulau-Pulau Karimata and between it and the Borneo coast.

6   The E of these passes, known as the "Inner Channel" leads between Pulau Panebangan (1°13′S, 109°15′E) and Tanjung Pasir (1°15′S, 109°24′E), with another passing between between Pulau-Pulau Pelapis (1°18′S, 109°10′E) and Pulau Panebangan. Both have a regular tide and convenient depths for anchoring and are therefore much frequented by vessels working through the strait, it being impossible to beat to windward through the main channel against a strong monsoon and a continuous current setting to leeward.

Selat Sunda ⇒ Selat Karimata
9.95

1   Pass E of Layang Layang (5°18′S, 106°04′E) and W of Batu Karang Permatan (5°24′S, 106°16′E) thence having passed S of Pulau Jagautara (5°12′S, 106°27′E), shape course to pass E of Gosong Mampango (3°35′S, 109°11′E). For cautions regarding oilfields in this area, see 9.94.

Passage north-bound through Selat Karimata
9.96

1   With a favourable wind, having passed E of Gosong Mampango steer N until past Pulau-Pulau Monparang

taking care to avoid the dangers stretching E from Terumbu Manggar, then alter course to the NW so as to pass between Karang Ontario and Pulau Serutu (1°43′S, 108°43′E), passing the light structure near the W end of Serutu at a distance of about 5 miles.

2   Proceeding N to the South China Sea, after passing between Karang Ontario and Pulau Serutu, keep approximately on the meridian of 108°, taking care to avoid the 5 m reef, lying about 28 miles NW of Karang Greig, then pass E of Pulau Pengikik Besar (0°15′N, 108°03′E). As an alternative to the to the main strait vessels may use the E channels, as described at 9.97.

3   If making for Selat Karimata from the E part of Jawa (7°00′S, 110°00′E), make for the E side of the S entrance to the straits, passing W of Gosong Aling (3°35′S, 110°15′E), Gosong Aruba (3°28′S, 110°12′E) and Gosong Jelai (3°24′S, 110°09′E), then steer NW so as to pass between Karang Ontario and Pulau Serutu.

North-bound through eastern channels
9.97

1   Vessels beating to windward through Selat Karimata may find it easier to take the E channels, described at 9.94, the most E of which is suitable for small vessels only. In these channels the sea is smoother and the current not so strong, it being wholly, or in part, overcome by the tidal stream and the indraught into the rivers on the W coast of Borneo; vessels also have the advantage of the change of wind at night and morning caused by the land breeze, which often brings it several more points to the E in both monsoons.

2   These channels have a convenient depth for anchoring, with a bottom of soft mud, but beating to windward through them is slow and tedious. Sounding gives good warning when approaching the Borneo side; vessels can pass fairly close to Pulau-Pulau Karimata.

3   Less water than charted has been reported throughout the "Inner Channel".

4   Coming from the SW, note the W limit for safe navigation, described at 9.94. When N of the dangers off Pulau Mangkut (3°04′S, 110°12′E) and off Tanjung Pagarantimun (2°15′S, 110°04′E), the Borneo coast may be approached to a depth of 15 m, and to 11 m in Teluk Sukadana (1°25′S, 109°43′E). The S group of Pulau-Pulau Layah (1°35′S, 109°20′E) should not be approached closer than 1 mile. Pass on either side of the N group of Pulau-Pulau Layah, noting that the depth quickly shoals to 9 m at 3 miles NE of Pulau Meledang, the most E of the group and at 2 miles N of Pulau Bulat, the most NE.

5   Between Pulau Krawang (1°44′S, 109°21′E) and the N group of Pulau-Pulau Layah the depths are from 22 m to 27 m, decreasing fairly regularly towards the Borneo coast. Greig Channel, between Pulau-Pulau Pelapis and Pulau Panebangan, is deep and bold either side.

Passage north-bound from Selat Karimata
9.98

1   Between Pulau Panebangan and Pulau Masatiga (0°56′S, 109°15′E) the Borneo coast may be approached to a depth of 11 m; but when about 8 miles NW of Pulau Masatiga, do not bring it to bear more than 135°, or stand into depths of less than 15 m, until off Sungai Padangtikar (0°40′S, 109°15′E), N of which the coast may approached to within 4 miles. Pulau Masatiga can usually be seen at a distance of 20 miles.

2   A vessel may stand off, or W of, Pulau-Pulau Leman (1°17′S, 108°54′E), observing that those islands kept bearing more than 140° lead E of Karang Twilight (1°02′S,

108°37′E), Karang Cina (0°57′S, 108°32′E) and Karang Greig Utara (0°52′S, 108°33′E).

3 Having cleared dangers as above, make good a course towards Pulau Datu (0°08′N, 108°36′E) or Pengikik Besar (0°15′N, 108°03′E).

Selat Sunda eastward ⇒ Banda Sea and Second Eastern Passage

Details
9.99

1 From November to February, vessels which have passed through Selat Sunda into the Java Sea, and find that the north-west monsoon in those waters and the north-east monsoon in the South China Sea, have already set in, are advised to make E at once and pick up the Second Eastern Passage (9.101) in the Ceram Sea N of the South Molukka Group.

2 Alternatively a vessel can join the First Eastern Passage (9.107) off the entrance to Selat Makasar (2°00′S, 118°00′E), although there is no advantage in doing so.

Directions
9.100

1 To join the Second Eastern Passage (9.101), the better recommended and more usual route is to stand NE from Selat Sunda and, having passed through the Java Sea, to pass through Selat Salayar (5°43′S, 120°30′E) and Alur Pelayaran Butung (5°20′S, 123°15′E) into the Banda Sea. With W winds, when coming from Selat Salayar, close Tanjung Batutoro (5°42′S, 122°47′E) to about 3 miles and keep along the coast as far as Tanjung Kassolanatumbi (5°16′N, 123°14′E) to avoid being set over towards Kepulauan Wakatohi (5°35′S, 124°00′E) in the light airs and S-going current which prevail offshore.

2 Alternatively, passage may be made N of Jawa, through Selat Sapudi (7°00′S, 114°15′E), N of Bali, Lombok and Sumbawa and through the Flores and Banda Seas to the Ceram Sea.

3 Directions for Ceram Sea, Selat Manipa, Selat Jailolo and Selat Dampier are given at 9.102, 9.103.

Second Eastern Passage

Details
9.101

1 **Archipelagic Sea Lanes** pass through the following sections of this passage:

   ASL IIIA (Part 1), through Banda Sea, Ceram Sea, Molucca Sea and to Pacific Ocean;

   ASL IIIA (Part 2), from Indian Ocean and through Savu Sea;

   ASL IIIB, through Selat Leti and Banda Sea;

   ASL IIID, from Indian Ocean, between Sawu and Roti and through Savu Sea;

   ASL IIIE through Molucca Sea to Celebes Sea.

   For details regarding Archipelagic Sea Lanes, see Appendix A.

2 For a summary of this passage, see 9.61. The passage from the South Indian Ocean to the South China Sea, through Selat Ombi (8°35′S, 125°00′E), is usually made during the season October to March. When proceeding to Singapore, the routes via Selat Sunda, summarised at 9.60, should be taken.

3 An alternative route in October and November is to pass through one of the central passages, Selat Bali (8°10′S, 114°25′E), Selat Lombok (8°47′S, 115°44′E) or Selat Alas

(8°40′S, 116°40′E) to join the passage from Selat Ombi in the Ceram Sea, see 9.62 for details.

4 December to April is the season of tropical storms, they may occur occasionally in November, see 6.16.

5 From Selat Ombi the route is either W of Pulau Buru (3°30′S, 126°30′E) or through Selat Manipa between Pulau Buru and Pulau Manipa (3°20′S, 127°35′E), into the Ceram Sea. Thence pass through Selat Jailolo (0°06′N, 129°10′E) or Selat Dampier (0°40′S, 130°30′E) into the Pacific Ocean. When in the Pacific Ocean, make to the E between the parallels of 1°30′N, and 3°00′N until able to pass E of Palau Islands (7°30′N, 134°30′E); however, between March and September pass W of these islands.

6 Having passed the Palau Islands, a variety of routes is available either through Surigao Strait (9°53′N, 125°21′E) and to the South China Sea via Mindoro Strait or Verde Island Passage (13°36′N, 121°00′E); or through San Bernadino Strait (12°33′N, 124°12′E) and Verde Island Passage. A more usual course would be keeping E of the Philippine Islands and to South China Sea via Balintang Channel (20°00′N, 122°20′E) or Bashi Channel (21°20′N, 121°00′E); to Shanghai between Sakishima Guntō (24°40′N, 124°45′E) and Okinawa Guntō (26°30′N, 128°00′E); to Japan by a route farther N in the full strength of the Japan Current (7.26).

7 For the sake of convenience the Second Eastern Passage can be divided into three parts:

   Selat Ombi to the Ceram Sea (9.102);

   Through the Ceram Sea to, and through, Selat Jailolo or Selat Dampier, or through Molucca Sea to Pacific Ocean (9.103);

   Continuation to South China Sea (9.104, 9.105 and 9.106).

Selat Ombi ⇒ Ceram Sea
9.102

1 Selat Ombi (8°35′S, 125°00′E) is the broad, deep passage separating the NW coast of Timor from Pulau-Pulau Alor (0°28′N, 104°18′E). From October to March it was frequently used by sailing vessels proceeding from Europe to China and Japan, and was also used by sailing vessels bound for the E part of Jawa from the South China Sea, from the middle of May to the end of June.

2 In the partially enclosed region N of Pulau-Pulau Sawu (10°30′S, 121°50′E) and Timor, known as the Savu Sea, especially in the E where it is continued E by Selat Ombi, the percentage of bright sky is greater than any other part of the archipelago and haziness is equally great when E winds blow. The rainfall is heaviest in December and January, but showers may occur with all W winds.

3 The south-east monsoon blows steadily between the middle of April and the end of September, from ESE to SE, the land breezes from Timor increase the force of the wind at night and the sea breezes diminish it by day. Similarly in the other season the wind will be most steady by day and unreliable at night.

4 In October and November the winds are from SE to SSW and, in December, from the SW quarter, accompanied by thunderstorms, but the north-west monsoon does not reach its full development, from W to WNW, until January and then begins to abate in February. Variable winds will then blow until April.

5 Proceeding NE through Selat Ombi, pass between Pulau Sumba (9°45′S, 120°00′E) and Pulau-Pulau Sawu (10°30′S, 121°50′) or between Pulau-Pulau Sawu and Pulau Roti (10°45′S, 123°00′E), if falling to leeward with a NW wind. Under the exceptional conditions of a strong NW wind and lee current, it may be desirable to pass W of Sumba and S of Flores.

6 Passage from Selat Ombi to the Ceram Sea may be made either W of Pulau Buru or through Selat Manipa (3°20'S, 127°22'E). If attempting to weather the W side of Buru, and falling to leeward, it is better to abandon the attempt and pass through Selat Manipa, which is a good, safe channel and conveniently situated for a call at Ambon (3°41'S, 128°10'E).

7 During the north-west monsoon, vessels making N should do so along the E coast of Buru, where adverse tidal stream is not so strong and the favourable tidal stream runs strongly. In the strength of the monsoons there may be a high sea running in Selat Manipa; if so consideration must be given to the use of Selat Kelang (3°15'S, 127°37'E), between Pulau Manipa and Pulau Kelang, but an adverse current prevails here during the north-west monsoon season.

Through Ceram Sea and Molucca Sea ⇒ Pacific Ocean.
9.103

1 Having entered the Ceram Sea, steer as directly as possible to pass through one of the channels between the chain of islands between Obi Mayor (1°25'S, 127°25'E) and Kofiau (1°10'S, 129°50'E) into the Halmahera Sea. The channel between Pulau Tobalai (1°38'S, 128°20'E) and Pulau Kekek (1°30'S, 128°38'E) is recommended in the north-west monsoon so as to keep well to windward.

2 Continue N through the Halmahera Sea and pass into the Pacific Ocean through Selat Jailolo or Selat Dampier. Selat Sagewin (0°55'S, 130°40'E), between Pulau Batanta and Pulau Salawati, should not be taken by sailing vessels, as there are frequent calms on account of the high land on either side and the rapid tidal streams, with strong eddies, are liable to make the vessel unmanageable. The only difficulty in Selat Jailolo arises from the strong tidal streams which cause whirlpools and tide-rips. The general directions for the passage of a sailing vessel through Selat Dampier are the same as those for a power vessel, which can be found in *Indonesia Pilot, Volume III*. If the wind is from the N, a sailing vessel, having passed through the narrows, should keep over towards Pulau Waigeo (0°12'S, 130°45'E), rather than Irian Jaya, to avoid being driven onto the Irian Jaya coast by the swell from the N. Great attention must be paid to the set of the currents.

3 Although the Molucca Sea is the principal passage for power vessels proceeding between the Celebes, Ceram, Banda and Arafura Seas, it is not recommended for a sailing vessel, working to the N during the north-west monsoon period, as the current sets with the wind at a rate of 16 to 24 miles a day. If obliged to pass through it, a sailing vessel would find it best to enter through Selat Peleng (1°00'S, 123°00'E), keeping along the Sulawesi coast.

Pacific Ocean ⇒ South China Sea by Surigao Strait
9.104

1 Surigao Strait (9°53'N, 125°21'E) is less frequented by sailing vessels than is San Bernadino Strait (12°33'N, 124°12'E), which is more to windward in the north-east monsoon. It is, however, more direct and safer than San Bernadino Strait, but it obliges sailing vessels that take it, if they are making for Manila, to work up the W coasts of Negros (10°30'N, 123°00'E) and Panay (11°00'N, 122°30'E) and the E coast of Mindoro (13°00'N, 121°10'E). It is of advantage to vessels going to the more S parts of

the Philippine Islands or to the Sulu Sea. Surigao Strait is safe and deep throughout its length and the shores of the islands that border it are steep-to.

2 At the entrance to Surigao Strait the north-east monsoon sets in towards the end of September and blows throughout October and November. In December NE winds alternate with N gales. In January winds blow from NE to ENE, accompanied by heavy rain. In February and March E winds prevail. In April, May and June the prevailing wind is SE, with occasional gales, called 'collas' from the S. In July, August and September, collas from the SW are frequent.

3 The NE winds, though strong, cease during the night, but winds from SE, S and SW will continue to blow. It generally rains with NNE and ENE winds but the rain ceases and the weather clears with E winds and more so with SE winds. With SW winds it remains clear unless a gale arises which sometimes brings rain.

4 In general there is no very bad weather in this part of the archipelago, except when a typhoon occurs. The season when a typhoon might occur is from the end of October to the beginning of January. They begin to blow from the NW and finish from the SE, having passed through either NE or SW. When they shift through NE they tend to blow stronger with more rain.

Pacific Ocean ⇒ South China Sea via San Bernadino Strait and Verde Island Passage.
9.105

1 When entering San Bernadino Strait (12°33'N, 124°12'E) from the E in the south-west monsoon, work to windward with the in-going stream and, when it loses its strength, make for the banks NW or W of Biri Island (12°42'N, 124°21'E), where anchorage can be had, on a sandy bottom, until the tide makes again.

2 On weighing, work according to the direction of the stream, so as to pass through Capul Pass, between Capul Island (12°26'N, 124°10'E) and Dalupiri Island (12°25'N, 124°15'E), or through Dalupiri Pass, between Dalupiri and Samar (12°00'N, 125°00'E). The latter is probably the safer, especially if coming from the S. If the tide should turn before a vessel has entered these passages, make for the open bay off Kinaguitman, S of Lipata Point (12°32'N, 124°16'E), in Samar Island. Anchorage can also be had, if necessary, in the channel on either side of Dalupiri Island, on a sandy bottom, strewn with big stones.

3 The only danger to guard against at this part is Diamante Rock (12°21'N, 124°12'E), once past this take either the passage between Naranjo Islands (12°21'N, 124°02'E) and Capul or between Naranjo Islands and Destacado Island (11°16'N, 124°06'E). The latter route is the better, shaping the course to round the N end of Ticao Islands (12°30'N, 123°42'E).

4 For information respecting winds, currents and passages with a fair wind through the strait, see *Philippine Islands Pilot*.

5 Verde Island Passage (13°36'N, 121°00'E) lies between the SW part of Luzon and the N coast of Mindoro. Verde Island (13°33'N, 121°05'E) divides the channel into N and S passages, both are safe but North Pass is preferred as South Pass is obstructed by Baco Islands (13°28'N, 121°10'E). It is a favourite route during the north-east monsoon for vessels coming from the S. Get to the N under the lee of Negros and Panay and, from the NW point of Panay, proceed between Mindoro and Tablas Island (12°25'N, 122°05'E) to Dumali Point (13°07'N, 121°33'E) and then on through Verde Island Passage and up the W

coast of Luzon, thus escaping the strong monsoon that is generally felt after clearing Lubang Islands.

Pacific Ocean ⇒ South China Sea passing north of Philippine Islands.
9.106

1 A vessel on passage E of the Philippine Islands will benefit from the increasing influence of the Japan Current (7.26) as far as the Balintang Channel (20°00′N, 122°20′E) or the Bashi Channel (21°20′N, 121°00′E) either of which can be used to enter the N part of the South China Sea. Details of these channels are described in *China Pilot, Volume III.*

First Eastern Passage

Details
9.107

1 **Archipelagic Sea Lane II** passes through sections of this passage, in particular Selat Lombok (8°47′S, 115°44′E) and Selat Makasar (2°00′S, 118°00′E), for details regarding Archipelagic Sea Lanes, see Appendix A.

2 As outlined at 9.62, the First Eastern Passage should be taken N-bound in October and November only. It is more suitable for S-bound vessels, but then only in May and from the South China Sea. It has little to recommend it on account of the adverse current setting to the S through Selat Makasar, often strongly, at all seasons. The winds are often boisterous and uncertain at the S end of the strait and light and variable at the N end, while navigation is difficult throughout almost the whole voyage to the open South China Sea.

3 The route runs from either Selat Lombok, Selat Bali (8°10′S, 114°25′E) or Selat Alas (8°40′S, 116°40′E), across the Java Sea into, and through, Selat Makasar to the Celebes Sea, thence to the Sulu Sea through the Basilan Strait (6°50′N, 122°00′E). It then passes up the W coasts of Mindanao, Negros and Panay and enters the South China Sea through Mindoro Strait (13°00′N, 120°00′E) or Verde Island Passage (13°36′N, 121°00′E).

Notes on Selat Bali, Selat Lombok and Selat Alas
9.108

1 Of these three straits, Selat Bali is the narrowest and the most difficult for sailing vessels. It was formerly preferred by them, due to the anchoring facilities it offered. Selat Lombok is the widest but Selat Alas is probably preferable as there are no dangers and anchorage can be obtained, if necessary, during the calms to which all these straits are more or less subject.

2 In Selat Bali, which is only 1 mile wide at its N end, the chief difficulty lies in the currents, and sailing vessels should only navigate this strait by day.

3 During the south-west monsoon, N of the area of the Trade Winds, the wind is mostly SSW and SSE to SE with a W-going stream. From July to September the wind can be very strong. In the north-west monsoon, a vessel N of the Trade Wind area may be set strongly to the E, both by wind and stream.

4 Selat Lombok is the most important passage between the Indian Ocean and Selat Makasar, mainly on account of its width and the ease with which it can be navigated. During the north-west monsoon, sailing ships average one day to make the N-bound passage; during the south-east monsoon this passage usually varies from one to three days.

5 Making the S-bound passage during the north-west monsoon takes at least one day, but usually more, in the south-east monsoon this passage is quick averaging 16 or

17 hours, however in April and October sailing vessels have experienced great difficulty in getting through the strait S-bound. February and March are the best months.

6 During the south-east monsoon, in Selat Lombok, calms are frequent from sunrise to noon, when a fresh S wind arises, turning to SE on the Bali side and to SSW on the Lombok side, blowing strong during the night.

7 During the north-west monsoon the winds are generally from NW, sometimes with violent squalls and a high sea in the N approach.

Directions for Selat Lombok
9.109

1 During the south-east monsoon, the South-east Trade Wind continues through the strait. When nearing the Strait keep E of the entrance and sight Lombok, taking into consideration that the vessel may be set W by the monsoon drift. Sail into the entrance close along the SW point of Lombok and then hold the Lombok side. At this season Pulau Nusapenida (8°44′S, 115°32′E) must never be approached, as in the event of calms, especially with a S-going stream, there is a danger of being set on to it.

2 In the transition months (March and the end of October and beginning of November), if W winds predominate, hold the Bali side, passing through Selat Badung (8°40′S, 115°20′E); if E winds predominate then hold the Lombok side. Selat Badung is always preferable as anchorage may be obtained there.

3 During the north-west monsoon, make for Bukit Badung and proceed through Selat Badung under the Bali shore.

Directions for Selat Alas
9.110

1 In the south-east monsoon the wind blows strongly from the S during the greater part of the day, but subsides towards evening, when the land breeze from Lombok begins. In the north-west monsoon variable and baffling S winds are often experienced in Selat Alas.

2 Approaching from the S, Selat Alas may be identified by the high rugged land of the SW part of Sumbawa, and the plateau forming the SE part of Lombok. From the N, Gunung Rinjani (8°25′S, 116°27′E) and the high NW part of Sumbawa are conspicuous and the islands, lying under the the coasts of Lombok and Sumbawa, will also be visible.

3 As all the straits E of Jawa are, more or less, subject to calms, sailing vessels proceeding through Selat Alas may find it necessary to anchor; it is therefore advisable to hold the Lombok side of the strait, where conditions for anchoring are more favourable.

Notes on passage through Java Sea and Selat Makasar ⇒ Celebes Sea
9.111

1 Having passed through Selat Alas, Selat Bali or Selat Lombok, as directed, steer to pass between Pulau-Pulau Kangean (6°50′S, 115°25′E) and Pulau-Pulau Tengah (7°20′S, 117°35′E), thence to enter Selat Makasar by one of the three channels into which the S entrance is divided. The middle of these three channels is to be preferred for entering the strait, though the most E channel is also frequently used, especially by vessels bound for Ujungpandang (5°07′S, 119°23′E). In the latter case Kepulauan Pabbiring, and its associated bank, rises so steeply from depths greater than 180 m that sounding will give no warning of a vessel's approach. The most W of these channels is seldom used, partly owing to the fact that no land is visible, making position fixing difficult and partly because no saving of distance is effected.

2 Some 150 miles N of these channels the strait is again divided into two channels by Pulau-Pulau Balabalagan (2°15′S, 117°30′E). The width of the W channel is 20 miles and that on the E side of the islands 45 miles. There are some dangers in the W channel, but it is never the less much frequented and, for some reasons preferred to the E, on account of the more moderate depths off the coast of Borneo, which permit anchoring in case of necessity, while the Sulawesi coast is steep-to in many places and destitute of anchorages.

3 Having passed N of Pulau-Pulau Balabalagan, there is no difficulty in navigating the remainder of Selat Makasar into the Celebes Sea.

Celebes Sea ⇒ South China Sea
9.112

4 This section of the passage passes from the Celebes Sea, through Basilan Strait (6°50′N, 122°00′E) into the Sulu Sea and thence by Mindoro Strait (13°00′N, 120°00′E) or Verde Island Passage (13°36′N, 121°00′E) to the South China Sea. Verde Island Passage (for directions see 9.105) is a favourite N-bound route during the north-east monsoon.

5 In Basilan Strait the channel N of Santa Cruz Islands (6°52′N, 122°04′E), although narrower than that on the S side of them, is generally preferred by sailing vessels for its better anchorage facilities.

6 The Sulu Sea is of great depth and offers no particular problems. For details of winds and currents in this area and its vicinity, see *Philippine Islands Pilot*.

PASSAGES SOUTH-BOUND THROUGH EASTERN ARCHIPELAGO

Charts:
> 5126 (1) to (12) Monthly Routeing Charts for Indian Ocean.
> 5309 Tracks followed by Sailing and Auxiliary Powered Vessels.
> 5310 World Surface Current Distribution.
> Diagram 9.3 World Sailing Ship Routes.

General information
9.113

1 There are three principal passages for vessels S-bound from the South China Sea through the Eastern Archipelago.

2 The Western Passage (9.114) passes through the South China Sea, W of the Philippine Islands and Borneo to Selat Sunda (6°00′S, 105°52′E), either direct or via Singapore.

3 The Eastern Passage (9.123) passes through the South China Sea, E of the Philippine Islands to Selat Jailolo (0°06′N, 129°10′E), and thence to Selat Ombi (8°35′S, 125°00′E) or to one of the central passages namely Selat Alas (8°40′S, 116°40′E), Selat Lombok (8°47′S, 115°44′E) or Selat Bali (8°10′S, 114°25′E).

4 The Central Passage (9.127) passes W of the Philippine Islands and E of Borneo, through Selat Makasar (2°00′S, 118°00′E) to one of the central passages.

5 Of these three passages the Western and Central are those used by vessels from ports in S China; the Central Passage is also used by vessels from Manila and ports in the S part of the Philippines or on the E side of Borneo. The Eastern Passage is used by vessels from ports in N China and Japan. In the strength of the south-west monsoon, vessels from ports in S China sometimes use the Eastern Passage.

Western Passage South-bound from South China Sea

Details
9.114

1 **Archipelagic Sea Lane I** passes through Selat Karimata (1°43′S, 108°34′E), for details regarding Archipelagic Sea Lanes, see Appendix A.

2 Passage may be taken either direct or via Singapore (1°12′N, 103°51′E), the latter being best during the north-west monsoon (October to April) and there are then two principal passages:
> One by Selat Riau (0°55′N, 104°20′E) and Selat Bangka (2°06′S, 105°03′E) or Selat Gelasa (2°53′S, 107°18′E), but in October by Selat Gelasa only, because light and baffling winds prevail, in that month, between Selat Riau and Selat Bangka;
> The other, known as the Inner Route, by Selat Durian (0°40′N, 103°40′E), Selat Berhala (0°50′S, 104°15′E) and Selat Bangka.

3 Selat Riau and Selat Berhala are particularly convenient for sailing vessels leaving Singapore for Europe in the north-west monsoon (north-east monsoon of the South China Sea). By using these routes the difficulties of heading E out of the Singapore Strait into the north-east monsoon of the South China Sea are avoided.

4 During the south-east monsoon, the ordinary route would be to beat out through the Singapore Strait to the E and work S by Selat Karimata (1°43′S, 108°34′E) or Selat Gelasa to Selat Sunda. At this time, vessels are frequently able to proceed much more quickly to the S by the Inner Route (9.77) than by the outer one. Convenient anchorage is always available in the straits for sailing vessels held up by wind or tidal streams.

5 To make the passage from the South China Sea to Selat Sunda, during the north-east monsoon of the South China Sea, a vessel having passed either E or W of Pulau-Pulau Anambas (3°00′N, 106°00′E) should proceed S through Selat Bangka or Selat Gelasa; but during October Selat Bangka should not be attempted, owing to calms and baffling winds which occur in its N approaches, at that time.

6 During the south-west monsoon, a vessel from Palawan Passage (10°20′N, 118°00′E), or one that has crossed to the Borneo coast from Mui Dinh (11°22′N, 109°01′E), should proceed by Selat Karimata or Selat Gelasa.

7 Directions for vessels S-bound through Selat Durian and Selat Berhala are the reverse of those given for the N-bound passage by the Inner Route described at 9.80 and 9.78. Directions for the other straits and channels follow.

Passage southward through Selat Riau
9.115

1 Vessels having a fair wind leaving Singapore at high water, or about the first quarter of the out-going (E-going) stream and, taking about 4 hours to reach the entrance to Selat Riau, will probably carry a fair tidal stream through both straits, but no dependence can be placed on it. See *Indonesia Pilot, Volume I*, for further details.

2 The directions given at 9.76 for coming N through Selat Riau, if reversed, will suffice for proceeding S. Deep-draught vessels should pass E of Karang Galang (1°09′N, 104°11′E).

3 At night, steer to pass ¾ mile E of the light on Karang Galang, from which position Pulau Tanjuk (0°57′N, 104°12′E) leading-lights will be in line bearing 180°. When Pulau Terkulai (0°57′N, 104°20′E) light is open S of Pulau Lobham (0°59′N, 104°15′E) bearing 100° and the vessel is

S of Karang Lolo (0°59′N, 104°13′E), shape to make about 135°, allowing for tide, until the light on Pulau Karas Kecil (0°44′N, 104°22′E) bears 154°, when it may be steered for on that bearing. Pass about ½ mile or more E of it and then keep it astern, bearing about 320° or less, as long as it is in sight, to lead in the fairway S of the strait.

Approach and passage southward through Selat Bangka
9.116

1   With a fair wind, when coming from the N, and having passed Pulau-Pulau Tuju (1°15′S, 105°15′E) and steering to the S to pass through Selat Bangka, there will be no difficulty, in clear weather, in determining the position. In such circumstances enter the strait E of Karang Ular (1°58′S, 104°57′E). In thick weather it often happens that no land can be seen until the vessel has arrived very near to the entrance to the strait, when it is important to locate the bank extending from the Sumatera coast and then proceed along its edge, in low water depths of 15 m to 11 m, carefully attending to soundings.

2   An offshore marine terminal has been established off the edge of the bank, 7½ miles N of Tanjung Ular (1°57′S, 105°08′E).

3   Sometimes Bukit Menumbing (2°07′S, 105°10′E) will be seen, but no other land, and in such case it would be prudent to proceed as before, keeping along the edge of the bank.

4   When working through Selat Bangka from the N, the passage W of Karang Ular is much to be preferred when the land is obscured and reliable bearings cannot be obtained; at other times the E channel is preferable. By reversing the directions given at 9.73 and 9.74, no difficulty should be experienced in leaving Selat Bangka.

Approach and passage southward through Selat Gelasa
9.117

1   Although the navigation of this strait is complicated by the many dangers in it, yet as the course is more direct, the prevailing winds more favourable and the distance less than by the safer route through Selat Bangka, many mariners prefer it, especially when S-bound from China late in the north-east monsoon.

2   In consequence of the N entrance to Selat Gelasa being so near the equator, the winds, even in the strength of the monsoons, are very uncertain, producing a corresponding uncertainty in the direction and force of the tidal streams and currents. A sailing vessel approaching from the N will, therefore, have to be principally guided by the winds and currents which are actually being encountered at the time rather than rely on the seasonal predictions given in *Indonesia Pilot, Volume I.*

3   In thick weather the greatest caution is necessary when approaching Selat Gelasa, for unless good observations or electronic positions can be obtained, there is no means of ascertaining an exact position. In such circumstances it is advisable to steer for Selat Bangka, where the soundings on the edge of the bank extending from the Sumatera coast may be a useful guide, as described at 9.116.

4   When approaching the NE coast of Bangka use every precaution not to become entangled among the outlying dangers when running S for Selat Gelasa in thick weather. Some of these dangers are over 40 miles from the shore, between Tanjung Tuing (1°37′S, 106°03′E) and Tanjung Berikat (2°34′S, 106°51′E).

5   Early in the north-east monsoon, when the wind is generally from N or NW, vessels intending to go through Selat Gelasa should pass between Pulau Toty (0°55′S, 105°46′E) and Pulau Dokan (0°58′S, 105°39′E), which lie off the N coast of Bangka. Later in the monsoon season the wind is more E and it is then better to pass from 10 to 20 miles E of Toty.

6   Cross bearings of the mountains on Bangka, in clear weather, will enable a vessel to clear Karang Iwan (1°40′S, 106°18′E) and Karang Severn (1°37′S, 106°31′E) which lie in the track to Selat Gelasa. Having passed N of Karang Severn, steer so as to get on the meridian of Pulau Gelasa (2°25′S, 107°04′E) before reaching the parallel of 1°50′S. Pulau Gelasa is visible, in clear weather, at a distance of over 30 miles, but it is not visible from Karang Lanrick (1°53′S, 106°57′E), the most N danger, for which a careful lookout is necessary. When Gelasa comes in sight bring it to bear 180°, which leads clear of all the dangers lying to the W. Then pass E of Pulau Gelasa and shape course for Selat Leplia (2°53′S, 106°58′E) which is the passage usually taken.

7   The directions given above, apply only to sailing vessels coming from China early in the north-east monsoon. Later in the monsoon, SE and E winds are often met with between Bangka and Pulau Belitung (3°00′S, 108°00′E) and it will be better to pass from 10 to 12 miles W of Pulau Pejantan (0°07′N, 107°12′E) and try, as soon as possible, to get on the meridian of Pulau Gelasa and when the island is sighted, bring it bear 180°, and proceed as above.

8   Late in the north-east monsoon also, SSW winds are often met in the S part of the South China Sea, obliging vessels to keep farther E towards the islands off Borneo. If this should happen in May or June, it would be tedious work getting to Selat Leplia, therefore steer for Pulau Langkuas (2°32′S, 107°37′E) off the NW point of Pulau Belitung and pass through Selat Baur.

9   Selat Gelasa can only be approached from the N, at night, by passing E of all the dangers lying N of it. Having passed well to the W of Karang Florence Adelaide (2°04′S, 108°05′E), shape course for the light on Pulau Langkuas and, when it comes in sight, alter course to pass about 4 miles W of Pulau Langir (2°48′S, 107°22′E).

Selat Leplia passage southward
9.118

1   In the early part of the north-east monsoon, N and NW winds prevail about the N entrance to Selat Gelasa and strong SE-going currents will generally be experienced between Pulau Gelasa and Pulau Liat (2°52′S, 107°03′E) especially near the N extremity of Pulau Liat. This current has been a frequent cause of accidents.

2   Vessels intending to proceed S-bound through Selat Leplia by night should take care to sight Tanjung Berikat (2°34′S, 106°51′E) during daylight.

3   Having passed from 1 to 2 miles E of Pulau Gelasa, steer to the SW until that island bears 014°, and then keep it on that bearing, astern, until the SE extreme of Pulau Kelapan (2°51′S, 106°50′E) bears 236° and the N point of Pulau Liat bears 125°. From this position keep in the fairway of the channel, steering about 185° to pass between the dangers off Pulau Celaka (2°52′S, 107°01′E) and Batu-Batu Discovery (2°53′S, 106°56′E), guarding against the effects of tidal streams and currents.

Selat Baur and Selat Limende passage southward
9.119

1   The directions given at 9.86 and 9.87 should be applied in reverse.

2   Selat Baur is the best channel to use for beating to windward against the south-east monsoon, since the currents in it are weaker than elsewhere.

Approach ⇒ and passage southward through Selat Karimata
9.120

1   If using the Main Channel during the north-west monsoon, take the channel E of Karang Ontario (2°00′S, 108°39′E). Approach Pulau Serutu (1°43′S, 108°43′E), with its summit bearing less than 152°, and then pass 4 or 5 miles W of the light-structure.

2   Then follow the directions given at 9.96 in reverse.

3   If using the E channels, then the directions given at 9.97 in reverse, are generally applicable. When the south-east monsoon is strong, smoother water with less current, will be found in these channels than in the main part of the strait.

General directions for passage from Selat Bangka, Selat Gelasa or Selat Karimata ⇒ Selat Sunda
9.121

1   **Cautions**. Areas of the Java Sea are being exploited for natural resources, see 7.38.2. These routes pass close to oilfields, with extensive areas of prohibited anchorage, it would be prudent to avoid these areas altogether wherever possible.

2   Dangerous shoals extend for about 35 miles to the S of Selat Gelasa, rendering great caution necessary when leaving it and making for Selat Sunda.

3   Having cleared the shoals S of Selat Gelasa and Selat Karimata, the route to Selat Sunda is the same as that from Selat Bangka, described below.

4   With a fair wind, after passing E of Five Fathom Banks (3°48′N, 106°29′E), in depths of 18 m to 22 m, steer to pass a prudent distance W of Karang Basa (5°12′S, 106°12′E) and Gosong-gosong Serdang (5°05′S, 106°15′E), from there follow the reverse of the directions given at 9.71.

5   When working S from a position W of Five Fathom Banks, follow the reverse of the directions given at 9.71, observing the caution to anchor at night if the position is at all doubtful. Gosong Sekopong (4°56′S, 106°03′E), with a least depth of 5 m, will be avoided by keeping Pulau Segama (5°10′S, 106°06′E) bearing more than 180°, and when S of them, the islands bearing less than 000° will lead E of Layang Layang (5°18′S, 106°04′E), with a depth of 8 m over it.

6   In the south-east monsoon, when the atmosphere is hazy and the coast rarely visible, great care is necessary when passing Pulau Segama, as from the N the islands appear as one.

Passage southward through Selat Sunda
9.122

1   The general description of Selat Sunda, together with the winds, sea and tidal streams to be expected there, is given in *Indonesia Pilot, Volume I*.

2   During the south-east monsoon (April to September), keep in the main fairway when the wind is favourable, but if proceeding through Selat Panaitan keep closer to the Jawa coast than to Pulau Panaitan. This route may also be taken at the beginning of the north-west monsoon, up to about the end of December, if conditions are favourable.

3   The monsoon is generally supposed to shift at about the beginning of October, but often is delayed for a month; the interval being filled with calms, light S winds and frequent heavy Sumatera squalls or south-westerlies. The squalls, at this season, generally take place at night, accompanied by heavy rain, thunder and lightning and are of short duration.

4   During the north-west monsoon (October to April), an alternative is offered between routes on the N and S sides of Selat Sunda.

9.122.1

1   **Northern route**. During the strength of the north-west monsoon, in January and February, the W channel between Pulau Sangian (5°57′S, 105°51′E) and Pulau-Pulau Sumur (5°52′S, 105°46′E) is recommended, giving the latter a berth of 1½ miles and thence working NW when winds are from the W.

2   If it is late in the day when Pulau-Pulau Sumur are sighted, with strong SW winds and an adverse stream, a vessel will do well to seek anchorage off the Sumatera coast or Tanjung Sumurbatu (5°50′S, 105°46′E) at the N end of the islands; or off Sindu, inshore of Palau Kandangbalak (5°53′S, 105°45′E), the SW island of the group. The vessel should be got aweigh immediately the stream turns, to take advantage of the morning land breeze.

3   Working through the passage between Pulau Sebuku (5°53′S, 105°31′E) and Sumatera, pass on whichever side of Pulau-Pulau Tiga (5°49′S, 105°32′E) the strong currents and hard squalls will allow, and thence N of Pulau Serdang (5°49′S, 105°23′E) and between it and Pulau Siuncal (5°48′S, 105°19′E). Alternatively pass N of Pulau Legundi (5°50′S, 105°16′E) and out through Selat Legundi, avoiding Karang Medusa (5°47′S, 105°16′E), 2 miles NE of Pulau Seserot, passing on either side of that island in mid-channel. In this manner a quick passage may be made through the strait if the wind is not too variable, besides having the advantage of anchorage being available on the E side of Pulau Sebuku (5°53′S, 105°31′E) or on the W side of Teluk Lampung (5°40′S, 105°20′E) if the current or wind prove too strong.

4   **Note** that Selat Legundi is 2 miles wide and is recommended to sailing vessels working out of Teluk Lampung in the north-west monsoon. The passages on either side of Pulau Seserot (5°48′S, 105°15′E) are equally good and, with contrary winds or current, there is anchorage on the E side of the island in depths of 18 m to 22 m, sand. Vessels drifting through the strait in a calm will be carried past the island by the off-set of the current. To the W of Selat Legundi is Teluk Kiluan (5°47′S, 105°06′E), where safe anchorage may be found, if required, by vessels with local knowledge only.

9.122.2

1   **Southern route**. This takes a sailing vessel through Selat Sunda, along the coast of Jawa and through Selat Panaitan, to the Indian Ocean. There are, on record, many instances of vessels having worked out of the strait during the north-west monsoon by taking this route with more ease and celerity than could have been effected by stretching into Teluk Lampung (5°40′S, 105°20′E), in consequence of the SW-going current from the Java Sea having developed its chief strength along the E side of the strait. This is, however, a lee shore and therefore dangerous at this season.

2   In spite of this, cases are on record in which vessels have worked through Selat Panaitan in a remarkably short time during a W gale, by carrying a heavy press of sail and tacking between squalls, when it was impossible for any vessel in the main channel to beat against the current and heavy seas. In this monsoon, particularly when working out, it is advisable to keep nearer the island shore, to obtain the help of the current, sometimes running to the

W, and to avoid being set upon the rocks about Tanjung Gedeh (6°46′S, 105°13′E) by the heavy swell. Near the Jawa shore, when outside anchorage depths, in a calm vessels would be in considerable danger.

3   **Conservation Areas** surround both Pulau Panaitan and Ujung Kulon, see *Indonesia Pilot, Volume I.*

Eastern Passage South-bound from South China Sea

Details
9.123

1   **Archipelagic Sea Lanes** pass through the following sections of this passage:

    ASL IIIA (Part 1), through Banda Sea, Ceram Sea, Molucca Sea and to Pacific Ocean;

    ASL IIIA (Part 2), from Indian Ocean and through Savu Sea;

    ASL IIIB, through Selat Leti and Banda Sea;

    ASL IIID, from Indian Ocean, between Sawu and Roti and through Savu Sea;

    ASL IIIE through Molucca Sea to Celebes Sea.

For details regarding Archipelagic Sea Lanes, see Appendix A.

2   The Eastern Route passes from the South China sea to Selat Ombi (8°35′S, 125°00′E), or to Selat Alas (8°40′S, 116°40′E), Selat Lombok (8°47′S, 115°44′E) or Selat Bali (8°10′S, 114°25′E). It is useful from the middle of May to the end of July.

3   During the strength of the south-west monsoon, the best route from Hong Kong (22°17′N, 114°10′E) and adjacent coast ports is to pass N of the Philippine Islands, through the Bashi Channel (21°20′N, 121°00′E) or Balintang Channel (20°00′N, 122°20′E) and then steer S along the E side of the Philippines, or SE towards Palau Islands (7°30′N, 134°30′E).

4   When the south-east monsoon is encountered, shape course to pass E of Halmahera, though Selat Jailolo (0°06′N, 129°10′E) or Selat Dampier (0°40′S, 130°30′E) to the Halmahera Sea, thence to the Ceram Sea; alternatively a vessel may pass to the Ceram Sea from the Pacific Ocean through the Molucca Sea.

5   From the Ceram Sea pass to the Banda Sea either through Selat Manipa, E of Pulau Buru (3°30′S, 126°30′E), or by passing W of Pulau Buru; then continue through the Banda Sea to Selat Ombi or pass through the Flores Sea to Selat Alas, Selat Lombok or Selat Bali. If bound to Selat Bali, the usual route is via Selat Sapudi (7°00′S, 114°15′E).

South China Sea ⇒ Pacific Ocean via Balintang Channel or Bashi Channel
9.124

1   Balintang Channel is reputed to be free of danger and frequently used by sailing vessels S-bound from ports in China. Bashi Channel is also used. Further details for these channels can be found in *China Sea Pilot, Volume III.*

Pacific Ocean ⇒ Ceram Sea
9.125

1   This part of the Eastern Route may be taken either via Selat Jailolo or Selat Dampier or through the Molucca Sea, as described at 9.123.

2   In Selat Jailolo, the deep channel lying between Pulau Muor (0°11′N, 128°57′E) and Pulau Gebe (0°06′S, 129°26′E) presents no difficulty, except from the strong

tidal streams, often accompanied by whirlpools and tide-rips, particularly of the NW extremity of Gebe.

3   Having passed through Selat Jailolo steer through the Halmahera Sea to enter the Ceram Sea through one of the channels between the chain of islands about 70 miles S of Selat Jailolo. The channel between Pulau Pisang (1°23′S, 128°55′E) and Kepulauan Boo (1°10′S, 129°25′E) is recommended for sailing vessels during the south-west monsoon.

4   When approaching Selat Dampier from the E, Tanjung Momfafa (0°18′S, 131°20′E) should be made out and a good berth given to the shoals, which extend about 7 miles ENE and which may be avoided by keeping Pulau Wayam (0°24′S, 131°15′E) bearing more than 245°; then proceed by following the reverse of the directions given at 9.103, further information can be found in *Indonesia Pilot, Volume III*. Having passed through Selat Dampier proceed directly to the Ceram Sea.

5   As for N-bound vessels (9.103), the Molucca Sea cannot be strongly recommended for vessels S-bound between the Pacific Ocean and the Ceram Sea. It is sometimes used by sailing vessels S-bound from China, with some advantage after September. It is, however, a tedious passage to beat through, as the currents set with the wind at rates of 16 to 24 miles a day. When it is difficult to get to the S by the channel between Pulau-Pulau Sula (1°50′S, 125°00′E) and Obi Mayor (1°25′S, 127°25′E), sailing vessels might try an alternative by keeping near the W coast of Halmahera and passing through Selat Patientie (0°25′S, 127°40′E), and thence through Selat Obi (1°00′S, 127°00′E) and Selat Tobalai (1°40′S, 128°15′E) to the Ceram Sea.

Ceram Sea ⇒ Indian Ocean via Selat Ombi, Selat Alas, Selat Lombok or Selat Bali
9.126

1   The recommended route is to pass through Selat Manipa (3°20′S, 127°22′E) into the Banda Sea and then to proceed as directly as possible to Selat Ombi or pass through the Flores Sea to Selat Alas, Selat Lombok or Selat Bali.

2   In passing S-bound through Selat Manipa during the south-east monsoon, keep towards the W side of Manipa, where the N-going current will not be so severely felt. See 9.102 and *Indonesia Pilot, Volumes II and III.*

3   Approaching Selat Alas from the N, Gunung Rinjani (8°25′S, 116°27′E) and the high NW part of Sumbawa are conspicuous. The 180 m depth contour from the S terminates about a mile from Tanjung Ringgit (8°52′S, 116°36′E); from the N it penetrates as far as a line running W from Pulau Belang (8°33′S, 116°47′E). The soundings between are deep but irregular. Selat Alas, in common with all the straits E of Jawa, is more or less subject to calms; it is therefore advisable for a sailing vessel to keep within soundings on the Lombok side, particularly as the currents are not so strong there as in the middle or on the E side.

4   When S-bound, it is advisable to get underway very early in the morning, in order to clear the strait, if possible, before the sea breeze sets in.

5   In Selat Lombok, S-bound during the south-east monsoon, with predominating SE winds, it is advisable to work up under the Bali shore, with a N-going stream, until the summit of Gunung Agung (8°20′S, 115°30′E) bears 270°; under these conditions, working to the S under the NW coast of Lombok is difficult, and the same applies to the Bali shore S of the parallel of Gunung Agung.

6   During the north-west monsoon, and in the transitional months, Selat Lombok from the N affords no particular difficulties, however, the remarks on the tidal streams in *Indonesia Pilot, Volume II* should be studied.

7   Selat Bali, described at 9.108, offers a safe passage to S-bound vessels during the north-west monsoon and, with the exception of Selat Alas, E of Lombok, is to be preferred to all the passages E of Jawa, as there is anchorage on both sides of the narrows in case passage could not be made in a single tide. For vessels coming from the N, the chief difficulty is the strength of the currents. Sailing vessels should only navigate this strait by day. During the north-west monsoon the water in the strait is smooth and the passage easy.

Central Passage South-bound from South China Sea

Details
9.127

1   **Archipelagic Sea Lane II** passes through sections of this passage, in particular Selat Lombok (8°47′S, 115°44′E) and Selat Makasar (2°00′S, 118°00′E), for details regarding Archipelagic Sea Lanes, see Appendix A.

2   The Central Passage runs from the South China Sea through Selat Makasar, to Selat Alas (8°40′S, 116°40°E), Selat Lombok or Selat Bali (8°10′S, 114°25′E). It is, in fact, the reverse of the First Eastern Passage (9.107). It is intended for vessels leaving China at the end of April or the beginning of May.

3   Summarising the route, a vessel should steer from the Macclesfield Bank (15°50′N, 114°30′E) to pass through Mindoro Strait (13°00′N, 120°00′E), thence across the Sulu Sea to the Celebes Sea through Basilan Strait (6°50′N, 122°00′E) or Sibutu Passage (4°55′N, 119°37′E). Basilan Strait is recommended for sailing vessels, though Sibutu Passage is sometimes used.

4   The voyage continues through the Celebes Sea and Selat Makasar into the Java Sea through which either of two routes may be taken, namely to Selat Sunda (6°00′S, 105°52′E) or to one of the central passages Selat Alas, Selat Lombok or Selat Bali. Vessels bound for Selat Bali usually pass through Selat Sapudi (7°00′S, 114°15′E).

5   If an alternative to Selat Makasar is desired then a vessel may pass from the Celebes Sea to the Banda Sea via the Molucca Sea, Ceram Sea and Selat Manipa (3°20′S, 127°22′E).

Passage through Mindoro Strait into Sulu Sea
9.128

1   The wide Mindoro Strait, separating the Calamian Group (12°00′N, 120°00′E) from Mindoro Island, is one of the most frequented channels for sailing vessels which leave Manila (14°32′N, 120°56′E) for the Indian Ocean towards the end of April and throughout the south-west monsoon

period; and by other vessels, throughout the year, from the ports in China to Australia. Land and sea breezes are felt on the coasts of the larger islands in the Mindoro Strait, mostly during the south-west monsoon and in the period between the monsoons, but they are not so regular during the north-east monsoon.

Passage through Sulu Sea ⇒ Selat Makasar
9.129

1   Making to the S through the Sulu Sea, it is best to keep on the E side, along the coast of Panay, and through the Basilan Strait. The more direct route from Mindoro Strait leading S through Sibutu Passage is not recommended and no special directions are available for it. For the Sulu Sea and Basilan Strait, see 9.112.

2   The passage from Basilan Strait or Sibutu Passage, across the Celebes Sea to Selat Makasar is as direct as possible.

Passage south-bound through Selat Makasar and Java Sea
9.130

1   In Selat Makasar the Borneo side provides anchorage in case of need, the coast of Sulawesi being steep-to. Although there are some dangers in the channel W of Pulau-Pulau Balabalagan (2°15′S, 117°30′E), it is nevertheless much frequented for the same reason.

2   On leaving Selat Makasar and entering the Java Sea, course must be shaped for Selat Sunda or for the N entrance to either Selat Alas, Selat Lombok or Selat Bali, see 9.108.

3   If bound for Selat Alas or Selat Lombok, steer to pass about 20 miles E of the dangers on that side of Pulau-Pulau Kangean (6°50′S, 115°25′E). If bound for Selat Bali the usual route is through Selat Sapudi, which is a good, safe channel with no dangers apart from Karang Tembaga (7°07′S, 114°08′E), which dry and Yacoba Elisabett Shoal (7°05′S, 114°11′E), lying on the W side. Selat Sapudi is preferable to both the channel W of Pulau Giliyang (7°00′S, 114°11′E) and Selat Raas (7°10′S, 114°27′E).

4   In Selat Sapudi and the passages farther E, and including Pulau-Pulau Kangean, the south-east monsoon prevails from April to October and the north-west monsoon from from November to March. In April and May all winds are S, in June the monsoon becomes dominant from SSE to SE and blows with greatest strength during July, August and September. In November winds are N, alternating with rain squalls from all points. In December N and NW winds last longer and squalls come from NW or WNW,; January and February are marked by very squally weather from NW to N, and in March it often continues to blow stiffly from W to WNW.

PASSAGES THROUGH RED SEA

Charts:
    5126 (1) to (12) Monthly Routeing Charts for Indian Ocean.
    5309 Tracks followed by Sailing and Auxiliary Powered Vessels.
    5310 World Surface Current Distribution.
    Diagram 9.3 World Sailing Ship Routes.

General notes
9.131

1   Sailing vessels, whether N-bound or S-bound, can experience great difficulties when working against the strong winds which, in the winter season, blow from either end of the Red Sea towards the centre and produce a short

hollow sea which, combined with the strong currents that often run with the wind, render the progress of such vessels very slow. In working to windward in the central channel, a vessel is recommended to favour the Arabian shore, but should not stand close in with a light wind or heavy swell. After dark a vessel should only stand towards the shore half of the distance that she stands out, and should never approach closer than 10 miles to the reefs, at night, in order to guard against the unexpected existence of a cross-current.

South-bound
9.132

1   For sailing vessels the most favourable part of the year for the S-bound passage is during the period of the south-west monsoon in the Arabian Sea (June to September), when N winds of variable strength prevail throughout the length of the Red Sea.

2   Particular attention should be paid to the currents in the Red Sea, this is particularly important in the narrower portions of the passage and their approaches. For further details, see *Red Sea and Gulf of Aden Pilot*.

3   On approaching the Straits of Bab el Mandeb (12°40′N, 43°30′E), choice must be made between using the Large Strait or the Small Strait. For ease of navigation, the Large Strait is recommended, while numerous accidents have occurred in the Small Strait, however anchorage in case of need can obtained there, although this is now restricted by a prohibited area which surrounds Mayyūn (12°39′N, 43°25′E).

North-bound
9.133

1   For the N-bound voyage, December, January and February are the best months, as the S winds often carry a vessel as far as the latitude of Jeddah (21°29′N, 39°11′E), and sometimes as far as that of Quṣeir (26°06′N, 34°17′E), even, at times, to Suez Bay (29°51′N, 32°33′E). After losing the S winds, a sailing vessel will have to beat against the N wind.

2   If as far N as Quṣeir, and bound for Suez Bay and a strong N wind is encountered, a vessel in the central channel of the Red Sea, or even on the W shore, ought to stand over to the Arabian coast, where she will probably reach Ḍubā (27°34′N, 35°30′E). Having worked up to 35 miles N of Ḍubā she may head over to Râs Muhammad (27°44′N, 34°15′E), leaving the Arabian coast at night. As she proceeds, the N winds will veer to NNE out of the Gulf of 'Aqaba, by sailing as close as possible, these will enable her to fetch Râs Muhammad.

PASSAGES FROM ADEN

Charts:
> 5126 (1) to (12) *Monthly Routeing Charts for Indian Ocean.*
> 5309 *Tracks followed by Sailing and Auxiliary Powered Vessels.*
> 5310 *World Surface Current Distribution.*
> Diagram 9.3 *World Sailing Ship Routes.*

Aden ⇒ Mumbai

Passage
9.134

1   During the south-west monsoon (April to September), take as direct a route as possible. Keep in the centre or rather towards the Arabian shore of the Gulf of Aden, to avoid the W-going current on the African coast. During the strength of the south-west monsoon, in June, July and August, when the weather is thick and heavy and observations very uncertain, steer direct for Kanhoji Angre (Khānderi) Island (18°42′N, 72°49′E) and watch the soundings carefully.

2   When steering for Mumbai (Bombay) (18°51′N, 72°50′E), from the middle of May until August, steady gales and clear weather will be experienced at times, until within 70 or 90 miles of the coast, but cloudy weather with rain and squalls may be expected as land is approached.

3   During the early part and strength of the south-west monsoon, great care must be observed not to get N of the entrance to the harbour, for then the N-going tidal stream, as well as the S swell, frequently sets along the bank towards the Gulf of Khambhāt (20°30′N, 72°00′E), and late in May, June and July it would be found difficult at times to work around Prongs Reef (18°53′N, 72°48′E). Therefore in these months a vessel should steer direct for Kanhoji Angre (Khānderi) Island, allowing for a N-going set of the tidal stream; although the prevailing current outside the depth of 55 m off the harbour, following the onset of the south-west monsoon is S-going. Endeavour should be made to make the island bearing between 090° and 135°, altering course as circumstances require, between those bearings, to carry a fair wind in entering the harbour.

4   If the wind is inclined to blow in squalls from the W to WNW, a vessel should not run too close inshore S of Kanhoji Angre (Khānderi) Island or even approach that island very closely, as there may be some difficulty in weathering it with these winds, which are sometimes experienced in June and July, but more frequently in August.

5   During the interval between the land and sea breezes in the forenoon a heavy smoky haze frequently hangs over the land, so great care should be exercised when approaching the land, shortly after daylight, between May and August. Occasionally this also occurs during the calm hours in the evening.

6   During the north-east monsoon (October to March), the passage from the Red Sea to India or the Persian Gulf is very tedious for sailing vessels and is seldom attempted. In former times, the passage between Aden and Mumbai, when taken at this season frequently took from 60 to 90 days.

7   If it is necessary to make the passage, work along the coast of Arabia, taking advantage of every shift of wind. Should the W-going current be strong inshore, stand out 60 to 80 miles from the land. If the wind is light, take advantage of the tides and land winds inshore, anchoring when necessary. When off Juzur al Ḥalāniyāt (Kuria Muria Islands) (17°30′N, 56°05′E), head towards Mumbai and, as progress to the E is made, the wind will draw to N or even W of N.

Aden ⇒ Sri Lanka and Bay of Bengal
Passage
9.135

1   During the south-west monsoon (April to October), pass N of Suquṭrá (12°30′N, 54°00′E) to avoid the heavy cross seas S of that island. It is at all times desirable to avoid passing S of Suquṭrá, if this means passing ʿAbd al Kūrī (12°10′N, 52°15′E) at night, as currents often set strongly N.

2   For Sri Lanka, proceed direct and thence to the Bay of Bengal, for directions in the Bay of Bengal, see 9.21 and 9.22.

3   During the north-east monsoon (October to March), keep along the Arabian coast to about 52°E; pass through the Eight Degree Channel (7°24′N, 73°00′E) or the Nine Degree Channel (9°00′N, 72°30′E), then steer round Sri Lanka and, having cleared the island, make E on the parallel of 5°N as far as the middle of the Bay of Bengal, then work N. From the meridian of 87°E, a vessel will probably reach Chennai (Madras) (13°07′N, 80°20′E). After mid-February round Sri Lanka at a distance of about 50 miles and then proceed direct.

4   The currents off the coast of Sri Lanka are strong and variable, see *West Coast of India Pilot* for further details.

Aden ⇒ Malacca Strait

Passage
9.136

1   During the south-west monsoon (April to October), pass N of Suquṭrá (12°30′N, 54°00′E), thence direct round the S end of Sri Lanka and across the Bay of Bengal, entering Malacca Strait S of Great Nicobar Island (7°00′N, 93°50′E).

2   During the north-east monsoon (October to March), work along the Arabian Coast as far as Ra's Fartak (15°38′N, 52°16′E), or just beyond it, and then head across the Arabian Sea, passing S of Minicoy (8°18′N, 73°02′E) and round the S end of Sri Lanka and across the Bay of Bengal. Pass close S of Great Nicobar Island, if the wind permits and then keep on the Malaysian side of the Malacca Strait, see 9.166 and 9.167 for directions through Malacca Strait.

3   Eight Degree Channel (7°24′N, 73°00′E) and Nine Degree Channel (9°00′N, 72°30′E) are separated by Minicoy. In Nine Degree Channel, the practice of steering to pass a few miles N of Minicoy, especially at night, is a dangerous one as the island is over 4 miles long, in a N/S direction, with the light on the SW side and the current, at times, setting strongly to the S. On the other hand, in Eight Degree Channel, a vessel should keep in the N part of the channel, nearer to Minicoy than the Maldives.

Aden ⇒ Selat Sunda

Low-powered vessel passage
9.137

1   Low-powered, or hampered, vessels may find the following routes more suitable than those described at 6.112 for full-powered vessels.

2   **October to April** the route is:
Round Raas Caseyr (11°50′N, 51°17′E), thence:
Direct to 1°00′S, 72°20′E in March and April, but via 3°00′N, 60°00′E from October to February, thence:
To 2°20′S, 76°30′E, thence:
To 3°00′S, 94°30′E, thence:
To Selat Sunda (6°00′S, 105°52′E).

3   **May to September** the route is:
As for fully-powered vessels (6.112) to landfall (5°45′N, 80°36′E), 10 miles S of Dondra Head, thence:
Across the equator at 96°30′E, thence:

Along the W coast of Sumatera by, either the outer route, W of all off-lying islands, or the middle route between the outer islands and those adjacent to the coast (see, *Malacca Strait and West Coast of Sumatera Pilot*), thence:
To Selat Sunda (6°00′S, 105°52′E).

4   Routes for low-powered vessels proceeding from Selat Sunda to Aden are described at 9.197.

Aden ⇒ Fremantle, Cape Leeuwin and southern Australia or New Zealand

Passages
9.138

1   During the south-west monsoon (April to October), when W winds prevail in the Gulf of Aden, proceed to the S of Sri Lanka, as described at 9.136.

2   After rounding the S point of Sri Lanka steer to the SE, to cross the equator in about 95°E; thence continue S across the South-east Trade Wind into the W winds and round Cape Leeuwin (34°23′S, 115°08′E), if not bound for Fremantle (32°02′S, 115°42′E).

3   During the north-east monsoon (October to March), proceed towards Sri Lanka, as directed at 9.136 for that season; thence with the north-east monsoon cross the equator in about 90°E into the north-west monsoon. Then proceed to the E, in that monsoon, as far as the E end of Jawa (7°00′S, 110°00′E); then stand across the South-east Trade Wind into the Westerlies and thence continue to Cape Leeuwin or Fremantle.

4   At all seasons, if bound to ports on the S or SE side of Australia, to Tasmania, or to New Zealand, continue S and SE in the Westerlies to join the appropriate part of the route from the Cape of Good Hope, described at 9.5.

Aden ⇒ Mauritius

Sailing ship passage
9.139

1   During the south-west monsoon (April to October), pass N of Suquṭrá (12°30′N, 54°00′E), then run through the south-west monsoon to cross the equator at about 72°E, or even to run through the One and Half Degree Channel (1°24′N, 73°20′E) and make to the S into the South-east Trade Winds, passing E of Chagos Archipelago (6°30′S, 72°00′E). Thence proceed direct to Mauritius (20°10′S, 57°30′E).

2   During the north-east monsoon (October to March), work along the Arabian coast until able to weather Ras Caseyr (11°50′N, 51°17′E), then run through the north-east and north-west monsoons, crossing the equator in about 64°E, and the parallel of 10°S in about 70°E. When the South-east Trade Winds have been picked up steer direct for Mauritius.

Low-powered vessel passage
9.140

1   Low-powered, or hampered, vessels may find the following routes more suitable than those described at 6.67 for full-powered vessels.

2   The choice of routes is influenced by monsoon conditions and the following seasonal routes are advised:
October to March:
From Ras Caseyr (11°50′N, 51°17′E), through the north-east monsoon, to cross the equator in about 64°E, thence:
S, passing E of Saya de Malha Bank (10°00′S, 61°00′E), into the South-east Trade Winds, thence:
To Port Louis (20°08′S, 57°28′E).

April to September:

From Ras Caseyr (11°50′N, 51°17′E), through the south-west monsoon, to cross the equator in about 72°E, or even through the One and Half Degree Channel (1°24′N, 73°20′E), thence:

S, passing E of Chagos Archipelago (6°30′S, 72°00′E) into the South-east Trade Winds, thence:

SW, through 12°S, 70°E, thence:

To Port Louis (20°08′S, 57°28′E).

3   The low-powered route from Mauritius to Aden is given at 9.43.

Aden ⇒ Cape of Good Hope

Passage
9.141

1   During the south-west monsoon (April to October), pass N of Suquṭrá (12°30′N, 54°00′E), then run through the south-west monsoon to cross the equator at about 72°E, or even to run through the One and Half Degree Channel (1°24′N, 73°20′E) and make to the S into the South-east Trade Winds, passing E of Chagos Archipelago (6°30′S, 72°00′E). Run through the South-east Trade Winds, passing S of Mauritius (20°10′S, 57°30′E) and about 100 miles S of Madagascar and make the African coast about 200 miles S of Durban (29°51′S, 31°06′E). Thence keep in the strength of the Agulhas Current (6.28) until abreast of Mossel Bay (34°11′S, 22°09′E), thence proceed direct round Cape Agulhas (34°50′S, 20°01′E). With W winds, after passing Algoa Bay (34°00′S, 26°00′E), keep within 40 or 50 miles of the coast. For further details, see 9.153.

2   **Caution**. Abnormal waves have occurred off the SE coast of South Africa, see 1.19 and 6.36 for details.

3   During the north-east monsoon (October to March), work along the Arabian coast until able to weather Ras Caseyr (11°50′N, 51°17′E), then run down the coast of Africa and through the Moçambique Channel, taking full advantage of the Moçambique and Agulhas Currents (6.28).

4   For further details, see 9.26 and 9.153.

Aden ⇒ Mombasa or Seychelles Group

Sailing ship passage
9.142

1   During the south-west monsoon (April to October), pass N of Suquṭrá (12°30′N, 54°00′E), then stand away to the SE on the starboard tack and cross the equator in about 70°E, or as far W as the monsoon permits.

2   The South-east Trade Wind will be met with, after passing through the Doldrums in about 2°S to 4°S, and having picked it up steer directly for the Seychelles (4°30′S, 55°30′E) if calling there, or towards Mombasa (4°05′S, 39°43′E). Allowance must be made for the probability of the wind heading and for the strong N-going current which will be encountered on nearing the African coast.

3   During the north-east monsoon (October to March), proceed as directed at 9.141, for that season, but heading for the desired port when it can be reached in the South-east Trade Wind.

Low-powered vessel passage
9.143

1   Low-powered, or hampered, vessels bound for the Seychelles, may find the following routes more suitable than those described at 6.79 for full-powered vessels.

2   **October to March**:

Along the Arabian coast until able to weather Raas Caseyr (11°50′N, 51°17′E), thence:

Direct to Mahé Island (4°35′S, 55°30′E).

3   **April to September**:

Round Raas Caseyr (11°50′N, 51°17′E), thence:

Through 3°N, 60°E, thence:

S across the equator and into the South-east Trade Wind, thence:

To Mahé Island (4°35′S, 55°30′E).

If the south-west monsoon is still blowing strongly in 3°N, 60°E, the SE course should be held until the monsoon's strength is lost before turning S.

4   Routes for low-powered vessels proceeding from Mahé Island to Aden are described at 9.170.

PASSAGES FROM THE WEST COAST OF INDIA AND SRI LANKA

Charts:

5126 (1) to (12) Monthly Routeing Charts for Indian Ocean.

5309 Tracks followed by Sailing and Auxiliary Powered Vessels.

5310 World Surface Current Distribution.

Diagram 9.3 World Sailing Ship Routes.

Karāchi ⇒ Mumbai

Passage
9.144

1   From Karāchi (24°46′N, 66°57′E) proceed direct, but in June, July and August get offshore, into depths of 27 m to 36 m, before standing S. Mumbai (18°51′N, 72°50′E) should be approached on the parallel of Kanhoji Angre (Khānderi) Island (18°42′N, 72°49′E), and the soundings should be carefully checked. There is considerable indraught into the Gulf of Kachchh (22°40′N, 69°30′E) from March to September.

Karāchi ⇒ Mombasa

Low-powered vessel passages (SW monsoon)
9.145

1   For low-powered, or hampered, vessels, in the full strength of the south-west monsoon, a more S route, may be preferred to those for full-powered vessels, described at 6.41.3.

2   After leaving the Indian coast, pass:

Through 12°50′N, 70°00′E, thence:

Through 6°00′N, 67°00′E, thence:

Direct to Mombasa (4°05′S, 39°43′E).

Karāchi ⇒ Moçambique Channel, Durban and Cape Town

Low-powered vessel passages (SW monsoon)
9.146

1   For low-powered, or hampered, vessels, in the full strength of the south-west monsoon, a more S route, may be preferred to those for full-powered vessels described at 6.42.3. Moçambique

2 After leaving the Indian coast, pass:

Through 12°50′N, 70°00′E, thence:

6°00′N, 67°00′E, thence:

Rejoining the route for full-powered vessels, described at 6.42.1, in 3°00′S, 54°00′E.

Karāchi or Mumbai ⇒ Mauritius

Low-powered vessel passages (SW monsoon)
9.147

1 For low-powered, or hampered, vessels, in the full strength of the south-west monsoon, more S routes, may be preferred to those for full-powered vessels, described at 6.68 and 6.70.

The routes from Karāchi (24°46′N, 66°57′E) or Mumbai (18°51′N, 72°50′E) pass:

E of Lakshadweep (12°00′N, 72°30′E), thence:

E of Maldives (4°00′N, 73°00′E), thence:

E of Chagos Archipelago (6°30′S, 72°00′E), thence:

Direct to Port Louis (20°08′S, 57°28′E).

Mumbai ⇒ Karāchi

Passage
9.148

1 In May and early June, on leaving Mumbai (18°51′N, 72°50′E), make to the W so as to be able weather Diu Head (20°41′N, 70°50′E) by 100 miles if bound into the Gulf of Kachchh (22°40′N, 69°30′E), or by 200 miles, if for Karāchi (24°46′N, 66°57′E). During June, July and August, when bound for Karāchi, be careful not to make the coasts of Sind and Kachchh before sighting Manora Point Light-structure (24°47′N, 66°59′E), as there is a SE-going set, and the wind is liable to lull occasionally inshore, leaving the vessel with a heavy swell and lee current.

2 In the first part of the south-west monsoon (May and June), the stream setting into the Gulf of Kachchh, during the flood, is greatly accelerated.

3 In September and October, also March and April, when NW winds are general, work directly towards Diu Head then along the coast. In November, December, January and February, work along the coast with the land and sea breezes, making due allowance for the tides, sighting the High Land of Saint John (Sunjan) (20°04′N, 72°50′E), or reaching the parallel of 20°N, before crossing to Diu Head, as the wind hangs much to N and NNE across the Gulf of Khambhāt (20°30′N, 72°00′E).

4 November is a calm month along the S coast of the peninsula separating the Gulfs of Kachchh and Khambhāt and it is frequently necessary to anchor on the in-going stream to avoid being swept into the Gulf of Khambhāt.

5 From November to January, when fresh NE winds blow outside the Gulf of Kachchh, and when working into it, anchor during the out-going stream off Dwārka (22°15′N, 68°58′E) or Kachchigadh (22°20′N, 68°58′E) and start, with the in-going stream, across the mouth of the gulf to make the Kachchh coast, where the water is smoother and a vessel can work to the E.

Mumbai, Cochin, Calicut or Malabar coast ⇒ Aden

Passage
9.149

1 From May to September, during the south-west monsoon, this passage is seldom taken; but in case of necessity it is given as follows, by what is known as the "Southern Passage".

2 Gain an offing from the Indian coast into depths of between 27 m and 36 m, or even to 75 m in the first part of south-west monsoon, as the wind then is more in the S quarter. Then steer down the coast, keeping in soundings of between 75 m to 90 m, this is advisable to keep clear of Lakshadweep (12°00′N, 72°30′E) in the thick, overcast, rainy weather that may be expected, and observations unobtainable. After passing Lakshadweep try to avoid being set farther E if possible. The wind will be from SW to WSW with hard W squalls and a SE-going current of 20 to 30 miles a day will be experienced.

3 Cross the equator and, when fairly in the South-east Trade Wind, run to the W, passing S of Chagos Archipelago (6°30′S, 72°00′E) and NE of the Seychelles (4°30′S, 55°30′E). Recross the equator in about 53°E or 54°E. Run through the south-west monsoon and make the African coast between Raas Xaafuun (10°26′N, 51°25′E) and Raas Caseyr (11°50′N, 51°17′E), due consideration being given the the strong NE-going current which will be experienced on nearing land. Pass close round Raas Caseyr and keep along the African coast up to Jasiired Maydh (11°14′N, 47°13′E), and then stand across the Gulf of Aden.

4 **Caution** is necessary when rounding Raas Caseyr from S or SE during the south-west monsoon, for more details, see 9.150.

5 From October to April, during the north-east monsoon, proceed direct but towards the end of the monsoon, in March and April, the winds are less constant in the Arabian Sea than in the four preceding months, and there are calms at times. In these months steer to pass S of Suquṭrá (12°30′N, 54°00′E); for, early in April, the north-east monsoon is nearly expended about this island and on the coast of Arabia and is succeeded by light breezes from SW and W with frequent calms. The current also begins to set strongly to the N about Suquṭrá, and between it and the coast of Africa. From about the end of March, therefore, it is advisable to pass about 50 miles S of the island in order to reach Raas Caseyr with the SW winds that may then be expected.

6 Leaving Mumbai (18°51′N, 72°50′E) late in April, shape a course to pass well S of Suquṭrá, in order to make the coast of Africa S of Raas Caseyr with the SW wind, which will probably be met with long before that shore is approached. The land may then be made anywhere between Raas Xaafuun and Raas Caseyr, and the remainder of the passage made, as directed above, for the south-west monsoon.

7 From November to February, sailing vessels bound to the Red Sea from Cochin (9°58′N, 76°14′E), Calicut (11°15′N, 75°46′E) and other ports on the Malabar coast, may steer directly W through the most convenient channel through Lakshadweep. Those from Cochin should pass through Nine Degree Channel (9°00′N, 72°30′E), but vessels from Mangalore (12°50′N, 74°50′E) or Cannanore (11°52′N, 75°22′E) should pass N of all the islands. In March and April, the prevailing winds between the Malabar coast and the African coast are from N to NW, so it is better to keep near the Malabar coast until N of Mount Dilli (12°01′N, 75°12′E) and to pass N of the islands. However, if the Nine Degree Channel is adopted, vessels should pass near Kalpeni (10°06′N, 73°39′E) and Suheli Par (10°03′N, 72°15′E), as the current sets S towards the Maldives (4°00′N, 73°00′E) in these months.

8 When W of Lakshadweep, in the period November to February, a course may be made to pass N of Suquṭrá but, in late March or early April, it is prudent to keep farther S, in 9°N or 10°N as the wind will permit. In May, when the south-west monsoon may be expected, it is advisable to keep well to the S.

Caution when approaching Raas Caseyr
9.150

1 Many vessels have been wrecked to the S of Raas Caseyr. Therefore the utmost caution is necessary when rounding this headland from the S or SE, during the south-west monsoon, which is when the weather is usually stormy, accompanied by a heavy sea and strong current, with the land generally obscured by a thick haze.

2 By day there is usually a gradual change in the colour of the water, from blue to dark green, as the land is approached; the sea decreases and the swell alters its direction to the E of S, when N and W of Raas Xaafuun. When the land cannot be clearly seen and recognised, extreme caution is necessary.

3 After rounding Raas Caseyr, keep towards the African shore until Jasiired Maydh (11°14′N, 47°13′E) is reached, then steer for Aden (12°45′N, 44°57′E). Beating along the African shore against strong W and WSW winds is sometimes tedious, but perseverance is more likely to succeed here than in the middle of the Gulf of Aden or the Arabian shore.

4 Well found sails and rigging are essential, for the wind frequently blows in severe gusts along the African coast.

Low-powered vessel passage
9.151

1 For low-powered, or hampered, vessels, in the full strength of the south-west monsoon, the following route, may be preferred to that for full-powered vessels described at 6.56.2.

From Mumbai the route is to:
   6°00′N, 67°00′E, thence:
   6°00′N, 60°00′E, thence:
   8°00′N, 52°40′E, thence:
   Between Raas Caseyr (11°50′N, 51°17′E) and ’Abd al Kŭrī (12°10′N, 52°15′E) (6.38.3), thence:
   Direct to Aden (12°45′N, 44°57′E).

Mumbai ⇒ Mombasa or Seychelles

Low-powered vessel passage
9.152

1 For low-powered, or hampered, vessels, in the full strength of the south-west monsoon, the following routes, may be preferred to those described for full-powered vessels at 6.52.1 or 6.80.

From Mumbai (18°51′N, 72°50′E), the route is:
   Through 6°00′N, 67°00′E, thence:
   Either direct to Mombasa (4°05′S, 39°43′E).
   Or cross the equator in 59°E, thence:
   To Mahé Island (4°35′S, 55°30′E).

Mumbai ⇒ Cape of Good Hope

Passage
9.153

1 From Mumbai (18°51′N, 72°50′E), from May to September, stand down the coast of India (for further details, see 9.155 and 9.156) and across the equator into the South-east Trade Wind, then steer to pass S of Mauritius (20°10′S, 57°30′E) and about 100 miles S of Madagascar and make the African coast about 200 miles S of Durban (29°51′S, 31°06′E). Thence keep in the strength of the Agulhas Current (6.28) until abreast of Mossel Bay (34°11′S, 22°09′E), thence proceed direct round Cape Agulhas (34°50′S, 20°01′E).

2 **Caution**. Abnormal waves have occurred off the SE coast of South Africa, see 1.19 and 6.36 for details.

3 In the early part of the monsoon (June and July), when the wind is more S than it is later, get offshore from Mumbai into about 90 m of water before standing down the coast, keeping in soundings of between 75 m to 90 m, to ensure being well inshore of Lakshadweep (12°00′N, 72°30′E).

4 In April and October the route is similar, but somewhat to the W and, in April, a considerable shortening can usually be effected by making a direct course from 15°S, 70°E to 30°S, 40°E, where the former course can be picked up.

5 From November to March, there are two routes for the first part of this passage, one leading E, and the other W, of the Comores (12°00′S, 44°30′E). The two routes rejoin in about 20°S and thence continue to Cape of Good Hope (34°21′S, 18°30′E).

6 To follow the route E of the Comores, proceed direct from Mumbai, W of the Seychelles (4°30′S, 55°30′E) and Amirante Isles (5°30′N, 53°20′E) and between Madagascar and the Comores on a rhumb line towards the African coast at Durban. Thence keep in the strength of the Moçambique Current and Agulhas Current. In rounding the Cape of Good Hope, if W winds prevail, keep on Agulhas Bank (35°30′S, 21°00′E) not more than 40 or 50 miles from the coast, where smoother water will be found.

7 A route passing W of the Comores is recommended by some navigators on account of a rather better current on the African side of the Moçambique Channel (17°00′S, 41°30′E). A vessel using this route would sail direct from Mumbai, as directed for the E route, above and, keeping on the African side of the channel, proceed S as for the E route.

8 When approaching the Moçambique Channel from the N, keep well off the land until up to Cabo Delgado (10°41′S, 40°38′E), as the wind sometimes hangs to the E or even S of E; thence stand down the coast, inside Saint Lazarus Bank (12°08′S, 41°22′E), keeping in the strength of the Moçambique and Agulhas Currents to reach Cape Agulhas (34°50′S, 20°01′E). A vessel will probably have to work to windward in the S part of the Moçambique Channel, as the prevailing winds there are S.

9 June, July and August are the worst months and January and February the best months for sailing vessels proceeding W-bound around the Cape of Good Hope. It should be borne in mind that the sea is less rough over the Agulhas Bank in depths of 110 m to 130 m, or less, during heavy gales than it is near its edge or S of it. If it is found necessary to heave-to, the port tack should be chosen as, with the exception of SE gales, any shift of wind is almost invariably counter-clockwise and the vessel will come up to the sea.

10 From October to April, E winds prevail as far S as the tail of Agulhas Bank in about 37°S, with variable, but chiefly W winds beyond.

11 Mariners should remember that off all parts of the S coast of Africa, especially off salient points, sunken wrecks or uncharted dangers may lie close inshore and it is not advisable, even for power-driven vessels, to approach this surf-beaten coast closer than 3 or 4 miles; sailing vessels should give Cape Agulhas a berth of 7 or 8 miles.

Low-powered vessel passage
9.154

1   For low-powered, or hampered, vessels, proceeding to Durban or Cape Town, in the full strength of the south-west monsoon, the following route, may be preferred to that described for full-powered vessels at 6.46.1.

From Mumbai, the route is to:

6°00′N, 67°00′E, thence:

Joining the direct route for full-powered vessels (6.46.1) in 3°00′S, 54°00′E, to the NW of the Seychelles Group (4°30′S, 55°30′E).

Mumbai ⇒ Colombo

Passage
9.155

1   At the outset of the south-west monsoon, when the wind sets to the SW, stand well offshore from Mumbai (18°51′N, 72°50′E) into depths of 70 m to 90 m of water before standing down the coast, keeping on the edge of the bank in soundings of between 70 m to 90 m, to keep clear of Lakshadweep (12°00′N, 72°30′E). On proceeding S the wind will generally become more favourable, veering to W or WNW, between Cochin (9°58′N, 76°14′E) and Cape Comorin (8°05′N, 77°33′E) S-going currents and WNW winds prevail from mid-July to mid-October.

2   October to May is the period of the north-east monsoon and of the land and sea breezes along the W coast of India. A summary of the weather that may be expected and advice for sailing vessels to make full use of these breezes is given in the following paragraph (9.156).

Land and sea breezes off the west coast of India
9.156

1   Except during the south-west monsoon, land and sea breeze effects are usually well developed near the coast, but the strength and duration of the land winds may be modified by the mountainous nature of the hinterland.

2   Off the Konkan coast, the south-west monsoon fails after the middle of September and is followed by light variable breezes, frequent calms, cloudy weather and occasional showers. This unsettled weather lasts for six to eight weeks, with prevailing winds from the NW but, occasionally from SW or S. On the Malabar coast there are occasional off-shore squalls.

3   Late in October, or early in November, a transitional storm may take place, with a high wind suddenly coming up from the S, blowing hard for several hours, accompanied by thunder and lightning. After this the north-east monsoon sets in with fine weather and land and sea breezes are experienced within 10 or 20 miles of the coast, these continue until March or April.

4   The sea breezes of the Malabar coast are fairly established throughout October, while the land winds are only occasional, light and uncertain. The sea breezes seldom fail, until they are merged into the south-west monsoon. The navigator, therefore, can calculate on sea breezes for eight months of the year, but for only half that period for regular land winds.

5   When sea and land breezes are regular, the sea breeze fails in the evening about sunset and is generally followed by a calm which continues until the land wind commences at about 2000 to 2200 hours.. At first it comes in fluctuating gentle breezes, but soon steadies from between NE and ESE and continues until 0900 or 1000 hours, it then begins to fail, decreasing to to a calm about mid-day. About this time, or soon after, the sea breeze sets in from WSW, W or NW and generally veers towards N in the evening, decreasing in strength.

6   In March and April, off the coast of Maharashtra, the land breezes are very light and uncertain, seldom starting until morning and continuing for so short a time that little advantage can be gained from them. It is therefore necessary to keep an offing in order to be ready for the sea breeze; which, at this time are usually NW between Mumbai and Cape Comorin, setting in about noon from WNW, veering gradually to NW and NNW in the evening, from which direction they continue during the first part of the night, declining afterwards to a calm about midnight, or early in the morning. A faint land breeze sometimes follows but, more frequently, light airs from the N or calms may be expected from just before midnight until the NW wind sets in about noon the following day.

7   In April the weather is mostly hazy and, at times, cloudy over the mountains in the evenings with light showers.

8   In May the prevailing winds along the coast, S of Mumbai, are from NW and W, but often variable and uncertain with cloudy threatening weather and light showers at times, accompanied by lightning from the SE. A gale from SW or S is liable to occur in this month and ships have made a speedy run along the coast to Mumbai; but it is more prudent to keep well out from the land and be prepared for bad weather, in order to avoid being driven on a lee shore if a storm should set in from the W. When NW winds prevail the weather is settled and clear of clouds although a little hazy; but it is cloudy and threatening when they blow from SE to SW. It sometimes happens that heavy clouds collect over the land in the evenings, producing a hard squall, with rain, about midnight. This has frequently been experienced between Mangalore and Shirali (14°00′N, 74°29′E) in May, and early in June, when these land squalls blow in sudden gusts through the gaps between the mountains.

9   The land and sea breezes described above require attention for sailing vessels to benefit from them to the full extent. During the night, with the land breeze, it is prudent to keep well inshore, if the wind permits without tacking, for it is stronger and steadier there than farther out. In the morning it is advisable to edge farther out to 15 or 20 miles offshore, or to soundings of 50 m to 55 m, before noon, ready for the sea breeze. In the evening it is desirable to be near the shore, before the land breeze starts; the coast may be approached to a depth 18 m in most places between Mumbai and Quilon (8°53′N, 76°35′E) and if close inshore before the land breeze starts short tacks should be made near the shore until it comes off. When calm its approach is frequently indicated by the noise of the surf on the beach, which can be heard at a considerable distance.

10   During the period of change, before the south-west monsoon has set in, the small coasting vessels run into the nearest river or place of shelter S of Mumbai in the afternoon, but large vessels should ensure adequate sea-room.

Mumbai ⇒ Bay of Bengal
Passage
9.157

1   Proceed first as for Colombo, described at 9.155, but so as to round the S coast of Sri Lanka. From June to mid-January make to the E to the middle of the Bay of Bengal, but during the other half of the year keep on the W side of the bay, thence to destination. See also 9.21 and 9.22 for further details.

Colombo ⇒ Mumbai and west coast of India

Passage
9.158

1　During the height of the north-west monsoon do not attempt to work N along this coast. At other times, between May and September, opportunities may exist to do so, see 9.156 for details and also *West Coast of India Pilot*.

2　In September and October, the N-bound passage is very tedious; on the S part of the coast a strong current sets strongly to the S and the wind is NW or variable, with frequent light airs; vessels often have to anchor to avoid drifting back. The weather is threatening at times with heavy showers. The land winds begin to blow about the beginning of October, S of Calicut (11°15′N, 75°46′E), but do not extend far offshore until November.

3　From December to February, regular land and sea breezes render navigation N-bound near the coast easy, as the sea is remarkably smooth and the sea breeze is at its strongest.

4　Where there are gaps in the mountain chain, as at the Pālghāt Gap, on the parallel of 10°45′N, the land winds in December and January continue sometimes to blow for more than a day, without any intervening sea breeze. This occurs also, but in a rather lesser degree, off Kārwār Head (14°48′N, 74°05′E) where the valley of the Sadāshivgarh River assumes a straight funnel shape in an E-W direction. In these months a sailing passage may sometimes be made from Cape Comorin to Mumbai in 6 to 8 days, and the return journey in 4 or 5 days. In November and early in December the sea breezes are weak, but become stronger afterwards. As February advances, the land breezes decrease in strength and duration and are not always regular.

5　In March and April, the land breezes will generally fail in strength and duration N of Mount Dilli (12°01′N, 75°12′E); make certain, therefore, particularly in April, to be well to seaward, in depths of 65 m to 75 m, at about noon so that a long stretch to the NNE or NE, with the NW wind, may be made. If near the shore early in the evening, with the wind at NW, make short tacks until the breeze veers to the N, which may be expected early in the night; then stretch to the NW or WNW to be ready for the sea breeze the following day.

6　When a strong NW wind sets in, it is liable to continue for two or three days, or longer, rendering it impracticable to gain any ground when working near the coast. At such times keep 60 miles or more from the land, where the winds are generally moderate and the sea smooth.

7　Late in April, and during May, keep well offshore towards Lakshadweep (12°00′N, 72°30′E), and when to the N of those islands keep even farther offshore, in case of a gale coming on. On the S part of the coast, S of Mount Dilli, when meeting headwinds in April and May, head off to the W of the islands, passing between Suheli Par (10°03′N, 72°15′E) and Minicoy (8°18′N, 73°02′E), or through any other channel through Lakshadweep, in order to benefit from the approaching W winds.

Colombo ⇒ Aden

Sailing vessel passage
9.159

1　The passage is hardly ever undertaken during the south-west monsoon (April to September). In case of necessity, however, the directions are to stand at once across the equator into the South-east Trade Wind, thence run W passing S of the Chagos Archipelago (6°30′S, 72°00′E) and NE of the Seychelles (4°30′S, 55°30′E). Recross the equator in about 53°E or 54°E and shape course to make the African coast at Raas Xaafuun (10°26′N, 51°25′E); round Raas Caseyr and keep along the African coast up to Jasiired Maydh (11°14′N, 47°13′E), and then stand across the Gulf of Aden.

2　**Caution** is necessary when rounding Raas Caseyr from S or SE during the south-west monsoon, for more details, see 9.150.

3　From October to March, pass through Nine Degree Channel (9°00′N, 72°30′E), then proceed direct for Aden (12°45′N, 44°57′E). After mid-March, pass S of Suquṭrá (12°30′N, 54°00′E), as light SW and W breezes may then be expected near this island. See 6.38.2 for further information regarding the approaches to Suquṭrá.

Low-powered vessel passages
9.160

1　For low-powered, or hampered, vessels, proceeding to Aden, in the full strength of the south-west monsoon, the following routes, may be preferred to those described for full-powered vessels at 6.58. The choice of a route depends largely on the power and sea-keeping qualities of a vessel.

For medium-powered vessels.
The route is:
　Through Eight Degree Channel (7°24′N, 73°00′E) on the parallel of 7°30′N, thence:
　Through 8°00′N, 60°00′E, thence:
Either:
　Through 8°00′N, 52°40′E and round Raas Caseyr (11°50′N, 51°17′E) (6.38.3), thence:
　Direct to Aden (12°45′N, 44°57′E).
Or:
　Through 13°00′N, 55°00′E and N of Suquṭrá, thence:
　Direct to Aden (12°45′N, 44°57′E).

Distances (S of Suquṭrá)
Colombo	2080 miles.
Dondra Head*	2140 miles.

Distances (N of Suquṭrá)
Colombo	2110 miles.
Dondra Head*	2170 miles.

* (10 miles S of).

For low-powered vessels.
There are alternative routes:
Either:
　Through Eight Degree Channel (7°24′N, 73°00′E), thence:
　Through 6°00′N, 67°00′E, thence:
　Through 6°00′N, 60°00′E, thence:
　Through 8°00′N, 52°40′E and round Raas Caseyr (11°50′N, 51°17′E) (6.38.3), thence:
　Direct to Aden (12°45′N, 44°57′E).
Or:
Provided that Olivelifuri (5°17′N, 73°35′E), the islet marking the N side of the entrance to Kaashidoo Channel, can be made between sunrise and noon, the route is:

Through Kaashidoo Channel (5°05′N, 73°24′E), thence:

Through 4°44′N, 60°00′E, thence:

Through 8°00′N, 52°40′E and round Raas Caseyr (11°50′N, 51°17′E) (6.38.3), thence:

Direct to Aden (12°45′N, 44°57′E).

For all small vessels.

The route is:

Through One and Half Degree Channel (1°24′N, 73°20′E), thence:

Through 2°00′N, 60°00′E, thence:

Through 8°00′N, 52°40′E and round Raas Caseyr (11°50′N, 51°17′E) (6.38.3), thence:

Direct to Aden (12°45′N, 44°57′E).

Colombo ⇒ Mombasa

Low-powered vessel passages
9.161

1 For low-powered, or hampered, vessels, proceeding to Mombasa (4°05′S, 39°43′E), in the full strength of the south-west monsoon, the following route, which crosses the equator into the South-east Trade Wind, may be preferred to those described for full-powered vessels at 6.54.

The route passes:

S of Diego Garcia (7°13′S, 72°23′E), thence:

As navigation permits to Mombasa (4°05′S, 39°43′E).

Colombo ⇒ Cape of Good Hope

Sailing ship passage
9.162

1 Pick up the route from Mumbai, described at 9.153, according to the time of year at the nearest available point; passing through Nine Degree Channel during the north-east monsoon (November to March), but directly to meet the May to September passage at about the equator.

Low-powered vessel passage
9.163

1 For low-powered, or hampered vessels, proceeding to Durban or Cape Town, in the full strength of the south-west monsoon, the following route may be preferred to those described for full-powered vessels at 6.48.

Masters of low-powered vessels, during the south-west monsoon, should consider a diversion to the S, by a route from the coast of Sri Lanka across the equator into the South-east Trade Wind, passing:

S of Diego Garcia (7°13′S, 72°23′E), thence:

To the N end of Moçambique Channel, thence:

As at 6.37 and 6.36 to destination.

Colombo ⇒ Fremantle, Cape Leeuwin and south and south-east Australia or New Zealand

Passage
9.164

1 From April to October, having rounded the S point of Sri Lanka, steer to the SE to cross the equator in about 95°E; then proceed S across the South-east Trade Wind into the W winds for a direct passage to Fremantle (32°02′S, 115°42′E) or to round Cape Leeuwin (34°23′S, 115°08′E). For passage to the E of Cape Leeuwin proceed as directed at 9.138.

2 From November to March, make as far to the E as possible in the north-west monsoon, and then proceed S across the South-east Trade Wind as above.

Colombo ⇒ Malacca Strait

Passage
9.165

1 In the south-west monsoon proceed direct to pass S of Great Nicobar Island (7°00′N, 93°50′E).

2 In the north-east monsoon, stand S to about 3°N, and then work NE towards the NW end of Sumatera, entering the Malacca Strait S of Great Nicobar Island.

Malacca Strait

References
9.166

1 Rules for vessels navigating through Malacca Strait and Singapore Strait are given in *Malacca Strait and West Coast of Sumatera Pilot* and Chart *5502 Mariners' Routeing Guide — Malacca and Singapore Straits* should also be consulted. See also 6.61 for details of the N approach channels to Malacca Strait.

Passage for sailing vessels
9.167

1 From April to October, after passing the NW end of Sumatera, the south-west monsoon will probably fail and it is advisable then to keep to the Malaysian side of the channel for better breeze and tidal streams. Sometimes a brisk wind will be carried as far as Pulau Pinang (5°20′N, 100°15′E) and, once the islands off the Malaysian coast have been sighted, there will be no difficulty in making to the S.

2 The winds on the E side of the Strait tend to be more favourable for a S-bound passage from October to March.

3 During the north-east monsoon (October to March), a sailing vessel N-bound should, after passing Pulau Pangkor (4°15′N, 100°35′E), keep near the edge of the mud flat that fronts the coast in order to avoid the strong wind and short sea likely to be encountered offshore near Pulau Pinang.

4 Directions for making Singapore (1°12′N, 103°51′E) are given at 9.83.

PASSAGES TO AND FROM SEYCHELLES GROUP

Charts:

5126 (1) to (12) Monthly Routeing Charts for Indian Ocean.

5309 Tracks followed by Sailing and Auxiliary Powered Vessels.

5310 World Surface Current Distribution.

Diagram 9.3 World Sailing Ship Routes.

Seychelles ⇐ ⇒ Mauritius

Low-powered vessel passage
9.168

1 For low-powered, or hampered vessels, proceeding between the Seychelles and Mauritius, the following routes

may be preferred to those described for full-powered vessels at 6.69.

Seychelles to Mauritius

November to March:
>E of Saya de Malha Bank (10°00′S, 61°00′E), thence:
>S until well into the South-east Trade Wind, thence:
>To Port Louis (20°08′S, 57°28′E).

April to October
>E to about 70°E, thence:
>S until well into the South-east Trade Winds, thence:
>Direct to Port Louis (20°08′S, 57°28′E).

Mauritius to Seychelles

November to March:
>W of the direct route until in the north-west monsoon.

April to October:
>As direct as navigation permits.

Seychelles ⇐ ⇒ Mombasa

Low-powered vessel passages
9.169

1   For low-powered, or hampered, vessels, proceeding between the Seychelles and Mombasa, the following routes may be preferred to those described for full-powered vessels at 6.78.

Mombasa to Seychelles

April to October:
>If unable to make Mahé Island (4°35′S, 55°30′E) by the direct route, the E course should be held until past the Seychelles Group (4°30′S, 55°30′E) and the island can be approached from the N.

October to April:
>The route is close N of the direct route until the north-west monsoon is picked up in about 45°E.

Seychelles to Mombasa

April to October:
>The route is direct, but allowance should be made for the probability of a head wind and the strong N-going current setting along the African coast.

October to April:
>The route is direct.

Seychelles ⇒ Aden

Low-powered vessel passages
9.170

1   For low-powered, or hampered, vessels, proceeding from the Seychelles to Aden from November to March, the following route may be preferred to those described for full-powered vessels at 6.79.

>Cross the equator in about 61°E, thence:
>>Into the north-east monsoon until able to weather Suquṭrá.

2   **Caution** must be exercised when approaching Suquṭrá, see 6.38.2 for further details.

3   Routes for low-powered vessels proceeding from Mahé Island to Aden are described at 9.143.

Seychelles ⇐ ⇒ Colombo

Low-powered vessel passages
9.171

1   For low-powered, or hampered, vessels, proceeding between the Seychelles and Colombo, the following routes may be preferred to those described for full-powered vessels at 6.81.

Seychelles to Colombo

November to March:
>Through 4°S, 70°E, thence:
>Across the equator in 80°E, thence:
>N to make the coast of Sri Lanka in 80°E, thence:
>To Colombo.

April to October:
>Through Eight Degree Channel (7°24′N, 73°00′E), or
>Through Kaashidoo Channel.

The latter is more direct but advisable by day only.

Colombo to Seychelles

October to April:
>Through Eight Degree Channel (7°24′N, 73°00′E), thence:
>SW to the equator in 54°E, thence:
>To Mahé Island (4°35′S, 55°30′E).

May to September:
>Establish a good offing from the coast of Sri Lanka, thence:
>S across the equator into the South-east Trade Winds, thence:
>S of Diego Garcia, thence:
>To Mahé Island (4°35′S, 55°30′E).

PASSAGES FROM PORTS IN THE BAY OF BENGAL

Charts:
>*5126 (1) to (12) Monthly Routeing Charts for Indian Ocean.*
>*5309 Tracks followed by Sailing and Auxiliary Powered Vessels.*
>*5310 World Surface Current Distribution.*
>*Diagram 9.3 World Sailing Ship Routes.*

Notes on navigation under sail in the Bay of Bengal
9.172

1   There is no difficulty proceeding from S to N or from W to E in the Bay of Bengal during the south-west monsoon; nor from N to S or from E to W during the north-east monsoon. See 9.21 for further details.

2   When the monsoon is contrary, a sailing vessel must work as necessary for the passage. At the change of the monsoon, voyages are usually tedious, for the light and variable winds, then prevalent, are as often adverse as favourable. Every change should be taken advantage of and the NE part of the Bay avoided, unless bound to or from one of the ports there.

3   **Cyclones**, as stated at 6.16, occur from May to November, with May, June October and November as the months of greatest frequency. They occur very occasionally in March, April and December and are almost unknown in January, and entirely so in February. See, *The Mariner's Handbook* and *Bay of Bengal Pilot* for further details.

4   If warning of a storm in the N part of the bay is given by E winds and a falling barometer between June and September; or by a squally E or NE wind driving low long-drawn masses of cloud before it; or a strong W-going current at the head of the bay in May, October or November, a vessel in the Hugli River (22°00′N, 88°00′E) or a port at the head of the bay, should remain in harbour until the weather moderates.

5 If at sea in the right-hand semicircle, the vessel should be hove-to on the starboard tack until the storm has passed; or if undoubtedly in the left-hand semicircle she should heave-to on the port tack if the wind is E of N, or run to the S keeping the wind on the starboard quarter when the wind is N or W of N.

6 Vessels lying in the roadsteads of the Coramandel coast, on the approach of a cyclonic storm, usually run in a S direction round the SW quadrant, and this is probably the only course open to sailing vessels.

Chennai ⇒ Kolkata

Passage
9.173

1 From April to August, proceed as directly as possible, making landfall about Bāvanapādu (18°34′N, 84°21′E).

2 In September and October, head over to North Andaman Island (13°15′N, 93°00′E) or Cape Negrais (16°03′N, 94°12′E); when 100 miles W of either, tack to the NW.

3 From November to January, make to the E across the bay, then towards the N on the E side or in the middle of it.

4 In February and March, steer direct, if possible; otherwise stand to the E across the bay as for November to January.

Chennai ⇒ Yangon, Moulmein or Mergui

Passage
9.174

1 During the south-west monsoon, sight Landfall Island (13°39′N, 93°01′E) if with a S wind, or Great Coco Island with a W wind. Pass through the Coco Channel and thence to the E, sighting Narcondam Island (13°25′N, 94°15′E). Then as directed by *Bay of Bengal Pilot* for Yangon (16°46′N, 96°11′E) or Moulmein (16°30′N, 97°38′E). If bound for Mergui (12°26′N, 98°36′E), pass S of Little Andaman Island (10°40′N, 92°30′E) and then steer for Tanangthayi Kyun (12°35′N, 97°51′E).

2 During the north-east monsoon, make to the N in the middle of the Bay and pass through Preparis North Channel (15°00′N, 93°40′E) or Preparis South Channel (14°33′N, 93°27′E), and then as directed by *Bay of Bengal Pilot* for Yangon or Moulmein, sounding continuously and allowing for tidal streams. If bound for Mergui, pass N of Andaman Islands and then work to the E to pass Mali Kyun (Tavoy Island) (13°05′N, 98°17′E) on either side. See also 9.21 and 9.22 for further details.

Bay of Bengal ⇒ Mumbai

Passage
9.175

1 This passage is seldom undertaken in the south-west monsoon. A vessel should first stand S across the equator into the South-east Trade Wind and then run W between 8°S and 9°S, passing S of Chagos Archipelago (6°30′S, 72°00′E). From 70°E, steer to recross the equator in 62°E or 63°E and sail to Mumbai (18°51′N, 72°50′E) direct.

2 During the north-east monsoon, steer as directed in 9.21 and round Sri Lanka at a convenient distance. After passing Cape Comorin (8°05′N, 77°33′E) keep the W coast of India in sight, so as to profit from the sea breezes described at 9.156.

Bay of Bengal ⇒ Aden

Passage
9.176

1 This passage is seldom undertaken in the south-west monsoon. A vessel should first stand S across the equator into the South-east Trade Wind and then run W between 8°S and 9°S, passing S of Chagos Archipelago (6°30′S, 72°00′E) and NE of the Seychelles (4°30′S, 55°30′E). Recross the equator in about 53°E or 54°E and shape course to make the African coast at Raas Xaafuun (10°26′N, 51°25′E); great caution is necessary in making the land. Round Raas Caseyr and keep along the African coast up to Jasiired Maydh (11°14′N, 47°13′E), and then stand across the Gulf for Aden (12°45′N, 44°57′E).

2 In the north-east monsoon, pass round Sri Lanka and through Nine Degree Channel (9°00′N, 72°30′E); thence steer to pass N of Suquṭrá (12°30′N, 54°00′E). After the middle of March pass S of Suquṭrá.

3 Directions for crossing the Arabian Sea and making landfall, see 9.149 and 9.150.

Bay of Bengal ⇒ Cape of Good Hope

Passage
9.177

1 From May to September, vessels leaving The Sandheads (20°54′N, 88°14′E) should make for the Orissa coast, sighting the land S of False Point (20°20′N, 86°44′E) and working to the SW along the shore; make short tacks during the day and long boards offshore during the night, leave the coast, when abeam of Kalingapatnam (18°19′N, 84°08′E), and stand down the bay. A comparatively smooth sea and a favourable current will be found near the shore and advantage may be taken of a veering wind in the squalls off the land.

2 When heading down the Bay of Bengal in the south-west monsoon, keep well W of the Andaman Islands, in order not be on a lee shore should a strong W gale set in. An alternative, and better, route passes through the Preparis North Channel (15°00′N, 93°40′E) or Preparis South Channel (14°33′N, 93°27′E) then work S in the comparatively smooth water E, and leeward, of the Andaman and Nicobar Islands. Fast sailing vessels from Kolkata, in the south-west monsoon, do beat down the Bay of Bengal, reaching 100 miles W of the Andaman Islands, but wear and tear is great and the time saved only slight.

3 From Chennai (13°07′N, 80°20′E) or the Coromandel coast, head directly across the equator into the South-east Trade Wind.

4 In either of the above cases and, from all parts of the Bay, head S so as to cross the equator in about 95°E, keeping on whichever tack provides greater S progress into the South-east Trade Wind.

5 Cross the meridian of 90°E in 10°S and, from this position, steer a direct course for Cape Agulhas (34°50′S, 20°01′E), passing about 200 miles S of Rodriguez Island (19°38′S, 63°25′E) and the same distance S of Madagascar. Make the African coast in about 33°S and keep in the strength of the Agulhas Current until abeam of Mossel Bay (34°11′S, 22°09′E), then round Cape Agulhas as described at 9.153.

6 From November to March, run straight down the Bay of Bengal to cross the equator in 86°E to 87°E and pick up the May to September route at about 15°S, thence direct to Cape Agulhas, as described above.

7 During October and April, run down the Bay on a line just E of 90°E, cross the equator at 90°E and pick up the

May to September route at about 15°S, thence direct to Cape Agulhas, as described above.

Bay of Bengal ⇒ Fremantle, Cape Leeuwin and south and south-east Australia or New Zealand

Passage
9.178

1 From March to October, having worked along the W shore of the Bay of Bengal, as described at 9.177, far enough to weather the Nicobar Islands and the islands fronting the SW coast of Sumatera, stand out of the Bay on the starboard tack, cross the south-east monsoon and the South-east Trade Wind. Having reached the prevailing W winds S of the Trade Wind, proceed E for Fremantle (32°02′S, 115°42′E) or round Cape Leeuwin (34°23′S, 115°08′E). The doldrum belt will be found to extend to about 4°S.

2 From November to April, stand down the middle of the Bay of Bengal and cross the equator into the north-west monsoon. Then make to the E, in the north-west monsoon, as far as Christmas Island (10°24′S, 105°43′E); then stand across the South-east Trade Wind into the Westerlies, and so to Fremantle or Cape Leeuwin.

3 For destinations to the E of Cape Leeuwin, proceed as directed at 9.138.

Kolkata ⇒ Chennai or Sri Lanka

Passage
9.179

1 During the south-west monsoon, make to the S without closing the E side of the Bay, as directed at 9.177. Steer for destination when 60 miles S of it.

2 If unable to work S, pass E of Andaman and Nicobar Islands and through Great Channel (6°10′N, 94°10′E).

Thence work across the Bay of Bengal. In June, July and August stand across the equator into the South-east Trade Wind, make to the W and recross the equator in about 83°E, if bound for Chennai (13°07′N, 80°20′E), then proceed direct. If bound for Colombo (6°58′N, 79°50′E), recross the equator in about 77°E.

3 During the north-east monsoon, steer direct. In September, with light S winds, work SW, keeping in soundings or stand out to sea. Keep off the coast in February and March as the current then runs N.

Kolkata ⇒ Yangon, Moulmein or Mergui

Passage
9.180

1 During the south-west monsoon steer to pass through Preparis South Channel (14°33′N, 93°27′E) and then as directed in *Bay of Bengal Pilot*, for Yangon (16°46′N, 96°11′E) or Moulmein (16°30′N, 97°38′E). If bound for Mergui (12°26′N, 98°36′E) pass either side of Coco Islands.

2 During the north-east monsoon, steer to pass round Alguada Reef (15°42′N, 94°10′E) and then work E sounding frequently and making full allowance for tidal streams

Kolkata ⇒ Singapore

Passage
9.181

1 During the south-west monsoon proceed direct through Preparis South Channel (14°33′N, 93°27′E) and the Malacca Strait. See 6.61, 9.166 and 9.167 for directions for passage of Malacca Strait.

2 During the north-east monsoon, proceed through one of the Preparis Channels, thence direct through Malacca Strait

3 For directions for Singapore Strait, see 9.81 and 9.82.

PASSAGES FROM PORTS IN BURMA (MYANMAR)

Charts:
> 5126 (1) to (12) Monthly Routeing Charts for Indian Ocean.
> 5309 Tracks followed by Sailing and Auxiliary Powered Vessels.
> 5310 World Surface Current Distribution.
> Diagram 9.3 World Sailing Ship Routes.

Yangon or Moulmein ⇒ Kolkata

Passage
9.182

1 Burma is known to the Burmese as Myanmar.

2 During the south-west monsoon, pass through one of the Preparis Channels, thence proceed as directly as possible.

3 During the north-east monsoon, pass S of Alguada Reef (15°42′N, 94°10′E) and then proceed N, about 30 miles off the Burma coast, before proceeding across to the NE; but after January, from Alguada Reef, stand into the middle of the Bay before heading N.

4 Vessels intending to leave Yangon (16°46′N, 96°11′E) or Moulmein (16°30′N, 97°38′E) during periods of strong NE winds, with a falling barometer, denoting the existence of a cyclonic storm E of the Andaman Islands, should wait until the storm has passed. This is indicated by rising barometer, and the wind shifting to E or S of E. See also 9.172.

Yangon or Moulmein ⇒ Chennai

Passage
9.183

1 During the south-west monsoon, keep well out to sea if the wind becomes W and endeavour to sight Narcondam Island (13°25′N, 94°15′E).

2 In working S keep W of, and at a moderate distance from, the Mergui Archipelago. Pass S of Great Nicobar Island (7°00′N, 93°50′E) and then work W to destination.

3 During the north-east monsoon, pass through the Preparis North Channel (15°00′N, 93°40′E) and thence proceed as directly as possible. After January, however, make the land S of destination on account of the N-going sets which occur off this coast after this month.

Yangon or Moulmein ⇒ Malacca Strait and Singapore

Passage
9.184

1 During the south-west monsoon, proceed as directed at 9.183, passing the S point of Ko Phuket (7°55′N, 98°20′E); thence proceed through Malacca Strait.

2 During the north-east monsoon, keep outside the Mergui Archipelago, sight the S point of Ko Phuket and proceed thence direct through the Malacca Strait. See 6.61 9.166 and 9.167 for directions for passage of Malacca Strait.

3   For directions for Singapore Strait, see 9.81 and 9.82.

Yangon or Moulmein ⇒ Cape of Good Hope

Passage
9.185

1   In both monsoons, head S, keeping E of the Andaman and Nicobar Islands, and pick up the May to September route from the Bay of Bengal (described at 9.177) at 15°S, following it to destination.

Mergui ⇒ Kolkata

Passage
9.186

1   During the south-west monsoon, work to the W and pass through Coco Channel (13°50′N, 93°10′E) or one of the Preparis Channels.

2   During the north-east monsoon, pass through any channel N of the Andaman Islands (12°30′N, 93°00′E), then proceed direct.

Mergui ⇒ Chennai

Passage
9.187

1   During the south-west monsoon,, after clearing the islands, work S to the N end of Sumatera; pass through Great Channel (6°10′N, 94°10′E) or S of Great Nicobar Island (7°00′N, 93°50′E) and proceed thence direct to Chennai (13°07′N, 80°20′E).

2   During the north-east monsoon, until the end of January, pass through any channel N of the Andaman Islands (12°30′N, 93°00′E), thence direct. After January pass S of Little Andaman Island (10°40′N, 92°30′E).

PASSAGES SOUTH-BOUND OR WEST-BOUND FROM SINGAPORE OR EASTERN ARCHIPELAGO

Charts:

> 5126 (1) to (12) *Monthly Routeing Charts for Indian Ocean.*
> 5309 *Tracks followed by Sailing and Auxiliary Powered Vessels.*
> 5310 *World Surface Current Distribution.*
> Diagram 9.3 *World Sailing Ship Routes.*

Singapore ⇒ Chennai

Passage
9.188

1   During the south-west monsoon, keep along the N coast of Sumatera, pass through Great Channel (6°10′N, 94°10′E) and work across the Bay of Bengal. In the height of the south-west monsoon (June, July and August), however, from the NW end of Sumatera cross the equator and make to the W in the South-east Trade Wind, recrossing the equator at about 83°E.

2   During the north-east monsoon, keep on the Malaysian coast until Ko Phuket (7°55′N, 98°20′E) is reached, then pass through either the Ten Degree Channel (10°00′N, 92°30′E) or Sombrero Channel (7°36′N, 93°33′E). In December and January make land N of Chennai (13°07′N, 80°20′E) on account of the S-going set on the Coromandel and Sri Lanka coasts.

Singapore ⇒ Colombo

Passage
9.189

1   During the south-west monsoon, keep along the N coast of Sumatera, pass through Great Channel (6°10′N, 94°10′E) thence cross the equator and make to the W in the South-east Trade Trade Wind, recrossing the equator at 77°E and proceeding as directly as possible to Colombo (6°58′N, 79°50′E).

2   During the north-east monsoon, pass on either side of Pulau Perak (5°42′N, 98°56′E) and between Pulau Rondo (6°04′N, 95°07′E) and Great Nicobar Island (7°00′N, 93°50′E); thence proceed direct, but if W winds are experienced near the N end of Sumatera, which is probable in October and November, keep to the N before altering course to the W.

Singapore ⇒ Kolkata

Passage
9.190

1   During the south-west monsoon, pass to the S of the Nicobar Islands (8°00′N, 94°00′E) and then steer directly for the Orissa coast (20°00′N, 86°00′E).

2   During the north-east monsoon, up to mid-January, pass E of Andaman Islands (12°30′N, 93°00′E). After mid-January, pass S of the Andaman Islands, or through Duncan Passage (11°05′N, 92°45′E), and work to the N in the middle of the Bay of Bengal, as NW and W winds are then found N of the Andaman Islands.

Singapore ⇒ Yangon or Moulmein

Passage
9.191

1   During the south-west monsoon, sight Narcondam Island (13°25′N, 94°15′E).

2   During the north-east monsoon, sight Great Western Torres Islands (11°48′N, 97°30′E), thence proceed as directed in *Bay of Bengal Pilot.*

Singapore ⇒ Darwin

Passage
9.192

1   From April to October, in the south-west monsoon in the South China Sea and the east monsoon on the N coast of Australia, when bound from Singapore (1°12′N, 103°51′E) to Darwin (12°25′S, to 130°47′E), proceed through Balabac Strait (7°34′N, 116°55′E), across Sulu Sea, through Basilan Strait (6°50′N, 122°00′E) and Selat Bangka (1°45′N, 125°05′E), as described at 10.62.2, thence through Selat Manipa (3°20′S, 127°22′E) for Darwin, as described at 10.84.1.

2   From November to April, during the north-east monsoon of the South China Sea, proceed through Selat Sunda (6°00′S, 105°52′E), as described at 9.64, thence for Darwin with the north-west monsoon. Alternatively, go through Selat Karimata (1°43′S, 108°34′E) and Selat Sapudi (7°00′S, 114°15′E) and into the Indian Ocean by Selat Lombok (8°47′S, 115°44′E) or Selat Alas (8°40′S, 116°40′E), as described at 9.109 and 9.110.

Singapore ⇒ Torres Strait

Passage
9.193

1   From April to October follow the route for Darwin, given above (9.192) for that season, as far as the Ceram Sea to join the route from Hong Kong to Torres Strait (10°36′S, 141°51′E), described at 10.83, or its alternative.

2   From November to April, proceed via Selat Karimata (1°43′S, 108°34′E), thence to the Arafura Sea either via Selat Lombok (8°47′S, 115°44′E) or Selat Alas (8°40′S, 116°40′E), as described at 9.109 and 9.110; or through the Java Sea and Flores Sea to join the April to October route described above.

Singapore ⇒ Fremantle or southern Australia

Passage
9.194

1   From April to October, proceed S through Selat Bangka (2°06′S, 105°03′E) and Selat Sunda (6°00′S, 105°52′E), as described at 9.64, thence across the South-east Trade Wind until in the region of W winds, whence a course may be shaped for Fremantle (32°02′S, 115°42′E) or Cape Leeuwin (34°23′S, 115°08′E). See also 10.61, for further details.

2   From November to April, if bound for one of the W or S ports of Australia, proceed through Selat Bangka, as described at 9.64, N of Jawa and through Selat Bali (8°10′S, 114°25′E) or Selat Lombok (8°47′S, 115°44′E), thence steer to the S into the South-west Trade Wind. Keep the ship close hauled on the port tack in the Trade Wind and, on losing the Trade Wind, steer to the S and SE into the W winds, thence proceed as directed for the Cape of Good Hope to Australia route (9.5 to 9.9).

Singapore ⇒ Selat Sunda and Cape of Good Hope

Passage
9.195

1   First proceed to the Indian Ocean, via Selat Sunda, by one of the routes described at 9.114.

2   From April to October, having cleared Selat Sunda, proceed directly to a position about 200 miles S of Rodriguez Island (19°38′S, 63°25′E) then pass the same distance S of Madagascar and as directed at 9.47.

3   From October to April, after clearing Selat Sunda, stand S into the South-east Trade Wind, passing through 16°S, 90°E; steer thence for a position 200 miles S of Rodriguez Island, thence as above. Note that this period is the cyclone season for the South Indian Ocean.

Singapore or Selat Sunda ⇒ Aden

Passage for sailing vessels
9.196

1   The route from Singapore (1°12′N, 103°51′E) may be taken either via Malacca Strait (described at 9.166 or 9.167) or via Selat Sunda (described at 9.64).

2   Having passed through Malacca Strait take departure as directed at 9.189.

3   From April to September follow the route into the South-east Trade Wind and then make to the W to pass S of Chagos Archipelago (6°30′S, 72°00′E) to join the Colombo to Aden route described at 7.159.

4   From October to March proceed S of Sri Lanka and through Nine Degree Channel (9°00′N, 72°30′E) as described at 7.159.

5   If the route through Selat Sunda is taken a similar procedure should be adopted, namely to join the Colombo to Aden route (7.159) S of Chagos Archipelago or in the Nine Degree Channel, according to season.

Passage for low-powered vessels
9.197

1   For low-powered, or hampered vessels, proceeding between the Selat Sunda and Aden, the following seasonal routes may be preferred to those described for full-powered vessels at 6.113.

   November to March the route is:
   Between Pulau-pulau Mentawai (1°35′S, 99°15′E) and Sumatera, thence:
   Through Selat Siberut (0°45′S, 98°28′E), thence:
   Across the equator at 97°00′E, thence:
   To 1°50′N, 95°00′E, thence:
   To 5°00′N, 90°00′E, thence:
   To 5°30′N, 85°00′E, thence:
   Through TSS off Dondra Head (5°55′N, 80°35′E), thence:
   As at 6.58.1 to Aden (12°45′N, 44°57′E).
   April, May, June and September routes are:
   Through 8°00′S, 68°00′E, passing close S of Chagos Archipelago (6°30′S, 72°00′E), thence:
   To 8°00′N, 52°40′E, thence:
   Round Raas Caseyr (11°50′N, 51°17′E) to Aden (12°45′N, 44°57′E).
   July and August routes are:
   Through 2°30′S, 65°00′E, thence:
   To 1°10′S, 61°30′E, thence:
   To 8°00′N, 52°40′E, thence:
   Round Raas Caseyr (11°50′N, 51°17′E) to Aden (12°45′N, 44°57′E).

2   Routes for low-powered vessels proceeding from Aden to Selat Sunda are given at 9.137.

Selat Sunda north-bound along the west coast of Sumatera

Passage
9.198

1   The three routes, Outer, Middle and Inner are described in detail in *Malacca Strait and West Coast of Sumatera Pilot*.

2   For a sailing vessel, the voyage in either direction at all seasons, is long and difficult on account of frequent calms, but it is generally more difficult to work N, rather than S, owing to the prevalence of SE-going currents, which continue to set even with, or after, a S wind. January and February are the best months for going N, while from September to November vessels will often be compelled to keep far out to sea in order to gain even a small distance N, working inshore during this period is almost impracticable.

3   The Outer Route, to the W of all the islands, is best of the three, especially for sailing vessels. SW and S winds often prevail here, whereas NW squalls, variable baffling winds, calms and S-going currents may be experienced close to the land.

4   The Middle Route, passes between the chain of large islands lying offshore and the small islands adjacent and interspersed along the coast. It should not be followed by sailing vessels when N-bound, nor at any other time, if it can be avoided. Although it is wide and may be used day or night by vessels of light draught, when the weather is clear and favourable, they are more at the mercy of currents when the winds are light and baffling and there is no anchorage. In some parts there are dangerous coral shoals where soundings will give no warning of approach.

5   The Inner Route, close along the coast, and between some of the islands and dangers off it, should also seldom be chosen by N-bound sailing vessels in either monsoon; but, as there are many places of moderate depth for anchoring, it is preferable to the Middle Route.

Selat Bali, Selat Lombok, Selat Alas or Selat Ombi ⇒ Cape of Good Hope

Passage
9.199

1   From Selat Bali (8°10′S, 114°25′E), Selat Lombok (8°47′S, 115°44′E) or Selat Alas (8°40′S, 116°40′E) head to the SW, during the south-east monsoon or direct to the S during the north-west monsoon to pick up the South-east Trade Wind at the nearest point, then make to cross 90°E at 22°S to 23°S. From this position head W along the parallel to the join the route from Singapore to Cape of Good Hope, described at 9.195, in about 23°S, 63°E.

2   From Selat Ombi (8°35′S, 125°00′E), pass through the Savu Sea and into the Indian Ocean between Timor (9°00′S, 125°00′E) and Sumba (9°45′S, 120°00′E). Thence steer to join the above route at the most convenient point, having regard to the prevailing wind at the time.

NORTHERN AUSTRALIA TO SYDNEY, INDIAN OCEAN AND SOUTH CHINA SEA

Charts:
　　5126 (1) to (12) Monthly Routeing Charts for Indian Ocean.
　　5309 Tracks followed by Sailing and Auxiliary Powered Vessels.
　　5310 World Surface Current Distribution.
　　Diagram 9.3 World Sailing Ship Routes.

Northern Australia ⇒ Sydney

Passage
9.200

1   During the south-east monsoon (April to October), head W to make North West Cape (21°47′S, 114°09′E) and beat S to round Cape Leeuwin (34°23′S, 115°08′E) and proceed by Bass Strait (40°00′S, 146°00′E) to Sydney (33°50′S, 151°19′E), described at 9.6 and 9.10.

2   During the north-west monsoon (November to April), proceed through the Torres Strait (10°36′S, 141°51′E) and stand into the Pacific Ocean until sufficiently far E to enable Sydney to be reached with the South-east Trade Wind. See 10.40 for details.

Northern Australia ⇒ Fremantle

Passage
9.201

1   During the north-west monsoon (November to April), short boards along the coast S of North West Cape (21°47′S, 114°09′E), will enable advantage to be taken of the land breezes. Only during the strength of the north-west monsoon should a sailing vessel proceed E-about via Torres Strait (10°36′S, 141°51′E), Bass Strait (40°00′S, 146°00′E) and Cape Leeuwin (34°23′S, 115°08′E).

Northern Australia ⇒ Cape of Good Hope

Passage
9.202

1   During the south-east monsoon (April to October), shape course through the Arafura Sea to join the route from the S part of the Eastern Archipelago, described at 9.199.

2   During the strength of the north-west monsoon proceed via Torres Strait (10°36′S, 141°51′E) and Bass Strait (40°00′S, 146°00′E). See 10.40 (to Sydney) and 9.225 (Sydney to Cape of Good Hope) for further details.

Northern Australia ⇒ Colombo

Passage
9.203

1   During the south-east monsoon (April to October), proceed W with the monsoon, crossing the equator in about 75°E, then steer as directly as possible to Colombo (6°58′N, 79°50′E) in the south-west monsoon.

2   From November to April, make to the N through Banda Sea and Molucca Sea, as described at 9.103, and round the N of Sulawesi. Then through the Basilan Strait (6°50′N, 122°00′E) into the Sulu Sea, which should be crossed, passing into the South China Sea through the Balabac Strait (7°34′N, 116°55′E), thence to Singapore (1°12′N, 103°51′E). Then proceed N, through the Malacca Strait, passing S of Great Nicobar Island (7°00′N, 93°50′E) to Colombo, in the north-east monsoon, as described at 9.189. For passages through the Eastern Archipelago and the approaches to Singapore, see 9.58 for details.

Northern Australia ⇒ Kolkata

Passage
9.204

1   During the south-east monsoon (April to October) on the N coast of Australia and the south-west monsoon in the Bay of Bengal, proceed as described at 9.203, for Colombo, but crossing the equator in about 82°E, thence steer E of Sri Lanka to the Sandheads (20°54′N, 88°14′E) at the mouth of the Hugli River.

2   Enough progress W should be made in the south-east monsoon before turning N to enter the limits of the south-west monsoon of the the the Indian Ocean and Bay of Bengal.

3   From November to April, during the north-west monsoon, follow the directions at 9.203 as far as Singapore, then as at 9.190 to Kolkata (22°30′N, 88°15′E).

Northern Australia ⇒ Singapore

Passage
9.205

1   From April to October, two routes are recommended.

2   The usual route passes N of Timor (9°00′S, 125°00′E), through Selat Wetar (8°16′S, 127°12′E) and Alur Pelayaran Wetar (8°00′S, 125°30′E) into the Flores Sea, continuing W along the N side of all the islands and through Selat

Sapudi to Selat Bangka (2°06'S, 105°03'E) or Selat Gelasa (2°53'S, 107°18'E), as described from 9.72 to 9.92.

3   An alternative route passes N or S of Timor and along the S side of all the islands, entering the Java Sea through Selat Sunda, as described from 9.67 to 9.70.2; thence to Singapore as directed at 9.59.

4   From November to April take the Colombo route, for that season, described at 9.203.

5   These routes are affected by Archipelagic Sea Lanes, see 9.57 and Appendix A for further details.

Northern Australia ⇒ Hong Kong

Passage
9.206

1   From April to October, proceed as directed at 9.205 for Singapore, at that season, but pass through Selat Gelasa (2°53'S, 107°18'E) or Selat Karimata (1°43'S, 108°34'E) and thence between Pulau-Pulau Anambas (3°00'N, 106°00'E) and Pulau-Pulau Natuna (4°00'N, 108°00'E) and into the South China Sea. Then steer between the Paracel Islands (16°30'N, 112°00'E) and Macclesfield Bank (15°50'N, 114°30'E), to Hong Kong (22°17'N, 114°10'E). In thick weather proceed through Selat Bangka (2°06'S, 105°03'E) in preference to Selat Gelasa or Selat Karimata.

2   From November to April the route is either by Bougainville Strait (6°40'S, 156°15'E) or by the Second Eastern Passage, described at 9.101. For the Bougainville Strait route, proceed E through Torres Strait (10°36'S, 141°51'E), thence E of Treasury Islands (7°25'S, 155°35'E), through Bougainville Strait and N of the Philippine Islands.

PASSAGES FROM SOUTH-WEST AND SOUTH AUSTRALIA

Charts:

     5126 (1) to (12) Monthly Routeing Charts for Indian Ocean.
     5309 Tracks followed by Sailing and Auxiliary Powered Vessels.
     5310 World Surface Current Distribution.
     Diagram 9.3 World Sailing Ship Routes.

Fremantle ⇒ Mauritius

Passage
9.207

1   In all seasons, steer NW from Fremantle (32°02'S, 115°42'E) into the strength of the South-east Trade Wind, which is generally found between 15°S and 20°S and where the Equatorial Current (6.28) sets to the W. Having reached 20°S, 90°E, in summer and two or three degrees nearer the equator, in the winter of the S hemisphere. Continue W to Mauritius (20°10'S, 57°30'E), passing about 50 miles S of Rodriguez Island (19°38'S, 63°25'E); though from November until April it is advisable to keep at a greater distance, as cyclones sometimes occur at this season, not only in this locality, but also in the area between these islands and the NW coast of Australia. After passing Rodriguez Island steer as directly as possible for Mauritius.

Fremantle ⇒ Cape of Good Hope

Passage
9.208

1   There are two routes to the Cape of Good Hope (34°21'S, 18°30'E), of which the Northern Route is available all year round while the Southern Route can only be used from December to March but is more direct.

2   For the Northern Route, proceed as for Mauritius described at 9.207, but pass 100 to 200 miles S of Rodriguez Island (19°38'S, 63°25'E) and thence about the same distance S of Madagascar, to make the African coast about 200 miles S of Durban (29°51'S, 31°06'E). Thence keep in the strength of the Agulhas Current until abreast of Mossel Bay (34°11'S, 22°09'E), then round Cape Agulhas as described at 9.153.

3   **Caution.** Abnormal waves have occurred off the SE coast of South Africa, see 1.19 and 6.36 for details.

4   For the Southern Route, steer for 30°S, 100°E, and thence make a nearly W course across the ocean to 40°E keeping between 27°S and 29°S, farther S in December and to the N in March. From 40°E steer towards the African coast to join the Northern Route to the E of Algoa Bay (34°00'S, 26°00'E). See notes on rounding Cape of Good Hope at 9.153.

Fremantle ⇒ Aden

Passage
9.209

1   From April to October, proceed direct to pass S of Chagos Archipelago (6°30'S, 72°00'E) to join the route from the Indian coast, as described at 9.149 and 9.150.

2   From November to April, follow a great circle track to 4°00'S, 73°30'E; thence proceed by rhumb line to round Raas Caseyr (11°50'N, 51°17'E).

Fremantle ⇒ Colombo

Passage
9.210

1   From April to October, with the South-east Trade Wind in the South Indian Ocean and the south-west monsoon N of the equator, cross the equator in 80°E, and thence proceed to Colombo (6°58'N, 79°50'E).

2   From November to April, with the north-west monsoon in the South Indian Ocean and the north-east monsoon in the Bay of Bengal, steer across the South-east Trade Wind to enter the north-west monsoon in about 10°S, 90°E. Thence continue N with the north-west monsoon crossing the equator in about 87°E, and with the north-east monsoon to Colombo.

Fremantle ⇒ Kolkata

Passage
9.211

1   From April to October, with the South-east Trade Wind in the South Indian Ocean and the south-west monsoon in the Bay of Bengal, proceed direct for the E coast of Sri Lanka and thence steer E of Sri Lanka to the Sandheads (20°54'N, 88°14'E) at the mouth of the Hugli River. See, 9.21 and 9.172 for further navigational notes.

2   From November to April, with the north-west monsoon in the South Indian Ocean, proceed direct to the equator, crossing it at about 93°E; and thence to make the land about the NW point of Sumatera. From there steer to pass W of the Nicobar Islands (8°00'N, 94°00'E) and thence steer N, close-hauled, and to the W of all the islands.

3   If the equator is crossed as late as March, keep well to the W in the Bay of Bengal as the current, at that time, runs N along the E coast of India and the winds will be found between SW and SE. In the middle of the Bay they are light and variable from NW to NE.

Fremantle ⇒ Singapore

Passage
9.212

1   From April to October, steer on a direct course for Selat Sunda (6°00′S, 105°52′E), taking care to make the land to the E of the strait as the W-going current is often strong near the S coast of Jawa (7°00′S, 110°00′E). Continue from Selat Sunda to Singapore (1°12′N, 103°51′E) as directed at 9.59.

2   From November to April, steer for 12°S, 102°E and then pass midway between Christmas Island (10°24′S, 105°43′E) and the Cocos or Keeling Islands (12°05′S, 96°52′E), there joining the route from Cape Town to Singapore described at 9.15.2. Alternatively, cross 20°S in about 110°E and then follow the Second Eastern Passage, described at 9.101, as far as Selat Manipa (3°20′S, 127°22′E). From this point proceed through Molucca Sea, Selat Bangka (2°06′S, 105°03′E), Celebes Sea, Basilan Strait (6°50′N, 122°00′E), Sulu Sea and Balabac Strait (7°34′N, 116°55′E) into the South China Sea. Thence proceed to Singapore in the north-east monsoon. This alternative route, though longer, will probably give a better passage.

Fremantle ⇒ Hong Kong

Passage
9.213

1   From April to October, proceed to Selat Sunda (6°00′S, 105°52′E), as directed at 9.212 and continue to the South China Sea as directed at 9.59.

2   From November to April, may either follow the seasonal directions given at 9.212 and pick up the the Second Eastern Passage, described at 9.101, or proceed to Singapore (1°12′N, 103°51′E) via Selat Sunda, as directed at 9.59; thence through Palawan Passage (10°20′N, 118°00′E) and along the coast of Luzon until able to stand across to Hong Kong (22°17′N, 114°10′E).

3   A vessel which, having passed Selat Sunda, finds that the north-west monsoon in the Java Sea and the north-east monsoon in the South China Sea have already begun, is advised to make to the E, as described at 9.99, to pick up the Second Eastern Passage.

Fremantle ⇒ south-east Australia or New Zealand

Passage
9.214

1   Stand S and proceed as directed from 9.8 to 9.12 for the voyage from Cape of Good Hope.

South-east Australia ⇒ Cape of Good Hope

Passage
9.215

1   There are two routes according to season.

2   The Northern Route is available from April to October, at the time of the year when the south-east monsoon of the Arafura Sea connects with the South-east Trade Wind of the Pacific Ocean and with the South-east Trade Wind of the Indian Ocean. Vessels using it should proceed first to the N, along the E coast of Australia and through Torres Strait (10°36′S, 141°51′E), thence through the Arafura Sea into the Indian Ocean and to the Cape of Good Hope (34°21′S, 18°30′E), as directed at 9.202.

3   Directions for passage through Bass Strait (40°00′S, 146°00′E) are given at 9.220.

4   The Southern Route should be used from December to April, when E winds are prevalent of the S coast of Australia. First proceed as directly as possible to round Cape Leeuwin (34°23′S, 115°08′E) at a safe distance, having regard to the weather prevailing at the time and the danger of being caught on a lee shore. From Cape Leeuwin stand to the NW into the South-east Trade Wind and join the route from Fremantle as directed at 9.208.

5   **Note.** Historically it was reported that masters of sailing vessels, bound for European ports from Adelaide, would often defer a decision whether to proceed E-about or W-about until they had ascertained the wind direction in the Australian Bight. Thus with a W wind they would sail E-about, as described at 10.10 for Cabo de Hornos. With an E wind they would take the Southern Route, described above, for the Cape of Good Hope. This consideration would apply equally today, when planning ocean passages under sail.

South-east Australia ⇒ Aden

Passage
9.216

1   From April to October, proceed as directed for the Northern Route, described at 9.215, to Torres Strait (10°36′S, 141°51′E), thence through the Arafura Sea. Having cleared all dangers in the Arafura Sea, steer to pass S of the Chagos Archipelago (6°30′S, 72°00′E) and as directed at 9.149.

2   From November to April, pass round Cape Leeuwin (34°23′S, 115°08′E) as directed for the Southern Route at 9.215 and thence steer NW to join the route from Fremantle to Aden described at 9.209.

South-east Australia ⇒ Colombo

Passage
9.217

1   From April to November, take the Northern Route, described at 9.215 to the Torres Strait (10°36′S, 141°51′E) and the Arafura Sea, whence course can be set for Colombo, as described at 9.203.

2   From December to April, when E winds are prevalent off the S coast of Australia, the Southern Route, described at 9.215, round Cape Leeuwin (34°23′S, 115°08′E), is taken. When clear of Cape Leeuwin stand to the NW into the South-east Trade Wind and enter the north-west monsoon in about 10°S, 90°E. Then steer N with the north-west monsoon across the equator in about 87°E and then with the north-east monsoon to Colombo (6°58′N, 79°50′E), bearing in mind that this is the cyclone season in the South Indian Ocean.

South-east Australia ⇒ Bay of Bengal

Passage
9.218

1   From April to November, proceed through the Torres Strait (10°36′S, 141°51′E) and the Arafura Sea, as described at 9.215. Keep in the South-east Trade Wind until 85°E is reached, then stand NW to cross the equator in about 80°E. From this point proceed direct to destination, allowing for the strong E-going current.

2   From December to April, pass round Cape Leeuwin (34°23′S, 115°08′E) and steer NW through the South-east Trade Wind so as to enter the north-west monsoon in about 85°E; then shape course towards the NW end of Sumatera and proceed W of Nicobar Islands (8°00′N, 94°00′E) and Andaman Islands (12°30′N, 93°00′E) to destination.

South-east Australia ⇒ Singapore

Passage
9.219

1   From April to November, three routes are available. N-about through the Torres Strait (10°36′S, 141°51′E), thence N of Timor (9°00′S, 125°00′E) and through the Java Sea; or S of the islands and through Selat Sunda (6°00′S, 105°52′E), as described at 9.205; or S-about round Cape Leeuwin (34°23′S, 115°08′E). The N-about routes are probably the best.

2   From December to April, in spite of the prevailing E winds to the S of Australia, a route S-about round Cape Leeuwin is not recommended for Singapore (1°12′N, 103°51′E) on account of the N winds and S-going currents prevalent between November and March in Selat Sunda, Selat Bangka, Selat Gelasa and Selat Karimata. A vessel has been known to take 30 days from Selat Sunda to Singapore, at this time of year, a distance of 500 miles. It is therefore advisable to proceed by the Outer Route, described at 10.39, to the E of Australia and through the Torres Strait, thence as directed at 9.203 for Colombo; or take the route E of Papua New Guinea, through Bougainville Strait (6°40′S, 156°15′E) and Surigao Strait (9°53′N, 125°21′E), described 9.104, into the South China Sea.

PASSAGES FROM SYDNEY TO PORTS IN THE INDIAN OCEAN

Charts:

    5126 (1) to (12) Monthly Routeing Charts for Indian Ocean.
    5309 Tracks followed by Sailing and Auxiliary Powered Vessels.
    5310 World Surface Current Distribution.
    Diagram 9.3 World Sailing Ship Routes.

Sydney ⇒ and through Bass Strait

Passage
9.220

1   There are two main routes, direct, and through Banks Strait (40°39′S, 148°05′E). By the direct route, in order to take advantage of the current as far as Cape Howe (37°30′S, 149°59′E), which appears to run strongest from November to March, keep along the outer edge of the charted 200 m depth contour, or at a distance of 15 to 18 miles from the coast, where the current runs stronger and with more regularity than elsewhere.

2   From about 15 miles E of Cape Howe, if the wind is S, do not steer a more W course than 212° until in 39°30′S on account of the danger to be apprehended from SE or S gales upon Ninety Mile Beach, between Cape Howe and Corner Inlet (38°50′S, 146°30′E). On reaching 39°30′S, steer to pass about 3 miles N of Wright Rock (39°36′S, 147°32′E) and the same distance S of the S point of Deal Island (39°29′S, 147°20′E), the SE of the Kent Group. Having passed the Kent Group steer to pass 2 or 3 miles S of Sugarloaf Rock (39°30′S, 146°39′E), and S of Judgement Rocks (39°31′S, 147°08′E).

3   From Sugarloaf Rock steer 15 or 20 miles to the N of King Island (39°50′S, 144°00′E), if the wind permits, but should the wind hold to the W of N, a course may be safely directed for the N point of Three Hummock Island (40°26′S, 144°55′E), taking care to avoid Mermaid Rock (40°23′S, 144°57′E) and Taniwha Rock (40°25′S, 144°59′E), passing afterwards N or S of King Island, as may be most favourable, but N is preferable.

Navigational notes
9.221

    An Area to be avoided encloses oil and gas fields extending between 20 miles SE and 45 miles S of Lakes Entrances (37°54′S, 147°59′E) on Ninety Mile Beach.

Submarine pipelines are laid between the fields and the shore. For details see *Australia Pilot, Volume II*.

1   Local experience has shown that with W and SW winds smoother water is inshore, off Ninety Mile Beach. As SW winds are the prevailing ones, mariners bound to the W may often take advantage of the smoother water, and an absence of danger, to approach the beach instead of avoiding it. A vessel inshore, when a E gale is threatened, should at once get offshore; these gales give warning signs.

2   Between December and March, as W gales veer to the S, it is advisable to stand towards the Tasmanian coast, to be ready to take advantage of the shift of wind.

3   Between April and November, and more particularly from September to November, the same course cannot be recommended, as in these months the wind tends to back to WNW.

4   The alternative route is via Banks Strait, which lies between Cape Barren Island (40°23′S, 148°15′E) and the N coast of Tasmania and offers an alternative entrance to the Bass Strait (40°00′S, 146°00′E). The chief dangers to be avoided on the S shore are the reefs and rocks off Swan Island (40°44′S, 148°06′E) and the foul ground and rocks N of Foster Islets (40°44′S, 147°58′E).

5   When working to the W, between December and March, when W gales are of short duration, it is advisable to stand towards the Tasmanian coast to take advantage of the shift of wind.

Sydney ⇒ Melbourne

Passage
9.222

1   Proceed as directed at 9.220 as far as Sugarloaf Rock (39°30′S, 146°39′E), thence to Port Phillip (38°20′S, 144°34′E) as directly as circumstances permit.

Bass Strait ⇒ Adelaide

Passage
9.223

1   In fine weather, from off Cape Otway (38°53′S, 143°31′E), steer to pass about 5 miles S of Cape Nelson (38°26′S, 141°32′E), 10 miles SW of Cape Northumberland (38°04′S, 140°40′E) and Cape Banks (37°54′S, 140°23′E), thence make direct course to Cape Willoughby (35°51′S, 138°08′E). Care must be taken at all times to guard against a set towards the land, but with S and W winds the coast

should be given a much greater berth, as a current of 1 kn sometimes sets towards it between Cape Otway and Cape Willoughby.

2   When entering the Gulf of Saint Vincent (35°00′S, 138°10′E), by the Backstairs Passage (35°45′S, 138°08′E), Young Rocks (36°23′S, 137°15′E) must be given a wide berth at night, but since they are above-water, they are not dangerous by day in clear weather. At times the sea breaks heavily off Cape Willoughby, see *Australia Pilot, Volume I*.

Bass Strait ⇒ Spencer Gulf

Passage
9.224

1   Proceed as directed at 9.223 to Cape Northumberland (38°04′S, 140°40′E).

2   Thence to Spencer Gulf (34°00′S, 137°00′E), giving a good berth to the SW Young Rock (36°24′S, 137°12′E), which is only 1·5 m high; except with strong SE winds, make allowance for the E-going set which usually prevails. From December to March, with SE winds, a current runs at about 1 kn to the NW.

3   In the event of threatening weather from the S and W, care must be taken to maintain a good distance offshore.

Sydney ⇒ Cape of Good Hope and all ports in the Indian Ocean

Passage
9.225

1   October and November are unsuitable months in which to start a passage from Sydney (33°50′S, 151°19′E) to the W either by Torres Strait (10°36′S, 141°51′E) or to the S of Australia.

2   From March to September, a route to the N of Australia should be taken, since the prevalence of strong W gales renders the S-about route very difficult, indeed, generally impracticable for sailing vessels during the whole period from April to November. The worst months for making the W-bound passage N of Australia are September to

November, for W-bound gales are then of frequent occurrence, the wind sometimes being from WSW to WNW and blowing very strong for more than a week at a time. From December to August, N winds are very common.

3   In these circumstances the best W bound route is via the Torres Strait and the Arafura Sea, taking, by preference, the Outer Route, described at 10.39, from Sydney, through the Coral Sea, to the Torres Strait.

4   From the Torres Strait directions are given as follows:

Kolkata	9.204
Cape of Good Hope	9.202
Colombo	9.203
Fremantle	9.201
Hong Kong	9.206
Singapore	9.205

5   From December to March a route S of Australia may be taken. During these months, proceed through the Bass Strait (40°00′S, 146°00′E), or S of Tasmania; E winds prevail in the Strait and along the S coast of Australia at that season and good passages have been made by keeping N of 40°S and passing round Cape Leeuwin (34°23′S, 115°08′E) into the South-east Trade Wind, which then extends well to the S. A vessel from Bass Strait bound round Cape Leeuwin is recommended, with a favourable wind, to shape a course which will lead about 150 miles S of the Cape.

6   In adopting this course advantage must be taken of every favourable change of wind, in order to make to the W. It is advisable not to approach too near the land as it would become a dangerous lee shore in SW gales, which are often experienced, even from December to March, and also contrary currents run strongest near the land.

7   After rounding Cape Leeuwin, stand to the NW into the South-east Trade Wind, and follow the directions given below:

Aden	9.209
Kolkata	9.211
Cape of Good Hope	9.208
Colombo	9.210
Mauritius	9.207

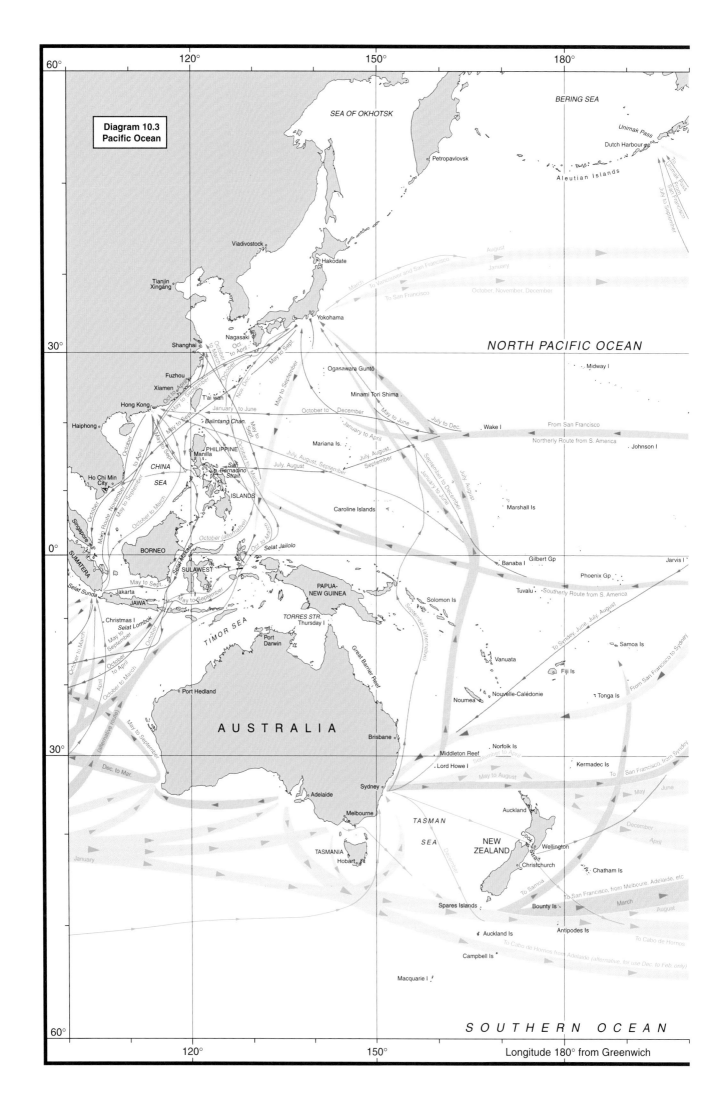

**Diagram 10.3
Pacific Ocean**

SEA OF OKHOTSK

BERING SEA

Petropavlovsk

Unimak Pass

Dutch Harbour

Aleutian Islands

Viadivostock

Hakodate

NORTH PACIFIC OCEAN

To Vancouver and San Francisco

To San Francisco

August

January

October, November, December

March

Tianjin
Xingang

Yokohama

Midway I

Nagasaki

Ogasawara Guntō

Oct
to April

Shanghai

October to March

May to Sept.

May to September

Nov.-Dec.

Fuzhou

Minami Tori Shima

Xiamen

T'ai wan

January to June

Hong Kong

May to Sept.

October to December

Balintang Chan.

May to June

July to Dec.

Wake I

From San Francisco

Haiphong

May to Sept.

January to April

Northerly Route from S. America

PHILIPPINE

Mariana Is.

Manilla

October to April

San
Bernadino
Strait

July, August, September

July, August,
September

Johnson I

Ho Chi Min
City

CHINA

SEA

May to September

ISLANDS

October to March

July, August

September to December

January to June

July, August

September to June

July, August

Caroline Islands

Marshall Is

Singapore

October

SUMATERA

Selat Sunda

May to Sept.

Selat Malaca

October (alternative)

BORNEO

SULAWESI

Oct. to May

Selat Jailolo

Banaba I

Gilbert Gp

Jarvis I

Phoenix Gp

Jakarta

JAWA

May to September

PAPUA-
NEW GUINEA

Tuvalu

Southerly Route from S. America

October to March

Christmas I

Selat Lombok

TORRES STR.

Thursday I

Solomon Is

Samoa Is

May to
September

October

TIMOR SEA

Port
Darwin

September (alternative)

Vanuata

Fiji Is

To Sydney, June, July, August

From San Francisco to Sydney

October to April

April

October to March

Great Barrier Reef

Nouvelle-Calédonie

Tonga Is

Noumea

May to September

Port Hedland

Dec. to Mar.

AUSTRALIA

Brisbane

Norfolk Is

September to April

Middleton Reef

Lord Howe I

Kermadec Is

To San Francisco, from Sydney

May to August

May

June

Adelaide

Sydney

To

Melbourne

December

Auckland

TASMAN

SEA

December

NEW
ZEALAND

Cook Strait

Wellington

April

TASMANIA

Christchurch

Chatham Is

January

To Samoa

To San Francisco, from Melboure, Adelaide, etc

Hobart

March

August

Spares Islands

Bounty Is

Auckland Is

Antipodes Is

To Cabo de Hornos

To Cabo de Hornos from Adelaide (alternative, for use Dec. to Feb. only)

Campbell Is

Macquarie I

SOUTHERN OCEAN

Longitude 180° from Greenwich

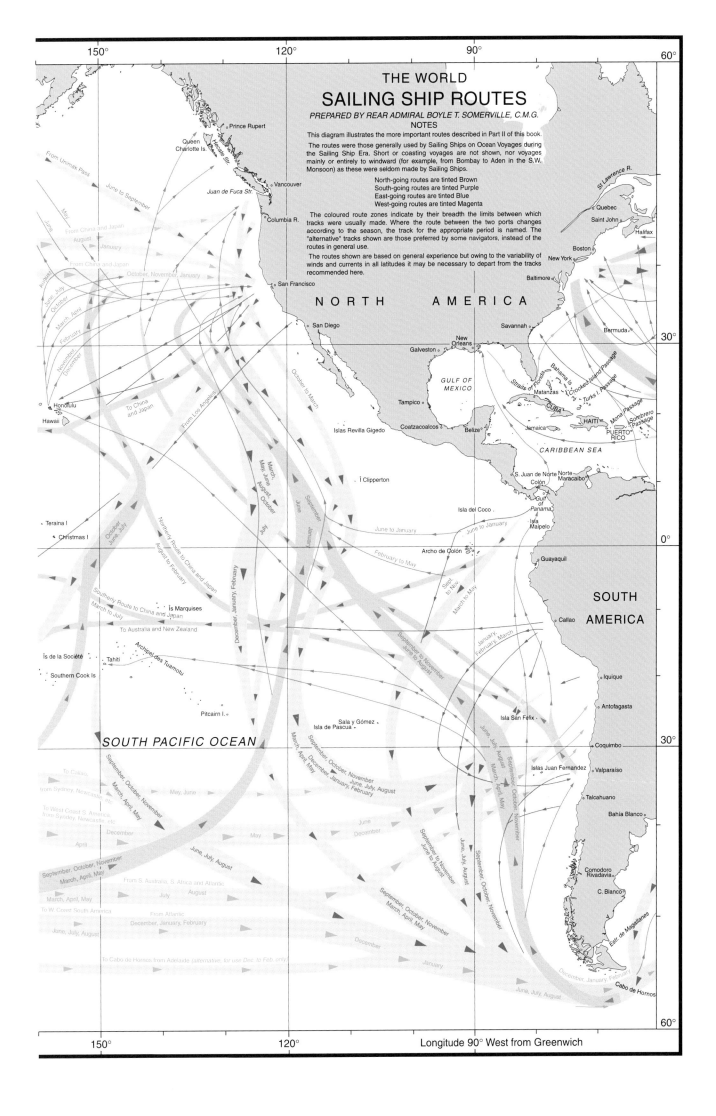

THE WORLD
SAILING SHIP ROUTES
PREPARED BY REAR ADMIRAL BOYLE T. SOMERVILLE, C.M.G.
NOTES

This diagram illustrates the more important routes described in Part II of this book.

The routes were those generally used by Sailing Ships on Ocean Voyages during the Sailing Ship Era. Short or coasting voyages are not shown, nor voyages mainly or entirely to windward (for example, from Bombay to Aden in the S.W. Monsoon) as these were seldom made by Sailing Ships.

North-going routes are tinted Brown
South-going routes are tinted Purple
East-going routes are tinted Blue
West-going routes are tinted Magenta

The coloured route zones indicate by their breadth the limits between which tracks were usually made. Where the route between the two ports changes according to the season, the track for the appropriate period is named. The "alternative" tracks shown are those preferred by some navigators, instead of the routes in general use.

The routes shown are based on general experience but owing to the variability of winds and currents in all latitudes it may be necessary to depart from the tracks recommended here.

Longitude 90° West from Greenwich

NOTES

CHAPTER 10

SAILING PASSAGES FOR PACIFIC OCEAN

GENERAL INFORMATION

COVERAGE

Chapter coverage
10.1

This chapter contains details of the following passages for sailing vessels and includes recommended routes for low-powered, or hampered, vessels:

Passages from South Africa and Southern Australia to Pacific Ocean ports (10.6).
Passages from and to Sydney (10.17).
Passages from New Zealand (10.42).
Passages from central South Pacific Island Groups (10.50).
Passages from Singapore and Eastern Archipelago (10.60).
Passages from Krung Thep or Ho Chi Minh City (10.74).
Passages from ports in China (10.79).
Passages from Manila (10.98).
Passages from Japan (10.105).
Passages from Islands in the North Pacific Ocean (excluding Hawaii) (10.117).
Passages from Honolulu (10.122).

Passages from Prince Rupert, Juan de Fuca Strait or Columbia River:
Sailing vessel routes (10.129).
Low powered vessel routes (10.132).
Passages from San Francisco (10.137).
Passages from Central America and Panama (10.150).
Passages from South American ports (10.161).
Passages from Cabo de Hornos (10.174).

Admiralty Publications
10.2

Relevant navigational publications should be consulted, in addition to this book, when planning and conducting passages. Details of these, and the areas covered, can be found, as follows:

Admiralty Sailing Directions; see 1.12.1
Admiralty List of Lights and Fog Signals; see 1.12.2
Admiralty Tide Tables; see 1.12.3
Admiralty List of Radio Signals; see 1.12.5
Admiralty Distance Tables; see 1.12.8

10.2.1

See also *The Mariner's Handbook* for notes on electronic products.

NAVIGATIONAL NOTES FOR THE PACIFIC OCEAN

Soundings and dangers
10.3

1   Very large areas of the Pacific Ocean are imperfectly surveyed, and many of the dangers are steep-to. See from 7.3 to 7.34 for notes on natural conditions, and 7.35 for further information on soundings. Also the notes on navigation in coral waters in *The Mariner's Handbook* and the appropriate volumes of *Admiralty Sailing Directions* should be consulted.

Currents
10.4

1   Particular attention to currents is required when navigating amongst the Pacific islands, see 7.36 for further details.

Navigation between the islands of the Pacific
10.5

1   Within the regions of the Trade Winds (7.6 and 7.15), there is no difficulty in travelling from E to W, the winds being fair.

2   From W to E, for short distances, a vessel may beat to windward, but for long distances, for instance from Fiji (18°00'S, 180°00') to Tahiti (17°35'S, 149°25'W), or from Tahiti to Pitcairn Island (25°04'S, 130°05'W), a vessel should head to the S through the Trade Winds into the Westerlies, then run to the E and re-enter the Trades Winds in about the meridian of the destination.

SOUTH AFRICA AND SOUTHERN AUSTRALIA TO PACIFIC OCEAN PORTS

Charts:
5127 (1) to (12) Monthly Routeing Charts for North Pacific Ocean.
5128 (1) to (12) Monthly Routeing Charts for South Pacific Ocean.
5309 Tracks followed by Sailing and Auxiliary Powered Vessels.
5310 World Surface Current Distribution.
Diagram 10.3 World Sailing Ship Routes.

South Africa ⇒ Cabo de Hornos
Indian Ocean passage
10.6

1   Follow the route described at 9.12, which passes S of Tasmania between 45°S and 47°S.

Icebergs
10.7

1   Icebergs are most numerous, near this route, midway between New Zealand and Cabo de Hornos (56°04'S,

67°15′W), but the periods of frequency vary considerably, and it may happen that while ships are meeting ice in the lower latitudes, those in higher latitudes may be free of it. See also 7.34 for further details.

Pacific Ocean usual passage
10.8

1   This passage, suitable throughout the year, passes S of New Zealand in about 48°30′S, or about 30 miles S of Snares Islands (48°01′S, 166°36′E). From this position a vessel should steer to the E between Bounty Islands (47°41′S, 179°03′E) and Antipodes Islands (49°35′S, 178°50′E), thence, inclining slightly S, the route follows a mean track of 51°S from about 150°W, across the ocean to 120°W. Keep at about 50°S from December to February, to be more clear of the ice; and at 52°S from June to August, but only if ice conditions are clear.

2   From 115°W, incline gradually S to round Islas Diego Ramírez (56°30′S, 68°43′W) and Cabo de Hornos. See 8.88 for directions for rounding the cape.

Pacific Ocean alternative passage
10.9

1   This passage is only recommended from December to February and follows a track farther S from the position S of Tasmania, described at 10.6, to pass between Auckland Islands (50°40′S, 166°00′E) and Campbell Island (52°33′S, 169°10′E) in about 52°S and to cross the Pacific Ocean between 54°S and 55°S.

2   This course would, if clear of ice and with favourable weather, doubtless ensure the quickest passage as being the shorter distance However, experience has shown that, at nearly all times of the year, a track, even as far N as 47°S, can be adopted with advantage. A more N track avoids loss of time at night and in thick weather, and the serious dangers incurred on account of the great quantities of ice normally met with in higher latitudes.

3   It is believed that a passage made between 47°S and 50°S will provide steadier winds, smoother water and less ice; a quicker passage may be expected in the better weather and with more security than in higher latitudes.

South-east Australia ⇒ Pacific Ocean

Adelaide to Cabo de Hornos
10.10

1   From Adelaide (34°48′S, 138°23′E), steer SE to join the main route, described at 10.6, in about 46°S, 146°E.

Melbourne to Cabo de Hornos
10.11

1   From Port Phillip (38°20′S, 144°34′E), in December to February, shape course to pass about 60 miles W of King Island (39°50′S, 144°00′E) and thence W of Tasmania to join the main route, described at 10.6, in about 46°S, 146°E. It is often necessary, and in heavy weather desirable, to make this passage at a considerable distance from the coast of Tasmania; namely from 120 to 250 miles from the W coast and round the S coast of the island.

2   For the rest of the year, and as an alternative to the above route, pass through the Bass Strait (40°00′S, 146°00′E) and steer to join the main route S of Snares Islands, as described at 10.8.

3   **Traffic Separation Schemes** have been established in the Bass Strait and SE of the Area to be Avoided (7.38.3) SE of Ninety Mile Beach. See also 8.4.

Hobart to Cabo de Hornos
10.12

1   Either join, the main route (10.8) S of Snares Islands; or the alternative route (10.9) between Auckland Islands and Campbell Island.

Adelaide, Melbourne or Hobart to Chilean ports
10.13

1   Proceed to 48°30′S, 166°30′E, S of Snares Islands as directed at 10.8, 10.10 or 10.12, thence make to the E across the Pacific Ocean, between 46°S and 48°S, being more to the S in March and more N in August, as far as 112°W. From this position steer as directly as possible for destination, bearing in mind the N-going current running up the whole W coast of South America.

Adelaide, Melbourne or Hobart to San Francisco or British Columbia
10.14

1   Proceed to 48°30′S, 166°30′E, S of Snares Islands as directed at 10.13, thence make for 41°S, 138°W, keeping about 60 miles N of the direct line in September and about 60 miles S of it in March. From 41°S, 138°W, proceed to 30°S, 124°W and from this position make nearly N through the South-east Trade Wind, crossing the equator in 116°W.

2   After picking up the North-east Trade Wind, in about 10°N, steer for 30°N, 131°W in November and for 30°N, 136°W in June and July. At other times cross 30°N between these positions.

3   From 30°N, proceed as direct to the destination as the prevailing W winds and the SE-going current will allow. The current crosses the track at a rate of 20 to 30 miles a day. See also 10.178, for details of the route from Cabo de Hornos to San Francisco, which joins this route soon after crossing the equator.

Melbourne to New Zealand
10.15

1   If the wind is W on departure from Port Phillip (38°20′S, 144°34′E), steer to pass Rodondo Island (39°14′S, 146°23′E) and then N of the Kent Group (39°30′S, 147°20′E). Then, for ports on the E side of South Island, steer S of Snares Islands (48°01′S, 166°36′E) and thence to destination.

2   For Wellington (41°22′S, 174°50′E), steer directly for Cook Strait (41°00′S, 174°30′E).

3   For Auckland (36°36′S, 174°49′E), steer for Three Kings Islands (34°10′S, 172°00′E) and thence round North Cape (34°25′S, 173°03′E) to destination.

4   **Traffic Separation Schemes** have been established in the Bass Strait and SE of the Area to be Avoided (7.38.3) SE of Ninety Mile Beach. See also 8.4.

5   If on leaving Port Phillip the wind is from E it may be desirable to run to the S, passing W of King Island (39°50′S, 144°00′E), then proceeding along the W coast of Tasmania, being prepared for the prevailing W or SW winds, when this coast becomes a dangerous lee shore, as described at 10.11. Having rounded the outlying dangers off the S coast of Tasmania, proceed to destination as above.

Adelaide or Melbourne to Sydney
10.16

1   If the wind is favourable for Bass Strait (40°00′S, 146°00′E), first steer for Rodondo Island (39°14′S, 146°23′E), passing about 20 miles S of Cape Otway (38°53′S, 143°31′E), if bound from Adelaide (34°48′S, 138°23′E).

2　　Having passed Rodondo Island and the Kent Group (39°30'S, 147°20'E), steer for a position about 20 miles SE of Rame Head (37°47'S, 149°30'E) and make Gabo Island (37°34'S, 149°55'E) or the land in the vicinity of Cape Howe (37°30'S, 149°59'E). If the wind is blowing hard from the S, a more E course should be steered to avoid Ninety Mile Beach, extending from Corner Inlet (38°50'S, 146°30'E) for 150 miles, nearly to Cape Howe, which would then become a dangerous lee shore.

3　　**Traffic Separation Schemes** have been established in the Bass Strait and SE of the Area to be Avoided (7.38.3) SE of Ninety Mile Beach. See also 8.4.

4　　From a position E of Cape Howe steer to the N along the E coast for Sydney (33°50'S, 151°19'E), at such a distance from the land as wind and weather conditions would suggest, bearing in mind that the current generally sets to the S at a distance of 20 to 60 miles from land.

5　　If, on leaving Adelaide or Melbourne, there should be an E wind it may be preferable to run to the S; if from Melbourne passing between Cape Otway and King Island (39°50'S, 144°00'E). Then proceed down the W coast of Tasmania, giving it a good berth, as directed at 10.11. Having cleared the outlying dangers off the S coast of Tasmania, proceed to the N as described above.

PASSAGES FROM AND TO SYDNEY

Charts:

5127 (1) to (12) Monthly Routeing Charts for North Pacific Ocean.

5128 (1) to (12) Monthly Routeing Charts for South Pacific Ocean.

5309 Tracks followed by Sailing and Auxiliary Powered Vessels.

5310 World Surface Current Distribution.

Diagram 10.3 World Sailing Ship Routes.

Sydney ⇒ Southern Australia and New Zealand

Sydney to Melbourne or Adelaide
10.17

1　　When proceeding S from Sydney (33°50'S, 151°19'E), keep at between 20 and 60 miles from the coast to take advantage of the S-going current.

2　　To make passage via Bass Strait (40°00'S, 146°00'E), follow the directions given at 9.220, if the wind permits.

3　　To make the passage S of Tasmania, reverse the directions for that route described at 10.16.

Sydney to Hobart
10.18

1　　Proceed to the S coast of Tasmania, as described at 10.17.

2　　On entering Storm Bay (43°15'S, 147°35'E) from the E, head over towards Cape Queen Elizabeth (43°15'S, 147°26'E) and steer along the E coast of North Bruny Island (43°10'S, 147°23'E) for the entrance to the River Derwent. In working against a NW wind work up along the same coast, to avoid the strong outset from Frederick Henry Bay (42°55'S, 147°35'E).

3　　If, when off Betsey Island (43°03'S, 147°29'E), the wind is blowing from the NW and preventing a vessel entering the River Derwent, good anchorage can be obtained either in Adventure Bay (43°18'S, 147°22'E) or Frederick Henry Bay. In calms, or light winds, vessels may anchor in Storm Bay, if necessary, until they get a breeze.

Sydney to Auckland
10.19

1　　There are two routes, according to the time of year, although it is sometimes possible to make a direct course to sight and pass 10 miles N of Three Kings Islands (34°10'S, 172°00'E) and thence around North Cape (34°25'S, 173°03'E). Sailing vessels which do not have a commanding breeze should not attempt to pass S of Three Kings Islands.

2　　From September to April, proceed to 30°S, 170°E and thence to Auckland (36°36'S, 174°49'E).

3　　From May to August, take a more S route through 35°S, 170°E.

Sydney to Wellington
10.20

1　　Take as direct a route as possible to Cook Strait (41°00'S, 174°30'E), noting that the best time of year for this passage is from October to February.

Sydney to Port Chalmers or adjacent ports
10.21

1　　Steer S of Snares Islands, as described at 10.22, then proceed as directly as possible to Port Chalmers (45°50'S, 170°30'E) or nearby ports.

2　　The passage round the SW end of South Island, and through Foveaux Strait (46°35'S, 168°00'E), is also possible but not recommended.

Sydney ⇒ South America

Sydney to Cabo de Hornos
10.22

1　　From Sydney (33°50'S, 151°19'E) at all seasons, and from whatever quarter the wind may blow, it is advisable to proceed to the S of New Zealand, rather than the N. Advantage should, therefore, be taken of the most favourable winds for either passing S of Snares Islands (48°01'S, 166°36'E) and Auckland Islands (50°40'S, 166°00'E) to join the route described at 10.8; or, if baffled by S winds and favoured by fine weather, the passage through Cook Strait (41°00'S, 174°30'E), may be taken, especially from October to February, joining the route, described at 10.45, to Cabo de Hornos (56°04'S, 67°15'W). See also 10.9 for an alternative route if passing S of New Zealand.

Sydney to ports on the west coast of South America
10.23

1　　Follow the directions at 10.19, according to season, as far as 170°E, thence proceed to destination as described below.

10.23.1

Sydney to ports between Talcahuano and Iquique.

After crossing 170°E, steer to cross the 180° meridian at about 35°S, then cross 150°W between 39°S and 43°S; to the N in November and December and to the S in April and May. Keep within these latitudes as far as 106°W and from that position curve the track gradually N towards the port of destination, making due allowance for the N-going current along the coast of South America. The winds will usually be from a S direction.

10.23.2
Sydney to ports between Iquique and Panama.

After crossing 170°E, steer to cross the 180° meridian between 33°S and 34°S and cross the ocean on a nearly E course, not going S of 36°S. On reaching 100°W, begin to make to the NE through the South-east Trade Wind to destination, making due allowance for the N-going current along the coast as far as the equator.

Sydney ⇒ San Francisco or British Columbia

Directions
10.24

1   From Sydney (33°50′S, 151°19′E), there are two routes, via Tahiti (17°35′S, 149°25′W) and via Fiji (18°00′S, 180°00′).

2   To make the passage via Tahiti, pass either N or S of New Zealand, or through Cook Strait (41°00′S, 174°30′E), according to the direction of the wind on leaving, but preferably through Cook Strait. Thence make to the NE so as to cross 30°S in about 160°W, and then N through the South-east Trade Wind, passing closely W of Îles de la Société.

3   From June to August, cross the equator in 148°W, but from October to February in 151°W, steering through the doldrums to 10°N, 143°W, where the North-east Trade Wind should be picked up. Head through the Trade Wind towards 30°N, 152°W, where the Westerlies should begin, when course can be directed to destination. From November to February the turn to the E can usually be made about 33°N, but in August continue to 40°N before turning towards the land. Allowance must be made for a current setting SE and S more strongly as the United States coast is approached. It is also felt off the coast of British Columbia, but there is complicated by tidal streams; see 10.105.1 for further details of these.

4   To make the passage via Fiji, take the Auckland route, described at 10.19, as far 170°E. Thence continue E, making nothing to N, as far as 176°E, when course may be altered towards the Fijian Islands. If not calling at Fiji, pass E of the group and thence steer due N to cross the equator and 18°N on the 180° meridian, thence head more to the E to 30°N, 172°W, from whence proceed, as directly as possible, to destination.

Sydney to, and among, South Pacific islands

Sydney to Tahiti
10.25

1   Follow the directions for the Tahiti route, described at 10.24.

Sydney to Fiji
10.26

1   Follow the directions for the Fiji route, described at 10.24.

Sydney to Nouméa
10.27

1   From Sydney (33°50′S, 151°19′E), pass between Lord Howe Island (31°32′S, 159°05′E) and Elizabeth Reef (29°55′S, 159°02′E), thence direct. The passage to Nouméa (22°18′S, 166°25′E) in a sailing vessel varies from 5 to 28 days, and is seldom made without encountering a gale.

Sydney to other South Pacific islands
10.28

1   When bound from the Australian coast to islands in the South Pacific Ocean, precise directions cannot be given on account of the irregularity of the wind. As a general rule, progress E must be made S of the Trade Wind limits, about 32°S. This is liable to interruption, however, especially between January and April. When on the longitude of the island to which bound, the Trade Wind may be entered and the ship sailed, running free, as the current will be found setting to windward, until near the islands.

2   For all practical purposes of navigation between the various groups of islands, it is important to note that they lie within the limits of the South-east Trade Wind (7.15) and the equatorial Current (7.27). For sailing vessels this means a favourable wind and current when proceeding from E to W, except when within the limits of the Equatorial Counter-current (7.28). Proceeding from W to E involves beating to windward, in a choppy sea against the current.

Sydney ⇒ Yokohama
Routeing details
10.29

1   From Sydney (33°50′S, 151°19′E) to Yokohama (35°26′N, 139°43′E), the route changes seasonally with the monsoons on both sides of the equator; for directions S of the equator see 10.30 and for those N of the equator see 10.31.

2   Alternative routes known as the Eastern, Middle and Western Routes may be taken, see 10.32 for details. They do not differ greatly from the other routes.

Passage south of the equator
10.30

1   During the north-west monsoon (November to March), pass between Lord Howe Island (31°32′S, 159°05′E) and Elizabeth Reef (29°55′S, 159°02′E), thence to the N between Récifs d'Entrecasteaux (18°20′S, 163°00′E) to the E and Bellona (21°30′S, 158°50′E) and Bampton (19°00′S, 158°40′E) Reefs to the W. From this position proceed between the Solomon (8°00′S, 158°00′E) and Santa Cruz (11°00′S, 166°15′E) Islands, crossing the equator in about 166°E, about 60 miles W of Nauru.

2   During the south-east monsoon (April to October), on leaving Sydney steer directly to the NE as far as 157°E, then to the N between Kenn (21°15′S, 155°48′E) and Bellona Reefs and E of Pocklington Reef (10°50′S, 155°45′E). Then either, through Bougainville Strait (6°40′S, 156°15′E) or through Pioneer Channel (4°40′S, 153°45′E), crossing the equator in about 155°E.

Passage north of the equator
10.31

1   From January to June, steer direct for Yokohama, passing E of the Caroline Islands (7°00′N, 150°00′E).

2   In July and August, take a more E track, passing about 100 miles W of the Marshall Islands (10°00′N, 170°00′E) and crossing 160°E in 18°N, thence direct to Yokohama.

3   From September to December, a track midway between these two is recommended.

Alternative routes
10.32

1   The Eastern Route is to Norfolk Island (29°02′S, 167°56′E), thence to Île Matthew (22°21′S, 171°21′E) and due N along 171°E to 11°S, crossing the equator in 166°E and through the E part of the Caroline Islands.

2   The Middle Route passes midway between Lord Howe Island and Elizabeth Reef, W of Nouvelle-Calédonie and between the Solomon and Santa Cruz Islands, to cross the equator in 153°E and through the middle of the Caroline Islands.

3   The Western Route is due N along 157°E as far as 11°S, thence through Bougainville Strait, and across the equator in 153°E, when a direct course may be steered for Yokohama.

Sydney ⇒ Hong Kong

Routeing details
10.33

1   From Sydney (33°50′S, 151°19′E) to Hong Kong (22°17′N, 114°10′E), there are three routes appropriate to the monsoon period.

North-west monsoon (October to March)
10.34

1   During this period, steer to pass midway between Lord Howe Island (31°32′S, 159°05′E) and Elizabeth Reef (29°55′S, 159°02′E). From this position pass N between Bellona Reefs (21°30′S, 158°50′E) and Nouvelle-Calédonie and thence between the Solomon (8°00′S, 158°00′E) and Santa Cruz (11°00′S, 166°15′E) Islands, to cross the equator in 159°E. Thence steer through the middle of the Caroline Islands (7°00′N, 150°00′E) and pass N of the Philippines.

2   The passage may be expected to be made in from 40 to 44 days.

First part of South-east monsoon (April to June)
10.35

1   During this period, steer NE as far as 157°E, and then due N as far as 11°S, thence through either Bougainville Strait (6°40′S, 156°15′E) or through Pioneer Channel (4°40′S, 153°45′E), crossing the equator in about 153°E. From this position steer to pass through the most W part of the Caroline Islands (7°00′N, 150°00′E) and through the Balintang Channel (20°00′N, 122°20′E) into the South China Sea.

Second part of South-east monsoon (July to September)
10.36

1   This route, through the Torres Strait (10°36′S, 141°51′E), is appropriate to the second part of the south-east monsoon, and the south-west monsoon of the South China Sea. It may be taken provided the vessel is through the Torres Strait before the end of September; if not then follow the directions given at 10.34 and 10.35. The Torres Strait Route may be expected to occupy about 40 days and, although not free of danger, may be navigated in safety by those experienced among coral reefs.

2   The passage from Sydney to the Torres Strait is described at 10.37. Directions for the straits and routes in the Eastern Archipelago are given in Chapter 9. In this area, there are several alternative routes between Torres Strait and Hong Kong

3   The first being to pass through Selat Wetar (8°16′S, 127°12′E) into the Flores Sea, thence along the N side of the islands and through Selat Sapudi (7°00′S, 114°15′E). From Selat Sapudi head N through Selat Karimata (1°43′S, 108°34′E), thence to Hong Kong.

4   The second alternative is to pass round the N end of Timor (9°00′S, 125°00′E), through Selat Ombi (8°35′S, 125°00′E), the Savu Sea and Selat Sumba (9°00′S, 119°30′E) to Selat Alas (8°40′S, 116°40′E), thence pass N to pass through Selat Karimata to the South China Sea. Otherwise a vessel may pass S of Timor and Sumba to Selat Alas, but this route leads through a part of the Arafura Sea with many known dangers, and probably many undiscovered ones.

5   The third alternative is to steer between Pulau-Pulau Aru (6°00′S, 134°30′E) and Pulau-Pulau Tanimbar (7°30′S, 131°30′E) to Selat Manipa (3°20′S, 127°22′E). Thence pass round the N end of Sulawesi and across the Celebes Sea, through Basilan Strait (6°50′N, 122°00′E) into the Sulu Sea. Then pass through Mindoro Strait (13°00′N, 120°00′E) into the South China Sea and Hong Kong.

Sydney ⇒ Torres Strait

Routeing details
10.37

1   From Sydney (33°50′S, 151°19′E) to the Torres Strait (10°36′S, 141°51′E), there are two routes. The Inner Route, which passes inshore of the Great Barrier Reef and the Outer Route which passes, seaward of the reefs, through the Coral Sea. The proper time for making either passage is from March to September, during the south-east monsoon. Large sailing vessels seldom take the Inner Route, but small vessels can do so without difficulty.

2   It is not desirable to reach the entrance to the Torres Strait before the beginning of April, in order to avoid the chance of an equinoctial gale and also to ensure that the south-east monsoon has begun in the Arafura Sea. Vessels have left Sydney as late as October and completed their passage, but generally it is much too late; for although the north-west monsoon does not blow until November, and sometimes later, the calms and light variable winds that precede it cause considerable delays.

Inner Route
10.38

1   Proceed as directly as possible, N along the coast, to Sandy Cape (24°42′S, 153°16′E).

2   The prevailing winds off the coast to Sandy Cape are NE from October to April and W from May to September. The mariner will have to contend with a strong S-going current running along the coast. The strength of this current is found on the edge of the charted 200 m depth contour, 20 to 30 miles from the coast, and will be avoided by keeping well outside this line.

3   Curtis Channel (24°20′S, 152°55′E) and Capricorn Channel (22°45′S, 152°00′E) are the only entrances to the Inner Route from SE and the latter is recommended. For details of these channels and of the Inner Route, see, *Australia Pilot, Volume III.*

Outer Route
10.39

1   On leaving Sydney avoid the S-going current by keeping within about 2 miles of the land until a direct course can be reasonably made to 24°S, 157°E. Thence, passing clear E of Cato Bank (23°15′S, 155°32′E) and Wreck Reefs (22°10′S, 155°20′E), proceed to 21°10′S, 156°35′E and continue on a NW course, to pass NE of Eastern Fields (10°10′S, 145°40′E) and Portlock Reefs (9°35′S, 144°50′E) to Bligh Entrance (9°12′S, 144°00′E).

Torres Strait ⇒ Sydney

Directions
10.40

1 The outer passage appears not to have been made very often, and, like the Inner Route from Torres Strait to Sydney, was formerly only considered practical during the north-west monsoon (November to February or March). The first objective after clearing Torres Strait (10°36′S, 141°51′E), in the north-west monsoon, will be to take advantage of W winds to make progress to the E, regarding immediate progress to the S as of secondary importance.

2 In the north-west monsoon, leave Torres Strait by Great North East Channel and, after clearing Eastern Fields (10°10′S, 145°40′E), take every advantages of the W breezes and try to reach a position in about 15°S, 156°E.

3 Having reached 156°E, and thus probably far enough E to take advantage of the South-east Trade Wind, proceed on a port tack and try to fetch Mellish Reef (17°25′S, 155°51′E). Great caution is necessary when in the neighbourhood of this reef and in the unexplored areas to the N of it, also there is generally a strong W-going set to contend with. If the wind permits, then pass between Kenn Reef (21°15′S, 155°48′E) and Wreck Reefs (22°10′S, 155°20′E), on the E side and Frederick Reef (21°00′S, 154°25′E) and Saumarez Reefs (21°50′S, 153°40′E) on the W side.

4 If unable to weather Frederick Reef, due to the South-east Trade Wind having too strong a S element, then pass W of it and between Saumarez Reefs and Swain Reefs (22°20′S, 152°42′E), where a S-going current will probably enable a vessel to weather Sandy Cape (24°42′S, 153°16′E), taking care to avoid Breaksea Spit (24°25′S, 153°10′E) and the shoals near its E edge.

5 As a rule a vessel should arrange to close the intermediate passage reefs in day-time, in order to check her position, as the current between Saumarez Reefs and Swain Reefs may seriously affect the vessels reckoning.

6 From Sandy Cape proceed for Sydney keeping the mainland coast in sight, to take advantage of the S-going current.

Sydney ⇒ Singapore

Directions
10.41

1 From March to September, the south-east monsoon period, the route is via the Torres Strait (10°36′S, 141°51′E), as described at 10.37. After passing through the strait, proceed by one of the two routes to Selat Karimata (1°43′S, 108°34′E), described at 10.36, thence through Selat Riau (0°55′N, 104°20′E) to Singapore (1°12′N, 103°51′E). This passage may also be made through Selat Gelasa (2°53′S, 107°18′E) or Selat Bangka (2°06′S, 105°03′E), instead of Selat Karimata.

2 A third route for March to September, from Torres Strait, is by the third alternative route given at 10.36, to Basilan Strait (6°50′N, 122°00′E) and the Sulu Sea; thence through Balabac Strait (7°34′N, 116°55′E) to the South China Sea and Singapore.

3 From November to February, during the north-west monsoon, steer to the N passing E of Papua New Guinea and, after meeting NE winds in about 5°N, pass S of Mindanao through Basilan Strait, Sulu Sea and Balabac Strait to the South China Sea and Singapore.

4 For routes through the Eastern Archipelago, see Chapter 9.

PASSAGES FROM NEW ZEALAND

Charts:
> *5127 (1) to (12) Monthly Routeing Charts for North Pacific Ocean.*
> *5128 (1) to (12) Monthly Routeing Charts for South Pacific Ocean.*
> *5309 Tracks followed by Sailing and Auxiliary Powered Vessels.*
> *5310 World Surface Current Distribution.*
> *Diagram 10.3 World Sailing Ship Routes.*

New Zealand ⇒ Australia

General information
10.42

1 In all cases steer to pass through Cook Strait (41°00′S, 174°30′E), or round the N end of North Island. These passages are always preferable to those round the S end of South Island where W winds prevail.

New Zealand ⇒ Sydney and ports northwards
10.43

1 Having cleared Cook Strait or the N end of North Island, proceed as directly as possible if bound for Sydney (33°50′S, 151°19′E) or Caloundra Head (26°48′S, 153°08′E), for Brisbane.

2 If bound for Torres Strait (10°36′S, 141°51′E), join the Outer Route, described at 10.39, at 24°S, 157°E.

3 If bound for ports N of Brisbane, join the Inner Route, described at 10.38, in Capricorn Channel.

New Zealand to southern Australia
10.44

1 Pass through Bass Strait (40°00′S, 146°00′E) if wind permits, otherwise pass S of Tasmania. Then join the routes from Sydney described from 9.220 to 9.224, and at 10.17.

New Zealand ⇒ South America

New Zealand to Cabo de Hornos
10.45

1 From Auckland (36°36′S, 174°49′E), join the trans-ocean route, described at 10.8, in about 51°S, 148°W. From Cook Strait (41°00′S, 174°30′E), join it in 170°W and from South Island ports, join it at the 180° meridian.

New Zealand to ports on the west coast of South America
10.46

1 From Auckland (36°36′S, 174°49′E), join the routes described at 10.23.1 or 10.23.2. From other departure points join the route described at 10.13. These trans-ocean routes should be joined at the nearest position.

New Zealand ⇒ San Francisco or British Columbia

Directions
10.47

1 Steer N so as to get through the Intertropical Convergence Zone, (see 7.4 and 7.13 for details), as quickly as possible; this particularly applies in July, August and September. At all times join the appropriate route from Sydney, described at 10.24, as soon as possible.

New Zealand ⇒ South Pacific Islands

Directions
10.48

1   Make to the E, keeping S of 40°S, as far as about 165°W if bound for Rarotonga (21°10′S, 159°47′W), or to 155°W if bound for Tahiti (17°35′S, 149°25′W), then head gradually N into the South-east Trade Wind and proceed direct.

2   The South-east Trade Wind is reasonably regular, among the Samoa, Tonga, Fiji and Nouvelle-Calédonie Islands, from April to October, but, from December to March, it is very light and uncertain and NW winds are frequent.

3   Cyclones sometimes pass over these localities from January to March.

New Zealand ⇒ South China Sea or Japan

Directions
10.49

1   Steer N to pick up, on 30°S, the appropriate route from Sydney at the following references:
　　　Yokohama (35°26′N, 139°43′E) at 10.29;
　　　Hong Kong (22°17′N, 114°10′E) at 10.33;
　　　Torres Strait (10°36′S, 141°51′E) at 10.37;
　　　Singapore (1°12′N, 103°51′E) at 10.41.

PASSAGES FROM ISLAND GROUPS IN THE CENTRAL SOUTH PACIFIC (BETWEEN NOUVELLE-CALÉDONIE AND ÎLES DE LA SOCIÉTÉ)

Charts:
　　　5127 (1) to (12) Monthly Routeing Charts for North Pacific Ocean.
　　　5128 (1) to (12) Monthly Routeing Charts for South Pacific Ocean.
　　　5309 Tracks followed by Sailing and Auxiliary Powered Vessels.
　　　5310 World Surface Current Distribution.
　　　Diagram 10.3 World Sailing Ship Routes.

Central South Pacific Islands ⇒ Sydney or southern Australia

Directions
10.50

1   Proceed W on about the latitude of the islands, taking full advantage of the Trade Wind and the favourable current. Pass about 150 miles S of Nouvelle-Calédonie (21°15′S, 165°20′E), then proceed, as directly as possible, to destination; or to the Bass Strait (40°00′S, 146°00′E) if bound for ports in southern Australia.

2   For passage through the Bass Strait to Port Phillip (38°20′S, 144°34′E) for Melbourne, or to Adelaide (34°48′S, 138°23′E), see from 9.220 to 9.224, and for passage from Bass Strait to Cape Leeuwin (34°23′S, 115°08′E), see 9.225.

3   For the route from Tahiti (17°35′S, 149°25′W), see 10.59.

Central South Pacific Islands ⇒ New Zealand

Directions
10.51

1   From the islands E of 170°W steer W in the Trade Wind, until that meridian is reached, then proceed as directly as possible to destination. Bear in mind that, when approaching New Zealand, a sailing vessel should, especially in winter, keep W, rather than E, of the direct route. Westerly winds are likely to be experienced S of the Trade Winds.

Central South Pacific Islands ⇒ South America

Central South Pacific Islands to Cabo de Hornos or Estrecho de Magallanes
10.52

1   From any of the Pacific island groups, head to the S through the Trade Winds and then into the W winds of the S hemisphere, then proceed by great circle to the W entrance to Estrecho de Magallanes (52°38′S, 74°46′W) or to round Cabo de Hornos (56°04′S, 67°15′W), but do not proceed S of the route from South Africa, described at 10.6.

Central South Pacific Islands ⇒ ports between Talcahuano and Panama
10.53

1   Head to the S, through the Trade Winds and into the Westerlies, to pick up the routes from Australia to the required destination in 160°W, or farther W, depending on starting point. The routes from Australia are described at 10.13, 10.23.1 and 10.23.2.

2   From July to October, however, a direct passage S of Archipel des Tuamoto (18°00′S, 141°00′W) can usually be made, passing Pitcairn Island (25°04′S, 130°05′W).

Central South Pacific Islands ⇒ San Francisco or British Columbia

Directions
10.54

1   Except from Tahiti, head to the S and join the route from Sydney, via Tahiti, described at 10.24.

Passages from Fiji

Fiji to Honolulu
10.55

1   From Fiji (18°00′S, 180°00′) head N through both Trade Winds and into the Westerlies, N of the North-east Trade Wind, thence making to the E to about 155°W, thence proceed direct to Honolulu (21°17′N, 157°53′W).

Fiji to Tahiti
10.56

1   From Fiji (18°00′S, 180°00′) head through the South-east Trade Wind into the Westerlies, then run down to the E, re-entering the Trade Wind in 150°W thence N to Tahiti (17°35′S, 149°25′W).

Passages east from Samoa

Directions
10.57

1   When sailing to the E from Samoa (14°00′S, 171°30′W), it will be found an advantage to keep on the S side of the group, where there is a favourable current and the winds are more regular and calms less frequent.

Passages from Tahiti

Tahiti to Honolulu
10.58

1   From Tahiti (17°35′S, 149°25′W), steer to the N to cross the equator in about 147°W and thence make the Hawaiian Islands (19°40′N, 155°30′W) from the E, to ensure the breeze:

2   The channel between Moorea (17°30′S, 149°50′W) and Tahiti (17°35′S, 149°25′W) should never be used by sailing ships except with steady winds from NE or SW, as these are the only winds that blow through the channel. When there is a fresh breeze from the E to the N of Tahiti. it is generally calm in this channel, and vessels have been becalmed for days, while a fresh breeze prevailed to seaward.

Tahiti to Australia or New Zealand
10.59

1   From Tahiti (17°35′S, 149°25′W) for Sydney (33°50′S, 151°19′E), run with the Trade Wind, steering to pass about 150 miles S of Nouvelle-Calédonie (21°15′S, 165°20′E), thence as directly as possible to Sydney. See 10.50 for directions to ports S of Sydney.

2   For Wellington (41°22′S, 174°50′E), run with the Trade Wind to about 170°, thence as directly as possible to Wellington. See also 10.51 for further details.

PASSAGES FROM SINGAPORE AND EASTERN ARCHIPELAGO

Charts:

> 5127 (1) to (12) Monthly Routeing Charts for North Pacific Ocean.
> 5128 (1) to (12) Monthly Routeing Charts for South Pacific Ocean.
> 5309 Tracks followed by Sailing and Auxiliary Powered Vessels.
> 5310 World Surface Current Distribution.
> Diagram 10.3 World Sailing Ship Routes.

Archipelagic Sea Lanes
10.60

1   **Archipelagic Sea Lanes (ASL)** (1.30) have been designated, by the Indonesian Authorities through the Eastern Archipelago, see 9.57 for details of these ASLs with regard to sailing vessels. Further details regarding Archipelagic Sea Lanes can be found at Appendix A and in *The Mariners' Handbook*.

Singapore ⇒ Sydney

Directions
10.61

1   From Singapore (1°12′N, 103°51′E), near and during the period of change from the south-east monsoon to the north-west monsoon (about October), sailing vessels may take five or six weeks to make the passage from Singapore to Selat Bangka (2°06′S, 105°03′E).

2   From mid-November to mid-February, the route is through the Torres Strait (10°36′S, 141°51′E). Leaving Singapore in the north-west monsoon, proceed through Selat Bangka or Selat Karimata (1°43′S, 108°34′E) and enter the Indian Ocean by Selat Lombok (8°47′S, 115°44′E) or Selat Alas (8°40′S, 116°40′E), after passing through Selat Sapudi (7°00′S, 114°15′E). Having reached the Indian Ocean, steer to pass N or S of Timor (9°00′S, 125°00′E) and thence to Torres Strait, see also 9.193. Continue as directed at 10.40.

3   From April to October, follow the directions given at 9.194, through Selat Sunda (6°00′S, 105°52′E) and into the Westerlies, then pass S of Australia and through the Bass Strait (40°00′S, 146°00′E) to Sydney. See also 9.6, 9.10 and 10.16 for further directions.

Singapore ⇒ Molucca Archipelago

Directions
10.62

1   From Singapore (1°12′N, 103°51′E), the passage should be made S of Borneo in the north-east monsoon and N of Borneo in the south-west monsoon.

10.62.1

1   From October to May, the north-east monsoon blows N of the equator and the north-west monsoon blows S of it. Proceed through Selat Karimata (1°43′S, 108°34′E), passing E of Karang Ontario (2°00′S, 108°39′E). On leaving the strait steer to pass 10 to 15 miles S of Gosong Gia (5°13′S, 113°17′E) and about 10 miles S of Masalembo Besar (5°33′S, 114°27′E), thence proceed to Selat Salayar (5°43′S, 120°30′E) as directly as possible.

2   After clearing Selat Salayar, Ambon (3°41′S, 128°10′E) is easily reached by passing S of Batu Ata (6°12′S, 122°42′E) and Pulau Binongko (5°57′S, 124°01′E); if bound for the Ceram Sea, first round the S point of Pulau Butung (5°00′S, 122°55′E) and then, after skirting the shore of that island and having passed Wangi-Wangi (5°20′S, 123°35′E), steer N as far as Pulau Wowoni (4°05′S, 123°05′E). Thence run for the S point of Pulau Sanana (2°15′S, 125°55′E) and thence into the Ceram Sea.

3   The currents, in this locality, set to the S and are very strong. If a vessel has been set to leeward of the N point of Pulau Buru (3°30′S, 126°30′E), it is best pass S of that island and thence through Selat Manipa (3°20′S, 127°22′E) to the Ceram Sea.

10.62.2

1   From May to September, the south-west monsoon blows N of the equator and the south-east monsoon blows S of it. Run S of Pulau-Pulau Anambas (3°00′N, 106°00′E) and then between Royal Charlotte Reef (6°56′N, 113°36′E) and Louisa Reef (6°20′N, 113°14′E), taking care to avoid the dangerous shoals bordering the Borneo coast and also of being set to leeward of Pulau Balambangan (7°18′N, 116°55′E) by the N-going current which prevails in the south-west monsoon. Having made Pulau Balambangan, round its N point and steer through Balabac Strait (7°34′N, 116°55′E) into the Sulu Sea. Cross the Sulu Sea to Sibutu Passage (4°55′N, 119°37′E), or one of the other passages through the Sulu Archipelago, and then cross the Celebes

Sea passing N of Sulawesi, thence work S through the Molucca Sea.

2 For an alternative route as far as Balabac Strait during the south-west monsoon, see 10.63.2.

3 Directions for the straits and channels in the Eastern Archipelago, and details of Archipelagic Sea Lanes, are given in Chapter 9, see from 9.56 to 9.58. See *China Sea Pilot, Volume II* for details of currents in the Palawan Passage.

Singapore ⇒ Sulu Sea

Directions
10.63

1 When E-bound in the Singapore Strait follow the reverse of the directions given at 9.82. The main route is via Balabac Strait (7°34′N, 116°55′E), which is approached by varied routes depending on season. The passage may also be made by proceeding S through Selat Karimata (1°43′S, 108°34′E), or Selat Gelasa (2°53′S, 107°18′E), through the Java Sea and then N through Selat Makasar (2°00′S, 118°00′E); but this route can only be recommended during October and November.

10.63.1

1 During the north-east monsoon (October to May), the route is via the Balabac Strait. During December to February, do not leave the Singapore Strait in strong NE winds, but find anchorage, under the N shore, as advised by the Port Authorities. In these months gales often occur with thick weather and rain lasting two or three days, while the SSE-going current outside the strait attains a rate of 2½ or 3 kn. A vessel leaving under these conditions would set bodily to leeward and, unable to reach Pulau Pejantan (0°07′N, 107°12′E), would have to work up the W coast of Borneo. Fine weather follows, with the wind backing to N or NW and the current offshore decreases to about 1¼ kn.

2 Having obtained the fine weather, the first objective should be to pass through the channel between Pulau-Pulau Natuna Besar (4°00′N, 108°00′E) and Pulau-Pulau Subi Besar (2°50′N, 108°50′E) or, if this proves impossible or difficult, use one of the passages to the S.

3 To pass through the channel between Pulau-Pulau Natuna Besar and Pulau-Pulau Subi Besar a vessel should leave the anchorage on the first of the out-going tidal stream, then steer to the NE, passing S of Pulau Midai (3°00′N, 107°47′E), then through the channel between Pulau-Pulau Natuna Besar and Pulau-Pulau Subi Kecil (3°03′N, 108°51′E), which is lit. This passage may be made in these months without much difficulty, especially at spring tides, when, it is reported, the wind after a few hours calm frequently shifts to the W, with squalls and rain, and then backs to SW and S, blowing moderately for 24 hours. Advantage can be taken of these conditions to weather Pulau-Pulau Subi Besar.

4 If after arriving in the vicinity of Pulau Midai, the wind continues E, steer to the N on the starboard tack, passing W of Pulau Midai, keeping at least 3 miles from its SW side to avoid shoal water extending 2½ miles from it. Pass about 5 miles W of Pulau Timau (3°18′N, 107°32′E) as coral reefs extend fully 3 miles from its SW side, with a least depth of 7 m. It is not recommended to pass between Pulau Timau and Pulau Midai on account of Karang-Karang Diana (3°03′N, 107°46′E). But, if the vessel's position can be verified and the wind permits, a vessel could pass through 4 or 5 miles S of Pulau Timau and Pulau Sedimin (3°24′N, 107°50′E), thus giving a

sufficiently wide berth to Karang-Karang Diana. However, the route S of Pulau Midai is to be preferred.

5 If passing W of Pulau Medai, when clear of Pulau Timau, there will be no difficulty in working towards the S point of Pulau-Pulau Natuna Besar as that island, when approached from the SW, gives shelter from the strong SW-going current of the monsoon. Off its S coast, at night in fine weather, the wind is off the land, but S and SE coasts should not be approached nearer than 6 or 7 miles on account of the off-lying dangers.

6 If being set to leeward of Pulau-Pulau Subi Besar, with a N wind, take one of three channels to the S of the group:

Alur Pelayaran Kota (2°40′N, 109°00′E), lies between Pulau Panjang and Pulau Serasan and its surrounding islands. The strait is clear apart from Karang Haynes (2°34′N, 108°51′E) lying in its W approaches.

Alur Pelayaran Serasan (2°24′N, 109°00′E), the sides of which are clear but the middle is encumbered by reefs.

Alur Pelayaran Api (2°00′N, 109°08′E), the currents in which are irregular and set in various directions and with considerable velocity to the SW for 16 to 19 hours at a time; for large vessels either of the other channels are preferable to this, as considerable caution and perseverance are required to work through. If using it keep to the Borneo side, in depths of 18 to 20 m, to avoid the current and to profit by the land winds. Directions for Alur Pelayaran Kota and Alur Pelayaran Api are given below.

7 When transiting Alur Pelayaran Kota give Pulau Panjang a good berth to avoid the reef which surrounds it and extends off its SW end. The winds among these islands, and as far E as Tanjung Sirik (2°47′N, 111°19′E), are generally from N to NNW. Once the passage is cleared proceed to the NE endeavouring, if not certain of longitude, to make Royal Charlotte Reef (6°56′N, 113°36′E), or Louisa Reef (6°20′N, 113°14′E), whichever is more accessible due to weather, by running on its parallel of latitude. The currents appear to be influenced by the prevailing winds and a set in the direction that it is blowing should be anticipated, the velocity of the current being in proportion to the force of the wind. Having reached either of these reefs, or having passed between them, steer E for about 100 miles towards Balabac Strait (7°34′N, 116°55′E) and through it to the Sulu Sea. See *China Sea Pilot, Volume II* and *Philippine Islands Pilot* for the W approaches to, and passage through, the Balabac Strait to the Sulu Sea.

8 Sailing vessels which have been forced to the S and having to work up the Borneo coast towards Alur Pelayaran Api may find favourable tidal streams near the shore, when a strong adverse current is running farther offshore, especially during the north-east monsoon. Pulau-Pulau Burung (0°45′N, 108°45′E) may be boldly approached from the W. Large sailing vessels are advised to pass outside the group, but smaller craft may, with some advantage, pass between them. For directions, see *China Sea Pilot, Volume II.*

10.63.2

1 In October and November, only, an alternative route via Selat Makasar (2°00′S, 118°00′E) may be taken. A vessel should proceed S to pass through Selat Gelasa (2°53′S, 107°18′E) or Selat Karimata (1°43′S, 108°34′E) and thence through the Java Sea to Selat Makasar, there joining the

First Eastern Passage, described at 9.107, and following it to the Sulu Sea.

10.63.3

1 During the south-west monsoon (May to September), first proceed to Alur Pelayaran Api as described below and thence, with a fair wind, parallel with the Borneo coast, as far as Balabac Strait, thence into the Sulu Sea. Directions from Singapore to Tanjung Api (1°57'N, 109°20'E), are as follows.

2 As far as the E entrance to the Singapore Strait the tidal streams are tolerably regular, but some miles offshore a current will be found setting about NNW in the south-west monsoon; its greatest strength will be encountered between Pulau Tioman (2°50'N, 104°10'E) and Pulau-Pulau Anambas (3°00'N, 106°00'E). To obviate the effect of this current, it is recommended to make good a course from the Singapore Strait to Pulau Mandariki (1°19'N, 107°02'E), by this route, should light airs prevail, there is the option of steering between Pulau Pengibu (1°35'N, 106°19'E) and Pulau Kayuara (1°32'N, 106°27'E) or S of Kayuara, thus avoiding Batu Acasta (1°39'N, 106°18'E). On passing Kayuara set course, allowing for a N-going set, to pass well S of Pulau Muri (1°54'N, 108°39'E), then keep Muri Light bearing 255°, which will lead about 2 miles S of Pulau Merundung (2°04'N, 109°06'E).

3 Banggi South Channel and Malawali Channel, between Pulau Banggi (7°15'N, 117°10'E) and Borneo are sometimes used by vessels navigating to ports on the NE coast of Borneo, but are intricate and demand careful navigation being for the greater part bounded by dangers. Balabac Main Channel (7°26'N, 117°14'E) is recommended, being much safer than either of these channels. See *China Sea Pilot, Volume II* and *Philippine Islands Pilot* for descriptions and directions for these channels.

Singapore ⇒ Manila

Passages for sailing vessels

10.64

1 From Singapore (1°12'N, 103°51'E) routes are seasonal depending on monsoon conditions.

10.64.1

1 During the north-east monsoon (October to May), follow the directions at 10.63.1 as far as the entrance to the Balabac Strait and continue thence to the N via the Palawan Passage (10°20'N, 118°00'E), between the charted 200 m depth contour W of Palawan and that of the off-lying foul ground. This channel is about 40 miles wide, except towards the S end where, between Royal Captain Shoal (9°03'N, 116°41'E) and Paragua Ridge (8°57'N, 117°21'E), it is 28 miles wide, which is the most dangerous part of the channel. From the N end of Palawan Passage, about 11°N, work to the N of Manila (14°32'N, 120°56'E), hugging the coast when possible. See *China Sea Pilot, Volume II* for details of currents in the Palawan Passage.

10.64.2

1 During the south-west monsoon (May to September), follow the directions at 10.63.3 to Tanjung Api, then proceed directly along the coast and through the Palawan Passage, with a fair wind, to Manila.

2 An alternative route is to pass S of Pulau-Pulau Anambas (3°00'N, 106°00'E) and Pulau-Pulau Natuna Besar (4°00'N, 108°00'E), then between Royal Charlotte Reef (6°56'N, 113°36'E) and Louisa Reef (6°20'N, 113°14'E), to pick up the Palawan Passage off Balabac Strait.

3 When working through the Palawan Passage, having conformed with the directions given for making the SW end of Palawan, in fine weather try to tack towards the shore in the afternoon, for the sun will then be astern and the patches near the edge of the bank will be more visible from aloft, enabling a tack offshore in ample time. In squally weather, and during heavy rain, these patches have been observed to impart a very distinct yellowish hue to the surface of the water. It is very desirable to get soundings before dark in order that a good departure may be made for the night. When on the inshore tack be prepared to tack offshore immediately on getting the first indication of the bank, which can shoal rapidly. See *China Sea Pilot, Volume II* for details of currents in the Palawan Passage.

4 When approaching the islands in the vicinity of Balabac Island (8°00'N, 117°00'E) and Palawan, if the wind is well to the S and the weather thick, Balabac Island may be approached near enough to obtain a good observation, but do not approach closer than 12 miles off, as extensive shoals lie off the W coast of the island. If the wind is W, with thick cloudy weather, Balabac Island should not be approached closer than 30 miles, as the wind usually forces a strong E-going current through the passage.

5 Off the SW end of Palawan, it is not unusual, particularly in squalls, for the wind to veer to the WNW, and sometimes NW, blowing with violence and placing the vessel on a lee shore with respect to the shoals inside the edge of the bank. This weather usually prevails off Palawan about September and October, rendering it uncertain and difficult to make the narrowest part of the channel, owing to the land being obscured.

Passage for low-powered vessels

10.65

Palawan Passage is useful for low-powered vessels N-bound during the north-east monsoon and may be more suitable than the routes given at 7.86 and 7.97 for full-powered vessels.

The routes between Singapore (1°12'N, 103°51'E) and Palawan Passage (10°20'N, 118°00'E) are given at 7.102. For cautions on position fixing, currents and directions for Palawan Passage, see *China Sea Pilot, Volume II*.

From the N end of the Palawan Passage, the route continues N to the W coast of Luzon and thence to Manila or onwards towards Hong Kong (22°17'N, 114°10'E) or the Taiwan Strait (24°00'N, 119°00'E).

Singapore ⇒ Hong Kong

Passages for sailing vessels

10.66

1 From Singapore (1°12'N, 103°51'E) routes are seasonal depending on monsoon conditions.

10.66.1

1 During the north-east monsoon (October to May), a route similar to the Main Route, described for powered vessels at 7.86.1, except that it passes initially between Pulau-Pulau Anambas (3°00'N, 106°00'E) and Pulau-Pulau Natuna Besar (4°00'N, 108°00'E), may be used, although it cannot be strongly recommended.

2 During the strength of the north-east monsoon use the Palawan Passage, as described at 10.64.1, as far as the N end of Palawan, then work up the coast of Luzon as far as Cape Bolinao (16°19'N, 119°47'E) or Cape Bojeador (18°30'N, 120°34'E). Strong winds and consequent high seas prevail between Cape Bolinao and Hong Kong (22°17'N, 114°10'E) at this time. Among the island groups

N of Luzon, winds are lighter and more irregular. See also notes on Pratas Reefs at 10.86.3 and Taiwan Strait at 10.89.

3 An alternative route through Selat Gelasa (2°53′S, 107°18′E) or Selat Karimata (1°43′S, 108°34′E) and thence through the Java Sea, Selat Salayar (5°43′S, 120°30′E) and the Banda Sea to join the Second Eastern Passage, described at 9.101, affords a leading or fair wind and favourable currents nearly throughout.

10.66.2

1 Between the monsoons, a route on the W side of the South China Sea is recommended, passing along the coast of Malaysia to Pulau Redang (5°47′N, 103°00′E), thence along the coast of Vietnam to 16°N, coastwise off the E side of Hainan Dao (19°10′N, 109°30′E) and inshore of Qizhou Liedao (19°58′N, 111°16′E) to make the mainland coast about Dafangi Dao (21°23′N, 111°12′E).

2 Note that NE and W gales blowing out of the Gulf of Tonkin, with dark weather and rain, have been experienced on this route, causing danger of being driven among the Paracel Islands (16°30′N, 112°00′E), but such gales are not frequent and the land should kept in sight, for smoother seas and the availability of anchorage.

3 Approaching Hong Kong, due to entry restrictions (see below), try to make a position about 8 miles S of Dawanshan Dao (21°57′N, 113°44′E), then steer S of Wenwei Zhou (21°49′N, 113°56′E), thence S of Jiapeng Liedao and S and E of Dangan Liedao to enter Sai Pok Liu Hoi Hap (West Lamma Channel) (22°13′N, 114°05′E) or Tung Pok Liu Hoi Hap (East Lamma Channel) (22°13′N, 114°10′E).

4 **Entry restricted**. The Chinese Authorities have declared that non-Chinese vessels are prohibited from navigation within Dangan Shuidao (Lema Channel) and within the area north of a line connecting the islands Xiaowanshan Dao, Dawanshan Dao, Jiapeng Liedao and Dangan Liedao. Exceptions are made for vessels entering Zhujiang Kou where they are required to follow the defined route. For details see *China Sea Pilot, Volume I.*

10.66.3

1 During the south-west monsoon (May to September), the Main Route, described for powered vessels at 7.86.1 is appropriate, except that it should pass between Pulau-Pulau Anambas (3°00′N, 106°00′E) and Pulau-Pulau Natuna Besar (4°00′N, 108°00′E), if the monsoon has not settled in.

2 The route on the W side of the South China Sea, described at 10.66.2, may also be used during the earlier part of the south-west monsoon.

3 During the latter part of the south-west monsoon, a route through the Palawan Passage is recommended, observing that, at this time of year, a N-bound route in the W part of the South China Sea is hampered by strong S-going currents in the vicinity of Îles Catwick (10°00′N, 109°00′E), with light winds, variable airs or calms. Steer N of Pulau Pengibu (1°35′N, 106°19′E) and between Pulau-Pulau Natuna Besar (4°00′N, 108°00′E) and Pulau-Pulau Subi Besar (2°50′N, 108°50′E) and thence for the Palawan Passage, as directed at 10.64.2.

Passage for low-powered vessels

10.67

1 Palawan Passage (10°20′N, 118°00′E) is useful for low-powered vessels N-bound during the north-east monsoon and may be more suitable than the routes given at 7.88 for full-powered vessels. See 10.65 for directions.

Singapore ⇒ ports north of Hong Kong

Passages for sailing vessels

10.68

1 From Singapore (1°12′N, 103°51′E) routes are seasonal depending on monsoon conditions.

10.68.1

1 During the north-east monsoon (October to May), proceed by the Palawan Passage, as described at 10.66.1, as far as Cape Bolinao (16°19′N, 119°47′E), then continue to work up the coast of Luzon and through the Balintang Channel (20°00′N, 122°20′E). Proceed up the E coast of T'ai-wan and to destination.

2 An alternative route during the north-east monsoon is via Selat Karimata and the Second Eastern Passage, as directed at 10.66.1 and 9.101.

10.68.2

1 Near the change of the monsoon, the inner route, described at 10.66.2 may be taken as far as Hong Kong (22°17′N, 114°10′E), except during the latter part of the south-west monsoon, thence along the coast of China to destination.

10.68.3

1 During the south-west monsoon (May to September), except the latter part, take one of the routes advised at 10.66.3 as far as Hong Kong and then continue along the coast of China to destination. In the latter part of this monsoon, use the Palawan Passage route.

Singapore ⇒ Ho Chi Minh City

Passages for sailing vessels

10.69

1 From Singapore (1°12′N, 103°51′E) to Ho Chi Minh City (10°50′N, 106°40′E), routes are seasonal depending on monsoon conditions.

10.69.1

1 During the south-west monsoon (May to September), the winds in the Singapore Strait are between SE and W and sailing vessels will have no difficulty getting through to the E.

2 Having cleared the Straits, steer to pass W of Con Son (8°38′N, 106°37′E), and thence along the edge of the bank fronting the mouths of the Mekong River, extending to the mouth of Song Sai Gon (10°18′N, 103°51′E).

3 Strong freshets run out of these rivers during south-west monsoon, to join the NE-going current, whereby vessels are required to keep the edge of the bank to avoid being set to leeward of Mui Vung Tau (10°19′N, 107°05′E). Keep sounding continuously whilst steering along the edge of the bank so as to remain in depths not less than 18 m. If the water begins to shoal, haul off to the E, when it will soon deepen as the depths are fairly regular. Continue along the edge of the bank in these depths until Mui Vung Tau bears less than 330°, when course may be set for the Song Sai Gon pilot.

10.69.2

1 During the north-east monsoon (October to May), steer as directed in 10.63.1 until clear of Pulau-Pulau Natuna Besar (4°00′N, 108°00′E) and then steer NE until reaching 112°E; after which head across the South China Sea to make Mui Vung Tau, or preferably the land to windward of the Cape, to avoid being set to leeward by the prevailing current.

2 From 7°N until 70 miles E of the mouths of the Mekong River, a strong current will be found setting to the SW, governed considerably by the prevailing wind, for when strong gales blow in the early part of this monsoon the SW-going current is stronger and often runs at a rate of

3 kn. The tidal streams are regular and set strongly near the Vietnamese coast during both monsoons.

3   In the latter part of March and April an E wind is often found E of Pulau-Pulau Anambas (3°00′N, 106°00′E) that will take a vessel to Con Son, from W of that island work to Mui Vung Tau, keeping towards the Vietnamese coast, which is very low and can seldom be seen at night.

4   From abreast the mouths of the Mekong River, the out-going stream will be found setting to windward, greatly assisting vessels standing inshore, but they should not stand close during the in-going stream, nor in less than 22 m of water during the night. Soundings should never be neglected when heading towards this low-lying land, which may be seen from a distance of about 10 miles in clear weather.

5   From December to February, and sometimes into March, the north-east monsoon often blows very strongly about the latitude of Îles Catwick (10°00′N, 109°00′E), and between the islands and the Vietnamese coast, continuing for two or three days with a heavy sea and strong current, the sky being generally thick and hazy throughout. A gradual rise in the barometer is a sure indication of an increase in the strength of the monsoon. If the monsoon proves too strong to contend with, head up for Con Son, where good shelter will be found, and anchor.

6   At about 90 miles from the coast, the wind, during settled weather, usually becomes ENE and E at about 1600 hrs, continuing fresh and blustery all night. This is the time to stand inshore, making allowance for the N-going set during the out-going tide, which should find the vessel to windward of Mui Ky Van (10°23′N, 107°16′E) in the morning in a good position to approach Song Sai Gon.

Passage for low-powered vessels
10.70

1   In July, during the strength of the south-west monsoon, or in December or January during the north-east monsoon, low-powered, or hampered, vessels may find the following routes more advantageous than those for full-powered vessels described at 7.87.

2   From Singapore to Ho Chi Minh City, in July, provided the ship's position is certain, the route passes W of Con Son (8°38′N, 106°37′E). In December and January, the best route is probably E of Pulau-pulau Natuna Besar (4°00′N, 108°00′E).

Singapore ⇒ Krung Thep

Passages for sailing vessels
10.71

1   From Singapore (1°12′N, 103°51′E) to Krung Thep (13°23′N, 100°35′E), routes are seasonal depending on monsoon conditions.

10.71.1

1   During the south-west monsoon (May to September), the winds in the Singapore Strait are between SE and W and sailing vessels will have no difficulty getting through to the E.

2   Having cleared the Straits, shape course for Pulau Redang (5°47′N, 103°00′E) and thence keep close to the W coast of the Gulf of Thailand, passing inside Ko Losin (7°20′N, 101°59′E) and Ko Kra (8°24′N, 100°44′E).

10.71.2

1   During the north-east monsoon (October to May), steer as directed in 10.63.1 until clear of Pulau-Pulau Natuna Besar (4°00′N, 108°00′E) and then steer NE until reaching 111°E or 112°E. This can be easily achieved, as the wind is invariably from N to NNW until reaching these meridians, E of which it generally veers to the NE when course should be set across the South China Sea to Hon Khoai (8°26′N, 104°49′E). Little or no current will be experienced until reaching 6°N or 7°N, when it will be found setting strongly to the SW, governed to a large extent by the prevailing winds.

2   In April and May, the best passages to the Gulf of Thailand are made by keeping close to the Malaysian coast but expect squalls, calms and rain, a weak current begins to set to the NE about this period.

Passage for low-powered vessels
10.72

1   In July, during the strength of the south-west monsoon, or in December or January during the north-east monsoon, low-powered, or hampered, vessels may find the following route more advantageous than those for full-powered vessels described at 7.87.

2   From Singapore to Krung Thep, in July, after passing Pulau Redang (5°47′N, 103°00′E), the route continues along the W shore of Gulf of Thailand. In December and January, the best route is probably E of Pulau-pulau Anambas (3°00′N, 106°00′E).

Eastern Archipelago ⇒ China

Passages for sailing vessels
10.73

1   Generally the various routes, according to season, are described from 9.57 to 9.130.

2   Passing N through the straits to the W of Borneo, usually from May to September, take the Main Route, described for powered vessels at 7.86.1, or the route on the W side of the South China Sea, described at 10.66.2.

3   Between November and April, the Second Eastern Passage, described at 9.101, is recommended. The First Eastern Passage, described at 9.107, can be considered but the disadvantages outweigh the advantages.

PASSAGES FROM KRUNG THEP OR HO CHI MINH CITY

Charts:
   5127 (1) to (12) Monthly Routeing Charts for North Pacific Ocean.
   5128 (1) to (12) Monthly Routeing Charts for South Pacific Ocean.
   5309 Tracks followed by Sailing and Auxiliary Powered Vessels.
   5310 World Surface Current Distribution.
   Diagram 10.3 World Sailing Ship Routes.

Krung Thep or Ho Chi Minh City ⇒ Hong Kong or ports northwards

Directions
10.74

1   From Krung Thep (13°23′N, 100°35′E) or Ho Chi Minh City Approaches (10°18′N, 107°04′E) to Hong Kong (22°17′N, 114°10′E), or ports farther northward, routes are seasonal depending on monsoon conditions.

10.74.1

1   During the north-east monsoon (November to April), any attempt to work N, from Krung Thep or Ho Chi Minh City, especially in the full strength of the monsoon, is so certain to be difficult that mariners are advised to head S to Pulau-Pulau Natuna Besar (4°00′N, 108°00′E). Then to join the Palawan Passage route, described at 10.66.1; or, if near the change of the monsoon, take the coastwise route, described at 10.66.2, if bound for Hong Kong.

2   If bound for ports N of Hong Kong either of these routes is possible, but consideration should be given to a route embodying, the Second Eastern Passage, described at 10.66.1.

10.74.2

1   During the south-west monsoon, if from Krung Thep, follow generally the directions given at 10.66.3 or 10.68.3.

Krung Thep ⇒ Singapore

Passage for sailing vessels
10.75

1   From Krung Thep (13°23′N, 100°35′E) to Singapore (1°12′N, 103°51′E), routes are seasonal depending on monsoon conditions.

10.75.1

1   During the north-east monsoon (November to April), the passage from Krung Thep, S through the Gulf of Thailand, will often be shortened by sighting Kaoh Kursrovie (11°07′N, 102°47′E) and passing inshore of Kaoh Tang (10°18′N, 103°08′E). Thence keep well E of Hon Panjang (9°18′N, 103°28′E) and steer well out to sea for a quick passage. Pass about 20 miles E of Pulau Tenggol (4°48′N, 103°41′E) and E of Pulau Aur (2°27′N, 104°31′E). See also 10.82.1.

10.75.2

1   During the south-west monsoon (May to September), keep close to the W shore of the Gulf of Thailand, passing inside Pulau-Pulau Perhentian (5°55′N, 102°45′E) and Pulau Redang (5°47′N, 103°00′E), Pulau Kapas (5°13′N, 103°16′E) and Pulau Tenggol. South of Pulau Kapas keep close inshore to avoid the current, passing inside Pulau Tioman (2°50′N, 104°10′E), Pulau Seri Buat (2°41′N, 103°55′E) and Pulau Sibu (2°13′N, 104°05′E). Then proceed to the Singapore Strait, taking advantage of the tidal streams and the land and sea breezes which prevail during settled weather in this monsoon.

2   The inshore channel extending from Pulau Sibu to Pulau Seri Buat, formed by a chain of islands and rocks parallel with the mainland, is a good safe one, having few hidden dangers and good anchorage all the way through.

Passage for low-powered vessels
10.76

1   In July, during the strength of the south-west monsoon, or in December or January during the north-east monsoon, low-powered, or hampered, vessels may find the following route more advantageous than those for full-powered vessels described at 7.87.

2   From Krung Thep to Singapore, in July, the route is along the W shore of Gulf of Thailand to Pulau Redang, thence inside Pulau Tenggol (4°48′N, 103°41′E), thence close inshore to Singapore Strait. In December and January the route is along the E shore of Gulf of Thailand, inshore of Kaoh Tang (10°18′N, 103°08′E) and Hon Panjang

(9°18′N, 103°28′E), (see *China Sea Pilot, Volume I*), thence to Singapore Strait.

Ho Chi Minh City ⇒ Singapore

Passage for sailing vessels
10.77

1   From Ho Chi Minh City Approaches (10°18′N, 107°04′E) to Singapore (1°12′N, 103°51′E), routes are seasonal depending on monsoon conditions.

10.77.1

1   During the north-east monsoon (November to April), from a position off Mui Vung Tau (10°19′N, 107°05′E), shape a course to pass E of Con Son (8°38′N, 106°37′E) and thence direct to Pulau Aur (2°27′N, 104°31′E). From Pulau Aur to Singapore proceed according to directions as from Hong Kong given at 10.82.1.

10.77.2

1   During the south-west monsoon (May to September), either the direct route, or a route E of Pulau-Pulau Natuna Besar (4°00′N, 108°00′E) may be taken. The latter is probably the better.

2   On the direct route, many good passages have been made by keeping close to the Vietnamese coast as far as Hong Trung Nho (8°35′N, 106°05′E), W of Con Son or Hon Khoai (8°26′N, 104°49′E), and then crossing the Gulf of Thailand with a strong NW wind until the Malaysian coast is reached. From Pulau Kapas (5°13′N, 103°16′E), follow the directions given at 10.75.2 for Krung Thep to Singapore.

3   Alternatively, the passage E of Pulau-Pulau Natuna Besar is considered, generally speaking, to be better, especially for large vessels.

4   After making departure from Mui Vung Tau, steer to the SW until the south-west monsoon forces the vessel off to a more SE course. This may be accomplished by taking every advantage of the N and NE winds, which frequently blow at night, and in some parts of the day, within a short distance of the coast. These local winds often carry vessels 40 or 50 miles SW of Con Son without any interruption.

5   While standing to the SE the full strength of the NE-going current will be met with in the neighbourhood of Charlotte Bank (7°08′N, 107°35′E); it gradually decreases and becomes slightly favourable when NE of Pulau-Pulau Natuna Besar. In this locality SE and E winds will generally be met with, and fast sailing vessels frequently pass through the channel between Pulau-Pulau Subi Besar (2°50′N, 108°50′E) and Pulau Midai (3°00′N, 107°47′E) and into the Singapore Strait. The channel is safe for all vessels and there are lights on Pulau Subi Kecil and Pulau Midai.

6   Strong W winds, with rain, frequently blow during the early part of this monsoon and may force vessels E to about 111°30′E. When this is the case make for Alur Pelayaran Api (2°00′N, 109°08′E), as described at 10.63.1, keeping close to the NW coast of Borneo from Tanjung Api (1°57′N, 109°20′E) to the S until Pulau-Pulau Burung (0°45′N, 108°45′E) is reached. This will be accomplished without difficulty, for strong land and sea breezes prevail, and the current is weaker nearer the coast. Do not leave the coast of Borneo too soon, as many vessels that have were not able to reach farther S than Pulau Aur (2°27′N, 104°31′E) or Pulau Tioman (2°50′N, 104°10′E).

7   Leaving Pulau-Pulau Burung, pass either N or S of Pulau-Pulau Tambelan (1°00′N, 107°34′E). If the wind is slight from the SW after leaving these islands, try to reach Pulau Mapor (1°00′N, 104°50′E) off the E side of Bintan.

8   The current offshore runs strongly to the N and through Alur Pelayaran Api. Vessels coming through this passage should keep to the N side, when possible, towards Pulau Merundung (2°04′N, 109°06′E), and should keep in depths of more than 24 m on the S side between Tanjung Datu (2°05′N, 109°38′E) and Tanjung Api. The latter point has steep-to shoals at 1½ miles off, but beyond that distance there is not less than 9 m between it and Tanjung Datu. Vessels should be ready to anchor in the passage or off any other part of the coast, as the tidal streams are greatly influenced by the current, which often changes without warning.

Passage for low-powered vessels
10.78

1   In July, during the strength of the south-west monsoon, or in December or January during the north-east monsoon, low-powered, or hampered, vessels may find the following route more advantageous than those for full-powered vessels described at 7.87.

2   From Ho Chi Minh City to Singapore, in July, the route is along the coast of Vietnam, thence across the mouth of the Gulf of Thailand to the Malaysian coast, passing inshore of Pulau Tioman (2°50′N, 104°10′E) and Pulau Sibu (2°13′N, 104°05′E), thence close inshore to Singapore. In December and January the route is E of Con Son, thence direct to Singapore.

PASSAGES FROM PORTS IN CHINA

Charts:

> 5127 (1) to (12) *Monthly Routeing Charts for North Pacific Ocean*.
> 5128 (1) to (12) *Monthly Routeing Charts for South Pacific Ocean*.
> 5309 *Tracks followed by Sailing and Auxiliary Powered Vessels*.
> 5310 *World Surface Current Distribution*.
> Diagram 10.3 *World Sailing Ship Routes*.

China or Japan ⇒ Indian Ocean
Summary of routes
10.79

1   Directions for the principal passages most frequently used by sailing vessels are given elsewhere in this book. References are as follows:

> Main routes at 9.114, 9.123, 9.127.
> Eastern Archipelago to Indian Ocean at 9.188 to 9.199.
> Hong Kong to Singapore at 10.82.
> Shanghai to Indian Ocean at 10.94.
> Hong Kong to Manila at 10.86.
> Manila to Indian Ocean and Australia at 10.104.

10.79.1

1   **The Western Route** (9.114) passes through the South China Sea to the W of the Philippine Islands and of Borneo to Selat Sunda through the Eastern Archipelago either direct, or via Singapore. The selection of which alternative to follow depends to a great degree on the final destination to be reached.

10.79.2

1   **The Eastern Route** (9.123) passes E of the Philippine Islands and then via Selat Jailolo, or the Molucca Sea, into the Ceram Sea and the Banda Sea. Thence it continues to Selat Ombi, or to one of the central straits (Selat Alas, Selat Lombok or Selat Bali). If bound for the Torres Strait the passage from the Banda Sea would be as described for the Hong Kong to Torres Strait route, described at 10.83, and the Hong Kong to Darwin route, described at 10.84.1.

10.79.3

1   **The Central Route** (9.127) passes W of the Philippine Islands, through the Sulu Sea and the Basilan Strait, E of Borneo through Selat Makasar and thence to one of the central straits (Selat Alas, Selat Lombok or Selat Bali). Alternatively a vessel, after leaving Selat Makasar, may head W through the Java Sea to enter the Indian Ocean through Selat Sunda.

Choice of route
10.80

1   Of the three principal routes, the Western and Central are used by vessels from the S coast of China. The Central Route is also used by vessels S-bound from Manila, the S part of the Philippines or the E part of Borneo. The Eastern Route is used by vessels from ports in the N part of China, from Japan and also from ports in the S part of China during the south-west monsoon.

Seasonal variation for routes from the south coast of China
10.81

1   During the north-east monsoon (September to February), pass between Macclesfield Bank (15°50′N, 114°30′E) and the Paracel Islands (16°30′N, 112°00′E), then about 60 miles E of Îles Catwick (10°00′N, 109°00′E), keeping to the E where the winds are more favourable. Then pass between Pulau-Pulau Anambas (3°00′N, 106°00′E) and Pulau-Pulau Natuna Besar (4°00′N, 108°00′E), to Selat Bangka (2°06′S, 105°03′E) or Selat Gelasa (2°53′S, 107°18′E), thence continue to Selat Sunda (6°00′S, 105°52′E).

2   In March, April and early May, after leaving the coast of China, head over to the coast of Luzon and proceed through the Palawan Passage (10°20′N, 118°00′E), along the coast of Borneo, through Alur Pelayaran Api (2°00′N, 109°08′E), past Pulau Pengiki Besar (0°15′N, 108°03′E) and through Selat Karimata (1°43′S, 108°34′E), thence close around Pulau Jagautara (5°12′S, 106°27′E) and direct to Selat Sunda. On this route E winds, without calms, but with fine weather and a smooth sea are likely to be experienced.

3   **Cautions**. Areas of the Java Sea are being exploited for natural resources, see 7.38.2. This route passes close to oilfields, with extensive areas of prohibited anchorage, it would be prudent to avoid these areas altogether wherever possible.

4   Alternatively, at the end of April or the beginning of May, head towards the Macclesfield Bank and then follow the Central Route, described at 10.79.3, by heading SE to join it at Verde Island Passage (13°36′N, 121°00′) or Mindoro Strait (13°00′N, 120°00′E).

5   From the middle of May to the end of July, cross the South China Sea and pass through Balintang Channel (20°00′N, 122°20′E) to join the Eastern Route, described at 10.79.2.

6   In August head towards Hainan Dao (19°10′N, 109°30′E), cross the Gulf of Tongkin and work down the coast of Vietnam, with the land and sea breezes, as far as Cap Varella (12°54′N, 109°28′E) or Mui Dinh (11°22′N, 109°01′E). Then cross to the coast of Borneo, keeping clear of any reefs, and work along that coast and through Selat Karimata, or Selat Gelasa, to Selat Sunda.

Hong Kong ⇒ Singapore

Directions
10.82

1   From Hong Kong (22°17′N, 114°10′E), to Singapore (1°12′N, 103°51′E), routes are seasonal depending on monsoon conditions.

10.82.1

1   During the north-east monsoon (October to March), steer to pass between Macclesfield Bank (15°50′N, 114°30′E) and the Paracel Islands (16°30′N, 112°00′E), and thence to pass E or W of Dao Phu Qui (10°32′N, 108°57′E) and Îles Catwick (10°00′N, 109°00′E). Thence passing W of Charlotte Bank (7°08′N, 107°35′E) and Pulau-Pulau Anambas (3°00′N, 106°00′E) steer to make Pulau Aur (2°27′N, 104°31′E).

2   Departing from Pulau Aur, bring it to bear about 000°, steering S until Horsburgh Light (1°20′N, 104°24′E) is sighted. When making for the entrance to Singapore Strait, steer for Horsburgh Light, making allowance for set, so as to pass 1 to 2 miles N of it.

3   **Traffic Separation Schemes** and **Prohibited Areas** have been extensively established throughout the Singapore Strait.

4   In slightly hazy weather, with Pulau Aur disappearing astern bearing 000° or less, steer to make a course between 192° and 204°, which may be required if the NE-going stream is setting out from the Singapore Strait. The depth will decrease regularly in steering to the S and low land will probably be seen to the W when in depths of from 33 m to 37 m; if so, coast along it at a distance of about 13 miles until Bukit Tautau (1°30′N, 104°15′E) is sighted. If in any doubt about the position, or if a depth of between 18 m and 22 m is obtained, haul off from the land or anchor.

5   Having made the entrance to the Singapore Strait, proceed as directed at 9.82.

6   In March, during the latter part of this monsoon, the winds are steady from the E, the weather is settled and the current is weak. In April, the prevailing winds are also from the E, but are much lighter and are accompanied by calms and squally weather. From the latter end of April until the middle of May, the monsoon gradually breaks up.

7   Caution is necessary if the weather is thick, with a fresh breeze, when near Pulau Aur. In these circumstances, heave-to, under its lee, and wait for a convenient time to proceed to the Straits. The current between Pulau Aur and the E point of Pulau Bintan (1°00′N, 104°30′E) sets about SSE, and a vessel, leaving Pulau Aur, and steering too much to the S can be set by this current, and the E-going stream from the Straits, so far to leeward of Bintan that they are obliged to proceed S of it and come up through Selat Riau (0°55′N, 104°20′E), as described at 9.76.

10.82.2

1   During the south-west monsoon, on leaving ports on the S coast of China, or from Manila (14°32′N, 120°56′E), during March to May, make to enter the Palawan Passage, at about 11°30′N, 118°30′E, and proceed SW through the passage, thence on a course along the NW coast of Borneo.

2   Pass through one of the passages through Pulau-Pulau Subi Besar (2°50′N, 108°50′E), then head across to the entrance to the Singapore Strait, and proceed as directed at 9.82.

3   When approaching the Singapore Strait be sure of the landfall. Keep well to the S before closing Pulau Bintan so as to allow for the current which sometimes runs to the N at a rate of 2 kn.

4   During the early part of the south-west monsoon, if the wind is in the SW on leaving Hong Kong, a good passage may occasionally be made by heading SE as far as 15°00′N, 115°30′E. Thence pass SW of Macclesfield Bank, to Îles Catwick or Mui Dinh (11°22′N, 109°01′E) and cross the Gulf of Thailand to Pulau Aur and thence to Singapore, as described above.

5   From August to October, after leaving Hong Kong, head towards Hainan Dao (19°10′N, 109°30′E), which will often be reached without tacking, as the wind frequently blows for days from SE or E in that part of the South China Sea. From thence cross the Gulf of Tongkin to the Vietnamese coast. Land and sea breezes and smooth water generally prevail close to that coast, for which reason it is best to work S as close to the shore as possible, taking advantage of every slant of wind, but being careful not to get too far off the land. It is sometimes possible to get as far S as Mui Dinh, in this way, but generally, after passing Cap Varella (12°54′N, 109°28′E), the monsoon is found to be blowing very fresh, out of the Gulf of Thailand, with frequent hard squalls rendering it impossible to progress much to windward. From Cap Varella, or from Mui Dinh if the vessel has been able to reach it, head away to the S, tacking if necessary, to weather the reefs or shoals, until the coast of Borneo is reached. Work along this coast and proceed W by one of the passages through Pulau-Pulau Subi Besar and to Singapore, as directed for March to May above.

Hong Kong ⇒ Torres Strait

Directions
10.83

1   From Hong Kong (22°17′N, 114°10′E), to Torres Strait (10°36′S, 141°51′E) the usual route crosses the the South China Sea to Lubang Island (13°45′N, 120°10′E) or Cape Calavite (13°27′N, 120°13′E) and enters the Sulu Sea through Mindoro Strait (13°00′N, 120°00′E), as described at 9.128; or by passing E of Lubang Island, through Verde Island Passage (13°36′N, 121°00′E) as described at 9.105 and Tablas Strait (13°00′N, 121°45′E).

2   In either case, having passed through Cuyo East Pass (10°30′N, 121°30′E), E of Sombrero Rocks (10°43′N, 121°33′E), proceed S through the Sulu Sea and through Basilan Strait (6°50′N, 122°00′E) into the Celebes Sea, as described at 9.112. Cross the Celebes Sea to Selat Bangka (1°45′N, 125°05′E), off the NW point of Sulawesi, passing through it into the Molucca Sea, and continue S to enter the Ceram Sea between Pulau-Pulau Sula (1°50′S, 125°00′E) and Obi Mayor (1°25′S, 127°25′E).

3   Cross the Ceram Sea and pass through Selat Manipa (3°20′S, 127°22′E), as described at 9.126 into the Banda Sea. Having cleared Selat Manipa, steer SE to pass between Pulau-Pulau Kai (5°40′S, 132°50′E) and Pulau-Pulau Tanimbar (7°30′S, 131°30′E), leaving Pulau Manuk (5°33′S, 130°18′E) to N or S as convenient.

4   Having passed Pulau-Pulau Tanimbar, the direct route to Torres Strait passes S of Pulau-Pulau Aru (6°00′S, 134°30′E) and past Ujung Salah (8°25′S, 137°39′E), but this route is not recommended owing to the dangers S and

SW of Pulau-Pulau Aru and to the chain of known and unexamined dangers lying W from Ujung Salah almost as far as 134°E. Instead a vessel is recommended to keep to the SE from Pulau-Pulau Tanimbar to cross 134°E at about 9°S, thence proceed to the Torres Strait.

5 In July and August, the alternative route for power driven vessels, described at 7.141.3, may prove useful to sailing vessels after passing Obi Mayor.

6 An alternative route for the whole passage, which can be used from April to October, is described at 10.84.2.

Hong Kong ⇒ Darwin

Directions
10.84

1 From Hong Kong (22°17′N, 114°10′E), to Darwin (12°25′S, 130°47′E), routes are seasonal.

10.84.1

1 From November to April, follow the directions given at 10.83, as far as Selat Manipa (3°20′S, 127°22′E), then steer to the SSE to pass E of Pulau-Pulau Penyu (5°22′S, 127°47′E) and between Pulau Damar (7°08′S, 128°36′E) and Pulau Teun (6°59′S, 129°08′E), thence pass between Pulau Sermata (8°12′S, 128°54′E) and Pulau Babar (7°55′S, 129°44′E) into the Arafura Sea. Proceed across the Arafura Sea to make Cape Fourcroy (11°48′S, 130°01′E), avoiding Flinders Shoal (9°53′S, 129°17′E) and other shoals near the route, thence proceed to Darwin. See *Australia Pilot, Volume V*, for a description of the dangers in the approaches to Darwin.

10.84.2

1 From April to October, steer across the South China Sea and pass through Balintang Channel (20°00′N, 122°20′E) into the Pacific Ocean. Thence proceed SE to pass either side of Palau Islands (7°30′N, 134°30′E); then make to the E to the Equatorial Counter-current, between 4°N and 8°N, until able to run through the Solomon Islands with the South-east Trade Wind, crossing the equator in about 158°E. After passing through the Solomon Islands, steer to the W, to enter the Torres Strait via the Bligh Entrance (9°12′S, 144°00′E), as described at 10.39, thence to Darwin.

2 From Palau Islands some navigators take the Saint George's Channel (4°20′S, 152°32′E), between New Ireland and New Britain, see *Pacific Islands Pilot, Volume I* for directions, instead of passing through the Solomon Islands; or, alternatively, the Pioneer Channel (4°40′S, 153°45′E), between New Ireland and the Solomon Islands, may be used.

Hong Kong ⇒ Sydney

Directions
10.85

1 From Hong Kong (22°17′N, 114°10′E), to Sydney (33°50′S, 151°19′E), routes are seasonal.

10.85.1

1 During the south-west monsoon (April to September), four routes are available. Two pass into the Pacific Ocean and run E of Australia; the other two pass through the Eastern Archipelago and W and S of Australia.

2 Directions for the route, which passes into the Pacific Ocean N of the Philippine Islands, are given at 10.84.2, as far as the Coral Sea. Thence steer S to join the route from the Torres Strait, described at 10.40, in about 15°S, 156°E.

3 For the Pacific Ocean route which passes S of the Philippine Islands, follow the directions given at 10.83, as far as the Celebes Sea. Then enter the Pacific Ocean S of Mindanao, between Sarangani Islands (5°27′N, 125°27′E) and Kepulauan Kawio (4°40′S, 125°25′E). Thence make to the E in the Equatorial Counter-current, as directed at 10.84.2, passing S of the Solomon Islands or through Saint George's Channel (4°20′S, 152°32′E), to join the route from the Torres Strait, described at 10.40.

4 For the routes passing W and S of Australia, pass through the Eastern Archipelago, either by the Eastern Route, described at 9.123, or by the Central Route, described at 9.127. On reaching the Indian Ocean proceed, as directed at 9.194, to round Cape Leeuwin and proceed as directed from 9.8 to 9.10.

10.85.2

1 During the north-east monsoon (October to February), proceed either via the Torres Strait, as directed at 10.83; or via Selat Sunda, described at 10.81, thence a passage W and S of Australia as directed at 9.194 and 9.8 to 9.10.

Hong Kong ⇒ Manila

Directions
10.86

1 From Hong Kong (22°17′N, 114°10′E) to Manila (14°32′N, 120°56′E), routes are seasonal.

10.86.1

1 In the north-east monsoon (October to April), make for the coast of Luzon, at about Piedra Point (16°19′N, 119°47′E). The current sets strongly to leeward, but decreases near Luzon. From Piedra Point, steer S for Manila Bay, giving the coastal dangers a wide berth.

10.86.2

1 During the south-west monsoon (May to September), take every advantage of the wind shifting to make to the S towards the Macclesfield Bank (15°50′N, 114°30′E), then steer direct for Manila Bay.

10.86.3

1 **Caution**. Pratas Reefs (20°40′N, 116°45′E), lies in the route between Manila and Hong Kong. It is a serious danger, especially in the north-east monsoon, when strong gales and thick clouds are sometimes prevalent for several weeks. As in this monsoon, vessels generally approach the reef from the SE, a great number of wrecks have occurred on this side. See *China Sea Pilot, Volume I*, for details.

Hong Kong ⇒ Yokohama

Directions
10.87

1 From Hong Kong (22°17′N, 114°10′E) to Yokohama (35°26′N, 139°43′E), routes are seasonal.

10.87.1

1 During the north-east monsoon (October to April), work up the coast of China as far as Shibeishan Jiăo (22°56′N, 116°30′E), taking advantage of the fact that the wind is N at night and E during the day. From Shibeishan Jiăo, head for the S end of T'ai-wan and work up on the E side of that island, a S-going set will be felt until reaching O-luan Pi (21°54N, 120°51′E), after passing which the Japan Current (7.26) will be experienced, setting N. Continue N to the W of Nansei Shotō (27°00′N, 129°00′E) as described below.

2 Towards the end of the north-east monsoon, head across the South China Sea until near the coast of Luzon, where the wind will be more E or even SE, then tack and head NNE along the E coast of T'ai-wan, and W of all the

islands of Nansei Shotō, with a generally favourable current. Then pass through one of the channels S of Ōsumi Kaikyō (30°55′N, 130°40′E), and from 50 to 80 miles off the S coast of Japan, in the strength of the Japan Current, to make land about Omae Saki (34°35′N, 138°12′E), thence coastwise to enter Uraga Suidō (35°10′N, 139°45′E).
10.87.2

1   During the south-west monsoon (May to September), run up the coast of China as far as Dongyin Dao (26°22′N, 120°30′E), thence steer to pass through Tokara Guntō, S of Akuseki Shima (29°27′N, 129°37′E), in preference to Ōsumi Kaikyō, (where dense fogs will probably be found, while farther seaward in the warm waters of the Japan Current the atmosphere is bright and clear). The course along the S coast of Japan is the same as for the north-east monsoon.

Hong Kong ⇒ Nagasaki

Directions
10.88

1   From Hong Kong (22°17′N, 114°10′E) to Nagasaki (32°42′N, 129°49′E), routes are seasonal.
10.88.1

1   During the north-east monsoon (October to April), follow the directions for the Yokohama route, described at 10.87.1, until N of T'ai-wan, after which continue as directly as navigation permits.
10.88.2

1   During the south-west monsoon (May to September), run up the coast of China as far as Dongyin Dao (26°22′N, 120°30′E), thence take a direct course to Nagasaki.

Hong Kong northwards to ports on the coast of China

General information for sailing vessels
10.89

1   From Hong Kong (22°17′N, 114°10′E), except when crossing the Taiwan Strait, there is no difficulty in making this passage in the south-west monsoon, but in the north-east monsoon, a sailing vessel should be prepared for meeting very rough weather.

2   The crossing of the Taiwan Strait is attended with considerable trouble at all times of the year, on account of strong, variable and sometimes opposite currents setting across the track. This is particularly noticeable at the change of the monsoon. In the S and W parts of the strait, a strong drift current setting to leeward, in both monsoons, must be allowed for. See, *China Sea Pilot, Volume III*, for further details.
10.89.1

1   During the north-east monsoon (October to April), make the passage either:
    E of T'ai-wan, so as to benefit from the Japan Current (7.26), the diminished strength of the monsoon and to avoid the short, heavy sea of the Taiwan Strait; or:
    Work up the coast of China, taking advantage of every favourable change of wind and tidal stream, anchoring whenever possible if conditions are unfavourable.

2   For the route E of T'ai-wan, work along the coast of China as far as Shibeishan Jiǎo (22°56′N, 116°30′E), to maintain as long as practicable the advantage of the land

wind at night, of smoother water and the E-going tidal stream out of the deep bays, which will generally be under the lee when on the starboard tack. There are numerous convenient anchorages should the wind blow too hard to make way. Keep within 10 miles of the land, to avoid being carried S by the monsoon drift current, whilst standing offshore; but as this cannot be done at night without risk, anchor, if possible in the evening and weigh between midnight and 0400 hours, when the wind being, generally, more off the land, will allow a good run on the offshore tack.

3   From Shibeishan Jiǎo head for the S end of T'ai-wan, as by passing E of that island, the heavy, short sea of the Taiwan Strait, and the constant S-going current, is avoided.

4   After rounding the S point of T'ai-wan, off which there is generally a troublesome sea, make short tacks as required to keep within the influence of the Japan Current.

5   The north-east monsoon does not blow with its full strength on the E coast of T'ai-wan, but strong gales are often experienced 20 miles to the E. If the wind declines in strength, with less sea, when on the inshore tack (particularly between 0900 and 1500 hours or up to sunset), it is advantageous to hug the coast as close as prudent. However, caution is required as the coast is mountainous and steep-to, so any sudden loss of wind, accompanied by swell, if followed by calm, could result in danger as there are no harbours. Stronger winds, with much rain, are encountered as progress is made to the E during the north-east monsoon. If an offshore course is maintained, whilst E of T'ai-wan, a constant succession of bad weather may be expected, with strong winds and a heavy sea.

6   Towards the close of the north-east monsoon, and still later, it is preferable to cross over towards Luzon than to beat up to Shibeishan Jiǎo against fresh NE breezes. Therefore head, on the port tack, towards the SE and pass through the SW-going current quickly On nearing Luzon, as the wind becomes more E, occasionally even SE, tack NNE, with a strong favourable current, and arrive E of T'ai-wan in less time than it would have taken to reach Shibeishan Jiǎo by keeping along the coast of China.

7   Having weathered the N end of T'ai-wan, it is still advisable to keep well to the E, and not approach the coast of China until 30°30′N is reached. In case of being driven to the W, take cautious advantage of the tidal streams through the S part of Zhoushan Qundao (30°22′N, 122°20′E).

8   Bound for Shantou (23°21′N, 116°40′E), Xiamen (24°27′N, 118°04′E) and the ports between that place and Minjiang Kou (26°05′N, 119°52′E), there is generally difficulty in getting round Shibeishan Jiǎo, for the tidal stream there is of no assistance. Advantage must therefore be taken of the wind, which will probably draw off the land after midnight, when by being inshore, good progress can be made and possibly Biao Jiǎo (23°14′N, 116°48′E) may be reached. For directions for Haimen Wan (23°08′E, 116°35′E) and Qiwang Wan (23°12′N, 116°43′E) anchorages, see *China Sea Pilot, Volume III*.

9   Having reached Biao Jiǎo, the NE-going stream assists a vessel to round it, and the out-going stream out of Han Jiang (23°20′N, 116°45′E) runs against the wind. If not passing inside Nan'ao Dao (23°26′N, 117°04′E) try to get along the S side of the island and anchor in the bay W of Nan Jiǎo (23°24′N, 117°06′E), should the weather be too bad to proceed. Both streams are strong off the bluff at the SE end of Nan'ao Dao and also off Shi Yu (23°35′N, 117°27′E), to round which take the first of the NE-going stream and the port tack.

10 Farther N, about Lishi Liedao (23°47′N, 117°43′E), the NE-going stream, with strong winds causes a very uneasy sea. Jiangjun Ao (24°01′N, 117°51′E) and Dingtai Wan (24°15′N, 118°05′E) are good stopping places for small vessels. The latter should be preferred, although at a loss of 2 or 3 miles, to anchorage in an exposed position in Xiamen harbour entrance, as when NE winds freshen there, during the in-going tide, they are generally accompanied by a mist, which obscures the entrance, and the tidal stream makes it difficult to proceed to sea.

11 Weitou Wan (24°30′N, 118°30′E), N of Xiamen, affords good shelter, while Shenhu Wan (24°39′N, 118°41′E) is not so good. The current during the monsoon overcomes the tidal streams and advantage must be taken of every slant of wind, bearing in mind that it is likely to draw off the land after midnight, and in the event of anchoring for shelter this is the time to get under way again, as waiting for daylight will lose the opportunity to make good progress offshore. The fog is, at times, thick and soundings must be taken, the bottom gradually changing from sand to mud as the shore is approached. There is fair anchorage under Dazuo Yan (24°53′N, 118°59′E), but it is not as good as that under Yang Yu (25°09′N, 119°30′E), and if the vessel is heading N, or anything E of it, the out-going tidal stream from Meizhou Wan (25°00′N, 119°05′E) is of assistance.

12 The most difficult part of the passage to Minjiang Kou (26°05′N, 119°52′E) is from Nanri Dao (25°13′N, 119°29′E) to Baiquan Leidao (25°58′N, 119°57′E). Vessels should keep outside Haitan Dao (25°30′N, 119°45′E), and head over towards the NW coast of T'ai-wan, to take advantage of a tidal stream running against the wind.

13 Off the coast of China N of Minjiang Kou, the indraught during the in-going tide must be considered.

14 There is good anchorage in a cove in the W island of Dongyin Dao (26°22′N, 120°30′E) but N of this, sailing vessels over 3·7 m draught, must keep off the coast in deep water. The tidal streams provide little assistance until Zhoushan Qundao (30°22′N, 122°20′E) is reached. The NE-going stream causes a disturbed sea in shallow water, while the SW-going stream causes too great a set to the S, unless the wind is well from the E. Nanjishan Liedao (27°28′N, 121°04′E) and Beijishan Liedao (27°38′N, 121°12′E) afford good shelter.

15 The route through the S channels of Zhoushan Qundao is not usually taken by sailing vessels. If working through the N part of this archipelago, advantage can be taken of the tidal streams.

16 The eddy tidal streams generally carry vessels clear of the larger islands, but caution is required to prevent being set in amongst detached rocks.

10.89.2

1 During the south-west monsoon (May to September), there is no difficulty in making the N-bound passage from Hong Kong, but the currents may be variable. See *China Sea Pilot, Volume III*, for further details.

Passage for low-powered vessels
10.90

1 During the north-east monsoon, low-powered, hampered, or small vessels are advised, if possible, to take advantage of the inshore passages described in *China Sea Pilot, Volumes I and II*, rather than that for full-powered vessels described at 7.90.1.

Hong Kong or Manila ⇒ North America and Panama

Directions
10.91

1 In all cases, from Hong Kong (22°17′N, 114°10′E) or Manila (14°32′N, 120°56′E), follow the routes for Yokohama, described at 10.87, and then follow the routes from Yokohama onwards described for:

Columbia River, Vancouver or Prince Rupert at 10.105;

San Francisco at 10.106;

Panama, at 10.106.1 as far as 150°W, thence direct.

Hong Kong or Manila ⇒ west coast of South America

Directions
10.92

1 From Hong Kong (22°17′N, 114°10′E) or Manila (14°32′N, 120°56′E), routes are seasonal.
10.92.1

1 During the south-west monsoon (May to September), the passage may be made either through the Bashi Channel (21°20′N, 121°00′E) or San Bernadino Strait (12°33′N, 124°12′E).

2 If proceeding by the Bashi Channel, continue as directed at 10.87.1 past Yokohama, then make to the E before heading S to join a suitable route from Sydney, described at 10.23, as convenient.

3 The Pacific Ocean may also be reached, in the south-west monsoon by the San Bernadino Strait, as described below. When clear of the strait, make to the E to join the route from Bashi Channel as convenient, or steer to the NE until in the Westerlies and then make to the E as above.

4 A vessel intending the passage of San Bernadino Strait should approach it through Verde Island Passage (13°36′N, 121°00′E) and then proceed to a position S of Marinduque Island (13°20′N, 122°00′E), to avoid being embayed with a SW wind in Nabasagan Bay (12°51′N, 123°12′E) on the W coast of Burias Island. A mid-channel course should be steered between Burias Island and Masbate Island (12°15′N, 123°30′E) and, when the SE point of Burias Island is passed, steer a NE course to pass N of Ticao Island (12°30′N, 123°42′E), giving San Miguel Island, off its N point, a good berth on account of the strength of the tidal stream near it.

5 If the wind is settled, steer for Naranjo Islands (12°21′N, 124°02′E), and thence pass midway between Capul Island (12°26′N, 124°10′E) and the islands off the SE point of Luzon, proceeding out of the strait by the channel S of San Bernadino Islets (12°46′N, 124°17′E).

6 If the SW wind is not settled, it is best to wait at anchor at San Jacinto, on the E side of Ticao Island, otherwise calms or light winds would leave the vessel subject to the effects of the tidal streams in the strait. The best time for leaving the anchorage is at the mid-point of the in-coming tide, for then the vessel is likely to get the first of the out-going tide when she is near Naranjo Islands.

7 If in danger of being carried near Calantas Rock (12°31′N, 124°05′E), it would be advisable to make for the coast of Luzon, where anchorage may be had, or to anchor on the bank in good time. The navigation of the strait requires great care and an anchor should always be ready to let go.

10.92.2

1   During the north-east monsoon (October to April), proceed, as directed at 10.83, as far as the Celebes Sea, and the either take the route direct to the Pacific Ocean or to the Coral Sea by the Torres Strait (10°36′S, 141°51′E).

2   For the Pacific Ocean route, cross the Celebes Sea to enter the Pacific Ocean S of Mindanao and then steer to pass N of New Guinea, and continue E between 2°N and 4°N as far as Kiribati (0°00′, 174°00′E). Thence steer SE into the Westerlies to join the appropriate route from Sydney, described at 10.23.

3   For the Torres Strait route, continue from the Celebes Sea to the Torres Strait, as directed at 10.83. After clearing the Torres Strait, continue along the S coast of New Guinea and the Louisade Archilpelago (11°00′S, 153°00′E), until far enough E to cross the Trade Winds into the Westerlies to join the appropriate route from Sydney, described at 10.23.

Shanghai south-bound coastwise

Directions
10.93

1   From Shanghai (31°03′N, 122°20′E), routes are seasonal.

10.93.1

1   During the north-east monsoon (October to April), after passing Maan Leidao (30°47′N, 122°42′E) and Dongfushan (30°08′N, 122°46′E), steer a good offshore course, passing outside the outer islands, giving them a good berth at night, closing the land to fix position by day, if necessary; for thick, hazy or rainy weather may always be expected.

10.93.2

1   During the south-west monsoon (May to September), although the constant adverse current makes this a tedious passage, a vessel of moderate sailing abilities can do it, as this monsoon is not steady in direction and land and sea breezes prevail.

2   Fog is frequent in the early part of the season, and caution is required; although it sometimes lifts near the land.

Shanghai ⇒ Indian Ocean

Directions
10.94

1   From Shanghai (31°03′N, 122°20′E), routes are seasonal.

10.94.1

1   During the north-east monsoon (October to April), take the coastwise route, described at 10.93.1, towards Hong Kong (22°17′N, 114°10′E) and pick up one of the routes to the Indian Ocean, listed at 10.79, proceeding either via Singapore or direct through the Eastern Archipelago.

10.94.2

1   During the south-west monsoon (May to September), steer direct for 15°N,132°E, to the E of the Philippine Islands, to pick up the route described at 10.109.2, Yokohama to the Indian Ocean.

Shanghai ⇒ Nagasaki

Directions
10.95

1   From Shanghai (31°03′N, 122°20′E) to Nagasaki (32°42′N, 129°49′E), routes are seasonal.

2   **Caution** is required in the vicinity of Socotra Rock (32°06′N, 125°11′E), which lies on the route. See, *China Sea Pilot, Volume III*, for further details.

10.95.1

1   During the north-east monsoon (October to March), with the wind E of N, make to the N at once, taking advantage of the tidal streams. As progress is made to the N, the wind usually backs through N to NW. Make allowance for the current which then sets to the SE or E.

10.95.2

1   In the south-west monsoon (March to September), during the E and SE winds which prevail from March to June, make to the E or S, even when a fair wind occurs, for it is sure to be of short duration. The tendency of the prevailing wind is to keep the vessel on the starboard tack and there is always a probability, during these months when the current sets to the NE, of being set towards the islands fringing the SW coast of Korea. If uncertain of the position when near Me Shima, in Danjo Guntō (32°02′N, 128°23′E), Tori Shima (32°15′N, 128°06′E) or Gotō Rettō (32°50′N, 129°00′E), arrange to approach these islands in daylight.

2   After June, with a steady south-west monsoon and a fair wind, steer from the estuary of Chang Jiang to pass between Danjo Guntō and Tori Shima.

3   The direct course from Jigu Jiăo (31°10′N, 122°23′E) leads midway between Tori Shima and Gotō Rettō, but it should not be taken, as the branch of the Japan Current (7.26) which sets through the Korea Strait, has to be crossed, and could set a vessel N of the S end of Gotō Rettō.

Shanghai ⇒ Yokohama

Directions
10.96

1   From Shanghai (31°03′N, 122°20′E) to Yokohama (35°26′N, 139°43′E), routes are seasonal.

10.96.1

1   During the north-east monsoon (October to March), if the wind, on departure, is E of N, as it frequently is in this monsoon, make to the N and when the wind backs to the NW, steer as directly as possible round the S end of Japan, and thence to destination in the strength of the Japan Current (7.26).

10.96.2

1   During the south-west monsoon (March to September), make to the E or S, as directed in 10.95.2 and then proceed round the S end of Japan, and thence to destination in the strength of the Japan Current (7.26).

Shanghai ⇒ North American ports

Directions
10.97

1   From Shanghai (31°03′N, 122°20′E) proceed, as directed at 10.96, to Yokohama (35°26′N, 139°43′E) and then follow the routes from Yokohama onwards described for:
    Columbia River, Vancouver or Prince Rupert at 10.105;
    San Francisco at 10.106.

PASSAGES FROM MANILA

Charts:

> *5127 (1) to (12) Monthly Routeing Charts for North Pacific Ocean.*
>
> *5128 (1) to (12) Monthly Routeing Charts for South Pacific Ocean.*
>
> *5309 Tracks followed by Sailing and Auxiliary Powered Vessels.*
>
> *5310 World Surface Current Distribution.*
>
> *Diagram 10.3 World Sailing Ship Routes.*

Manila ⇒ North and South American ports

Directions
10.98

1   From Manila (14°32′N, 120°56′E), the routes to ports in North and South America are described at 10.91 and 10.92, respectively.

Manila ⇒ Singapore

Directions
10.99

1   In all seasons, from Manila (14°32′N, 120°56′E) to Singapore (1°12′N, 103°51′E), steer to pass N of the central dangers of the South China Sea to Îles Catwick (10°00′N, 109°00′E), thence proceed direct to Pulau Aur (2°27′N, 104°31′E) and to Singapore. For more detail, see the directions for Hong Kong to Singapore at 10.82.

Manila ⇒ Ho Chi Minh City

Directions
10.100

1   From Manila (14°32′N, 120°56′E) to Ho Chi Minh City Approaches (10°18′N, 107°04′E), routes are seasonal.

10.100.1

1   During the north-east monsoon (October to March), take a direct passage across the South China Sea, allowing for the current which sets with the wind.

10.100.2

1   During the south-west monsoon (April to September), sailing vessels will find the voyage long and tedious, whichever route they adopt. The following route has been recommended.

2   On leaving Manila Bay take the Verde Island Passage (13°36′N, 121°00′E) and pass down the E side of Mindoro and the W coast of Panay, thence crossing the Sulu Sea to pass out through the Balabac Strait (7°34′N, 116°55′E) and work down the NW coast of Borneo in order to make to the W and cross the South China Sea, passing E of Pulau-Pulau Natuna Besar (4°00′N, 108°00′E).

Manila ⇒ Hong Kong or Xiamen

Directions
10.101

1   From Manila (14°32′N, 120°56′E) to Hong Kong (22°17′N, 114°10′E) or Xiamen (24°27′N, 118°04′E), routes are seasonal.

10.101.1

1   During the north-east monsoon (October to March), there is a choice between two routes, W or E of T'ai-wan, if bound for Xiamen. For Hong Kong work up the coast of Luzon, then proceed as directed at 10.66.1.

2   For the passage W of T'ai-wan, keep near the coast to Cape Bojeador (18°30′N, 120°34′E) and then work N to O-luan Pi (21°54N, 120°51′E) and thence along the SW and W coasts of T'ai-wan until able to stand across the the Taiwan Strait to Xiamen.

3   For the passage E of T'ai-wan, if the monsoon is well set in, it might be advisable to to stand to the E, passing N of Luzon and work to the N with the benefit of the Japan Current (7.26), passing E of, and round the N of, T'ai-wan. Thence, allowing for the current, steer to make the China coast N of the destination.

10.101.2

1   During the south-west monsoon (April to September), proceed direct, making allowance for a lee current.

Manila ⇒ Iloilo

Directions
10.102

1   From Manila (14°32′N, 120°56′E) to Iloilo (10°42′N, 122°36′E), routes are seasonal.

10.102.1

1   During the north-east monsoon (October to March), pass through Verde Island Passage (13°36′N, 121°00′E) and Tablas Strait (13°00′N, 121°45′E) and continue S along the W coast of, and round the S end of, Panay to Iloilo.

10.102.2

1   During the south-west monsoon (April to September), as above as far as Dumali Point (13°07′N, 121°33′E), then steer to pass S of Simara Island (12°49′N, 122°03′E) and between Tablas Island (12°25′N, 122°05′E) and Romblon Island (12°30′N, 122°15′E). Thence pass through Jintotolo Channel between Jintotolo Island (11°51′N, 123°07′E) and Zapato Islands (11°45′N, 123°01′E) and then, turning S, proceed along the E coast of Panay to Iloilo.

Manila ⇒ Cebu

Directions
10.103

2   At all seasons, from Manila (14°32′N, 120°56′E), take the south-west monsoon route for Iloilo, described at 10.102.2, as far as Jintotolo Island and then proceed to Malapascua Island (11°20′N, 124°07′E) and thence S to Cebu (10°18′N, 123°55′E).

Manila ⇒ Indian Ocean and Australia

Directions
10.104

1   From Manila (14°32′N, 120°56′E) the routes are seasonal.

10.104.1

1   During the north-east monsoon (October to March), follow the route for Singapore, described at 10.99, and then take the appropriate route onward, described between 9.188 and 9.199.

2   If not calling at Singapore and bound to the S through the Eastern Archipelago, proceed as described in 10.99 after passing Îles Catwick (10°00′N, 109°00′E), between Pulau-Pulau Anambas (3°00′N, 106°00′E) and Pulau-Pulau Natuna Besar (4°00′N, 108°00′E), thence as directed between 9.113 and 9.130, joining the route from Singapore for the passage onwards through the Indian Ocean.

10.104.2

1   During the south-west monsoon (April to September), proceed as directed at 10.83 as far as the Celebes Sea. Thence either continue on the Central Route from the South China Sea to the Indian Ocean, described at 10.79.3; or cross the Celebes Sea to pass through Selat Bangka

(1°45′N, 125°05′E) into the Molucca Sea. Thence continue S to the Ceram Sea and through Selat Manipa (3°20′S, 127°22′E) and the Banda Sea to Selat Ombi (8°35′S, 125°00′E) and the Indian Ocean.

2 In both cases join the route described at 9.199, if bound for the Cape of Good Hope. If bound for other ports join, or steer to join, as directly as possible the appropriate route from Singapore, described between 9.188 and 9.198.

3 For the Torres Strait (10°36′S, 141°51′E) follow the directions given at 10.83, from Verde Island Passage (13°36′N, 121°00′E) onwards.

4 An alternative route to the E coast of Australia is to pass into the Pacific Ocean through the San Bernadino Strait (12°33′N, 124°12′E), as described at 10.92.1. Thence proceed SE, making to cross the equator in about 158°E, passing through the Solomon Islands and continuing onwards to the S to join the route, described at 10.40, from Torres Strait to Sydney in 15°S, 156°E.

PASSAGES FROM JAPAN

Charts:

 5127 (1) to (12) Monthly Routeing Charts for North Pacific Ocean.
 5128 (1) to (12) Monthly Routeing Charts for South Pacific Ocean.
 5309 Tracks followed by Sailing and Auxiliary Powered Vessels.
 5310 World Surface Current Distribution.
 Diagram 10.3 World Sailing Ship Routes.

Yokohama ⇒ Columbia River, Vancouver or Prince Rupert

Directions
10.105

1 From Yokohama (35°26′N, 139°43′E) to the Columbia River (46°15′N, 124°05′W), Vancouver (49°17′N, 123°19′W) or Prince Rupert (54°19′N, 130°20′W), cross 167°E in about 42′N, being 30 miles N of that position in August and the same distance S of it in January. From this position steer almost due E, with a fair wind and favourable current, so as to cross 150°W in about 44°N, keeping a little to the N throughout the voyage in summer and to the S in winter. From 150°W proceed direct to destination, still with a fair wind.

10.105.1

1 **Caution**. The tidal streams, on the approach to Vancouver Island (49°30′N, 125°00′W), cause a general set towards the land and an indraught into all sounds during the in-going tide. Sailing vessels, therefore, when making the Juan de Fuca Strait (48°30′N, 124°47′W) during the winter, especially during November and December, and experiencing E and SE winds, should try and hold a position SW of Tatoosh Island (48°24′N, 124°44′W), and on no account open up the entrance to the strait until able to proceed well inside. See, *British Columbia Pilot, Volume I*, for further details.

Yokohama ⇒ San Francisco

Directions
10.106

1 From Yokohama (35°26′N, 139°43′E) to San Francisco (37°45′N, 122°40′W), the routes are seasonal.

10.106.1

1 From April to September, follow the directions given at 10.105 as far as 44°N, 150°W and thence proceed as directly as possible to San Francisco.

10.106.2

1 From October to March, winter conditions require a longer route farther to the S. First steer to cross 165°E in 40°N, thence along that latitude as far as 140°W or 135°W, thence proceed directly to San Francisco.

Yokohama ⇒ Honolulu

Directions
10.107

1 From Yokohama (35°26′N, 139°43′E), steer to cross 160°E at 41°30′N, then cross 180° in 43°30′N, thence proceed directly to Honolulu (21°17′N, 157°53′W), making allowance, on approaching the land, for a W-going current running at a rate of about 1 kn.

Yokohama ⇒ Singapore

Directions
10.108

1 From Yokohama (35°26′N, 139°43′E) to Singapore (1°12′N, 103°51′E), the routes are seasonal.

10.108.1

1 From October to April, proceed first to pass S of Tanega Shima (30°35′N, 131°00′E) and through Tokara Kaikyō (30°12′N, 130°15′E), thence steer SW to join the coastwise route from Shanghai southward, as described at 10.93.1, in about 28°N.

10.108.2

1 From May to September, two routes are appropriate, W or E of the Philippines.

2 For the passage W of the Philippines, pass E of all the island groups of Nansei Shotō (27°00′N, 129°00′E) and thence through the Bashi Channel (21°20′N, 121°00′E). From thence make for Îles Catwick (10°00′N, 109°00′E) and thence to Pulau Aur (2°27′N, 104°31′E). The passage from Pulau Aur to Singapore is described at 10.82.1.

3 For the passage E of the Philippines, first steer to the S passing E of Nanpō Shotō (30°00′N, 141°00′E), the chain of islands lying S of the SE point of Honshū. Thence make to the SSW for Selat Jailolo (0°06′N, 129°10′E), passing about 300 miles E of the Philippines. Thence pass S through the Ceram Sea and Selat Manipa (3°20′S, 127°22′E) into the Banda Sea, thence through the Flores Sea and the Java Sea, finally through on of the straits between Sumatera and Borneo to Singapore.

4 Directions for the straits and seas, through the Eastern Archipelago, are given between 9.56 and 9.130.

Yokohama ⇒ Indian Ocean

Directions
10.109

1 From Yokohama (35°26′N, 139°43′E) to the Indian Ocean, the routes are seasonal.

2 Directions for the appropriate routes S-bound, through the Eastern Archipelago, are given between 9.113 and 9.130.

10.109.1

1 From October to April, follow the directions at 10.108.1, to Singapore, then proceed to the Indian Ocean through

either Selat Sunda (6°00′S, 105°52′E) or the Malacca Strait (4°00′N, 100°00′E).

2   If not calling at Singapore, proceed as above but after passing Îles Catwick (10°00′N, 109°00′E), pass between Pulau-Pulau Anambas (3°00′N, 106°00′E) and Pulau-Pulau Natuna Besar (4°00′N, 108°00′E), to Selat Gelasa (2°53′S, 107°18′E) or Selat Karimata (1°43′S, 108°34′E) and thence to Selat Sunda and the Indian Ocean.

10.109.2

1   From May to September, there are two alternative routes.

2   For the first take the route W of the Philippines, for October to April, as described at 10.108.1, towards Singapore thence through one of the straits.

3   The second route follows that passing E of the Philippines, as described at 10.108.2, leaving as necessary to enter the Indian Ocean through one of the straits between Selat Ombi (8°35′S, 125°00′E) and Selat Sunda.

Yokohama ⇒ Sydney

Direct routes
10.110

1   The direct routes from Yokohama (35°26′N, 139°43′E) to Sydney (33°50′S, 151°19′E), are seasonal.

10.110.1

1   During the north-east monsoon (October to April), steer to cross 160°E in 20°N,, thence to cross the equator at 168°E. Then steer to pass E of Vanuatu (17°00′S, 168°00′E) and Nouvelle-Calédonie (21°15′S, 165°20′E), thence direct to Sydney, passing N of Middleton Reef (29°28′S, 159°04′E).

2   Alternatively, after crossing the equator, pass between the Solomon Islands (8°00′S, 158°00′E) and Santa Cruz Islands (11°00′S, 166°15′E), then W of Bampton Reefs (19°00′S, 158°40′E), thence to Sydney, making the Australian coast S of Sandy Cape (24°42′S, 153°16′E) and thence continue S along the coast.

10.110.2

1   During the south-west monsoon (May to September), first make to the E, keeping N of 35°N, until in about 170°E, then head S through the North-east Trade Wind to cross the equator in 173°E. Then pass E of Vanuatu and of Nouvelle-Calédonie, thence to Sydney, as described at 10.110.1.

Routes via Guam
10.111

1   From Yokohama (35°26′N, 139°43′E), if intending to call at Guam (13°27′N, 144°35′E) on passage to Sydney (33°50′S, 151°19′E), pass E of Nanpō Shotō (30°00′N, 141°00′E), the chain of islands lying S of the SE point of Honshū, and E or W of the Mariana Islands (17°00′N, 146°00′E), according to conditions prevailing at the time. From Guam to Sydney the routes are seasonal.

10.111.1

1   During the north-east monsoon (October to April), steer to the S with the North-east Trade Wind, and pass through Bougainville Strait (6°40′S, 156°15′E), or between the Solomon Islands and Santa Cruz Islands.

2   From Bougainville Strait proceed to a position in about 15°N, 156°E to join the route from Torres Strait to Sydney, described at 10.40.

3   From the position E of the Solomon Islands proceed as for the alternative route, described at 10.110.1.

10.111.2

1   During the south-east monsoon (May to September), in the S hemisphere, pass the Solomon Islands as described at 10.111.1, and make enough progress to the E to ensure a clear run across the Coral Sea, making the Australian coast S of Sandy Cape (24°42′S, 153°16′E) and thence continue S along the coast, where the prevailing wind will be W at this time of year.

Yokohama ⇒ Hong Kong or Xiamen

Directions
10.112

1   From Yokohama (35°26′N, 139°43′E) to Hong Kong (22°17′N, 114°10′E) or Xiamen (24°27′N, 118°04′E), routes are seasonal.

10.112.1

1   During the north-east monsoon (October to March), head to the SW across the Japan Current (7.26) as far as 28°N, 135°E, thence N of Tokuno Shima (22°47′N, 128°58′E) and, after passing Iō Tori Shima (27°52′N, 128°14′E), steer for Dongyin Dao (26°22′N, 120°30′E) and down the coast of China, as described at 10.93.1.

10.112.2

1   During the south-west monsoon (April to September), steer SE from Uraga Suidō (35°10′N, 139°45′E) to cross 30°N in about 145°E. Thence, passing E of Ogasawara Guntō (27°00′N, 142°00′E), and E and S of Kazan Rettō (25°00′N, 141°20′E), cross 140°E in about 21°N. Thence a direct course to pass N of Luzon and straight to Hong Kong, making allowance for the NE-going set in the South China Sea.

Yokohama ⇒ Shanghai

Directions
10.113

1   From Yokohama (35°26′N, 139°43′E) to Shanghai (31°03′N, 122°20′E), it was formerly recommended that the best sailing route was through Seto Naikai (the Japanese Inland Sea), avoiding the strength of the Japan Current (7.26) by keeping near the coast between Yokohama and Kii Suidō (34°00′N, 134°50′E) and sailing as directly as possible after passing through Kanmon Kaikyō (33°58′N, 130°52′E) and the Korea Strait.

2   Owing to traffic, and other factors, the route through Seto Naikai is probably no longer feasible without detailed local knowledge. Either a coastwise route S of Japan and through Ōsumi Kaikyō (30°55′N, 130°40′E), taking advantage of the local counter-currents, see *Japan Pilot, Volume II* for further details; or an ocean route S of the strongest part of the Japan Current, seem preferable.

Yokohama ⇒ Hakodate

Directions for coastal routes
10.114

1   From Yokohama (35°26′N, 139°43′E) to Hakodate (41°46′N, 140°41′E), the routes are seasonal.

10.114.1

1   In winter (November to March), make the passage as close inshore as safety will allow, as the wind is usually off the land and there is smooth water near the coast. In the event of encountering a NE gale, the best course is to make for the nearest sheltered anchorage, if available. Frequent snowstorms often obscure the land, and the currents are irregular, which make it necessary to use every precaution when navigating this part of the coast.

10.114.2

1   In summer (May to September) keep offshore and take advantage of the Japan Current (7.26). Fog will usually be encountered as far N as Kinkasan To (38°18′N, 141°35′E). Close the land to the S of Shiriya Saki (41°26′N, 141°28′E) and round that promontory at a distance of not less then 2 miles to avoid Ō Ne (41°27′N, 141°28′E).

2   In thick weather, if the land about Shiriya Saki has not been sighted, a rise in water temperature, floating debris such as plants trees and driftwood or heavy tide-rips, may help in determining that the vessel is to the N of Shiriya Saki and in the influence of the E-going current through Tsugaru Kaikyō (41°39′N, 140°48′E).

3   If proceeding direct to Hakodate from the E entrance to Tsugaru Kaikyō, a vessel may, after passing not less then 2 miles off Shiriya Saki, steer for Esan Misaki (41°49′N, 141°11′E), so as to take advantage of the cold W-going stream along the S coast of Hokkaidō, remembering that the NE-going current is sometimes found close inshore near Shiokubi Misaki (41°42′N, 140°58′E).

Directions for Tsugaru Kaikyō
10.115

1   Approaching Tsugaru Kaikyō from E, the adverse current will be avoided by keeping near the shore, keeping clear of Ō Ne and the dangers off Ōma Saki (41°33′N, 140°55′E).

2   Make Shiriya Saki bearing about 310°, passing it at a distance of not less then 2 miles and, when N of it, keep towards the S shore to avoid the current and to be in a position to anchor if becalmed. By keeping towards this shore a vessel may possibly be set a considerable distance by the W-going stream, while the NE-going current is running strongly in the middle of the strait.

3   Wait at anchor SE of Ōma Saki for a favourable opportunity to cross the strait and, as the winds during summer are generally light from the SW for a considerable period, freshening a little when the W-going stream makes, this is the proper time to weigh anchor.

4   Proceeding from Hakodate to the W, against SW winds, keep near the shore when N of Yagoshi Misaki (41°31′N, 140°25′E) and if unable to round it, anchor about 2 miles NE of it, weighing again when the next W-going stream makes.

5   With a light wind a sailing vessel might not clear the strait in one tide, in which case it would be better to wait at anchor, E of Shirakami Misaki (41°24′N, 140°12′E), and take the whole of the following tide to get sufficiently to the W rather than run any risk of being swept back through the strait by the current.

6   Approaching Tsugaru Kaikyō from SW during foggy weather, guard against being carried to the N, by the current, and past the entrance. If the weather is clear when nearing Nyūdō Saki (40°00′N, 139°42′E), it might be as well to sight it.

7   If the weather thickens when nearing Nyūdō Saki, good, though open, anchorage over a sandy bottom will be found to the S of it; while to the N, the bottom is rocky, though anchorage is still possible.

8   Sailing vessels, passing through Tsugaru Kaikyō, particularly to the W, should be prepared to anchor immediately and keep close to the shore.

9   For details of currents, tidal streams and ice, see *Japan Pilot, Volume I*.

Nagasaki ⇒ China coast

Directions
10.116

1   From Nagasaki (32°42′N, 129°49′E) to Shanghai (31°03′N, 122°20′E), steer as direct a course as circumstances will allow, keeping rather to windward of the course as, except near the coast of Japan, the drift of the current is usually to leeward.

2   **Caution** is required in the vicinity of Socotra Rock (32°06′N, 125°11′E), which lies on the route. See, *China Sea Pilot, Volume III*, for further details.

3   For Hong Kong (22°17′N, 114°10′E), Xiamen (24°27′N, 118°04′E) and ports in the vicinity, routes are seasonal.

10.116.1

1   During the north-east monsoon (October to April), steer to make the coast of China a little to the S of Zoushan Qundao (30°1;22′N, 122°20′E) and thence sail coastwise, as described at 10.93.1.

10.116.2

1   During the south-west monsoon (May to September),, first head across to the coast of China and then make to the S coastwise.

PASSAGES FROM ISLANDS IN THE NORTH PACIFIC OCEAN (EXCLUDING HAWAII)

Charts:

  5127 (1) to (12) Monthly Routeing Charts for North Pacific Ocean.

  5128 (1) to (12) Monthly Routeing Charts for South Pacific Ocean.

  5309 Tracks followed by Sailing and Auxiliary Powered Vessels.

  5310 World Surface Current Distribution.

  Diagram 10.3 World Sailing Ship Routes.

General notes and cautions
10.117

1   In this chapter only the routes from Honolulu are given in detail. From other islands in the North Pacific Ocean, the most favourable route to be taken can be ascertained by consulting the charts mentioned above.

North Pacific Islands ⇒
Asia or North and South America

Directions
10.118

1   *Diagram 10.3 World Sailing Ship Routes* shows that little difficulty will be experienced in deciding on the most suitable route for a vessel W-bound to any port in Asia, the Eastern Archipelago, the Indian Ocean or Japan. A number of routes from North or South America and from Australia pass near the islands and can be joined at a convenient position.

2   For a vessel E-bound, the general principle is to head N, or S, through the Trade Winds to reach the "Westerlies", described at 7.10 and 7.17; a favourable current may be experienced in the area of the Westerlies.

3   Passages, E-bound across the Pacific Ocean, that may conveniently be joined, are given at 10.105 and 10.106

from Yokohama; at 10.22, 10.23 and 10.24 from Sydney; or making S to join the route from South Africa described at 10.8.

North Pacific Islands ⇒
other North Pacific Islands

Directions
10.119

1   For a sailing vessel W-bound, no difficulty should be experienced as the wind is favourable and, except in the Equatorial Counter-current between 4°N and 8°N, a favourable current should also assist the passage.

2   Proceeding E-bound to Honolulu (21°17′N, 157°53′W), head N through the Trade Winds as far as about 40°N or until the W winds are met. Cross 180° meridian in about 43°N and 160°W in 40°N; thence head SE to a position in about 35°N, 153°W and thence proceed directly to Honolulu. See also the directions for Yokohama to Honolulu, given at 10.107.

3   To the other North Pacific islands the direct mean course can be steered over short distances, but this usually means working to the E against the North Equatorial Current (7.26).

4   In most cases it is probably best to head S or SE into the Equatorial Counter-current (7.28), and then work E until able to fetch the destination, making allowance for the W-going set as the vessel makes to the N.

North Pacific Islands ⇒
South Pacific Islands or Sydney

Directions
10.120

1   The longitude of most of the principal island groups of the South Pacific Ocean is E, or little or nothing to the W, of that of similar groups in the North Pacific Ocean, except for the Hawaiian Islands. Therefore the first objective, in all passages, must be to make to the E while still N of the equator, which is usually crossed between 168°E and 173°E. Probably the most advantageous passage to reach this objective is to head S, or as much to the SE as can be made on the port tack, until the region of the Equatorial Counter-current (7.28) is reached, between 4°N and 8°N, then work E until able to cross the equator, as described above. From the equator proceed as follows:

> For the Solomon Islands (8°00′S, 158°00′E), Vanuatu (17°00′S, 168°00′E) or Nouvelle-Calédonie (21°15′S, 165°20′E), proceed direct.
> For Sydney (33°50′S, 151°19′E), follow the relevant directions at 10.110 from Yokohama.
> For Fiji (18°00′S, 180°00′), pass down the W side of Tuvalu (8°00′S, 179°00′E) and thence direct.
> For Samoa (14°00′S, 171°30′W), stand S as far as the latitude of the Fijian Islands, weathering them, if possible, until able to make Samoa on the starboard tack.
> For the islands E of Samoa, it is best to head S through the Trade Winds into the Westerlies; then run down making eastward until the longitude of the destination is reached. Then re-enter the Trade Winds and proceed to destination.

North Pacific Islands ⇒ Torres Strait

Directions
10.121

1   At all times of the year, the route through the Solomon Islands (8°00′S, 158°00′E) may be taken, by following the general directions at 10.84.2 as far as the equator, modified as necessary for the point of departure.

2   During the north-east monsoon (October to April), an equally good, or even better passage may be made by steering direct to pass through Selat Jailolo (0°06′N, 129°10′E), and joining the route from Hong Kong, described at 10.83, in the Ceram Sea.

PASSAGES FROM HONOLULU

Charts:
> *5127 (1) to (12) Monthly Routeing Charts for North Pacific Ocean.*
> *5128 (1) to (12) Monthly Routeing Charts for South Pacific Ocean.*
> *5309 Tracks followed by Sailing and Auxiliary Powered Vessels.*
> *5310 World Surface Current Distribution.*
> *Diagram 10.3 World Sailing Ship Routes.*

General remarks on winds, currents and sailing passages around the Hawaiian Islands
10.122

1   With regard to winds, the North-east Trade Wind seems to divide at Cape Kumukahi (19°31′N, 154°49′W), part following the coast to the NW around Upolu Point (20°16′N, 155°51′W), where it loses its force; the other part following the SE coast around Ka Lae (18°55′N, 155°41′W), where it also loses its force.

2   On the W coast of Hawaii the sea breeze sets in about 0900 hrs and continues until after sunset, when the land breeze springs up.

3   Sailing vessels coming from the W, bound to the ports on the windward, or SE, side of Hawaii, should pass close to Upolu Point and keep near the coast, as the wind is generally much lighter than offshore. Those from the W, bound for ports on the the E side of Hawaii should keep well to the N until clear of Alenuihaha Channel (20°20′N, 156°00′W).

4   On account of the current, which nearly always sets to the N along the W coast of Hawaii, it is advisable for sailing vessels to make the land S of their destination as, during calms or light airs, a vessel is liable to drift to the N.

5   With regard to navigation, Alenuihaha Channel is 26 miles wide and clear of dangers. The North-east Trade Wind, which predominates throughout the year, frequently blows through the channel with great strength and there is also a strong current setting W, but, during calms there is, at times, an E-going set of about 1 kn, which during "kona" winds (the reverse of the Trade Wind) may increase to 2 or 3 kn.

6   Vessels from any of the W ports of Hawaii are therefore recommended to keep close in under the lee of the island until reaching Upolu Point, when they will be able to head across to the Alalakeiki Channel (20°35′N, 156°30′W). Those from the N, bound for Hilo (19°44′N, 155°04′W), will probably find it impossible to weather Upolu Point from the W side of Maui (20°45′N, 156°20′W), but on getting under the lee of Hawaii, the Trade Wind fails until reaching the S point of the island, when they will have to beat against wind and current along the SE coast.

Honolulu ⇒ Tahiti

Directions
10.123

1 From Honolulu (21°17′N, 157°53′W), head first to the N of the Hawaiian Islands and then make to the E in the North-east Trade Wind, cross the equator, well to the E, then proceed SW in the South-east Trade Wind to Tahiti (17°35′S, 149°25′W).

Honolulu ⇒ Fiji, Australia and New Zealand

Directions
10.124

1 From Honolulu (21°17′N, 157°53′W) for Fiji (18°00′S, 180°00′), proceed as directly as navigation permits, with a fair Trade Wind.

2 For Australia and New Zealand, take the route to Fiji, as described above, as far as about 170°W to 175°W, then follow the directions given at 10.50 and 10.51 and proceed direct.

3 Due to the prevalence of W winds off the coast of New Zealand, except when bound for Auckland (36°36′S, 174°49′E), it is best to pass down the W coast of New Zealand. Thence through the Cook Strait (41°00′S, 174°30′E) for ports in North Island, if conditions are favourable; or round South Island and N along the E coast.

Honolulu ⇒ China, Japan or Philippine Islands

Directions
10.125

1 The routes described from 10.140 to 10.143, from San Francisco (37°45′N, 122°40′W) to these destinations, pass close S of Hawaii and, from Honolulu (21°17′N, 157°53′W), should be picked up at any position between 160°W and 170°W.

Honolulu ⇒ San Francisco

Directions
10.126

1 From Honolulu (21°17′N, 157°53′W) to San Francisco (37°45′N, 122°40′W), throughout the year, first steer due N before turning E on reaching steady W winds. The turning point varies in latitude, being farthest N in August and farthest S in November and December. The onward routes are seasonal, as follows:

> August, turn E in approximately 40°N and steer along that parallel to 150°W. Thence proceed directly to destination.

> June and July, turn to the NE between 35°N and 36°N, and steer on a slightly curving course to cross 150°W in approximately 39°N. Thence proceed directly to destination.

> May, September and October, turn to the NE in about 30°N and steer on a curving course to cross 150°W between 37°30′N and 38°00′N. Thence proceed directly to destination.

> March and April, turn to the NE between 26°N and 27°N and steer on a curving course to cross 150°W in about 36°30′N. Thence proceed directly to destination.

> January and February, turn to the NE between 25°N and 26°N and steer on a curving course to cross 150°W in about 33°N. Thence proceed directly to destination.

> November and December, turn to the NE in about 24°N and steer on a curving course to cross 150°W between 32°00′N and 32°30′N. Thence proceed directly to destination.

2 **Note** that the curving course referred to above can best be understood by reference to *Diagram 10.3 World Sailing Ship Routes.*

Honolulu ⇒ ports between San Francisco and Panama

Directions
10.127

1 From Honolulu (21°17′N, 157°53′W) proceed N, as directed at 10.126, but turn E instead of NE. Make E as directly as possible, gradually altering course to ESE after reaching between 150°W and 140°W, depending on destination, the latter to the more S ports. Join the route from San Francisco, described at 10.146, at a convenient position.

Honolulu ⇒ ports between Panama and Cabo de Hornos

Directions
10.128

1 From Honolulu (21°17′N, 157°53′W), the most important objective must be to make E progress as soon as possible so as to be able to stand SE to join one of the routes from San Francisco:

> To South American ports, described at 10.147 and 10.148.

> To Cabo de Hornos (56°04′S, 67°15′W) described at 10.149.

2 In all cases it appears advisable to join these routes N of the equator, where the Equatorial Counter-current will be advantageous if getting too far W on the passage S.

PASSAGES FROM PRINCE RUPERT, JUAN DE FUCA STRAIT OR COLUMBIA RIVER

SAILING VESSEL ROUTES

Charts:
> 5127 (1) to (12) Monthly Routeing Charts for North Pacific Ocean.
> 5128 (1) to (12) Monthly Routeing Charts for South Pacific Ocean.
> 5309 Tracks followed by Sailing and Auxiliary Powered Vessels.
> 5310 World Surface Current Distribution.
> Diagram 10.3 World Sailing Ship Routes.

Prince Rupert, Juan de Fuca Strait or Columbia River ⇒ Honolulu and Yokohama

Directions
10.129

1 From Prince Rupert (54°19′N, 130°20′W), head S through Hecate Strait (53°00′N, 131°00′E) and from Juan de Fuca Strait (48°30′N, 124°47′W) or the Columbia River (46°15′N, 124°05′W), head seaward to a safe distance offshore, but keeping as close inshore as prudence dictates, to avoid the heavy seas to be experienced farther out.

Proceed S until within about 300 miles NW of San Francisco (37°45′N, 122°40′W), and thence directly to Honolulu (21°17′N, 157°53′W). The route W of Honolulu is seasonal.

2 From May to November, proceed to 20°N and head due W to the 180° meridian, thence to 25°N, 160°E, thence to Yokohama (35°26′N, 139°43′E), allowing for the NE-going set of the Japan Current (7.26).

3 During the winter, from December to April, a vessel may have to keep farther S to get the strength of the Trade Winds for the run to the W, after leaving Honolulu. The directions from San Francisco, given at 10.140, should be followed, according to date, after running to the W.

4 An alternative route for all seasons is to make SW from Honolulu to join one of the seasonal routes from San Francisco as described at 10.140.

Prince Rupert, Juan de Fuca Strait or Columbia River ⇒ Sydney

Directions
10.130

1 From Prince Rupert (54°19′N, 130°20′W), head S until reaching the North-east Trade Wind, passing on either side of Queen Charlotte Islands (53°10′N, 132°10′W).

2 From Juan de Fuca Strait (48°30′N, 124°47′W) or the Columbia River (46°15′N, 124°05′W), head SW at once to pick up the Trade Wind.

3 Then, from June to August, proceed as directly as possible, crossing the equator at about 170°W and passing W of the Fiji Islands (18°00′S, 180°00′) and SE of Nouvelle-Calédonie (21°15′S, 165°20′E). At other times of the year, cross the equator between 150°W and 155°W, passing S of the Tonga Islands (20°00′S, 174°30′W) and Fiji Islands.

4 Make the Australian coast S of Sandy Cape (24°42′S, 153°16′E) as described at 10.110.1.

Prince Rupert, Juan de Fuca Strait or Columbia River ⇒ San Francisco and South America

Directions
10.131

1 The following routes vary in some degree from those recommended for San Francisco to South American ports, by the Unites States Naval Oceanographic Office, and described between 10.147 to 10.149.

2 From Prince Rupert (54°19′N, 130°20′W), Juan de Fuca Strait (48°30′N, 124°47′W) or the Columbia River (46°15′N, 124°05′W) to San Francisco (37°45′N, 122°40′W), at all seasons, keep as near the shore as is prudent, in order to avoid the heavy sea felt farther out.

10.131.1

1 During October to March, if bound from Juan de Fuca Strait to Valparaíso (33°02′S, 71°37′W), head down the coast, keeping about 100 miles from it until near the latitude of San Francisco and from thence pass W of, and in sight of, Isla de Guadalupe (29°00′N, 118°17′W), where in all probability the North-east Trade Wind will be met with. Then steer to sight Île Clipperton (10°18′N, 109°13′W), and pass W of it, where the Trade Wind will be lost.

2 The belt of variable wind and calms will then be entered, which at this season, on 120°W, is 250 to 350 miles wide, and it may not be possible to cross the equator much to windward of 118°W. Every effort should be made not to cross farther W than that or the vessel would not be able to weather Henderson Island (24°22′S, 128°19′W) or Pitcairn Island (25°04′S, 130°05′E), in the vicinity of which light baffling winds from S to SE would be experienced.

3 In all probability, at this time of year, the South-east Trade Wind will be met with between 5°N (November and December) and 3°N (towards March), when the vessel should make, as nearly as the wind will permit, a course of 180°.

4 In about 6°S the Trade Winds generally become more E in direction, sometimes backing N of E, and a vessel should make the following positions:
> 20°S, 124°W, thence;
> 35°S, 120°W, thence;
> 39°S, 110°W, thence;
> 40°S, 100°W, thence;
> 39°S, 90°W, thence proceed direct to Valparaíso, passing S of Isla Robinson Crusoe (33°37′S, 78°52′W).

5 Calms and variable winds will experienced in the vicinity of 30°S, settling into the NW quarter. From Valparaíso to Callao (12°02′S, 77°14′W), steer N along the coast. See also directions at 10.147.
10.131.2

1 During April to September, vessels bound for South American ports should adopt a more W course, passing the latitude of San Francisco in about 129°W. Thence keeping farther from the land to avoid the calms and light variable winds experienced at this season along the coast of lower California and the Gulf of Panama. After meeting the North-east Trade Wind in about 30°N, 127°W, head S on, or near, the longitude of 125°W, which avoids, not only the calms, but also the hurricanes that can be met E of 125°W, during August and September. Occasionally, but rarely, these storms can be encountered W of 125°W.

2 The North-east Trade Wind will be lost at this season in 11°N or 12°N, and the belt of Doldrums will be found to be not so wide as during the winter. The South-east Trade Wind, at this time of year, should be met with in about 8°N but if, as is most likely to be the case at first, the wind is more S, then head more E to recover some of the ground lost by keeping farther W in the North-east Trade Wind. Try to cross the equator between 118°W and 120°W and when, soon after crossing it, the wind backs more to the E head to the S to weather Ducie Island (24°40′S, 124°48′W). Reach 40°S before making to the E, so as to meet the NW winds, as calms and variable winds are met with N of 40°S. After passing 90°W head up for Isla Robinson Crusoe and Valparaíso.

LOW POWERED VESSEL ROUTES

Charts:
> *5127 (1) to (12) Monthly Routeing Charts for North Pacific Ocean.*
> *5128 (1) to (12) Monthly Routeing Charts for South Pacific Ocean.*
> *5309 Tracks followed by Sailing and Auxiliary Powered Vessels.*
> *5310 World Surface Current Distribution.*
> *Diagram 10.3 World Sailing Ship Routes.*

Prince Rupert and Juan de Fuca Strait ⇒ Guam or Yap

Directions
10.132

1   For low-powered, or hampered vessels, W-bound in winter, the following routes may be preferred to those given at 7.271, in order to avoid the worst of the weather and adverse current:

From Juan de Fuca Strait (48°30′N, 124°47′W):
Rhumb line to 32°00′N, 145°00′W, thence:
Along the parallel of 32°N to 180°00′, thence:
Great circle to Guam (13°27′N, 144°35′E) or Yap (9°28′N, 138°09′E).

From Prince Rupert (54°19′N, 130°20′W):
Through Hecate Strait (53°00′N, 131°00′W), thence:
Rhumb line to join the route from Juan de Fuca Strait in 32°00′N, 145°00′W.

Juan de Fuca Strait ⇒ Manila or Singapore

Directions
10.133

1   For low-powered, or hampered, vessels one of the following alternatives to the N great circle routes, described at 7.300, may be preferred:

April, May and October:
Rhumb line to 35°00′N, 160°00′W, thence:
Along the parallel of 35°00′N to 170°00W, thence:
Great circle to San Bernadino Strait (12°33′N, 124°12′E), thence:
As described in *Philippine Islands Pilot* to Manila (14°32′N, 120°56′E) or as described at 7.97 to Singapore (1°12′N, 103°51′E).

November to March:
By rhumb line to 32°00′N, 145°00′W, thence:
Along the parallel of 32°00′N to 175°00W, thence:
Great circle to San Bernadino Strait (12°33′N, 124°12′E), thence:
As described in *Philippine Islands Pilot* to Manila (14°32′N, 120°56′E) or as described at 7.97 to Singapore (1°12′N, 103°51′E).

If weather conditions are unfavourable on the parallel of 32°00′N, course can be held to reach 30°00′N and the ocean crossing can be made on that parallel.

Juan de Fuca Strait ⇒ ports in China

Directions
10.134

1   For low-powered, or hampered, vessels one of the following alternatives to the N great circle routes, described at 7.301, may be preferred:

April, May and October:
Rhumb line to 35°00′N, 160°00′W, thence along the parallel of 35°00′N as follows:
For Shanghai (31°03′N, 122°20′E):
To the vicinity of Nojima Saki (34°54′N, 139°53′E), thence:
Through Ōsumi Kaikyō (30°55′N, 130°40′E), thence:
As navigation permits to destination.
For Hong Kong (22°17′N, 114°10′E):
To 165°00′E, thence:
Great circle to Bashi Channel (21°20′N, 121°00′E), thence:
As navigation permits to destination.

November to March:
Rhumb line to 32°00′N, 145°00′W, thence along the parallel of 32°00′N as follows:
For Shanghai (31°03′N, 122°20′E):
To 150°00′E, thence:
Through Ōsumi Kaikyō (30°55′N, 130°40′E), thence:
As navigation permits to destination.
For Hong Kong (22°17′N, 114°10′E):
To 165°00′E, thence:
Great circle to Bashi Channel (21°20′N, 121°00′E), thence:
As navigation permits to destination.

If weather conditions are unfavourable on the parallel of 32°00′N, course can be held to reach 30°00′N and the ocean crossing can be made on that parallel.

Prince Rupert ⇒ Manila or Singapore

Directions
10.135

1   For low-powered, or hampered, vessels one of the following alternatives to the N great circle routes, described at 7.305, may be preferred:

April, May and October:
From Dixon Entrance (54°30′N, 132°30′W) by rhumb line to 35°00′N, 160°00′W, thence:
Along the parallel of 35°00′N to 175°00′W, thence:
Great circle to San Bernadino Strait (12°33′N, 124°12′E), thence:
As described in *Philippine Islands Pilot* to Manila (14°32′N, 120°56′E), or as described at 7.97 to Singapore (1°12′N, 103°51′E).

November to March:
From Hecate Strait (53°00′N, 131°00′W) by rhumb line to 32°00′N, 145°00′W, thence:
Along the parallel of 35°00′N to 175°00′W, thence:
Great circle to San Bernadino Strait (12°33′N, 124°12′E), thence:
As described in *Philippine Islands Pilot* to Manila (14°32′N, 120°56′E), or as described at 7.97 to Singapore (1°12′N, 103°51′E).

If weather conditions are unfavourable on the parallel of 32°00′N, course can be held to reach 30°00′N and the ocean crossing can be made on that parallel.

Prince Rupert ⇒ ports in China

Directions
10.136

1   For low-powered, or hampered, vessels one of the following alternatives to the N great circle routes, described at 7.306, may be preferred:

April, May and October:
From Dixon Entrance (54°30′N, 132°30′W) by rhumb line to 35°00′N, 160°00′W, thence:
Along the parallel of 35°00′N as follows:
For Shanghai (31°03′N, 122°20′E):
To the vicinity of Nojima Saki (34°54′N, 139°53′E), thence:
Through Ōsumi Kaikyō (30°55′N, 130°40′E), thence:
As navigation permits to destination.
For Hong Kong (22°17′N, 114°10′E):
To 165°00′E, thence:
Great circle to Bashi Channel (21°20′N, 121°00′E), thence:
As navigation permits to destination.

November to March:
From Hecate Strait (53°00′N, 131°00′W) by rhumb line to 32°00′N, 145°00′W, thence:

Along the parallel of 32°00′N as follows:
For Shanghai (31°03′N, 122°20′E):
To 150°00′E, thence:
Through Ōsumi Kaikyō (30°55′N, 130°40′E), thence:
As navigation permits to destination.
For Hong Kong (22°17′N, 114°10′E):
To 165°00′E, thence:

Great circle to Bashi Channel (21°20′N, 121°00′E),
thence:
As navigation permits to destination.
If weather conditions are unfavourable on the parallel of 32°00′N, course can be held to reach 30°00′N and the ocean crossing can be made on that parallel.

PASSAGES FROM SAN FRANCISCO

Charts:
5127 (1) to (12) Monthly Routeing Charts for North Pacific Ocean.
5128 (1) to (12) Monthly Routeing Charts for South Pacific Ocean.
5309 Tracks followed by Sailing and Auxiliary Powered Vessels.
5310 World Surface Current Distribution.
Diagram 10.3 World Sailing Ship Routes.

San Francisco ⇒ Prince Rupert, Juan de Fuca Strait or Columbia River

Directions
10.137

1 The routes from San Francisco (37°45′N, 122°40′W) to Prince Rupert (54°19′N, 130°20′W), Juan de Fuca Strait (48°30′N, 124°47′W) or the Columbia River (46°15′N, 124°05′W) are seasonal.

10.137.1

1 From November to April, during the bad weather season, the vessels should, at once, be taken well out to sea. This will be easy as the wind is usually NW. When far enough offshore to have nothing to fear from SW or NW winds, make as much progress to the N as possible. To the N of the latitude of Cape Mendocino (40°26′N, 124°25′W), SW winds prevail, enabling a vessel to complete the voyage without difficulty, but land should be made 20 to 30 miles S of the port.

10.137.2

1 From April to November, the fine weather season, the wind almost invariably blows from between NW and NE. After leaving San Francisco run about 200 miles offshore and then make to the N, profiting from every shift of wind and always standing on the most favourable tack. It would be well not to approach the land until up to the latitude of the destination unless the vessel can reach the port, or nearly so, without tacking. If bound for Prince Rupert it would be well not to approach the land until nearly abreast of Langara Island (54°15′N, 133°03′W) at the NW extremity of Queen Charlotte Island (53°10′N, 132°10′W). Hecate Strait (53°00′N, 131°00′W) may also be taken.

San Francisco ⇐ ⇒ Unimak Pass, Aleutian Islands

Directions
10.138

1 From San Francisco (37°45′N, 122°40′W) to Unimak Pass (54°15′N, 164°30′W), the tracks for sailing vessels, recommended by the United States Pilot charts, for May until October are as follows:

May and June, make W from San Francisco to 145°W and thence proceed direct to Unimak Pass.
July, August and September continue to 155°W before turning N.
October, continue to 158°W before turning N.

2 From Unimak Pass to San Francisco, proceed as directly as possible.

San Francisco ⇒ Honolulu

Directions
10.139

1 At all seasons, the route from San Francisco (37°45′N, 122°40′W) to China and Japan passes close S of the Hawaiian Islands, and is therefore nearly direct for Honolulu (21°17′N, 157°53′W). On leaving San Francisco, run to the SW for the North-east Trade Wind. From June to December, clear the coast as soon as possible, steering about 266° to avoid the calms E of 128°W. Near the Hawaiian Islands the Trade Wind may possibly veer to E or even SE, particularly from October to May. Approach the land from ENE, when all local winds will be fair. When making a landfall remember that the currents often run at a rate of 20 miles per day, and that calms and baffling winds are common to leeward of the islands.

2 See also remarks on winds, currents and sailing passages around the Hawaiian Islands, at 10.122.

San Francisco ⇒ Yokohama

Directions
10.140

1 From San Francisco (37°45′N, 122°40′W) to Yokohama (35°26′N, 139°43′E), proceed as directed at 10.139, but pass S of Hawaii. Then stand to the W between 15°N and 20°N, being to the N in summer and S in winter. On reaching 160°E, proceed as follows:
January to April, head farther W on the previous course, to 150°E and thence curve round to the WNW, to NW and finally N; passing about 60 miles W of Ogasawara Guntō (27°00′N, 142°00′E) and W of the other islands of Nanpō Shotō (30°00′N, 141°00′E), steering N to Yokohama.
May and June, make to the WNW at once, so as to cross 150°E between 23°N and 24°N, then proceed direct to Yokohama passing about 200 miles E of Ogasawara Guntō.
July to December, leave the track across the Pacific Ocean in 163°E, instead of 160°E, and set course as directly as possible to Yokohama.

2 Alternatively, some navigators recommend heading for Yokohama on reaching the 180° meridian, but this is not a very usual practice.

3   At all times of the year, allowance must be made for the Japan Current (7.26) setting across the track during the latter part of the voyage.

San Francisco ⇒ northerly ports
of the South China Sea

Directions
10.141

1   From San Francisco (37°45′N, 122°40′W), follow the directions given at 10.140 as far as 160°E, and then stand slightly to the N to clear the most N of the Mariana Islands (17°00′N, 146°00′E), and then pass through the Bashi Channel (21°20′N, 121°00′E) to destination.

2   For an alternative route in the north-east monsoon, see 10.101.1. See also 10.143 for a passage via Manila (14°32′N, 120°56′E).

San Francisco ⇒ Shanghai or Nagasaki

Directions
10.142

1   From San Francisco (37°45′N, 122°40′W) to Shanghai (31°03′N, 122°20′E) or Nagasaki (32°42′N, 129°49′E), follow the directions given at 10.141 across the Pacific Ocean but, on arriving in about 135°E, makes as directly as possible for either destination.

San Francisco ⇒ southerly ports
of the South China Sea

Directions
10.143

1   From San Francisco (37°45′N, 122°40′W) to ports in the S part of the South China Sea, the routes are seasonal.
10.143.1

1   During the north-east monsoon (October to March), follow the directions given at 10.141 as far as the Bashi Channel (21°20′N, 121°00′E) and then proceed S along the W coast of Luzon if bound for Manila (14°32′N, 120°56′E). For Ho Chi Minh City Approaches (10°18′N, 107°04′E), proceed directly across the South China Sea, allowing for the current which sets with the wind. For Singapore (1°12′N, 103°51′E), proceed as directed at 10.99.
10.143.2

1   During the south-west monsoon (April to September), leave the W-bound track across the Pacific Ocean, described at 10.140, in 160°E and steer for 15°N, 150°E. Thence, passing S of the Mariana Islands (17°00′N, 146°00′E), head directly for the San Bernadino Strait (12°33′N, 124°12′E) or for the Surigao Strait (9°53′N, 125°21′E) and thence through the Philippine Islands for Manila, Iloilo (10°42′N, 122°36′E) or other Philippine ports.
10.143.3

1   From Mindoro Strait (13°00′N, 120°00′E) or Verde Island Passage (13°36′N, 121°00′E) proceed as directed:
    For Singapore at 10.99;
    For Ho Chi Minh City at 10.100;
    For Hong Kong (22°17′N, 114°10′E) or Xiamen (24°27′N, 118°04′E) at 10.101.

San Francisco ⇒ Australian ports
south of Bribane

Directions
10.144

1   The routes usually followed are seasonal, after taking a direct route from San Francisco (37°45′N, 122°40′W), through the North-east Trade Wind, to about 10°N, 145°W.
10.144.1

1   From 10°N, 145°W, in June to August, steer a direct course, passing N of the Fiji Islands (18°00′S, 180°00′) and S of Nouvelle-Calédonie (21°15′S, 165°20′E), to Caloundra Head (26°48′S, 153°08′E), for Brisbane; or making the coast S of Sandy Cape (24°42′S, 153°16′E) for Sydney (33°50′S, 151°19′E). From Sydney continue S, as directed from 9.220 to 9.224 and from 10.17 to 10.18.
10.144.2

1   From 10°N, 145°W, in September to May, steer a direct course to cross the equator:
    in 152°W, in December to February;
    in 150°W from March to June, and:
    in 152°W to 153°W from September to November.

2   At whatever point the equator is crossed, steer to cross 10°S near 155°W and thence pass S of the Tonga Islands (20°00′S, 174°30′W), cross the 180° meridian in 24°S to 25°S and 160°E in 26°S to 27°S. Thence proceed to destination, passing N of Middleton Reef (29°28′S, 159°04′E), if bound for Sydney making the coast S of Sandy Cape.
10.144.3

1   Alternative seasonal routes were recommended by the French Authorities as follows:
    From January to July, cross 10°N in 143°W and the equator in 148°W. From January to March, no area of calms will be found between the North-east and South-east Trade Winds. From April to June the chance of calms will only be 2%. From the equator steer for position 10°S, 155°W, and continue thence as directed at 10.144.2.
    From July to September, steer to 10°N, 148°W and cross the equator between 150°W and 153°W. In this season there will only be 2% to 3% of calms between 10°N and the equator, provided the vessel does not get E of the prescribed course. Thence proceed as directed above for January to July.
    From October to January, steer to 10°N, 138°W and cross the equator in 143°W. By following this route there will only be from 2% to 3% of calms between the two Trade Wind regions. Farther to the W at this season, more calms are likely. Then proceed as directed above for January to July.

San Francisco ⇒ Pacific Islands

Directions
10.145

1   From San Francisco (37°45′N, 122°40′W), if bound for the North Pacific Islands, head to the SW into the Trade Wind and the North Equatorial Current (7.26) and then head W on 15°N as far as 170°W, thence steer, as directly as navigation permits, to destination.

2   For Tahiti (17°35′S, 149°25′W), steer SW on nearly the direct course to cross the equator in 140°W, or a little to the E, and then direct, allowing for the set of the South Equatorial Current (7.27).

3　For Samoa (14°00′S, 171°30′W), Fiji (18°00′S, 180°00′) and islands to the W, steer from 10°N, 145°W as directed at 10.144.2 until arriving on the latitude of the island of destination and then run W on this latitude.

San Francisco ⇒ Panama

Directions
10.146

1　From San Francisco (37°45′N, 122°40′W) to Panama, between December and May, when the prevailing winds on the W coast of Mexico are from the N, and the current is favourable, first head well offshore and head S, keeping about 100 miles off the coast of California, and about 150 miles off that of Mexico, shaping a course to make Isla Jicarita (7°13′N, 81°48′W), which is 55 miles W of Punta Mariato, which is a good landfall for vessels bound for Panama (Balboa) (8°53′N, 79°30′W) from the W.

2　Between June and November, when calms, variable winds and, sometimes, hurricanes prevail on the W coast of Mexico, head well out to sea after leaving San Francisco and then make course to cross the equator in about 104°W and then head on S until sure of reaching Panama on the starboard tack.

3　Bound for Panama from the N, try to make Isla Jicarita and then try to keep under the land as far as Punta Mala (7°28′N, 80°00′W). If unable to achieve this, head across to the other side of the Gulf of Panama, where the current will be found favourable. On getting E of Punta Mala, shape a course for Isla Galera (8°12′N, 78°47′W) and use the passage E of Archipiélago de las Perlas (8°20′N, 79°00′W) with caution. See *South America Pilot, Volume III* for further details. If, however, tempted up the gulf by a fair wind, try to get on the W side of Archipiélago de las Perlas, where anchorage and less current will be found, should the wind fail.

4　Off the coast N of Punta Guascama (2°37′N, 78°25′W), the winds become more favourable and rains more frequent. This coast is very wet and the rains are abundant all year round, with very few fair days. This weather is found as far as Cabo Corrientes (5°29′N, 77°33′W), the prevailing wind is SW, but NE winds are not uncommon. Offshore in this zone, between 2°N and 5°N, the winds are baffling, especially from March to May.

5　Between Cabo Corrientes and Panama, the prevailing winds are from the N and W, with frequent squalls and wet weather from the SW between June and October.

6　Within 60 miles of the coast there is a constant current to the N. After passing Punta Mala it meets the Mexico Current from the WNW and thus causes numerous ripplings and a short uneasy sea so often met with at the entrance to the Gulf of Panama. This troubled water will be found

more or less to the S, according to the strength of the contending streams.

San Francisco ⇒ Callao or Iquique

Directions
10.147

1　From San Francisco (37°45′N, 122°40′W), proceed 300 miles to the SW, then working round to S so as to cross 30°N in about 127°W. From this position make a straight course to the SSE, roughly parallel with the coast to a position in 5°N, 110°W, where the South-east Trade Wind should be met. Head through the Trade Wind, on the port tack, to 20°S, 118°W.

2　From this position a course gradually approaching the coast may be made, as the Trade Wind is lost and the S winds are felt. The positions reached will be about 34°S, 110°W from September to November, about 37°S, 110°W from December to May and intermediate latitudes from June to August.

3　As the coast is neared, S winds and a N-going current will be obtained, by which Iquique (20°12′S, 70°10′W) or Callao (12°02′S, 77°14′W) can be reached. In both cases make the port well to the S to allow for the N-going current which runs the whole length of the South American coast.

San Francisco ⇒ Coquimbo, Valparaíso and Coronel

Directions
10.148

1　From San Francisco (37°45′N, 122°40′W) to Coquimbo (29°57′S, 71°21′W), Valparaíso (33°02′S, 71°37′W) or Coronel (37°06′S, 73°10′W), proceed as directed at 10.147, but do not attempt to make much progress to the E after arriving at 20°S, 118°W, until well S of 35°S.

2　Make coast well S of the destination, in order to allow for the N-going current which runs the whole length of the South American coast.

San Francisco ⇒ Cabo de Hornos

Directions
10.149

1　From San Francisco (37°45′N, 122°40′W), proceed 300 miles offshore, then head nearly due S to 5°N, at about 126°W from December to February and between 120°W and 122°W during the rest of the year, being farthest E in March.

2　When the South-east Trade Wind is met, head S to cross 30°S at about 124°W. As soon as the Trade Wind is lost and the Westerlies picked up, at about 35°S make as direct a course as possible to round Cabo de Hornos (56°04′S, 67°15′W) as directed in 8.88.

PASSAGES FROM CENTRAL AMERICA AND PANAMA

Charts:
　　5127 (1) to (12) Monthly Routeing Charts for North Pacific Ocean.
　　5128 (1) to (12) Monthly Routeing Charts for South Pacific Ocean.
　　5309 Tracks followed by Sailing and Auxiliary Powered Vessels.
　　5310 World Surface Current Distribution.
　　Diagram 10.3 World Sailing Ship Routes.

PASSAGES FROM BAJA CALIFORNIA

Baja California ⇒ North American ports
Directions
10.150

1　On account of the contrary S-going current, the only way to make passage from any port, on the coast of Baja California (28°00′N, 113°30′W), is to proceed W on the starboard tack until the variable winds are reached, in about

130°W, and then head N, as directed at 10.137. Between July and January, vessels may have to head W as far as 140°W.

2   In the past, lumber vessels bound for Juan de Fuca Strait (48°30′N, 124°47′W) found it advantageous to keep as near the land as practicable, in order to take advantage of the SE storms, which veer to become SW. Rapid passages have been made in this manner.

Baja California ⇒ Pacific Ocean ports

Directions
10.151

1   From ports on the coast of Baja California (28°00′N, 113°30′W), to Honolulu (21°17′N, 157°53′W) and the North Pacific Islands proceed directly when in the North-east Trade Wind, see 10.139 and 10.145 for details.

2   For Sydney (33°50′S, 151°19′E), steer to join the route from San Francisco, described at 10.144, at 10°N, 145°W from June to August and at the equator at other times.

3   For other Pacific Ocean destinations, steer SW to join the appropriate route from San Francisco in a convenient position.

PASSAGES FROM PANAMA

Notes on passage out of the Gulf of Panama
10.152

1   From Balboa (8°53′N, 79°30′W), bound in any direction, the chief difficulty is the passage out of the Gulf of Panama, as light and baffling winds or calms are met with at all seasons.

2   From October to April, the prevailing wind in the gulf is from the N, for the remainder of the year the wind tends more to the W, and land and sea breezes are felt, varied by calms and occasional squalls from the SW.

3   North of 5°N, between 80°W and 110°W, is a region of calms and light winds, varied by squalls of wind and rain. S of 5°N and W of 80°W, between the mainland and Archipiélago de Colón (0°00′, 90°00′W), the wind is between S and W all the year round and, except from February to June, is fairly strong. Whether bound N or S from Panama, head to the S and gain the South-east Trade Wind. By doing so the doldrums and irregular winds will not only be avoided but more suitable weather will be found.

Panama ⇒ Central America

Directions
10.153

1   From Balboa (8°53′N, 79°30′W), the passage to ports along the coast of Central America is slow and troublesome for sailing vessels and advantage must be taken of every shift of wind to get to the NW. The currents will be favourable as far as the Golfo de Fonseca (13°10′N, 87°45′W), but if bound for Acapulco (16°50′N, 99°55′W) or Mazatlán (23°11′N, 106°26′W), the passage may be better made by standing off from the coast after reaching Golfo de Fonseca.

2   If a "Norther" (2.5) is blowing in Golfo de Tehuantepec (16°00′N, 95°00′W) and sail can be carried, it is advisable to run well to the W, without trying to make any progress N. If obliged to heave-to, then two to four days of heavy weather may be expected, with a high, short sea, clear sky overhead and a dense red haze near the horizon.

3   It is reported that if the summits of Sierra Chimalapa, N of Laguna Inferior (16°15′N, 94°45′E), are hidden, about sunset, by a slate-coloured vapour then a Norther will blow the following day. If similar mists are seen on the ocean horizon at sunset, then a SSW wind will blow the next day.

Panama ⇒ San Francisco or Juan de Fuca Strait

Directions
10.154

1   From June to January, having left the Gulf of Panama, as described at 10.152, steer to pass N of Archipiélago de Colón (0°00′, 90°00′W), in order to gain the South-east Trade Wind, thence keeping on 2°N until 105°W is reached. Then alter course NW to pass W of Île Clipperton (10°18′N, 109°13′W), in the neighbourhood of which the North-east Trade Wind will be met, then continue NW to 20°N, 120°W and, if bound for San Francisco (37°45′N, 122°40′W), continue to 35°N, 135°W. If bound for Juan de Fuca Strait (48°30′N, 124°47′W), keep on NW until reaching 40°N, 138°W, then head in for the coast as the wind allows, remembering always to make land to windward of the destination.

2   From February to May, cross the equator between Archipiélago de Colón and the mainland and head W until past 105°W, then alter course NW to pass W of Île Clipperton and proceed as described above for June to January.

Panama ⇒ Australia or New Zealand

Directions
10.155

1   From the Gulf of Panama, cross the equator, passing S of Archipiélago de Colón (0°00′, 90°00′W) into the South-east Trade Wind, as described at 10.152 and 10.153. When in the Trade Wind, run SW to cross 120°W between 11°S to 12°S, thence heading W to pass S of Îles Marquises (9°00′S, 140°00′W) and N of Archipel des Tuamoto (18°00′S, 141°00′W) and join the route from San Francisco, described at 10.144, in 14°S or 15°S on the meridian of 160°W. See also *Diagram 10.3 World Sailing Ship Routes*.

2   If bound for New Zealand, leave the route for Sydney (33°50′S, 151°19′E) in about 170°W and proceed direct, noting that it is advisable, except when bound for Auckland (36°36′S, 174°49′E), to pass down the W coast, and round the S of New Zealand, owing to the prevalence of W winds.

3   If conditions are favourable, when off the Cook Strait (41°00′S, 174°30′E), approach Wellington (41°22′S, 174°50′E) through it.

Panama ⇒ South American ports

General information
10.156

1   From the Gulf of Panama, passages S are all slow and difficult for a sailing vessel, on account of the contrary coastal current, which sets N throughout the year, and the equally contrary light, but persistently S winds.

2   The general opinion appears to be that, if bound for ports N of Callao (12°02′S, 77°14′W) then it is better to beat down the coast. But if bound for ports farther S, such as Matarani (16°59′S, 72°07′W), Iquique (20°12′S, 70°10′W), Antofagasta (23°38′S, 70°26′W), it is better to get offshore into the Trade Wind then reach the coast by the Westerlies, S of 30°S, and run N with a fair wind and current to the destination.

Panama to Golfo de Guayaquil
10.157

1   From the Gulf of Panama, make the best way S until between 5°N and the equator and try, if possible, to keep near the meridian of 80°W, then make a SW course if the winds will allow it. If the wind is SW head to the S, but if it is SSW, then head W if it is a good steady breeze. If the wind is light and baffling, with rain, then the vessel is in the doldrums and should get to the S as soon as possible, taking advantage of every slant of wind to Golfo de Guayaquil (3°00′S, 80°00′W).

Panama to Callao
10.158

1   From the Gulf of Panama, follow the directions at 10.157, as far as Golfo de Guayaquil, then work close inshore as far as Islas Lobos de Afuera (6°57′S, 80°42′W). Approach these islands with care, see *South America Pilot, Volume III* for further details. Try always to be close to the land soon after the sun has set, so that advantage may be taken of the land breeze which, however light, usually begins about this time. This will frequently enable a vessel to make way along the shore throughout the night, and be in a good position for the start of the sea breeze.

2   After passing Islas Lobos de Afuera it would be advisable to work S until the latitude of Callao is approached, then head in. If it cannot be reached then continue working S along the coast, as directed above, remembering that the wind becomes E on leaving the coast. Some navigators attempt to make this passage by standing off for several days, hoping to come in on the opposite tack, but this will generally be found to be fruitless, owing to the N-going current.

Panama to ports between Matarani and Valparaíso
10.159

1   From the Gulf of Panama to ports between Matarani (16°59′S, 72°07′W) and Valparaíso (33°02′S, 71°37′W)

follow the directions for leaving the Gulf of Panama, according to season, given at 10.152.

        From June to January, head W after crossing 2°N and pass N of Archipiélago de Colón (0°00′, 90°00′W), taking care to keep S of 5°N. Winds from S and SSW will persist as far as 85°W, after passing which the wind will be S and the vessel can be considered to be in the Trade Wind.

        From February to May, it is better to cross the equator between Archipiélago de Colón and the coast, before proceeding to the W. This could take a week, but is far preferable to encountering the contrary weather N of Archipiélago de Colón at this season. On this route it must be remembered that S of 1°N the wind becomes more E as the vessel leaves the coast and in 83°W is frequently found E of S.

2   The seasonal routes from Panama, passing N or S of Archipiélago de Colón, given above, meet at about 20°S, 100°W. On reaching this position begin, if possible, to make S and E towards the coast, crossing 30°S at about 95°W. Thence, as the W winds and the N-going current begin to be felt, and eventually the SW and S coastal winds, gradually head up towards the destination, making land to the S, to allow for the N set.

Panama to Cabo de Hornos
10.160

1   From the Gulf of Panama to Cabo de Hornos (56°04′S, 67°15′W), proceed as directed at 10.159 to 20°S, 100°W, and then continue heading to the S, crossing 30°S between 102°W and 103°W. From this position, or on reaching the Westerlies, gradually curve towards the SE, to 50°S, 90°W, being N of the track from September to November and S of it from June to August. Round Cabo de Hornos as directed at 8.88.

PASSAGES FROM SOUTH AMERICAN PORTS

Charts:
> *5127 (1) to (12) Monthly Routeing Charts for North Pacific Ocean.*
> *5128 (1) to (12) Monthly Routeing Charts for South Pacific Ocean.*
> *5309 Tracks followed by Sailing and Auxiliary Powered Vessels.*
> *5310 World Surface Current Distribution.*
> *Diagram 10.3 World Sailing Ship Routes.*

Callao ⇒ Panama, Central America and Mexico

Directions
10.161

1   From Callao (12°02′S, 77°14′W), head N along the coast with a favourable current and a S wind. See notes on winds, weather and currents at 10.146.

2   To ports on the coast of South America, N of the Gulf of Panama, follow the general directions given at 10.153.

Callao ⇒ San Francisco or Juan de Fuca Strait

Directions
10.162

1   From Callao (12°02′S, 77°14′W), head out from the coast to pick up the South-east Trade Wind, then steer NW to cross the equator between 112°W and 115°W, and 5°N and 7°N in 115°W to 118°W, to join the route from Panama, described at 10.154.

Callao ⇒ Australia or New Zealand

Directions
10.163

1   From Callao (12°02′S, 77°14′W), steer W in the South-east Trade Wind to join the route from Panama, described at 10.155, in about 12°S, 122°W.

Callao ⇒ China, Philippine Islands and Japan

Directions
10.164

1   From Callao (12°02′S, 77°14′W), steer W in the South-east Trade Wind to join the route from Valparaíso, described at 10.171, in about 12°S, 122°W.

Callao ⇒ ports as far south as 27°S

Directions
10.165

1 From Callao (12°02′S, 77°14′W) all ports, to 27°S, lie within the South-east Trade Wind, therefore it is recommended normally to work along the shore as far as Isla San Gallán (13°51′N, 76°28′W), where the coast tends more to the E, so that a long leg and a short leg may be made, with the land just in sight, to Rada de Arica (18°30′S, 70°20′W) or to any of the ports between it and Bahía Pisco (13°42′S, 76°15′W).

2 When proceeding from Callao to Bahía Pisco it is recommended to stand off the land at night, and towards it during the day, until S of 13°S, when it is advisable to keep within 4 or 5 miles of the shore down to Bahía Pisco. For details of currents, see *South America Pilot, Volume III*.

3 As an alternative, a sailing vessel of poor performance, might do better running through the Trade Wind belt and making to the S, well offshore, so as to return to the N along the coast, with the current, rather than attempting to work to windward against a Trade Wind that never varies more than a few points.

4 Care is necessary when approaching Punta Caldera (27°03′S, 70°52′W) in the very light winds, as the current will tend to set the vessel on the rocks N of Punta Francisco (27°02′S, 70°50′W).

Callao ⇒ ports south of 30°S

Directions
10.166

1 From Callao (12°02′S, 77°14′W) to ports S of 30°S, there is no doubt that, by standing offshore, a quicker passage will be made than by working along the coast. Therefore, on leaving Bahía del Callao, head well out to the SW through the South-east Trade Wind, and:

> From January to March, cross 18°S in 90°W and then 30°S in 95°W. From this position, as soon as the Westerlies begin to be felt steer E for destination, making to S of the desired port on account of the N-going current.

> From April to December, a lesser distance offshore from the coast will suffice and, on leaving Bahía del Callao, steer so as to cross 18°S in 85°W and 30°S in 90°W, and thence steering E for destination as described above.

Callao ⇒ Cabo de Hornos

Directions
10.167

1 From Callao (12°02′S, 77°14′W) to Cabo de Hornos (56°04′S, 67°15′W), follow the directions given in 10.166, according to season, but on reaching 30°S, 90°W, continue to the S to cross 50°S between 85°W and 90°W, being to the E of the track from September to November and to the W of it from June to August. Then steer to round Cabo de Hornos, as described at 8.88.

Valparaíso ⇒ South American ports northwards

Directions
10.168

1 From Valparaíso (33°02′S, 71°37′W), steer N along the coast. Calms and light winds may be expected in the vicinity of 30°S, but S winds, and a N-going current, will be experienced throughout the remainder of the voyage.

Valparaíso ⇒ Panama, Central America and Mexico

Directions
10.169

1 From Valparaíso (33°02′S, 71°37′W), head to the NW, crossing 30°S in about 77°W and then head N until on the parallel of Callao (12°02′S, 77°14′W), from whence keep about 150 miles from the land until reaching the Gulf of Panama.

2 For ports in Central America and Mexico, N of Panama, proceed as directed at 10.153.

Valparaíso ⇒ San Francisco, Juan de Fuca Strait or Prince Rupert

Directions
10.170

1 From Valparaíso (33°02′S, 71°37′W), the best route to follow is the same at all times of the year. head to the NW, passing E of Isla San Félix (26°19′S, 80°04′W) and crossing 17°S in 90°W. With the South-east Trade Wind, steer to cross the equator in about 118°W. Continue NW into the North-east Trade Wind and cross 20°N in 138°W; 30°N in 142°W and 40°N in 140°W.

2 In May and June, the North-east Trade Wind is often very weak to the N of 20°N, and frequently a belt of calm exists between 20°N and 30°N.

3 For San Francisco (37°45′N, 122°40′W), after losing the North-east Trade Wind, make to the E as soon as the W winds are met with, which will be from about 33°N during the winter to 40°N in the summer, up to the end of August, making allowance for the SE-going current.

4 For Juan de Fuca Strait (48°30′N, 124°47′W), or Prince Rupert (54°19′N, 130°20′W), on reaching 40°N, at all times of the year, cross 47°N in 130°W before steering directly for the destination. Prince Rupert may be approached by passing W of Queen Charlotte Island (53°10′N, 132°10′W) or via Hecate Strait (53°00′N, 131°00′W). Allowance must be made for a SE-going current, setting across the track and attention is called to the cautionary statements in 10.105.1.

Valparaíso ⇒ China, Philippine Islands and Japan

Directions
10.171

1 From Valparaíso (33°02′S, 71°37′W), the passage may be made by using either the North-east or South-east Trade Winds. These two routes are described hereunder as the Northerly or Southerly Route, respectively.

10.171.1

1 The Northerly Route, for departures between August and February, passes through 12°S, 122°W, where the route from Callao (10.164) joins it, and continues through the South-east Trade Wind to cross the equator in about 138°W and 10°N in about 143°W. The North-east Trade Wind will be found near this parallel, thence continue, as from North America, passing S of the Hawaiian Islands (19°40′N, 155°30′W), and joining the appropriate route from San Francisco to the destination, as listed below for:

> The Philippine Islands at 10.143;
> Hong Kong (22°17′N, 114°10′E) at 10.141 or 10.143;
> Shanghai (31°03′N, 122°20′E) at 10.142;
> Singapore (1°12′N, 103°51′E) at 10.143;

Nagasaki (32°42′N, 129°49′E) at 10.142;
Yokohama (35°26′N, 139°43′E) at 10.140.

10.171.2

1   For the Southerly Route, recommended for departures between March and July, on leaving Valparaíso, steer NW into the South-east Trade Wind. Having picked up the Trade Wind, pass S of Îles Marquises (9°00′S, 140°00′W), S of Kiribati (0°00′, 174°00′E) and N of the Caroline Islands (7°00′N, 150°00′E) to a position in about 13°N, 130°E.

2   At this position, join the Second Eastern Passage, described at 9.101, from the Eastern Archipelago to China and Japan, during the north-east monsoon (October to March), E of the Philippines and, from April to September pass through San Bernadino Strait (12°33′N, 124°12′E) for Manila (14°32′N, 120°56′E) and the South China Sea.

3   Bound to Yokohama (35°26′N, 139°43′E), leave the route when S of Kiribati, cross the equator in about 168°E and join the appropriate route from Sydney, described from 10.29 to 10.32, soon afterwards. See also *Diagram 8.3 World Sailing Ship Routes.*

4   See also details of the routes from San Francisco, described from 10.141 to 10.143.

5   If there is a N wind on leaving Valparaíso, head W for as long as it lasts and then NW into the South-east Trade Wind. In the latitudes of Valparaíso, from June to August, N winds occasionally extend far across the Pacific Ocean.

Valparaíso ⇒ Australia or New Zealand

Directions

10.172

1   From Valparaíso (33°02′S, 71°37′W), steer to the NW to join the route from Panama, described at 10.155, between 120°W and 130°W and 10°S and 12°S.

2   See also the remarks at 10.171.2, regarding leaving Valparaíso with a N wind.

Valparaíso ⇒ Cabo de Hornos

Directions

10.173

1   From Valparaíso (33°02′S, 71°37′W), the same rule applies for rounding Cabo de Hornos (56°04′S, 67°15′W), as that from Callao, described at 10.167, namely first to make offshore for 500 or 600 miles to the SW until the Westerlies are steady and certain, and the strength of the NE-going current is lost. This rule also applies to any port on the W coast of South America.

2   From Valparaíso or Talcahuano (36°41′S, 73°06′W), the position to make for is about 40°S, 84°W. Thence head nearly S crossing 50°S in about 85°W. From this position alter course gradually to the SE and E to round Cabo de Hornos, as directed at 8.88.

PASSAGES FROM CABO DE HORNOS

Charts:

> 5127 (1) to (12) Monthly Routeing Charts for North Pacific Ocean.
> 5128 (1) to (12) Monthly Routeing Charts for South Pacific Ocean.
> 5309 Tracks followed by Sailing and Auxiliary Powered Vessels.
> 5310 World Surface Current Distribution.
> Diagram 10.3 World Sailing Ship Routes.

Reference

10.174

1   For directions for rounding Cabo de Hornos (56°04′S, 67°15′W) W-bound, see 8.23.

2   For directions for rounding Cabo de Hornos E-bound, see 8.88.

Cabo de Hornos ⇒ Valparaíso

Directions

10.175

1   After passing 70°W, in about 57°S, as described at 8.23, for rounding Cabo de Hornos (56°04′S, 67°15′W), head NW and then N, keeping at a distance of about 150 miles from the land. Begin to close the land at about 40°S, W winds and a favourable current will be found from about 48°S. Make the landfall S of Valparaíso (33°02′S, 71°37′W).

2   Some navigators prefer to head farther to the NW to about 50°S, 80°W, before turning N towards destination.

Cabo de Hornos ⇒ South American ports northwards of Valparaíso

Directions

10.176

1   Vessels bound for South American ports N of Valparaíso (33°02′S, 71°37′W), should round Cabo de Hornos (56°04′S, 67°15′W), as described at 8.23, and when W of the meridian of Cabo Pilar (52°43′S, 74°41′W) take every opportunity to make W progress as far as 82°W to 84°W. Thence steer as direct to destination, as is consistent with, making use of the steady winds prevailing offshore. Be careful not to get to leeward of the destination when approaching the land.

Cabo de Hornos ⇒ Panama, Central America and Mexico

Directions

10.177

1   Proceed as directed at 10.176 until reaching 82°W to 84°W, then steer N to close the land to a distance of about 60 miles off, or just N of Golfo de Guayaquil (3°00′S, 80°00′W). After crossing the equator steer for Isla Galera (8°12′N, 78°47′W), taking care, especially in the dry season, to stand inshore with the first N winds. By doing so, a vessel will most probably have a favourable current along the coast, whereas by keeping in the centre or on the W side of the gulf, a strong S-going set will be experienced.

2   After making Isla Galera and clearing Banco San José (8°08′N, 78°39′W), navigation towards Panama between Archipiélago de las Perlas (8°20′N, 79°00′W) and the mainland is clear and easy, with the advantage of being able to anchor during adverse conditions of wind and tide. As a rule the passage E of the islands should be taken, but with a strong S wind the navigator is tempted to run up the gulf, in which case he should keep to the W side of Archipiélago de las Perlas, where anchorage and less current will be found, if the wind should fail, which can always be expected in these regions.

3　If bound for ports in Central America or Mexico, N of the Gulf of Panama, proceed generally as directed in 10.153.

Cabo de Hornos ⇒ San Francisco and ports northwards

Directions
10.178

1　Having rounded Cabo de Hornos (56°04′S, 67°15′W), as described at 8.23, and bound for San Francisco (37°45′N, 122°40′W), head to the NW so as to cross 50°S between 80°W and 85°W, and thence due N to 30°S.

2　Then proceed NW again, running through the South-east Trade Wind to cross the equator between 112°W and 115°W. Throughout the whole voyage from Cabo de Hornos, keep to the E from September to November, and to the W from June to August.

3　After crossing the equator, steer to cross 120°W between 13°N and 15°N, where the route divides into two branches, according to season.

4　From March to October, make for 30°N, 137°W and turn towards the land when the Westerlies are reached, about 35°N, allowing for the SE-going current setting across the track.

5　From November to February, make for 30°N, 132°W, and from that position, when the Westerlies are met, curve gradually round to make to the N of San Francisco, allowing for the SE-going current setting across the track.

6　For Columbia River (46°15′N, 124°05′W), Juan de Fuca Strait (48°30′N, 124°47′W), or Prince Rupert (54°19′N, 130°20′W), follow the routes given above as far as 30°N, then continue to the NW, curving to the E on reaching, or nearing, 45°N to make the desired destination, allowing for the SE-going current setting across the track.

Cabo de Hornos ⇒ Honolulu

Directions
10.179

1　From Cabo de Hornos (56°04′S, 67°15′W), follow the directions given at 10.178 as far as 30°S, or if necessary a little farther N, to enter the South-east Trade Wind, then proceed as directly as possible to Honolulu (21°17′N, 157°53′W), crossing the equator between 120°W and 125°W.

Cabo de Hornos ⇒ Philippine Islands, China, Japan, Australia or New Zealand

Directions
10.180

1　From Cabo de Hornos (56°04′S, 67°15′W), follow the directions given at 10.178 as far as 30°S, and then run in the South-east Trade Winds to about 12°S, 122°W to join:
　　The route from Valparaíso, described at 10.171, for the Philippine Islands, China or Japan, and;
　　The route from Panama, described at 10.155, for Australia or New Zealand.

Cabo de Hornos ⇒ Pacific Islands

Directions
10.181

1　From Cabo de Hornos (56°04′S, 67°15′W), follow the directions given at 10.178 as far as 30°S, 85°W.

2　Then, if bound for Tahiti (17°35′S, 149°25′W), run WNW in the South-east Trade Wind, either passing through Archipel des Tuamotu (18°00′S, 141°00′W), by way of Passe de Fakarava (16°00′S, 145°50′W). If without modern navigational equipment it would be better to pass S of Pitcairn Island (25°04′S, 130°05′E). For directions for Archipel des Tuamotu, see *Pacific Islands Pilot, Volume III.*

3　For islands in the W part of the North Pacific Ocean, after crossing 30°S, run in the South-east Trade Wind to about 12°S, 122°W, to join the most N route from Valparaíso, described at 10.171.1, for the Philippine Islands, China or Japan. Leave this route in about 175°E if bound for the Marshall Islands (10°00′N, 170°00′E) or in 160°E or 165°E, if bound for the Caroline Islands (7°00′N, 150°00′E) or farther W, and proceed direct to destination. The route to Honolulu is given at 10.179.

4　For islands in the W part of the South Pacific Ocean, proceed either via Tahiti as described above, and thence having reached the latitude of destination, head direct. Alternatively, the route described at 10.155 could be followed as far as 160°W, thence to destination.

TABLE A

BEAUFORT SCALE OF WIND

Beaufort Number	Descriptive Term	Mean velocity		SPECIFICATIONS	Probable wave height* in metres	Probable wave height* in feet
		Knots	m/s			
0	Calm	>1	0–0·2	Sea like a mirror	—	—
1	Light air	1–3	0·3–1·5	Ripples with the appearance of scales are formed, but without foam crests	0·1 (0·1)	¼ (¼)
2	Light breeze	4–6	1·6–3·3	Small wavelets, still short but more pronounced; crests have a glassy appearance and do not break	0·2 (0·3)	½ (1)
3	Gentle breeze	7–10	3·4–5·4	Large wavelets; crests begin to break; foam of glassy appearance; perhaps scattered white horses	0·6 (1)	2 (3)
4	Moderate breeze	11–16	5·5–7·9	Small waves, becoming longer; fairly frequent white horses	1 (1·5)	3½ (5)
5	Fresh breeze	17–21	8·0–10·7	Moderate waves, taking a more pronounced long form; many white horses are formed (chance of some spray)	2 (2·5)	6 (8½)
6	Strong breeze	22–27	10·8–13·8	Large waves begin to form; the white foam crests are more extensive everywhere (probably some spray)	3 (4)	9½ (13)
7	Near gale	28–33	13·9–17·1	Sea heaps up and white foam from breaking waves begins to be blown in streaks along the direction of the wind	4 (5·5)	13½ (19)
8	Gale	34–40	17·2–20·7	Moderately high waves of greater length; edges of crests begin to break into the spindrift; the foam is blown in well-marked streaks along the direction of the wind	5·5 (7·5)	18 (25)
9	Strong gale	41–47	20·8–24·4	High waves; dense streaks of foam along the direction of the wind; crests of waves begin to topple, tumble and roll over; spray may affect visibility	7 (10)	23 (32)
10	Storm	48–55	24·5–28·4	Very high waves with long overhanging crests; the resulting foam, in great patches, is blown in dense white streaks along the direction of the wind; on the whole, the surface of the sea takes a white appearance; the tumbling of the sea becomes heavy and shock-like; visibility affected	9 (12·5)	29 (41)
11	Violent storm	56–63	28·5–32·6	Exceptionally high waves (small and medium-sized ships might be for a time lost to view behind the waves); the sea is completely covered with long white patches of foam lying along the direction of the wind; everywhere the edges of the wave crests are blown into froth; visibility affected	11·5 (16)	37 (52)
12	Hurricane	64 and over	32·7 and over	The air is filled with foam and spray; sea completely white with driving spray; visibility very seriously affected	14 (—)	45 (—)

* This table is only intended as a guide to show roughly what may be expected in the open sea, remote from land. It should never be used in the reverse way; ie. for logging or reporting the state of the sea. In enclosed waters, or when near land, with an off-shore wind, wave heights will be smaller and the waves steeper. Figures in brackets indicate the probable/maximum height of waves.

Table B

Table showing principal areas affected and months in which seasonal winds normally occur

Hemisphere	Area	General Wind Direction	Jan	Feb	Mar	Apr	May	Jun	Jul	Aug	Sep	Oct	Nov	Dec
Northern Hemisphere	South China Sea	NE	5-6	4-5	4								5-6	5-6
Northern Hemisphere	Eastern China Sea	NE-N	5	5	5							4-5	5	5
Northern Hemisphere	Yellow Sea	N-NW	5	5	5							4-5	5	5
Northern Hemisphere	Japan Sea	N-NW	5	5	5							4-5	5	5
Northern Hemisphere	North Indian Ocean	NE	4	4	4								4	4
Northern Hemisphere	South China Sea	SW					3	4	4	4-5				
Northern Hemisphere	Eastern China Sea	SW-S						3-4	3-4					
Northern Hemisphere	Yellow Sea	SW-SE						3-4	3-4					
Northern Hemisphere	Japan Sea	SW-S-E						3-4	3-4					
Northern Hemisphere	North Indian Ocean	SW						5-6	6	6	5			
South Hemisphere	Indonesian waters	W-NW	3	3	3									3
South Hemisphere	Arafura Sea	NW	5	5	3-4									3-4
South Hemisphere	N and NW Australian Waters	W-NW	4-5	4-5										
South Hemisphere	Indonesian waters	SE					4-5	4-5	4-5	4-5	4-5			
South Hemisphere	Arafura Sea	SE					4	4	4	4	4			
South Hemisphere	N and NW Australian Waters	SE-E				3-4	4-5	4-5	4-5	4-5	3-4			

Seasonal Winds - normal periods
Seasonal Winds - variable periods at onset and termination
Figures indicate typical wind force (Beaufort)

Seasonal Wind/Monsoon Table - West Pacific and Indian Ocean

369

Table C

Table showing principal areas affected and months in which tropical storms normally occur

Area & Local name	Jan	Feb	Mar	Apr	May	Jun	Jul	Aug	Sep	Oct	Nov	Dec	A	B
North Atlantic, West Indies region (hurricane)													10	5
North-East Pacific (hurricane)													15	7
North-West Pacific (typhoon)													25-30	15-20
North Indian Ocean Bay of Bengal (cyclone)													2-5	1-2
North Indian Ocean Arabian Sea (cyclone)													1-2	1
South Indian Ocean W of 80°E (cyclone)													5-7	2
Australia W, NW, N coasts & Queensland coast (hurricane)													2-3	1
Fiji, Somoa, New Zealand (North Island) (hurricane)													7	2

Start/Finish of season

Period of greatest activity

Period affected when season early/late

Column A: Approximate average frequency of tropical storms each year
Column B: Approximate average frequency of tropical storms each year which develop Force 12 winds or stronger

Tropical Storm Table

Table D

STANDARD TIME ZONE CHART OF THE WORLD

Boundaries shown on this chart are approximate.
See also UKHO Chart 5006 - The World Time Zone Chart

All islands in the Line and Phoenix Groups within the Republic of Kiribati observe the same date as the islands in the Gilbert Group, even though they are positioned on opposite sides of the International Date Line.

INTERNATIONAL DATE LINE

EUROPE and N. AFRICA
(see larger scale)

See Note

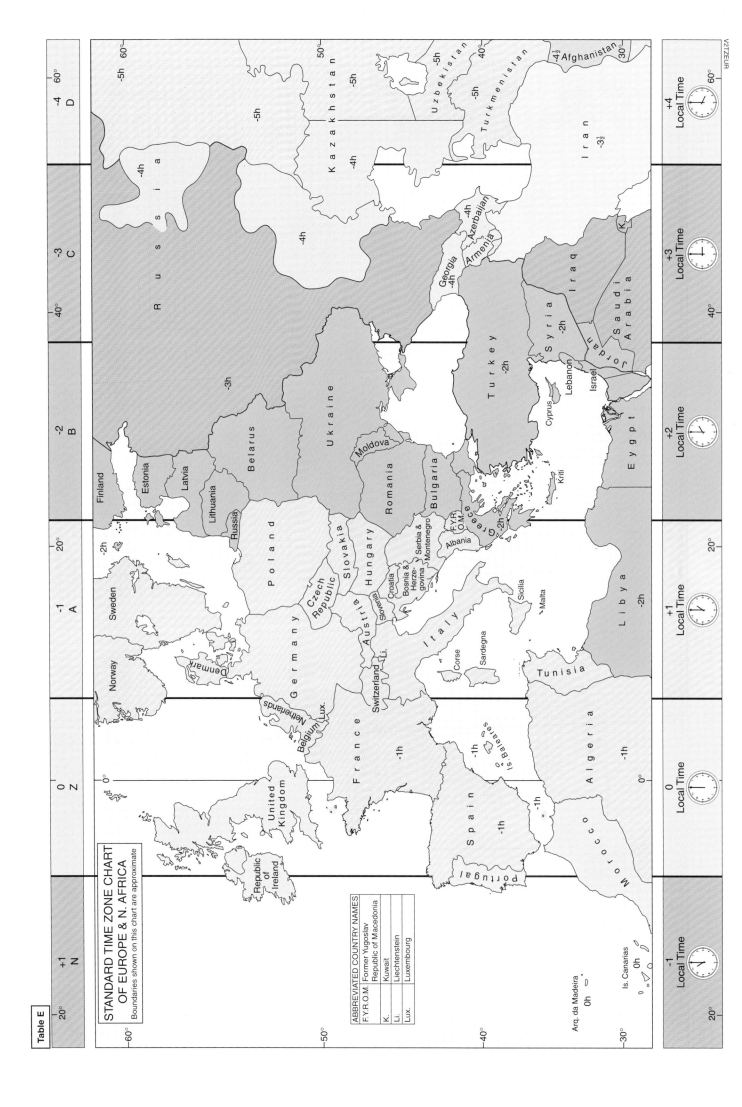

Eastern Archipelago
Archipelagic Sea Lanes

APPENDIX A

ARCHIPELAGIC SEA LANES

Establishment of Archipelagic Sea Lanes

The Indonesian Government has established Archipelagic Sea Lanes (ASLs) for passages through the Eastern Archipelago.

This Appendix explains their purpose, use and the rights and obligations of foreign vessels exercising right of innocent passage through Indonesian Waters.

The relevant Indonesian Government Regulations, numbers 36, 37 and 38 of 2002 are also given.

Definition

Archipelagic sea lanes and air routes are routes through and above the territorial sea and archipelagic waters of an archipelagic State from one part of the high seas or an exclusive economic zone to another part of the high seas or an exclusive economic zone. They are defined by a series of continuous axis lines from the entry points of passage routes to the exit points. The axis lines are delimited by a series of geographic co-ordinates of latitude and longitude, referred to a geodetic datum. Ships and aircraft exercising archipelagic sea lanes passage shall not deviate more than 25 miles to either side of the axis lines, provided that such ships and aircraft shall not navigate closer to the coast than 10% of the distance between the axis line and the nearest points on islands bordering the sea lanes.

Purpose

An archipelagic State may designate sea lanes and air routes thereabove, suitable for the continuous and expeditious and unobstructed transit of foreign ships and aircraft through or over its archipelagic waters and adjacent territorial sea between one part of the high seas or an exclusive economic zone and another part of the high seas or an exclusive economic zone. All ships and aircraft enjoy the right of archipelagic sea lanes passage in such sea lanes and air routes in their normal mode.

Archipelagic Sea Lanes adopted by the IMO

When an archipelagic State submits proposed archipelagic sea lanes to the IMO, the recognised competent international organisation, the IMO will ensure that the proposed sea lanes are in conformity with the relevant provisions of UNCLOS. The IMO will also determine whether the submission is a full or partial sea lanes proposal. It should be noted that within archipelagic sea lanes traffic is not separated except in traffic separation schemes. It should also be noted that the axis of the archipelagic sea lane does not indicate the deepest water, nor any recommended route or track. The first partial system of archipelagic sea lanes in Indonesian archipelagic waters was adopted in 1998 and came into force in December 2002.

Archipelagic Sea Lanes not adopted by the IMO

If an archipelagic State only proposes a partial system of archipelagic sea lanes, or where it decides not to designate archipelagic sea lanes, archipelagic sea lanes passage is available for ships and aircraft through and above all routes normally used for international navigation and overflight.

Charting of Archipelagic Sea Lanes

Admiralty charts show all adopted archipelagic sea lanes, including the axis lines and the lateral limits of the sea lanes.

Archipelagic Sea Lanes through Eastern Archipelago

The following archipelagic sea lanes, defined by their respective axis lines, have been designated by the Indonesian Government in Indonesian Government Regulation Number 37 of 2002:

ASL I — South China Sea — Selat Karimata — Western Java Sea — Selat Sunda — Indian Ocean

Co-ordinates	Reference number in Regulation 37-2002
3°35'·0N, 108°51'·0E	I - 1
3°00'·0N, 108°10'·0E	I - 2
0°50'·0N, 106°16'·3E	I - 3
0°12'·3S, 106°44'·0E	I - 4
2°01'·0S, 108°27'·0E	I - 5
2°16'·0S, 109°19'·5E	I - 6
2°45'·0S, 109°33'·0E	I - 7
3°46'·8S, 109°33'·0E	I - 8
5°12'·5S, 106°54'·5E	I - 9
5°17'·3S, 106°44'·5E	I - 10
5°17'·3S, 106°27'·5E	I - 11
5°15'·0S, 106°12'·5E	I - 12
5°57'·3S, 105°46'·3E	I - 13
6°18'·5S, 105°33'·3E	I - 14
6°24'·8S, 104°41'·4E	I - 15

ASL IA — Northeast of Pulau Bintan

Co-ordinates	Reference number in Regulation 37-2002
1°52'·0N, 104°55'·0E	IA - 1
0°50'·0N, 106°16'·3E	I - 3

ASL II — Celebes Sea — Selat Makasar — Selat Lombok — Indian Ocean

Co-ordinates	Reference number in Regulation 37-2002
0°57'·0N, 119°33'·0E	II - 1
0°00'·0N, 119°00'·0E	II - 2
2°40'·0S, 118°17'·0E	II - 3
3°45'·0S, 118°17'·0E	II - 4
5°28'·0S, 117°05'·0E	II - 5
7°00'·0S, 116°50'·0E	II - 6
8°00'·0S, 116°00'·0E	II - 7
9°01'·0S, 115°36'·0E	II - 8

ASL IIIA (part 1) — Pacific Ocean — Molucca Sea — Ceram Sea — Banda Sea

Co-ordinates	Reference number in Regulation 37-2002
3°27'·0N, 127°40'·5E	IIIA - 1
1°40'·0N, 126°57'·5E	IIIA - 2
1°12'·0N, 126°54'·0E	IIIA - 3
0°09'·0N, 126°20'·0E	IIIA - 4

Co-ordinates	Reference number in Regulation 37-2002
1°53'·0S, 127°02'·0E	IIIA - 5
2°37'·0S, 126°30'·0E	IIIA - 6
2°53'·0S, 125°30'·0E	IIIA - 7
3°20'·0S, 125°30'·0E	IIIA - 8
7°50'·0S, 125°21'·2E	

ASL IIIA (part 2) — Savu Sea — Indian Ocean

Co-ordinates	Reference number in Regulation 37-2002
8°52'·0S, 124°05'·0E	
9°03'·0S, 123°34'·0E	IIIA - 10
9°23'·0S, 122°55'·0E	IIIA - 11
10°12'·0S, 121°18'·0E	IIIA - 12
10°44'·5S, 120°45'·8E	IIIA - 13

ASL IIIB — Banda Sea — Selat Leti

Co-ordinates	Reference number in Regulation 37-2002
3°20'·0S, 125°30'·0E	IIIA - 8
4°00'·0S, 125°40'·0E	IIIB - 1
8°03'·0S, 127°21'·2E	IIIB - 2

ASL IIIC — Banda Sea — Aru Sea

Co-ordinates	Reference number in Regulation 37-2002
3°20'·0S, 125°30'·0E	IIIA - 8
4°00'·0S, 125°40'·0E	IIIB - 1
6°10'·0S, 131°45'·0E	IIIC - 1
6°44'·0S, 132°35'·0E	IIIC - 2

ASL IIID — Savu Sea —Between Sawu and Roti — Indian Ocean

Co-ordinates	Reference number in Regulation 37-2002
9°23'·0S, 122°55'·0E	IIIA - 11
10°58'·0S, 122°11'·0E	IIID - 1

ASL IIIE — Celebes Sea — Molucca Sea

Co-ordinates	Reference number in Regulation 37-2002
4°32'·2N, 125°10'·4E	IIIE - 2
4°12'·1N, 126°01'·0E	IIIE - 2
1°40'·0N, 126°57'·5E	IIIA - 2

INDONESIAN GOVERNMENT REGULATIONS

The following Indonesian Government Regulations:
Number 36 of 2002, Rights and obligations of foreign vessels when exercising innocent passage via Indonesian Waters:
Number 37 of 2002, Rights and obligations of foreign ships and aircraft when exercising right of Archipelagic Sea Lane passage via the established Archipelagic Sea Lanes:
Number 38 of 2002, Relating to List of coordinates of geographical points of Indonesian Archipelago baseline:
are detailed below.

NUMBER 36 OF 2002

RELATING TO RIGHTS AND OBLIGATIONS OF FOREIGN VESSELS WHEN EXERCISING AN INNOCENT PASSAGE VIA THE INDONESIAN WATERS

THE PRESIDENT OF THE REPUBLIC OF INDONESIA

Considering:
a) that Law Number 6 of 1996 concerning the Indonesian Waters, which constituted the addition to the United Nations Convention concerning Law of the Sea of 1982, which amongst others, stipulates that the rights and obligations of foreign vessels in the conduct of Innocent Passage and be further determined in Government Regulations;
b) that based upon the consideration as referred to letter a, it is necessary to stipulate a Government Regulation on the Rights and Obligations of Foreign Vessels when exercising Out Innocent Passage via the Indonesian Waters.

In view of:
1. Article 5 paragraph (2) of the 1945 Constitution as amended with the Third Amendment of the 1945 Constitution;
2. Law Number 6 of 1996 concerning the Indonesian Waters (State Gazette of the Republic of Indonesia of 1996 Number 73; Supplementary State Gazette Number 3647);

DECIDED

To Stipulate:

GOVERNMENT REGULATION RE THE RIGHTS AND OBLIGATIONS OF FOREIGN VESSELS WHEN EXERCISING AN INNOCENT PASSAGE VIA THE INDONESIAN WATERS

CHAPTER I

GENERAL PROVISIONS

Article 1
In this Government Regulation:
(1) Law shall refer to Law Number 6 of 1996 concerning the Indonesian Waters.
(2) Passage shall refer to the definition as provided in Article 11 paragraph (2) of the Law;
(3) Innocent Passage shall refer to the definition as provided in Article 12 of the Law.
(4) The Indonesian Waters shall refer to the definition as provided in Article 1 point 4 of the Law.
(5) Territorial Seas shall mean the sea lines as referred to the definition as provided in Article 3 paragraph (2) of the Law.
(6) Indonesian Archipelago Waters shall refer to the definition as provided in Article 3 paragraph (3) of the Law.

(7) Indonesian Interior Waters shall refer to the definition as provided in Article 3 paragraph (4) of the Law.

(8) Sea Lanes shall refer to the shipping lanes commonly used for shipping purposes determined as the lanes for safe, continuous and fast sailing.

(9) Passage Separating Scheme shall refer to the passage separation arrangement for the sailing safety via Sea Lanes.

(10) Navigation Charts shall refer to the sea maps compiled and used for the navigation purposes in sea in compliance to the international standard in respect to the sailing safety.

(11) Convention shall refer to the United Nations Convention concerning the Law of the Sea of 1982 as referred to Article 1 point 9 of the Law.

CHAPTER II

IMPLEMENTATION OF INNOCENT PASSAGE IN THE TERRITORIAL WATERS AND INDONESIAN ARCHIPELAGO WATERS

First Part

Rights and Obligations of Foreign Vessels

Article 2

(1) All foreign vessels may exercise the right to Innocent Passage via the Territorial Waters and Archipelago Waters for the purpose of passing from a part of the open sea or exclusive economic zone to the other part of the open sea or exclusive economic zone without going into the Interior Waters or drop anchor in the middle of the sea, or port facilities outside the Interior Waters for the purpose of passing from the open sea or exclusive economic zone to pass to or from the Interior Waters or drop anchor in the middle of the sea, or port facilities outside the Interior Waters.

(2) The exercise of Innocent Passage Right as referred to paragraph (1) shall be carried out by the using the Sea Lanes commonly used for international sailing in compliance to Article 11 and adhering to the sailing guidelines issued by the competent authorities in charge of sailing safety.

Article 3

(1) Any foreign vessel from a part of open sea or exclusive economic zone exercising Innocent Passage crossing the Territorial Seas and Archipelago Waters to the other part of open sea or exclusive economic zone must use the Sea Lanes according to the port of origin and destination of its sailing.

(2) Any foreign vessel from a part of open sea or exclusive economic zone en route to the Interior Waters or one of the ports or the other way around exercising Innocent Passage crossing the Territorial Seas and Archipelago Waters to must use the Sea Lanes according to the origin and destination.

(3) Any foreign vessel exercising Innocent Passage must be within the limits of reasonable sailing lanes with the speed and direction according to the normal navigation en route to the sailing destination.

(4) In exercising the Innocent Passage as referred to paragraph (1), paragraph (2), and paragraph (3), foreign vessels are prohibited from dropping their anchors, stop, loitering, unless such is necessary due to force majeure, or disaster or saving humans, vessels or aircrafts suffering from a disaster.

Article 4

(1) In exercising Innocent Passage via the Territorial Seas and Archipelago Waters, no foreign vessels are allowed to perform of the following activities:

a). intimidating or use of force towards the sovereignty, territorial integrity, country's coastal political freedom, or any other way that violates the international law principles as contained in the United Nations Charter;

b). exercising or practicing using any type of weapon;

c). gathering information that will be detrimental to the country's defense and security;

d). waging any propaganda aimed at inducing the country's defense and security;

e). launching, landing or flying any aircraft from or to the vessel;

f). launching, landing, or loading any military equipment and instrument from or to the vessel; or

g). loitering in the Territorial Seas and Archipelago Waters or performing any other activities not directly related to the passage.

Article 5

(1) In exercising Innocent Passage via the Territorial Seas and Archipelago Waters, no foreign vessel is allowed to perform the following activities:

a). unloading or loading any commodity, currency, or passenger contradictory with the laws and regulations of the customs, fiscal, immigration, or sanitary;

b). fishing;

c). research or survey;

d). activity aimed at interfering any communication system, facility, or other communication facility;

e). deliberate polluting and causing serious pollution.

(2) In exercising Innocent Passage via the Territorial Seas and Archipelago Waters, no foreign vessel is allowed to perform the following activities:

a). damaging or disrupting navigation instruments and means, and other navigation facilities or installations;

b). damaging biodiversity sources; or

c). ruining or disrupting sea cables and pipelines.

Article 6

In exercising Innocent Passage via the Territorial Seas and Archipelago Waters in narrow straits, foreign vessels in sailing in Sea Lanes as designated are prohibited from sailing and approaching the coastline less than 10% (ten percent) of the width of such straits.

Article 7

(1) Foreign vessels used for fishing purposes in exercising Innocent Passage via Territorial Seas and Archipelago Waters must sail in the Sea Lanes as referred to Article 11.

(2) In exercising Innocent Passage, the foreign vessels as referred to in paragraph (1) must keep their fishing equipment in the vessels' holds.

Article 8

(1) Any foreign vessel deployed for a marine research or survey in exercising Innocent Passage via the Territorial Seas and Archipelago Waters must sail in the Sea Lanes as referred to in Article 11.

(2) In exercising Innocent Passage, the foreign vessels as referred to in paragraph (1) must keep their research and survey equipment that are not part of the navigation equipment in non-operating mode.

Article 9

Any foreign vessel sailing in the Sea Lanes must:

a). keep tuning to the Indonesia's Seamen Message Radio;

b). keep monitoring the sailing activities of vessels involved in inter-Archipelago sailing.

Article 10

(1) Foreign vessels must fully pay any levy imposed on them with regard to special services rendered to them while exercising Innocent Passage via Territorial Seas and Archipelago Waters.

(2) Any foreign vessel that fails to meet their obligations as referred to paragraph (1) are subject to execution pursuant to the prevailing provisions of civil law of procedure.

Second Part

Sea Lanes and Separating Scheme

Article 11

(1) Foreign tankers, foreign fishing boats, marine research vessels or foreign hydrographical survey vessels, and foreign nuclear-powered ships or foreign vessels laden with nuclear materials or other hazardous or poisonous substances in exercising Innocent Passage for the purpose of passing from one part of open sea or exclusive economic zone to the other part of open sea and exclusive economic zone via the Indonesian Waters must use the sea lanes commonly designated for international sailing as referred to paragraph (2).

(2)

a). For sailing from the South China Sea to the Indian Ocean and vice versa, those vessels may use Sea Lanes that are commonly used for international sailing, namely via Natuna Sea, Karimata Strait, Java Sea and Sunda Strait.

b). For sailing from Sulawesi Sea to the Indian Ocean or vice versa, those vessels may use Sea Lanes that are commonly used for international sailing, namely via Makassar Strait, Flores Sea, and Lombok Strait.

c). For sailing from the Pacific Ocean to the Indian Ocean or vice versa, those vessels may use Sea Lanes that are commonly used for international sailing, namely via Maluku Sea, Seram Sea, and Banda Sea, Ombai Strait and Sawu Sea.

d). For sailing from the Pacific Ocean to Timor Sea or Arafura Sea or vice versa, those vessels may use Sea Lanes that are commonly used for international sailing, namely via Maluku Sea, Seram Sea, and Banda Sea.

(3) The Sea Lanes as referred to paragraph (2) are mentioned in navigation charts or sea guidelines that are exclusively published for sailing safety.

Article 12

(1) For sailing safety in Sea Lanes as referred to Article 11, the Government may determine passage separating scheme.

(2) Any foreign vessel sailing in the Sea Lanes in which passage separating scheme is determined as referred to paragraph (1) must comply with the use of such passage separating scheme.

Article 13

(1) For sailing safety in a strait used for international sailing, the Minister, whose scope of duties and responsibilities cover transportation, may determine the Sea Lanes in Archipelago Waters to be used as part of the passage separating scheme in respect to the implementation of transit passage via such straits.

(2).The sailing of a foreign vessel using the Sea Lanes as referred to in paragraph (1) shall be exercised by adhering to the provisions of Innocent Passage in Archipelago Waters.

Third Part

Suspension of Innocent Passage

Article 14

(1) Temporary suspension of Innocent Passage for foreign vessels in certain areas in Territorial Seas and Archipelago Waters as urgently required for security reasons or for battle exercises shall be announced by the Chief of the Indonesian Armed Forces.

(2) Temporary Suspension of Innocent Passage for foreign vessels in certain areas in Territorial Seas and Archipelago Waters as referred to in paragraph (1), shall be notified by the Ministry of Foreign Affairs to the foreign countries through the diplomatic channels and announced through the Indonesia's Seamen Message following its stipulation by the Chief of the Indonesian Armed Forces regarding the period and the areas affected by such temporary suspension.

(3) Innocent Passage for foreign vessels in certain areas in Territorial Seas and Archipelago Waters as referred to paragraph (1) shall take into effect at the latest 7 (seven) days following its notice and announcement as referred to paragraph (2).

CHAPTER III

CLOSING PROVISIONS

Article 15

Upon the enactment of this Government Regulation, Government Regulation Number 8 of 1962 concerning Innocent Passage for Foreign Vessels in Indonesian Waters (State Gazette of the Republic of Indonesia Number 36, Supplementary State Gazette Number 2466) is declared null and void.

Article 16

This Government Regulation shall take into effect from its enactment date.

For the public cognizance, this Government Regulation shall be promulgated in the State Gazette of the Republic of Indonesia.
Stipulated in Jakarta
on June 28, 2002
PRESIDENT OF THE REPUBLIC OF INDONESIA
signed
MEGAWATI SOEKARNOPUTRI

Enacted in Jakarta
on June 28, 2002
signed
BAMBANG KESOWO

NUMBER 37 OF 2002

RELATING TO RIGHTS AND OBLIGATIONS OF FOREIGN SHIPS AND AIRCRAFT WHEN EXERCISING RIGHT OF ARCHIPELAGIC SEA LANE PASSAGE VIA THE ESTABLISHED ARCHIPELAGIC SEA LANES

THE PRESIDENT OF THE REPUBLIC OF INDONESIA,

Considering:
a). that, the provisions in Law Number 6 of 1996 concerning Indonesian Waters, which constituted an addition to the United Nations Convention concerning Law of the Sea 1982, which, amongst others, stipulated that the rights and obligations of foreign ships and aircraft in the conduct of Archipelagic Sea Lane Passage will be further determined in Governmental Regulations;

b). that, Law Number 6 of 1996 concerning Indonesian Waters, also stipulated that the Government, by publishing the axes on nautical charts, will determine the most suitable sea lanes, including the flight paths above the sea lanes, for the conduct of archipelagic sea lane passage;

c). that, during 69th session of the International Maritime Organization in 1998, the Maritime Safety Committee, with resolution MSC.72 (69), accepted the Indonesian submission concerning Indonesian Archipelagic Sea Lanes;

d). that, based on the developments in a, b and c, there is a requirement for the establishment of Government Regulations concerning the Rights and Obligations of Foreign Ships and Aircraft when exercising Right of Archipelagic Sea Lane Passage via the established Sea Lanes;

Referring to:
(1) Article 5 paragraph (2) of the 1945 Constitution as amended in the Third Amendment of the 1945 Constitution;

(2) Law Number 6 of 1996, concerning Indonesian Waters (State Gazette 1996 Number 73, Supplementary State Gazette Number 3647);

DECIDES

To Stipulate:

GOVERNMENT REGULATION CONCERNING THE RIGHTS AND OBLIGATIONS OF FOREIGN SHIPS AND AIRCRAFT WHEN EXERCISING RIGHT OF ARCHIPELAGIC SEA LANE PASSAGE VIA THE ESTABLISHED ARCHIPELAGIC SEA LANES.

CHAPTER 1

GENERAL PROVISIONS

Article 1

In this regulation:
(1) An Archipelagic Sea Lane is a sea lane as defined in Article 1 section 8 of the relevant law in which it is described as a lane for the conduct of Right of Passage of Archipelagic Sea Lanes.

(2) The law referred to is number 6 of 1996 relating to Indonesian Waters.

(3) Right of Archipelagic Sea Lane Passage is the right of foreign ships and aircraft to transit, as defined in paragraph 18 paragraph (1) and paragraph (2) of the law.

(4) Right of Peaceful Passage is the right of a foreign ship to transit as defined in Article 11 of the law.

(5) Territorial Sea is the Territorial Sea as defined in Article 3 paragraph (2) of the law.

(6) Archipelagic Waters are waters as described in Article 3 paragraph (3) of the law.

(7) Convention is the convention as defined in Article 1 number 9 of the law.

CHAPTER II

RIGHTS AND OBLIGATIONS OF FOREIGN SHIPS AND AIRCRAFT EXERCISING RIGHT OF ARCHIPELAGIC SEA LANE PASSAGE

Article 2

Foreign Ships and aircraft can exercise right of Archipelagic Sea Lane Passage in order to sail or fly from one sea or exclusive economic zone to another sea or exclusive economic zone via Indonesian territorial seas and archipelagic waters.

Article 3

(1) The Right of Archipelagic Sea Lane passage as described in paragraph 2 can be exercised via a sea lane or via the air above a sea lane which has been described as an archipelagic sea lane for the purpose of the Right of Archipelagic Sea Lane Passage as described in paragraph 11.

(2) In accordance with this regulation, the Right of Archipelagic Sea Lane Passage can be exercised in other Indonesian waters after those waters have been established as archipelagic sea lanes which can be used for the purpose of Right of Archipelagic Sea Lane Passage.

Article 4

(1) Foreign ships and aircraft exercising Right of Archipelagic Sea Lane Passage must transit via or above the archipelagic sea lane as quickly as possible using the normal method, only for the purposes of a direct, continuous and uninterrupted transit.

(2) Whilst foreign ships and aircraft exercise archipelagic sea lane passage, they must not stray more than 25 nautical miles to either side of the axis of the sea lane, with the provision that the vessel or aircraft must not approach the coast to a distance less than 10 % the distance between the coastline and the point at which the sea lane borders the coast line.

(3) When exercising right of Archipelagic Sea Lane Passage, foreign ships and aircraft must not pose a threat or use violence towards the sovereignty, regional totality or political independence of the Republic of Indonesia or in any other way contravene basic International Law as laid down in the United Nations Charter.

(4) When exercising right of Archipelagic Sea Lane Passage, foreign military aircraft and warships must not conduct military exercises or exercise any type of weapons with ammunition.

(5) With the exception of force majeure or in the event of a disaster, aircraft exercising Right Archipelagic Sea Lane Passage must not land in Indonesia.

(6) With the exception of force majeure, in the event of a disaster or in order to render assistance to people or other vessels experiencing a disaster, all foreign ships exercising Right of Archipelagic Sea Lane Passage must not stop, anchor or loiter.

(7) Foreign ships and aircraft exercising Right of Archipelagic Sea Lane Passage must not conduct illegal transmissions or interfere with telecommunications systems and must not communicate directly with people or unlawful groups in Indonesia.

Article 5

Whilst exercising Right of Archipelagic Sea Lane Passage, foreign ships or aircraft, including research or hydrographic vessels, must not conduct oceanographic research or hydrographic survey either with the use of detection equipment or sample gathering equipment, except when approved to do so.

Article 6

(1) Whilst exercising Right of Archipelagic Sea Lane Passage, foreign vessels, including fishing vessels, must not conduct fishing operations.

(2) Whilst exercising Right of Archipelagic Sea Lane Passage, foreign fishing vessels, as well as fulfilling the obligations in paragraph (1), must store all fishing equipment within the hold.

(3) Whilst exercising Right of Archipelagic Sea Lane Passage, foreign ships and aircraft must not embark or disembark people, goods or currency not in accordance with customs, immigration, fiscal and health laws except in the accordance with force majeure or in the event of a disaster.

Article 7

(1) Whilst exercising Right of Archipelagic Sea Lane Passage, foreign vessels are obligated to observe the generally accepted regulations, procedures and international practices concerning international shipping, including the regulations relating to collision avoidance at sea.

(2) Whilst exercising Right of Archipelagic Sea Lane Passage in a sea lane which has been established as a sea lane for the "Rule of the Road/Right of Way", foreign ships are obliged to observe the "Rule of the Road/Right of Way".

(3) Whilst exercising Right of Archipelagic Sea Lane Passage, foreign ships must not damage or disrupt navigation facilities, as well as submarine cables and pipes.

(4) Whilst exercising Right of Archipelagic Sea Lane Passage in a sea lane with natural resource exploitation or exploration facilities, foreign ships must not sail within 500 meters of the prohibited zone as detailed around the facility.

Article 8

(1) Foreign civil aircraft exercising Right of Archipelagic Sea Lane Passage must:
   a). observe the aviation regulations as established by the International Civil Aviation Organization concerning flight safety;
   b). continuously monitor the radio frequencies as directed by the responsible air traffic authority or the appropriate international emergency radio frequency.

(2) Foreign national aircraft exercising Right of Archipelagic Sea Lane Passage must:
   a). respect the regulations concerning flight safety as detailed in paragraph (1) a;
   b). observe the obligations as detailed in paragraph (1) b.

Article 9

(1) Foreign ships exercising Right of Archipelagic Sea Lane Passage must not expel oil, oily waste and other dangerous good into the marine environment, or conduct other activities in contention with standard international regulations to prevent, reduce and control marine pollution originating from the ship.

(2) Foreign ships exercising Right of Archipelagic Sea Lane Passage are forbidden to dump waste in Indonesian waters.

(3) Foreign ships exercising Right of Archipelagic Sea Lane Passage which are nuclear powered, transporting nuclear goods, other dangerous goods or poison must carry documentation and advise of special preventative measures as stipulated by international agreements for such ships.

Article 10

(1) The person or legal bodies responsible for the operation of foreign cargo ships and aircraft, or foreign government owned ships and aircraft used for commercial trade, must take responsibility for any loss or damage suffered by Indonesia as a result of any ship or aircraft not adhering to the provisions in paragraphs 7, 8 and 9 whilst exercising Right of Archipelagic Sea Lane Passage.

(2) Nationally flagged vessels or nationally listed aircraft bear the international responsibility for loss and

damage suffered by Indonesia as a result of any war ship or aircraft not adhering to the provisions in paragraphs 7, 8 and 9 whilst exercising Right of Archipelagic Sea Lane Passage.

CHAPTER III

THE DETERMINATION OF ARCHIPELAGIC SEA LANES WHICH CAN BE USED FOR THE RIGHT OF ARCHIPELAGIC SEA LANE PASSAGE

Article 11

(1) Archipelagic Sea Lane I is to be used to exercise Right of Archipelagic Sea Lane Passage between the South China Sea and the Indian Ocean, or reverse, and transits the Natuna Sea, Karimata Strait, Java Sea and Sunda Strait. This constitutes the axis which connects points I-1 through to I-15 as detailed in the Coordinate Table and explained in Article 12, paragraph (2).

(2) Archipelagic Sea Lane I, as outlined in paragraph (1), contains Archipelagic Sea Lane Branch IA which joins Archipelagic Sea Lane I at point I-3 for sailing from the Singapore Strait via the Natuna Sea, or reverse. This constitutes the axis which connects points IA-1 and 1-3 as detailed in the Coordinate Table and explained in Article 12, paragraph (2).

(3) Archipelagic Sea Lane II is to be used to exercise Right of Archipelagic Sea Lane Passage from the Sulawesi Sea to the Indian Ocean, or reverse, via the Makassar Strait, Flores Sea and Lombok Strait. This constitutes the axis which connects points II-1 through to II-8 as detailed in the Coordinate Table and explained in Article 12, paragraph (2).

(4) Archipelagic Sea Lane IIIA I to be used to exercise Right of Archipelagic Sea Lane Passage from the Pacific Ocean to the Indian Ocean, or reverse, via the Maluku Sea, Seram Sea, Banda Sea, Ombai Strait and Sawu Sea. This constitutes the axis which connects points IIIA-1 through to IIIA-13 as detailed in the Coordinate Table and explained in Article 12, paragraph (2).

(5) Archipelagic Sea Lane III-A as detailed in paragraph (4) includes:

a). Archipelagic Sea Lane Branch IIIB joins Archipelagic Sea Lane IIIA at point IIIA-8 for sailing from the Pacific Ocean to the Indian Ocean, and reverse, via the Maluku Sea, Seram Sea, Banda Sea and Leti Strait. This constitutes the axis which connects points IIIA-8, IIIB-1 and IIIB-2 as detailed in the Coordinate Table and explained in Article 12, paragraph (2).

b). Archipelagic Sea Lane Branch IIIC joins Archipelagic Sea Lane Branch IIIB at point IIIB-1 for sailing from the Pacific Ocean to the Arafura Sea, or reverse, via the Maluku Sea, Seram Sea and Banda Sea. This constitutes the axis which connects points IIIB-1, IIIC-1 and IIIC-2 as detailed in the Coordinate Table and explained in Article 12, paragraph (2).

c). Archipelagic Sea Lane Branch IIID joins Archipelagic Sea Lane IIIA at point IIIA-11 for sailing from the Pacific Ocean to the Indian Ocean, or reverse, via the Maluku Sea, Seram Sea, Banda Sea, Ombai Strait and Sawu Sea.

This constitutes the axis which connects points IIIA-11 and IIID-1 as detailed in the Coordinates Table and explained in Article 12, paragraph (2).

d). Archipelagic Sea Lane Branch IIIE joins Archipelagic Sea Lane IIIA at point IIIA-2 for sailing from the Indian Ocean to the Sulawesi Sea, and reverse, via the Sawu Sea, Ombai Strait, Banda Sea, Seram Sea and Maluku Sea or, for sailing from the Timor Sea to the Sulawesi Sea, or reverse, via the Leti Strait, Banda Sea, Seram Sea and Maluku Sea or, for sailing from the Arafura Sea to the Sulawesi Sea, or reverse, via the Banda Strait, Seram Sea and Maluku Sea. This constitutes the axis which connects points IIIA-2, IIIE-1 and IIIE-2 as detailed in the Coordinate Table and explained in Article 12, paragraph (2).

Article 12

(1) The archipelagic sea lane axis and connecting points mentioned in Article 11 are included on navigation charts which are published as required.

(2) Geographic coordinates for the archipelagic sea lane connecting points mentioned in Article 11 are summarised in the Coordinate Tables at the beginning of this appendix.

(3) The positions for connecting points I-1, I-15, IA-1, II-1, II-8, IIIA-1, IIIA-13, IIIB-2, IIIC-2, IIID-1 and IIIE-2, as connecting points outside the archipelagic sea lane axis are included in the Geographic Coordinate Table and explained in Article 12, paragraph (2) are located where the archipelagic sea lane axis intersects territorial sea border.

(4) If, as a result of a natural change, the connecting points mentioned above are not in the position as detailed in the Geographic Coordinate Table mentioned in Article 12, paragraph (2) then the geographic position must be decided by the situation in the field.

(5) An illustrated Chart displaying the axis and the connecting points as detailed in Article 11 is at the beginning of this appendix.

CHAPTER IV

OTHER PROVISIONS

Article 13

Provisions in this Government Regulation do not lessen right of foreign vessels to exercise right of peaceful passage in archipelagic sea lanes.

Article 14

The provisions in this Government Regulation concerning Indonesian Archipelagic Sea Lanes and Indonesian Archipelagic Sea Lane Passage do not take effect in the Leti Strait and part of the Ombai Strait which border East Timor. With the changed status of East Timor, the waters in that area are no longer part of Indonesian Archipelagic Waters.

Article 15

Within 6 months of this Government Regulation taking effect, foreign ships and aircraft may only exercise Right of Archipelagic Sea Lane Passage via the archipelagic waters detailed in this Government regulation.

CHAPTER V

Article 16

This Government Regulation comes into force with effect the date of law. Let it be known that this law has been written into the Republic of Indonesia State Gazette.

Stipulated in Jakarta
June 28, 2002
PRESIDENT OF THE REPUBLIC OF INDONESIA
Signed
MEGAWATI SOEKARNOPUTRI

Enacted in Jakarta
June 28, 2002
Republic of Indonesia State Secretary
Signed
BAMBANG KESOWO

NUMBER 38 OF 2002

RELATING TO LIST OF COORDINATE OF GEOGRAPHICAL POINTS OF INDONESIAN ARCHIPELAGO BASELINE

THE PRESIDENT OF THE REPUBLIC OF INDONESIA

Considering:

a). that Law Number 6 of 1996 re the Indonesian Waters determined as the follow up of the ratification of the United Nations Convention re Marine Law of 1982 determines further the rights and obligations of foreign vessels in carrying out Peaceful Passage and be regulated further in a Government Regulation;

b). that apart from the adequate scale as required for determining the territorial borders of Indonesian Waters, List of Coordinate of Geographical Points of Indonesian Archipelago Baseline describing the territorial borders of the Indonesian waters may be immediately determined to meet the need.

c). that based upon the consideration as referred to in letter a and letter b, it is necessary to stipulate Government Regulation on LIST COORDINATE OF GEOGRAPHICAL POINTS OF INDONESIAN ARCHIPELAGO BASE LINE;

In view of:

(1). Article 5 paragraph (2) of the 1945 Constitution as amended with the Third Amendment of the 1945 Constitution;

(2). Law Number 6 of 1996 re the Indonesian Waters (Official Gazette of the Republic of Indonesia of 1996 Number 73, Supplementary Official Gazette Number 3647;

DECIDED

To Stipulate:

GOVERNMENT REGULATION RE THE LIST COORDINATE OF GEOGRAPHICAL POINTS OF INDONESIAN ARCHIPELAGO BASE LINE;

CHAPTER I

GENERAL PROVISIONS

Article 1

In this Government Regulation the following terms shall refer to:

(1). Geographical Coordinate shall refer to the coordinate of which its scale is expressed in degree, minute, second and angle second in the latitude axis and geographical longitude.

(2). Low Water Line shall refer to hydrographical datum of navigation map determined at the average position of the nebtide Low Water Line.

(3). Hydrographical Datum shall refer to the chart subside surface constituting a reference to the sea surface used to reduce the sea depth figures on the navigation Charts.

(4). Navigation Charts shall refer to the sea Charts compiled for navigation purposes in sea in compliance to the international standard in respect to sailing safety.

(5). Geodetic Datum shall refer to mathematical reference to determine the geographical coordinate or for hydrographical Charts.

(6). Coastal general directions shall refer to the average directions as pointed by the coastal line directions having the same general equation in a given location.

(7). Archipelago general configuration shall refer to the layout of islands or the most outer island groups or most outer dry reefs and most outer low ebb elevation of which each of them form a certain configuration.

(8). Latitude and longitude shall refer to reference system of geographical coordinate axis of the earth surface.

(9). Nautical mile shall refer to the geographical mile comprising 1/60 (one sixtieth) latitude degree.

Article 2

(1) The Government has drawn the Archipelago Base Line to determine territorial sea width.

(2) The drawing of the Archipelago Base Line as referred to in paragraph (1) has been carried out based on:

a). Archipelago Straight Base Line;
b). Ordinary Base Line;
c). Straight Baseline
d). Closing Line of Gulfs;
e). Closing Lines of River Mouths, Canals and Estuaries; and
f). Closing Line of Seaports.

CHAPTER II

THE DRAWING OF ARCHIPELAGO BASE LINE

First Part

Archipelago Straight Base Line

Article 3

(1) Among the most outer islands, and the most outer dry reefs of the Indonesian archipelago, the base line used to measure the territorial sea width is the Archipelago Straight Base Line.

(2) The Archipelago Straight Base Line as referred to paragraph (1) is the straight line connecting the most outer points of the Low Water Line on the most outer line of the most outer islands, and the most outer dry reefs on the same most outer point on the Low Water Line on the most outer point of the most outer island, other most outer dry reefs lying side by side.

(3) The length of Archipelago Straight Base Line as referred to paragraph (2) may not exceed 100 (one hundred) nautical miles, except that 3% (three percent) of the total length of Archipelago Straight Base Line may exceed such length, up to 125 (one hundred and twenty five) nautical miles maximum.

(4) The drawing of Archipelago Straight Base Line as referred to paragraph (2) and paragraph (3) is carried out by not deviating too far from the archipelago general configuration.

(5) The drawing of Archipelago Straight Base Line as referred to paragraph (2) may be carried out by utilizing the most outer points of the Low Water Line at each low ebb elevation of which on top of it lies a lighthouse or similar installation that permanently lies on the water surface or low ebb elevation that partly or wholly lies within the distance not exceeding the territorial sea width of the Low Water Line of the nearest island.

(6) The waters lie on the side of the Archipelago Straight Base Line as referred to paragraph (1) are the Archipelago Waters, and the waters located at the outer side of Archipelago Straight Base Line are Territorial Seas.

Second Part

Ordinary Base Line

Article 4

(1) In case the geographical shape of the most outer island's coast is normal, with the exception as stated in Article 5, Article 6, Article 7, and Article 8, the Base Line used to measure the Territorial Sea width is the Ordinary Base Line.

(2) The Ordinary Base Line as referred to paragraph (1) is the Low Water Line along the coast as determined pursuant to the prevailing Hydrographical Datum.

(3) On the most outer island situated in an atoll or the most outer island with reefs around it, the Base Line used to measure the Territorial Sea width is the Ordinary Base Line comprising Low Water Line on the sides of the atoll or the most outer reefs towards the sea.

(4) Low Water Line as referred to paragraph (2) and paragraph (3) appear on the large-scale Navigation Charts as officially issued by the Government navigation chart making body.

(5) The waters situated in the sides of the Ordinary Base Line as referred to paragraph (1) and paragraph (2) are the Interior Waters and the waters situated on the outer side of such Ordinary Base Line is the Territorial Sea.

Third Part

Straight Base Line

Article 5

(1) On the coast having a sharp-pointed hollow, the base line used to measure the Territorial Sea is the Straight Base Line.

(2) Straight Base Line as referred to in paragraph (1) is the straight line drawn between the between the most outer points on the protruding Low Water Line adjacent the hollow mouth of such coast.

(3) On the coast due to its delta or other natural condition, its coastal line is unstable, the base line used to measure the territorial sea width is the Straight Base Line.

(4) Straight Base Line as referred to in paragraph (3) is the straight line drawn between the between the most outer Points on the most protruding Low Water Line towards the se on a delta or such other natural condition.

(5) The waters situated on the sides of the Straight Base Line as referred to paragraph (1) and paragraph (3) are interior waters and waters situated on the outer side of Straight Base Line is the Territorial Sea.

Fourth Part

Gulf Closing Line

Article 6

(1) On the coastal hollow comprising a gulf, the base line used to measure the territorial sea width is the Gulf Closing Line.

(2) The Gulf Closing Line as referred to paragraph (1) is the straight line drawn between the most outer points of the most protruding Low Water Line adjacent the mouth of such gulf.

(3) The Gulf Closing Line as referred to paragraph (1) may only be drawn if the size of the gulf is or larger than ½ (half) of a circle of which its diameter is the closing line drawn on the mouth of such gulf.

(4) In case in the gulf there are islands forming more than one gulf mouth, the length of the Gulf Closing Line of several gulf mouths is 24 (twenty four) nautical miles maximum.

(5) The waters situated on the side of the Gulf Closing Line as referred to paragraph (1) is the Interior Waters and the waters situated on the outer side of such Gulf Closing Line is the Territorial Sea.

Fifth Part

Closing Line of River Mouth, Canal and Estuary

Article 7

(1) On a River Mouth or Canal, the base line used to measure the Territorial Sea width is the Base Line as the closing line of such river mouth or canal.

(2) The straight line as referred to paragraph (1) is drawn between the most outer points on the protruding and opposite Low Water Line.

(3) In case Straight Line as referred to paragraph (1) is not applicable due to the Estuary in the river mouth, the estuary closing line uses straight lines connecting the estuary points and the most outer points on the Low Water Line of the mouth river side.

(4) The waters situated on the interior side of the closing line as referred to paragraph (1) and paragraph (3) are the Interior Waters and the waters situated on the outer side of such closing line is the Territorial Sea.

Sixth Part

Seaport Closing Line

Article 8

(1) In the seaport area, the base line used to measure the Territorial Sea width are the straight lines as the seaport closing area, comprising the most outer permanent building constituting inseparable part of the seaport system as part of the coast.

(2) The straight line as referred to paragraph (1) is drawn between the most outer points on the coastal Low Water Line and the most outer points of the most outer permanent building constituting inseparable part of the seaport system.

(3) The waters situated at the side of the closing lines of the seaport area as referred to paragraph (1) are the Interior Waters and waters situated at the side of the such closing line is the Territorial Sea.

CHAPTER III

LIST OF COORDINATE OF GEOGRAPHICAL POINTS OF INDONESIAN ARCHIPELAGO BASE LINE

Article 9

(1) The positions of the most outer points of archipelago base lines used to determine the Territorial Sea width as referred to Article 3, Article 4, Article 5, Article 6, Article 7, and Article 8 are determined in the Geographical Coordinate along with the Geodetic Datum reference used.

(2) The Geographical Coordinate from the most outer points of the archipelago base line to determine the Territorial Sea width as referred to Article 3, Article 4, Article 5, Article 6, Article 7, and Article 8 is the one contained in the List of Geographical Coordinate forming an enclosure of

this Government Regulation, but not included in this book.

(3) The Geographical Coordinate from the most outer points as referred to paragraph (2) contains the positions of geographical points mentioned in the Latitude and Longitude and accompanied with the description of the waters where such points are located, data on the field, type of base line between the most outer points, reference maps with their scale information and Geodetic Datum used.

(4) The enclosure as referred to as referred to paragraph (2) constitutes inseparable part of this Government Regulation.

Article 10

In the event that on the Indonesian Waters section, the data of Geographical Coordinate of the Most Outer Points are not included in the enclosure as referred to Article 9 paragraph (2) or in case due to some natural changes the Geographical Coordinate of such Outer Points are not considered being in the position as contained in such enclosure, the Geographical Coordinate of the Most Outer Points used are the actual Geographical Coordinate of the Most Outer Points as found in the field.

CHAPTER IV

SUPERVISION AND ENHANCEMENT

Article 11

(1) The Government shall update on a regular basis to improve and complete the provision on Geographical Coordinate of the Most Outer Points to draw the Archipelago Base Lines as referred to Article 3, Article 4, Article 5, Article 6, Article 7, and Article 8.

(2) In case in the future there are evidently the most outer islands, atoll, dry reefs, most outer low ebb elevation, then the gulf, river mouth, canal or estuary and seaport that may be used to determine the most outer points of the Archipelago Base Lines are not included in the enclosure as referred to Article 9 paragraph (2), changes shall be made to such enclosure according to the most recent data available.

(3) In case in the future, the Geographical Coordinate of the Most Outer Points, most outer islands, atoll, most outer dry reef, most outer low ebb elevation, the gulf, river mouth, canal or estuary and seaport change, the enclosure shall undergo some revisions as referred to Article 9 paragraph (2).

CHAPTER V

DETERMINING THE INTERIOR WATERS BORDERS IN ARCHIPELAGO WATERS

Article 12

(1) The determination of interior waters borders in archipelago waters shall be carried out using the Ordinary Base Line, Straight Base Line, and Closing Line in River Mouth, Canal, or Estuary, in the Gulfs and Seaports situated in coast of the islands facing the archipelago waters.

(2) The ruling on the determining the borders of Interior Waters as referred to in paragraph (1) shall be further regulated in a separate Government Regulation.

CHAPTER VI

CLOSING PROVISIONS

Article 13

Upon the enactment of this Government Regulation, Government Regulation Number 61 of 1996 re the List of Geographical Coordinate of Indonesian Archipelago Base Line Points in Natuna Sea (Official Gazette of the Republic of Indonesia Number 3768) shall be declared null and void.

Article 14

This Government Regulation shall take into effect on its enactment date.

For the public cognizance, this Government Regulation shall be promulgated in the Official Gazette of the Republic of Indonesia.

Stipulated in Jakarta
on June 28, 2002
PRESIDENT OF THE REPUBLIC OF INDONESIA
signed
MEGAWATI SOEKARNOPUTRI

Enacted in Jakarta
on June 28, 2002
signed
BAMBANG KESOWO

GAZETTEER

The approximate geographical positions of places mentioned in the text of this volume are listed in the Gazetteer.
Obsolete names are given in brackets after the new names.
Asterisks (*) follow locations which are Arrival and Departure Positions (1.4).

Abaiang Atoll	1°58′ N, 172°50′E
Abang Kecil	0°33′N, 104°14′E
Abang, Selat	0°32′N, 104°15′E
'Abd al Kūrī	12°10′N, 52°15′E
Abrolhos, Canal dos	17°50′S, 38°45′W
Abrolhos, Archipelago dos	18°00′S, 38°40′W
Acapulco	16°50′N, 99°55′W
Acasta, Batu	1°39′N, 106°18′E
Acebuche, Punta del	36°02′N, 5°27′W
Açores, Arquipélago dos	38°00′N, 27°00′W
Ad Dakhla	23°42′N, 15°56′W
Adelaide*	34°48′S, 138°23′E
Aden*	12°45′N, 44°57′E
Adolphus Channel	10°40′S, 142°35′E
Adventure Bay	43°18′S, 147°22′E
Agalega Islands	10°26′S, 56°39′E
Agulhas Bank	35°30′S, 21°00′E
Agulhas, Cape	34°50′S, 20°00′E
(15 miles S of)*	35°05′S, 20°00′E
Agung, Gunung	8°20′S, 115°30′E
Ahunui	19°40′S, 140°25′W
Aitutaki	18°54′S, 159°47′W
Akbar, Beting	2°39′S, 107°15′E
Akuseki Shima	29°27′N, 129°37′E
Akutan Pass	54°05′N, 166°00′W
Al=The (definite article)	see proper name
Alalakeiki Channel	20°35′N, 156°30′W
Alas, Selat	8°40′S, 116°40′E
Albardão, Banco do	33°10′S, 52°25′W
Albatross Islet	40°23′S, 144°39′E
Alborán, Isla de	35°56′N, 3°02′W
Aldabra Group	9°25′S, 46°20′E
Alenuihaha Channel	20°20′N, 156°00′W
Algeciras	36°08′N, 5°27′W
Algiers*	36°46′N, 3°05′E
Algoa Bay	34°00′S, 26°00′E
Alguada Reef	15°42′N, 94°10′E
Alice Shoal	16°05′N, 79°20′W
Alijos, Rocas	24°57′N, 115°45′W
Aling, Gosong	3°35′S, 110°15′E
Almina, Peninsula de la	35°54′N, 5°17′W
Alor, Pulau–Pulau: Kep. Lingga	0°28′N, 104°18′E
Alor, Pulau–Pulau: Selat Alor	8°15′S, 124°30′E
Alor, Selat	8°15′S, 123°55′E

Alphard Banks	35°03′S, 20°54′E
Alta Vela	17°28′N, 71°39′W
Alur Pelayaran=Passage	see proper name
Amami Guntō	28°00′N, 129°05′E
Amaya, Khawr al	29°25′N, 49°06′E
Amazonas, Rio	1°16′N, 49°30′W
Ambon*	3°41′S, 128°10′E
Amchitka Pass	51°20′N, 180°00′
Amirante Isles	5°30′N, 53°20′E
Amsterdam, Île	37°50′S, 77°35′E
Anadyrskiy Zaliv	64°00′N, 178°00′W
Anambas, Pulau-Pulau	3°00′N, 106°00′E
Andaman Islands	12°30′N, 93°00′E
Aniva, Mys	46°02′N, 143°25′E
Anjouan, Île	12°15′S, 44°30′E
Anna, Pulo	4°40′N, 131°58′E
Anser Group	39°18′S, 146°30′E
Antigua	17°05′N, 61°45′W
Antiope Reef	18°14′S, 168°20′W
Antipodes Islands	49°35′S, 178°50′E
Antofagasta	23°38′S, 70°26′W
Anyer Lor	6°03′S, 105°55′E
Api, Alur Pelayaran	2°00′N, 109°08′E
Api, Tanjung	1°57′N, 109°20′E
Apia: Samoa*	13°47′S, 171°45′W
Apo East Pass	12°40′N, 120°45′E
Apolima Strait	13°48′S, 172°10′W
Ara, Gosong	0°48′N, 104°57′E
Archipel, Arcipielago, Archipiélago, Arquipelago, or Arquipiélago= Archipelago	see proper name
Arica, Rada de	18°30′S, 70°20′W
Arorae Island	2°39′S, 176°54′E
Arrecife=Reef	see proper name
Aru, Pulau–Pulau	6°00′S, 134°30′E
Aruba Island	12°26′N, 69°56′W
Aruba, Gosong	3°28′S, 110°12′E
Arus, Tanjung	1°53′N, 125°05′E
Ascension Island	7°55′S, 14°25′W
Assumption Island	9°45′S, 46°30′E
Astoria	46°12′N, 123°51′W
Astove Island	10°04′S, 47°43′E
Ata, Batu	6°12′S, 122°42′E
Atalia, Ponta da	0°35′S, 47°21′W
Atkin, Karang	0°33′S, 104°02′E
Attu Island	52°55′N, 173°00′E

Auckland Islands	50°40'S, 166°00'E
Auckland*	36°36'S, 174°49'E
Aur, Pulau: Selat Gelasa	3°00'S, 107°13'E
Aur, Pulau: Selat Karimata	1°14'S, 109°18'E
Aur, Pulau: South China Sea	2°27'N, 104°31'E
Australes, Îles	23°00'S, 151°00'W
Ava=Passage	see proper name
Avalon Peninsula	47°20'N, 53°00'W
Awal, Gosong	3°24'S, 107°36'E
Ayam, Tanjung	1°21'N, 104°12'E
Ayerlancur, Tanjung	2°53'S, 107°20'E
Bab el Mandeb, Straits of	12°40'N, 43°30'E
Babar, Pulau	7°55'S, 129°44'E
Babuyan Channel	18°40'N, 122°00'E
Bacan	0°30'S, 127°30'E
Backstairs Passage	35°45'S, 138°08'E
Baco Islands	13°28'N, 121°10'E
Badung, Bukit	8°50'S, 115°10'E
Badung, Selat	8°40'S, 115°20'E
Baffin Bay	73°00'N, 70°00'W
Baginda	3°04'S, 106°43'E
Baginda, Tanjung	3°05'S, 106°44'E
Baginda, Karang-Karang	3°07'S, 107°05'E
Bahamas, The	25°00'N, 75°00'W
Bahia, Bahía=Bay	see proper name
Bahía Blanca*	39°10'S, 61°45'W
Baia=Bay	see proper name
Baiquan Liedao	25°58'N, 119°57'E
Baja California	28°00'N, 113°30'W
Bajo=Shoal	see proper name
Bakau, Pulau	3°02'S, 107°09'E
Bakau, Tanjung	0°21'S, 103°47'E
Bakung	2°57'S, 106°53'E
Balabac Island	8°00'N, 117°00'E
Balabac Main Channel	7°26'N, 117°14'E
Balabac Strait	7°34'N, 116°55'E
Balabalagan, Pulau-Pulau	2°15'S, 117°30'E
Balambangan, Pulau	7°18'N, 116°55'E
Balboa*	8°53'N, 79°30'W
Baleares, Islas	39°45'N, 3°00'E
Bali, Selat	8°10'S, 114°25'E
Balikpapan*	1°21'S, 116°56'E
Balintang Channel*	20°00'N, 122°20'E
Ballard Bank	46°40'N, 52°50'W
Ball's Pyramid	31°45'S, 159°15'E
Balmoral Reef	15°40'S, 175°50'E
Bampton Reefs	19°00'S, 158°40'E
Banaba (Kiribati)	0°50'S, 169°35'E
Banco=Bank	see proper name
Banda, Pulau-Pulau	4°35'S, 129°45'E
Banggi, Pulau	7°15'N, 117°10'E
Banggi Selatan, Selat	7°05'N, 117°06'E
Bangka, Pulau	2°00'S, 106°00'E
Bangka Passage: Celebes Sea	2°00'N, 125°15'E
Bangka, Selat: Celebes Sea	1°45'N, 125°05'E
Bangka, Selat: Java Sea	2°06'S, 105°03'E
Bangkok Bar; see Krung Thep*	13°23'N, 100°35'E
Banjul (Bathurst) Bar*	13°32'N, 16°54'W
Banks Islands	14°00'S, 167°30'E
Banks Strait	40°39'S, 148°05'E
Banks, Cape	37°54'S, 140°23'E
Banteng, Karang (Buffalo Rock)	1°09'N, 103°49'E
Baram, Tanjung	4°36'N, 113°58'E
Barbados: Bridgetown*	13°06'N, 59°39'W
Barbuda	17°40'N, 61°45'W
Barcelona*	41°20'N, 2°10'E
Barquero, Ria del	43°45'N, 7°38'W
Basa, Karang	5°12'S, 106°12'E
Bashi Channel	21°20'N, 121°00'E
Basilan Strait*	6°50'N, 122°00'E
Bass Strait*	40°00'S, 146°00'E
Bassas da India	21°28'S, 39°46'E
Bassas de Pedro	13°00'N, 72°25'E
Batam, Pulau	1°07'N, 104°07'E
Batanme, Pulau	1°55'S, 130°05'E
Batanta, Pulau	0°53'S, 130°40'E
Bathurst Island	11°40'S, 130°20'E
Bathurst; see Banjul	13°32'N, 16°54'W
Batu=Rock	see proper name
Batubelayar	0°25'N, 104°16'E
Batutoro, Tanjung	5°42'S, 122°47'E
Baur, Selat	3°00'S, 107°18'E
Bāvanapādu	18°34'N, 84°21'E
Bawean, Pulau	5°50'S, 112°40'E
Beijishan Liedao	27°38'N, 121°12'E
Beira	19°52'S, 34°58'E
Beirut*	33°55'N, 35°31'E
Belang, Pulau	8°33'S, 116°47'E
Belem	1°27'S, 48°30'E
Belitung, Pulau	3°00'S, 108°00'E
Belize*	17°20'N, 88°01'W
Bell Reef	40°23'S, 144°05'E
Belle Isle, Straits of*	51°44'N, 56°00'W
Bellona Reefs	21°30'N, 158°50'E
Belukar, Pulau	0°51'N, 103°39'
Beluru, Bukit	3°10'S, 107°40'E
Belvedere, Karang	2°12'S, 107°02'E
Ben Sekka (Enghela), Ras	37°21'N, 9°45'E

Benan, Pulau	0°28′N, 104°27′E
Benggala, Selat	5°50′N, 95°07′E
Benguela, Porto de	12°35′S, 13°24′E
Benin, Bight of	5°00′N, 3°00′E
Benin, River	5°45′N, 5°00′E
Bergen*	60°24′N, 5°18′E
Berhala, Pulau	0°52′S, 104°24′E
Berhala, Selat	0°50′S, 104°15′E
Berikat, Pulau	2°28′S, 106°58′E
Berikat, Tanjung	2°34′S, 106°51′E
Berlenga, Ilha	39°25′N, 9°30′W
Bermuda*	32°23′N, 64°38′W
Besar, Pulau	2°53′S, 106°08′E
Betata, Terumbu	1°11′N, 104°09′E
Beting=Reef	see proper name
Betsey Island	43°03′S, 147°29′E
Beveridge Reef	20°00′S, 167°47′W
Biafra, Bight of	3°00′N, 7°00′E
Biao Jião	23°14′N, 116°48′E
Biaro, Pulau	2°06′N, 125°23′E
Bijagós, Arquipélago dos	11°15′N, 16°00′W
(75 miles SW of)*	10°40′N, 17°40′W
Bikar Atoll	12°15′N, 170°06′E
Bikini Atoll	11°36′N, 165°23′E
Bilbao	43°21′N, 3°02′W
Binongko, Pulau	5°57′S, 124°01′E
Bintan, Pulau	1°00′N, 104°30′E
Bintan Besar, Gunung	1°05′N, 104°27′E
Bioko	3°30′N, 8°45′E
Biri Island	12°42′N, 124°21′E
Bishop Rock (5 miles S of)*	49°47′N, 6°27′W
Black Pyramid	40°29′S, 144°20′E
Blanco, Cabo	47°12′S, 65°45′W
Bligh Entrance	9°12′S, 144°00′E
Blis, Karang	3°16′S, 107°12′E
Bluff*	46°38′S, 168°21′E
Bo Hai	38°30′N, 119°30′E
Boavista, Ilha da	16°05′N, 22°55′W
Bojador, Cabo	26°07′N, 14°30′W
Bojeador, Cape	18°30′N, 120°34′E
Boleng, Selat	8°25′S, 123°20′E
Bolinao, Cape	16°19′N, 119°47′E
Bomatu Point	8°23′S, 151°07′E
Bombay; see Mumbai*	18°51′N, 72°50′E
Bombay Reef	16°02′N, 112°30′E
Bon, Cap	37°05′N, 11°03′E
Bonny River*	4°13′N, 7°01′E
Bonvouloir Islands	10°21′S, 151°52′E
Boo, Kepulauan	1°10′S, 129°25′E

Booby Island	10°36′S, 141°55′E
Borda, Cape	35°45′S, 136°35′E
Bordeaux	44°50′N, 0°30′W
Boston*	42°20′N, 70°46′W
Bougainville Island	6°00′S, 155°15′E
Bougainville Strait*	6°40′S, 156°15′E
Bounty Islands	47°41′S, 179°03′E
Brava, Ilha	14°50′N, 24°42′W
Breaksea Spit	24°25′S, 153°10′E
Brest	48°22′N, 4°30′W
Brett, Cape	35°10′S, 174°20′E
Brindisi	40°40′N, 17°59′E
Brisbane	27°20′S, 153°12′E
Caloundra Head for Brisbane	26°49′S, 153°10′E
Bristol Bay	57°30′N, 160°30′W
Brooks Banks	24°10′N, 167°00′W
Browse Island	14°06′S, 123°33′E
Buau, Pulau	1°03′N, 104°13′E
Buen Suceso, Bahía	54°50′S, 65°15′W
Buffalo Rock; see	
Banteng, Karang	1°09′N, 103°49′E
Bukit=Hill	see proper name
Bulat, Pulau	1°28′S, 109°24′E
Bungaran, Pulau	4°00′N, 108°10′E
Bungin, Tanjung	4°32′S, 105°54′E
Bungo Suidō	33°00′N, 132°15′E
Buntunga, Batu	4°40′S, 117°10′E
Bûr Sa'îd (Port Said)*	31°21′N, 32°33′E
Burias, Island	12°50′N, 123°15′E
Buru, Pulau: Ceram Sea	3°30′S, 126°30′E
Buru, Pulau: Selat Durian	0°52′N, 103°30′E
Burung, Pulau-Pulau	0°45′N, 108°45′E
Butung, Pulau	5°00′S, 122°55′E
Butung, Alur Pelayaran	5°20′S, 123°15′E
Butung, Selat	4°56′S, 122°47′E
Butunga, Batu	4°40′S, 117°00′E
Cabezos, Bajo de Los	36°01′N, 5°43′W
Cabo Verde, Arquipélago de	17°00′N, 24°00′W
Cabo=Cape	see proper name
Cabot Strait*	47°32′N, 59°30′W
Cagliari	39°12′N, 9°06′E
Caicos Passage*	22°15′N, 72°20′W
Cakang, Tanjung	0°37′N, 104°17′E
Calabar River Entrance	4°24′N, 8°18′E
Calamian Group	12°00′N, 120°00′E
Calantas Rock	12°31′N, 124°05′E
Calavite, Cape (5 miles W of)*	13°27′N, 120°13′E
Calcanhar, Cabo	5°10′S, 35°29′W
(60 miles ENE of)*	4°40′S, 34°35′W

Calcutta; see Kolkata . . . . 22°30′N, 88°15′E
  Approach* . . . . 21°00′N, 88°13′E
Caldera, Punta . . . . 27°03′S, 70°52′W
Calicut . . . . 11°15′N, 75°46′E
Callao* . . . . 12°02′S, 77°14′W
Caloundra Head* . . . . 26°48′S, 153°08′E
Caluula, Raas . . . . 11°59′N, 50°47′E
Campbell Island . . . . 52°33′S, 169°10′E
Campeche, Banco de . . . . 22°30′N, 90°00′W
Çanakkale Boğazı* . . . . 40°09′N, 26°24′E
Canarias, Islas . . . . 29°00′N, 15°00′W
Cani, Îles . . . . 37°21′N, 10°07′E
Cannanore . . . . 11°52′N, 75°22′E
Cap, Capo=Cape . . . . see proper name
Cape Barren Island . . . . 40°23′S, 148°15′E
Cape Breton Island . . . . 46°00′N, 60°30′W
Cape Coast Castle . . . . 5°06′N, 1°14′W
Cape Point . . . . 34°21′S, 18°30′E
Cape Town* . . . . 33°53′S, 18°26′E
Capel Banc . . . . 25°15′S, 159°40′E
Capricorn Channel . . . . 22°45′S, 152°00′E
Capul Island . . . . 12°26′N, 124°10′E
Capul Pass . . . . 12°25′N, 124°12′E
Cargados Carajos Shoals . . . . 16°35′S, 59°40′E
Caringin . . . . 6°21′S, 105°49′E
Caroline Islands . . . . 7°00′N, 150°00′E
Carpentaria Shoals . . . . 10°45′S, 141°03′E
Cartier Islet . . . . 12°32′S, 123°33′E
Carvoeiro, Cabo . . . . 39°21′N, 9°24′W
Casablanca* . . . . 33°38′N, 7°35′W
Caseyr, Raas (Cap Guardafui) . . . . 11°50′N, 51°17′E
Cato Island . . . . 23°15′S, 155°32′E
Catoche, Cabo . . . . 21°36′N, 87°04′W
Catwick, Îles . . . . 10°00′N, 109°00′E
Cay Sal Bank . . . . 23°40′N, 80°00′W
Cayenne . . . . 5°02′N, 52°19′W
Cayo, Cayos=Cay, Cays . . . . see proper name
Cebu* . . . . 10°18′N, 123°55′E
Celaka, Pulau . . . . 2°52′S, 107°01′E
Cemara, Gosong . . . . 0°49′N, 104°14′E
Ceram; see Seram, Pulau . . . . 3°00′S, 129°00′E
Ceva-i-Ra . . . . 21°44′S, 174°38′E
Chagos Archipelago . . . . 6°30′S, 72°00′E
Chang Jiang (Yangtze River) . . . . 31°03′N, 122°20′E
Charlotte Bank: Pacific Ocean . . . . 11°47′S, 173°13′E
Charlotte Bank: South China Sea . . . . . . 7°08′N, 107°35′E
Chatham Islands . . . . 44°00′S, 176°30′W
Chaussée=Bank, Causeway . . . . see proper name
Che Kamat, Pulau . . . . 1°21′N, 104°14′E

Chennai (Madras)* . . . . 13°07′N, 80°20′E
Chenal=Channel . . . . see proper name
Chesapeake Bay* . . . . 36°56′N, 75°58′W
Chesterfield Reefs . . . . 19°55′S, 158°20′E
Chetvertyy Kuril'skiy Proliv . . . . 50°00′N, 155°00′E
Chiloé, Isla . . . . 42°40′S, 74°00′W
China Strait . . . . 10°35′S, 150°40′E
Christmas Island: Indian Ocean . . . . 10°24′S, 105°43′E
Christmas Island: Pacific Ocean;
  see Kiritimati Atoll . . . . 1°55′N, 157°25′W
Cikoneng, Tanjung . . . . 6°04′S, 105°53′E
Cina, Karang . . . . 0°57′S, 108°32′E
Cires, Punta . . . . 35°55′N, 5°29′W
Clarence Strait . . . . 12°00′S, 131°00′E
Clarion, Isla . . . . 18°20′N, 114°45′W
Clipperton, Île . . . . 10°18′N, 109°13′W
Clyde, River . . . . 55°45′N, 4°58′W
Cochin . . . . 9°58′N, 76°14′E
Coco Channel . . . . 13°50′N, 93°10′E
Coco, Isla del . . . . 5°32′N, 87°04′W
Cocos (Keeling) Islands . . . . 12°05′S, 96°52′E
Coiba, Isla . . . . 7°25′N, 81°45′W
Colombo* . . . . 6°58′N, 79°50′E
Colón* . . . . 9°23′N, 79°55′W
Colón, Archipiélago de . . . . 0°00′, 90°00′W
Columbia River . . . . 46°15′N, 124°05′W
Comodoro Rivadavia* . . . . 45°51′S, 67°26′W
Comorin, Cape . . . . 8°05′N, 77°33′E
Comores (Comoro Islands) . . . . 12°00′S, 44°30′E
Con Son . . . . 8°38′N, 106°37′E
Conception, Point . . . . 34°27′N, 120°28′W
Congalton Skar . . . . 1°22′N, 104°19′E
Congo, River . . . . 6°00′S, 12°30′E
Cook Strait . . . . 41°00′S, 174°30′E
Cooper, Karang . . . . 3°22′S, 107°35′E
Coquimbo . . . . 29°57′S, 71°21′W
Corner Inlet . . . . 38°50′S, 146°30′E
Coromandel coast . . . . 15°00′N, 80°00′E
Coronel . . . . 37°06′S, 73°10′W
Corrientes,Cabo: Argentina . . . . 38°01′S, 57°32′W
Corrientes, Cabo: Colombia . . . . 5°29′N, 77°33′W
Corse (Corsica) . . . . 42°00′N, 9°00′E
Corse, Cap . . . . 43°01′N, 9°22′E
Coruña, La . . . . 43°22′N, 8°24′W
Corvo, Ilha do . . . . 39°40′N, 31°05′W
Cosmoledo Group . . . . 9°45′S, 47°35′E
Creus, Cabo . . . . 42°19′N, 3°19′E
Crocodile Shoal . . . . 1°11′N, 104°16′E
Crooked Island Passage* . . . . 23°15′N, 74°25′W
Cruz, Cabo . . . . 19°51′N, 77°44′W

Cu Lao Ré	15°23′N, 109°07′E
Cuba	22°30′N, 80°00′W
Cukubalimbing, Tanjung	5°56′S, 104°33′E
Cumberland, Cape; see Nahoï	14°37′S, 166°37′E
Cunene, Rio	17°15′S, 11°45′E
Curaçao: Willemstad*	12°06′N, 68°56′W
Curacoa Reef	15°29′S, 173°37′W
Curtis Channel	24°20′S, 152°55′E
Cuyo East Pass	10°30′N, 121°30′E
Dafangji Dao	21°23′N, 111°12′E
Dakar*	14°41′N, 17°24′W
Dalupiri Island	12°25′N, 124°15′E
Dalupiri Pass	12°25′N, 124°18′E
Damar, Pulau	7°08′S, 128°36′E
Dampier, Selat	0°40′S, 130°30′E
Dangan Liedao	22°02′N, 114°13′E
Dangan Suidao (Lema Channel)	22°07′N, 114°15′E
Danjo Guntō	32°02′N, 128°23′E
Dao Phu Qui	10°32′N, 108°57′E
Dapur, Pulau	3°08′S, 106°31′E
Darwin*	12°25′S, 130°47′E
Dasseneiland (Dassen Island)	33°25′S, 18°05′E
Dato, Tanjung	0°00′, 103°48′E
Datu, Pulau	0°08′N, 108°36′E
Datu, Tanjung	2°05′N, 109°38′E
Davis Strait	60°00′N, 56°00′W
Dawanshan Dao	21°57′N, 113°44′E
Daxingshan Jiao	22°33′N, 114°55′E
Dazuo Yan	24°53′N, 118°59′E
Deal Island	39°29′S, 147°20′E
Dedap, Pulau	0°30′N, 104°16′E
Degong, Pulau	0°47′N, 103°32′E
Delaware Bay*	38°48′N, 75°02′W
Delgado, Cabo	10°41′S, 40°38′E
Dempo, Pulau	0°36′N, 104°18′E
Dempo, Selat	0°36′N, 104°15′E
Denmark Strait	66°00′N, 29°00′W
Derwent, River	43°00′S, 147°22′E
Désappointment, Îles du	14°10′S, 141°20′W
Destacado Island	12°16′N, 124°06′E
Deux Fréres, Les; see Hong Trung Nho	8°35′N, 106°05′E
Dhióriga Korínthou	37°57′N, 22°57′E
Diamante Rock	12°21′N, 124°12′E
Diamond Passage	17°30′S, 151°15′E
Diamond Shoal Light-buoy	35°09′N, 75°18′W
Diana, Karang-Karang	3°03′N, 107°46′E
Diego Garcia	7°13′S, 72°23′E
Diego Ramírez, Islas	56°30′S, 68°43′W
Dilli, Mount	12°01′N, 75°12′E

Dingtai Wan	24°15′N, 118°05′E
Dinh, Mui	11°22′N, 109°01′E
Direction Bank	18°20′N, 72°15′E
Discovery, Batu-Batu	2°53′S, 106°56′E
Diu Head	20°41′N, 70°50′E
Dixon Entrance	54°30′N, 132°30′W
Dokan, Pulau	0°58′S, 105°39′E
Dominican Republic	19°00′N, 71°00′W
Dondra Head	5°55′N, 80°35′E
(10 miles S of)*	5°45′N, 80°36′E
Dongfushan	30°08′N, 122°46′E
Dongyin Dao	26°22′N, 120°30′E
Douala*	3°54′N, 9°32′E
Double–headed Shot Cays	23°57′N, 80°28′W
Drake Passage	60°00′S, 67°00′W
Dry Tortugas	24°04′N, 82°55′W
N–bound from*	24°25′N, 83°00′W
S–bound to*	24°30′N, 83°05′W
Ḍubā	27°34′N, 35°30′E
Duc de Gloucester, Groupe d'Îles	20°38′S, 143°17′W
Ducie Island	24°40′S, 124°48′W
Duddel Shoal	9°57′S, 136°00′E
Dumali Point	13°07′N, 121°33′E
Duncan Passage	11°05′N, 92°45′E
Durai, Pulau	0°32′N, 103°36′E
Durban*	29°51′S, 31°06′E
Durham Shoal	16°07′S, 173°49′W
Durian Besar, Pulau	0°43′N, 103°43′E
Durian Kecil, Pulau	0°44′N, 103°40′E
Durian, Selat	0°40′N, 103°40′E
Dutch Harbour*	53°56′N, 166°29′W
Dwārka	22°15′N, 68°58′E
East Lamma Channel; see Tung Pak Liu Hoi Hap	22°13′N, 114°10′E
East London	33°02′S, 27°57′E
Eastern Fields	10°10′S, 145°40′E
Ebon Atoll	4°38′N, 168°43′E
Eddystone Rock	43°52′S, 146°59′E
Eight Degree Channel	7°24′N, 73°00′E
El=The (definite article)	see proper name
Elafonísou, Stenó*	36°25′N, 22°57′E
Elizabeth Reef	29°55′S, 159°02′E
Ellice Island Group; see Tuvalu	8°00′S, 179°00′E
Enewetak Atoll	11°30′N, 162°15′E
Engaño, Cape	18°35′N, 122°08′E
Enggano, Pulau	5°30′S, 102°20′E
Enghela, Ras; see Ras Ben Sekka	37°21′N, 9°45′E
English Channel	50°00′N, 3°00′W
Ensenada=Bay	see proper name
Entrecasteaux, Récifs d'	18°20′S, 163°00′E

Erimo Misaki	41°56′N, 143°14′E
Ernest Legouvé Reef	35°12′S, 150°40′W
Esan Misaki	41°49′N, 141°11′E
Espartel, Cabo	35°47′N, 5°56′W
Espiritu Santo Island	15°20′S, 166°50′E
Estados, Isla de los	54°50′S, 64°10′W
Estrecho=Strait	see proper name
Eua Island	21°21′S, 174°56′W
Europa Point	36°06′N, 5°21′W
Europa, Île	22°20′S, 40°20′E
Færoe Islands; see Føroyar	62°00′N, 7°00′W
Fair Isle	59°30′N, 1°40′W
Fakarava, Passe de	16°00′S, 145°50′W
Falkland Islands: Stanley*	51°40′S, 57°49′W
Falloden Hall Shoal	1°21′N, 104°19′E
False Bay; see Valsbaai	34°15′S, 18°40′E
False Point, Orissa coast	20°20′N, 86°44′E
Fanning Island; see Tabuaeran	3°51′N, 159°22′W
Farewell Cape: New Zealand	40°30′S, 173°41′E
Farilhoēs, Os	39°29′N, 9°33′W
Farquar Group	10°10′S, 51°10′E
Farrall Rocks	15°50′N, 82°19′W
Fartak, Ras	15°38′N, 52°16′E
Farvel Kap	59°46′N, 43°55′W
(75 miles S of)*	58°30′N, 44°00′W
Fastnet Rock (5 miles S of)*	51°18′N, 9°36′W
Fernando de Noronha, Ilha de	3°50′S, 32°27′W
Ferrol, Ria de El	43°28′N, 8°20′W
Fifty Fathoms Flat, The	18°15′N, 71°25′E
Fiji Islands	18°00′S, 180°00′
Filippo Reef	5°31′S, 151°40′W
Finisterre, Cabo	42°53′N, 9°16′W
Five Fathom Banks	3°48′S, 106°29′E
Fleurieu Group	40°30′S, 144°45′E
Flinders Shoal	9°53′S, 129°17′E
Flint Island	46°11′N, 59°46′W
Florence Adelaide, Karang	2°04′S, 108°05′E
Flores: Eastern Archipelago	8°40′S, 121°30′E
Flores: Arquipélago dos Açores	39°25′N, 31°15′W
Flores, Selat	8°20′S, 123°00′E
Florida Reefs	25°00′N, 80°30′W
Florida, Straits of*	27°00′N, 79°49′W
Fogo, Ilha do	14°55′N, 24°25′W
Fonseca, Golfo de	13°10′N, 87°45′W
Føroyar (Færoe Islands)	62°00′N, 7°00′W
Fortaleza	3°43′S, 38°31′W
Foster Islets	40°44′S, 147°58′E
Four, Chenal de	48°27′N, 4°51′W
Fourcroy, Cape	11°48′S, 130°01′E

Foveaux Strait	46°35′S, 168°00′E
Fowey Rocks	25°35′N, 80°06′W
Francisco, Punta	27°02′N, 70°50′W
Fratelli, Les	37°18′N, 9°24′E
Frederick Hendrick, Cape	45°52′S, 147°59′E
Frederick Henry Bay	42°55′S, 147°35′E
Frederick Reef	21°00′S, 154°25′E
Freetown*	8°30′N, 13°19′W
Fremantle*	32°02′S, 115°42′E
French Frigate Shoals	23°45′N, 166°10′W
Frio, Cabo: Brazil	23°01′S, 42°00′W
Frio, Cape: Namibia	18°26′S, 12°00′E
Fuerteventura, Isla de	28°20′N, 14°00′W
Furneaux Group	40°00′S, 148°15′E
Futuna, Île	14°20′S, 178°05′W
Fuzhou	26°05′N, 119°18′E
Gabo Island	37°34′S, 149°55′E
Gaferut Island	9°14′N, 145°23′E
Galangbaru, Pulau	0°40′N, 104°15′E
Galang, Karang	1°09′N, 104°11′E
Galera, Isla	8°12′N, 78°47′W
Galite, Canal de la	37°18′M, 9°00′E
Galite, Îles de la	37°30′N, 8°55′E
Galle, Point de	6°01′N, 80°12′E
Galleons Passage*	10°57′N, 60°55′W
Gambia, The	13°32′N, 16°55′W
Gannet Passage	10°35′S, 141°50′E
Gardner Pinnacles	25°00′N, 168°00′W
Gardafui, Cap; see Raas Caseyr	11°50′N, 51°17′E
Gata, Cabo de	36°43′N, 2°11′W
Gebe, Pulau	0°06′S, 129°26′E
Gedeh, Tanjung: Jawa	6°46′S, 105°13′E
Gedeh, Tanjung: Sumatera	2°20′S, 104°55′E
Gelasa, Pulau	2°25′S, 107°04′E
Gelasa, Batu	2°25′S, 107°03′E
Gelasa, Selat	2°53′S, 107°18′E
Genova*	44°23′N, 8°53′E
Genting, Karang: Selat Durian	0°40′N, 103°43′E
Genting, Karang: Selat Gelasa	3°34′S, 107°41′E
Genting, Tanjung	3°14′S, 107°36′E
Geographe Reef	34°18′S, 114°59′E
Georges Bank	41°40′N, 67°44′W
Georgetown: Guyana	6°50′N, 58°10′W
Geresik, Pulau	3°00′S, 107°16′E
Germaine Bank	5°09′N, 107°35′W
Geyser, Récif du	12°22′S, 46°25′E
Ghana	7°00′N, 1°00′W
Għawdex (Gozo)	36°03′N, 14°15′E
Gia, Gosong	5°13′S, 113°17′E
Gibraltar, Strait of (6 miles S of Europa Point)*	36°00′N, 5°21′W

390

Gilbert Group: Kiribati	0°00′, 174°00E
Giliyang, Pulau	7°00′S, 114°11′E
Gironde, La*	45°40′N, 1°28′W
Gladstone	
North Channel	23°44′S, 151°21′E
South Channel	23°52′S, 151°33′E
Glorieuses, Îles	11°35′S, 47°20′E
Golfo=Gulf	see proper name
Golovnina, Proliv	48°12′N, 153°15′E
Good Hope, Cape of	34°21′S, 18°30′E
(145 miles S of)*	36°45′S, 19°00′E
Gosol, Terumbu	5°53′S, 105°54′E
Gosong=Shoal, sandbank	see proper name
Gotō Rettō	32°50′N, 129°00′E
Gough Island	40°20′S, 9°55′W
Gozo; see Għawdex	36°03′N, 14°15′E
Gracias á Dias, Cabo	15°00′N, 83°10′W
Graham Land	66°00′S, 64°00W
Gran Canaria, Isla de	28°00′N, 15°30′W
Grand Banks of Newfoundland	45°30′N, 52°00′W
Grand Cayman	19°20′N, 81°15′W
Gravois, Pointe de	18°01′N, 73°54′W
Great Abaco Island	26°25′N, 77°05′W
Great Bahama Bank	23°30′N, 78°00′W
Great Coco Island	14°07′N, 93°23′E
Great Fish Point	33°32′S, 27°07′E
Great Inagua Island	21°00′N, 73°20′W
Great Nicobar Island	7°00′N, 93°50′E
Great North East Channel	9°20′S, 144°00′E
Great Western Torres Islands	11°48′N, 97°30′E
Green Point	33°54′S, 18°24′E
Greig Channel: Selat Karimata	1°15′S, 109°13′E
Greig Utara, Karang	0°52′S, 108°33′E
Greig Shoals	0°54′S, 108°32′E
Griphølen	63°15′N, 7°37′E
Grupo=Group of islands	see proper name
Guadalcanal	9°40′S, 160°00′E
Guadalupe, Isla de	29°00′N, 118°17′W
Guam*	13°27′N, 144°35′E
Guang'ao Wan	23°12′N, 116°43′E
Guanos, Punta	23°09′N, 81°39′W
Guardafui, Cap; see Raas Caseyr	11°50′N, 51°17′E
Guardian Bank	9°30′N, 87°30′W
Guascama, Punta	2°37′N, 78°25′W
Guayaquil, Golfo de	3°00′S, 80°00′W
Guinea, Gulf of	2°00′N, 3°00′E
Gundul, Pasir	6°27′S, 105°42′E
Gunong, Gunung=Mountain	see proper name
Gusung=Shoal, sandbank	see proper name

Guyana	5°00′N, 60°00′W
Guyane Française	4°00′N, 53°00′W
Haaian, Karang	3°29′S, 107°07′E
Habana*	23°10′N, 82°22′W
Hadd, Ra's al	22°30′N, 59°48′E
Haimen Wan	23°08′N, 116°35′E
Hainan Dao	19°10′N, 109°30′E
Haitan Dao	25°30′N, 119°45′E
Haïti	19°00′N, 72°30′W
Hakodate	41°46′N, 140°41′E
Ḥalāniyāt, Juzur al	
(Kuria Muria Islands)	17°30′N, 56°05′E
Halifax*	44°31′N, 63°30′W
Halmahera	0°30′N, 128°00′E
Hambre, Puerto del	53°40′S, 70°55′W
Hampton Harbor	42°54′N, 70°49′W
Han Jiang: Shantou	23°20′N, 116°45′E
Hancock, Karang	3°34′S, 107°05′E
Hangklip, Cape	34°23′S, 18°50′E
Harans Reef	21°32′S, 168°54′W
Hatteras, Cape	35°14′N, 75°31′W
Hawaii	19°40′N, 155°30′W
Haymet Rocks	27°11′S, 160°13′W
Haynes, Karang	2°34′N, 108°51′E
Hecate Strait	53°00′N, 131°00′W
Helen Mar Reef	1°07′N, 103°46′E
Helen Shoal	19°12′N, 113°53′E
Heluputan, Karang	0°38′N, 105°08′E
Henderson Island	24°22′S, 128°19′W
Hereheretue Atoll	19°52′S, 145°00′W
Hervey Islands; see Manuae	19°15′S, 158°55′W
High Land of Saint John	20°04′N, 72°50′E
Hilo	19°44′N, 155°04′W
Hinatuan Passage	9°52′N, 125°28′E
Hispaniola	19°00′N, 71°00′W
Ho Chi Minh City (Sai Gon)	10°50′N, 106°40′W
Song Sai Gon approaches*	10°18′N, 107°04′W
Hobart*	42°58′S, 147°23′E
Hog Island; see Shirāli	14°00′N, 74°29′E
Hokkaidō	43°00′N, 143°00′E
Hon=Island	see proper name
Hong Kong*	22°17′N, 114°10′E
Hong Trung Nho	
(Les Deux Freres)	8°35′S, 106°05′E
Honolulu*	21°17′N, 157°53′W
Honshū	36°00′N, 138°00′E
Hormuz, Strait of*	26°27′N, 56°32′E
Hornos, Cabo de (5 miles S of)*	56°04′S, 67°15′W
Horsburgh Light	1°20′N, 104°24′E
Horta	38°32′N, 28°36′W

Hout Bay	34°04'S, 18°21'E
Houtman Abrolhas	28°35'S, 113°45'E
Howe, Cape	37°30'S, 149°59'E
Hudson Bay	60°00'N, 85°00'W
Hudson Strait	61°00'N, 64°50'W
Hugli River	22°00'N, 88°00'E
Hunter, Île	22°24'S, 172°05'E
Hydrographer's Passage	20°38'N, 150°21'E
Hyères, Îles d'	43°00'N, 6°20'E
Île=Island	see proper name
Ilha=Island	see proper name
Iloilo*	10°42'N, 122°36'E
Ince Point	10°30'S, 142°19'E
Indispensable Reefs	12°30'S, 160°25'E
Indespensible Strait	9°00'S, 160°30'E
Indreleia:	
North entrance	71°10'N, 25°47'E
South entrance	58°58'N, 5°45'E
Inishtrahull (5 miles N of)*	55°31'N, 7°15'W
Inubo Saki	35°42'N, 140°52'E
Investigator Strait	35°30'S, 137°00'E
Iō Tori Shima: Amami Guntō	27°52'N, 128°14'E
Iquique*	20°12'S, 70°10'W
Isla, Isola=Island	see proper name
İstanbul	41°01'N, 29°00'E
Isumrud Strait	4°45'S, 145°50'E
Iwan, Karang	1°40'S, 106°18'E
İzmir*	38°25'N, 27°00'E
Jabung, Tanjung	1°01'S, 104°22'E
Jaguatara, Pulau	5°12'S, 106°27'E
Jailolo, Selat*	0°06'N, 129°10'E
Jakarta*	6°03'S, 106°53'E
Outer Channel	5°50'S, 106°33'E
Jamaica	18°00'N, 77°30'W
Jan Mayen	71°00'N, 8°30'W
Jang, Tanjung	0°18'S, 105°00'E
Jarum, Malang	1°06'N, 104°13'E
Jarvis Island	0°23'S, 160°01'W
Jati, Tanjung	2°58'S, 106°03'E
Jawa	7°00'S, 110°00'E
Jazirat=Island	see proper name
Jelai, Gosong	3°24'S, 110°09'E
Jiangjun Ao	24°01'N, 117°51'E
Jiapeng Liedao	21°53'N, 114°03'E
Jicarita, Isla	7°13'N, 81°48'W
Jeddah	21°29'N, 39°11'E
Jigu Jiao	31°10'N, 122°23'E
Jintotlo Channel	11°48'N, 123°05'E

Jintotlo Island	11°51'N, 123°07'E
Johor Shoal	1°19'N, 104°04'E
Jomard Entrance	11°15'S, 152°06'E
Jora, Bukit	0°43'N, 103°43'E
Juan Fernández, Archipiélago de	33°40'S, 78°50'W
Juan de Fuca Strait*	48°30'N, 124°47'W
Juan de Nova, Île	17°03'S, 42°42'E
Judgement Rocks	39°31'S, 147°08'E
Jupiter Inlet	26°57'N, 80°04'W
Ka Lae	18°55'N, 155°41'W
Kaashidoo Channel; see Kardiva	5°05'N, 73°24'E
Kachchh, Gulf of	22°40'N, 69°30'E
Kachchigadh	22°20'N, 68°58'E
Kadavu Passage	18°45'S, 178°00'E
Kahoolawe Island	20°33'N, 156°36'W
Kai, Pulau-Pulau	5°40'S, 132°50'E
Kait, Karang	3°27'S, 107°00'E
Kaiwi Channel	21°12'N, 157°40'W
Kalangbahu, Pulau	3°02'S, 107°10'E
Kalao, Pulau	7°17'S, 120°55'E
Kalb, R'as al	14°02'N, 48°41'E
Kalingapatnam	18°19'N, 84°08'E
Kalpeni Island	10°06'N, 73°39'E
Kalukalukuang, Pulau	5°10'S, 117°39'E
Kameleon, Batu	0°31'N, 104°07'E
Kandangbalak, Pulau	5°53'S, 105°45'E
Kangean, Pulau-Pulau	6°50'S, 115°25'E
Kanhoji Angre (Khānderi) Island	18°42'N, 72°49'E
Kanmon Kaikyō	33°58'N, 130°52'E
Kaoh Kusrovie	11°07'N, 102°47'E
Kaoh Tang	10°18'N, 103°08'E
Kap=Cape	see proper name
Kapas, Pulau	5°13'N, 103°16'E
Karāchi*	24°46'N, 66°57'E
Karang=Coral, Reef	see proper name
Karangjajar, Pulau-Pulau	6°41'S, 105°11'E
Karas Besar, Pulau	0°45'N, 104°20'E
Karas Kecil, Pulau	0°44'N, 104°22'E
Kardiva (Kaashidoo) Channel	5°05'N, 73°24'E
Karimata, Pulau-Pulau	1°43'S, 108°54'E
Karimata, Selat	1°43'S, 108°34'E
Karimun Besar, Pulau	1°05'N, 103°20'E
Karimun Kecil, Pulau	1°09'N, 103°23'E
Karimunjawa, Pulau-Pulau	5°54'S, 110°15'E
Karpáthou, Stenó	35°55'N, 27°30'E
Kārwār Head	14°48'N, 74°05'E
Kas Tang	10°18'N, 103°08'E
Kasenga, Pulau	3°03'S, 107°21'E
Kassolanatumbi, Tanjung	5°16'S, 123°14'E
Kata, Karang	0°46'N, 104°25'E

Katang Lingga, Pulau	0°30′N, 104°25′E
Kateman, Pulau	0°16′N, 103°40′E
Katimabongko, Tanjung	2°20′S, 105°14′E
Kauai Channel	21°50′N, 158°45′W
Kawio, Kepulauan	4°40′N, 125°25′E
Kayuara, Pulau	1°32′N, 106°27′E
Kazan Rettō	25°00′N, 141°20′E
Kebatu, Pulau	3°48′S, 108°04′E
Kebua, Karang	6°22′S, 105°48′E
Keeling Island; see Cocos Islands	12°05′S, 96°52′E
Keith Reef	37°49′N, 10°55′E
Kekek, Pulau	1°30′S, 128°38′E
Keladi, Bukit	2°54′S, 107°03′E
Kelang, Pulau	3°13′S, 127°44′E
Kelang, Selat	3°15′S, 127°37′E
Kelapan, Pulau	2°51′S, 106°50′E
Kelemar, ...	2°58′S, 107°13′E
Kelian, Tanjung	2°05′S, 105°08′E
Kelso Banc	24°00′S, 159°20′E
Kembung, Karang	2°51′S, 107°20′E
Kenn Reef	21°15′S, 155°48′E
Kent Group	39°30′S, 147°20′E
Kepulauan=Group of islands	see proper name
Kerguelen, Îles	49°00′S, 69°30′E
Kermedec Islands	30°30′S, 178°30W
Keuchenius Reef; see Pinang, Karang	2°02′S, 106°37′E
Khambhāt, Gulf of	20°30′N, 72°00′E
Khānderi Island; see Kanhoji Angre Island	18°42N, 72°49′E
Khoai, Hon	8°26′N, 104°49′E
Kii Suidō	34°00′N, 134°50′E
Kiluan, Teluk	5°47′S, 105°06′E
King Island	39°50′S, 144°00′E
Kingston: Jamaica*	17°54′N, 76°44′W
Kinkasan Tō	38°18′N, 141°35′E
Kiribati:	
Gilbert Group	0°00′ 174°00′E
Phoenix Group	4°00′S 173°00′W
Kiritimati Atoll (Christmas Island)	1°55′N, 157°25′W
Ko=Island	see proper name
Kofiau Island	1°10′S, 129°50′E
Koliot, Terumbu	5°55′S, 105°49′E
Kolkata (Calcutta)	22°30′N, 88°15′E
Approach*	21°00′N, 88°13′E
Konkan Coast	17°00′N, 73°00′E
Korean Strait	34°30′N, 129°30′E
Korek Rapat	0°41′N, 104°20′E
Kosciusko Bank	10°26′S, 179°30′E
Kosrae Island	5°20′N, 163°00′E
Kota, Alur Pelayaran	2°40′N, 109°00′E

Kra, Ko	8°24′N, 100°44′E
Krawang, Pulau	1°44′S, 109°21′E
Krung Thep (Bangkok) Bar*	13°23′N, 100°35′E
Kualacenaku, Teluk	0°10′S, 103°45′E
Kueel, Pulau	2°59′S, 107°08′E
Kula, Ko	10°15′N, 99°15′E
Kulon, Ujung	6°45′S, 105°20′E
Kumukahi, Cape	19°31′N, 154°49′W
Kuria Muria Islands; see Ḥalāniyāt, Juzur al	17°30′N, 56°05′E
Kuril'skiye Ostrova	48°00′N, 153°00′E
Ky Van, Mui	10°23′N, 107°16′E
Kyūshū	33°00′N, 131°00′E
La Guaira, Puerto	10°36′N, 66°56′W
La Pérouse Strait	45°45′N, 142°00′E
La, Le, Les=The	see proper name
Labu, Tanjung: Selat Bangka	2°58′S, 106°20′E
Labu, Tanjung: Selat Leplia	2°57′S, 106°55′E
Laccadive Islands; see Lakshadweep	12°00′N, 72°30′E
Lae	6°45′S, 147°00′E
Lagoon Reef	9°27′S, 144°54′E
Lagos*	6°23′N, 3°24′E
Laguna Inferior	16°15′N, 94°45′W
Lahi, Ava	21°00′S, 175°10′W
Lakeba Passage	17°50′S, 178°30′W
Lakes Entrance	37°54′S, 147°59′E
Lakshadweep (Laccadive Islands)	12°00′N, 72°30′E
Lamakera, Selat	8°30′S, 123°10′E
Lampung, Teluk	5°40′S, 105°20′E
Lances, Playa de los	36°01′N, 5°37′W
Landfall Island	13°39′N, 93°01′E
Langara Island	54°15′N, 133°03′W
Langir, Pulau	2°48′S, 107°22′E
Langkuas, Pulau	2°32′S, 107°37′E
Lanrick, Karang	1°53′S, 106°57′E
Larabe, Karang	3°32′S, 107°10′E
Las Palmas*	28°07′N, 15°24′W
Laughlan Islands	9°17′S, 153°41′E
Laurot, Pulau-Pulau	4°50′S, 115°45′E
Laut, Pulau	3°40′S, 116°15′E
Layah, Pulau-Pulau	1°35′S, 109°20′E
Layang Layang	5°18′S, 106°04′E
Layaran, Pasir	7°47′S, 122°18′E
Le Cher, Gosong	8°29′S, 136°17′E
Le Maire, Estrecho de	54°50′S, 65°00′W
Leeuwin, Cape	34°23′S, 115°08′E
(20 miles SW of)*	34°28′S, 114°45′E
Leeward Islands	16°00′N, 62°00′W
Legundi, Pulau	5°50′S, 105°16′E
Legundi, Selat	5°49′S, 105°13′E

Leixões*	41°10′N, 8°42′W
Lelari, Tanjung	2°49′S, 105°57′E
Lema Channel; see Dangan Suidao	22°07′N, 114°15′E
Leman, Pulau–Pulau	1°17′S, 108°54′E
Leman, Karang	0°28′N, 104°28′E
Lepar, Pulau	2°57′S, 106°48′E
Leplia, Selat	2°53′S, 106°58′E
Les Deux Freres; see Hong Trung Nho	8°35′S, 106°05′E
L'Esperence Rock	31°21′S, 178°50′W
Lesser Antilles	15°00′N, 61°00′W
Liaodong Wan	40°00′N, 121°00′E
Liat, Pulau	2°52′S, 107°03′E
Liberia	6°00′N, 10°00′W
Libreville*	0°25′N, 9°17′E
Lihou Reef	17°20′S, 152°00′E
Lima, Pulau–Pulau: Selat Gelasa	3°03′S, 107°24′E
Lima, Pulau–Pulau: Selat Makasar	5°05′S, 117°04′E
Limende, Selat	3°00′S, 107°12′E
Linapacan Strait	11°37′N, 119°57′E
Lindesnes (4 miles S of)*	57°55′N, 7°03′E
Lingga, Pulau-Pulau	0°00′, 104°30′E
Lingga, Pulau	0°10′S, 104°35′E
Lion's Head	33°56′S, 18°23′E
Lions, Gulf of	43°00′N, 4°00′E
Lipata Point	12°32′N, 124°16′E
Lisboa*	38°36′N, 9°24′W
Lishi Liedao	23°47′N, 117°43′E
Litke, Proliv	59°00′N, 163°30′E
Little Andaman Island	10°40′N, 92°30′E
Little Bahama Bank	27°00′N, 78°30′W
Liverpool	53°25′N, 3°00′W
Livorno	43°33′N, 10°19′E
Lizard Point	49°58′N, 5°12′W
Lobam, Pulau	0°59′N, 104°15′E
Lobam Kecil, Pulau	0°59′N, 104°13′E
Lobito*	12°19′S, 13°35′E
Lobos de Afuera, Islas	6°57′S, 80°42′W
Lolo, Karang	0°59′N, 104°13′E
Lombok	8°30′S, 116°30′E
Lombok, Selat*	8°47′S, 115°44′E
London Reefs	8°51′N, 112°13′E
Long Island	40°50′N, 73°00′W
Lopez, Cap	0°37′S, 8°43′E
Lord Howe Island	31°32′S, 159°05′E
Losin Ko	7°20′N, 101°59′E
Louisa Reef	6°20′N, 113°14′E
Louisade Archipelago	11°00′S, 153°00′E
Luanda, Porto de	8°47′S, 13°16′E
Lubang Island	13°45′N, 120°10′E

Ludai, Bukit	3°09′S, 107°44′E
Luzon	16°00′N, 121°00′E
Luzon Strait	20°00′N, 121°00′E
Lyra Reef	1°50′S, 153°25′E
Lyttleton	43°36′S, 172°49′E
Maan Liedao	30°47′N, 122°42′E
Maatsuyker Islands	43°36′S, 146°20′E
Mabut Laut	0°49′N, 104°18′E
Macauley Island	30°14′S, 178°26′W
Macclesfield Bank	15°50′N, 114°30′E
Madeira, Ilha de	32°40′N, 17°00′W
Madras; see Chennai*	13°07′N, 80°20′E
Madura, Pulau	7°00′S, 113°30′E
Maestro de Campo Island	12°56′N, 121°43′E
Maéwo Island	15°10′S, 168°05′E
Magallanes, Estrecho de:	
E entrance (3 miles S of Dungeness)*	52°27′S, 68°26′W
W entrance (5 miles NNW of Cabo Pilar)*	52°38′S, 74°46′W
Magdalena, Karang	2°02′S, 107°00′E
Mahé Island*	4°35′S, 55°30′E
Mahia Peninsula	39°10′S, 177°55′E
Makasar, Selat	2°00′S, 118°00′E
Makassar; see Ujungpandang	5°08′S, 119°22′E
Mala, Bahía	36°10′N, 5°20′W
Mala, Punta	7°28′N, 80°00′W
Malabata, Punta	35°49′N, 5°45′W
Malacca Strait	4°00′N, 100°00′E
Malaita Island	9°00′S, 161°00′E
Malang=Reef, shoal	see proper name
Malang, Batu	3°15′S, 107°28′E
Malapascua Island	11°20′N, 124°07′E
Malawali, Selat	7°00′N, 117°21′E
Malden Island	4°00′S, 154°55′W
Maldives	4°00′N 73°00′E
Mali Kyun (Tavoy Island)	13°05′N, 98°17′E
Malpelo, Isla	4°00′N, 81°35′W
Malta*	35°55′N, 14°33′E
Mampango, Gosong	3°35′S, 109°11′E
Mandariki, Pulau	1°19′N, 107°02′E
Mangaia Island	21°53′S, 157°55′W
Mangalore	12°50′N, 74°50′E
Manggar, Terumbu	2°54′S, 108°57′E
Mangkut, Pulau	3°04′S, 110°12′E
Manila*	14°32′N, 120°56′E
Manilang, Bukit	0°37′N, 103°43′E
Manipa, Pulau	3°20′S, 127°35′E
Manipa, Selat*	3°20′S, 127°22′E
Manoel Luis, Recif	0°50′S, 44°15′W

Manora Point	24°47′N, 66°59′E
Mantang, Pulau	0°47′N, 104°33′E
Manua Islands	14°13′S, 169°35′W
Manuae Island: Îles Sous–le–Vent	16°30′S, 154°40′W
Manuae (Hervey) Island: Southern Cook Islands	19°15′S, 158°55′W
Manuk, Pulau	5°33′S, 130°18′E
Manzanillo	19°05′N, 104°20′W
Mapor, Pulau	1°00′N, 104°50′E
Maputo, Baia de	26°00′S, 32°50′E
Marangbolo, Tanjung	3°15′S, 107°31′E
Marettimo, Isola	37°58′N, 12°03′E
Margarita, Canal de	10°51′N, 64°05′W
Maria Theresa, Reef	36°50′S, 136°39′W
Maria Van Dieman, Cape	34°29′S, 172°38′E
Maria, Îles	21°48′S, 154°42′W
Mariana Islands	17°00′N, 146°00′E
Mariato, Punta	7°12′N, 80°53′W
Marinduque Island	13°20′N, 122°00′E
Marquises, Îles	9°00′S, 140°00′W
Marseille*	43°18′N, 5°20′E
Marshall Islands	10°00′N, 170°00′E
Martin Vaz, Ilhas	20°31′S, 28°51′W
Marutea, Sud	21°30′S, 135°35′W
Masalembo, Besar	5°33′S, 114°27′E
Masalembo, Kecil	5°25′S, 114°25′E
Masatiga, Pulau	0°56′S, 109°15′E
Masbate Island	12°15′N, 123°30′E
Maşīrah, Kalīj	19°30′N, 58°00′E
Maspari, Pulau	3°13′S, 106°13′E
Maspari, Alur Pelayaran	3°08′S, 106°06′E
Mataiva	12°50′S, 148°45′W
Matanzas, Puerto de	23°03′N, 81°34′W
Matarani	16°59′S, 72°07′W
Matthew, Île	22°21′S, 171°21′E
Maui	20°45′N, 156°20′W
Mauke	20°08′S, 157°23′W
Mauritania	20°00′N, 16°00′E
Mauritius	20°10′S, 57°30′E
Maya, Pulau	1°08′S, 109°35′E
Mayaguana Passage	22°30′N, 73°20′W
Maydh, Jasiired	11°14′N, 47°13′E
Mayotte, Île	12°50′S, 45°10′E
Mayyūn	12°39′N, 43°25′E
Mazatlán	23°11′N, 106°26′W
Me Shima	32°00′N, 128°21′E
Medang, Pulau	8°09′S, 117°23′E
Medang, Karang	3°22′S, 106°56′E
Medusa, Karang	5°47′S, 105°16′E
Meizhou Wan	25°00′N, 119°05′E

Mekong River	9°50′N, 106°48′E
Melbourne:	
Port of	37°50′S, 144°54′E
Port Phillip*	38°20′S, 144°34′E
Meledang, Pulau	1°29′S, 109°23′E
Mellish Reef	17°25′S, 155°51′E
Melvill, Gosong	3°02′S, 106°15′E
Melvill, Karang	0°52′N, 103°37′E
Melville Island	11°30′S, 131°00′E
Mendanau, Pulau	2°52′S, 107°25′E
Mendocino, Cape	40°26′N, 124°25′W
Menjangang, Tanjung	3°50′S, 105°57′E
Menjangang, Gosong	3°45′S, 106°12′E
Mentawai, Pulau-Pulau	1°35′S, 99°15′E
Menumbing, Bukit	2°07′S, 105°10′E
Merak Besar, Pulau	5°56′S, 105°59′E
Merapas, Pulau	0°56′N, 104°55′E
Mergui	12°26′N, 98°36′E
Mermaid Rock	40°23′S, 144°57′E
Merundung, Pulau	2°04′N, 109°06′E
Mesanak, Pulau	0°25′N, 104°32′E
Messina, Stretto di*	38°12′N, 15°36′E
Mewstone	43°45′S, 146°22′E
Midai, Pulau	3°00′N, 107°47′E
Middleton Reef	29°28′S, 159°04′E
Midway Islands	28°13′N, 177°21′W
Mikomoto Shima	34°35′N, 138°57′E
Mindanao	8°00′N, 125°00E
Mindoro	13°00′N, 121°10′E
Mindoro Strait	13°00′N, 120°00′E
Minerva Reefs	23°45′S, 179°00′W
Minicoy	8°18′N, 73°02′E
Minjiang Kou	26°05′N, 119°52′E
Miskam, Teluk	6°30′S, 105°45′E
Miskito Channel	14°20′N, 83°10′W
Mississippi River:	
Gulf Outlet Channel*	29°25′N, 88°58′W
South Pass*	28°59′N, 89°07′W
South-west Pass	28°52′N, 89°25′W
Misteriosa Bank	19°00′N, 83°43′W
Moçambique Channel	17°00′S, 41°30′E
Moçambique, Porto de	15°05′S, 40°43′E
Moçamedes, Baía de; see Baía Namibe	15°10′S, 12°07′E
Mombasa*	4°05′S, 39°43′E
Momfafa, Tanjung	0°18′S, 131°20′E
Momparang, Pulau-Pulau	2°35′S, 108°45′E
Mona Passage*	18°20′N, 67°50′W
Monrovia*	6°21′N, 10°50′W
Montague Island	36°15′S, 150°14′E
Montevideo, Puerto de	34°54′S, 56°15′W

Montreal	45°30′N, 73°40′W
Moonlight Head	38°46′S, 143°14′E
Moorea	17°30′S, 149°50′W
Morane	23°10′S, 137°08′W
Morant Cays	17°25′N, 76°00′W
Morotiri Islands	27°55′S, 143°30′W
Mossel Bay	34°11′S, 22°09′E
Moulmein	16°30′N, 97°38′E
Mubut Laut	0°49′N, 104°18′E
Muci, Pulau	0°32′S, 104°02′E
Muhammad, Râs	27°44′N, 34°15′E
Mui=Cape	see proper name
Mujeres, Isla	21°12′N, 86°43′W
Mukoshima Rettō	27°40′N, 142°10′E
Mumbai (Bombay)*	18°51′N, 72°50′E
Mungging, Pulau	1°22′N, 104°18′E
Muor, Pulau	0°11′N, 128°57′E
Muri, Pulau	1°54′N, 108°39′E
Murung, Tanjung	3°02′S, 106°54′E
Musa, Jbel	35°54′N, 5°25′W
Muscaṭ	23°37′N, 58°35′E
Muwaylih, Al	27°40′N, 35°29′E
Nabasagan Bay	12°51′N, 123°12′E
Naga, Karang	3°27′S, 107°37′E
Nagasaki*	32°42′N, 129°49′E
Nahoī (Cumberland), Cape	14°37′S, 166°37′E
Nakhodka*	42°48′N, 132°57′E
Namibe, Baía (Moçamedes, Baía de)	15°10′S, 12°07′E
Namonuito Islands	8°45′N, 150°00′E
Nan Jiāo	23°24′N, 117°04′E
Nangka, Pulau-Pulau	2°25′S, 105°48′E
Nanjishan Liedao	27°28′N, 121°04′E
Nanpeng Liedao	23°16′N, 117°17′E
Nanpō Shotō	30°00′N, 141°00′E
Nanri Dao	25°13′N, 119°29′E
Nansei Shotō	27°00′N, 129°00′E
Nantucket TSS	40°30′N, 69°15′W
Nanuku Passage	16°40′S, 179°10′W
Nanumea Atoll	5°40′S, 176°07′E
Napoli*	40°50′N, 14°17′E
Naranjo Islands	12°21′N, 124°02′E
Narcondam Island	13°25′N, 94°15′E
Natuna Besar, Pulau-Pulau	4°00′N, 108°00′E
Naturaliste, Cape	33°32′S, 115°01′E
Naturaliste, Reefs	33°13′S, 115°02′E
Nauru	0°31′S, 166°56′E
Navassa Island	18°24′N, 75°01′W
Nazareth Bank	14°30′S, 60°40′E
Near Islands	52°40′N, 173°00′E

Necker Island	23°35′N, 164°45′W
Negrais, Cape	16°03′N, 94°12′E
Negro, Cabo	15°40′S, 11°55′E
Negros	10°30′N, 123°00′E
Nelson, Cape	38°26′S, 141°32′E
Netscher Shoal	1°09′N, 104°15′E
New Britain	5°40′S, 151°00′E
New Caledonia; see Nouvelle–Calédonie	21°15′S, 165°20′E
New Hebrides; see Vanuatu	17°00′S, 168°00′E
New Ireland	3°20′S, 152°00′E
New Orleans	29°58′N, 90°05′W
(see also Mississippi River).	
New Bank; see Bajo Nuevo: Carribbean Sea	15°54′N, 78°40′W
New Shoal; see Bajo Nuevo: Gulf of Campeche	21°50′N, 92°05′W
New York* (Ambrose Channel)	40°28′N, 73°50′W
Newport Rock Channel	29°52′N, 32°33′E
Ngatik Atoll	5°50′N, 157°11′E
Nginang, Pulau	1°01′N, 104°10′E
Nicholas Channel	23°15′N, 80°00′W
Nicobar Islands	8°00′N, 94°00′E
Nigeria	7°00′N, 6°00′E
Nihoa	23°00′N, 162°00′W
Nine Degree Channel	9°00′N, 72°30′E
Ninety Mile Beach	38°15′S, 147°23′E
Niua Fo'ou	15°36′S, 175°38′W
Niuatoputapa Group	16°00′S, 173°45′W
Niue Island	19°03′S, 169°51′E
Niur, Kuala	1°00′S, 103°48′E
Nojima Saki	34°54′N, 139°53′E
Nordkapp (5 miles N of)*	71°15′N, 25°40′E
Norfolk Island	29°02′S, 167°56′E
Normanby Island	10°00′S, 151°00′E
North Andaman Island	13°15′N, 93°00′E
North Balabac Strait	8°15′N, 117°00′E
North Bruny Island	43°10′S, 147°23′E
North Cape: New Zealand	34°25′S, 173°03′E
North Danger Reef	11°25′N, 114°21′E
North Elbow Cay	23°56′N, 80°29′W
North Sahul Passage	10°10′S, 127°05′E
North–East Providence Channel*	25°50′N, 77°00′W
North–West Providence Channel	26°10′N, 78°35′W
North West Cape: Australia	21°47′S, 114°09′E
Northumberland, Cape	38°04′S, 140°40′E
Nouméa*	22°18′S, 166°25′E
Nouvelle–Calédonie (New Caledonia)	21°15′S, 165°20′E
Nova Scotia	45°00′N, 64°00′W
Novoya Zemlaya	74°00′N, 57°00′E
Nuevo, Bajo: Gulf of Campeche	21°50′N, 92°05′W

Nuevo, Bajo: Caribbean Sea . . . . 15°54′N, 78°40′W

Nukumanu Islands (Tasman Islands) . . . . . 4°35′S, 159°25′E

Nukusogea . . . . 19°13′S, 178°20′W

Nuku'alofa* . . . . 21°00′S, 175°10′W

Nukulaelae Atoll . . . . 9°22′S, 179°52′E

Nusapenida, Pulau . . . . 8°44′S, 115°32′E

Nyera, Karang . . . . 3°12′S, 107°28′E

Nyūdō Saki . . . . 40°00′N, 139°42′E

Ō Shima: Sagami Nada . . . . 34°45′N, 139°22′E

Obi Mayor . . . . 1°25′S, 127°25′E

Obi, Selat . . . . 1°00′S, 127°00′E

Ogasawara Guntō . . . . 27°00′N, 142°00′E

Ogea Driki . . . . 19°12′S, 178°24′W

Okinawa Guntō . . . . 26°30′N, 128°00′E

Old Bahama Channel: E entrance . . . . . . 21°30′N, 75°45′W

Olinda, Ponta de . . . . 8°01′S, 34°50′W

Olivelifuri . . . . 5°17′N, 73°35′E

O–luan Pi . . . . 21°54′N, 120°51′E

Olyutorskiy, Zaliv . . . . 60°00′N, 169°00′E

Ōma Saki . . . . 41°33′N, 140°55′E

Omae Saki . . . . 34°35′N, 138°12′E

Ombi, Selat . . . . 8°35′S, 125°00′E

Ombak, Karang . . . . 3°25′S, 107°10′E

One and Half Degree Channel . . . . 1°24′N, 73°20′E

Ontario, Karang . . . . 2°00′S, 108°39′E

Ontong Java Group . . . . 5°30′S, 159°30′E

Orkney Islands . . . . 59°00′N, 3°00′W

Oroluk Lagoon . . . . 7°30′N, 155°20′E

Ortegal, Cabo . . . . 43°46′N, 7°52′W

Os=The (definite article) . . . . see proper name

Osborn Passage . . . . 12°40′S, 124°00′E

Ōsumi Kaikyō . . . . 30°55′N, 130°40′E

Otago . . . . 45°45′S, 170°44′E

Otway, Cape . . . . 38°53′S, 143°31′E

Ouessant, Île d' . . . . 48°28′N, 5°05′W

   (10 miles W of)* . . . . 48°28′N, 5°23′W

Padangtikar, Sungai . . . . 0°40′S, 109°15′E

Pagalu, Isla de . . . . 1°26′S, 5°37′E

Pagarantimun, Tanjung . . . . 2°15′S, 110°04′E

Palau Islands . . . . 7°30′N, 134°30′E

Palawan . . . . 10°00′N, 118°40′E

Palawan Passage . . . . 10°20′N, 118°00′E

Palermo . . . . 38°08′N, 13°24′E

Palmas, Cape . . . . 4°22′N, 7°42′W

   (20 miles SSW of)* . . . . 4°06′N, 7°54′W

Palmeirinhas, Ponta das . . . . 9°06′S, 13°00′E

Palmerston Atoll . . . . 18°05′S, 163°15′W

Palmyra Island . . . . 5°53′N, 162°05′W

Palos, Cabo de . . . . 37°58′N, 0°41′W

Panaitan, Pulau . . . . 6°35′S, 105°13′E

Panaitan, Selat . . . . 6°40′S, 105°15′E

Panama Canal:

   Colón* . . . . 9°23′N, 79°55′W

   Balboa* . . . . 8°53′N, 79°30′W

Panama, Gulf of . . . . 8°00′N, 79°00′W

Panay . . . . 11°00′N, 122°30′E

Pandan, Karang . . . . 2°53′S, 107°12′E

Panebangan, Pulau . . . . 1°13′S, 109°15′E

Pangkil, Pulau . . . . 0°50′N, 104°22′E

Pangkor, Pulau . . . . 4°15′N, 100°35′E

Panjang, Gosong . . . . 6°25′S, 105°47′E

Panjang, Hon . . . . 9°18′N, 103°28′E

Panjang, Pulau . . . . 2°45′N, 108°55′E

Pantar, Selat . . . . 8°20′S, 124°20′E

Pantelleria, Isola di . . . . 36°47′N, 12°00′E

Papeete* . . . . 17°30′S, 149°36′W

Pará, Río: Salinópolis Pilot Stn* . . . . 0°30′S, 47°23′W

Paracel Islands . . . . 16°30′N, 112°00′E

Paradip* . . . . 20°15′N, 86°42′E

Paragua Ridge . . . . 8°57′N, 117°12′E

Paramaribo . . . . 5°50′N, 55°10′W

Pasangtenang . . . . 6°09′S, 105°51′E

Pasir, Karang . . . . 3°29′S, 107°10′E

Pasir, Tanjung . . . . 1°15′S, 109°24′E

Pasir=Sand, sandy beach . . . . see proper name

Passe=Passage . . . . see proper name

Passo Karang . . . . 1°08′N, 104°10′E

Patientie, Selat . . . . 0°25′S, 127°40′E

Payung Besar, Pulau . . . . 5°49′S, 106°33′E

Pearl Bank . . . . 5°48′N, 119°42′E

Pedra Branca . . . . 43°52′S, 146°58′E

Pedro Bank . . . . 17°10′N, 78°45′W

Pejantan, Pulau . . . . 0°07′N, 107°12′E

Pelali, Bukit . . . . 1°25′N, 104°12′E

Pelanduk Subang Mas . . . . 0°57′N, 104°10′E

Pelangkat, Pulau . . . . 0°45′N, 103°35′E

Pelapis, Pulau-Pulau . . . . 1°18′S, 109°10′E

Peleng, Selat . . . . 1°00′S, 123°00′E

Pelorus Reef . . . . 22°51′S, 176°26′W

Pematan, Karang . . . . 5°24′S, 106°16′E

Peña, Punta de la . . . . 36°03′N, 5°40′W

Penang, Pulau; see Pinang, Pulau . . . . . . 5°20′N, 100°15′E

Peñas, Golfo de . . . . 47°20′S, 75°00′W

Pencaras, Pulau . . . . 0°58′N, 104°10′E

Penedos de São Pedro e São Paulo . . . . . . 0°55′N, 29°21′W

P'eng–chia Yü . . . . 25°38′N, 122°04′E

Pengelap, Pulau . . . . 0°30′N, 104°17′E

Pengelap, Selat . . . . 0°30′N, 104°20′E

P'eng–hu Ch'ün–tao . . . . 23°30′N, 119°30′E

P'eng–hu Kang–tao . . . . 23°30′N, 119°53′E

Pengibu, Pulau . . . . 1°35′N, 106°19′E

Pengikik Besar, Pulau . . . . 0°15′N, 108°03′E

Penguin Bank . . . . 11°28′S, 175°30′E

Penmarc'h, Pointe de . . . . 47°48′N, 4°23′W

Penrhyn Island (Tongareva) . . . . 9°00′S, 158°03′W

Penyu, Pulau-Pulau . . . . 5°22′S, 127°47′E

Penzhinskiy Zaliv . . . . 60°00′N, 158°00′E

Perak, Pulau . . . . 5°42′N, 98°56′E

Perasi Besar, Pulau . . . . 0°43′N, 103°39′E

Perasi Kecil, Pulau . . . . 0°46′N, 103°38′E

Perhentian Besar, Pulau . . . . 5°55′N, 102°45′E

Perla, La . . . . 36°03′N, 5°26′W

Perlas, Archipiélago de las . . . . 8°20′N, 79°00′W

Permatan, Karang . . . . 5°24′S, 106°16′E

Permisan, Pegunungan . . . . 2°36′S, 105°57′E

Petong, Pulau . . . . 0°37′N, 104°05′E

Petropavlovsk* . . . . 53°00′N, 158°38′E

Phoenix Group: Kiribati . . . . 4°00′S 173°00′W

Phuket, Ko . . . . 7°55′N, 98°20′E

Piedra Point . . . . 16°19′N, 119°47′E

Pilar, Cabo . . . . 52°43′S, 74°41′W

Pillar, Cape . . . . 43°14′S, 148°02′E

Pinang, Karang (Keuchenius Reef) . . . . . . 2°02′S, 106°37′E

Pinang, Pulau (Penang, Pulau) . . . . 5°20′N, 100°15′E

Pine, Cape . . . . 46°37′N, 53°32′W

Pingelap Atoll . . . . 6°12′N, 160°40′E

Pioneer Channel* . . . . 4°40′S, 153°45′E

Piraiévs* . . . . 37°56′N, 23°37′E

Pisang, Pulau . . . . 1°23′S, 128°55′E

Pisco, Bahia . . . . 13°42′S, 76°15′W

Pitcairn Island . . . . 25°04′S, 130°05′W

Placentia Bay . . . . 47°00′N, 54°30′W

Planier, Île de . . . . 43°12′N, 5°14′E

Plasit, Karang . . . . 1°01′N, 104°13′E

Playa=Beach . . . . see proper name

Pocklington Reef . . . . 10°50′S, 155°45′E

Pointe Noire* . . . . 4°46′S, 11°49′E

Poluostrov Kamchatskiy . . . . 56°30′N, 159°00′E

Ponta Delgada* . . . . 37°44′N, 25°39′W

Pontianak . . . . 0°01′S, 109°20′E

Popole, Pulau . . . . 6°24′S, 105°48′E

Port Chalmers . . . . 45°50′S, 170°30′E

Port Dalrymple . . . . 41°02′S, 146°45′E

Port Elizabeth . . . . 33°57′S, 25°40′E

Port Hedland* . . . . 20°12′S, 118°32′E

Port Louis: Mauritius* . . . . 20°08′S, 57°28′E

Port Moresby . . . . 9°29′S, 147°07′E

Port Phillip . . . . 38°20′S, 144°34′E

Port Said; see Bûr Sa'îd* . . . . 31°21′N, 32°33′E

Portlock Reefs . . . . 9°35′S, 144°50′E

Porto Grande:
   Arquipélago de Cabo Verde* . . . . 16°54′N, 25°02′W

Poulo=Island . . . . see proper name

Pratas Reef . . . . 20°40′N, 116°45′E

Preparis North Channel . . . . 15°00′N, 93°40′E

Preparis South Channel . . . . 14°33′N, 93°27′E

Prince Consort Bank . . . . 7°53′N, 110°00′E

Prince Rupert* . . . . 54°19′N, 130°20′W

Prince of Wales Bank . . . . 8°10′N, 110°30′E

Prince of Wales Channel . . . . 10°30′S, 142°15′E

Proliv=Channel, strait . . . . see proper name

Prongs Reef . . . . 18°53′N, 72°48′E

Providencia, Isla de . . . . 13°23′N, 81°22′W

Puerto Rico . . . . 18°15′N, 66°30′W

Pulau=Island . . . . see proper name

Pulau-Pulau=Group of islands . . . . see proper name

Punggai, Tanjung . . . . 1°26′N, 104°18′E

Punggung, Tanjung . . . . 0°45′N, 104°31E

Punta=Point . . . . see proper name

Qiwang Wan . . . . 23°12′N, 116°43′E

Qizhou Liedao . . . . 19°58′N, 111°16′E

Quinaguitman . . . . 12°31′N, 124°17′E

Queen Charlotte Islands . . . . 53°10′N, 132°10′W

Queen Elizabeth, Cape . . . . 43°15′S, 147°26′E

Quilon . . . . 8°53′N, 76°35′E

Quṣeir . . . . 26°06′N, 34°17′E

Raas, Ras, R'as=Cape . . . . see proper name

Raas, Selat . . . . 7°10′S, 114°27′E

Race, Cape . . . . 46°39′N, 53°04′W

Rada=Roadstead . . . . see proper name

Radressa, Ras; see Rhiy di–Irīsal . . . . 12°35′N, 54°29′E

Raffles Light . . . . 1°10′N, 103°44′E

Raja, Tanjung . . . . 1°54′S, 106°11′E

Rakahanga . . . . 10°03′S, 161°06′W

Rakata, Pulau . . . . 6°09′S, 105°26′E

Rame Head . . . . 37°47′S, 149°30′E

Rangas, Tanjung . . . . 2°37′S, 118°49′E

Ranggas, Pulau . . . . 0°45′N, 104°29′E

Rangoon; see Yangon:

   Port of . . . . 16°46′N, 96°11′E

   River entrance* . . . . 16°09′N, 96°17′E

Raoul Island . . . . 29°15′S, 177°54′W

Rapa, Île . . . . 27°37′S, 144°17′W

Rarotonga . . . . 21°10′S, 159°47′W

Rat Island . . . . 51°48′N, 178°20′E

Ray, Cape . . . . 47°37′N, 59°18′W

Raya, Gosong	2°40′S, 106°53′E
Raz=Race, violent tidal stream	see proper name
Recife=Reef	see proper name
Recife: Brazil:	
Port of*	8°04′S, 34°51′W
Landfall off*	8°00′S, 34°40′W
Recife, Cape: South Africa	34°02′S, 25°42′E
Redang, Pulau	5°47′N, 103°00′E
Reid Rocks	40°15′S, 144°10′E
Reinga, Cape	34°25′S, 172°40′E
Remunia Shoals	1°27′N, 103°39′E
Réunion, Île de la	20°55′S, 55°15′E
Revillagigedos, Islas	19°25′N, 110°30′W
Rhiy di–Irīsal (Radressa, Ras)	12°35′N, 54°29′E
Riau, Selat	0°55′N, 104°20′E
Richardson, Karang	0°37′N, 103°43′E
Ringgit, Tanjung	8°52′S, 116°36′E
Rinjani, Gunung	8°25′S 116°27′E
Rio de Janiero*	22°59′S, 43°10′W
Río de La Plata*	35°10′S, 56°15′W
Rio=River	see proper name
Robbie Bank	11°03′S, 176°57′W
Robinson Crusoe, Isla	33°37′S, 78°52′W
Roca, Cabo da	38°46′N, 9°30′W
Roca=Rock	see proper name
Rocas, Atol das	3°52′S, 33°49′W
Rodondo Island	39°14′S, 146°23′E
Rodriguez Island	19°38′S, 63°25′E
Romblon Island	12°30′N, 122°15′E
Romblon Pass	12°43′N, 122°12′E
Rondo, Pulau	6°04′N, 95°07′E
Rosas, Golfo de	42°15′N, 3°09′E
Rose Island	14°33′S, 168°09′E
Rossel Spit	11°27′S, 154°24′E
Roti, Pulau	10°45′S, 123°00′E
Roti, Selat	10°25′S, 123°30′E
Rottnest Island	32°00′S, 115°30′E
Rotumah Shoals	13°30′S, 179°12′W
Royal Captain Shoal	9°03′N, 116°41′E
Royal Charlotte Reef	6°56′N, 113°36′E
Rukan, Pulau-Pulau	0°35′N, 103°45′E
Rukan Selatan, Pulau	0°33′N, 103°46′E
Rukan Tengah, Pulau	0°35′N, 103°46′E
Rukan Utara, Pulau	0°37′N, 103°45′E
Sabalana, Pulau-Pulau	7°00′S, 118°30′E
Sabang	5°53′N, 95°18′E
Sable Island	43°55′N, 60°00′W
Sagewin, Selat	0°55′S, 130°40′E
Sagoweel, Bukit	2°54′S, 107°22′E

Sahul Banks	11°30′S, 125°00′E
Sai Gon; see Ho Chi Minh City	10°50′N, 106°40′W
Sai Gon, Song (Approaches)*	10°18′N, 107°04′W
Sai Pok Liu Hoi Hap	
(West Lamma Channel)	22°13′N, 114°05′E
Saint Ann Shoals	8°00′N, 13°30′W
Saint Francis, Cape	34°13′S, 24°50′E
Saint George's Channel*	4°20′S, 152°32′E
Saint Helena	15°57′S, 5°43′W
Saint John's Island;	
see Pulau Sakijang Bendera	1°13′N, 103°51′E
Saint Lawrence, Gulf of	48°00′N, 61°00W
Saint Lazarus Bank	12°08′S, 41°22′E
Saint Lucia	14°02′N, 61°01′W
Saint Lucia/Saint Vincent Channel*	13°30′N, 61°00′W
Saint Mary's, Cape	46°49′N, 54°12′W
Saint–Paul, Île	38°43′S, 77°32′E
Saint–Pierre, Île	46°47′N, 56°10′E
Saint Vincent	13°15′N 61°15′W
Saint Vincent, Gulf of	35°00′S, 138°10′E
Sainte–Marie, Cap	25°35′S, 45°08′E
Sakhalin, Ostrov	50°00′N, 143°00′E
Sakijang Bendera, Pulau	
(Saint John's Island)	1°13′N, 103°51′E
Sakishima Guntō	24°40′N, 124°45′E
Sakunci, Gosong	7°50′S, 117°15′E
Sal, Ilha do	16°45′N, 22°55′W
Sala y Gomez, Isla	26°28′S, 105°28′W
Salah Ujung	8°25′S, 137°39′E
Salawati, Pulau	1°05′S, 130°50′E
Salayar, Pulau	6°10′S, 120°29′E
Salayar, Selat	5°43′S, 120°30′E
Saldanha Bay	33°05′S, 17°55′E
Salinopolis	0°37′S, 47°22′W
Salvador*	13°00′S, 38°32′W
Samadang, Tanjung	6°38′S, 105°14′E
Samar	12°00′N, 125°00′E
Samoa	14°00′S, 171°30′E
San Agustin, Cape	6°16′N, 126°11′E
(18 miles E of)*	6°15′N, 126°30′E
San Ambrosio, Isla	26°20′S, 79°52′W
San Antonio, Cabo: Cuba	21°52′N, 84°57′W
San Antonio, Cabo: Spain	38°48′N, 0°12′E
San Bernadino Islands	12°46′N, 124°17′E
San Bernadino Strait*	12°33′N, 124°12′E
San Cristóbal Island	10°30′S, 161°45′E
San Diego*	32°38′N, 117°15′W
San Diego, Cabo	54°39′S, 65°07′W
San Félix, Isla	26°19′S, 80°04′W
San Francisco*	37°45′N, 122°40′W
San Gallán, Isla	13°51′S, 76°28′W

San Jacinto, Puerto	12°34′N, 123°44′E
San José, Banco	8°08′N, 78°39′W
San Juan, Cabo	54°43′S, 63°49′W
San Juan del Norte	10°56′N, 83°43′W
San Miguel Island	12°43′N, 123°35′E
San Salvador Island	24°00′N, 74°30′W
San Sebastian, Cabo de	41°53′N, 3°12′E
Sanana, Pulau	2°15′S, 125°55′E
Sand Key	24°27′N, 81°53′W
Sandakan*	5°49′N, 118°07′E
Sandheads, The	20°54′N, 88°14′E
Sandy Cape	24°42′S, 153°16′E
Sangian, Pulau	5°57′S, 105°51′E
Sangihe, Pulau-Pulau	3°30′N, 125°35′E
Sangkarang Pulau-Pulau	5°00′S, 119°15′E
Sanglang Besar, Pulau	0°37′N, 103°41′E
Santa Barbara Channel	34°15′N, 120°00′W
Santa Cruz Islands: Basilan Strait	6°52′N, 122°04′E
Santa Cruz Islands: Solomon Islands	11°00′S, 166°15′E
Santanilla, Islas (Swan Islands)	17°25′N, 83°52′W
Santiago, Ilha de	15°06′N, 23°40′W
Santo Antão, Ilha de	17°00′N, 25°10′W
São Miguel, Ilha de	37°45′N, 25°30′W
São Roque, Cabo de	5°29′S, 35°16′E
São Tomé, Isla de	0°12′N, 6°37′E
São Tomé, Cabo de	22°02′S, 41°03′W
São Vicente, Cabo de: Portugal	37°01′N, 9°00′W
São Vicente, Canal de: Arquipélago de Cabo Verde	16°56′N, 25°06′W
Sape, Selat	8°30′S, 119°20′E
Sapudi, Pulau	7°07′S, 114°20′E
Sapudi, Selat	7°00′S, 114°15′E
Sarangani Islands	5°27′N, 125°27′E
Sardegna	40°00′N, 9°00′E
Sarmiento, Banco	52°30′S, 68°04′W
Satawal Island	7°20′N, 147°02′E
Sau, Pulau	1°04′N, 104°11′E
Saumarez Reefs	21°50′S, 153°40′E
Savai'i Island	13°35′S, 172°30′W
Sawu, Pulau-Pulau	10°30′S, 121°50′E
Saya de Malha Bank	10°00′S, 61°00′E
Saya, Pulau	0°47′S, 104°56′E
Sebesi, Pulau	5°57′S, 105°29′E
Sebong, Tanjung	1°07′N, 104°14′E
Sebuku, Pulau	5°53′S, 105°31′E
Sedimin, Pulau	3°24′N, 107°50′E
Segama, Pulau	5°10′S, 106°06′E
Seguci, Karang	0°43′N, 104°22′E
Sein, Chaussée de	48°00′N, 5°00′W
Sein, Raz de	48°01′N, 4°45′W
Sekopong, Tanjung	4°56′S, 105°54′E
Sekopong, Gosong	4°56′S, 106°03′E
Selanga, Pulau-Pulau	0°30′N, 104°21′E
Sele, Selat	1°00′S, 131°00′E
Selemar, Pulau	2°59′S, 107°06′E
Selfridge Bank	20°56′S, 157°05′E
Seliu, Pulau	3°13′S, 107°32′E
Semadang, Tanjung	6°38′S, 105°14′E
Sembulang, Tanjung	0°52′N, 104°16′E
Sendara, Karang	0°41′N, 104°37′E
Senegal, Fleuve	16°00′N, 16°30′W
Senyavin Islands	6°55′N, 158°10′E
Seram, Pulau	3°00′S, 129°00′E
Serasan	2°31′N, 109°04′E
Serasan, Alur Pelayaran	2°24′N, 109°00′E
Serdang, Pulau	5°49′S, 105°23′E
Serdang, Gosong-Gosong	5°05′S, 106°15′E
Serdang, Tanjung	4°27′S, 105°54′E
Seri Buat, Pulau	2°41′N, 103°55′E
Sermata, Pulau	8°12′S, 128°54′E
Serrana Bank	14°17′N, 80°24′W
Serrat, Cap	37°14′N, 9°13′E
Serupi, Beting	3°58′N, 112°16′E
Serutu, Pulau	1°43′S, 108°43′E
Seserot, Pulau	5°48′S, 105°15′E
Setapa, Tanjung	1°20′N, 104°08′E
Seto Nakai (Inland Sea of Japan)	34°10′N, 133°30′E
Severn, Karang	1°37′S, 106°31′E
Seychelles Bank	5°00′S, 56°00′E
Seychelles Group	4°30′S, 55°30′E
Shandong Bandao	37°00′N, 122°00′E
Shanghai*	31°03′N, 122°20′E
Shannaqiif, Raas	11°41′N, 51°15′E
Shantou	23°21′N, 116°40′E
Shenhu Wan	24°39′N, 118°41′E
Shetland Islands	60°30′N, 1°00′W
Shibeishan Jiāo	22°56′N, 116°30′E
Shiokubi Misaki	41°42′N, 140°58′E
Shirakami Misaki	41°24′N, 140°12′E
Shirāli (Hog Island)	14°00′N, 74°29′E
Shiriya Saki	41°26′N, 141°28′E
Shi Yu	23°35′N, 117°27′E
Sibbalds, Gosong	5°46′S, 117°07′E
Siberut, Selat	0°45′S, 98°28′E
Sibu, Palau	2°13′N, 104°05′E
Sibutu Passage	4°55′N, 119°37′E
Sicilia	37°30′N, 14°00′E
Sidmouth Rock	43°51′S, 147°01′E
Simara Island	12°49′N, 122°03′E
Simedang, Pulau	3°19′S, 107°12′E

Simedang Kecil, Pulau	3°18′S, 107°13′E
Simons Bay	34°11′S, 18°26′E
Sind	23°30′N, 69°00′E
Singapore, Port of*	1°12′N, 103°51E
Singapore Strait	1°10′N, 103°50′E
South Channel	1°16′N, 104°24′E
Middle Channel	1°22′N, 104°24′E
North Channel	1°25′N, 104°22′E
Singkep	0°30′S, 104°40′E
Singkeplaut, Pulau–Pulau	0°42′S, 104°28′E
Sirik, Tanjung	2°47′N, 111°19′E
Sisal, Arrecife	21°20′N, 90°10′W
Siulung, Pulau	0°47′N, 104°36′E
Siuncal, Pulau	5°48′S, 105°19′E
Skagerrak	57°30′N, 8°30′E
Snares Islands*	48°01′S, 166°36′E
Société, Îles de la	16°00′S, 152°00′W
Socotra Rock	32°06′N, 125°11′E
Sofala, Banco de	20°30′S, 35°30′E
Sōfu Gan: Nanpō Shotō	29°47′N, 140°21′E
Solander Island	46°34′S, 166°50′E
Solomon Islands	8°00′S, 158°00′E
Sombrero Channel	7°36′N, 93°33′E
Sombrero Key	24°38′N, 81°07′W
Sombrero Passage*	18°25′N, 63°45′W
Sombrero Rocks	10°43′N, 121°33′E
Sonsorol Islands	5°20′N, 132°13′E
Soreh, Karang	0°53′N, 104°22′E
Sorol Atoll	8°10′N, 140°25′E
South East Point: Victoria, Australia	39°08′S, 146°26′E
South East Cape: Tasmania, Australia	43°39′S, 146°50′E
South Georgia	54°30′S, 36°30′W
South Head: Port Jackson	33°50′S, 15°17′E
South Luconia Shoal	5°00′N, 112°35′E
South West Cape: Tasmania, Australia	43°35′S, 146°03′E
Southern Cook Islands	20°00′S, 158°15′W
Spartivento, Capo	38°53′N, 8°51′E
Speke, Karang	0°37′S, 104°06′E
Spencer Gulf	34°00′S, 137°00′E
Spitsbergen	79°00′N, 15°00′E
Split Point	38°28′S, 144°06′E
Sri Lanka	8°00′N, 81°00′E
Stanley Harbour: Falkland Islands	51°40′S, 57°49′W
St John's Harbour: Newfoundland*	47°34′N, 52°38′W
Stanton, Alur Pelayaran	3°00′S, 106°15′E
Starbuck Island	5°37′S, 155°55′W
Stenó=Strait	see proper name
Stewart Island	47°00′S, 167°45′E
Stewart Islands	8°25′S, 162°52′E
Storm Bay	43°15′S, 147°35′E
Subar Laut, Pulau	1°13′N, 103°50′E
Subi Besar, Pulau-Pulau	2°50′N, 108°50′E
Subi Kecil, Pulau	3°03′N, 108°51′E
Suez Bay (Bahr el Qulzum)	29°51′N, 32°33′E
Sugarloaf Point	32°27′S, 152°32′E
Sugarloaf Rock	39°30′S, 146°39′E
Suheli Par	10°03′N, 72°15′E
Suji, Karang-Karang	3°34′S, 106°55′E
Sukadana, Teluk	1°25′S, 109°43′E
Sula, Kepulauan	1°50′S, 125°00′E
Sulawesi	1°00′S, 120°00′E
Sultan Shoal	1°14′N, 103°39′E
Sulu Archipelago	6°00′N, 121°00′E
Sumba	9°45′S, 120°00′E
Sumba, Selat	9°00′S, 119°30′E
Sumbawa	8°30′S, 118°00′E
Sumburgh Head	59°51′N, 1°16′W
Sumur, Pulau-Pulau	5°52′S, 105°46′E
Sumurbatu, Tanjung	5°50′S, 105°46′E
Sunda, Selat*	6°00′S, 105°52′E
Sungai=River	see proper name
Suquṭrá	12°30′N, 54°00′E
Surabaya*	7°12′S, 112°44′E
Surigao Strait*	9°53′N, 125°21′E
Surinam	4°00′N, 56°00′W
Suva*	18°11′S, 178°24′E
Suvarov; see Suwarrow Islands	13°15′S, 163°05′W
Suwarrow (Suvarov) Islands	13°15′S, 163°05′W
Swain Reefs	22°20′S, 152°42′E
Swains Island	11°00′S, 171°05′W
Swan Islands (Islas Santanilla)	17°25′N, 83°54′W
Swan Island	40°44′S, 148°06′E
Syahbandar, Gosong	5°07′S, 105°56′E
Sybrandi, Karang	5°41′S, 105°51′E
Sydney*	33°50′S, 151°19′E
Tablas Island	12°25′N, 122°05′E
Tablas Strait	13°00′N, 121°45′E
Table Bay	33°50′S, 18°30′E
Tabuaeran (Fanning) Island	3°51′N, 159°22′W
Tagula Island	11°20′S, 153°10′E
Tahiti	17°35′S, 149°25′W
Tail of the Bank: Newfoundland	43°00′N, 50°00′W
T'ai–wan	24°00′N, 121°00′E
Taiwan Banks	23°00′N, 118°30′E
Taiwan Strait	24°00′N, 119°00′E
Taizhou Liedao	28°25′N, 121°55′E
Takarewataya, Karang	6°04′S, 118°55′E
Takoradi*	4°53′N, 1°44′W

Talcahuano	36°41'S, 73°06'W
Talisei Island	1°50'N, 125°00'E
Taloh, Tanjung	1°01'N, 104°14'E
Tamana Island	2°30'S, 176°00'E
Tambelan, Pulau-Pulau	1°00'N, 107°34'E
Tampico	22°17'N, 97°44'W
Tanangthayi Kyun	12°35'N, 97°51'E
Tanega Shima	30°35'N, 131°00'E
Tangier	35°47'N, 5°47'W
Tanimbar, Pulau-Pulau	7°30'S, 131°30'E
Taniwha Rock	40°25'S, 144°59'E
Tanjuk, Pulau	0°57'N, 104°12'E
Tanjong, Tanjung=Cape	see proper name
Tanjungsau, Pulau	1°03'N, 104°11'E
Taongi Atoll	14°38'N, 168°59'E
Tapai, Pulau–Pulau	0°05'N, 104°29'E
Tarābulus*	32°56'N, 13°12'E
Tarakan*	3°15'N, 117°54'E
Tarifa, Isla de	36°00'N, 5°37'W
Tasman Island	43°14'S, 148°00'E
Tasman Islands; see Nukumanu Islands	4°35'S, 159°25'E
Tatoosh Island	48°24'N, 124°44'W
Tautau, Bukit	1°30'N, 104°15'E
Tauu Islands	4°45'S, 157°00'E
Tavoy Island; see Mali Kyun	13°05'N, 98°17'E
Tahuantepec, Golfo de	16°00'N, 95°00'W
Telang Kecil, Pulau	0°42'N, 104°36'E
Teluk=Bay	see proper name
Tembaga, Karang: Selat Banka	2°41'S, 105°53'E
Tembaga, Karang: Selat Sapudi	7°07'S, 114°08'E
Tempurung, Pulau	5°54'S, 105°56'E
Ten Degree Channel	10°00'N, 92°30'E
Tenang, Karang	2°31'S, 108°55'E
Tenerife, Isla de	28°20'N, 16°35'W
Tengah, Karang	0°51'N, 103°34'E
Tengah, Pulau-Pulau	7°20'S, 117°35'E
Tenggol, Pulau	4°48'N, 103°41'E
Tepoto, Île	14°06'S, 141°27'W
Teraina (Washington) Island	4°43'N, 160°25'W
Terceira, Ilha	38°48'N, 27°15'W
Terkulai, Pulau	0°57'N, 104°20'E
Terumbu=Rock awash at low water	see proper name
Tétouan, Ensenada de	35°35'N, 5°15'W
Teun, Pulau	6°59'S, 129°08'E
Thessaloníki*	40°37'N, 22°56'E
Tho Chau, Hon	9°18'N, 103°28'E
Three Hummock Island	40°26'S, 144°55'E
Three Kings Islands	34°10'S, 172°00'E
Thunder Knoll	16°30'N, 81°20'W

Tianjin Xingang*	38°56'N, 117°51'E
Ticao Island	12°30'N, 123°42'E
Tunda, Pulau	5°49'S, 106°17'E
Tierra del Fuego	54°30'S, 68°00'W
Tiga, Karang	3°14'S, 107°27'E
Tiga, Pulau-Pulau	5°49'S, 105°32'E
Timau, Pulau	3°18'N, 107°32'E
Timor	9°00'S, 125°00'E
Timur, Gosong-Gosong	0°40'S, 103°50'E
Tioman, Pulau	2°50'N, 104°10'E
Tiung Reef	2°15'S, 107°00'E
Tobago	11°15'N, 60°35'W
(15 miles N of North Point)*	11°35'N, 60°35'W
Tobalai, Pulau	1°38'S, 128°20'E
Tobalai, Selat	1°40'S, 128°15'E
Tokara Guntō	29°20'N, 129°20'E
Tokara Kaikyō	30°12'N, 130°15'E
Tokelau Group	9°00'S, 172°00'W
Tokuno Shima	22°47'N, 128°58'E
Tōkyō Wan	35°20'N, 139°45'E
Tondang, Tanjung	1°14'N, 104°19'E
Tonga Islands	20°00'S, 174°30'W
Tongareva; see Penrhyn Island	9°00'S, 158°03'W
Tongatapu	20°59'S, 175°10'W
Tonkin, Gulf of	19°30'N, 107°00'E
Tori Shima: Nanpō Shotō	30°28'N, 140°20'E
Tori Shima: S of Gotō Rettō	32°15'N, 128°06'E
Torres Islands	13°15'S, 166°35'E
Torres Strait*	
(3½ miles W of Booby Island)	10°36'S, 141°51'E
Toty, Pulau	0°55'S, 105°46'E
Trafalgar, Cabo	36°11'N, 6°02'W
Treasury Islands	7°25'S, 155°35'E
Triangulo Oeste	20°58'N, 92°18'W
Trieste*	45°39'N, 13°44'E
Trinidad; Port of Spain*	10°38'N, 61°34'W
Trinidade, Ilha da	20°30'S, 29°20'W
Tristan da Cunha Group	37°00'S, 12°20'W
Trondheim*	63°27'N, 10°23'E
Tsugaru Kaikyō*	41°39'N, 140°48'E
Tsushima	34°25'N, 129°20'E
Tuamotu, Archipel des	18°00'S, 141°00'W
Tubu, Pulau	1°05'N, 104°10'E
Tuing, Tanjung	1°37'S, 106°03'E
Tuju, Pulau-Pulau	1°15'S, 105°15'E
Tula, Gosong	0°58'N, 104°15'E
Tung Pak Liu Hoi Hap (East Lamma Channel)	22°13'N, 114°10'E
Tunjuk, Pulau	0°57'N, 104°12'E
Tuntungkalik, Tanjung	5°48'S, 105°05'E
Tureia Atoll	20°50'S, 138°35'W

Turks Island Passage*	21°48′N, 71°16′W
Turks Islands	21°31′N, 71°08′W
Tutuila Island	14°18′S, 170°42′W
Tuvalu (Ellice Island Group)	8°00′S, 179°00′E
Twilight, Karang	1°02′S, 108°37′E
Twin Island	10°27′S, 142°26′E
Uban, Tanjung	1°04′N, 104°12′E
Udiep, Pulau	0°32′N, 104°18′E
Ujelang Atoll	9°50′N, 160°54′E
Ujungpandang (Makassar)*	5°08′S, 119°22′E
Ujung=Point	see proper name
Ular, Karang	1°58′S, 104°57′E
Ular, Tanjung	1°57′S, 105°08′E
Ulawa Island	9°45′S, 162°00′E
Ulithi Atoll	10°00′N, 139°40′E
Ulul Island	8°35′N, 149°40′E
Unimak Pass	54°15′N, 164°30′W
Upolu Island	13°55′S, 171°45′W
Upolu Point	20°16′N, 155°51′W
Uraga Suido	35°10′N, 139°45′E
Valdes Peninsula	42°30′S, 64°00′W
Valparaíso*	33°02′S, 71°37′W
Valsbaai (False Bay)	34°15′S, 18°40′E
Van Dieman, Cape	11°10′S, 130°23′E
Van Sittard Reef	2°12′S, 106°45′E
Vancouver	49°00′N, 123°22′W
Vancouver Island	49°30′N, 125°00′W
Vanguard Bank	7°28′N, 109°37′E
Vanikolo	11°40′S, 166°50′E
Vanuatu (New Hebrides)	17°00′S, 168°00′E
Varella, Cap	12°54′N, 109°28′E
Vatoa Island	19°50′S, 178°13′W
Vema Seamount	31°48′S, 8°20′E
Venezia*	45°24′N, 12°29′E
Venezuela	10°00′N, 67°00′W
Vera Cruz	19°12′N, 96°08′W
Verde Island	13°33′N, 121°05′E
Verde Island Passage*	13°36′N, 121°00′E
Vereker Banks	21°05′N, 116°00′E
Vermelha, Ponta	5°40′S, 12°10′E
Vert, Cap	14°43′N, 17°30′W
Vigo*	42°13′N, 8°50′W
Villano, Cabo	43°10′N, 9°13′W
Virgenes, Cabo	52°20′S, 68°21′W
Virgin Rocks	46°29′N, 50°46′W
Vitiaz Strait*	5°51′S, 147°30′E
Vivorillo, Cayos	15°51′N, 83°18′W
Vivero, Ria de	43°43′N, 7°35′W

Vladivostok	43°02′N, 131°58′E
Volsella Shoal	9°54′S, 136°14′E
Vostok Island	10°06′S, 152°23′W
Vung Tau, Mui	10°19′N, 107°05′E
Waigeo, Pulau	0°12′S, 130°45′E
Wailingding Dao	22°06′N, 114°02′E
Wakatohi, Kepulauan	5°35′S, 124°00′E
Wake Island	19°17′N, 166°39′E
Walvis Bay	22°54′S, 14°30′E
Wanganella Bank	32°31′S, 167°24′E
Wangi-Wangi	5°20′S, 123°35′E
Wangkang, Malang	2°48′S, 107°21′E
Warren Hastings, Karang–Karang	2°21′S, 106°56′E
Washington Island; see Teraina Island	4°43′N, 160°25′W
Wayam, Pulau	0°24′S, 131°15′E
We, Pulau	5°50′N, 95°18′E
Weitou Wan	24°30′N, 118°30′E
Wellington*	41°22′S, 174°50′E
Wenwei Zhou	21°49′N, 113°56′E
Wessel, Cape	11°00′S, 136°45′E
West Cape Howe	35°05′S, 117°35′E
West Fayu Island	8°04′N, 146°42′E
West Lamma Channel; see Sai Pok Liu Hoi Hap	22°13′N, 114°05′E
West Reef: London Reefs	8°51′N, 112°13′E
Wetar, Alur Pelayaran	8°00′S, 125°30′E
Wetar, Selat*	8°16′S, 127°12′E
Wilder Shoal	8°16′N, 173°26′W
Willoughby, Cape	35°51′S, 138°08′E
Windward Islands	13°00′N, 61°00′W
Windward Passage	20°00′N, 74°00′W
Wizard Reef	8°50′S, 51°04′E
Woleai Atoll	7°20′N, 143°50′E
Wowoni, Pulau	4°05′S, 123°05′E
Wrath, Cape (5 miles N of)*	58°43′N, 5°00′W
Wreck Reefs	22°10′S, 155°20′E
Wright Rock	39°36′S, 147°32′E
Wuqiu Yu	25°00′N, 119°27′E
Xaafuun, Raas	10°26′N, 51°25′E
Xiamen, Dao	24°27′N, 118°04′E
Xiaoban Men	30°12′N, 122°36′E
Yacoba Elisabett Shoal	7°05′S, 114°11′E
Yagoshi Misaki	41°31′N, 140°25′E
Yampi Sound	16°10′S, 123°30′E
Yangon (Rangoon):	
Port of	16°46′N, 96°11′E
River entrance*	16°09′N, 96°17′E
Yang Yu	25°09′N, 119°30′E
Yangtze River; see Chang Jiang	31°03′N, 122°20′E

Yap* .... 9°28′N, 138°09′E

Yingkou .... 40°41′N, 122°14′E

Yokohama* .... 35°26′N, 139°43′E

Young Rocks .... 36°23′S, 137°15′E

Yucatan, Peninsula de .... 21°00′N, 88°00′W

Yucatan Channel .... 21°30′N, 85°40′W

Zanzibar Island .... 6°10′S, 39°18′E

Zapato Islands .... 11°45′N, 123°01′E

Zephyr Bank .... 15°55′S, 176°50′E

Zhoushan Dao .... 30°03′N, 122°10′E

Zhoushan Qundao .... 30°22′N, 122°20′E

INDEX

NOTES

NOTES

NOTES

NOTES

NOTES

NOTES

PUBLICATIONS OF THE
UNITED KINGDOM HYDROGRAPHIC OFFICE

A complete list of Sailing Directions, Charts and other works published by the Hydrographer of the Navy, together with a list of Agents for their sale, is contained in the "Catalogue of Admiralty Charts and Publications", published annually. The list of Admiralty Distributors is also promulgated in Admiralty Notice to Mariners No 2 of each year, or it can be obtained from:

The United Kingdom Hydrographic Office,
Admiralty Way,
Taunton, Somerset
TA1 2DN

Produced in the United Kingdom
for the UKHO by Pindar plc

NOTE:- The zones, areas and seasonal periods shown on this chart relate to THE MERCHANT SHIPPING (LOAD LINE) RULES 1998.

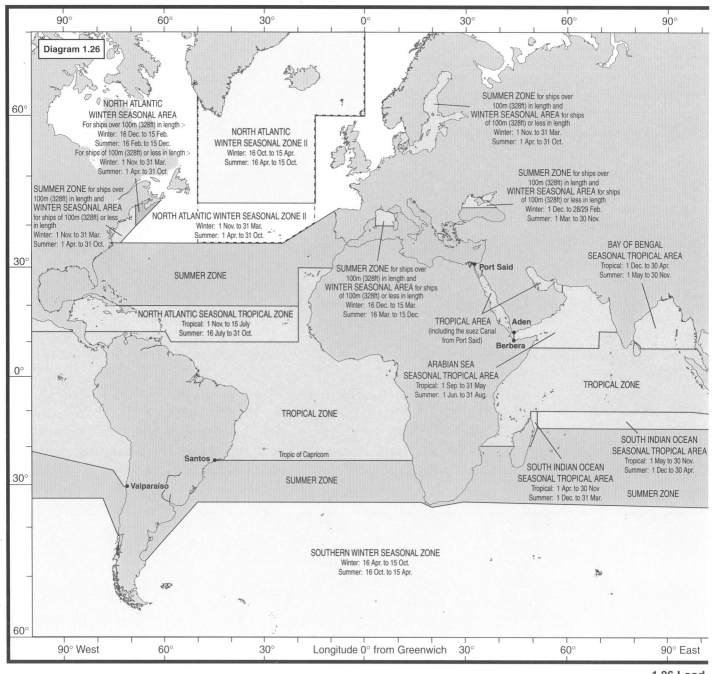

Diagram 1.26

NORTH ATLANTIC WINTER SEASONAL AREA
For ships over 100m (328ft) in length :-
Winter: 16 Dec. to 15 Feb.
Summer: 16 Feb. to 15 Dec.
For ships of 100m (328ft) or less in length :-
Winter: 1 Nov. to 31 Mar.
Summer: 1 Apr. to 31 Oct.

NORTH ATLANTIC WINTER SEASONAL ZONE II
Winter: 16 Oct. to 15 Apr.
Summer: 16 Apr. to 15 Oct.

SUMMER ZONE for ships over 100m (328ft) in length and WINTER SEASONAL AREA for ships of 100m (328ft) or less in length
Winter: 1 Nov. to 31 Mar.
Summer: 1 Apr. to 31 Oct.

SUMMER ZONE for ships over 100m (328ft) in length and WINTER SEASONAL AREA for ships of 100m (328ft) or less in length
Winter: 1 Dec. to 28/29 Feb.
Summer: 1 Mar. to 30 Nov.

SUMMER ZONE for ships over 100m (328ft) in length and WINTER SEASONAL AREA for ships of 100m (328ft) or less in length
Winter: 1 Nov. to 31 Mar.
Summer: 1 Apr. to 31 Oct.

NORTH ATLANTIC WINTER SEASONAL ZONE II
Winter: 1 Nov. to 31 Mar.
Summer: 1 Apr. to 31 Oct.

BAY OF BENGAL SEASONAL TROPICAL AREA
Tropical: 1 Dec. to 30 Apr.
Summer: 1 May to 30 Nov.

SUMMER ZONE

NORTH ATLANTIC SEASONAL TROPICAL ZONE
Tropical: 1 Nov. to 15 July
Summer: 16 July to 31 Oct.

SUMMER ZONE for ships over 100m (328ft) in length and WINTER SEASONAL AREA for ships of 100m (328ft) or less in length
Winter: 16 Dec. to 15 Mar.
Summer: 16 Mar. to 15 Dec.

Port Said

TROPICAL AREA (including the suez Canal from Port Said)

Aden
Berbera

ARABIAN SEA SEASONAL TROPICAL AREA
Tropical: 1 Sep. to 31 May
Summer: 1 Jun. to 31 Aug.

TROPICAL ZONE

TROPICAL ZONE

SOUTH INDIAN OCEAN SEASONAL TROPICAL AREA
Tropical: 1 May to 30 Nov.
Summer: 1 Dec to 30 Apr.

Santos

Tropic of Capricorn

SUMMER ZONE

Valparaíso

SOUTH INDIAN OCEAN SEASONAL TROPICAL AREA
Tropical: 1 Apr. to 30 Nov
Summer: 1 Dec. to 31 Mar.

SUMMER ZONE

SOUTHERN WINTER SEASONAL ZONE
Winter: 16 Apr. to 15 Oct.
Summer: 16 Oct. to 15 Apr.

90° West   60°   30°   Longitude 0° from Greenwich   30°   60°   90° East

1.26 Load